WA 1094952 6

D1426493

Muscles Alive

Their Functions Revealed by Electromyography

FIFTH EDITION

Muscles Alive

Their Functions Revealed by Electromyography

FIFTH EDITION

JOHN V. BASMAJIAN, M.D., F.A.C.A., F.R.C.P.(C.)

Director
Rehabilitation Centre
Chedoke-McMaster Hospitals and
Professor of Medicine and Anatomy
McMaster University
Hamilton, Ontario, Canada

CARLO J. DE LUCA, Ph.D.

Director
NeuroMuscular Research Center and
Professor of Biomedical Engineering
Boston University
Boston, Massachusetts

Williams & Wilkins

BALTIMORE • PHILADELPHIA • HONG KONG
LONDON • MUNICH • SYDNEY • TOKYO

A WAVERLY COMPANY

Editor: John Butler
Associate Editor: Carol Eckhart
Copy Editor: Stephen Siegforth
Design: JoAnne Janowiak
Illustration Planning: Lorraine Wrzosek
Production: Anne G. Seitz

Williams & Wilkins
428 E. Preston Street
Baltimore, Md. 21202, U.S.A.

Accurate indications, adverse reactions, and dosage schedules for drugs are provided in this book, but it is possible that they may change. The reader is urged to review the package information data of the manufacturers of the medications mentioned.

Made in the United States of America

First Edition, 1962
Reprinted, 1966
Second Edition, 1967
Reprinted, 1968, 1969
Italian Edition, 1971
Reprinted, 1972
Third Edition, 1974
Reprinted, 1974
Spanish Edition, 1976
Fourth Edition, 1979
Reprinted, 1979

Library of Congress Cataloging in Publication Data

Basmajian, John V., 1921-
Muscles alive, their functions revealed by electromyography.
Bibliography: p.
Includes index.
1. Muscles. 2. Electromyography. I. De Luca, Carlo J. II. Title. [DNLM: 1. Electromyography. 2. Muscles—physiology. WE 500 B315m]
QP321.B33 1985 612′.74 84-13102
ISBN O-683-00414-X

94 95 96 97 2 3 4 5 6 7 8 9 10
11 12 13 14 15

Preface to the Fifth Edition

In the life of every successful book there comes a period when simple expansion and updating is not enough. Now is that time for *Muscles Alive.* Although in its fourth edition it still enjoys a gratifying success requiring reprintings and translations to meet the demand; cosmetic changes must be avoided; and who better than my former graduate student, now a distinguished scientist, to join me in this fresh approach.

Dr. De Luca has become a co-author with wide powers to rewrite great sections of the book, particularly the important chapters that might be called basic EMG, Kinesiology, and Neurophysiology. Retaining such earlier material as is still significant, he has woven a text that is bright and very useful to all who rely on this book for authoritative facts and opinions.

Changes have not been confined to the basic EMG sections. I have revised and interwoven new materials (including hundreds of new citations) into the sections on regional kinesiology. Thus *Muscles Alive* enters its twenty-third year of life as a vigorous young adult and not a jaded old warrior. My hope is that this renewal made possible by Dr. De Luca's joining me will not only serve its community of scholars well today, but that in the future it will grow even more useful and be very much *alive.*

John V. Basmajian

Hamilton, 1985

Preface to the First Edition

ALMOST a hundred years ago, Duchenne concluded the preface of his epoch-making *Physiology of Motion* with the remark that after ten years of continuous research his task was completed with the writing of his book. After a similar period of some ten years of intense research on muscle function by a modern electronic technique that Duchenne himself would have enthusiastically employed, I have brought together into this book my own findings and those of many others from all over the world. Pleased as I am to be completing my present toil, my main feeling is one of renewed and profound admiration for the old master who, single-handed and against considerable opposition, produced a work that has stoutly withstood a century's buffeting. True, many modifications of his teaching have been dictated by clinical observations (particularly by Beevor) and by the electromyographic findings to be described in this volume. What impresses me most, however, is that this book will in no way replace or even subordinate his. On the contrary, it will complement and illuminate it, as did Beevor's Croonian Lectures *On Muscular Movement* to the Royal College of Physicians of London in 1903. I would like to hope, then, that this work will be accepted kindly as the direct and vigorous lineal descendant of the works of Duchenne and Beevor.

In particular, this book is intended for all those who deal with living muscles and movement, and I have consciously tried to make it indispensable for such workers. These include physiologists, zoologists, and anatomists, and their students; orthopedic surgeons, kinesiologists, physical medicine specialists and therapists; neurologists; and physical educationists. Indeed, it is difficult to stop the list there because the chapter on normal pharyngeal and laryngeal muscles and another on eye muscles will be of special interest to specialists in those fields. Such chapters were especially meant to be comprehensible to ordinary scientific readers as well.

Finally, it is my fond hope that the reader will soon discover that this book is not just another treatise on standard kinesiology, a subject that is already quite adequately dealt with by an impressive (and sometimes oppressive) series of books. Nonetheless, the informed reader will soon detect that it includes a great deal of both old and new information that rightfully belongs in standard kinesiology textbooks but which has not been generally available to their authors.

Acknowledgments

The list of people who have contributed to the success of this book over more than two decades grows longer each year. In previous editions special mention was made of those who acted as midwives at its birth: Otto Mortensen, J. C. B. Grant, R. G. MacKenzie, Phillippe Bauwens, W. A. Hawke, W. T. Mustard, A. W. Ham, A. N. Mitchell, and Glen Shine. A long list of others may be found in the previous edition of students and associates who helped the senior author through four editions.

We now add our warmest thanks to our wives, Dora and Chris, who have tolerated the diversion of our time and thoughts to the production of the fifth edition. There is no question that writing a book compares with a minor marital infidelity. Therefore we crave forgiveness, too. We also thank our secretaries, Jane Eaid in Hamilton and Kimberly Hood in Boston, for significant help with production. The junior author wishes to acknowledge the varied contributions of associates and past students. He is indebted to Allen L. Cudworth and Liberty Mutual Insurance Co. for the support which made possible the realization of some of the concepts reported in this book. To all, many thanks; we hope you will be pleased with the results.

John V. Basmajian
Carlo J. De Luca

Contents

CHAPTER 1

Introduction

Electromyography is the study of muscle function through the inquiry of the electrical signal the muscles emanate.

Inherent movement is the prime sign of animal life. For this and many other reasons, man has shown a perpetual curiosity about the organs of locomotion in his own body and in those of other creatures. Indeed, some of the earliest scientific experiments known to us concerned muscle and its functions.

With the reawakening of science during the Renaissance, interest in muscles was inevitable. Leonardo da Vinci, for example, devoted much of his thought to the analysis of muscles and their functions. So, too, did the acknowledged "father" of modern anatomy, Andreas Vesalius, whose influence through his monumental work, the *Fabrica*, extends down to this day. In one sense, however, the heritage of Vesalius was unfortunate because it stressed the appearance and the geography of dead muscles rather than their dynamics (Fig. 1.1).

During subsequent years, a series of scientists gave life back to the muscles. The first logical deduction of muscle-generated electricity was documented by Italian Francesco Redi in 1666. He suspected that the shock of the electric ray fish was muscular in origin and wrote, "It appeared to me as if the painful action of the electric ray was located in these two sickle-shaped bodies, or muscles, more than any other part" (Biederman, 1898). The relationship between electricity and muscle contraction was first observed by Luiggi Galvani in 1791. In his epoch-making experiments, he depolarized the muscles of a frog's legs by touching them with metal rods (see Fig. 1.2). His concept of "animal electricity" was enthusiastically received throughout Europe. Galvani's original book, *De Viribus Electricitatis*, has been translated into English by Green (1953). This discovery is generally acknowledged as representing the birth of neurophysiology, thereby making Galvani the father of this field which continues to expand rapidly.

Many rushed to confirm Galvani's results and praise his discovery. Among them was Alessandro Volta, who initially embraced the discovery and in retrospect wrote "it contains one of the most beautiful and surprising discoveries and the germ of many others" (Volta, 1816). But, within two years, in 1793, Volta questioned Galvani's findings by proving that dissimilar metals in contact with an electrolyte (such as those present in body tissues) would generate an electric current. In the following year Galvani reaffirmed his concept when he found that a muscle contraction

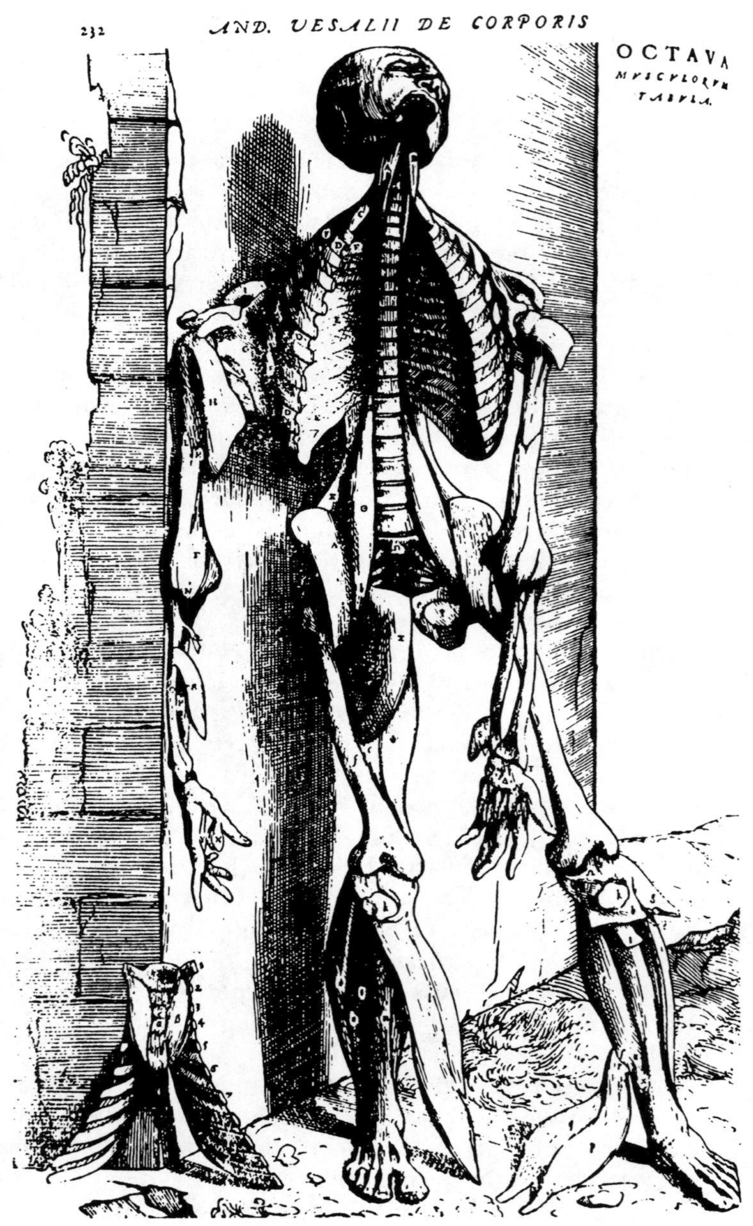

Figure 1.1. A "muscle-man" from Vesalius' *Fabrica*. (From a rare 1555 edition in the Library of Queen's University.)

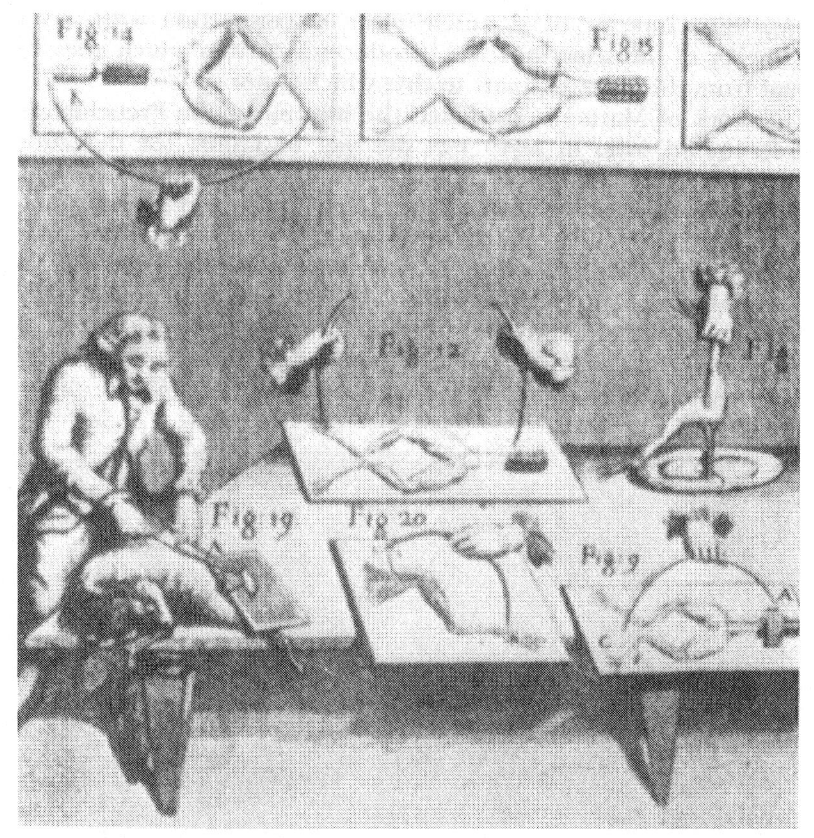

Figure 1.2. Galvani's demonstrations of the effects of electricity on muscles of frogs and sheep. (From Fulton's reproduction of a plate in Galvani's *De Viribus Electricitatis in Motu Musculari Commentarius,* 1792.)

could be elicited by placing the free end of a nerve across a muscle without the intervention of metals.

However, Volta's blow was so powerful that the concept of animal electricity was not discussed meaningfully for four decades. The engineering development of Volta, in providing a device for generating electrical currents and stimulating muscles conveniently, was not matched by a comparable engineering development in providing equipment and techniques to detect the electrical current in the muscles. In 1820, Schweigger built the first practical galvanometer based on Oersted's discoveries on magnetism. Five years later Nobili improved the sensitivity by compensating for the torque of the earth's magnetic field. Using this improved galvanometer, Carlo Matteucci in 1838 finally proved that electrical currents did originate in muscles. In 1844 he wrote, "The interior of a muscle place in connection with any part whatsoever of the same muscle. . .produces a current which goes in the animal from the muscular part to that which is not so."

The work of Matteucci attracted the interest of the Frenchman DuBois-Reymond, who in 1849 was the first to report the detection of voluntarily elicited electrical signals from human muscles. DuBois-Reymond's achievement was an example of ingenuity and stalwartness. He devised a surface electrode which consisted of a wire attached to a blotting paper immersed in a jar of saline solution. Figure 1.3 is a reproduction of Figure 147 in his book. This diagram demonstrates the recording apparatus. He found that when the fingers were immersed in the saline solution, and the arms and hand were contracted (as shown in the figure) the deflection on the galvanometer was minute (approxi-

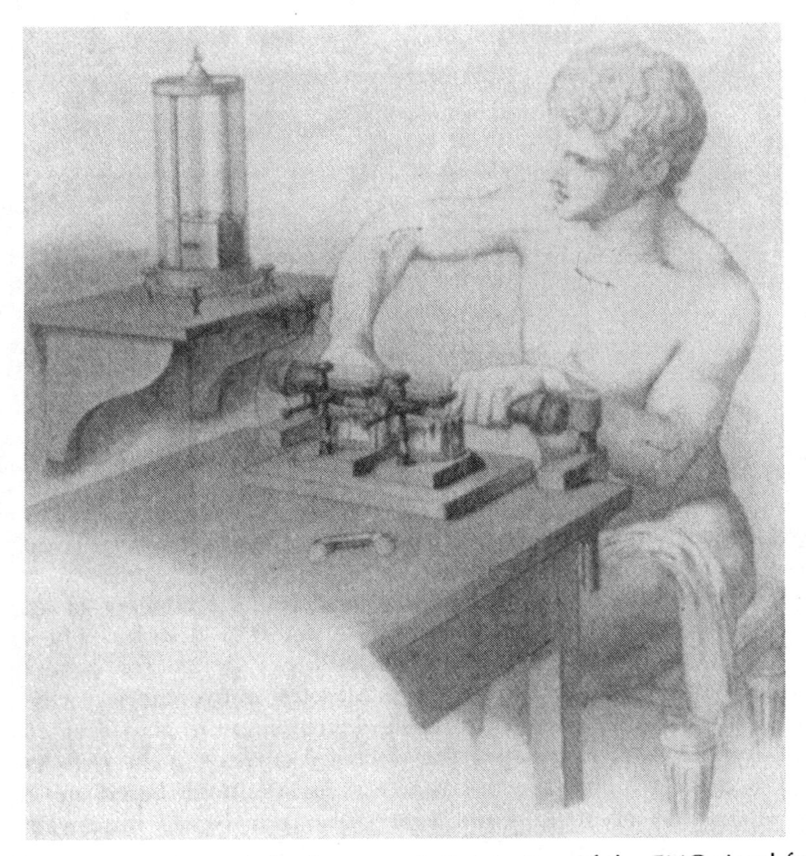

Figure 1.3. Depiction of the first recorded detection of the EMG signal from human muscles during voluntary contraction. (From Figure 147 of the book *"Uber Thierische Elektricitat"* by Du Bois-Reymond published in 1849.)

mately 2 to 3°). He realized that the impedance of the skin reduced the current which he could detect to drive the galvanometer. He circumvented this problem by inducing a blister in each forearm. Then he removed the skin and placed the open wounds in contact with the saline solution of the electrode. Upon contraction he measured a sizable deflection (65°) on his galvanometer. He repeated the contraction three times for each arm and always obtained similar results. To remove doubt, he repeated the whole experiment several weeks later, after the original wounds had healed. He obtained the same results.

However, measurements from human musculature remained unwieldy until the metal surface electrode was employed by the German Piper (1907). The detection techniques were further simplified with the advent of the cathode ray tube, which was invented by Braun (1897) and was first used to amplify action potentials in conjunction with a string galvanometer by Forbes and Thacher (1920). Two years later, Gasser and Erlanger (1922) used a cathode ray oscilloscope in place of the galvanometer, which up to this time was the sole apparatus able to "show" the signals from the muscles. This application, along with their wise interpretation of the action potentials which they were able to "see," earned Gasser and Erlanger a Nobel Prize in 1944.

During the 19th century, the capability of detecting the electromyographic or myoelectric signal from a human remained a sophisticated and delicate venture. As such, it was mastered only by a few, and useful achievements were slow in coming. On the other hand, the task of electrically stimulating a muscle by applying current through the skin was relatively simple and gained wide attention. In fact, many unqualified charlatans rushed to exploit its novelty by claiming that a "properly" applied dose of electricity could perform miraculous cures on a wide variety of ailments ranging from tic douloureux to chronic functional disorders (Holbrook, 1959).

However, among this morass of deceptions flowered the work of few individuals who remained true to the scientific inquiry. Towering above them was the Frenchman Duchenne, who in the middle of the past century, skillfully applied electrical stimulation to investigate systematically the dynamics and functions of intact skeletal muscles (Figure 1.4). His immortal work, *Physiologie des Mouvements*, is now available in an English edition (translated by E.B. Kaplan). No one before or after has contributed so much to our understanding of muscular function, although Beevor's (1903) work cannot be ignored.

It was only natural that the business of detecting electric signals from, and applying electric currents to, muscles should attract the attention of an electrical engineer. Such was the case for the Englishman Baines, who published his works in 1918. He argued that appropriate technical considerations should be administered before obtaining or interpreting

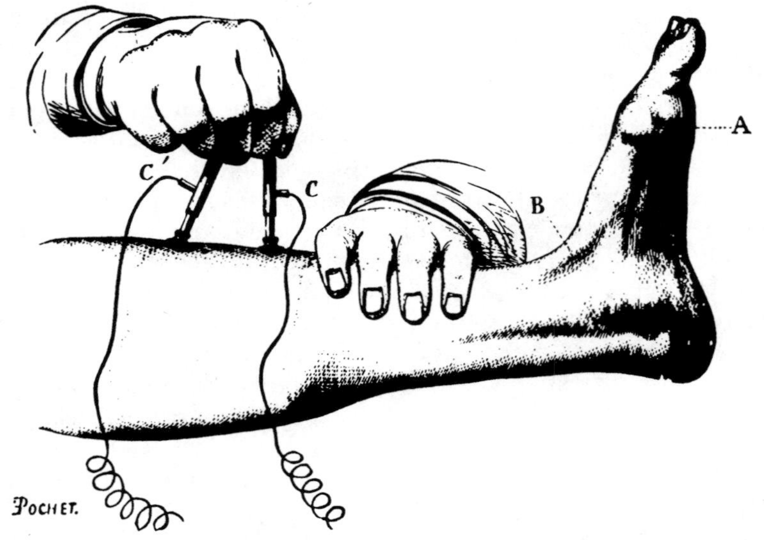

Figure 1.4. Duchenne's illustration of electrical stimulation of muscles.

data related to electrophysiological phenomena. This call still echoes among the numerous abuses that have been promulgated throughout the past seven decades. Baines was the first to formalize the analogy between the propagation of pulses in a nerve trunk and an electric cable. This approach subsequently became known as the cable theory. He initiated the concept of modeling parts of the nervous system with electrical circuits in attempting to explain their behavior (see Fig. 1.5). It may be said that he was the first biomedical engineer.

With the introduction of vacuum tube amplifiers, the task of detecting the electromyographic signal was greatly simplified. Soon the new "art" of electromyography was put to practical usages in the clinical environment. The first successful attempt at detecting a signal from a dysfunctional muscle was made by Proebster (1928) who obtained "tracings" from a muscle with peripheral nerve paralysis.

However, the impact on the clinical community occurred after the introduction of the needle electrode by Adrian and Bronk (1929). This approach, for the first time, enabled us to observe the electrical activity associated with individual muscle fibers (or small groups of muscle fibers). Although Adrian and Bronk's motivation was the investigation of the motor control schemes which acted on the muscle, the immediate impact was on the clinical community. The use of the needle electrode was methodologically exploited by Buchthal and his colleagues during the 1950s and 1960s.

As the quality and availability of electronics apparatus improved;

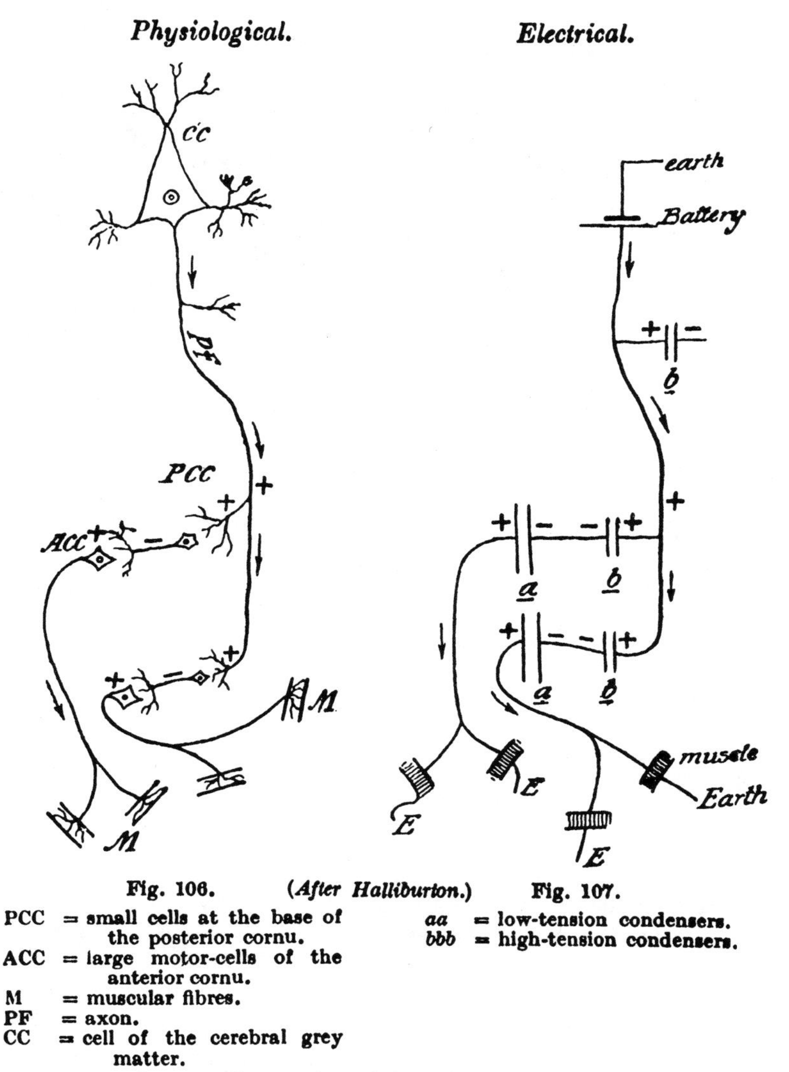

Figure 1.5. First known electrical model of the nervous system. Published by Arthur E. Baines in 1918. He was possibly the first biomedical engineer. (Reproduced from Figures 106 and 107 of his book.)

anatomists, kinesiologists, and orthopedic surgeons began to make increasing use of electromyography. The first study that gained wide acceptance was that of Inman et al (1944), who reported their work on the movement of the shoulder region. However, kinesiological studies proliferated only after the appearance of the electrically stable silver-

silver chloride surface electrode and the nonobtrusive inserted wire-electrode which appeared on the scene circa 1960.

In 1960, a group of Russian engineers led by Kobrinsky revealed the design of a hand prosthesis controlled by myoelectric signals detected from the forearm muscles. This demonstration excited the engineering and rehabilitation community. Engineers in Canada, England, Sweden, Austria, and the USA rushed to explore and exploit this new area. Their interest was fueled by funds made available to provide replacement limbs for numerous children who were born with undeveloped limbs due to the ingestion of the drug thalidomide by their mothers during pregnancy. Although the work of the Russians generated the excitement, it should be mentioned that the earliest recorded effort to employ electromyographic signals for controlling prostheses should be accorded to Reinhold Reiter, who applied for a patent describing the concept in 1945 in Germany.

Thus, we have seen that during the past two centuries a wide variety of individuals of diverse training have been fascinated by and have contributed to the understanding of the electrical phenomenon which is associated with a muscle contraction. The contents of this book were chosen to describe the more recent developments in electromyography which are the heirs of all the efforts that have preceded.

TERMINOLOGY

Sure signs of a progressive and developing field are the evolution and modification of definitions and standards. Two groups have offered such guidelines: the IFSECN at its Second International Congress (Guld et al, 1970), and the Second Congress of the International Society of Electrophysiological Kinesiology (ISEK) in 1972. The latter group has revised its definitions and has published a manual (Winter et al, 1980).

Some of the following terminology and set of definitions are an abridged version of those found in the ISEK manual "Units, Terms and Standards in the Reporting of EMG Research" (1980). They have been extracted with permission.

Alpha-motoneuron—The neural structure whose cell body is located in the anterior horn of the spinal cord and which, through its relatively large diameter axon and terminal branches, innervates a group of muscle fibers.

Motor unit (MU)—The term used to describe the single smallest controllable muscular unit. The motor unit consists of a single alpha-motoneuron, its neuromuscular junction, and the muscle fibers it innervates (as few as 3, as many as 2000).

Muscle fiber action potential or motor action potential (MAP)—The name given to the detected waveform resulting from the depolarization wave as it propagates in both directions along each muscle fiber from its motor end plate.

Motor unit action potential (MUAP)—The name given to the detected waveform consisting of the spatiotemporal summation of individual muscle fiber action potentials originating from muscle fibers in the vicinity of a given electrode or electrode pair.

Motor unit action potential train (MUAPT)—The name given to a repetitive sequence of MUAPs from a given motor unit.

Interpulse interval (IPI)—The time between adjacent discharges of a motor unit. It is a semirandom quantity.

Instantaneous firing rate—The parameter which represents the inverse value of the interpulse interval.

Average firing rate—The average firing rate of a motor unit over a given period of time. It is measured in units of pulses per second.

Synchronization—The term which describes the tendency for a motor unit to discharge at or near the time that another motor unit discharges. It therefore describes the interdependence or entrainment of two or more motor units.

Electromyographic (EMG) signal—The name given to the total signal detected by an electrode. It is the algebraic summation of all MUAPTs from all active motor units within the pick-up area of the electrode.

Myoelectric signal—An alternative nomenclature for the electromyographic signal.

Amplitude—That quantity which expresses the level of the signal activity.

Time duration—The amount of time over which a waveform presents detectable energy.

Phase—In electromyography, the net excursion of the amplitude of a signal in either the positive or negative direction.

Shape—The characteristics of a signal which remain unaltered with linear scaling in either the amplitude or time domains. An example of such characteristics are the phases of an action potential.

Waveform—The term which describes all aspects of the excursion of the potential, voltage, or current associated with a signal as a function of time. It incorporates all the notions of shape, amplitude, and time duration.

Decomposition—The process whereby individual MUAPs are extracted from the electromyographic signal.

Electrode—A device or unit through which an electrical current enters or leaves an electrolyte, gas, or vacuums.

Detection surface—The portion of the electrode which is in direct contact with the medium which is being sensed.

Unipolar electrode—One which consists of one detection surface.

Bipolar electrode—One which consists of two detection surfaces.

Concentric electrode—A unipolar electrode in which the detection surface is located in the center of a metallic shield (typically, the cannula of a needle), which in turn is connected to ground.

Detection—The process of sensing the signal by the electrode.

Recording—The process which creates a record of the detected signal on any media (CRT, paper, magnetic tape, etc.)

Isometric contraction—One during which the length of the contracting muscle remains constant. Generally, the muscle length is assessed by monitoring the angle of the joint being affected.

Anisometric contraction—One during which the length of the contracting muscle may vary.

Ballistic contraction—One that is executed with the greatest speed physiologically possible.

Maximal voluntary contraction (MVC)—The greatest amount of effort that an individual may exert. Usually, the effort is concentrated on one muscle or on one joint. It is generally measured by monitoring the force or torque output.

Agonist muscle—One which initiates a contraction.

Antagonist muscle—One which actively provides a negative contribution to a particular function during a contraction.

Synergist muscle—One which actively provides an additive contribution to a particular function during a contraction.

We would like to draw attention to the appellation of two terms: *firing rate* and *interpulse interval*. These terms are used to describe parameters which in the past have been referred to as firing frequency and interspike interval. Such terminology has its origin in the early days of electrophysiology when the available electronics apparatus was considerably more limited than today's versions and could not provide either detailed or correct representation of the behavior of the MUAPT. Thus, earlier investigators could only observe the presence of spikes (deflections) in a noisy signal. Modern technology enables us to observe the individual action potentials as distinguishable pulses. Thus, it is more proper to describe the time interval as the interpulse interval. The use of the term firing frequency is improper because the concept of frequency implies periodicity. It is now clear that the motor unit does not discharge in a periodic fashion, but rather in a semirandom fashion. It is, therefore, more correct to use the terminology of stochastic processes, i.e., firing rate. The concept of rate provides information concerning the number of firings per unit time without any restriction on the temporal regularity of the discharges.

THE MOTOR UNIT

The reader must have a clear knowledge of the structural and functional units in striated muscles to appreciate fully much of the literature in electromyography. The structural unit of contraction is, as everyone knows, the muscle cell or muscle fiber (Fig. 1.6). Best described as a very fine thread, this muscle fiber has a length ranging from a few millimeters to 30 cm and a diameter of 10 to 100 μm. On contracting it will shorten to about 57% of its resting length (Haines, 1932, 1934).

By looking at the intact normal muscle during contraction, one would believe, quite erroneously, that all the muscle fibers were in some sort of continuous smooth shortening. In fact, this is not true; instead, there is a virtual buzzing of asynchronous activity in which the fibers are undergoing very rapid contractions and relaxations.

In normal mammalian skeletal muscle, the fibers probably never contract as individuals. Instead, small groups of them contract in concert. On investigation, one finds that all the members of each of these groups of muscle fibers are supplied by the terminal branches of one nerve fiber or axon whose cell body is in the anterior horn of the spinal grey matter. Now, this nerve cell body, plus the long axon running down the motor

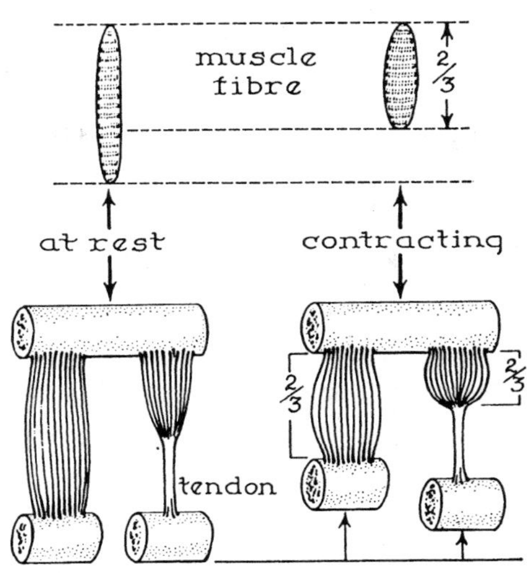

Figure 1.6. The structural unit of contraction is the muscle fiber. The greatest amount a whole muscle can actively shorten is dependent on the maximal contraction of its contractile units. (From Basmajian, © 1970, Williams & Wilkins, Baltimore.)

nerve, plus its terminal branches and all the muscle fibers supplied by these branches, together constitute a *motor unit* (Fig. 1.7). The motor unit is, then, the functional unit of striated muscle, since an impulse descending the motoneuron causes all the muscle fibers in one motor unit to contract almost simultaneously. The disparity in the time activation of different muscle fibers of the same motor unit has two causes. One is the variable delay introduced by the length and diameter of the individual axon branches innervating individual muscle fibers. This delay is fixed for each muscle fiber. The other delay is introduced by the random discharge of acetylcholine packets released at each neuromuscular junction. Because this is a random process, the excitation of each muscle fiber of a motor unit is a random function of time. This random excitation of each muscle fiber appears as a *jitter* when the electrical discharges of the individual muscle fibers are monitored. This phenomenon was first observed by Ekstedt (1964), who coined the name. In normal individuals the standard deviation of the jitter is about 20 μs. The chief use of this discovery has been in clinical diagnosis.

The termination of the axon of the muscle fiber defines an area known as the endplate region. These endplates (neuromuscular junctions) are usually, but not always, located near the middle of the muscle fibers (Fig. 1.8). This has been shown by Coërs and Woolf (1959) in human skeletal muscle, by Gurkow and Bast (1958) in the trapezius and sternomastoid of the hamster, by Jarcho et al (1952) in the gracilis of the rat, and by

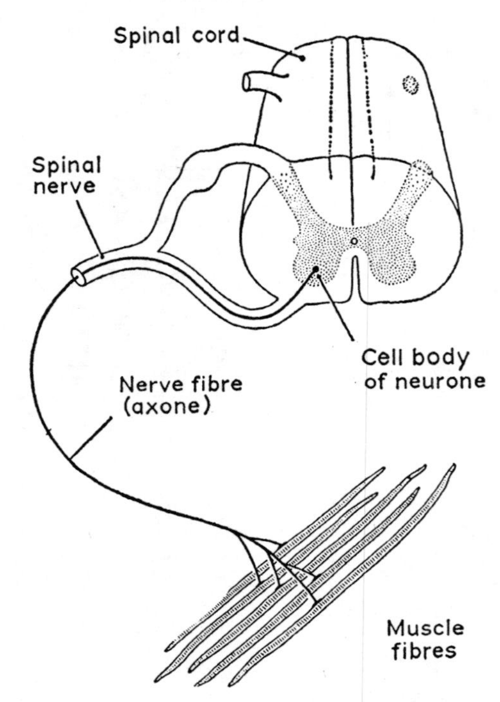

Figure 1.7. Scheme of a motor unit. (Modified from Basmajian, 1955a.)

Dutta and Basmajian (1960) in the pharyngeal constrictors of the rabbit. However, our own observations indicate that in the case of the human tibialis anterior, the endplate region is proximal to the middle of the muscle. In some muscles of some individuals, two aggregations of motor endplates or motor points may exist in a single muscle. Such is often the case in the long head of the human biceps brachii.

The number of muscle fibers that are served by one axon, i.e., the number in a motor unit varies widely, but certain rules have been established in recent years. Generally, it has been agreed that muscles controlling fine movements and adjustments (such as those attached to the ossicles of the ear and to the eyeball and the larynx) have the smallest number of muscle fibers per motor unit. On the other hand, large coarse-acting muscles, i.e., those in the limbs, have larger motor units. The muscles that move the eye have small motor units with less than 10 fibers/unit, as do the human tensor tympani muscle of the middle ear, the laryngeal muscles and the pharyngeal muscles. These are all rather small delicate muscles which apparently control fine or delicate movements.

Krnjević and Miledi (1958) report 7 to 17 fibers/motor unit in the rat

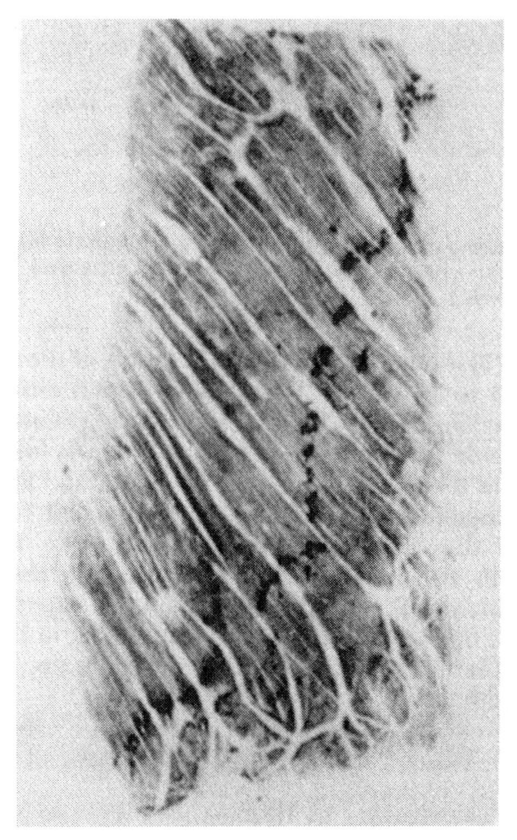

Figure 1.8. Bundle of parallel muscle fibers with endplates (*dark dots*) stained by cholinesterase technique (From Coërs and Woolf, © 1959, Blackwell Scientific Publications, Oxford, and Charles C Thomas, Springfield, IL.)

diaphragm, which suggests that this muscle too has a fine or delicate control. The size of motor units in the rabbit pharyngeal muscles is also quite small, ranging from as few as 2 to a maximum of only 6 (Dutta and Basmajian, 1960). The size of the motor units in our study was determined by tracing the individual nerve fibers along their final distribution to the muscle fibers (Figs. 1.9 and 1.10). Other observers have calculated the total number of muscle fibers in a muscle and the total number of nerve fibers in its motor nerve. Then, by dividing the former by the latter figure, they have calculated the size of the motor units. The latter method is rather questionable because it involves sympathetic fibers as well as motor fibers (Fig. 1.11). Nonetheless, it is a method that does produce reasonable approximations.

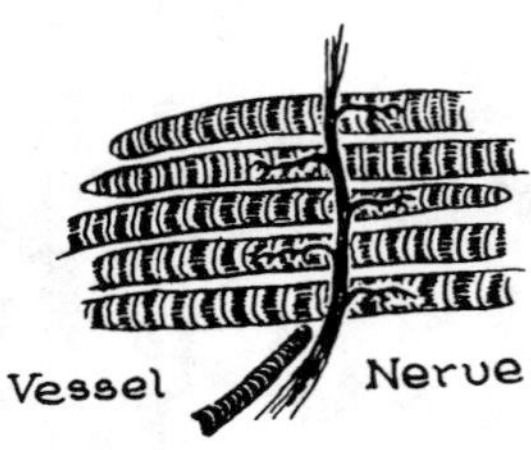

Figure 1.9. Drawing of a nerve bundle ending on muscle fiber-teased specimen (low power, phase contrast microscope). (From Dutta and Basmajian, © 1960, *Anatomical Records*.)

Tergast (1873) estimated that the motor units of the sheep extraocular muscles have 3 to 10 muscle fibers; Bors (1926) estimated 5 to 6 for human extraocular muscles. More particularly, Feinstein et al (1955) reported 9 muscle fibers/motor unit in the human lateral rectus, 25 in platysma, 108 in the first lumbrical of the hand, and 2000 in the medial head of gatrocnemius. Christensen (1959) reported 770 muscle fibers/motor unit in the biceps brachii of infants. Van Harreveld (1947) reported 100 to 125 muscle fibers/motor unit in the sartorius of the rabbit; Berlendis and De Caro (1955), 27 in the stapedius and 30 in the tensor tympani of the rabbit; Wersall (1958), 10 in the human tensor tympani; and Ruedi (1959), 2 to 3 muscle fibers/motor unit in the human laryngeal muscles.

The above innervation ratios represent average values. In fact within a muscle, there exists a hierarchical arrangement of motor unit sizes.

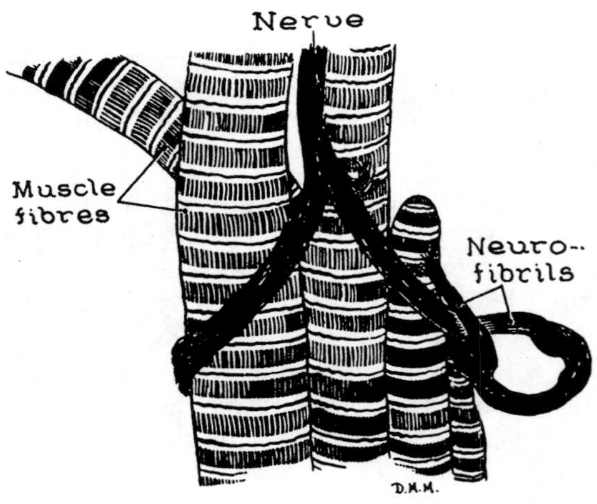

Figure 1.10. Drawing of a photograph of nerve fibers ending on muscle fibers. (Magnification, about ×500.) (From Dutta and Basmajian, © 1960, *Anatomical Records*.)

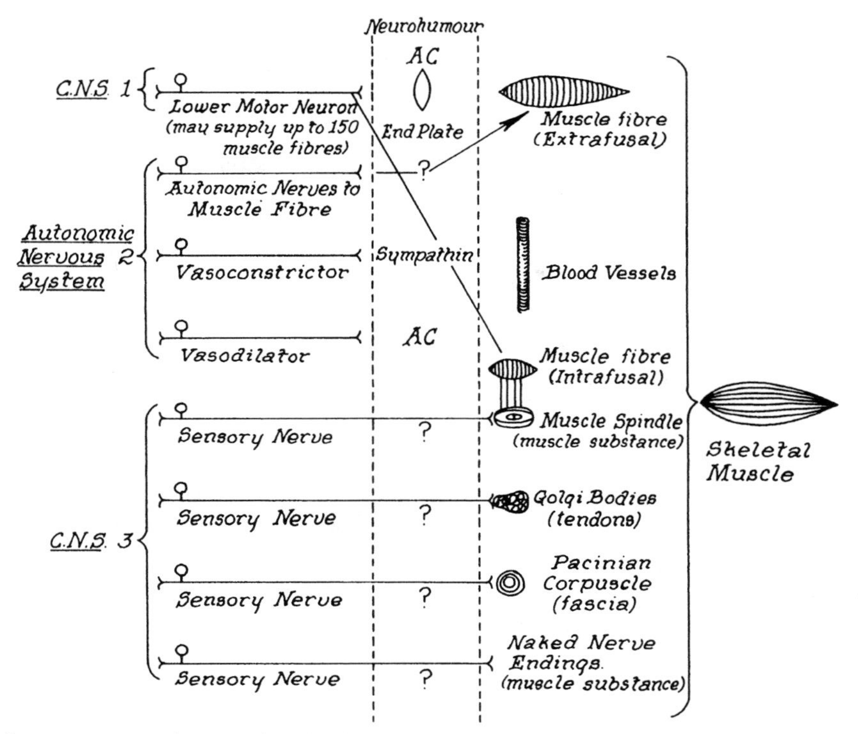

Figure 1.11. Scheme of multiple innervation of skeletal muscle. (After Solandt, from Dutta and Basmajian, © 1960.)

The motor units with a smaller number of muscle fibers are innervated by the smaller alpha motoneuron and are excited earlier during a contraction requiring a progressively increasing force. Larger motor units are innervated by larger alpha motoneurons and become activated at progressively higher force levels.

Van Harreveld (1946, 1947), working with the rabbit's sartorius, concluded that the fibers in a motor unit may be scattered and intermingled with fibers of other units. Thus, the individual muscle bundles one sees in cross-section in routine histological preparations of normal striated muscles rarely, if ever, correspond to individual motor units as such. Norris and Irwin (1961) went further with their conclusion (supported by excellent evidence) that in rat muscle the fibers of a motor unit are widely scattered.

Buchthal et al (1957), using an elegant 12-lead multielectrode technique, finally demonstrated quite conclusively that (in the human biceps brachii) the fibers of each motor unit were localized in an approximately circular region with an average diameter of 5 mm, but in some cases reaching a spread of 20 mm. As a rule of thumb, the motor unit territory may be considered to be approximately one-third of the cross-sectional area of the muscle. Buchthal and his colleagues also showed that fibers of up to 30 different motor units may be located within the territory occupied by any one motor unit.

MUSCLE FIBER TYPE

It is beyond the scope of this book to entertain a detailed discussion on the various classifications of muscle fibers, but some basic knowledge is necessary to appreciate better the structure of muscles.

During the past several decades there has been a growing tendency to establish definable, meaningful categories describing the physiological and biochemical properties of muscle fibers. During the decade of the 1970s, numerous investigations were performed attempting to establish criteria for the distinction of the muscle fiber categories. Although such distinctions are undeniably useful, they may be self-defeating in some cases. For example, overemphasis on the differences among some populations of fiber types tends to make us forget that the physiological and biochemical properties of muscle fibers are a continuum. Depending on which segment of the distribution of the continuum we look at, we see distinctions from other segments. If the segments are sufficiently far apart, the characteristics of the fiber population described therein will be definably different. Most of the work on fiber type categorization has been performed on cats, with some relevant work on guinea pigs and rabbits. In these animals the typing distinction appears to be more clearly apparent than in human muscle fibers.

Muscles of higher-order mammals (including the human) consist of muscle fibers which vary widely in their physiological, morphological, and biochemical properties. Within any one animal, different muscles contain varying amounts of the different fiber types. For related details on 36 human muscles, the reader is referred to the work of Polgar et al (1973) and Johnson et al (1973). However, the muscle fibers belonging to one motor unit show a remarkable homogeneity in their properties.

For several decades muscle fibers have been categorized according to their appearance. Some fibers have a pinkish or reddish visual appearance due to the relatively large amount of blood that is supplied to them by their large vascularization; other fibers appear much paler in coloring, reflecting a less prolific vascularization. These two categories have been referred to as *red* and *white* fibers. Numerous investigations have shown that when the motoneuron of a motor unit consisting of red fibers is stimulated, the resulting force twitch is slower rising and longer lasting than the force twitch which results when a motor unit consisting of white fiber is stimulated. Thus, we have a situation in which red fibers are slow twitch and white fibers are fast twitch.

During the 1960s and 1970s these basic subdivisions of muscle fiber types were subjected to more detailed analysis, and additional subdivisions have been defined and proposed. Engel (1962, 1974) proposed that the fibers be identified as type I and type II. Histochemical tests for key enzymes which correspond closely to the physiological properties of muscle fibers have been performed. The myosin ATPase affinity is an indicator of the fibers' contractile speed (Bárány, 1967). Based on this fact, Brooke and Kaiser (1970) suggested that the muscle fibers be categorized according to the different sensitivities of the myosin ATPase. The type I fibers are acid stable and alkaline labile, whereas the converse is true for the type II fibers. The intensity of staining for specific enzymes of the glycogenolytic and glycolytic metabolic pathways provides an indication of the fibers' capacity to perform work (usually in short bursts) in the absence of oxygen. This is known as the anaerobic capacity. Conversely, oxidative enzymes provide information concerning the capability of the contractile mechanism to use oxygen as its fuel. This is known as aerobic capacity. A high aerobic capacity indicates that a muscle fiber is resistant to fatigue as long as oxygen can be supplied to it via its vascularization. It also follows that a muscle fiber with a high aerobic capacity would have a reddish appearance.

The terminology of Peter et al (1972) is commonly accepted to describe these biochemical-histological properties. Thus, FG stands for fast (high myosin ATPase) glycolytic (high anaerobic and low aerobic capacity); FI stands for fast with high anaerobic and "intermediate" aerobic capacity; FOG stands for fast oxidative (high aerobic capacity) and glycolytic (high anaerobic capacity); SO stands for slow (low myosin ATPase) with oxidative capacity.

An alternative approach for subdividing the categories of muscle fibers has been proposed by Burke et al (1971). They suggested the following classification based on the mechanical response of all the muscle fibers of a motor unit when their motoneuron was stimulated by a single electrical pulse (contractile response) and to a sustained train of stimuli (contractile fatigue response). They expanded the classification of the slow and fast twitch by introducing the evaluation of contractile fatigue, which may be measured by observing the time at which the amplitude of the twitch response declines and/or the rate at which it declines. They denoted FF for fast (contracting), quickly fatigable units; F(Int) for fast, intermediate fatigable units; FR for fast, fatigue-resistant units; and S for slow (contracting) fatigue-resistant units.

It should be noted that these histochemical distinctions may be made much more easily in cat soleus and gastrocnemius muscles than in human muscles. Saltin and Gollnick (1981) argue for the distinction among human muscle fiber types to be made solely on the basis of the myofibrillar ATPase stability to low pH. They point out that the oxidative capacity of fast twitch human muscle fibers is not discrete. Furthermore, the contractile fatigue measures which have been described apply to artificial stimulus train, which provides a similar excitation to all the motor units. Natural, voluntary stimulus trains are not similar for the various motor units involved in generating a muscle contraction. For example, the firing rate of latter recruited motor units is lower than that of earlier recruited motor units. Thus, the proposed classification provides a description of the mosaic of the motor unit fatigability but may not provide a description of the motor unit fatigability during a voluntary contraction.

CHAPTER 2

Apparatus, Detection, and Recording Techniques

In this chapter we will discuss details concerning electrode types and configurations, as well as associated instrumentation that has a bearing on the quality of the EMG signal that is detected and subsequently displayed, recorded, or processed. The process of sensing the signal by the electrode is referred to as *detection*. The word *recording* is reserved for describing the process that creates a record on any media (CRT, paper, magnetic tape, etc.).

Before beginning a substantive discussion on electrodes, it is necessary to ensure some minimal knowledge concerning the concepts of "impedances" and "filter functions."

CONCEPT OF IMPEDANCE AND FILTER FUNCTIONS

All forms of matter present an impedance to the transmission of an electric current. The impedance function is a vector quantity, hence it is expressed in terms of complex numbers, the real part of which denotes the resistance and the imaginary part of which denotes the susceptance. This latter part exists due to the presence of capacitance and/or inductance, two basic electrical properties of matter. In media such as muscle tissues, fatty tissue, and skin, the inductance is essentially unmeasurable. However, the capacitance is present in a significant amount and cannot be overlooked.

One of the simplest expressions of an impedance function, which is useful for conceptualizing the electrical characteristics of electrodes and tissue, is the impedance of a resistance in series with a capacitor presented in Figure 2.1. In this configuration the impedance function is expressed as a vector

$$Z(\omega) = R + \frac{1}{j\omega C}, \quad \text{where } j \text{ is } \sqrt{-1}, \text{ an imaginary quantity}$$

R = the resistance (ohms), C = the capacitance (farads), $\omega = 2\pi f$ and f = the frequency (Hz).

A vector may also be expressed in terms of its magnitude and phase (direction). The magnitude is the square root of the sum of the squared real part and the squared imaginary part.

$$|Z(\omega)| = \frac{(1 + \omega^2 C^2 R^2)^{1/2}}{\omega C}$$

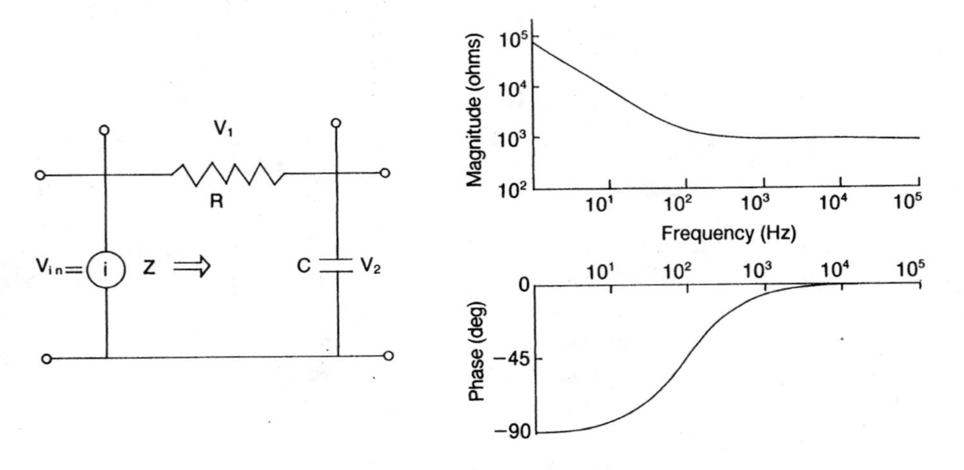

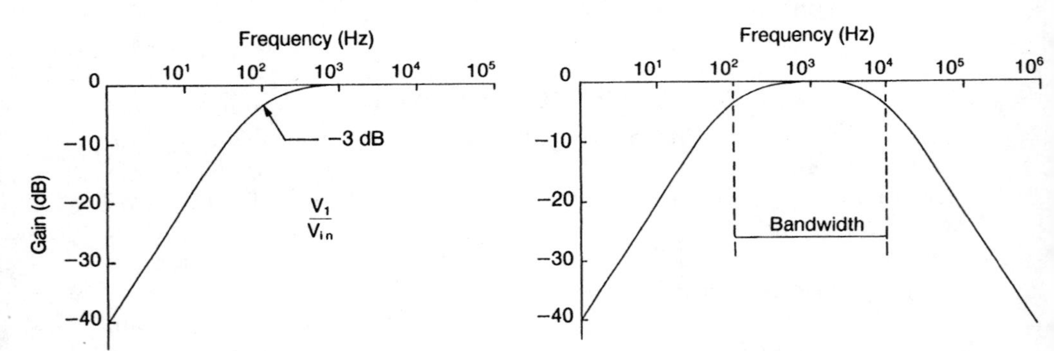

Figure 2.1. (*Top left*) A simple electrical circuit consisting of a resistor and capacitor. (*Top right*) The magnitude and phase of the impedance function of the electrical circuit, with $R = 1k\Omega$ and $C = 1.59\mu f$. Note that when the magnitude decreases by a factor of 0.707, the phase angle is 45°. (*Bottom left*) The filter function of the ratio of V_1/V_{in}. This is a high-pass filter. (*Bottom right*) A band-pass filter. The bandwidth is defined as the range of frequency between the high and low 3 dB points.

The phase is the inverse tangent of the ratio of the imaginary part to the real part.

$$\Phi Z(\omega) = -tan^{-1}\left(\frac{1}{\omega CR}\right)$$

The magnitude is measured in units of ohms and the phase in units of degrees. The above two functions are plotted on the top right-hand quadrant of Figure 2.1, with $R = 1k\Omega$ and $C = 1.59\mu f$. Note that as $f \rightarrow 0$, $|Z(\omega)| \rightarrow \infty$, and as $f \rightarrow \infty$, $|Z(\omega)| \rightarrow R$.

The impedance of a circuit describes the relationship between the voltage and current, $V = ZI$. Thus an alternative way of looking at the frequency-dependent characteristics of the impedance function is as a filter between the current and the voltage. In the specific example provided here, the impedance behaves as a high-pass filter because at higher frequencies the impedance is less. This filtering concept will be useful in describing the filtering properties of the electrode-electrolyte junction in latter portions of this chapter.

A more straightforward description of a filter function is expressed by the ratio of an output voltage to the input voltage. In our example

$$\left| \frac{V_1}{V_{in}} \right| = (1 + 1/\omega^2 R^2 C^2)^{-1/2}$$

$$\left| \frac{V_2}{V_{in}} \right| = (1 + \omega^2 R^2 C^2)^{-1/2}$$

Note that the ratio will be unitless and is therefore measured as a gain. The gain of $|V_1/V_{in}|$ will be greater at higher frequencies; thus it is referred to as a high-pass filter, whereas $|V_2/V_{in}|$ is a low-pass filter. The filter function for the high-pass filter is presented in the bottom left-hand quadrant. Note that in both examples of the two types of filters, when $f = 1/2\pi RC$, the magnitude and gain decrease by $\sqrt{1/2} = 0.707$. It is common to measure the magnitude and gain in the decibel (dB) scale:

$$\text{gain} = 20 \log \left| \frac{V_{out}}{V_{in}} \right|$$

and a gain of 0.707 = 3 dB. The value of the frequency where the gain decreases by 3 dB is referred to by various terms such as the corner frequency, the break frequency, the cutoff frequency, and the 3 *dB* point. When the magnitude function is plotted on a log-log scale, as is the case in Figure 2.1, the slope of the function is 6 dB/octave or 20 dB/decade. An octave represents a doubling of the frequency and a decade indicates an order of magnitude. Note that 20 dB is equivalent to a factor of 10.

The combination of a low-pass and a high-pass filter yields a band-pass filter, as demonstrated in Figure 2.1 (*bottom*). The width of the "frequency-window" of a band-pass filter is measured by its bandwidth. This is the frequency range between the lower and upper 3 dB points. It should now be apparent that the bandwidth may be increased by either increasing the slope of the rolloff or by shifting the 3 dB points. As was the case in the impedance function, at the cutoff frequency, the phase angle distortion is −45°, a relatively large amount. Hence, if the signal processed by such a filter were to contain considerable energy at this frequency, considerable distortion of the signal would occur. This difficulty may be overcome by choosing a cutoff frequency that is 10 times the greatest signal frequency of interest, in which case the filter would pass the signal with negligible amplitude reduction and a maximal phase shift of about 5°.

ELECTRODES

The electrodes used in electromyography could well be, and actually are, of a wide variety of types and construction. Their use depends on the first principle that they must be relatively harmless and must be brought close enough to the muscle under study to pick up the current generated by the ionic movement. The segment of the electrode which makes direct electrical contact with the tissue will be referred to as the *detection surface.* In electromyography these are used either singularly or in pairs. These configurations will be referred to as *monopolar* and *bipolar.*

The two main types of electrodes used for the study of muscle behavior are surface (or skin) electrodes and inserted (wire and needle) electrodes. Each has its advantages and its limitations, and they will now be described.

Surface Electrodes

Surface electrodes can be constructed as either passive or active. In the passive configuration, the electrode consists of a detection surface that senses the current on the skin through its skin-electrode interface. In the active configuration, the input impedance of the electrodes is greatly increased, rendering it less sensitive to the impedance (and therefore quality) of the electrode-skin interface.

One of the earliest, if not the earliest, reported usages of surface electrodes specifically for the purpose of detecting EMG signals from a human muscle was by Piper (1912). He used a metal plate. The design of passive surface electrodes has not changed much since Piper's days; conceptually, the metal electrodes used for this purpose today function similarly.

Often one finds that the simple silver discs used widely in electroencephalography are also used as passive surface electrodes in electromyography (Fig. 2.2). Their advantages revolve around one point: convenience. For example, they are readily obtained from supply houses; they can be applied to the skin after very little training and with reasonable success (within the limitations to be discussed); and they give little discomfort to the subject.

Since a poor contact must be avoided, continued pressure is important. The pressure provided by the adhesive strips or collars used to secure the electrodes is usually adequate. Electrical contact is greatly improved by the use of a saline gel or paste; this is retained between electrode and skin by making the silver disc slightly concave on the aspect to be applied to the skin. The dead surface layer of the skin, along with its protective oils, must be removed to lower the electrical impedance. This is best

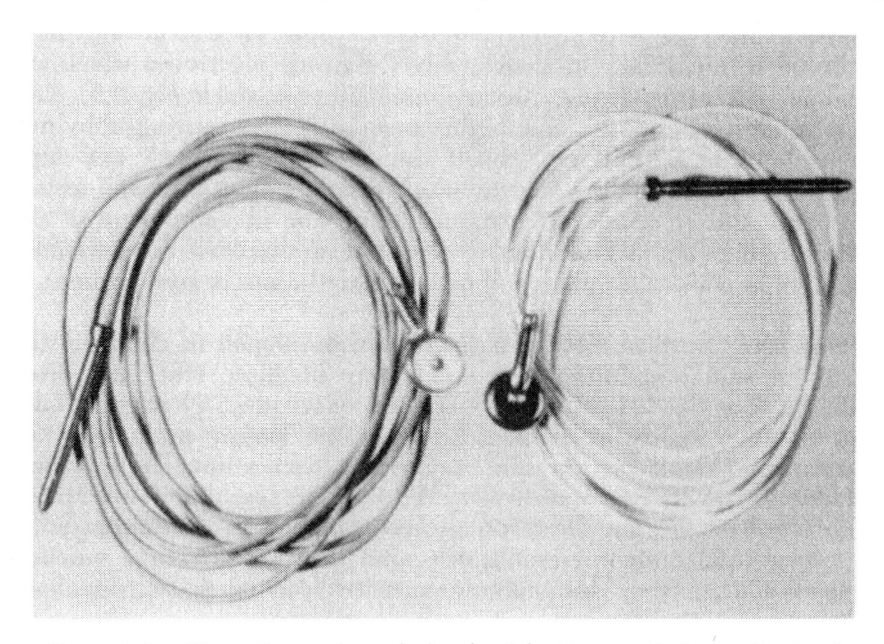

Figure 2.2. Silver disc surface electrodes (electroencephalographic type).

done by light abrasion of the skin at the site chosen for electrode application. In recent years, we have found that it is best produced by "rubbing in" those types of electrode gels that have powdered abrasives included in their formula. For additional details concerning the advantages of skin abrasion, the reader is referred to Tam and Webster (1977) and Burbank and Webster (1978).

In attempting to reduce the mass of the electrode, silver-metal films have been painted on the skin. These electrodes have been described by Rositano (1970). Although they may be convenient for special applications, such as detecting perioral muscle activity (Allen et al, 1972) or for long-lasting recording sessions such as space flights, they generally provide an inferior performance, as compared to that of conventional passive surface electrodes (Konopacki and Cole, 1982).

The lack of chemical equilibrium at the metal-electrolyte junction sets up a polarization potential that may vary with temperature fluctuations, sweat accumulation, changes in electrolyte concentration of the paste or gel, relative movement of the metal and skin, and the amount of current flowing into the electrode. Various construction designs have been implemented attempting to stabilize the polarization potential (Boter et al, 1966; Girton and Kamiya, 1974, among others). It is important to note that the polarization potential has both a dc and an ac component. The ac component is greatly reduced by providing a reversible chloride exchange interface with the metal of the electrode. Such an arrangement is found in the widely used silver-silver chloride electrodes which are commercially available (e.g., Beckman miniature model in Fig. 2.3). This type of electrode has become highly popular in electromyography due to its light mass (250 mg), small size (11-mm diameter), and high reliability and durability. The diminished polarization potential associated with this electrode is a major benefit. The dc component of the polarization potential is nullified by electronic means when the electrodes are used in pairs. This point will be elaborated upon in later sections of this chapter.

The active surface electrodes have been developed to eliminate the need for skin preparation and conducting medium. They are often referred to as "dry" electrodes or "pasteless" electrodes. These electrodes can be either resistively coupled (Lewes, 1965; Bergey et al, 1971) or capacitively coupled to the skin (Lopez and Richardson, 1969; Potter and Menke, 1970; Betts and Brown, 1976). In the case of the capacitively coupled electrode, the detection surface is coated with a thin layer of dielectric (nonconducting) substance, and the skin electrode junction behaves as a capacitor. Although the capacitively coupled electrodes have

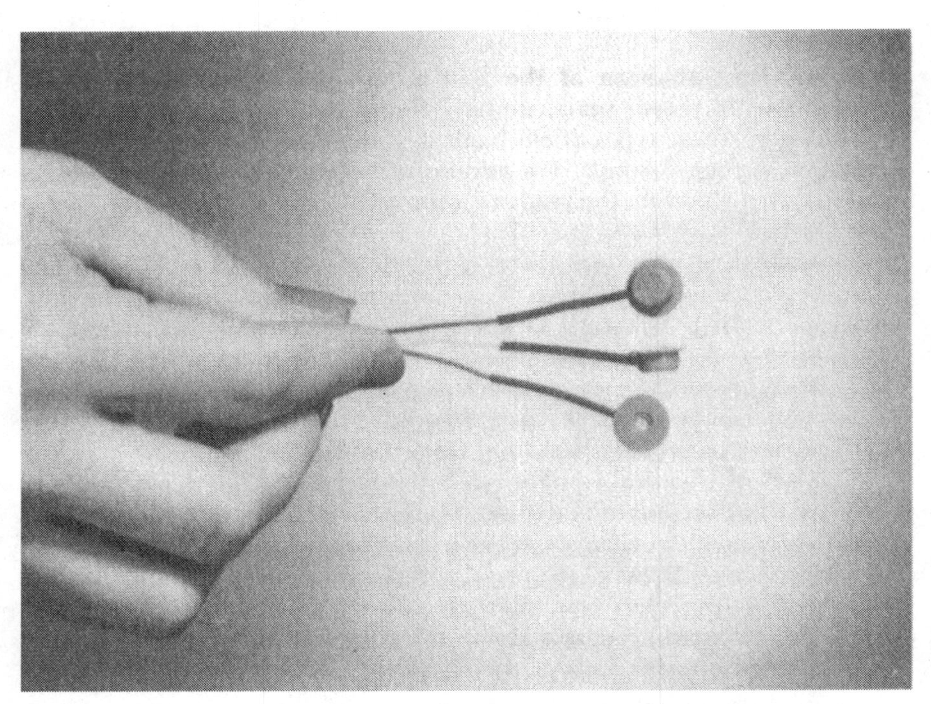

Figure 2.3. Miniature silver-silver chloride electrodes (Beckman type).

the advantage of not requiring a conductive medium, they have a higher inherent noise level (Potter and Menke, 1970). Also, these electrodes do not have long-term reliability because their dielectric properties are susceptible to change with the presence of perspirtion and the erosion of the dielectric substance. For these reasons they have not yet found a place in electromyography.

An adequately large input impedance is achieved when the resistance is in the order of 10^{12} ohm and the capacitance is small (typically, 3 or 4 pF). The advent of JFET microelectronics has made possible the construction of amplifiers housed in integrated circuitry which have the required input impedance and associated necessary characteristics. However, the physical construction of the active electrode remains important because the input capacitance from the metal surfaces to the input of the active circuitry is to be minimized. Two examples of such electrodes are presented in Figure 2.4. These electrodes were conceptualized and designed by De Luca and were constructed in the NeuroMuscular Research Laboratory of the Liberty Mutual Research Center in Hopkinton, MA. They each have two detection surfaces and associated electronics circuitry within their housing; the circular one contains a stainless steel ring around its perimeter which serves as a ground. These electrodes

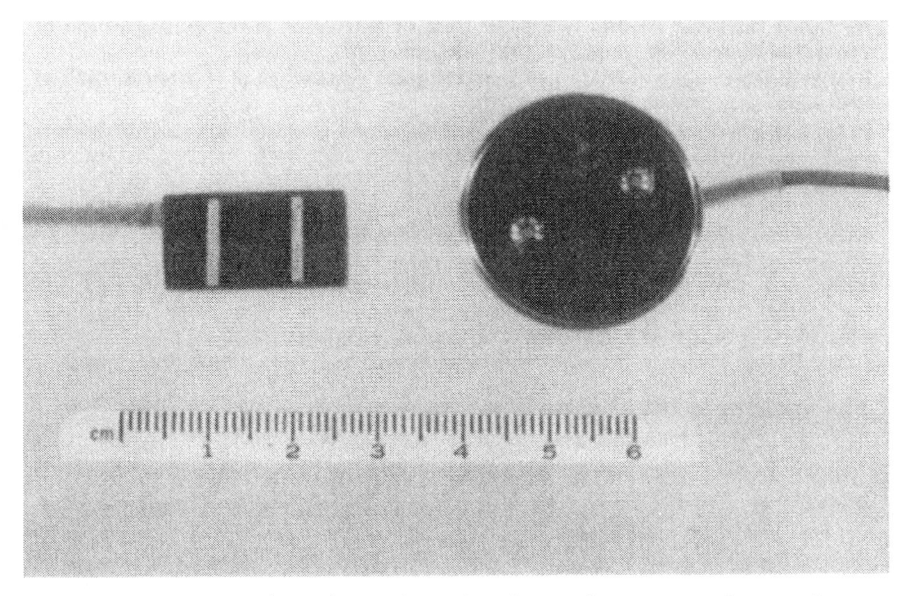

Figure 2.4. Active surface electrodes in bipolar configurations. The circular unit contains a ground ring around the perimeter of the electrode (patent pending). These electrodes do not require any skin preparation or conductive paste or gels. (Conceptualized and designed by C.J. De Luca.)

are an outgrowth of several years of research in the NeuroMuscular Laboratory, and their precursor was reported by De Luca et al (1979). The details of the electronics circuit design considerations will be discussed in a later section in this chapter.

The active surface electrodes are preferable not only because they provide an EMG signal of greater fidelity, but also because they are convenient to use. The simplicity and speed with which they may be applied to the skin is rapidly making them the electrode of choice for pragmatic applications as in busy clinical environments and myoelectrically controlled prosthetics. Within a few years, they will inevitably be the preferred type of surface electrode in the research environment. In our laboratories, active surface electrodes have been in common usage for the past 4 years.

The chief disadvantages of surface electrodes are that they may be used effectively only with superficial muscles and that they cannot be used to detect signals selectively from small muscles. In the latter case, the detection of "cross-talk" signals from other adjacent muscles becomes a concern. These limitations are often outweighed by their advantages in the following circumstances:

1. When representation of the EMG signal corresponding to a substantial part of the muscle is required.
2. In motor behavior studies when the time of activation and the magnitude of the signal contain the required information.
3. In psychophysiological studies of general gross relaxation of tenseness, such as in biofeedback research and therapy.
4. In the detection of EMG signals for the purpose of controlling external devices such as myoelectrically controlled prostheses and other like aids for the handicapped population.
5. In clinical environments where a relatively simple assessment of the muscle involvement is required, e.g., in physical therapy evaluations and sports medicine evaluations.
6. Where the simultaneous activity or interplay of activity is being studied in a fairly large group of muscles under conditions where palpation is impractical, e.g., in the muscles of the lower limb during walking.
7. In studies on children or other individuals who object to needle insertions.

The surface electrodes are commonly used to detect gross EMG signals consisting of the electrical activity from numerous individual motor units within the pickup area of the detection surfaces. However, we have often used such electrodes to detect MUAPTs during low level muscle contractions. Gydikov and Kosarov (1972) have reported similar accomplishments. The reader is cautioned that this detection technique requires practice and may be frustrating.

Needle Electrodes

By far the most common indwelling electrode is the needle electrode. As such, a wide variety of needle electrodes are now commercially available. The most common needle electrode, in turn, is the "concentric" electrode, first described and used by Adrian and Bronk (1929) and now used widely by clinical and research electromyographers. The monopolar configuration contains one insulated wire in the cannula (Fig. 2.5). The tip of the wire is bared and acts as a detection surface. The bipolar configuration contains a second wire in the cannula and provides a second detection surface. The needle electrode has two main advantages. One is that its relatively small pickup area enables the electrode to detect individual motor unit action potentials conveniently, especially during relatively low-force contractions. The other is that they may be conveniently repositioned within the muscle (after insertion) so that new territories may be explored or the signal quality may be improved. These amenities have naturally led to the development of various specialized versions to study particular properties or aspects of the motor unit.

Buchthal and his colleagues used a multifilament electrode with 12 individually insulated wires located longitudinally along the length of the cannula. With this electrode, they were able to ascertain the size of the territory of the motor unit (Buchthal et al, 1957a) and measure the voltage decrease as a function of distance in muscle tissues (Buchthal et al, 1957b).

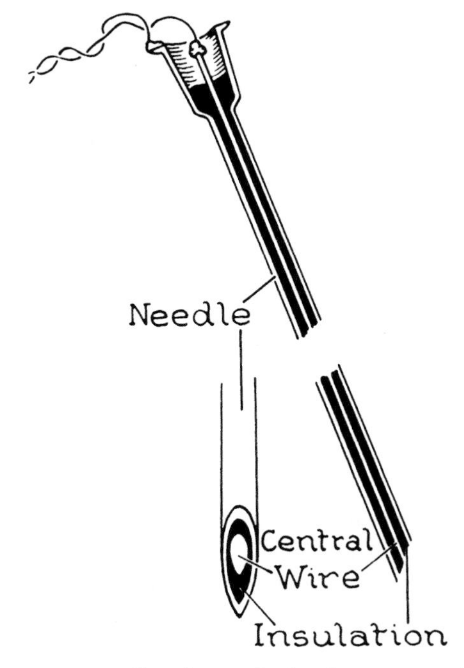

Figure 2.5. Concentric needle electrode in the monopolar configuration. A bipolar configuration may be realized by locating two insulated wires within the cannula of the needle.

In 1972, De Luca and Forrest described the design of a four-filament needle electrode for simultaneously detecting more than one electrical representation of the activity of the motor units within the pickup area of the electrode. This electrode contained four detection surfaces consisting of the cross-sectional area of the 25-μm diameter wires located at the corner of a square in the tip of the cannula. With this electrode, De Luca and Forrest demonstrated that multichannel detection of EMG signals is essential to eliminate ambiguity in the identification of individual MUAPs. Recently, in collaboration with our colleagues Broman and Mambrito, we produced an improved version of this electrode. This latter version, along with the preferred detection configuration, may be seen in Figure 2.6. This version houses somewhat larger (75-μm diameter) detection surfaces on the side of the cannula. This particular architecture of the electrode has proven to provide the necessary pickup selectivity while still detecting several MUAPTs.

Another productive exploit of the architecture of the needle electrode was originally reported by Ekstedt and Stålberg (1973) and has been subsequently popularized by Stålberg and his colleagues in Uppsala. This type of electrode consists of a modified monopolar needle electrode whose central wire is relatively small in diameter (25 μm) and is exposed on the side of the cannula. By detecting from this wire with respect to a

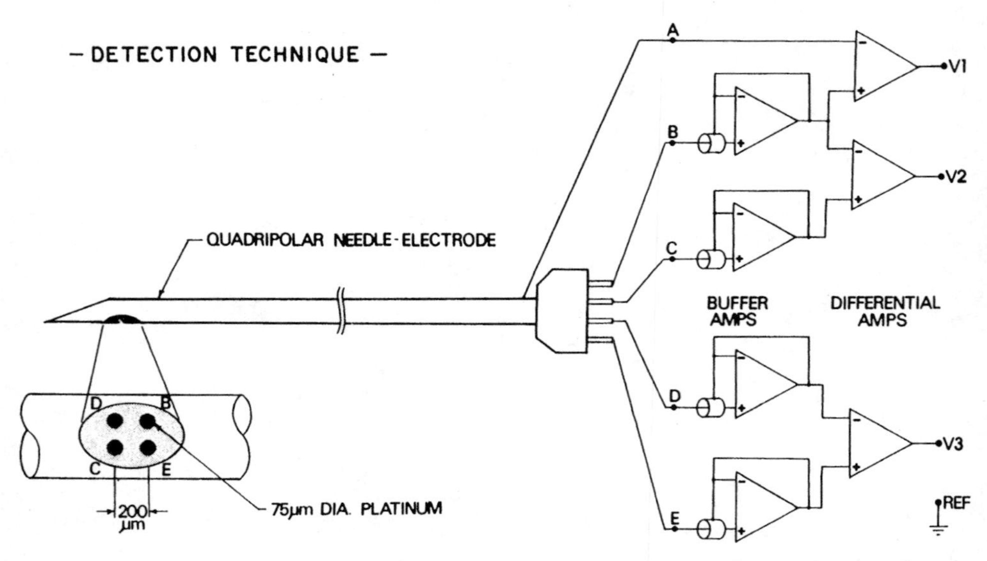

Figure 2.6. A schematic representation of a lightweight quadripolar needle electrode configured to detect three independent channels of EMG signals. Any two of the detection surfaces (cross-sectional areas of the wires) may be used, as a bipolar pair.

distant electrode it is possible to obtain the signal activity of only one or two muscle fibers of a motor unit in most muscles. This type of electrode is presented in Figure 2.7.

In 1980, Stålberg modified the single fiber electrode and constructed a "macro" electrode which detects the MUAP from many fibers of the motor unit. This feat is accomplished by insulating the cannula of the needle to within 15 mm of its tip and using the exposed part of the cannula as the main electrode contact (refer to Fig. 2.7). This uncommonly large detection surface will obviously detect MUAPs from numerous motor units whose fibers are scattered throughout the 15-mm length of the cannula. A chosen MUAP is recovered by locating it with the single fiber electrode (on the same needle) and using this MUAP to trigger average the signal detected by the exposed part of the cannula. The process of trigger averaging consists of using a clearly detectable signal to denote the time occurrence of a related event in a noisier signal. This is accomplished by repetitively averaging a segment of the noisy signal, beginning at a time denoted by the clearly detectable signal or the "trigger" signal. If the noise is random in nature it will cancel out, while the signal of the related event will be enhanced.

Some clinicians prefer the monopolar type of electrode, which is an outgrowth of the electrode introduced by Jasper and Ballem (1949).

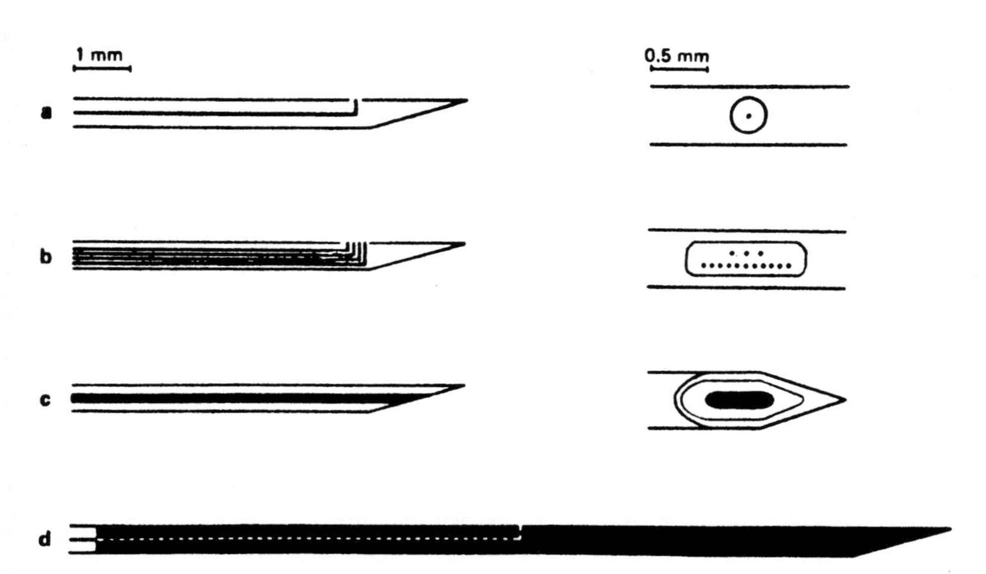

Figure 2.7. Examples of various needle electrodes. (*a*) Single fiber electrode with one detection surface. (*b*) Multipolar electrode. (*c*) Concentric needle electrode. (*d*) Macroneedle electrode. (*Black area* is noninsulated.) (From E. Stålberg, © 1980, *Journal of Neurology, Neurosurgery, and Psychiatry*.)

Unlike the concentric monopolar electrode, this electrode consists of a metallic needle that is insulated throughout, except for the tip, which is the detection surface. This concept has now reached a refinement that has proven useful for recording directly from nerve fibers. This microneurography detection was originated by Vallbo et al (1979). The electrode consists of an insulated tungsten filament approximately 150 to 200 μm in diameter and exposed for 5 to 10 μm at the tip. This tungsten filament is inserted as a monopolar electrode in the nerve, and the neuroelectric signal is detected with respect to a reference electrode located on the surface of the skin.

Wire Electrodes

Since 1961 this type of electrode has been popularized by Basmajian and his associates (see Basmajian and Stecko, 1962). Similar electrodes that differ only in the details of their construction have been independently developed by a number of other researchers, notably Close et al. (1960) and Long and Brown (1962). Wire electrodes have proven a boon to kinesiological studies because they are extremely fine, and therefore painless, and are easily implanted and withdrawn.

Wire electrodes may be made from any small diameter, highly non-oxidizing, stiff wire with insulation. Metals such as platinum alloys, silver, and nickel-chromium alloys are preferable. Insulations such as nylon, polyurethane, and Teflon are conveniently available. Such wires are available under the trade names of Karma (Wilber B. Drive Co., Harrison, NJ), Stablohm (Johnson Matthey Metals Ltd., London, England, and California Fine Wire Co., Grover City, CA), and various other manufacturers. Wire diameters in various sizes are available, the smallest being 25 μm. At least one distributor, A-M Systems, Inc. (Toledo, Ohio), has multistrand fine wire available. Of the available metals and alloys which are used to draw the fine wires, we prefer the 90% platinum-10% iridium alloy. It offers the appropriate combination of chemical inertness, mechanical strength, and stiffness. The Teflon or nylon insulators are preferred because they add some mechanical rigidity to the wires, making them easier to handle. Also, these insulators are less apt to crack, thus exposing the wires in unplanned locations and providing additional detection surfaces that may compromise the detection strategy.

The steps for making an electrode are displayed in Figure 2.8 and may be described as follows: (1) A double strand of insulated fine wire is passed through the cannula of a hypodermic needle. A small loop is left distally, and 5 to 7 cm of wire are left proximally. (2) A small amount of the insulation at the distal tip and 3 to 4 cm of each strand proximally is removed, either by burning it off with an alcohol flame or (depending on the type of insulation) etching it off with chemical solvents. (3) The loop is cut, leaving 1 to 2 mm of bared wire distally on each strand. These bare ends are staggered so that they will not come in contact.

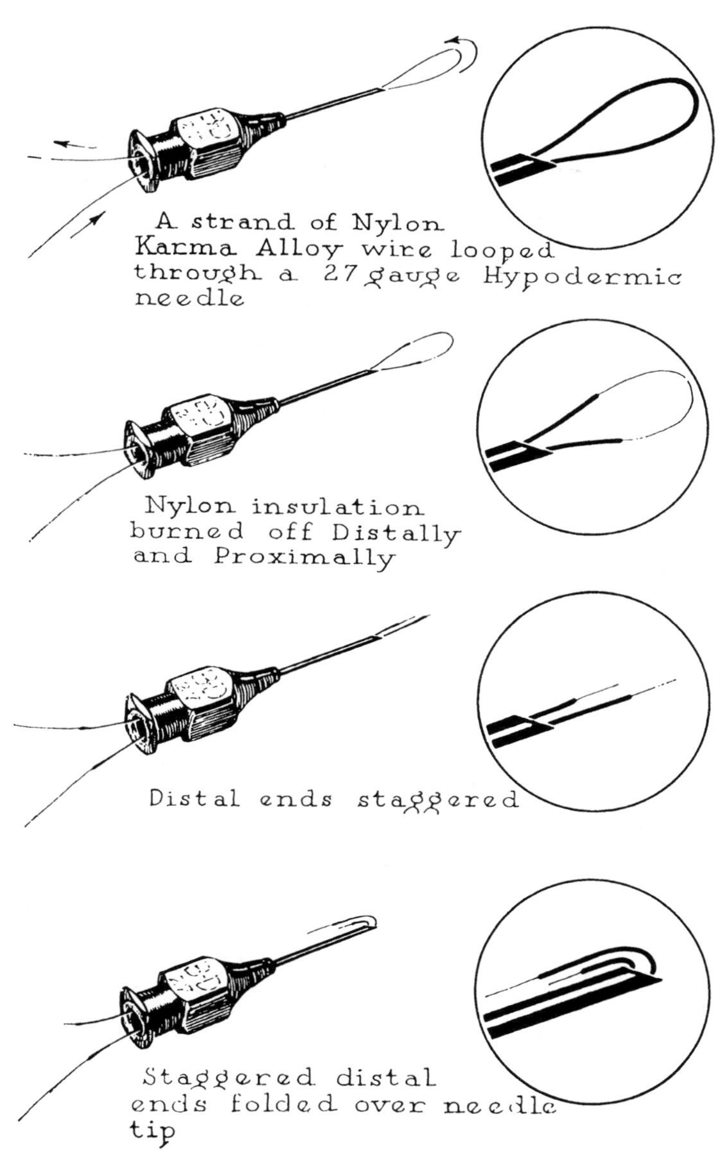

Figure 2.8. Steps in making a bipolar wire electrode with its carrier needle used for insertion. (From Basmajian and Stecko, © 1962, © *Journal of Applied Physiology*.)

They are then bent sharply back to lie against the needle shaft for a short distance. If preferred, the wires may be twisted together.

Such electrode assemblies may be driven easily into a muscle without anesthesia, and the attendant pain is the usual pain resulting from the needle puncture. If fresh, sharp, 27-gauge needles are used, the pain is minimal and transitory. The needle withdraws easily, and its removal only rarely dislodges the electrodes, for they are retained by the hooks at their ends. The electrodes are taped to the skin at the site of emergence to ensure that an accidental tug does not remove them. At the end of an experiment a gentle pull brings the electrodes out painlessly, for each wire is so pliable that the barb straightens out on traction and offers little if any palpable resistance. We have had no accidental breakage in many thousands of uses; nor would we be disturbed if we had, because the fine wire is innocuous. Jonsson and Bagge (1968) have described the various deformations and dislocations of fine-wire electrodes and report that very vigorous exercise may break 25-μm wires. Fortunately, the wire is available in 50- and 75-μm sizes also, and these have been shown to be much tougher.

If one wishes to insert individual monopolar wire electrodes, the modification of our technique by Scott (1965) is satisfactory. The hypodermic needle is used to insert the electrode and also acts as a cutting instrument once the electrode is deep in place. A single strand of fine wire is passed through the hypodermic needle, and a long loop is turned back from the needle tip. The wire is cut by pulling on its free ends, one of which emerges from the skin alongside the needle and one out of the needle. Both the needle and its contained wire are then withdrawn, leaving one unbarbed wire deep in place.

After a period of use, we found only one tedious step or complication with wire electrodes, i.e., the connection of the almost invisible filament to the input of the amplifiers. Others have met and overcome this problem in different ways. For example, Long and his colleagues at Highland View Hospital in Cleveland relied on preconnection of the fine-wires to standard wires to produce a unit. However, when their needles are withdrawn, they cannot be discarded because they cannot be drawn over the connections. Because we prefer disposable needles and electrodes that are used only once and discarded, we make connections after the needle is entirely removed. Making the connections can be a tedious procedure when it must be done a dozen times for one experimental setup. Therefore, being lazy, we developed a simple method.

After finding that soldering, microwelding, and miniature alligator clamps all have drawbacks, we devised a spring-wire coil connector (Basmajian et al, 1966) which has proved itself in 20 years of extended usage. It is a brass spring (about 4 mm in diameter by 12 mm in length) soldered permanently to the free ends of each amplifier lead-in wire. The spring is tightly wound from a resilient 22 "spring-brass" wire which gives considerable pinch between adjacent coils. This type of hard-brass wire is available through ordinary commercial channels. To make connection to the wire electrodes after they have been inserted and the needle has been discarded, the spring is bent slightly between the thumb and index finger. This spreads the coils and allows the bared end of the electrode wire to be slipped between one or two pairs of coils. Released, the spring clamps the fine wire and gives good electrical connection instantly. Wrapping a bit of adhesive tape around the connection (for protection and insulation) completes a procedure that takes only moments and saves many tedious minutes required by other methods.

The wide acceptance of wire electrodes has led to a variety of innovations exploiting the versatility of this technique. Leifer (1969) used a four-filament electrode ensemble consisting of 25-μm stainless steel wires cemented together with epoxy. The four wires were exposed at their tips and were arranged in a straight line. Hennerz (1974) used a three-filament (100-μm silver wire) arrangement. In the same year, Shiavi (1974) decribed the construction of a wire electrode consisting of three staggered pairs of wires which provided three independent channels of EMG signals. This wire cluster was approximately 250 μm in diameter and was inserted into the muscle by the standard method. All these multifilament fine wire arrangements have, as their main goal, the detection of EMG signals which present a clearer view of the individual MUAPTs. It has been our experience that with the proper combination of wire diameter, staggering arrangement of the wire, and interwire spacing, progress towards the desired intention may be made. However, some trial and error is involved, and the beginner should not easily abandon the task.

Two other innovative users of wire electrodes have been proposed. Caldwell and Reswick (1975) described an electrode that could be inserted in a muscle and would remain within the muscle for months. The electrodes consisted of fine wires wound into a coil filled with silicone rubber. These coils, when inserted via a hypodermic needle, anchored themselves to the muscle fibers and thus resisted dislodgement.

A truly innovative concept has been proposed by Andreassen and A. Rosenfalck (1978). They described a "side-hole" electrode which consists of two 75-μm insulated fine wires which are twisted. Unlike all previous cases, the detection surfaces were two holes (10 to 25 μm) burned into the side of the wires. The electrode was inserted by a curved cannula which entered the skin, passed through the muscle, and exited through the skin. In this fashion, the location of the detection surfaces could be positioned throughout the muscle by pulling on either end of the wires. This approach overcomes the main disadvantages of wire electrodes; the inability to be repositioned after insertion. This interesting technique should be used with caution and, advisably, in an aseptic environment. Continual and excessive movement of the protruding wires may provide a bacterial conduit.

In kinesiological studies in which the main purpose of using wire electrodes is to record a signal that is proportional to the contraction level of muscle, repositioning of the electrode is not important. But, for other applications, such as recording distinguishable MUAPTs, this limitation is damaging. Some of us have used the term "poke-and-hope" to describe the standard wire electrode technique for this particular application. Another limitation of the wire electrode is its tendency to migrate after it has been inserted, especially during the first few contractions of the muscle. In fact, we suggest that the muscle with the electrode be contracted and relaxed at least one-half dozen times before any measurements are taken. The pumping action of the muscle will tend to lodge the barbs at the end of the electrode into the tissue. It will also draw some wire subcutaneously, providing excess wire to be released if the external tension on the wire increases. The reproducibility of the EMG signal detected with wire electrodes discussed by Jonsson and Reichmann (1968), Komi and Burskirk (1970) and, more recently, by Gans and Gorniak (1980) applies to parameters of the total EMG signal (such as amplitude), not to the more sensitive parameters of the individual MUAPs.

ELECTRODE TREATMENT

Cleaning

It is good practice to clean needle electrodes with a swab soaked with 70% ethanol and 30% distilled water solution after every use of the electrode. This procedure will remove the debris (such as skin particles, coagulated blood, muscle tissue, etc.) which tends to accumulate near the detection surfaces. Alternatively, the needle electrodes may be cleaned by placing them in an ultrasonic vibrator, of the type that is used to clean small pieces of jewelry. These vibrators may be purchased as small receptacles which contain a cleansing fluid and conveniently hold several needles simultaneously. We have been very pleased with our experience with such devices. Regardless of the technique used, it is imperative to remove the attached debris from the tip of the needles. Their accumulation will change the impedance properties of the electrode, which will result in a deterioration of the quality of the detected signal.

Surface electrodes should also be cleaned after each application. If the electrodes are of the type which require a conductive paste or gel, then any residual paste or gel should be removed by wiping them with gauze dampened in distilled water before the conductive material hardens. Once the conductive material hardens (usually within 30 minutes of exposure to the air), it may still be removed by more vigorous wiping action with gauze dampened with distilled water. The use of solvents or cleansing agents is greatly discouraged. If the surface electrode is of the type which does not require conductive paste or gel, it is recommended that the metallic contacts of the electrode be regularly cleaned in order to remove any oxide layer which may accumulate on the detection surface. This should be done by swabbing them with 70% ethanol-30% distilled water mixture. This mixture will evaporate without leaving a film.

Impedance Reduction

Needle electrodes, especially those that have detection surfaces with small surface areas, tend to have a high impedance due to the high resistance of the exposed area and low capacitance of the metal-electrolyte interface. The impedance of the electrode may be effectively reduced by an electrolytic treatment originally described by Buchthal et al (1957a). A modified procedure was described by De Luca and Forrest (1972). This procedure consists of placing the needle electrode in a receptacle containing a 1 N saline solution (0.9% NaCl) and a platinum metal strip. The terminals of the electrode are connected to the negative terminals of a variable power supply, and the platinum strip is connected to the positive terminal. A dc current of 1 mA is passed through the needle electrode until small bubbles are seen forming and effervescing from the detection surfaces of the electrodes. Continue to supply the current until the bubbles rise to the surface in a continuous stream. If the bubbles do not form, increase the current gradually until the bubbles are in evidence. The reader is cautioned that an excessive amount of current will damage the detection surfaces of the electrode.

This electroplating procedure deposits a rough layer of salts on the metal of the contact area and thus increases the surface area. This results in a decreased impedance. The procedure also neutralizes some of the corrosive processes which degrade the metallic surfaces as a function of time, and, in the case of bipolar electrodes, it will tend to balance the impedances of the two detection surfaces. The usefulness of this latter development will become apparent in subsequent sections of this chapter. This approach of balancing input impedances of the detection surfaces has also been applied after the electrode has been inserted. Basmajian (1973) described such usage for the purpose of eliminating occasional high frequency artifacts from indwelling wire electrodes.

Sterilization

Prior to piercing any electrode through the skin and into a muscle, it is manditory to sterilize it. This may be accomplished with dry heat, boiling water, or steam. Of these three approaches, autoclaving at 15 lb/inch or 10 newtons/m^2 pressure for 30 minutes is preferred. If this method is not available in your environment, or if it proves to be inconvenient, then we suggest dry heat. In this approach, simply securely wrap the electrode(s) in a paper folder and place the package in an oven for 60 minutes at a temperature of 130°C. Both these approaches are appropriate for needle and wire electrodes. For needle electrodes, it is sometimes convenient to sterilize them in boiling water that is under more than 1 atmosphere of pressure.

Antibacterial chemical baths are not recommended for sterilizing electrodes.

During sterilization, caution should be exerted to ensure that the temperature does not damage the insulation of the wires and the adhesive used to bond the wires to the needles. The commonly used materials for this purpose may be continuously exposed to the following temperatures without mechanical damage: Teflon, 150 to 200°C; nylon, 80 to 120°C; and expoxy, ~ 175°C. Continual exposure to higher temperature may structurally damage needle and wire electrodes.

HOW TO CHOOSE THE PROPER ELECTRODE

The specifics of the type of electrode that is chosen to detect the EMG signal depend on the particular application and the convenience of use. The application refers to the information that is expected to be obtained from the signal, for example, obtaining individual MUAPs or the gross EMG signal reflecting the activity of many muscle fibers. The convenience aspect refers to the time and effort that the investigator wishes to devote to the disposition of the subject or the patient. Children, for example, are commonly resistant to having needles inserted in their muscles.

We recommend the following electrode usage. However, the reader should keep in mind that crossover applications are always possible for specific circumstances.

- Surface electrodes
  - Time-force relationship of EMG signals
  - Kinesiological studies of surface muscles
  - Neurophysiological studies of surface muscles
  - Psychophysiological studies
  - Interfacing an individual with external electromechanical devices
- Needle electrode
  - MUAP characteristics
  - Control properties of motor units (firing rate, recruitment, etc.)
  - Exploratory clinical electromyography
- Wire electrodes
  - Kinesiological studies of deep muscles
  - Neurophysiological studies of deep muscles
  - Limited studies on motor unit properties
  - Comfortable recording procedure from deep muscles

Each of these suggested categories often has specific applications that require special properties of the electrode. We will now discuss some of these properties. They are: electrode configuration; voltage decrement function of muscle tissue; electrode pickup area and cross-talk electrical properties of electrode-electrolyte interface; and filtering properties of bipolar configuration. The novice in electronics should review the section on impedance and filter functions at the beginning of this chapter prior to reading the following material.

Electrode Configuration

The electrical activity inside a muscle or on the surface of the skin outside a muscle may be easily acquired by placing an electrode with only one detection surface in either environment and detecting the electrical potential at this point with respect to a "reference" electrode located in an environment which is either electrically quiet or contains electrical signals which are unrelated to those being detected. (By unrelated, it is meant that the two signals have minimal physiological and anatomical association.) A surface electrode is commonly used as the reference electrode. Such an arrangement is called monopolar and is at times used in clinical environments because of historical precedents and its relative technical simplicity. In the early days of electromyography (circa 1940) electronics amplifiers were considerably inferior to and much more limited than today's versions. Schematic arrangement of the monopolar detection configuration may be seen in Figure 2.9.

The monopolar configuration has the drawback that it will detect all the electrical signals in the vicinity of the detection surface; that includes unwanted electrical signals from sources other than the muscle being investigated.

The bipolar detection configuration overcomes this limitation. This configuration is also displayed in Figure 2.9. In this case, two detection surfaces are used to detect two potentials in the muscle tissue of interest, each with respect to the reference electrode. The two signals are then fed to a differential amplifier which amplifies the difference of the two signals, thus eliminating any "common mode" components in the two signals. Signals emanating from the muscle tissue of interest near the detection surface will be dissimilar at each detection surface due to the localized electrochemical events occurring in contracting muscle fibers. Whereas "ac noise" signals originating from a more distant source (such as 50- or 60-Hz electromagnetic signals radiating from power cords, outlets, and electrical devices) as well as "dc noise" signals (such as polarization potentials in the metal-electrolyte junction) will be detected with an essentially similar amplitude at both detection surfaces and, therefore, will be subtracted prior to being amplified. This idealized behavior of the differential amplifier cannot be achieved with present day electronics. The measure of the ability of the differential amplifier to eliminate the common mode signal is called the common mode rejection ratio.

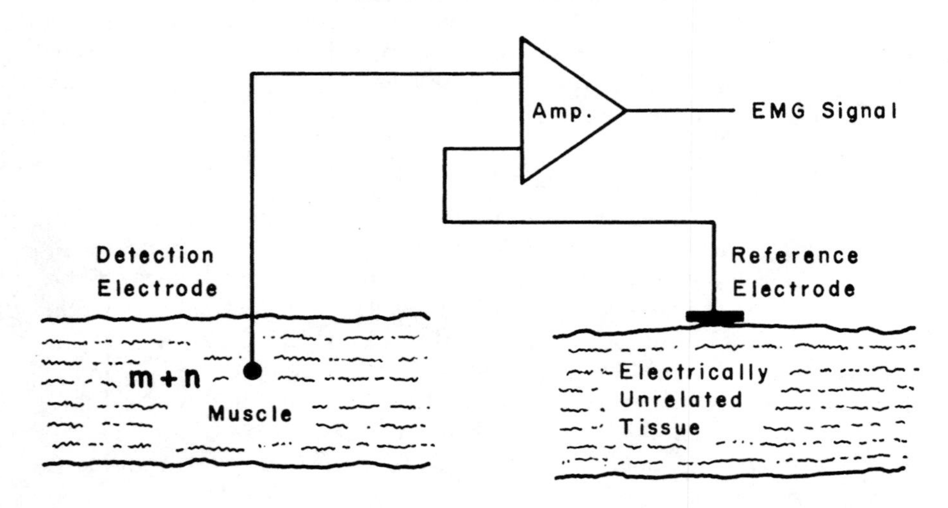

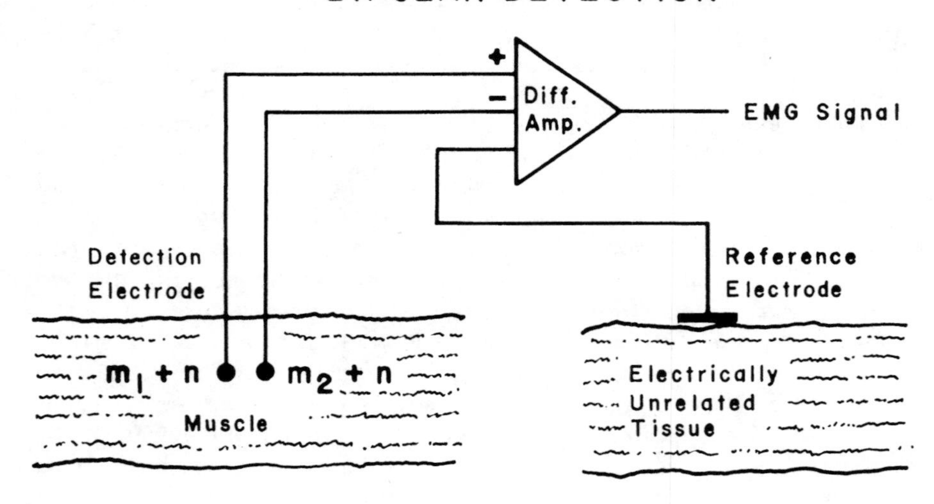

Figure 2.9. (*Top*) Monopolar detection arrangement. (*Bottom*) Bipolar detection arrangement. Note that in the bipolar detection arrangement, the EMG signals are considered to be different, whereas the noise is similar.

Decrement Function of Muscle Tissue

Muscle tissue presents an internal impedance to the propagation of electric currents. The impedance is frequency dependent; it is less for lower frequencies than for higher frequencies. It will also be a function of the distance between the sources of the EMG signal and the detection surfaces of the electrode. Thus, the muscle and adjacent tissues may be considered as a distance-dependent filter. Lindström (1970), through elaborate mathematical modeling, was able to calculate the tissue filter functions. His results, which are a simplified representation of the real environment, are nonetheless very helpful in providing guidance and insight in the behavior of tissue properties. They are presented in Figure 2.10. These curves represent the tissue filter properties as a function of distance perpendicular to the muscle fiber.

In reality, the impedance of muscle tissue is not isotropic, i.e., similar in all directions. In fact, it is highly direction dependent, i.e., anisotropic. The anisotropy is due to the nonhomogeneity of the anatomical construction of a muscle; muscle fibers are normally arranged lengthwise, and the surrounding extracellular fluid forms lengthwise channels parallel to the muscle fibers. These "channels" of lower impedance branching throughout the muscle make it very difficult to define precisely the current distribution within a muscle. In fact, the situation is considerably aggravated when the signal propagates through the fatty tissue and the skin to reach the surface of the skin, where it may be detected by surface electrodes. The considerably different electrical properties of the muscle tissue, fatty tissue, and skin cause inflections in the current field.

This anisotropic property of the muscle tissue impedance has been known since the earliest attempts were made to measure it. Hermann (1871), using the crude instruments available in the last century, found that the magnitude of the impedance in a direction perpendicular to the muscle fibers was considerably greater than that in a direction parallel to the muscle fibers. Recently, Epstein and Foster (1983) have reported detailed measurements which indicate that the magnitude of the impedance in the perpendicular direction is 7 to 10 times greater than that in the longitudinal direction. These results are consistent with other reported measurements of a similar nature.

Referring to Figure 2.10, it can be seen that at higher frequencies, the signal amplitude will decline sharply near the surface of the muscle fiber ($D = 0$) and then gradually diminish. This measure is known as the "decrement function." This function is typically obtained by plotting the peak-to-peak amplitude of a muscle fiber action potential observed as the detecting electrode is moved away from the active muscle fiber along a perpendicular direction. (Note that the peaks of the action potential contain high frequency components. It is for this reason that the high frequency region of Figure 2.10 is used to provide a comparison in the

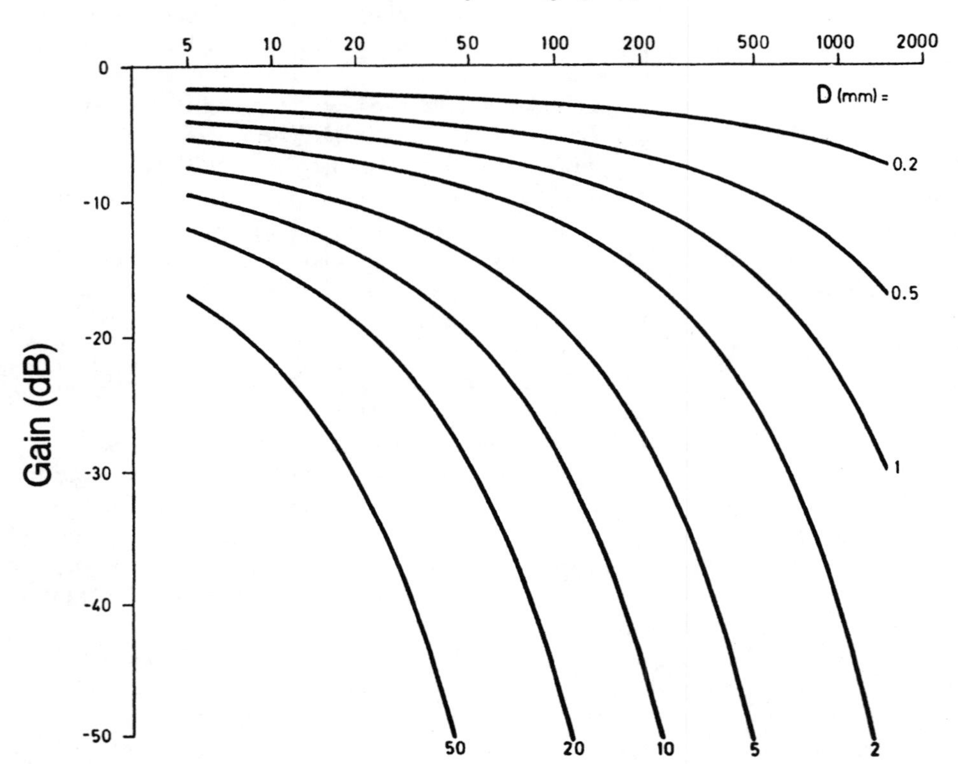

Figure 2.10. Representation of the tissue filter function. The parameter (*D*) indicates the distance from an active fiber to the detection electrode. These curves were obtained by designating the conduction velocity along the muscle fiber to be 4 m/s and the diameter of the muscle fiber to be 100 μm. (From L. R. Lindström, © 1970, *Technical Report, Chalmers University of Technology, Sweden.*)

frequency domain for the peak-to-peak decrease in the time domain.) Figure 2.11 which is taken from the theoretical work of Andreassen and A. Rosenfalck (1978), clearly demonstrates the dramatic decrease in the peak-to-peak amplitude for different electrode configurations and orientations. Gath and Stålberg (1978) reported empirical results supporting the theory.

The information in Figures 2.10 and 2.11 indicates that small displacements of the electrode with respect to the active fibers, when the electrode is near the surface of the active fiber, cause drastic changes in the waveform of the detected signal. If the electrode is moved 100 μm from the surface of a fiber, the peak-to-peak amplitude decreases by approximately 75%. It is this sharp radial decline which accounts for the sometimes drastic modifications of MUAP waveforms during muscle contractions. Even attempted constant-force isometric contractions may provide sufficient relative movement between the electrode and active muscle fibers to seriously disturb the amplitude and shape of the signal. At times this disturbance may be sufficient to render impossible the identification of MUAPs belonging to the same MUAPT.

Figure 2.11 also demonstrates that *rotation* of the detection surfaces of a needle electrode will greatly modify the amplitude of the signal, especially if they are near the active muscle fiber. Empirical observations of this behavior have been reported by Andreassen and A. Rosenfalck (1978) for indwelling electrodes and by Vigreux et al (1979) for surface electrodes. This aspect of the detecting techniques provides both advantages and disadvantages. The wise investigator will exploit this property

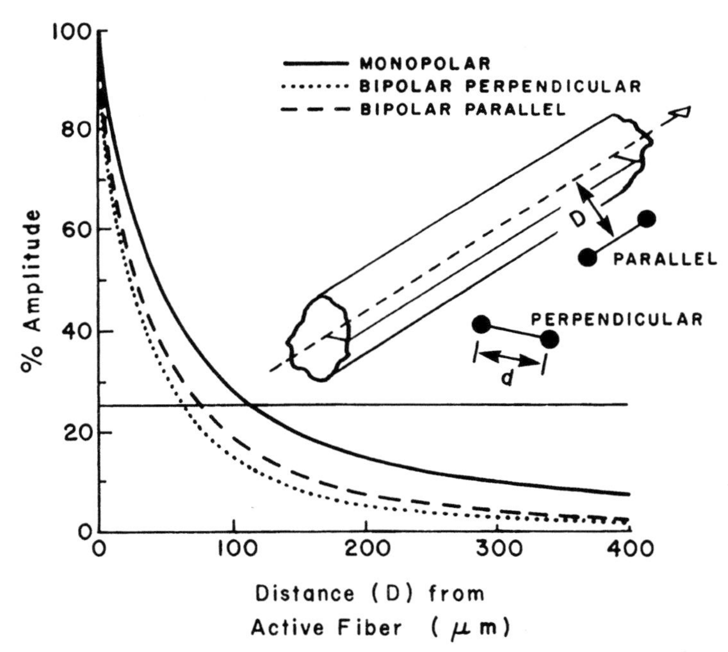

Figure 2.11. Decline of the peak-to-peak amplitude of the muscle fiber action potential for a monopolar and bipolar electrode placed perpendicularly or parallel to the direction of the muscle fibers. The *insert* describes the orientations. The *ordinate* represents the amplitude in terms of percentage of the amplitude at distance 0 from the fiber surface. The *abscissa* presents the distance from the fiber surface to the nearest detection surface. The distance at which the amplitude declines to 25% is expressed for the three electrode arrangements by the *horizontal line*: monopolar, 116 μm; bipolar perpendicular, 63 μm; bipolar parallel, 76 μm. (From S. Andreassen and A. Rosenfalck, © 1978, *IEEE Transactions in Biomedical Engineering.*)

to improve the quality of the detected signal. It also follows that caution must be taken to prevent the needle electrode from unwanted rotations during a contraction, including attempted isometric contractions.

Electrode Selectivity or Pick-Up Area and Cross-Talk

When an electrical current propagates in a volume conductor, it is theoretically possible to detect a potential at any location throughout the medium. But, as is evidenced in Figure 2.11, the voltage gradient decreases quickly. Therefore, if an electrode is placed more than 2 or 3 mm from the surface of an active muscle fiber the detected signal will have a very low amplitude, possibly lower than that of the extraneous unwanted signals, and thus will provide no useful information. It is therefore necessary to establish an arbitrary demarcation value that will define the pickup area. Several approaches have been proposed. Pollak (1971) suggested that the demarcation point be where the peak-to-peak amplitude of the action potential diminishes to 10%. Andreassen and A. Rosenfalck (1978) suggested a decrease to 25%, whereas Gath and Stålberg (1978) suggested that it be designated by the distance where the amplitude of the action potential diminishes to 200 μV. The definitions relating to the percentage decrease are more direct than those referring to an absolute value. In the latter case, the diameter of the muscle fibers will effect the measure because it is related to the amplitude of the action potential (P. Rosenfalck, 1969).

The selectivity of an electrode will depend on the area of the detection surface, and in the case of bipolar electrodes, on the distance between the two detection surfaces. By using mathematical derivations based on the earlier work of P. Rosenfalck (1969), Andreasson and A. Rosenfalck (1978) were able to determine the selectivity of different types of electrodes. Pursuing their definitions of the pickup area, they obtained values for monopolar electrodes, bipolar electrodes oriented perpendicularly to the muscle fibers, and bipolar electrodes oriented in parallel to the muscle fibers. Figure 2.12 presents the borders of the pickup area (decrease to 25%) for electrodes whose detection surface has a diameter of 25 μm. It is interesting to note that the monopolar configuration is less sensitive (larger pickup area) than the bipolar configuration; and in the latter case the selectivity increases when the detection surfaces are oriented perpendicularly to the muscle fibers.

The voltage decrement functions for the bipolar configurations presented in Figure 2.11 are obtained by performing the operation depicted in Figure 2.13. In this figure, *d* represents the distance between the two detection surfaces. Take the amplitude of the monopolar decrement function beginning at a distance, *d*, and subtract it from the decrement function of the other detection surface. The resultant describes the voltage decrement function of the differential bipolar arrangement.

In their report, Andreassen and A. Rosenfalck (1978) estimated that

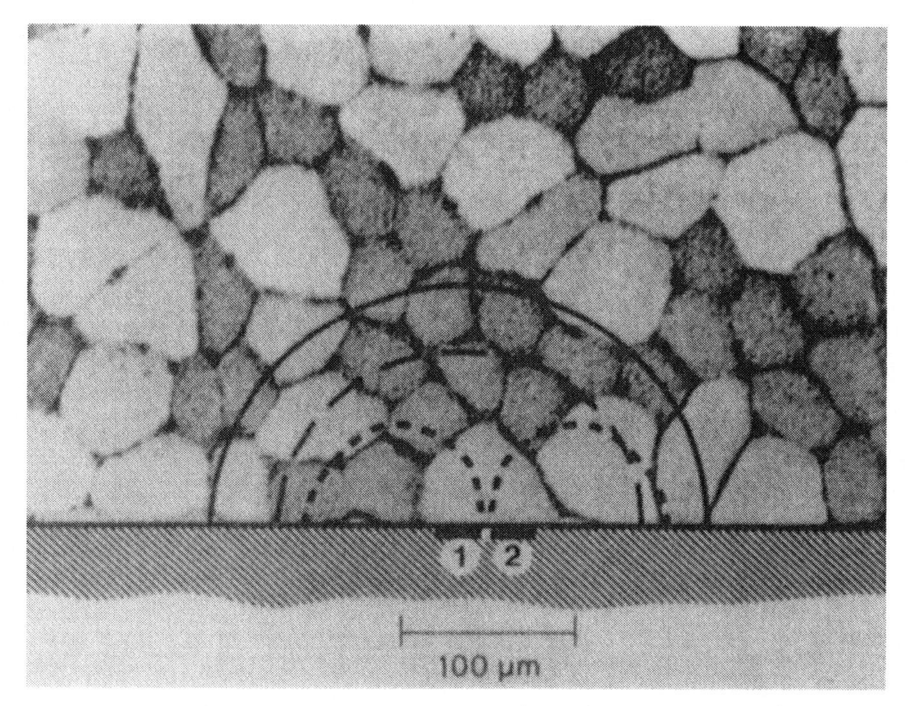

Figure 2.12. Territories of pickup areas (decline of 75% of the amplitude) for: monopolar electrode (full semicircle) with detection surface at position 1; a bipolar perpendicular electrode (*dashed line* with two lobes) with detection surfaces of positions 1 and 2; and bipolar parallel electrode (*dashed semicircle*) with detection surfaces at position 1 and 25 μm above or below position 1. At various positions on the micrograph, the number of fibers within the pickup area ranges from 9 to 17 for the monopolar electrode, 2 to 7 for the bipolar perpendicular electrode, 5 to 9 for the bipolar parallel electrode. (From S. Andreassen and A. Rosenfalck, © 1978, *IEEE Transactions in Biomedical Engineering.)*

for the monopolar configuration, 9 to 17 muscle fibers are located in the pickup area. For a bipolar configuration (25-μm diameter wires spaced 50 μm apart), 2 to 9 muscle fibers would be circumscribed by the pickup area if the detection surfaces were oriented perpendicularly to the fibers, 5 to 9 if oriented parallel to the muscle fibers. The reader is cautioned that these numbers would be higher if a more generous measure of the pickup area were used, say, a decrease to 10% of the amplitude. In any case, it is apparent that the most selective electrode is the bipolar electrode, which is constructed with the smallest detection surfaces and with the smallest separation between the detection surfaces. The selectivity is further accentuated by orienting the detection surfaces in a direction perpendicular to that of the muscle fibers.

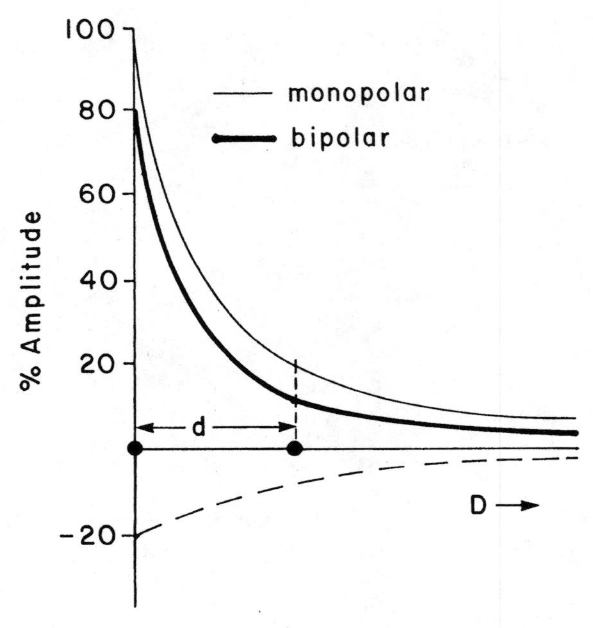

Figure 2.13. Method for calculating the decrement function of bipolar electrodes from the decrement function of a monopolar electrode. The detection surfaces of the bipolar electrode are identical and similar to those of the monopolar electrode. The distance (*D*) describes the center-to-center separation between the detection surfaces of the bipolar electrode. The operation is performed by taking the segment of the monopolar curve, beginning at *d* to infinity; inverting (as required by differential arrangement); shifting it back to $d = 0$; and subtracting its values from the original monopolar curve.

The reader is reminded that during submaximal contractions, not all the muscle fibers will be active and that adjacent fibers commonly belong to different motor units. Therefore, by judiciously placing a highly selective monopolar or bipolar needle electrode in the muscle, it is possible to detect extracellular action potentials from single muscle fibers during submaximal contractions. This is the basis of single fiber electromyography.

The high selectivity, although a blessing for some applications, may be a nightmare for others. Consider the situation where there is some relative movement between the active muscle fibers and electrode. Minor relative movements in the order of 100 μm will locate the pickup area of a selective electrode in an entirely different field of muscle fibers which, given the random intermingling of muscle fibers of different motor units, may belong to different motor units. Such minute relative movements may occur even during attempted constant-force isometric contractions.

Wire electrodes, which generally have larger detection surfaces and are usually spaced 1 to 2 mm apart, therefore will have a larger pickup area. However, unlike needle electrodes they may move within the muscle without external indications. Such hidden relative movements between the detecting surfaces and the active fibers may cast serious ambiguity on the reliability of data relating to properties of individual motor units.

Surface electrodes, even the ones which have relatively small (less than 2 mm) detection surfaces, are not generally considered to be selective. In fact, an efficient design of surface electrodes is directed at obtaining as much activity as possible from one muscle. However, such attempts must be counterbalanced with the discrimination of EMG signals from adjacent muscles, including muscles deep to the one of interest. This "interference" of EMG signals from muscles other than the one(s) under the electrode is referred to as *cross-talk*. There is no fixed solution for guarding against cross-talk. Each electrode configuration and anatomical architecture of the adjacent musculature requires a specific solution to the design of the surface electrode.

Surface electrodes are generally used in the bipolar configuration. The differential amplification arrangement is essential to remove the unwanted "noise" signals on the surface of the skin which are generally present in most environments. (It should be mentioned that monopolar surface electrodes may be used successfully if the reference electrode is large, makes good electrical contact with the skin, and is judiciously located.) The size of the detecting surfaces is not highly critical. Although, ideally they should be as large as possible, the advantages of increasing their dimension quickly disappears above a diameter of 5 mm. Therefore, the major question with respect to selectivity is how far apart the detection surfaces should be located. This question may easily be answered by performing the calculation represented in Figures 2.13. Lynn et al (1978) proposed a rule-of-thumb to simplify the above calculation: If the electrical characteristics of the tissue(s) beneath the electrode are reasonably homogeneous, the distance between the detecting surfaces corresponds roughly to the distance from which muscle fiber will contribute meaningfully to the EMG signal. The reader is cautioned that the anisotropy of the tissue(s) beneath the electrode may considerably alter the direction of the current path and render this simplification meaningless.

We recommend that a standard interdetection-surface spacing of 1 cm be used in surface electrodes. This spacing is compatible with the anatomical architecture of most muscles in the human body. In the following section, discussing signal bandwidth considerations, it will be seen that this spacing has other advantages.

Filtering Properties of Electrode-Electrolyte Interface

All electrode-electrolyte (including skin) interfaces have an impedance consisting of resistances and capacitances. This impedance is frequency dependent. Given that the signal generated by the muscle fibers is a current source, the electrode-electrolyte interface can be thought of in terms of a filter for the voltage generated on the electrode. Most electrodes that are used in electromyography may be conveniently considered to be a high-pass filter. The characteristics of the high-pass filter vary with electrode type and with the characteristics of the electrode-electrolyte junction. This behavior is particularly sensitive for needle electrodes having small detection surfaces (see De Luca and Forrest, 1972). The reader who is interested in more details on this topic is referred to the book by Geddes (1972). Figure 2.14 presents the magnitude of the impedance function for monopolar and bipolar needle electrodes. The detection surfaces were formed by the circular cross-section of the 25-μm diameter nickel-chromium alloy wires spaced 50 μm apart (in the bipolar configuration). The figure also shows the effect of the electrolysis procedure mentioned earlier in this chapter. These impedance values are typical of extremely selective (small detection surface) electrodes. The *dashed lines* represent Bode representations of the impedance function. The equations derived from these representations are provided for each case of Figure 2.14. Each additional term in the equations indicates and increases in the complexity of the resistor-capacitor model required to describe the behavior of the magnitude of the impedance.

For wire and surface electrodes, the magnitude of the impedance is considerably less. Antonelli et al (1982) have reported measurements for typical surface and wire electrode bipolar configurations. Their data is presented in Figure 2.15. Note that the magnitude of the impedance is generally higher for wire electrodes and for smaller detection surfaces.

Filtering Property of Bipolar Configurations

This is a property of bipolar electrodes when used in a differential amplification arrangement.

As demonstrated in Figure 2.9, the two input signals are subtracted and then amplified. Now, if the two detection surfaces are placed parallel to the muscle fibers, the action potential waveform will reach one detecting surface before the other. The differences in the times of arrival will be a function of the conduction velocity of the muscle fiber(s) and the interdetection surface separation. It follows that the frequency components of the propagated signal whose wavelength is equal to the interdetection surface separation (d) will cancel out (refer to signal 1 in Fig. 2.16*A* for a graphic explanation). In a similar fashion, it may be argued that the frequencies component whose wavelength is equal to $2d$ will be amplified with no loss. (refer to *signal 2* in Fig. 2.16A). This pattern will be repeated for every multiple integral frequency value. The

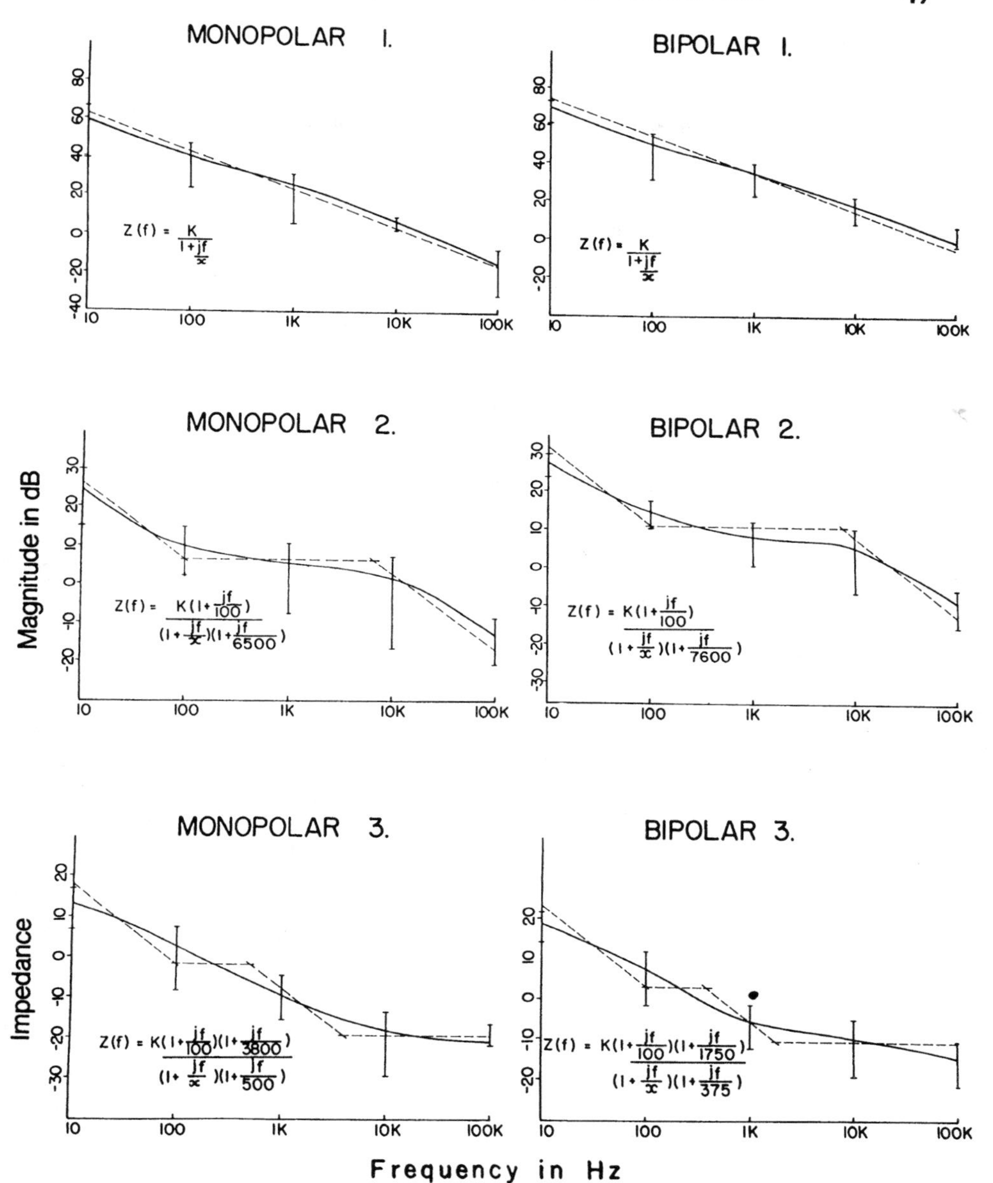

Figure 2.14. Magnitude of the impedance of a monopolar and bipolar needle electrode: (1) before electrolysis; (2) 72 hr after electrolysis and (3) 10 minutes after electrolysis. These values represent the average and standard deviations of 12 electrodes built with 25 μm wires and spaced approximately 50 μm apart. The *dashed lines* represent Bode representations which describe the impedance function. The equations derived from these representations are provided for each case. (From C.J. De Luca and W.J. Forrest, © 1972, *IEEE Transactions in Biomedical Engineering.*)

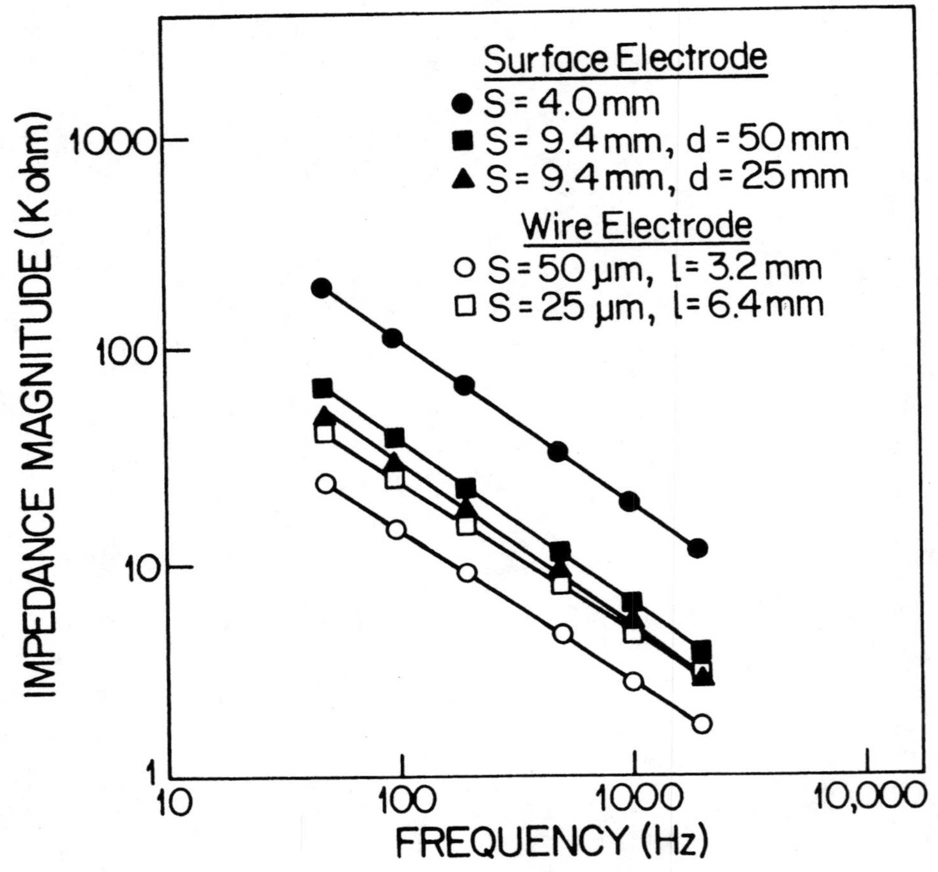

Figure 2.15. Typical values of the magnitude of the impedance of surface and wire electrodes. The *filled circle* represents a monopolar arrangement; the rest are all bipolar. *S*, diameter of detection surface; *D*, interdetection surface spacing; *L*, exposed tip length. (From D. Antonelli et al, © 1982, Pathokinesiology Lab, Rancho Los Amigos Hospital, California.)

corresponding cancellation frequency and pass frequency values may be expressed as:

$$f_{cancellation} = \frac{nv}{d} \qquad n = 1, 2, 3, \cdots$$

$$f_{pass} = \frac{nv}{2d} \qquad n = 1, 3, 5, \cdots$$

where v is conduction velocity along muscle fibers and d is interdetection surface separation. It further stands to reason that frequencies having other values will be subject to some attentuation.

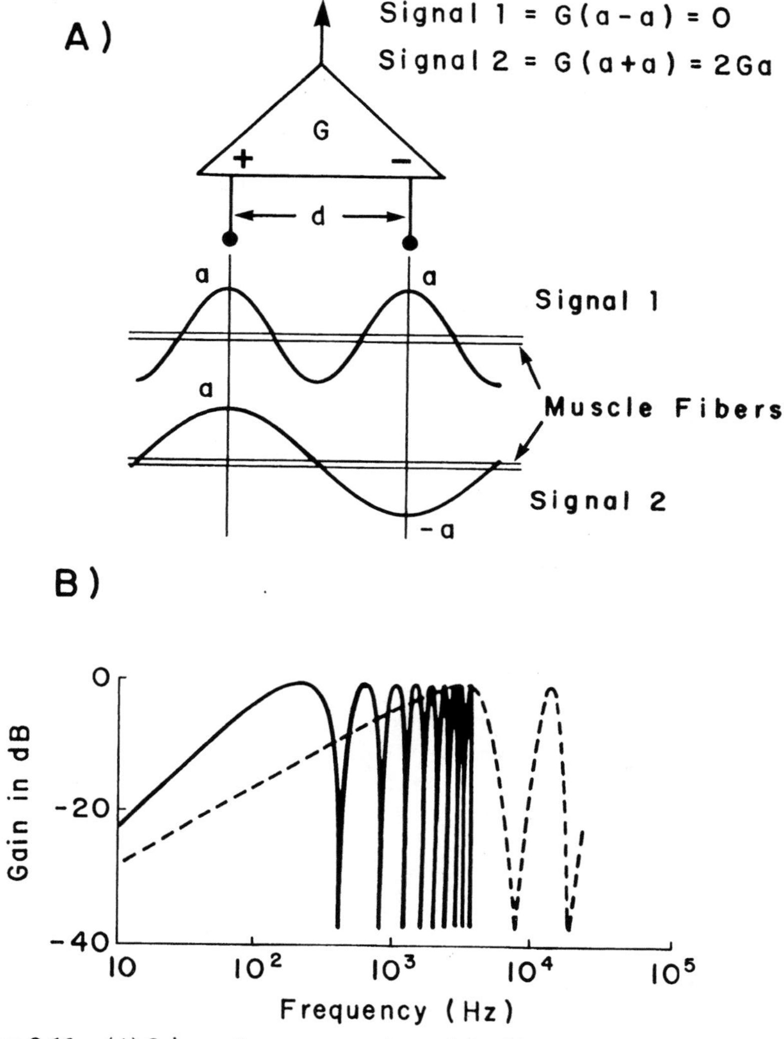

Figure 2.16. (*A*) Schematic representation of the filtering aspects of the differentially amplified bipolar detection. As the signal travels along a muscle fiber at its conduction velocity, it will pass by both detection surfaces sequentially, with a delay proportional to the interdetection surface spacing (*d*). Some of the frequency components of the signal will have wavelengths which are multiples of the distance *d* (cancellation frequency); these will cancel out when amplified differentially. When the wavelength is equal to *2d* (as in signal 2), the signal will add (pass frequency). (*B*) Alternating behavior of the filter function and cancellation frequencies. The *solid line* represents the filter function of a surface electrode and is calculated for an interdetection surface spacing of 1 cm and a conduction velocity of 4 m/s. The *dashed line* represents the filter function of a typical needle electrode and is calculated for an interdetection surface spacing of 0.5 mm and a conduction velocity of 4 m/s. It is apparent that the bipolar filter function is of concern for surface electrodes and has minor relevance in needle electrodes.

By applying elegant mathematical modeling, Lindström (1970) was able to derive the complete differential filter function, whose magnitude may be expressed as:

$$R(\omega,d) = K \sin^2 \left(\frac{\omega d}{2v}\right)$$

where K is a scaling factor representing the various gain factors of the electrode-electrolyte interface. This function is displayed in Figure 2.16*B*. This particular example was calculated by letting $d = 1$ cm and $v = 4$ m/s.

In the case of needle or wire electrodes, where the interdetection surface spacing is considerably smaller, the first cancellation frequency would occur at considerably higher values. Let us take, for example, a needle electrode with an interdetection surface separation of 0.5 mm. For a conduction velocity of 4 m/s, the first cancellation point would occur at 8 kHz, and the first pass frequency would be at 4 kHz. These values are located well into the higher end of the EMG signal spectrum. Hence, only the first and possibly the second cycles will influence the EMG signal. The shape of the latter filter function is also shown in Figure 2.16*B*.

Now it must be pointed out that these functions have been calculated for one muscle fiber which is modeled as an infinitely long cylinder in an unbounded medium, a clearly unrealistic representation which, nonetheless, does provide a useful expression of the filtering characteristics of the filter. Of greater concern in the environment of real muscle fibers is that they are not all necessarily oriented parallel to each other and, therefore, parallel to the electrode, and that they do not all have identical conduction velocities. Thus, the resulting filter function of the bipolar electrodes to the EMG signal, which may be thought of as a summation of the individual filter functions of the action potential, in all likelihood will not be so well defined, nor will it have such well-defined "dips."

The above equation indicates that the bandwidth of the electrode filter function increases as the interdetection surface distance d decreases. Empirical verification of this fact has been reported by Parker and Scott (1973), Zipp (1978), and Lynn et al (1978), all of whom presented evidence that the bandwidth of the detected EMG signal increases as the interdetection surface distance decreases.

ELECTRONICS CONSIDERATIONS

It is now apparent that the EMG signal is filtered by the tissue and the electrode in the process of being detected. Before the signal may be observed, it is necessary to amplify it. This latter procedure may also modify the frequency characteristics of the signal. In order to describe this process, it is necessary to describe some properties and parameters of electronic amplifiers. They are:

(a) Noise characteristics
(b) Signal-to-noise ratio
(c) Gain
(d) Common mode-rejection ratio
(e) Input impedance and input bias current
(f) Bandwidth

For this purpose it is useful to refer to Figure 2.9 and define the following terms:

G = gain of the amplifier
m = the detected wanted signal (the EMG signal)
n = the detected unwanted signal (the noise)

Noise

This term can be defined as any unwanted signal which is detected together with the wanted signal. Our environment is inundated with myriad electrostatic and electromagnetic fields. The presence of electrostatic fields has been completely overlooked in electromyogaphy. This oversight has been mainly due to the fact that the equipment used generally filters out DC signals. However, the presence of "static electricity" on the surface of a subject may reach proportions which may damage the electrode characteristics and, possibly, the amplifiers. High levels of static electricity are often present when a subject wears polyester clothing and the humidity level of the air is low. Electromagnetic fields are ever present in a variety of forms such as 50 or 60 Hz from power lines and electrical devices which operate on line current, radio signals, television signals, and communications signals, to name a few.

To these, also add electrical noise generated by the very equipment which we employ to detect and record the EMG signal. These are: (1) the "thermal noise" generated by the electrodes; this physical property of metals is proportional to the square root of the resistance of the detection surface and cannot be eliminated but may be reduced to the point that it is not a factor of concern by cleaning the electrode contacts, as described in a previous section in this chapter; (2) the noise generated by the first stage of the amplifiers; this is a physical property of semiconductors and cannot be completely eliminated, but may be reduced by the continual advances that are being made in semiconductor physics. The user has no recourse but to choose (or construct) an amplifier that has low noise.

To the above sources we should add another which assumes particular importance in electromyography, that is, *motion artifact.* This disturbance may occur in two locations; at the electrode-tissue interface or at the wire leads connecting the electrodes to the amplifier. The prior source has two origins. One is any relative movement of the electrode and tissue. As described previously in this chapter, when any two materials having dissimilar electrical properties come in contact with each other, there is a lack of chemical equilibrium at the junction, which in turn generates a polarization potential. Any relative movement at the junction modulates the polarization potential and generates an AC current which generates the noise signal. The other is the "skin potential." Under normal conditions, a voltage of approximately 20 mV exists across the skin layers. It is generally believed that this potential is originated by "injury currents" of the dead cells as they migrate to the surface of the skin. In any case, the voltage varies as the skin is stretched, as is the case when the muscles underneath it contract or as a limb is displaced. It is of interest to point out that abrasion of the skin reduces this component of motion artifact because as the skin is pierced, the voltage across the skin is shorted out (refer to Burbank and Webster, 1978, for details). The noise resulting from the leads movement is caused by the natural phenomenon that is used to create current in a generator. That is, a metallic wire (the lead) is moved through electromagnetic fields (which as described before are pervasive in our environment). The voltage resulting from these mechanical artifacts may be large (several millivolts) so that they seriously contaminate the EMG signals. This problem is accentuated when the input impedance of the amplifier is high because a small current passing through a high impedance may generate a high voltage. The reader is reminded of the basic relation ($V = IZ$) between voltage (V), impedance (Z), and current (I). It should also be pointed out that because these forms of electrical noise are induced by movements of the body tissues, they will be limited to frequency components that are less than 30 Hz. The human body, or parts of it, will not oscillate with any easily measurable energy or higher frequencies.

All these noise sources are the critical ones because they are added to the wanted signal prior to amplification; therefore any amplification will increase both the wanted and unwanted signals.

Signal-to-Noise Ratio

In any scheme for detecting, amplifying, or recording signals, the ratio of the wanted signal to the unwanted signal is the single most important factor to be considered. It is the factor which measures the quality of the signal.

Gain

Referring to Figure 2.9, we can describe the amplification of the detected signal in the following fashion:

Monopolar amplified signal $= G(m + n)$

Bipolar amplified signal $= G[(m_1 + n) - (m_2 + n)]$
$= G(m_1 - m_2)$

The advantage of the bipolar configuration is now apparent. Ideally, the noise component is removed. This idealized representation for the differential amplification associated with the bipolar detection configuration indicates that if the noise signal fed to the amplifier is similar in all respects (amplitude, phase and frequency components), then it will be totally eliminated. This perfect cancellation does not occur in real differential amplifiers for two reasons. First, the amplifiers cannot subtract perfectly. The measure of how well the differential amplifier subtracts (reject) the common mode signal is called the "common mode rejection ratio," and will be addressed in the following section. The second reason is that the noise signal reaching the two input stages of the differential amplifier is not necessarily common mode. This is particularly true if the tissue media is anisotropic.

Another point that should be mentioned concerns the amount of amplification required to observe or record the EMG signal. It is apparent in the above formulation that the bipolar configuration will require greater amplification. However, this is of no concern because the values of the gains required in both cases are well within the capabilities of ordinary electronics amplifiers.

Common Mode-Rejection Ratio

In practice, the performance of differential amplifier circuits departs from the ideal characteristics mentioned above. Gain imbalance and nonlinearities in the amplifier's differential input stages cause errors in the subtraction process. As a result, signals common to each input (E_{cm}) do not cancel completely and produce an undesirable common-mode error voltage at the amplifier output (E_e). The ratio between the common-mode voltage (E_{cm}) of the amplifier and its common-mode error voltage (E_e) is defined as the common-mode rejection ratio (CMRR)

$$CMRR = \frac{E_{cm}}{E_e}$$

The importance of the CMRR becomes apparent when dealing with the effects of external fields such as power line-induced interference radiating from the environment. Referring to Figure 2.17, we can model the effect of an external signal field acting on the tissue media as two current sources (i_n) in parallel with their respective tissue impedances (Z_{tn}). If the tissue media impedance (Z_n) is isotropic, and the external field gradient across the tissue media is constant, then the fields induced currents (i_n) at each input are equal and will cancel. Obviously, the higher the CMRR of the amplifier, the better the cancellation of these undesirable currents.

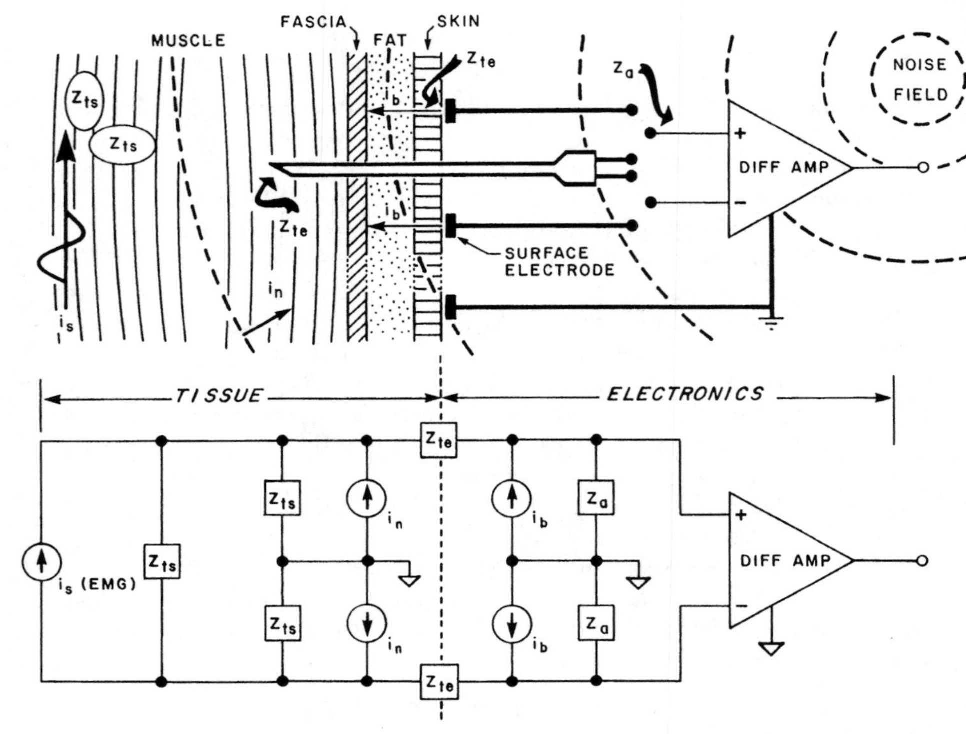

Figure 2.17. (*Top*) Diagramatic representation of impedances and currents in the tissues, electrode, and amplifier:

i_s = source current from EMG signal
i_n = common current from noise
i_b = input bias current from amplifier
Z_{ts} = tissue impedance seen by source current
Z_{tn} = tissue impedance seen by noise current
Z_{te} = tissue electrode impedance of the metal electrolyte interface
Z_a = input impedance of amplifier

(*Bottom*) The schematic diagram representing the electrical interaction of the EMG signal, extraneous noise, electrode, and amplifier.

Input Impedance and Input Bias Current

In order to accurately measure the amplitude and waveshape parameters of the EMG signal, it is necessary to understand how the input impedance and input bias current of the differential amplifier can influence these parameters. The concept of impedance has been discussed at the beginning of this chapter. The input bias current may be thought of as the minimal constant current required to keep the amplifier active. Since the differential amplifier is not ideal, it has a finite impedance at each input and nonzero input bias current. As demonstrated in Figure

2.17, the bias current flows out of the amplifier. Thus, it stands to reason that any signal which has a current less than the bias current will not be amplified. In modern amplifiers this current is considerably small (< 100 pA) so that it does not present any danger to the subject when the electrode is on the skin or in a skeletal muscle. However, in needle electrodes which have small detection surfaces (<100 μm), the current may be sufficient to alter the chemical structure of the surface layer over repeated applications. This, in turn, will alter the metal-electrolyte filtering characteristics of the electrode. (Referring to Fig. 2.17, we can model the input of the amplifier with an impedance (Z_a) and current source (i_b) from each input to ground reference. When an EMG signal source is connected to the inputs, these bias currents flow through the signal source. The output of the signal source is likewise shunted across the input impedance (Z_a). The amount of signal shunting depends on Z_a and the distributed impedance of the EMG signal source.

The distributed impedance of the EMG signal source is determined by the impedance characteristics of the tissue (Z_{ts}) and the tissue-electrode interface, (Z_{te}). As discussed in previous sections, these impedances have both resistive and reactive components due to the capacitive effects of tissue media and electrode interface. The value of the distributed source impedance can vary greatly, depending on the impedance of the needle or surface electrode interface configuration (Z_{te}) and the amount of intervening tissue (Z_{ts}), typically 10^4 to 10^6 ohms at 1 kHz. *To minimize waveshape distortion and attentuation of the signal source due to the shunting by the amplifier, the input impedance (Z_a) should be much larger (10^{12} in parallel with 5 pf) than the distributed source impedance.*

Bandwidth

All amplifiers have limits on the range of frequency over which they operate. In fact, limitations of amplifiers are commonly measured in gain-bandwidth quotient. The value of the quotient is defined by the type of semiconductor components used. This limitation does not present a problem in electromyography because amplifiers providing the required gain over the necessary bandwidth are commonly available and easy to design. The bandwidth of an amplifier may be conceptualized as a window in the frequency domain. The frequencies of a signal that coincide between the borders of the window, i.e., the bandwidth, will pass with minimal, if any, diminution; whereas, frequency components outside the bandwidth will be suppressed or eliminated. In this sense, an amplifier can be considered as a filter with gain (refer to Fig. 2.1 and the discussion at the beginning of the chapter for more details).

Some amplifiers are specially designed to amplify dc signals. Such amplifiers require special circuitry to eliminate "drift" as a function of temperature. They generally have low bandwidths, usually up to 100

Hz. They are typically used to amplify signals such as those generated by force, pressure, and temperature transducers.

In electromyography, it is highly advisable not to DC couple the electrodes to the amplifiers for the following reasons: (1) The dc polarization potential present at the electrode-electrolyte interface may be as large as the EMG signal being detected. (2) Motion artifacts in the lead wire will generally have low frequencies (<20 Hz), and thus they would also be amplified. (3) The frequency components of the EMG signal below 20 Hz are unstable and fluctuate with a considerable amount of unpredictability. This point will be elaborated in Chapter 3. For these reasons, it is recommended that for general applications the low-frequency 3 dB point be set at 20 Hz. The high frequency 3 dB point should be set to a value slightly higher than the highest frequency components of the wanted signal. As will be seen shortly, this value is dependent on the type of electrode used to detect the signal. Any noise signal having frequency components greater than the high 3 dB point will be attentuated, thus increasing the signal-to-noise ratio of the amplified signal.

RECORDING ASPECTS

Having detected and amplified the signal, it is necessary to display it in some fashion so that it may be employed. A variety of devices and media may be used for this purpose. The most common approach is to display the amplified signal on an oscilloscope. If a permanent record is required, it is useful to record the signal on a frequency-modulated (FM) tape recorder. FM tape recorders are used because they have a bandpass which includes dc. The great advantage of storing the signal on FM tape is that it may be replayed and transferred to other media. Permanent visual records are commonly produced on strip-chart recorders, of which there are at least three types. One type prints on paper via an electromechanical pen, another via a galvanographic or oscillographic process on ultraviolet sensitive paper, and the third prints on paper via an ink jet spray.

The primary concern of recording the signal is that the bandwidth of the device be greater than the bandwidth of the amplified signal. If the signal has been amplified with a proper gain (yielding a ±1 V peak-to-peak value), the noise considerations become minimal at this stage. For oscilloscopes, this is a trivial issue, because they all have much wider bandwidth specifications. FM magnetic tape recorders have bandwidths whose upper 3 dB point is linearly related to the speed of the tape. All brand-name commercially available FM tape recorders have bandwidths which are required for EMG signal storage. It is only necessary to ensure that the appropriate speed is used when recording the signal.

Strip chart recorders present a more restrictive barrier. The type which prints with an electrical-mechanical pen has severe bandwidth limitations, typically 0 to 60 Hz. Clearly any attempt to record the EMG signal directly on such devices sacrifices most of the energy in the signal and, thus, most of the information. However, such devices are obviously convenient. They may be properly employed if the signal is first recorded on FM tape at a relatively high speed, and is subsequently played back to the strip-chart recorder at a lower speed. The ratio of the high to low speeds effectively multiples the bandwidth of the strip-chart recorder. So, if a speed reduction ratio of 32 is used, the strip-chart recorder will have an effective bandwidth of approximately 0 to 2000 Hz. Such a bandwidth is acceptable for surface electrodes and adequate for wire and needle electrodes. This procedure has the disadvantage that it cannot be performed in real time. The ultraviolet sensitive strip-chart recorders, as well as some ink jet recorders which are now appearing on the market, have bandwidths that are considerably higher, typically 0 to 750 Hz. These may be used to record EMG signals detected with surface electrodes directly without sacrificing a significant part of the signal. They may be (and are often) used to record signals directly from indwelling electrodes. Although they provide a visually proper representation of signals obtained by indwelling electrodes, it is wise to remember that the higher frequency components have been removed. Therefore, measurements of the amplitude of peaks and peak-to-peak slopes will misrepresent the actual (postdetected, prerecorded) values. However, these devices may be used in real time, and as such may inherently provide some advantages, especially in clinical environments. Nonetheless, the images traced on the strip-chart must be interpreted with the forewarned caution.

With the exploding popularity of digital computers in laboratories and clinics, it has become possible to record the EMG signals directly on digital storage media (computer memory, disks, or digital magnetic tape). This is accomplished by sampling the signal at regular intervals and expressing the amplitude value at each point as a binary value (power of 2) and storing this value. This operation is known as digitization. In this operation, the sampling rate is an important factor. A minimal requirement to preserve the frequency information of the signal is that the sampling rate be at least twice the value of the highest frequency component of the signal. This is known as the Nyquist frequency. Other considerations such as signal amplitude, prefiltering, aliasing, etc., must also be addressed when digitizing a signal. Although computer programs for performing this operation are readily available, the novice is strongly advised to obtain assistance from competent individuals prior to attempting this operation.

OVERVIEW OF DETECTION AND RECORDING PROCEDURES

It is now apparent that the entire procedure of acquiring an observable EMG signal consists of a catenation of filtering processes, each of which modifies, in some respects, the amplitude and frequency characteristics of the observed signal.

It is important to remember that the characteristics of the observed EMG signal are a function of the apparatus used to acquire the signal as well as the electrical current which is generated by the membrane of the muscle fibers.

For the sake of convenience, the block diagram in Figure 2.18 presents all the individual filtering steps which have been discussed. Some minor filtering effects due to the size of the detection surfaces, location of the electrode with respect to the neuromuscular junction and tendons, etc. have not been discussed. Their effects are difficult to calculate and do not alter the signal characteristics in any considerable amount.

Practical Considerations

It is always desirable to detect and record the EMG signal with minimal distortion and extraneous contamination. We will now list some practical considerations which will assist in accomplishing the task. These pointers essentially represent a summary of the previous discussions in this chapter, the details of which should now be familiar to the reader.

Tissue Filtering

1. The voltage decrement function decreases rapidly with distance; therefore, inserted electrodes will only detect signals from nearby muscle fibers. The amplitude of action potentials decreases to 25% within 100 μm.

2. The filtering characteristics of the muscle tissues is a function of the

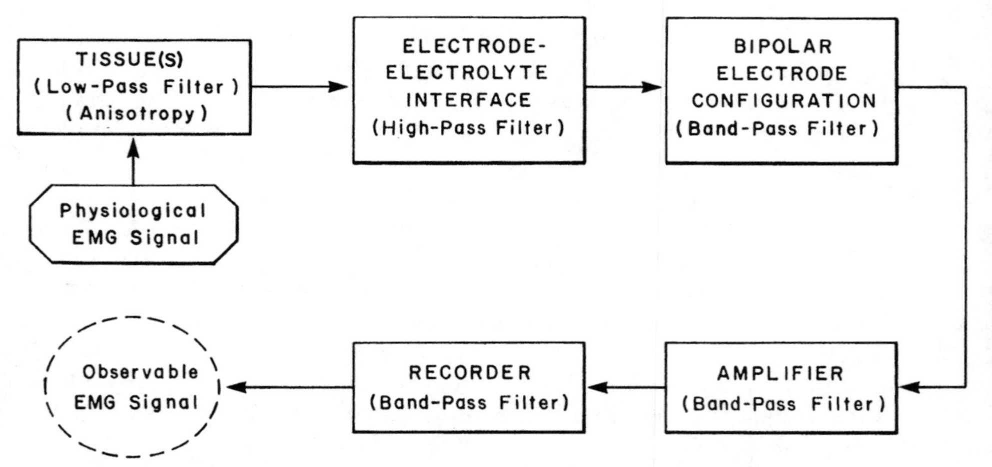

Figure 2.18. Block diagram of all the major aspects of the signal acquisition procedure. Note the variety of physical properties that act as filters to the EMG signal before it can be observed. The term physiological EMG signal refers to the collection of signals which emanate from the surface of the muscle fibers. They are not observable.

distance between the active muscle fibers and the detection surface(s) of the electrode. In the case of surface electrodes, the thickness of the fatty and skin tissues must also be considered. The tissue(s) behaves as a low-pass filter whose bandwidth and gain decreases as the distance increases.

3. The muscle tissue is highly anisotropic. Therefore, the orientation of the detection surfaces of the electrode with respect to the length of the muscle fibers is critical.

Electrode-Electrolyte Interface

1. This electrochemical junction also behaves as a high-pass filter.

2. The gain and bandwidth will be a function of the area of the detection surfaces, electrolytic treatment of the surfaces, and any chemical-electrical alteration of the junction.

3. The detection surfaces should always be kept clean.

Bipolar Electrode Configuration

1. This property ideally behaves as a band-pass filter. However, this is only true if the inputs to the amplifier are balanced and if the filtering aspects of the electrode-electrolyte junctions are equivalent.

2. A larger interdetection surface spacing will render a lower bandwidth. This aspect is particularly significant for surface electrodes.

3. The greater the interdetection surface spacing, the greater the susceptibility of the electrode to detecting measurable amplitudes of EMG signals from adjacent and deep muscles. Again, this aspect is particularly significant for surface electrodes. A rule of thumb is that the electrodes will detect measurable signals from a distance equal to the interdetection surfaces spacing. However, the anisotropy of the tissues beneath the electrode may augment the sensitivity of the electrodes along the surface of the muscle creating cross-talk.

4. An interdetection surface spacing of 1.0 cm is recommended for surface electrodes.

Amplifier Characteristics

1. These should be designed and/or set for values which will minimally distort the EMG signal detected by the electrodes.

2. The length of the leads to the input of the amplifier (actually, the first stage of the amplification) should be as short as possible and should not be susceptible to movement. This may be accomplished by building the first stage of the amplifier (the preamplifier) in a small configuration which may be physically located near (within 10 cm) of the electrode. The necessity of this precaution is accentuated when amplifiers with high input impedance ($>10^7$ ohm) are used.

3. Typical settings and characteristics are:

a. Gain: such that it renders the output with an amplitude of approximately ±1 V.

b. Input impedance > 10^{12} ohms resistance in parallel with 5 pf capacitance.
c. Common mode rejection ratio: >100 dB
d. Input bias current: as low as possible (typically less than 50 pA)
e. Noise < 5μV rms
f. Bandwidth (3 dB points for 12 dB/octave rolloff):

Surface electrodes	20–500 Hz
Wire electrodes	20–1000 Hz
Monopolar and bipolar needle electrodes for general use	20–1000 Hz
Needle electrodes for signal decomposition	1000–10,000 Hz
Single fiber electrode	20–10,000 Hz
Macroelectrode	20–10,000 Hz

Recording Characteristics

The effective or actual bandwidth of the device or algorithm that is used to record or store the signal must be greater than that of the amplifiers.

Other Considerations

1. It is preferable to have the subject, the electrode, and the recording equipment in an electromagnetically quiet environment. However, if all the procedures and cautions discussed in this chapter are followed and heeded, high quality recordings will be obtained in the electromagnetic environments found in most institutions, including hospitals.

2. When using indwelling electrodes, great caution should be taken to minimize (eliminate, if possible) any relative movement between the detection surfaces of the electrodes and the muscle fibers. Relative movements of 0.1 mm may dramatically alter the characteristics of the detected EMG signal and may possibly cause the electrode to detect a different motor unit population.

WHERE TO LOCATE THE ELECTRODE

Prior to entertaining a discussion on the preferred location of electrodes in a muscle or on the surface of a muscle, we would like to dispel a common fallacy which, to our knowledge, has been prevalent for many years among electromyographers, that is, the common practice of locating the detection electrode in the vicinity of the motor point of the muscle. (This location is the surface projection of the anatomical center of the innervation zone of the muscle.) The reasoning presumably is that the motor point is to some extent an electrically and anatomically definable point. Another fallacious rationale is at times advanced by the belief that the motor point, being the location where an externally applied electrical current causes the maximal excitation of the muscle, should therefore be the location that provides the greatest signal amplitude. This is not so. Even if it were so, the availability of larger amplitude is not an important consideration. However, there are three important considerations. They are: (1) signal-to-noise ratio; (2) signal stability (reliability); and (3) cross-talk from adjacent muscles. The importance of the signal-to-noise ratio has been discussed already. The stability consideration addresses the issue of the modulation of the signal amplitude due to relative movement of the active fibers with respect to the detection surfaces of the electrode. The issue of the cross-talk concerns the detection by the electrode of signals emanating from adjacent muscles.

For most configurations of needle electrodes the question of cross-talk is of minor concern, because the electrode is so selective that it only detects from nearby muscle fibers (see Fig. 2.12). Due to the fact that the muscle fibers of different motor units are scattered in a semirandom fashion throughout the muscle, the location of the electrode becomes irrelevant from the point of view of signal quality and information content. The stability of the signal will not necessarily be improved in any one location. It is, nonetheless, wise to stay clear of the innervation zone so as to reduce the probability of irritating a nerve ending. And of course, one should stay clear of suspected locations of blood vessels. These two latter precautions should be taken, not only as a courtesy to the subject, but also for technical considerations. If a nerve ending is irritated, "spontaneous" signals unrelated to the motor aspects of muscles will be detected. If a blood vessel is ruptured, the resulting pool of blood in the interstitial spaces will "short" the input of the electrode by decreasing the impedance between the detection surfaces.

For wire electrodes, all the considerations which have been discussed for needle electrodes also apply; and in this case, any complication will be unforgiving, in that the electrode may not be relocated. Because the wire electrodes have a larger pickup area, a concern arises with respect to how the location of the insertion effects the stability of the signal. This question is even more dramatic in the case of surface electrodes and will be discussed in that context.

The precautions required to reduce cross-talk with surface electrodes has already been discussed. In this section we will discuss the susceptibility of signal to the location on the muscle. In order to measure this susceptibility, it is necessary to devise a procedure whereby the excitation to the muscle remains constant while the location of the electrode is varied. Gerbino, Gilmore, and De Luca satisfied this condition by supramaximally stimulating, with 0.5-ms current pulses, the tibialis nerve of rabbits via implanted stimulation electrodes. In this fashion, a compound action potential of consistent amplitude and shape was generated in the gastrocnemius muscle. The detection electrode was always arranged with its detection surfaces along the length of the fibers of the gastrocnemius muscle and was located in four distinct locations: near the Achilles tendon; halfway between the tendon and the center of the innervation zone; on the innervation zone; and between the innervation zone and the origin. While the electrode was maintained in each location, the

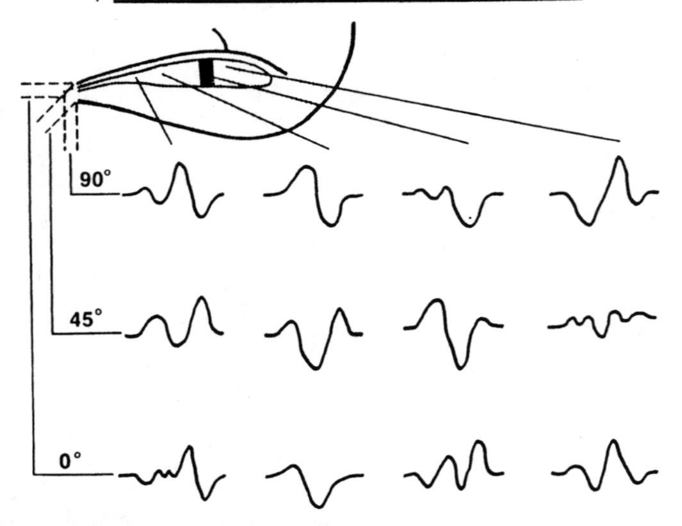

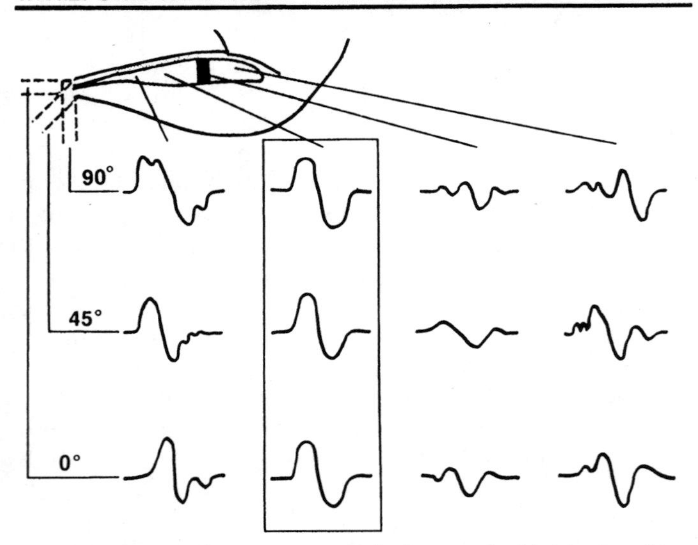

Figure 2.19. Compound action potentials detected with bipolar electrodes from the surface of the skin above the gastrocnemius muscle of a rabbit. A stimulation electrode was implanted around the tibialis nerve, and supramaximal electrical stimulation was applied. The detection electrode was located in four positions: near the Achilles tendon; 1 halfway between the tendon and the center of the innervation zone (*double-shaded* band); the center of the innervation zone; and between the innervation zone and the origin. The angle of the ankle joint was held at 0, 45, and 90. (*Top panel*) The electrode was attached to the skin in the normal fashion. (*Bottom panel*) the electrode was held firmly against the skin and underlying muscle, reducing any relative movement between the muscle and the electrode. Note the more consistent waveforms in the latter case. Also note that the preferred location is the area between the insertion tendon and the innervation zone.

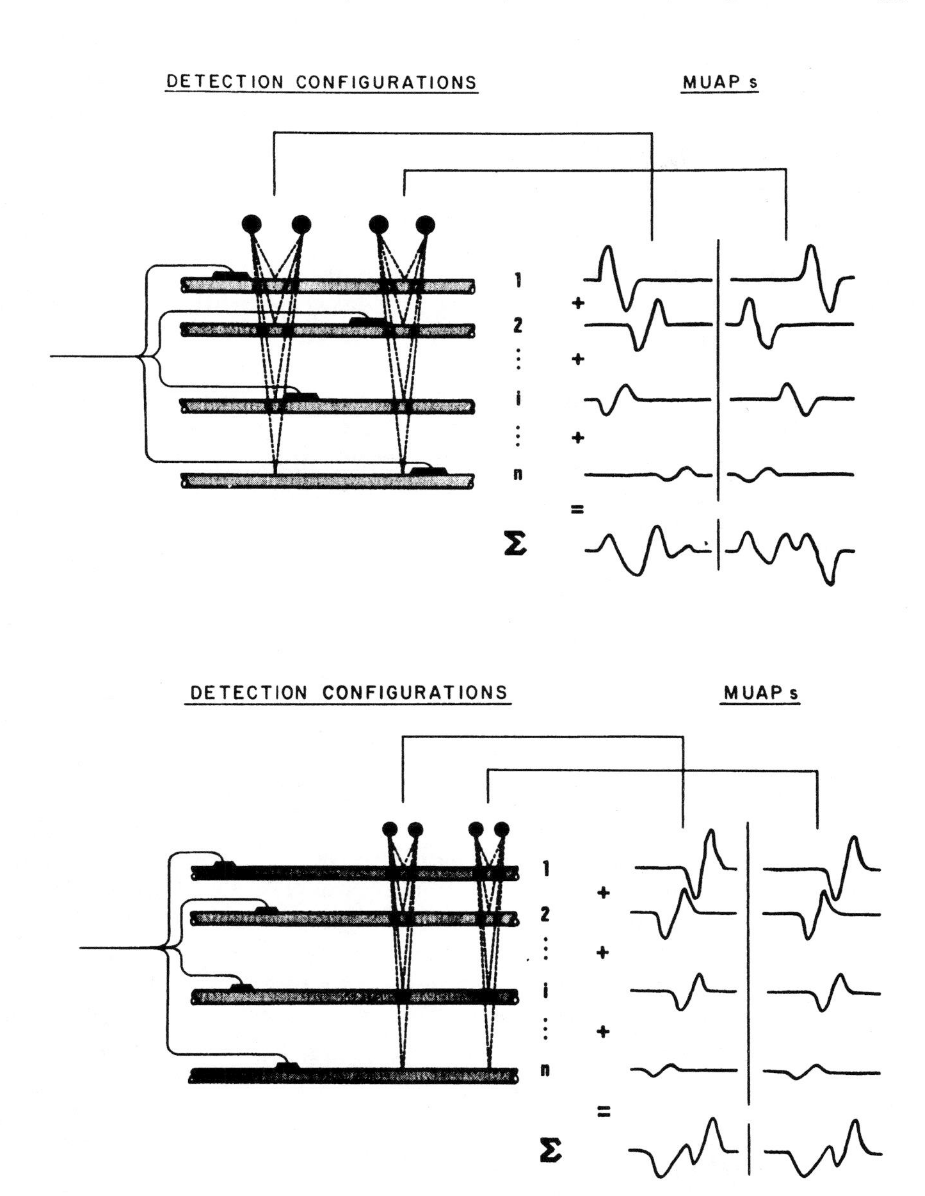

Figure 2.20. Schematic diagram explaining the susceptibility of the motor unit action potential (MUAP) waveform to electrode displacement. *Top panel* represents two detection locations in the innervation zone and the associated modifications in the detected MUAP. *Bottom panel* presents the similar data for two comparable detection locations well away from the innervation zone.

ankle was fixed at 0°, 45°, or 90°, thus simulating the muscle length changes during shortening contractions. The compound action potential was detected for each electrode location and for each ankle position. Two sets of measurements were taken. In one set, the electrode was attached to the skin in a normal fashion; in the other set, the electrode was held firmly against the skin with pressure applied against the muscle, thereby reducing the relative movement of the electrode and skin with respect to the muscle fibers.

A typical example of the results is displayed in Figure 2.19. Note that the shape of the compound action potential is susceptible to the location of the electrode along the muscle, and in the case where no pressure was applied to stabilize the electrode, the shape was altered by the position of the ankle. In both cases, the location which provided the most unreliable results was the one corresponding to the innervation zone.

These observations may be conveniently explained by using a schematic approach for the formulation of the MUAP, which in some sense has a comparable genesis to the compound action potential. (The reader is referred to Chapter 3 for a detailed explanation of this concept.) This schematic representation is presented in Figure 2.20. The MUAP is formed by linearly superimposing the action potentials from all the active muscle fibers in the vicinity of the detection electrode. In the *top panel*, the electrode is located near the innervation zone, in the *bottom panel* away from the zone. It is apparent in the top panel that a slight displacement of the electrode results in a drastically different waveform for the MUAP, whereas in the bottom panel an equivalent displacement does not result in such drastic alterations of the waveform.

It is also apparent from Figure 2.20, and may be noted in the empirical results of Figure 2.19, that the relative movement of the detection electrode with respect to the active fibers (as may easily occur when a muscle shortens and the electrode remains stationary on the skin) may, by itself, alter the characteristics of the detected signal. This modification of the EMG signal would not be related to physiological aspects of the contracting muscle. A situation of this nature is found in investigations which acquire EMG signals during anisometric contraction, such as gait. The modulation of the amplitude of the EMG signals obtained from lower limb muscles during gait should be interpreted with great caution.

We suggest that the preferred location of an electrode is in the region halfway between the center of the innervation zone and the further tendon.

CHAPTER 3

Description and Analysis of the EMG Signal

The EMG signal is the electrical manifestation of the neuromuscular activation associated with a contracting muscle. It is an exceedingly complicated signal which is affected by the anatomical and physiological properties of muscles, the control scheme of the peripheral nervous system, as well as the characteristics of the instrumentation that is used to detect and observe it. Most of the relationships between the EMG signal and the properties of a contracting muscle which are presently employed have evolved serendipitously. The lack of a proper description of the EMG signal is probably the greatest single factor which has hampered the development of electromyography into a precise discipline.

This chapter will present two main concepts. The first is a discussion of a structured approach for interpreting the information content of the EMG signal. The mathematical model which is developed is based on current knowledge of the properties of contracting human muscles. These properties are discussed in Chapter 5. The extent to which the model contributes to the understanding of the signal is restricted to the limited amount of physiological knowledge currently available. However, even in its present form, the modeling approach supplies an enlightening insight into the composition of the EMG signal.

The second concept in this chapter concerns a discussion of methodologies that are useful for processing and analyzing the signal.

REVIEW OF NOMENCLATURE

Throughout this chapter, specialized terms will be used to describe distinct aspects of signals. These terms will be defined now so as to eliminate possible confusion.

Waveform—The term which describes all aspects of the excursion of the potential, voltage, or current associated with a signal or a function of time. It incorporates all the notions of shape, amplitude, and time duration.
Amplitude—That quantity which expresses the level of signal activity.
Time duration—The amount of time over which a waveform presents detectable energy.
Phase—In electromyography, this term refers to the net excursion of the amplitude of a signal in either the positive or negative direction.
Shape—The characteristics of a signal which remains unaltered with linear scaling in either the amplitude or time domains. An example of such characteristics is the phases of an action potential.

The distinction between the concept of shape and waveform is depicted in Figure 3.1. In Figure 3.1*A* the amplitude is scaled linearly by 2 and by 0.5, but the shape remains unaltered. In Figure 3.1*B* the time is scaled linearly by 2 and by 0.5, but the shape remains unaltered. In Figure 3.1*C* the amplitude and time are scaled nonlinearly, and the number of phases is altered. In this latter case the shape changes.

THE MOTOR UNIT ACTION POTENTIAL

Under normal conditions, an action potential propagating down a motoneuron activates all the branches of the motoneuron; these in turn activate all the muscle fibers of a motor unit (Krnjević and Miledi, 1958a;

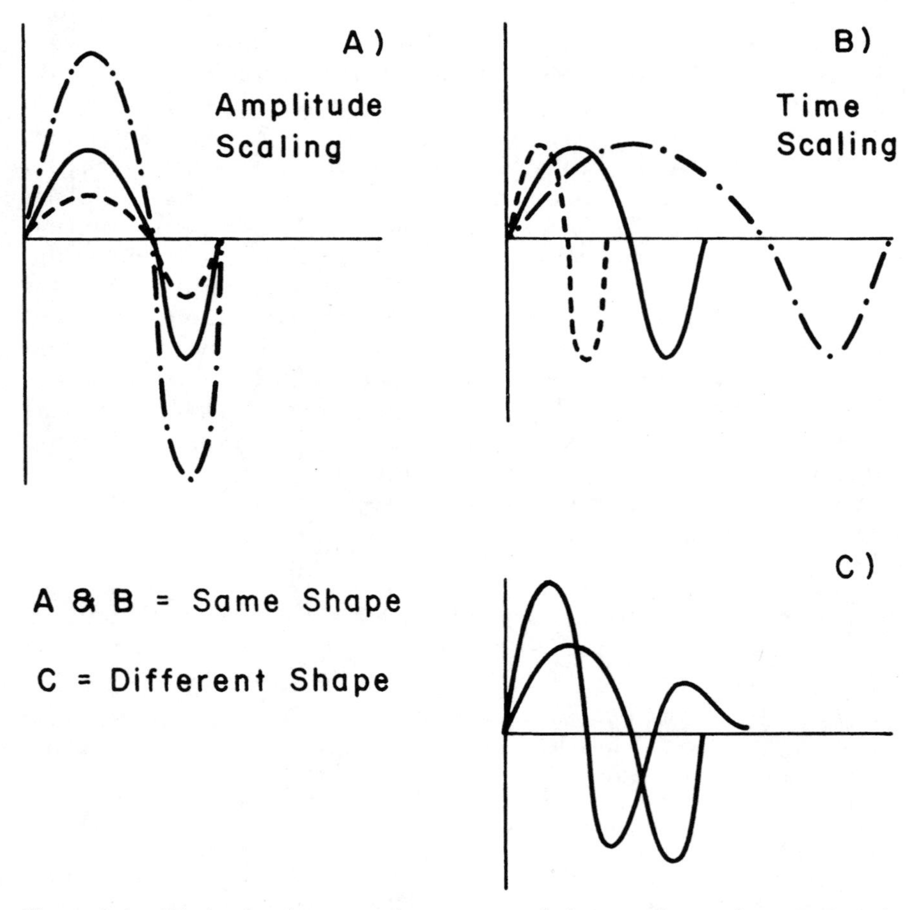

Figure 3.1. Distinction between the concept of shape and waveform. In *A* and *B* the shape remains unaltered through linear scaling (0.5 and 2) of the amplitude and time. In *C* the shape is modified through nonlinear scaling and an additional phase. In all three examples, the waveform has been altered.

Paton and Wand, 1967). When the postsynaptic membrane of a muscle fiber is depolarized, the depolarization propagates in both directions along the fiber. The membrane depolarization, accompanied by a movement of ions, generates an electromagnetic field in the vicinity of the muscle fibers. An electrode located in this field will detect the potential or voltage (with respect to ground), whose time excursion is known as an action potential. A schematic representation of this situation is presented in Figure 3.2. In the diagram, the integer n represents the total number of muscle fibers of one motor unit that are sufficiently near the recording electrode for their action potentials to be detected by the electrode. For the sake of simplicity, only the muscle fibers from one motor unit are depicted. The action potentials associated with each muscle fiber are presented on the right side of Figure 3.2. The individual muscle fiber action potentials represent the contribution that each active muscle fiber makes to the signal detected at the electrode site.

For technical reasons, the detection electrode is typically bipolar, and the signal is amplified differentially. The waveform of the observed action potential will depend on the orientation of the detection electrode contacts with respect to the active fibers. For simplicity, in Figure 3.2 the detection surfaces of the electrode are aligned parallel to the muscle fibers. With this arrangement, the observed action potentials of the muscle fibers will have a biphasic shape, and the sign of the phases will depend upon the direction from which the muscle membrane depolarization approaches the detection site. To clarify the relative position of the neuromuscular junction of each muscle fiber and the recording site in Figure 3.2, lines have been drawn between the nearest point on each muscle fiber and the detection surfaces. In the diagram, a depolarization approaching from the right side is reflected as a negative phase in the action potential and *vice versa.* Note that when the depolarization of the muscle fiber membranes reaches the point marked by the two lines, the corresponding muscle fiber action potential will have a zero interphasic value.

In human muscle tissue, the amplitude of the action potentials is dependent on the diameter of the muscle fiber, the distance between the active muscle fiber and the detection site, and the filtering properties of the electrode. The amplitude increases as $V = ka^{1.7}$, where a is the radius of the muscle fiber and k is a constant (Rosenfalck, 1969); it decreases approximately inversely proportional to the distance between the active fiber and the detection site. The filtering properties of a bipolar electrode are a function of the size of the detection surfaces, the distance between the contacts, and the chemical properties of the metal-electrolyte interface. For details see discussion in previous chapter.

The duration of the action potentials will be inversely related to the

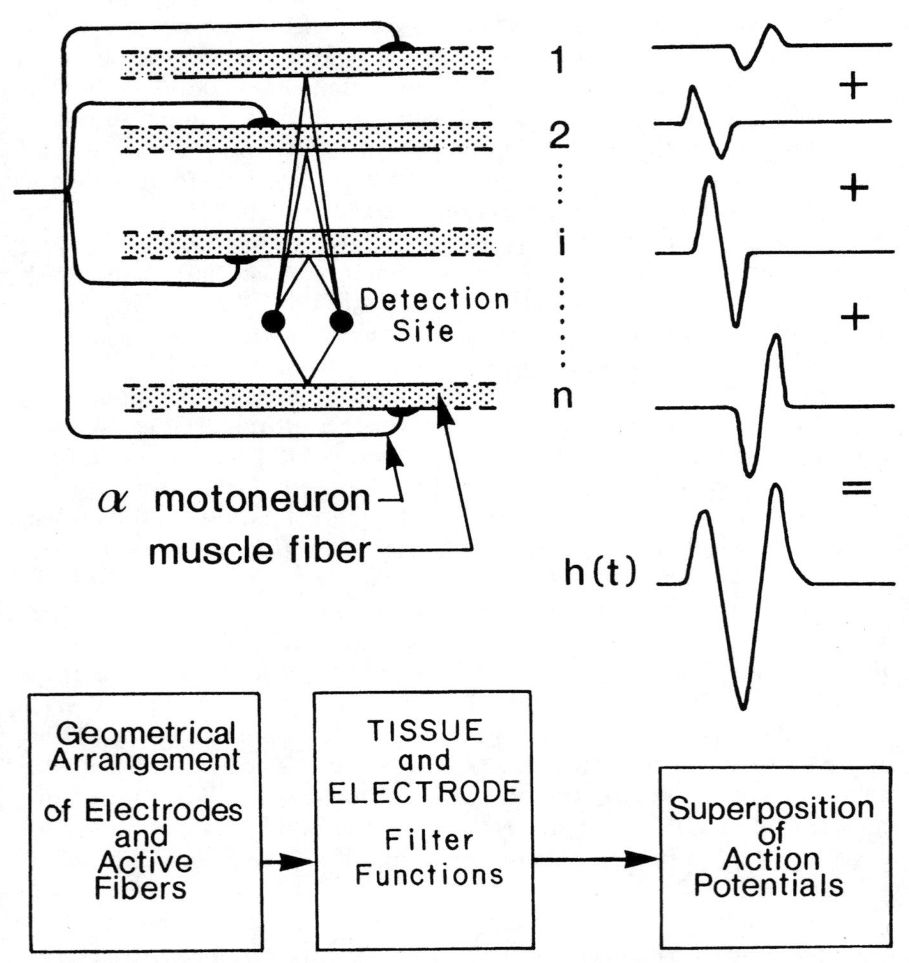

Figure 3.2. Schematic representation of the generation of the motor unit action potential.

conduction velocity of the muscle fiber, which ranges from 3 to 6 m/s. The relative time of initiation of each action potential is directly proportional to the difference in the length of the nerve branches, and the distance the depolarizations must propagate along the muscle fibers before they approach the detectable range (pickup area) of the electrode. This relative time of initiation is also inversely proportional to the conduction velocities of the nerve branch and the muscle fiber. The time delay caused by propagation along the muscle fibers is an order of magnitude greater than that caused by the nerve branch because of the much faster alpha-motoneuron conduction velocity (in the order of 50 to 90 m/s).

The waveform and, therefore, the frequency spectrum of the action potentials will be affected by the tissue beween the muscle fiber and the detection site. As described in the previous chapter, the presence of this tissue creates a low-pass filtering effect whose bandwidth decreases as the distance increases. This filtering effect of the tissue is much more pronounced for surface electrode recordings than for indwelling electrode recordings because indwelling electrodes are located closer to the active muscle fibers.

Thus far, muscle fiber action potentials have been considered as distinguishable individual events. However, since the depolarizations of the muscle fibers of one motor unit overlap in time, the resultant signal present at the detection site will constitute a spatial-temporal superposition of the contributions of the individual action potentials. The resultant signal is called the motor unit action potential (MUAP) and will be designated as $h(t)$. A graphic representation of the superposition is shown on the *right side* of Figure 3.2. This particular example presents a triphasic MUAP. The shape and the amplitude of the MUAP are dependent on the geometric arrangement of the active muscle fibers with respect to the electrode site as well as all the previously mentioned factors which affect the action potentials (Refer to discussion in previous chapter for details.)

If muscle fibers belonging to other motor units in the detectable vicinity of the electrode are excited, their MUAPs will also be detected. However, the shape of each MUAP will generally vary due to the unique geometric arrangement of the fibers of each motor unit with respect to the detection site. MUAPs from different motor units may have similar amplitude and shape when the muscle fibers of each motor unit in the detectable vicinity of the electrode have a similar spatial arrangement. Even slight movements of indwelling electrodes will significantly alter the geometric arrangement and, consequently, the amplitude and shape of the MUAP.

Given the various factors that affect the shape of an observed MUAP, it is not surprising to find variations in the amplitude, number of phases, and duration of MUAPs detected by one electrode, and even larger variations if MUAPs are detected with different electrodes. In normal muscle, the peak-to-peak amplitude of a MUAP detected with indwelling electrodes (needle or wire) may range from a few microvolts to 5 mV, with a typical value of 500 μV. According to Buchthal et al (1954), the number of phases of MUAPs detected with bipolar needle electrodes may range from one to four with the following distribution: 3% monophasic, 49% biphasic, 37% triphasic, and 11% quadriphasic. MUAPs having more than four phases are rare in normal muscle tissue but do appear in abnormal muscle tissue (Marinacci, 1968). The time duration of MUAPs may also vary greatly, ranging from less than 1 to 13 ms (Stålberg et al, 1975; Basmajian and Cross, 1971).

Petersén and Kugelberg (1949) first reported a slight prolongation of the MUAP with advancing age. Later, Sacco et al (1962) proved, in a systematic study of abductor digiti quinti, biceps brachii, and tibialis anterior of normal infants (3 months of age) and adults, that the duration of the MUAPs was significantly shorter in the muscles of infants (Fig. 3.3). This they explained in terms of the increase in width of the endplate zone with growth. In persons from 20 to 70 years of age, the mean duration of the MUAPs increased 25% further in the biceps brachii, but they remained unaltered in the abductor digiti quinti. Whenever the duration of the MUAPs increased with age, there was an increase in mean amplitude. The Russian physiologists Fudel-Osipura and Grishko (1962) reported that in aged persons the amplitude of the MUAP decreased and the time duration increased. A similar but more detailed

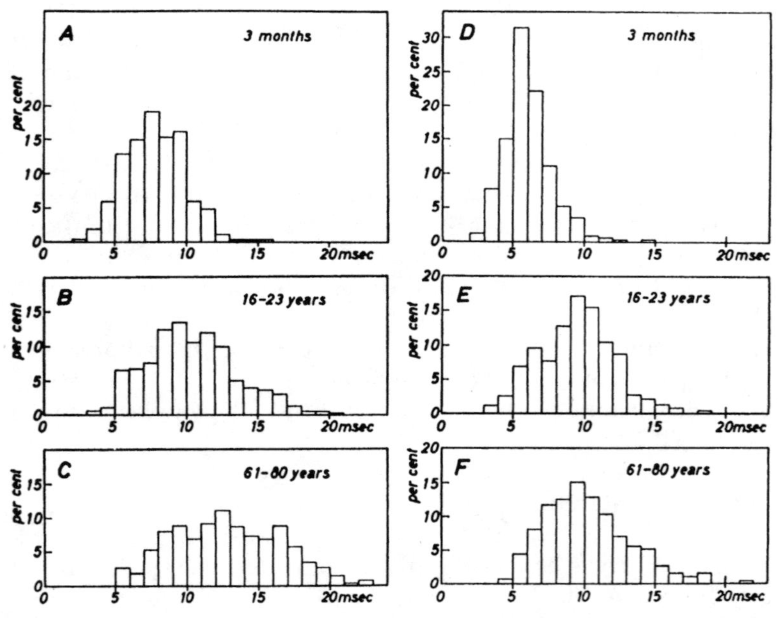

Figure 3.3. Histograms of duration of motor unit potentials recorded with concentric electrodes from biceps brachii (*A–C*) and abductor digiti quinti (*D–F*) of subjects of different ages. (From G. Sacco et al, © 1962.)

study was reported by Carlson et al (1964). They found a considerable number of highly complex and long duration MUAPs in more than half of the older age group (Fig. 3.4), suggesting that such a deviation from normal motor unit activity is a characteristic of aged skeletal muscle. (No polyphasic or long-duration potentials were noted in the young normal individuals who made up their control group.) In view of the absence of denervation (fibrillation) potentials in all aged subjects and the finding of normal motoneuron conduction velocities, Carlson et al could not relate the presence of complex potentials to a neurogenic disturbance. They thought this could be explained on the basis of physiological alteration of the muscle fiber with an associated delay in fiber response.

It should be emphasized that the amplitude and shape of an observed MUAP are a function of the geometrical properties of the motor unit, muscle tissue, and detection electrode properties. The filtering properties of the electrode (and possibly the cable connecting the electrode to the preamplifiers, as well as the preamplifiers themselves) can cause the observed MUAPs to

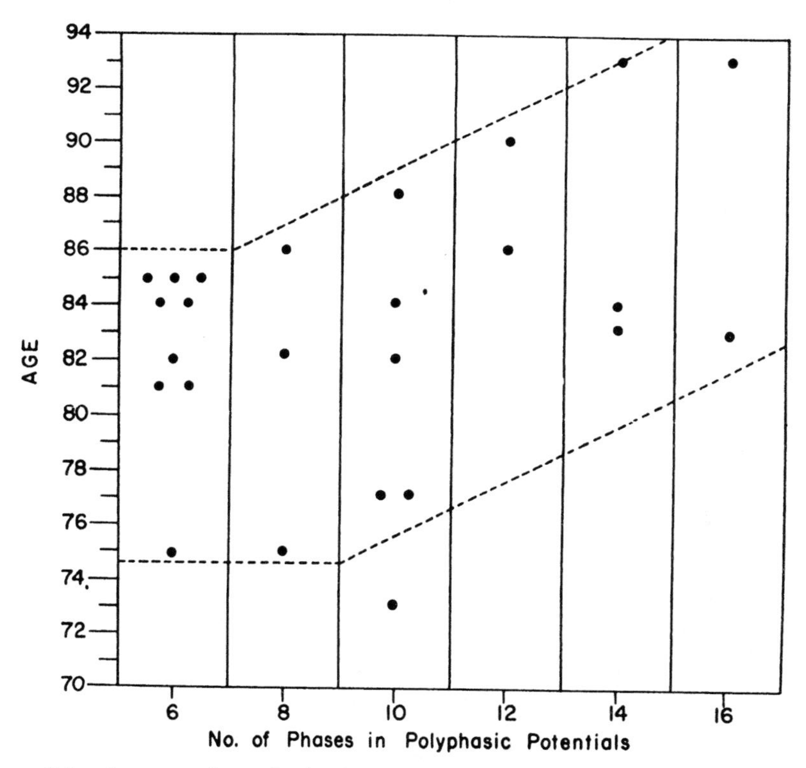

Figure 3.4. Increase in polyphasic activity as a function of age. (From K.E. Carlson et al, 1964, *American Journal of Physical Medicine*)

have additional phases and/or longer durations. This is an inevitable behavior of most filter networks. These effects were known empirically when Petersén and Kugelberg (1949) reported that the configuration of the electrode affected the duration and amplitude of the detected action potentials.

THE MOTOR UNIT ACTION POTENTIAL TRAINS

The electrical mainfestation of a MUAP is accompanied by a twitch of the muscle fibers. In order to sustain a muscle contraction, the motor units must be repeatedly activated. The resulting sequence of MUAPs is called a motor unit action potential train (MUAPT). The waveform of the MUAPs within a MUAPT will remain constant if the geometric relationship between the electrode and the active muscle fibers remains constant, if the properties of the recording electrode do not change, and if there are no significant biochemical changes in the muscle tissue. Biochemical changes within the muscle could affect the conduction velocity of the muscle fiber and filtering properties of the muscle tissue.

The muscle fibers of a motor unit are randomly distributed throughout a subsection of the normal muscle and are intermingled with fibers belonging to different motor units. Evidence for this anatomical arrangement in the rat and cat has been presented by Edström and Kugelberg (1968), Doyle and Mayer (1969) and Burke and Tsairis (1973). There is also indirect electromyographic evidence suggesting that a similar arrangement occurs in human muscle (Stålberg and Ekstedt, 1973; Stålberg et al, 1976). The cross-sectional area of a motor unit territory ranges from 10 to 30 times the cross-sectional area of the muscle fibers of the motor unit (Buchthal et al, 1959; Brandstater and Lambert, 1973). This admixture implies that any portion of the muscle may contain fibers belonging to 20 to 50 motor units. Therefore, a single MUAPT is observed when the fibers of only one motor unit in the vicinity of the electrode are active. Such a situation occurs only during a very weak muscle contraction. As the force output of a muscle increases, motor units having fibers in the vicinity of the electrode become activated, and several MUAPTs will be detected simultaneously. This is the case even for highly selective electrodes which detect action potentials of single muscle fibers. As the number of simultaneously detected MUAPTs increases, it becomes more difficult to identify all the MUAPs of any particular MUAPT due to the increasing probability of overlap between MUAPs of different MUAPTs.

A Model for the Motor Unit Action Potential Train

Several investigators have attempted to formulate mathematical expressions for the MUAPT (Bernshtein, 1967; Libkind, 1968; De Luca, 1968; Coggshall and Bekey, 1970; Stern, 1971; Brody et al, 1974; Gath, 1974; De Luca, 1975). Of these investigators, only Libkind (1968), De

Luca (1975), and Gath (1974) have employed empirically derived information to construct the model.

The MUAPT may be completely described by its IPIs and the waveform of the MUAP. From a mathematical point of view, it is convenient to describe the MUAPT as a random process in which the waveform of the MUAP is present at random intervals of time. Considerable support for this approach is presented in the remainder of this chapter. This approach requires that we describe mathematically the waveform and the rate of occurrence of the MUAPs in the train, i.e., the firing rate.

For the purpose of our discussion, the firing rate will be considered to be only a function of time (t) and force (F) and will be denoted as $\lambda(t, F)$. This restriction in the notation is adopted for convenience. However, it should be clearly understood that the derivations which follow apply for any general description of the firing rate. If future investigations reveal definitive relationships between the firing rate and the force rate, velocity, and acceleration of a contraction, they can be readily incorporated into the ensuing model with no loss of generality. A systematic way of obtaining a mathematical expression for $\lambda(t, F)$ is to fit the IPI histogram with a probability distribution function, $p_x(x, t, F)$. The inverse of the mean value of $p_x(x, t, F)$ will be the firing rate, or

$$\lambda(t, F) = \left[\int_{-\infty}^{\infty} x p_x(x, t, F)\, dx\right]^{-1}$$

Alternatively, a mathematical expression for $\lambda(t, F)$ could be obtained by performing a regression analysis of the IPIs as a function of time and force.

On the other hand, it would be extremely difficult to give a unique mathematical description of the MUAP because there are many possible shapes. However, if a MUAPT is isolated and the MUAP can be identified, it would be possible to make a piecewise approximation of the shape. Refer to De Luca (1975) for additional information on the mathematical representations.

It is further convenient to decompose the MUAPT into a sequence of Dirac delta impules, $\delta_i(t)$, which are passed through a filter (black box) whose impulse response is $h_i(t)$. The impulse response of the filter may be constructed to be time variant in order to reflect any change in the waveform of the MUAP during a sustained contraction. This concept is not included in this part of the development of the model so that the equations do not become unnecessarily cumbersome.

If each Dirac delta impulse marks the time occurrence of a MUAP in a MUAPT, the output of the filter will be the MUAPT or $u_i(t)$. The integer i denotes a particular MUAPT. This decomposition, shown in Figure 3.5, allows us to treat the two characteristics of the MUAPT separately.

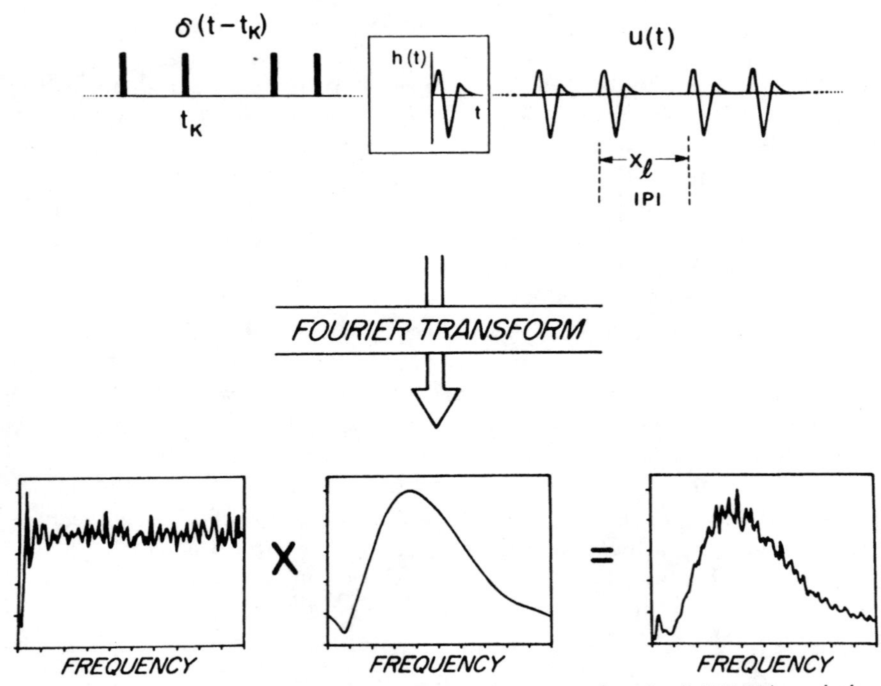

Figure 3.5. Model for a motor unit action potential train (MUAPT) and the corresponding Fourier transform of the interpulse intervals (IPIs), the motor unit action potentials (MUAP), and the MUAPT.

The Dirac delta impulse train can be described by

$$\delta_i(t) = \sum_{k=1}^{n} \delta(t - t_k)$$

It follows that the MUAPT, $u_i(t)$, can be expressed as

$$u_i(t) = \sum_{k=1}^{n} h_i(t - t_k)$$

where $t_k = \sum_{l=1}^{k} x_l$ for $k, l = 1, 2, 3, \ldots, n$. In the above expressions, t is a real continuous random variable, t_k represents the time locations of the MUAPs. x represents the IPIs, n is the total number of IPIs in a MUAPT, and i, k, and l are integers which denote specific events.

Reports have appeared in the literature which argue in favor of the existence of minimal (if any) dependence among the IPIs of a particular MUAPT (Masland et al, 1969; De Luca and Forrest, 1973a; Kranz and Baumgartner, 1974; and others). It could be argued that these reports do not firmly conclude that the IPIs of a MUAPT are statistically independent. In fact, it is easy to envisage specific circumstances and situations where the Ia-alpha motoneuron reflex loop may inject some dependence. However, the overwhelming majority of the available data supports an approach for modeling the IPI train as a renewal process which provides considerable mathematical and practical convenience.

It is now possible to write expressions for two time-domain parameters of signals, i.e., the mean rectified value and the mean-squared value (whose square root is the root-mean-squared value, or rms) by invoking one restriction, that is, the waveform of the MUAP remains invariant throughout the train. Then, and only then, it follows that:

$$\text{Mean rectified value} = E\{|u_i(t, F)|\} = \int_0^\infty \lambda_i(\hat{t}, F)|h_i(t - \hat{t})|\, d\hat{t}$$

$$\text{Mean-squared value} = MS\{u_i(t, F)\} = \int_0^\infty \lambda_i(\hat{t}, F)\, h_i^2(t - \hat{t})\, d\hat{t}$$

where $\hat{t}$ is a dummy variable and E is the mathematical symbol for the expectation or the mean. Although the above equations can be solved, the computation requires the execution of a convolution. De Luca (1975) has shown that since $\lambda(t, F)$ is slowly time varying, the above expressions can be greatly simplified to:

$$E\{|u_i(t, F)|\} = \underline{|h_i(t)|}\, \lambda_i(t, F)$$

$$MS\{u_i(t, F)\} = \underline{h_i^2(t)}\lambda_i(t, F)$$

In most cases, this approximation introduces an error of less than 0.001%. The bar denotes an integration from zero to infinity as a function of time. These mathematical computations are displayed in Figure 3.6. It should be noted that the first term on the right side of the equations is now a scaling value and is independent of time. Hence, these MUAPT parameters are reduced to the expression of the firing rate multiplied by a scaling factor.

To compute the expression for the power density spectrum (frequency content) of a MUAPT it is necessary to consider additional statistics of the IPIs and the actual MUAP waveform. The IPIs can be described as a real, continuous, random variable. Only minimal (if any) dependence exists among the IPIs of a particular MUAPT. Therefore, the MUAPT may be represented as a renewal pulse process. A renewal pulse process is one in which each IPI is independent of all the other IPIs.

The power density spectrum of a MUAPT was derived from the above formulation by Le Fever and De Luca (1976) and independently by Lago and Jones (1977). It can be expressed as:

$$S_{u_i}(\omega, t, F) = S_{\delta_i}(\omega, t, F)|H_i(j\omega)|^2$$

$$= \frac{\lambda_i(t, F)\cdot\{1 - |M(j\omega, t, F)|^2\}}{1 - 2\cdot\text{Real}\,\{M(j\omega, t, F)\} + |M(j\omega, t, F)|^2}\{|H_i(j\omega)|^2\}$$

$$\text{for} \quad \omega \neq 0$$

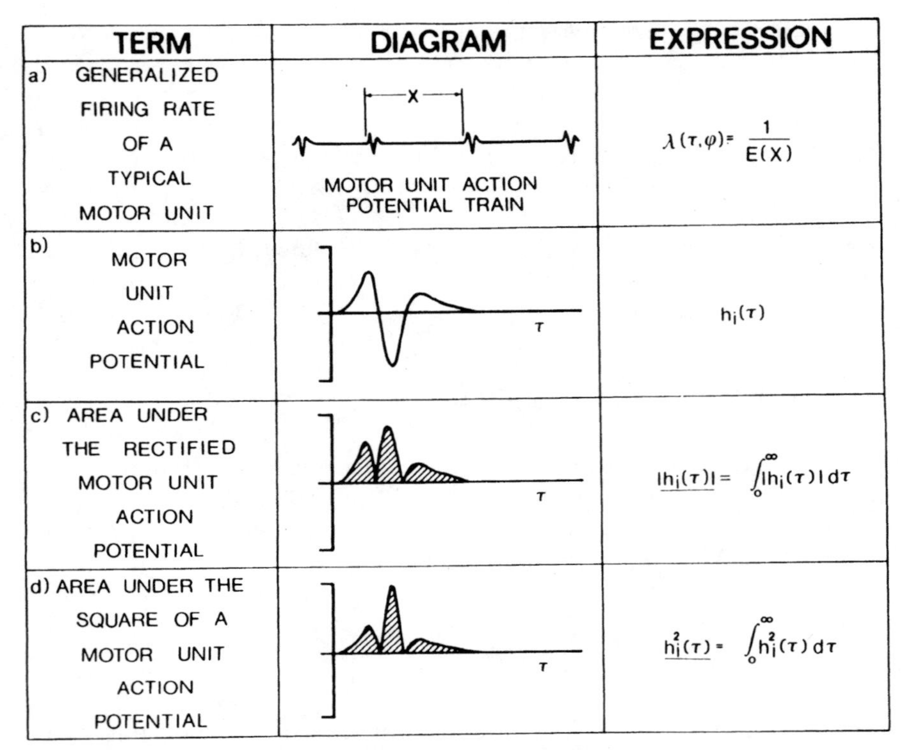

TERM
DIAGRAM
EXPRESSION
a) GENERALIZED FIRING RATE OF A TYPICAL MOTOR UNIT
X
MOTOR UNIT ACTION POTENTIAL TRAIN
λ(τ,φ)= 1/E(X)
b) MOTOR UNIT ACTION POTENTIAL
τ
h_i(τ)
c) AREA UNDER THE RECTIFIED MOTOR UNIT ACTION POTENTIAL
τ
|h_i(τ)| = ∫_0^∞ |h_i(τ)| dτ
d) AREA UNDER THE SQUARE OF A MOTOR UNIT ACTION POTENTIAL
τ
h_i^2(τ) = ∫_0^∞ h_i^2(τ) dτ

Figure 3.6. Explanation of some of the terms in the expressions in the text.

where

ω = the frequency in radians
$H_i(j)$ = the Fourier transform of $h_i(t)$
$M(j\omega, t, F)$ = the Fourier transform of the probability distribution function, $p_x(x, t, F)$ of the IPIs.

More recently, Blinowska et al (1979) have derived similar expressions. The above expressions clearly indicate that the power density spectrum may be constructed from the energy spectra of the MUAPs of the motor units and the firing statistics of the motor unit. In addition, the contribution of the firing statistics has a multiplicative effect on the energy spectrum of the MUAP.

By representing $h_i(t)$ by a Fourier series, LeFever and De Luca (1976) were able to show that in the frequency range of 0 to 40 Hz the power density spectrum is affected primarily by the IPI statistics. A noticeable peak appears in the power density spectrum at the frequency corresponding to the firing rate and progressively lower peaks at harmonics of the firing rate. The amplitude of the peaks increases as the IPIs become more regular. Beyond 40 Hz, the power density spectrum is essentially determined by the shape of $h_i(t)$.

This point has been verified empirically. By using the computer-assisted decomposition technique developed by LeFever and De Luca (1982) and Mambrito and De Luca (1983), it is possible to obtain highly accurate IPI measurements of MUAPT records many seconds long. The details of this technique are described in Chapter 4. The Fourier transform of the IPIs may then be computed directly. Figure 3.7 presents the magnitude of such Fourier transforms for MUAPTs that were all detected during two separate isometric constant-force contractions maintained at 50% of maximal voluntary force in the first dorsal interosseous muscle. The time duration of the MUAPT segment that was analyzed was 5 s. The function with the solid line represents the average. The histograms present the IPI distribution of each motor unit, the one on the left corresponding to the function with the broken line and the middle histogram to the function with the dash-dot line. Some statistics of the IPIs are presented in Table 3.1. The coefficient of variation, which is the ratio of the standard deviation to the mean value, is a measure of the regularity with which the motor unit is discharging. The smaller the coefficient of variation, the sharper and higher will be the peak corresponding to the firing rate in the magnitude of the Fourier transform. When the coefficient of variation is higher (0.26, 0.28), then the peak is less sharp and has lower amplitude (see Figure 3.7*b*).

Concepts of Normalization and Generalized Firing Rate

A generalized representation of the EMG signal must contain a formulation which allows a comparison of the signal between different muscles and individuals. This is not a problem in some contractions, such as those involving ballistic movements. However, it is a requirement in isometric and anisometric contractions. The formulation for comparison may be obtained by normalizing the variables of the EMG signal with respect to their maximal measurable value in the particular experimental procedure. For example, in a constant-force isometric contraction, the time is normalized with respect to the duration that the individual can maintain the designated force level. The contraction force is normalized with respect to the force value of a MVC. The *normalized contraction-time* will be denoted by τ, the *normalized force* by ϕ, and their maximal value is 1.

In the model, expressions for the parameters of the EMG signal are formed by a superposition of the equations of the MUAPT derived in the previous section. Such an approach requires that the mathematical relationship of the firing rates of all the individual MUAPTs be first definable and then known. Such information is difficult to obtain. Even if it could be obtained it may prove to be useless due to the difficulty in

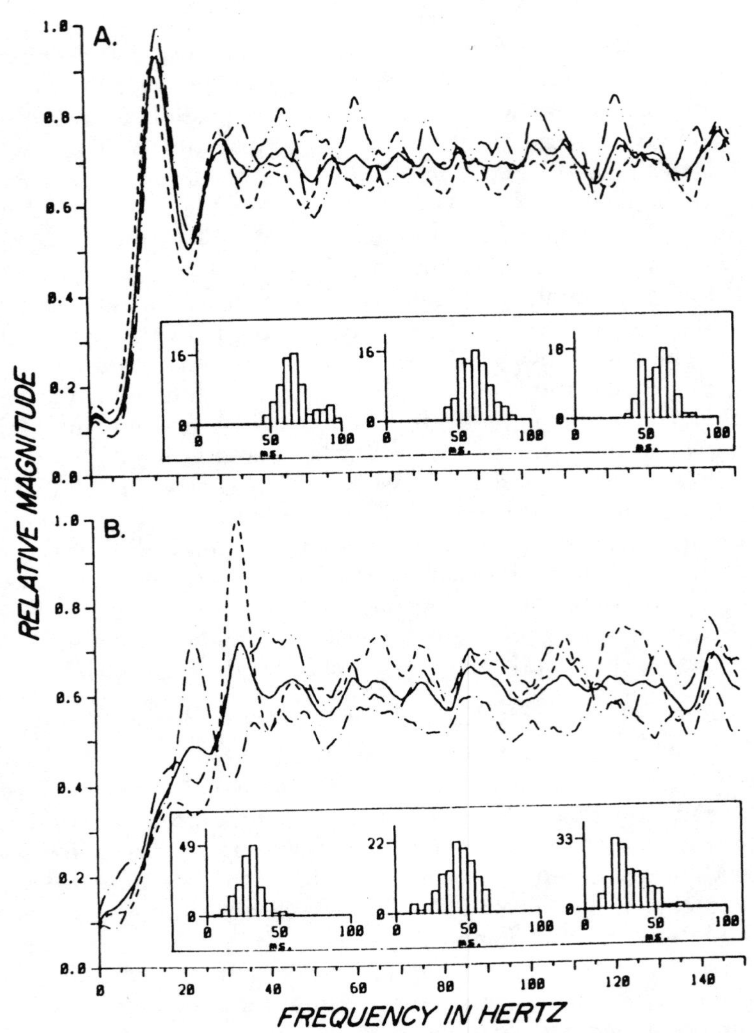

Figure 3.7. Magnitude of the Fourier transforms and histograms of the interpulse intervals (IPI) of the motor unit action potential trains obtained from the first dorsal interosseous muscle during two separate constant-force isometric contractions performed at 50% maximal force. The function plotted with the *solid line* represents the average. The histograms present the IPI distribution, with the one on the *left* corresponding to the function with the *broken line* and with the middle histogram corresponding to the function with the *dash-dot* line. Note that the peak corresponding to the average firing rate value is present in some functions.

Table 3.1
Interpulse Statistics of MUAPTs Obtained During Isometric Constant-Force Contractions Sustained at 50% of Maximal Force in the First-Dorsal Interosseous Muscle[a]

	μ[b] (ms)	SD (ms)	CV
	58.5	9.3	0.20
Fig. 3.7*A*	61.8	10.0	0.16
	69.3	13.7	0.16
	29.7	8.3	0.28
Fig. 3.7*B*	31.3	11.8	0.38
	43.4	11.2	0.26

[a] This is the detailed data of the functions and histograms plotted in Figure 3.5.
[b] μ, mean; SD, standard deviation; CV, coefficient of variation.

defining this as a function of time and force. There is considerable evidence (which will be described in Chapter 5) that the firing rate is not necessarily a monotonic function of either time or force, and the relationship between firing rate and force is also dependent on the particular muscle. To overcome this barrier, De Luca (1968) introduced the concept of the *generalized firing rate* and defined it as the mean value of the firing rates of the MUAPTs detected during a contraction. For a detailed description of the calculation of the generalized firing rate, refer to De Luca and Forrest (1973a).

It must be emphasized that the generalized firing rate is a mathematical concept which may only be properly mathematically described with extreme difficulty. It will be dependent on many factors which may effect the occurrence of the IPIs, such as recruitment of a new motor unit, minute force perturbations in the intended force output of a muscle, reflex activity, etc. One example of the mathematical formulation has been described by De Luca and Forrest (1973a) and is presented here:

$$\lambda(\tau, \phi) = \frac{1000}{\beta(\tau, \phi)\Gamma[1 + 1/\kappa(\tau, \phi)] + \alpha} \text{ pulses per second}$$

$$\kappa(\tau, \phi) = 1.16 - 0.19\tau + 0.18\phi$$
$$\beta(\tau, \phi) = \exp\,(4.60 + 0.67\tau - 1.16\phi) \text{ ms}$$
$$\alpha = 3.9 \text{ ms}$$
$$\text{for } 0 < \tau < 1,\ 0 < \phi < 1$$

The above values are valid for the middle fibers of the deltoid muscle during a constant force isometric contraction. Other relationships will exist for other muscles. The above equation is plotted in Figure 3.8.

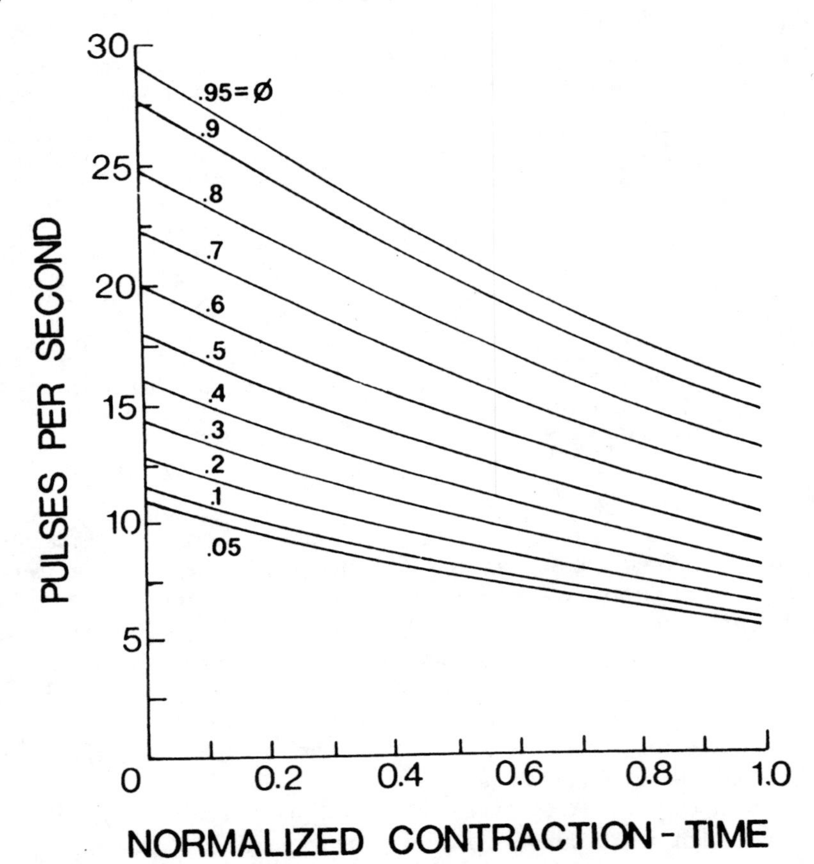

Figure 3.8. Generalized firing rate of motor unit action potential trains as a function of normalized contraction-time at various normalized constant-force levels. The force was normalized with respect to the maximal isometric contraction.

A Model for the EMG Signal

The EMG signal may be synthesized by linearly summing the MUAPTs as they exist when they are detected by the electrode. This approach is expressed in the following equation:

$$m(t, F) = \sum_{i=1}^{p} u_i(t, F)$$

and is displayed in Figure 3.9, where 25 mathematically generated MUAPTs are added to yield the signal at the bottom. This composite signal bears striking similarity to real EMG signals.

Biro and Partridge (1971) obtained empirical evidence which justified this approach. The modeling approach was later expanded by De Luca

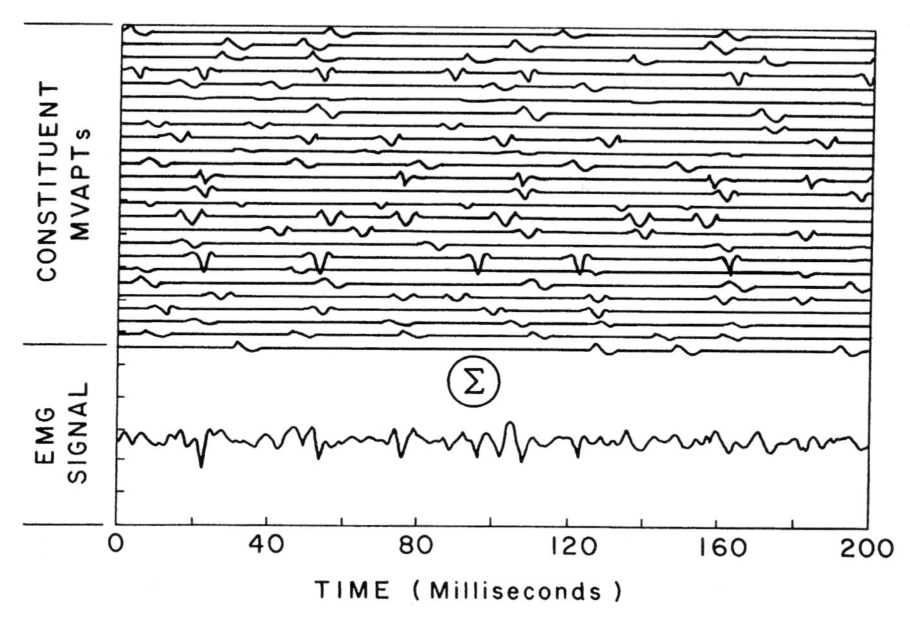

Figure 3.9. An EMG signal formed by adding (superimposing) 25 mathematically generated MUAPTs.

and van Dyk (1975) and Meijers et al (1976). A schematic representation of the model is shown in Figure 3.10. The interger p represents the total number of MUAPTs which contribute to the potential field at the recording site. Each of the MUAPTs can be modeled according to the approaches presented in Figure 3.2 and 3.5. The superposition at the recording site forms the physiological EMG signal, $m_p(t, F)$. This signal is not observable. When the signal is detected, an electrical noise, $n(t)$, is introduced. The detected signal will also be affected by the filtering properties of the recording electrode, $r(t)$, and possibly other instrumentation. The resulting signal, $m(t, F)$, is the observable EMG signal. The location of the recording site with respect to the active motor units determines the waveform of $h(t)$, as described at the beginning of this chapter.

A Comparison of the Time-Dependent Parameters of the EMG Signal

From this concept it is possible to derive expressions for the mean rectified value, the root-mean-squared value, and the variance of the rectified EMG signal. The expressions are presented in Figure 3.11. Their derivation can be found in the article by De Luca and van Dyk (1975). In Figure 3.11, each of the terms of the expressions are associated with five physiological correlates which affect the properties of the EMG signal.

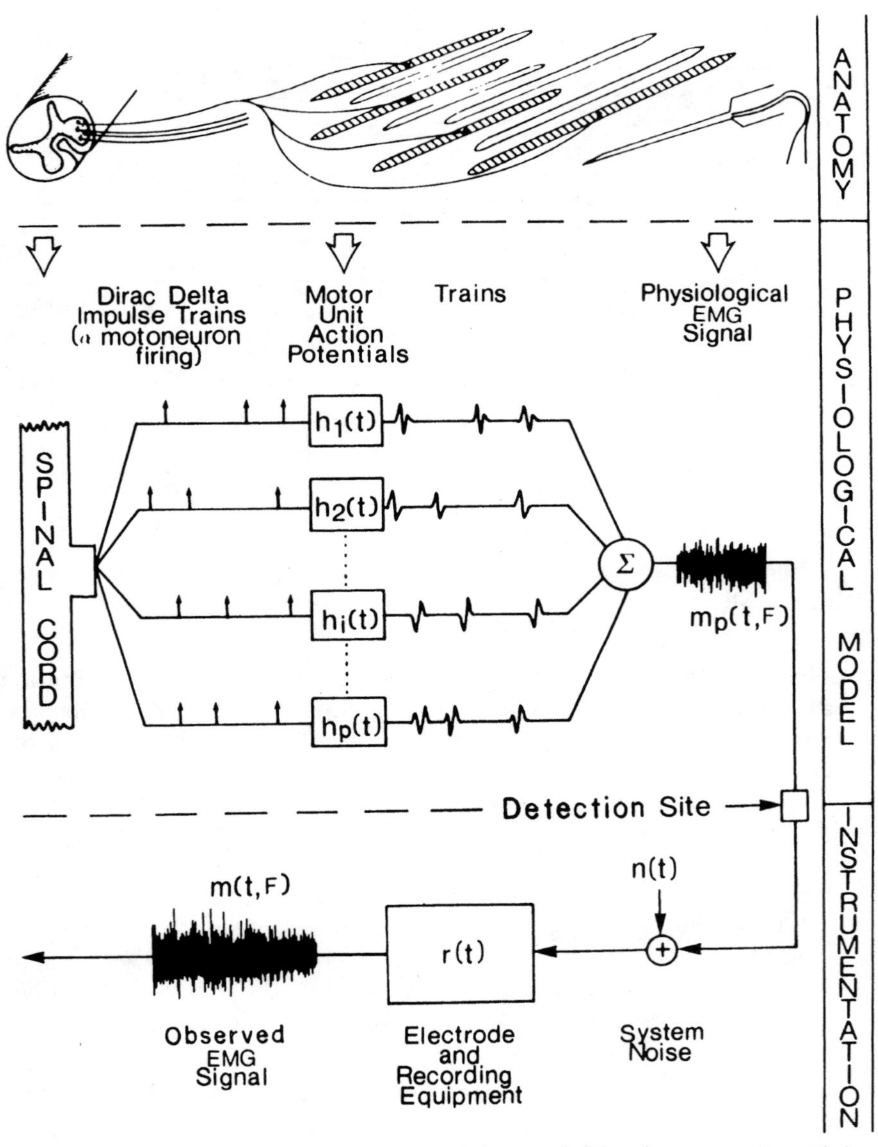

Figure 3.10. Schematic representation of the model for the generation of the EMG signal.

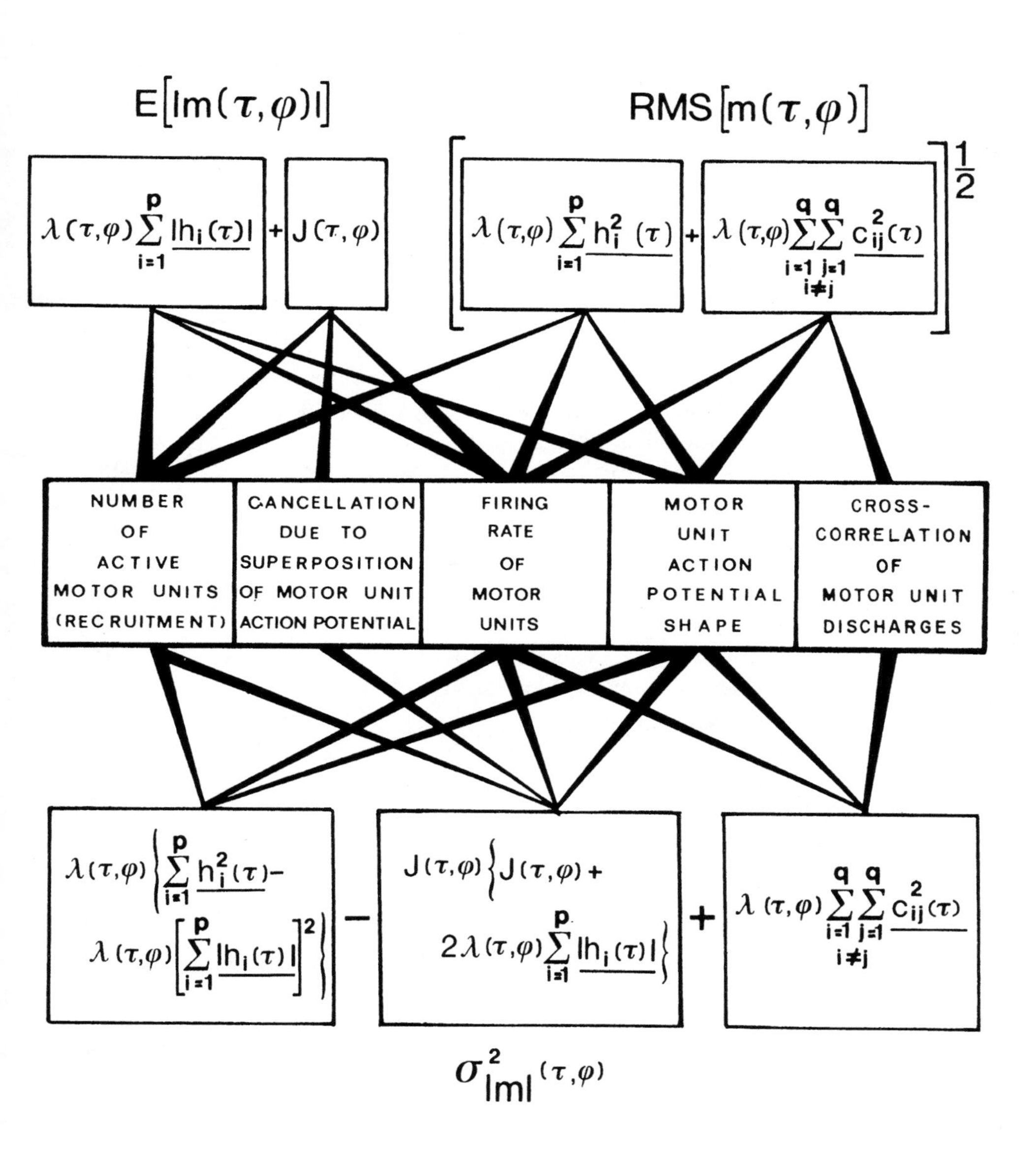

Figure 3.11. Theoretical expressions for parameters of the EMG signal and their relation to physiological correlates of a contracting muscle.

In the equation of the mean rectified value, the term $J(t, F)$ is a nonpositive term which accounts for the cancellation in the signal due to the superposition of opposite phases of the MUAPTs. In a sense, the *superposition term* represents the EMG activity which is generated by the muscle but is not available in the observed EMG signal. The expression for the mean rectified value confirms that this parameter of the EMG signal is dependent on the number and firing rates of the MUAPTs detected by the electrode, the area of the MUAPs, and the amount of cancellation occurring from the superposition of the MUAPTs.

The integral of the mean rectified value is a commonly used parameter in electromyography. By definition, it will be dependent on the same physiological correlates as the mean rectified value. It is often used to obtain a relationship between the EMG signal and the force output of the muscle. A linear relationship has been reported often. Considering all the physiological correlates involved, the nonlinearity of the motor unit behavior, and the viscoelastic properties of muscle tissue, a linear relationhip should be considered as accidental.

The root-mean-squared value is also dependent on the number and firing rates of the MUAPTs and the area of the MUAPTs but is not affected by the cancellation due to the MUAPT superposition. However, it is affected by the cross-correlation between the MUAPTs, represented by the $c_{ij}^2(\tau)$ terms. In the corresponding equation in Figure 3.11, the integer v denotes the number of MUAPTs that are cross-correlated. Note that any two MUAPTs can still be synchronized even if their cross-correlation term is zero, since the lack of cross-correlation is not sufficient to prove independence. Under this condition, their synchronization has no effect on the root-mean-squared value.

The expression for the variance of the rectified signal is more complicated, containing all the terms which are present in the previous two parameters. Therefore, it reflects the combined effect of all the physiological correlates. This parameter represents the AC power of the rectified EMG signal and should prove to be useful in analyzing the EMG signal. However, it has been used sparingly. Most of the past investigations have dealt with the DC level of the EMG signal.

The approach used thus far has been directed at relating the measurable parameters of the EMG signal to the behavior of the individual MUAPTs. However, when the electrode detects a large number of MUAPTs (greater than 15), such as would typically be the case for a surface electrode, the law of large numbers can be invoked to consider a simpler, more limited approach. In such cases, the EMG signal can be effectively represented as a band-limited signal with a Gaussian distributed amplitude. It is also possible to represent the amplitude by a carrier signal consisting of a random signal whose statistical properties are those of the signal during an isometric contraction, and a modulation signal reflecting the force and time-dependent properties of the signal. Such a representation was first publicized by Kreifeldt and Yao (1974) and was later expanded by Shwedyk et al (1977) and Hogan and Mann (1980).

PROPERTIES OF THE POWER DENSITY SPECTRUM OF THE EMG SIGNAL

According to the model in Figures 3.10 and 3.11, the power density spectrum of the EMG signal may be formed by summing all the auto- and cross-spectra of the individual MUAPTs, as indicated in this expression:

$$S_m(\omega) = \sum_{i=1}^{p} S_{\mu_i}(\omega) + \sum_{\substack{i,j=1 \\ i \neq j}}^{q} S_{u_i u_j}(\omega)$$

where $S_{u_i}(\omega)$ = the power density of the MUAPT, $u_i(t)$; and $S_{u_i u_j}(\omega)$ = the cross-power density spectrum of MUAPTs $u_i(t)$ and $u_j(t)$. This spectrum will be nonzero if the firing rates of any two active motor units are correlated. Finally, p = the total number of MUAPTs that comprise the signal; q = the number of MUAPTs with correlated discharges. For details of this mathematical approach, refer to De Luca and van Dyk (1975).

De Luca et al (1982b) have shown that many of the concurrently active motor units have, during an isometric muscle contraction, firing rates which are greatly correlated. It is not yet possible to state that all concurrently active motor units are correlated (although all our observations to date support this point). Therefore, q is not necessarily equal to p, which represents the total number of MUAPTs in the EMG signal. The above equation may be expanded to consider the following facts:

1. During a sustained contraction, the characteristics of the MUAP shape may change as a function of time (t). For example, De Luca and Forrest (1973a), Broman (1973, 1977), Kranz et al (1981), and Mills (1982) have all reported an increase in the time duration of the MUAP.
2. The number of MUAPTs present in the EMG signal will be dependent on the force of the contraction (F).
3. The detected EMG signal will be filtered by the electrode before it can be observed. This electrode filtering function will be represented by $R(\omega, d)$, where d is the distance between the detection surfaces of a bipolar electrode.

Note that the recruitment of motor units as a function of time during a constant force has not been considered; however, the required modification to the equation is trivial, and the concept may easily be accommodated. The concept of "motor unit rotation" during a constant force contraction (i.e., newly recruited motor units replacing previously active motor units) which has, at times, been speculated to exist, has also not been included. No account may be found in the literature which has provided evidence of this phenomenon by definitively excluding the likelihood that the indwelling electrode has moved relative to the active muscle fibers and, in fact, records from a new motor unit territory in the muscle. For a review of these details consult the material in Chapter 5. Hence:

$$S_m(\omega, t, F) = R(\omega, d)\left[\sum_{i=1}^{p(F)} S_{u_i}(\omega, t) + \sum_{\substack{i,j=1 \\ i\neq j}}^{q(F)} S_{u_iu_j}(\omega, t)\right]$$

The notation used in this equation may appear to be awkward in that it contains functions of time and frequency, when the frequency variable in turn represents a transformation from the time domain. This notation is used to describe functions that have time-dependent frequency properties. From a practical point of view, such equations are useful to describe the properties of "slow" nonstationary stochastic processes.

It is apparent from the examples presented in Figure 3.7 that the individual MUAPT power density function $S_{u_i}(\omega, t, F)$ may or may not have a peak at the frequency value corresponding to the firing rate, depending on the regularity with which the motor unit discharges, i.e., the coefficient of variation of the IPIs. When more than one MUAPT is present, the presence of a peak will also depend on the amount of separation between the mean of the individual IPI histograms. (This latter parameter determines how the peaks and valleys in the initial part of the Fourier transform superimpose and cancel out; note the different results of Figures 3.7*A* and *B*.) In general, when many MUAPTs having a wide range of individual coefficients of variation are present, the peaks will be less pronounced, and the effect of the IPI statistics on the magnitude of the Fourier transform of the EMG signal will be negligible above 30 Hz. No peaks will be distinct in the low frequency region. Empirical evidence of this behavior has been reported by De Luca (1968) and Hogan (1976), who noted large peaks occurring between 8 and 30 Hz. Hogan (1976) further showed that as successively more motor units are detected during increasing force level contractions, the amplitude of the peak diminishes with respect to the remainder of the spectrum. Hogan's data are presented in Figure 3.12.

It should be noted that the two parameters which have been identified as affecting the presence of the firing rate peak are both related to synchronization. That is, the smaller the coefficient of variation and the closer the average firing rate values, the greater the probability of two or more motor units discharging during a specific time interval. It should also be immediately added that physiological events may also occur that may render the MUAPTs dependent and thereby introduce a third parameter to the concept of synchronization.

Now, a major question arises concerning the cross-power density term. Does it really vary as a function of time during a sustained contraction

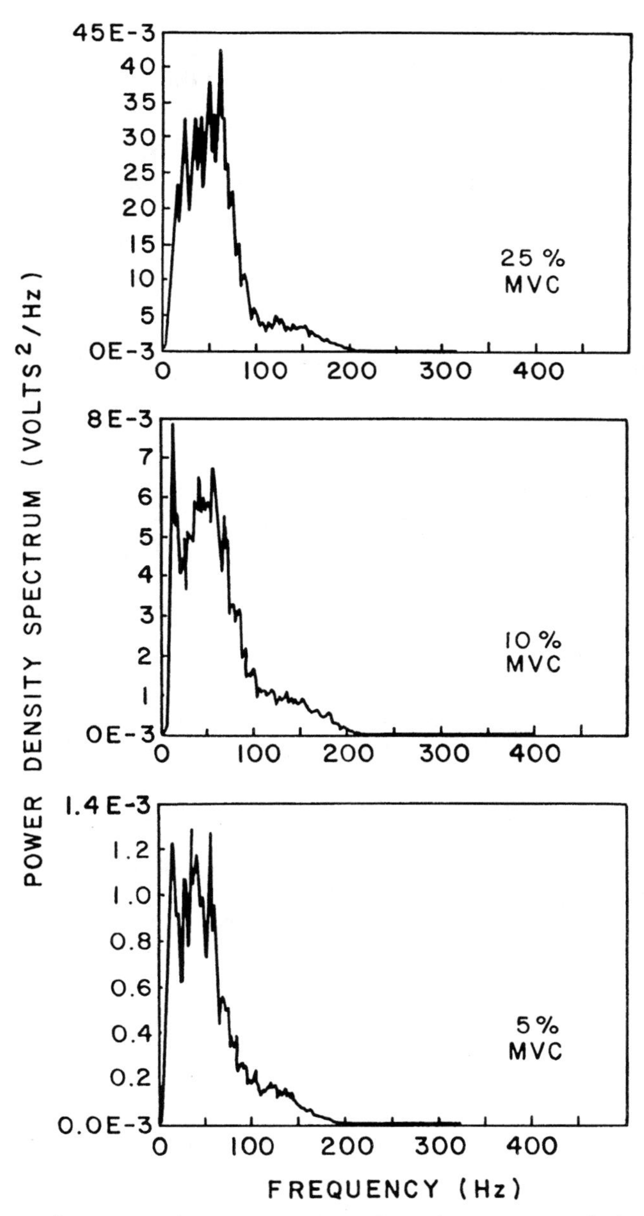

Figure 3.12. The power density spectrum when the EMG signal obtained from the biceps brachii during isometric constant-force contractions which are performed at 5%, 10%, and 25% of maximal level. Note that the peaks below 30 Hz are more pronounced during the 5% and 10% maximal level contractions. (From N.J. Hogan, © 1976, M.I.T., Cambridge, MA.)

so as to singly influence the power density spectrum of the EMG signal? There are three eventualities that may influence its time dependency: (1) the characteristics of the shape of the MUAP $u_i(t)$ and $u_j(t)$ change as a function of time; (2) the number of MUAPTs which are correlated varies as a function of time; (3) the degree of cross-correlation among the correlated MUAPTs varies. A change in the shape of the MUAP of $u_i(t)$ and $u_j(t)$ would not only cause an alteration in the cross-power density term but also would cause a more pronounced modification in the respective autopower density spectra. Hence, the power density spectrum of the EMG signal would be altered regardless of the modifications of the individual cross-power density spectra of the MUAPTs. There is to date no direct evidence to support the other two points. In fact, De Luca et al (1982*a* and *b*) have presented data which indicate that the cross-correlation of the firing rates of the concurrently active motor units does not appear to depend on either time during, or force of a contraction. This apparent lack of time-dependent cross-correlation of the firing rates is not inconsistent with previously mentioned observations, indicating that the synchronization of the motor unit discharges tends to increase with contraction time. These two properties can be unrelated.

Up to this point, the modeling approach has provided an explanation of the following aspects and behavior of the power density spectrum:

1. The amplitude increases with additionally recruited MUAPTs.
2. The IPI firing statistics influence the shape of the spectrum below 40 Hz, although this effect is not necessarily consistent, and is less evident at higher force when an increasing number of motor units are active.
3. The tendency for motor units to "synchronize" will affect the spectral characteristics but will be limited to the low frequency components.
4. Modification in the waveform of MUAPs within the duration of a train will effect most of the spectrum of the EMG signal. This is particularly worrisome in signals that are obtained during contractions that are anisometric, because in such cases the waveform of the MUAP may change in response to the modification of the relative distance between the active muscle fibers and the detection electrode.

The above associations do not fully explain the now well-documented property of the EMG signal, which manifests itself as a shift towards the low frequency end of the frequency spectrum during sustained contractions. It is apparent that modifications in the total spectral representation of the MUAPs can only result from a modification in the characteristics of the shape of the MUAP *per se*. During attempted isometric contraction, such modifications have their root cause in events that occur locally within the muscle. Broman (1973) and De Luca and Forrest (1973a) were the first to present evidence that the MUAP increases in time duration during a sustained contraction. More recently, Kranz et al (1981) and Mills (1982) have provided further support.

This approach was pursued by Lindström (1970), who derived a different expression for the power density spectrum of the EMG signal detected with bipolar electrodes. He approached the problem by representing the muscle fibers as cylinders in which a charge is propagated along their length. The simplified version of his results may be expressed as:

$$S_m(\omega) = R(\omega, d)\left[\frac{1}{v^2} G\left(\frac{\omega d}{2v}\right)\right]$$

where v = the average conduction velocity of active muscle fibers contributing to the EMG signal; G = the shape function which is implicitly dependent on many anatomical, physiological, and experimental factors; and d = the distance between the detection surfaces of the bipolar electrode.

One of the factors which is incorporated in the G function is the filtering effect of the tissue between the source (active muscle fibers) and the detection electrode. Note that this distance depends on the location (depth) of the muscle fibers within the muscle, plus the thickness of the fatty and skin tissues beneath the electrode. The form of this expression consists of modified Bessel functions of the second kind (Lindström, 1970). A plot of the expression for the "distance" or "tissue" filtering function has been presented in Figure 2.10. It has the characteristics of a low-pass filter with a cutoff frequency that is inversely related to the distance between the recording electrode and active muscle fibers.

The above expression may be generalized by introducing the time- and force-dependent effects on the EMG signal; then

$$S_m(\omega, t, F) = R(w, d)\left[\frac{1}{v^2(t, F)} G\left(\frac{\omega d}{2v(t, F)}\right)\right]$$

This representation of the power density spectrum explicitly denotes the interconnection between the spectrum of the EMG signal and the conduction velocity of the muscle fibers. Such a relationship is implicit in the previously presented modeling approach because any change in the conduction velocity would directly manifest itself in a change in the time duration of $h(t)$ as seen by the two detection surfaces of a stationary bipolar electrode. If the conduction velocity were to decrease, the depolarization current would require more time to traverse the fixed distance along the fibers in the vicinity of the detection surfaces and the detected MUAPs would have longer time durations. Hence, the frequency spectrum of the MUAPs and the EMG signal, which they comprise, will have a relative increase in the lower-frequency components and a decrease in the higher-frequency components. In other words, a shift toward the low-frequency end would occur. The above equation also explains the amplitude increase of the EMG signal as the conduction velocity decreases.

As discussed in the previous chapter, Lindström (1970) was able to show that for a bipolar electrode located between the innervation zone and the tendon of a muscle, the magnitude of the electrode filter function,

$$R(\omega, d) = K sin^2(\omega d/2v)$$

where d is the distance between the detection surfaces and v is the conduction velocity along the muscle fibers. The sin^2 term has a zero value at regular frequency intervals. These frequency values are related to the interdetection surface spacing and the conduction velocity of the muscle fibers. (Refer to Chapter 2 for details.) This function has a multiplicative effect on the power density spectrum, therefore, it too will be zero at the particular frequency values. Thus, if the interdetection surface spacing is known, it may be possible to calculate the average conduction velocity of the muscle fibers whose action potentials contribute to the detected signal.

These "dips" in the power density spectrum have been reported by several investigators (Lindström et al, 1970; Broman, 1973; Agarwal and Gottlieb, 1975; Lindström and Magnusson, 1977; Trusgnich et al, 1979; Hogan and Mann, 1980; and others), but their presence is not always prominent. Consider the case in which the detected signal consists of several MUAPTs whose respective MUAPs all contribute substantial energy to the power density spectrum but whose muscle fibers have different conduction velocities. Then a filter function somewhat similar to that shown in Figure 2.16 would apply for every MUAP, but the location of the zero points of the function would be different in accordance with the filter equation. Hence, the sum of the electrode filter functions would constitute a function with no clearly demonstrable dips, and the power density spectrum would not contain a discernible dip.

The opposite is true when there is one MUAPT in the EMG signal which contributes considerably more energy to the EMG signal or when most of the motor units which contribute to the EMG signal have nearly similar conduction velocities.

In a sense, the presence of the dips in analogous to the presence of the "peaks" at the low frequency end of the spectrum. Both will be present when individual MUAPTs in the EMG signal contribute considerable energy to the signal. But, it is possible for only one of these indicators to be present, depending on the firing statistics of the concurrently active motor units that contribute to the detected EMG signal.

At this point, it is necessary to recollect some consistently observed properties of the motor units and the EMG signal as a function of time and force during a contraction. As the force output of the muscle increases, the number of active motor units increases, and the firing rates of all the active motor units generally increase or remain nearly constant. As the time of a sustained contraction increases, the time duration of MUAPs increases, and the spectrum of the ME signal shifts towards the low frequency end. The ramifications of these factors interpreted through the explanations provided by the mathematical models are all graphically illustrated in Figure 3.13.

A Test for the Time Domain Parameters of the Model

Before testing the parameters derived by the model, it is necessary to comment that the neuromuscular system is an extraordinary actuator. It is capable of generating and modulating force under a wide variety of

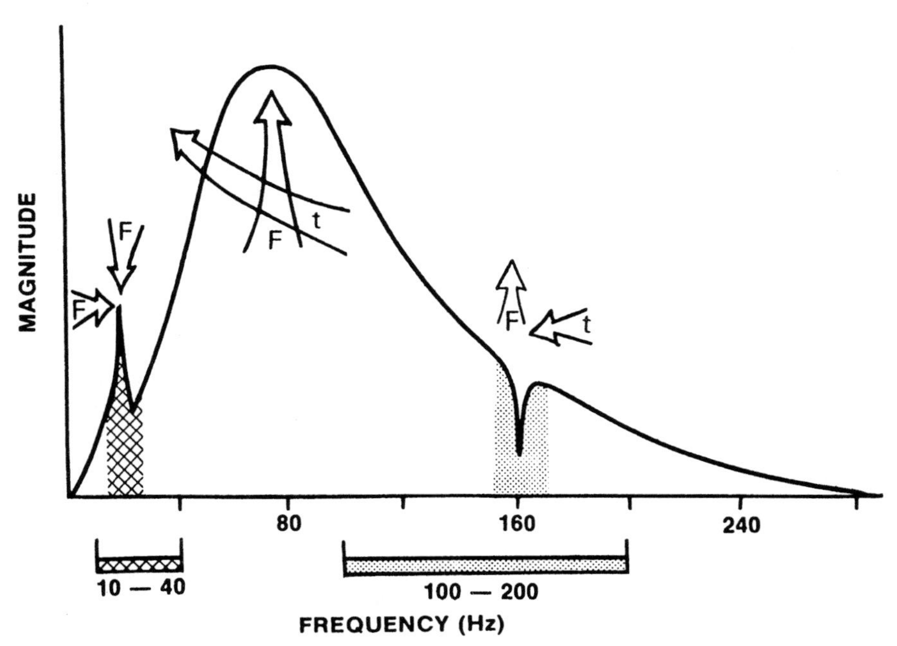

Figure 3.13. Diagramatic representation of the frequency spectrum of the differentially detected EMG signal with a graphic representation of the modifications which occur as a function of time and force of contraction. The shape has been purposefully exaggerated so as to accentuate interesting segments. The direction of the *arrows* indicates the direction of the modification on a particular segment of the spectrum caused by increasing either force or time. (For example, an arrow with the letter *F* indicates that as the force increases, the segment to which it is pointing will be modified in the direction of the *arrow*). The peak in the low frequency components is associated with the firing rates of motor units; the dip in the high frequency components is associated with the conduction velocity along the muscle fibers. The bars on the frequency axis indicate the range over which the peaks and dips may occur.

static (isometric) and dynamic (velocity, acceleration) conditions. It is known that the behavior of the MUAPTs varies for different types of contractions. However, it is possible to test the model for the EMG signal recorded during constant-force isometric contractions.

In a study performed by Stulen and De Luca (1978), EMG signals were simultaneously recorded differentially with bipolar surface and needle electrodes from the deltoid muscle while 11 subjects performed sustained constant-force isometric contractions at 25, 50, and 75% of maximal voluntary contraction (MVC). The empirical values of the parameters corresponding to those derived previously were calculated and compared. Let us consider the empirical root-mean-squared parameter which is plotted in Figure 3.14. The *solid lines* represent the average value for the 11 subjects. The *vertical lines* indicate 1 SD about the average. For convenience, the magnitude of the values has been normalized with respect to the largest value of the average.

Note that the amplitude of the root-mean-squared parameter increases as a function of time when the EMG signal is detected with surface electrodes, and decreases when detected with indwelling electrodes. Why? These signals were recorded simultaneously from the same area of the same muscles. To explain this apparent paradox we must turn our attention to the model.

It is possible to solve the equation for the root-mean-squared parameter in Figure 3.14 with the following restrictions: (1) no recruitment occurs during a constant-force contraction, (2) the areas of the MUAPs do not change; and (3) the MUAPTs are not cross-correlated. With these assumptions, the root-mean-squared parameter is directly proportional to the square root of the generalized firing rate. In fact, if the generalized firing rate of Figure 3.8 is normalized, it provides an exceptionally good fit to the mean value of the root-mean-squared parameter of the signal recorded with indwelling electrodes at 25 and 59% MVC but not at 75% MVC. It appears that the decrease in the signal from contractions executed at less than 50% MVC is due to the decrease in the firing rates of the motor units and that recruitment and synchronization do not play a significant role. But apparently at 75% MVC, other physiological correlates affect the EMG signal. Synchronization is a likely candidate to explain the different behavior of the root-mean-squared curves at 75% MVC. This indication is also implied in the behavior of the empirical mean rectified parameter and the variance of the rectified signal. However, it is necessary to emphasize that this result does not provide direct evidence of synchronization.

During a muscle contraction maintained at a constant force, if the firing rate of the motor units decreases and there is no significant recruitment, a complementary mechanism must occur to maintain the constant force output. One possible mechanism is the potentiation of

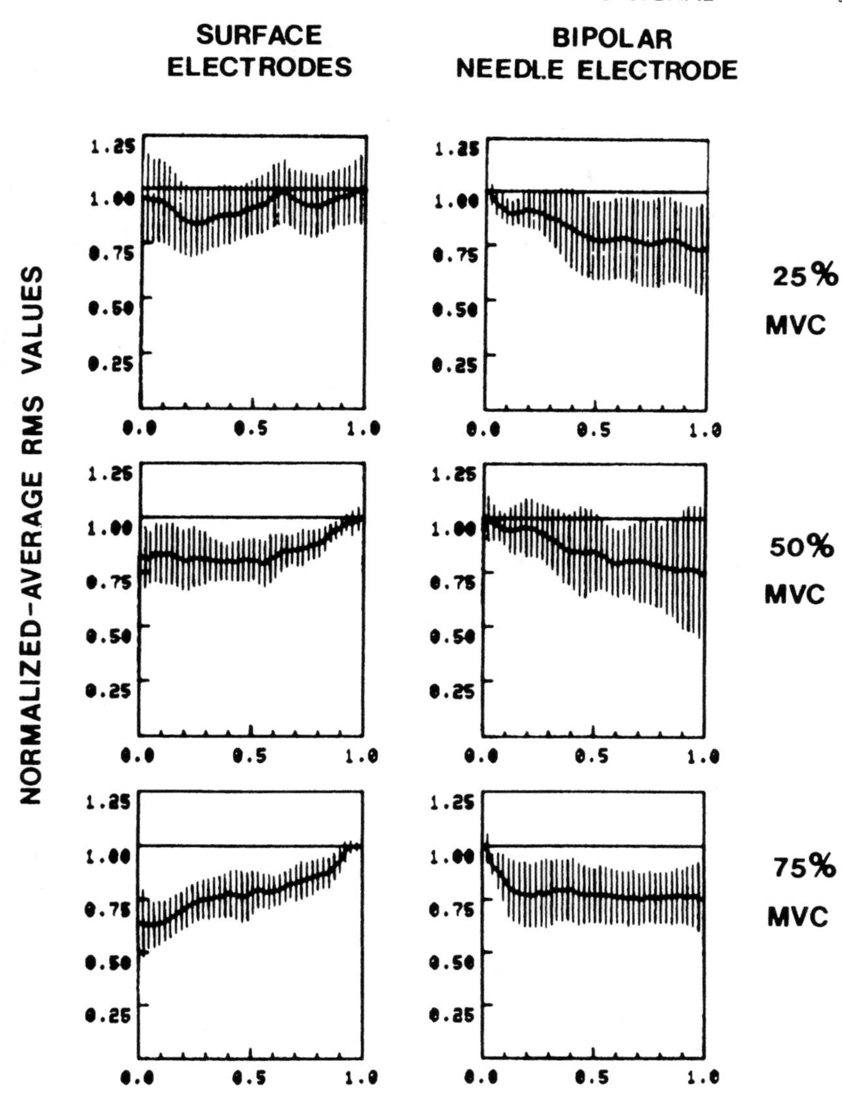

Figure 3.14. Average of the normalized root-mean-squared values from all subjects, plotted as a function of contraction time. Both the amplitude and time duration are normalized to their respective maxima. The *vertical lines* indicate 1 SD about the average.

twitch tension of the motor units as a contraction progresses. Evidence for potentiation of twitch tension caused by sustained repetitive stimulation has been presented by Gurfinkel' and Levik (1976) *in situ* in the human forearm flexors and by Burke et al (1976) *in vivo* in the cat gastrocnemius.

Now, let us consider the EMG signals recorded with surface electrodes. Why is the root-mean-squared value increasing during all three force levels when the firing rate is decreasing? The behavior of the firing rate is not affected by the type of electrodes used to record the signal. One possible explanation is as follows. During a sustained contraction, the conduction velocity along the muscle fibers decreases (Stålberg, 1966; Lindström et al, 1970). As a result, the time duration of the MUAPs increases (De Luca and Forrest, 1973a; Broman, 1973). This change in the shape of the MUAPs is reflected in the power density spectrum as a shift toward the lower frequencies which has been well documented. Hence, more signal energy passes through the tissue between the active fibers and the surface electrodes. Lindström et al (1970) have made theoretical calculations which show that the muscle tissue and differential electrodes act as low-pass filters. (Refer to Figs. 2.10 and 2.16 in the previous chapter for details.) As the distance between the active fibers and the electrodes increases, the bandwidth of the tissue-filter decreases. The increasing effect on the signal, due to the tissue and electrode filtering, overrides the simultaneously decreasing effect of the firing rate. The filtering effect is not seen in the signal recorded with the indwelling electrodes because the active fibers are much closer to the detection electrode.

Hence, the apparent paradox in the behavior of the EMG signal recorded with surface and indwelling electrodes is resolved by considering the detection arrangement.

It has been shown that the modeling approach is in agreement with the empirical results that could be tested. The model has also helped to resolve some ambiguities as well as to give insight into the information contained in the EMG signal. The limited discussion and arguments based on the describable known MUAPT behavior that have been presented are not sufficient to establish the generality of the model. However, as additional information describing the behavior of MUAPTs becomes available (especially the force dependence), the model should prove to be more useful and revealing.

TIME DOMAIN ANALYSIS OF THE EMG SIGNAL

Most of the parameters and techniques that will be discussed in this section have found wide use over the past years. Some of the approaches are relatively new and have only received acclaim among some groups.

As may be seen from the previous sections, the EMG signal is a time-

and force (and possibly other parameters)-dependent signal whose amplitude varies in a random nature above and below the zero value. Typically, the signal is recorded with AC coupled amplifiers. This guarantees that the average value (in this case the DC level) will be zero. This situation will prevail even if the amplifers are DC coupled, if the DC polarization potential is zero. In any case, it is apparent that simple averaging of the signal directly will not provide any useful information.

Rectification

A simple method that is commonly used to overcome the above restriction is rectifying the signal before performing more pertinent analysis. The process of rectification involves the concept of rendering only positive deflections of the signal. This may be accomplished either by eliminating the negative values (half-wave rectification) or by inverting the negative values (full-wave rectification). The latter is the preferred procedure because it retains all the energy of the signal.

Smoothing of the Rectified Signal

The rectified signal still expresses the random nature of the amplitude of the signal. A useful approach for extracting amplitude-related information from the signal is to smoothen the rectified signal. This procedure may be accomplished either by analog means or digital means. The concept of smoothing involves the suppression of the high-frequency fluctuations from a signal so that its deflections appear smoother. This may now be recognized as low-pass filtering procedure that has been discussed at the beginning of Chapter 2. The amount of smoothing performed on the signal will depend on the bandwidth of the low-pass filter that is used; the smaller the bandwidth, the greater the smoothing.

Averages or Means of Rectified Signals

The equivalent operation to smoothing, in a digital sense, is averaging. By taking the average of randomly varying values of a signal, the larger fluctuations are removed, thus achieving the same results as the analog smoothing operation. The mathematical expression for the average or mean of the rectified EMG signal is:

$$\overline{|m(t)|}_{t_j-t_i} = \frac{1}{t_j - t_i} \int_{t_i}^{t_j} |m(t)| \, dt$$

t_i and t_j are the points in time over which the integration and, hence, the averaging is performed. The shorter this time interval, the less smooth this averaged value will be.

The above expression will provide only one value over the time window $T = t_j - t_i$. In order to obtain the time-varying average of a complete record of a signal, it is necessary to move the time window T duration along the record; this operation is referred to as *moving average*. This may be accomplished in a variety of ways. For example, the window T may be shifted forward one, or two, or any number of digitized time intervals up to the time (T) equivalent to the width of the window. In the latter case, each average value that is produced from the integral operation is unbiased, that is, calculated from data not common to the previous value. When the time window is shifted for a time less than its width, the calculated average values will be biased, that is, each average value will have been derived from some data that is common to the data used to calculate the previous average value. The moving average operation may be expressed as:

$$\overline{|m(t)|} = \frac{1}{T} \int_{t}^{t+T} |m(t)| \, dt$$

Like the equivalent operation in the analog sense, this operation introduces a lag, that is, T time must pass before the value of the average of the T time interval may be obtained. In most cases this outcome does not present a serious restriction, especially if the value of T is chosen wisely. For typical applications we suggest values ranging from 100 to 200 ms. However, it should be noted that the smaller the time window T, the less smooth will be the time-dependent average (mean) of the rectified signal. The lag may be removed by calculating the average for the middle of the window.

$$\overline{|m(t)|} = \frac{1}{T} \int_{t-T/2}^{t+T/2} |m(t)| \, dt$$

But, in this case, fringe problems occur at the beginning and the end of a record when either side of the window is less than $T/2$.

Integration

The most commonly used and abused data reduction procedure in electromyography is the concept of integration. One of the earliest (if not the earliest) references to this parameter was by Inman et al (1952). They erroneously applied the term. Their analysis procedure used a linear envelope detector to follow the envelope of the EMG signal as the force output of the muscle was varied. It is not surprising that with such an incorrect introduction the term should become improperly applied. The literature of the past three decades is swamped with improper usage of this term although, happily, within the past decade it is possible to find an increasing number of proper usages.

The term *integration*, when applied to a procedure for processing a signal, has a well-defined meaning which is expressed in a mathematical sense. It applies to a calculation which obtains the area under a signal or a curve. The units of this parameter are V·s or mV·ms.

It is apparent that a signal, such as the observed EMG signal which has an average value of zero, will also have a total area (the integrated value) of zero. Therefore, the concept of integration may only be applied to the rectified value of the EMG signal. The operation is expressed as:

$$I\{|m(t)|\} = \int_{o}^{t} |m(t)|\, dt$$

Note that the operation is a subset of the procedure of obtaining the average rectified value. Since the rectified value is always positive, the integrated rectified value will increase continuously as a function of time. The only difference between the integrated rectified value and the average rectified value is that in the latter case the value is divided by T, the time over which the average is calculated. *There is no additional information in the integrated rectified value.*

As in the case with the average rectified value, the integrated rectified value may be more usefully applied by integrating over fixed time periods, thereby indicating any time-dependent modifications of the signal. In such cases, the operation may be expressed as

$$I\{|m(t)|\} = \int_{t}^{t+T} |m(t)|\, dt$$

We suspect that there are two principal reasons which account for the wide spread usage of this operation. The first is simply its historical precedent. Regardless of the improper usage of the operation, its application has been continuously employed over the past three decades. The second is that if a sufficiently long integration time T is chosen, the integrated rectified value will provide a smoothly varying measure of the signal as a function of time.

The Root-Mean-Square (RMS) Value

The derivations of the mathematical expressions for the time- and force-dependent parameters presented in the earlier parts of this chapter indicated that the rms value provided more information than the previously described parameters. (Refer to Figure 3.11 for more details.) However, its use in electromyography has been sparse in the past. The recent increase in its use is possibly due to the availability of analog chips that perform the rms operation and to the increased awareness for technical competence in electromyography. The rms value is obtained by performing the operations described by the term, in reverse order, that is

$$RMS\{m(t)\} = \left(\frac{1}{T}\int_{t}^{t+T} m^2(t)\, dt\right)^{1/2}$$

We recommend the use of this parameter above the others.

A comparison of the analysis techniques operating on the same signal is presented in Figure 3.15. In this case the signal was obtained with wire electrodes in the biceps brachii during an isometric contraction.

Zero Crossings and Turns Counting

This method consists of counting the number of times per unit time that the amplitude of the signal contains either a peak or crosses the zero

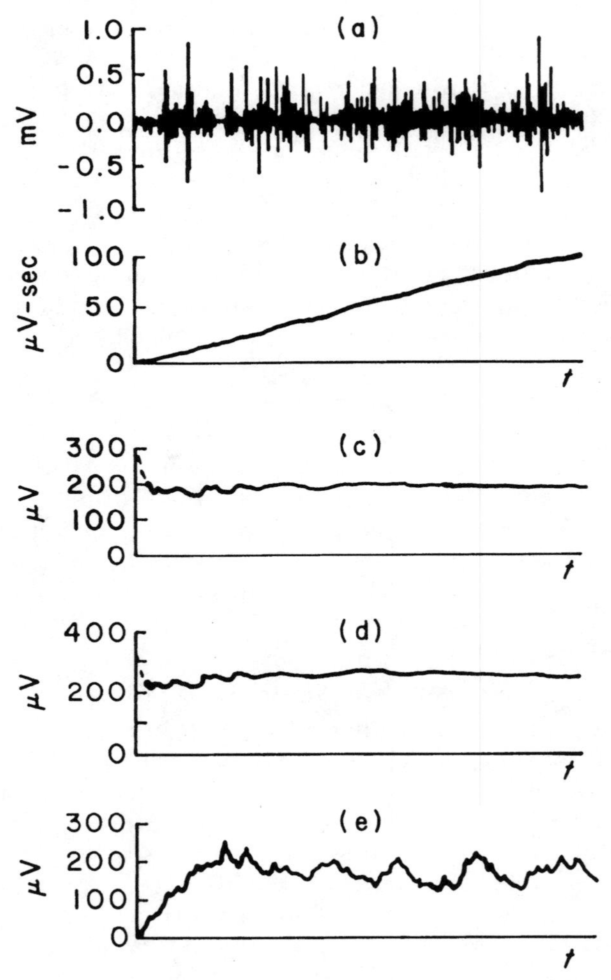

Figure 3.15. Comparison of four data reduction techniques: (*a*) raw EMG signal obtained with wire electrodes from biceps brachii during a constant-force isometric contraction; (*b*) the integrated rectified signal, (c) the average rectified signal; (*d*) the root-mean square signal; (e) smoothed rectified signal. In the latter case, the signal was processed by passing the rectified EMG signal through a simple RC filter having a time constant of 25 ms. The time base for each plot is 0.5 ms. (From J.V. Basmajian et al, 1975.)

value of the signal. It was popularized in electromyography by Willison (1963). The relative ease with which these measurements could be obtained quickly made this technique popular among clinicians. Extensive clinical applications have been reported, some indicating that a discrimination may be made between myopathic and normal muscle. However, such distinctions are usually drawn on a statistical basis.

This technique is not recommended for measuring the behavior of the signal as a function of force (when recruitment or decruitment of motor units occurs) or as a function of time during a sustained contraction. Lindström et al (1973) have shown that the relationship between the turns or zeros and the number of MUAPTs is linear for low level contractions. But, as the contraction level increases, the additionally recruited motor units contribute MUAPTs to the EMG signal. When the signal amplitude attains the character of Gaussian random noise, the linear proportionality no longer holds.

FREQUENCY DOMAIN ANALYSIS OF THE EMG SIGNAL

The analysis of the EMG signal in the frequency domain involves measurements and parameters which describe specific aspects of the frequency spectrum of the signal. Fast Fourier transform techniques are commonly available and are convenient for obtaining the power density spectrum of the signal.

An idealized version of the power density spectrum of the EMG signal, along with various parameters of interest, is presented in Figure 3.16.

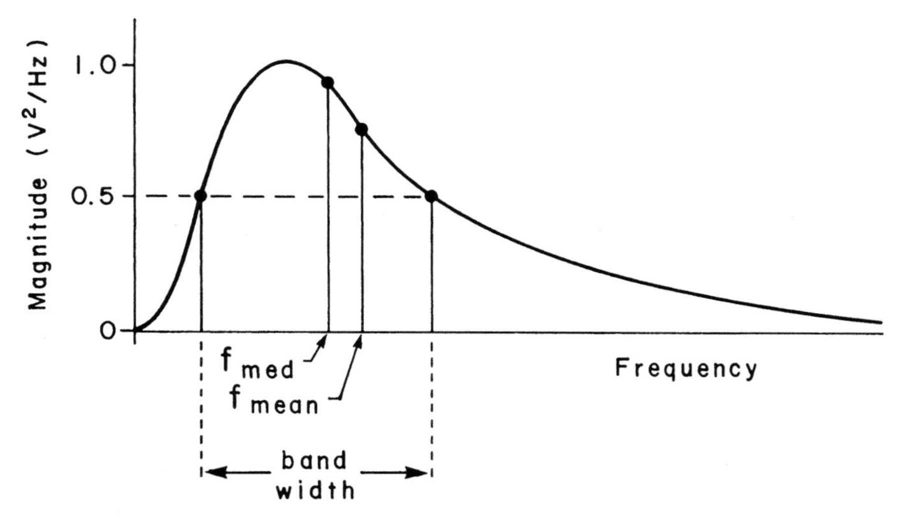

Figure 3.16. An idealized version of the frequency spectrum of the EMG signals. Three convenient and useful parameters: the median frequency, f_{med}; the mean frequency, f_{mean}; and the bandwidth are indicated.

Note that this plot has linear scales, because in our opinion such a representation provides a more direct expression of the power distribution. A logarithmic scale, which is the scale of preference in other disciplines such as acoustics, would compress the spectrum and unnecessarily distort the distribution.

Three parameters of the power density spectrum may be conveniently used to provide useful measures of the spectrum. They are: the median frequency, the mean frequency, and the bandwidth of the spectrum. Other parameters such as the mode frequency and ratios of segments of the power density spectrum have been used by some investigators but are not considered reliable measures, given the inevitably noisy nature of the spectrum (refer to Fig. 3.12). The bandwidth measure has been discussed in the previous chapter. Note that because the amplitude scale is in V/Hz, the 3 dB points or the corner frequencies are defined by a decrease of a factor of 0.5. The median frequency and the mean frequency are defined by the following equations:

$$\int_{o}^{f_{med}} S_m(f)\, df = \int_{f_{med}}^{\infty} S_m(f)\, df$$

$$f_{mean} = \frac{\int_{o}^{f} f\, S_m(f)\, df}{\int_{o}^{f} S_m(f)\, df}$$

where $S_m(f)$ is the power density spectrum of the EMG signal. Stulen and De Luca (1981) performed a mathematical analysis to investigate the restrictions in estimating various parameters of the power density spectrum. The median and mean frequency parameters were found to be the most reliable, and of these two, the median frequency was found to be less sensitive to noise. This quality is particularly useful when the signal is obtained during low level contractions when the signal to noise ratio may be less than 6.

CHAPTER 4

Decomposition of the EMG Signal and Analysis of the Motor Unit Action Potential Trains

Decomposition of the EMG signal is the procedure by which the EMG signal is separated into its constituent motor unit action potential trains. This concept is illustrated in Figure 4.1. The development of a system to accomplish such a decomposition will be beneficial both to researchers interested in understanding motor unit properties and behavior, and to clinicians interested in assessing and monitoring the state of a muscle.

In the clinical environment, measurements of some characteristics of the motor unit action potential (MUAP) waveform (for example, shape, amplitude, and time duration) are currently used to assess the severity of a neuromuscular disease or, in some cases, to assist in making a diagnosis. Thus, the decomposition of the EMG signal is useful in two ways. First, a partial decomposition must be implicitly performed by the clinical investigator to ensure that what is actually observed is a MUAP and not a superposition of two or more MUAPs or some other ephemeral artifact. Second, averaging the MUAP waveforms present in the same train will produce a low noise representation of the MUAP and, hence, provide a more faithful representation of the events occurring within the muscle. Any decomposition scheme devised for such application (i.e., to extract only MUAP shape and amplitude) will have weak constraints on its performance. A useful technique should allow detection of some, but not necessarily all, firings of a single unit in a particular record.

For physiological investigation, both the statistic of the interpulse intervals (IPI, time between two successive firings of the same motor unit) and the MUAP waveform characteristics are used to study motor unit properties and motor control mechanisms of muscles. In these conditions, much stronger constraints are imposed on the performance of a decomposition technique. It is desirable, in fact, to monitor the simultaneous activities of as many motor units as possible. Furthermore, all the firings of the observed motor units should be detected. Shiavi and Negin (1973) showed that an error of 1% in the detection of a motor unit firing prevented the observation of some relevant motor unit behavioral phenomena. Statistical analysis of IPI also implies acquisition and processing of relatively long EMG signal records (in the order of dozens of seconds), thus increasing the time required for the decomposition.

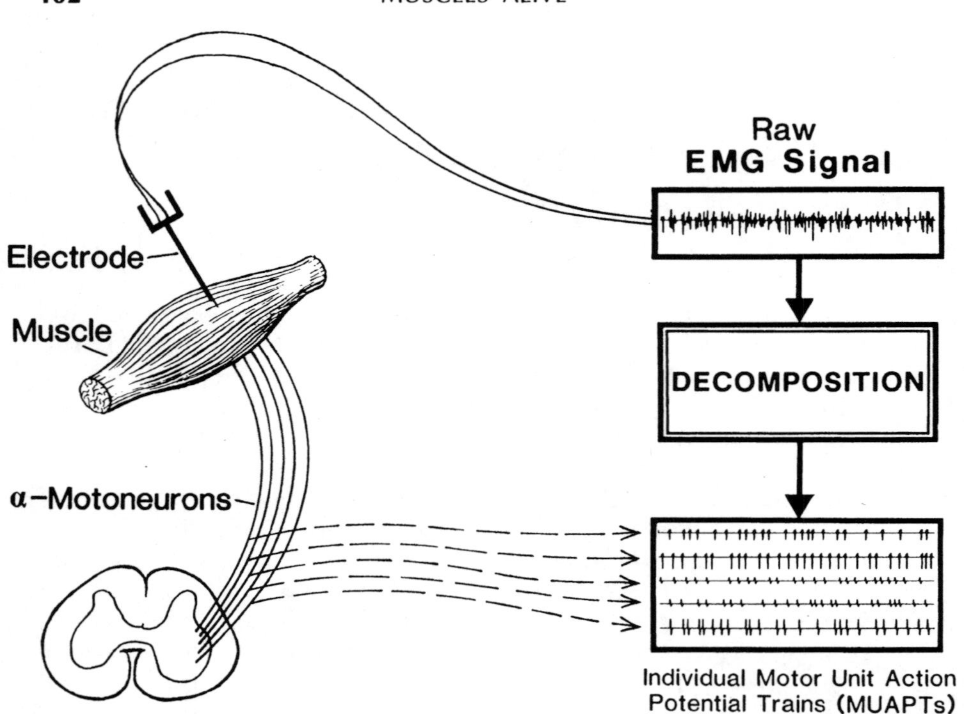

Figure 4.1. A schematic representation of the decomposition of the EMG signal into its constituent motor unit action potential trains (From C.J. De Luca et al, © 1982, *Journal of Physiology*.)

It is apparent that a decomposition technique satisfying the requirements for physiological investigations will also provide all the information currently used in clinical studies. The additional information on the temporal behavior of motor unit firing may also provide useful information which clinicians may exploit in the future.

Given the above requirements and constraints, any design approach to EMG signal decomposition must address two major issues: convenience of use and accuracy. The first point implies that any suitable technique should work on EMG signals that are routinely and repetitively acquired by a specific and convenient method. The second issue is more important since it validates the results. Therefore, it is essential that any method used for decomposing the EMG signal should be able to provide a measure of its accuracy.

Due to the novel approach presented by this system, it is useful to clarify some points concerning the capabilities and applicability of this system:

1. Not all the EMG signals acquired with this technique can be decomposed with

a 100% accuracy. There are many factors which determine the suitability of any particular EMG signal record. Force level of the muscle contraction is not necessarily a major hindrance; EMG signals detected at near maximal force levels have been decomposed successfully. Far more important are the dissimilarity of the MUAP waveforms belonging to different motor units, the number of MUAPTs present, and the stability of the MUAP waveform during the record.

2. The decomposition algorithm of the system described in the following pages may be used in a variety of modes, ranging from fully automatic to highly human-operator interactive. The chosen mode of operation will determine the tradeoff between the accuracy of the data and the amount of time required to perform a decomposition. For a record containing six MUAPTs, the time required to decompose the signal with 100% accuracy will range from 15 s to 15 min for 1 s of data, depending on the quality of the data. The same data may be decomposed in a fully automatic mode, requiring from 1 to 15 s for 1 s of data, but the accuracy of the decomposition would be approximately 65%.
3. As many as 11 MUAPTs have been decomposed accurately from an EMG signal. To date, the longest EMG signal record that has been decomposed accurately was 144 s long and contained approximately 7000 discharges of four motor units.
4. The signal conditioning which is performed in various phases of the system modifies the waveform of the MUAPs. Therefore, standard measurements of the MUAP waveform such as amplitude, time duration, and number of phases, may not be compared to those of conventionally acquired and recorded signals. However, it is important to note that such information may be easily made available by using the cannula of the needle or one of the wire surfaces for acquiring EMG signals in a conventional manner, and using the event timing from the decomposed MUAPTs to trigger average the conventionally obtained signal. This suggested procedure is similar to the "macro" EMG signal technique described by Stålberg (1980) with the additional advantage of recovering the waveform of many MUAPs, other than only one, as in the case for Stålberg's technique.
5. This chapter will focus on the decomposition aspects.

BACKGROUND

In the past, several investigators have devised techniques to identify MUAPs from each motor unit action potential train contained in the EMG signal. The different techniques that have been employed may be generally categorized as either visual identification by the investigators (Clamann, 1970; De Luca and Forrest, 1972, 1973; Desmedt and Godaux, 1977; Gurfinkel' et al, 1970; Gurfinkel' et al., 1964; Hannerz, 1974; Kranz and Baumgartner, 1974; Masland et al, 1969; Maton and Bouisset, 1972; Person and Kudina, 1971; and others) or automatic identification by electronic apparatus (Andreassen, 1977; Dill et al, 1972; Friedman, 1968; Gerstein and Clark, 1964; Glaser and Marks, 1966; Keehn, 1966; Leifer, 1969; McCann and Ray, 1966; Mishelevich, 1970; Schmidt, 1971; Schmidt and Stromberg, 1969; Shiavi, 1972; Simon, 1965; and others). Procedures that consist exclusively of visual analysis limit the scope and accuracy, as well as requiring a tremendous amount of time for performing the MUAP identifications and firing time measurements. The criteria upon which automatic identifications are based may be categorized as either feature extraction (peak amplitude, rise time, area, or some other characteristic of the MUAP waveform) or signal space representation (usually referred to as correlation, matched filter, template, or square-error separation techniques). One of the major problems with most automatic detection schemes is the inability to resolve waveforms produced by superposition of two or more simultaneously occurring MUAPs. Most automatic detection schemes also cannot accommodate a slow change in a MUAP waveform's shape or amplitude throughout a contraction. This latter consideration is important because the relative position of the recording electrode and active muscle fibers is subject to variation during a muscle contraction.

The system described in this chapter overcomes some of the limitations in the previous approaches and satisfies the requirements for physiological investigation as specified above. The initial description of this concept dates back to LeFever and De Luca (1978). A detailed description of the precursor system may be found in LeFever (1980), with a shortened description being given by LeFever and De Luca (1982) and LeFever et al. (1982). The subsequent modifications, some of which are described in this chapter may be found in their entirety in Mambrito (1983).

The major features of the system are:

1. Multiple channel recording of the EMG signal to increase discrimination power among MUAPs.
2. Recording bandwidth of 1–10 kHz.
3. Highly computer-assisted recording and decomposition techniques.
4. Slow variations in the shape of the MUAP waveforms and IPI statistics are allowed.
5. MUAP superposition can be decomposed in most cases.
6. Means for on-line checking of the EMG signal quality in terms of decomposition suitability.
7. Means for verifying the validity of the results.

The major limitations are:

1. Only records derived from attempted isometric contractions have been decomposed.
2. To date, a number of 9 MUAPTs simultaneously present in the EMG signal have been found to constitute a practical limit in the number of decomposable units from one record.
3. The technique requires interaction with a highly trained operator.

SIGNAL ACQUISITION

The EMG signal acquisition and quality verification system is depicted in Figure 4.2. The system requires the capability of recording multiple

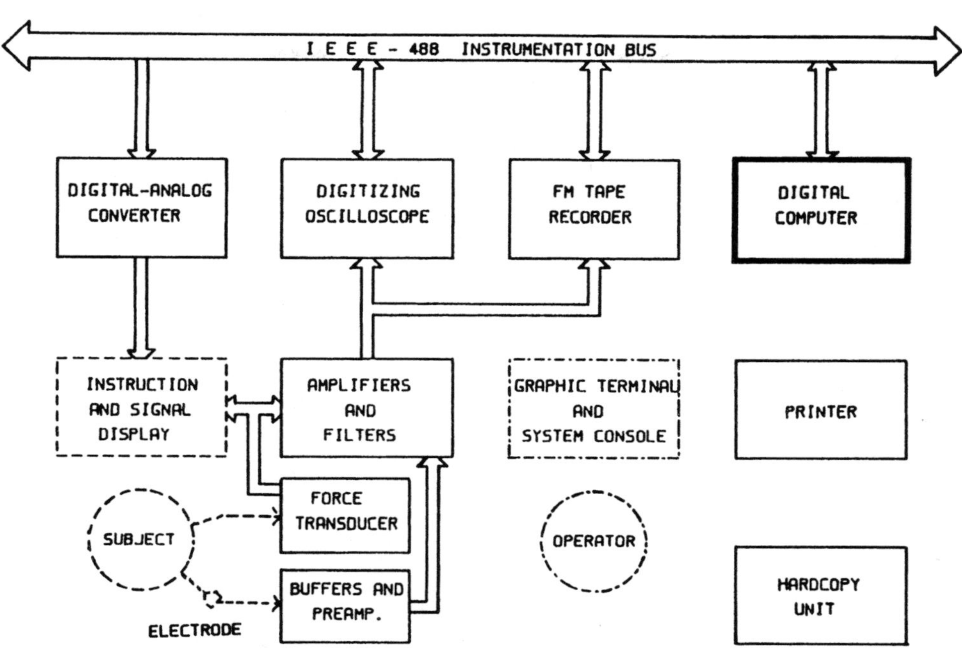

Figure 4.2. EMG signal acquisition, quality verification, and decomposition system.

independent channels of EMG signal. A special electrode to accomplish this task has been constructed based on the design of an electrode reported in an earlier study (De Luca and Forrest, 1972). A schematic of the new lightweight quadripolar electrode may be seen in Figure 4.3. It consists of 25-gauge stainless steel tubing having an opening in the wall of the shaft approximately 2 mm from the proximal edge of the tip. In this opening are exposed the cross-sectional areas of four 75-μm diameter insulated wires (90% platinum-10% iridium), located at the corners of a square and spaced approximately 200 μm apart. This geometrical arrangement was chosen so that the activity from four or five motor units would be consistently detected in most muscles. The four wires (pick up areas) terminate on four male pin connectors mounted on an insulated base that is epoxied to the shaft. The shaft itself makes electrical contact with another pin. The five pins on the electrode may be connected to form a variety of differential recording arrangements, each providing a channel of EMG signal.

Figure 4.3 presents one of several possible combinations providing three differential channels of EMG signal. The lines A, B, C, D, and E are individually shielded and fed into five high input-impedance front end buffers (10^{12} ohms input resistance and 25 pA bias current) and are

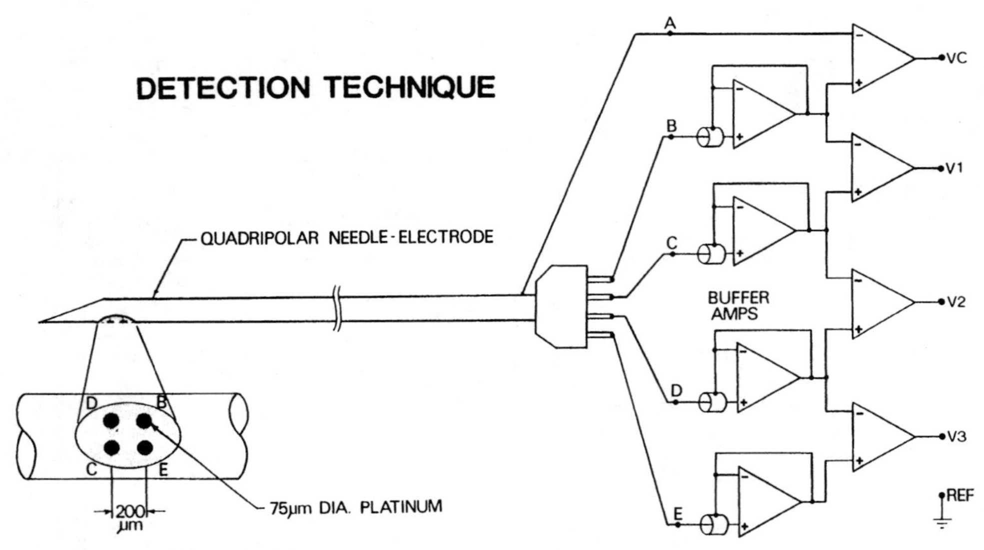

Figure 4.3. A schematic representation of the lightweight quadripolar needle electrode configured to detect three independent channels of EMG signal (V1, V2, V3) for the purpose of decomposition. A fourth channel VC is also displayed for the purpose of simultaneously recording one conventional EMG signal channel.

successively fed into a set of four differential amplifiers. For the purpose of decomposition, the differential amplifier outputs V1, V2, and V3 of three channels are band-pass filtered using differential amplifiers with low and high frequency 3 dB points set at 1 and 10 kHz. (The fourth channel, VC, may be differentially amplified with a bandwidth of 20 Hz to 10 kHz, providing a conventional EMG signal from which the conventional waveforms of the MUAP may be recovered by trigger-averaging from the decomposed MUAPTs.) The procedure of setting the lower 3-dB point at 1 kHz rather than at a lower frequency is consistently observed to reduce the amplitude of the slower rise-time MUAP waveforms produced by muscle fibers distant from the recording site. As indicated in Figure 4.2, the outputs of this last stage of amplification and filtering (the block indicating the amplifiers and filters) are viewed on an oscilloscope. When the oscilloscope is triggered by a MUAP arrival, 20-ms long segments of the signal are transmitted to the digital computer. These segments can then be plotted sequentially on the graphic terminal, and decomposition attempts can be made. These operations enable the operator to assess the spatial discrimination among MUAP waveforms and the stability of the recording, i.e., to make a judgment on how convenient it is to decompose that particular EMG signal. If sufficiently high quality EMG signals are detected, the data collection may proceed, otherwise the electrode(s) should be repositioned. During an experiment, the output of the last stage of amplification and filtering is recorded on an FM tape recorder at a sufficient speed to provide a bandwidth up to 20 kHz. With this arrangement, it is possible to obtain MUAP with peak to peak rise times as short as 100 μs.

The main advantage of multiple channel recording is to increase the discrimination power among different MUAPs. This fact is absolutely essential for performing a correct decomposition. The necessity of this feature is dramatically illustrated in Figure 4.4, which contains segments of three channels of simultaneously detected signals with MUAPs from five motor units. Note that in channel 1, MUAPs 4 and 5 have similar waveforms; such is the case for MUAPs 1 and 2. On channel 2, MUAPs 3 and 4 have similar representation whereas, MUAPs 1 and 5 have similar waveshapes on channels 2 and 3. Finally, MUAPs 1, 3, and 5

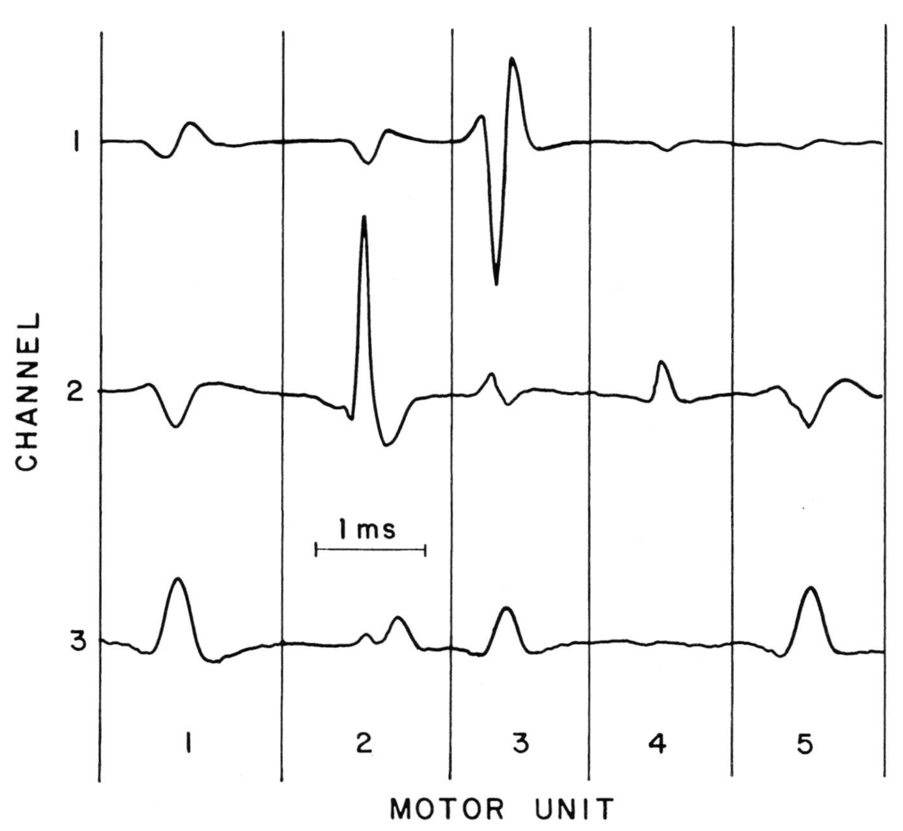

Figure 4.4. Three-channel representation of action potentials from five different motor units. The three channels represent the same electrical event (MUAP) as seen from three different geometrical perspectives.

have similar representation on channel 3. It is apparent that any identification and decomposition technique attempting to discriminate among several simultaneously active motor units using only one channel of information will not be accurate.

DATA SAMPLING AND COMPRESSION

The analog signals are transferred off-line to digital storage. As will be clarified later, due to the signal processing method recommended to detect firings, a sampling rate several times higher than the Nyquist frequency (which is twice the maximal signal frequency in this case) must be used.

A sampling rate of 50 kHz is recommended since some of the MUAP waveforms obtained using the wideband recording technique described before have frequency spectra that range up to 10 kHz. The high sampling rate may be conveniently achieved by playing back the EMG signal 32 times slower than it is recorded and sampling at a rate of 1.5625 kHz. The computer program that samples data stores only those segments of data containing positive or negative peaks above a preset threshold. This threshold is selected by the operator dependent upon the level of background noise in the data. The portions of data intervals between stored segments are stored only as a number of skipped samples. This method reduces the storage requirements from 5 to 20 times less than uncompressed storage. An example of compressed EMG signal is shown in Figure 4.5, where the numbers near the *vertical bars* indicate number of milliseconds skipped between sampled waveforms.

SIGNAL CONDITIONING

The analog high pass filtering at 1 kHz is effective in substantially reducing both amplitude and the time duration of slow rise-time MUAP waveforms recorded from fibers distant from the electrode. However, it is sometimes useful to filter the record further to reduce the degree of superposition among MUAPs by further shortening their time duration. In such cases, a symmetric Hamming window, finite impulse response digital filter is used. This type of filter has no phase distortion which could add undesirable extraphases to the MUAP waveforms. The parameters of the filter (high- and low-cutoff frequency and rolloff) can be chosen specifically for each record using the power spectrum of specific MUAP waveforms in the record as a guide.

THE DECOMPOSITION ALGORITHM

The detection of the occurrence of a particular motor unit firing is based upon the *maximum a posteriori* probability receiver theory, which has found wide applications in the field of communications (Van Trees, 1968). The theoretical computations have been derived under a set of assumptions, none of which in practice is exactly appropriate for the

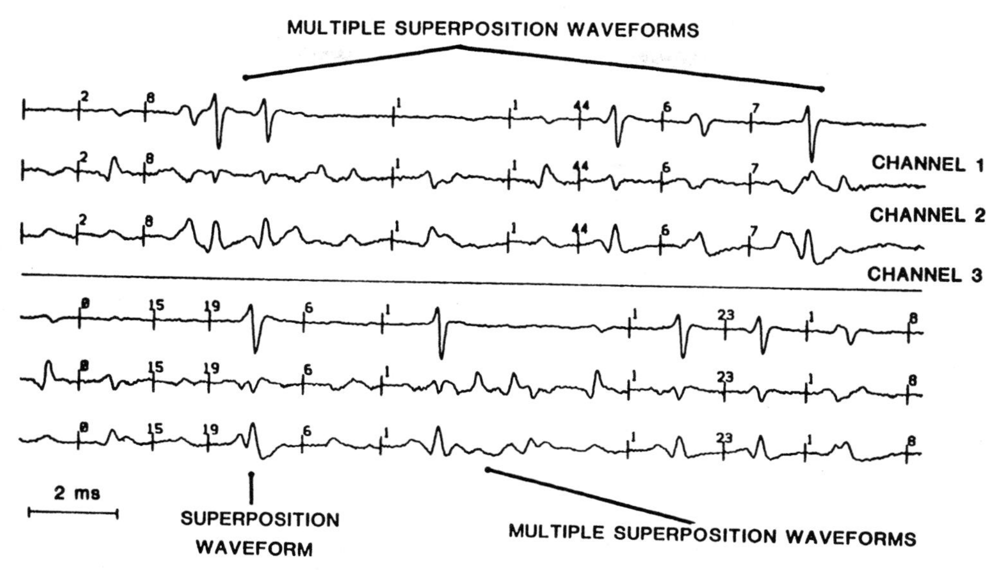

Figure 4.5. An example of three channels of a real, filtered, and time-compressed EMG signal. The *numbers* above the *vertical separating lines* (skipped interval markers) represent the time in milliseconds which contained no useful information and was removed.

EMG signal. The *maximum a posteriori* probability receiver theory assumptions and EMG signal characteristics differ in the following manner:

1. One, and only one, of a set of M signals is present from time t to time $t+T$. The number of M signals present in the set, and the time t of possible occurrence of any one signal in the set are known. (What is unknown is which one of the M signals will actually occur.) In the case of the EMG signal, more than one MUAP may be present in the same time interval t to $t+T$. The time t of possible occurrence, and the number M of motor units firing at a certain time are unknown.
2. The exact waveform or template of each signal (or the waveform in absence of perturbing noise) is known, and all the templates have the same time duration T. In the case of the EMG signal, different MUAPs have different durations; the exact template of each MUAP is unknown; and the waveform shape of the same MUAP may change in time during a contraction.
3. The *a priori* probability of occurrence of any of the M signals in any interval t to $t+T$ is known. Obviously, this is not the case for the EMG signal.
4. The signal is perturbed only by additive, zero mean, Gaussian distributed, white noise with known variance. In the case of the EMG signal, the perturbing noise consists mainly of low amplitude MUAPs which cannot be detected with reliability, thus, the perturbing noise is not white and its variance is unknown.

Theoretical Computations in the Maximum A Posteriori Probability Receiver Algorithm. The *maximum a posteriori* probability receiver algorithm uses the following decision criteria when an unknown waveform occurs. The most likely template is chosen, given the characteristics of the particular occurrence (i.e., the template with the *maximum a posteriori* probability is chosen as the one corresponding to the unknown waveform). It can be shown that this is equivalent to making the decision with the minimum probability of error.

In order to perform this task, the following computations must be made. Upon occurrence of an unknown waveform, the difference signals between unknown waveform and *M* templates are computed. Then the energies in the difference signals are also computed. These energy values, or squared errors, are modified using a weighting factor derived from the probability values described above in no. 4, and the variance of the perturbing noise. In particular, squared errors resulting from matching less likely templates are increased while squared errors resulting from matching more likely templates are decreased. The template which gives lowest final value of modified squared error is chosen as the one corresponding to the unknown waveform. The variance of the perturbing noise controls the degree to which the probability weighting factors effect the decision. For low variance values, the decision is more affected by the energy term (i.e., by the similarity between "unknown" waveform and template), while for high variance values, the probability weighting factor (i.e., the probability of occurrence of each template) dominates the decision. If the signal has multiple channel representations (as in the case of the EMG signal), all the energy computations are performed for each channel; the total squared error (before the modification with the probability weighting factor) is obtained by summing the squared errors for each channel.

The original *maximum a posteriori* probability receiver decision technique has been extensively modified to take into account the wide difference between the theoretical *maximum a posteriori* probability receiver case and practical EMG signal case. In the following parts of this section we will describe the modification to the *maximum a posteriori* probability receiver algorithm and its implementation.

The Motor Unit Templates. The first problem is how to obtain the set of templates. Each motor unit template is an estimate of the MUAP waveform (amplitude and shape on all three channels). Prior to the analysis of an EMG signal record, both the number of different motor units (whose firing can be detected) and their corresponding MUAP waveforms are unknown. Therefore, all templates must be created during the decomposition. When the operator has decided that a waveform in the EMG signal is produced by an action potential of a "new motor unit," a new template is created using the waveform itself. This template may also be updated at each successive detection of the motor unit firing by averaging the template with the detected waveform in the EMG signal. This operation will improve the estimate of the MUAP waveform by reducing the amount of perturbing noise in the template and will also compensate for slow variations in template shape. (The term new motor unit is used nonrigorously.) The new MUAP waveform may be from an already firing, although previously undetected, motor unit which has increased in amplitude (due to electrode movement) so that it now exceeds the sampling threshold. If the amplitude of this MUAP waveform is significantly greater than the sampling threshold, a newly recruited motor unit has probably been detected. If not, subsequent analysis of the firing pattern will permit the distinction between a newly recruited and an already firing motor unit to be made.

The A Priori Probabilities of Firing. The second problem is how to obtain the *a priori* probability of occurrence of all the MUAPs whose templates are available in any interval t to $t+T$ of the record. This may be accomplished by measuring the interpulse interval (IPI) between adjacent firings of a motor unit. The mean and the variance of the IPIs of the detected motor unit can be recursively obtained by using expressions that have been reported by LeFever and De Luca (1982). The use of a running recursive expression during the decomposition allows slow time variations in the IPI mean and variance. If the time sequence of firing of each motor unit is modeled as a renewal process with Gaussian distributed IPIs, the relative probability of occurrence of a MUAP for each motor unit can be approximated from IPI mean and variance. Such probability is continually updated at each firing detection.

Detection of a Motor Unit Firing. At this point, it would be possible to compute the squared error as described in the theoretical *maximum a posteriori* probability receiver computations. However, it has been found that the absolute change in a MUAP waveform from one firing to the next is roughly proportional to the waveform amplitude. Hence, the ratio between squared error and template energy should be used as the decision criterion instead of simply the squared error.

Because a motor unit firing can occur at any unknown instant of time, the problem arises as to how to align the templates with the "unknown" detected waveform to compute the squared error. To achieve this, the peak (greatest absolute value) of each motor unit template is aligned with the peak of the unknown waveform and shifted back and forth to achieve the minimal squared error. During this operation, an alignment error may occur which, at most, is equal to one half the sampling period. It is this alignment error that poses the requirement for the high sampling rate reported in the data sampling and compression section. For expressions on the alignment error, see LeFever and De Luca (1982).

Perturbing Noise. The last dissimilarity between theoretical and practical conditions relates to the nature of the perturbing noise and its variance, which is used to modify the squared error value. In the decomposition algorithm, the noise is still considered white; rather than estimating the value of the variance from the EMG signal, the value is set by the operator to control the degree to which the probability weighting factor affects the decision.

While the theoretical *maximum a posteriori* probability receiver algorithm always automatically leads to a choice of one among the *M* available templates, this cannot be allowed for the EMG signal. In fact, the random superposition of two or more differing MUAP waveforms may produce a complex waveform quite similar to some other previously identified MUAP not actually present. The peak that has been detected may also have arisen from only the background noise. Alternatively, the MUAP detected may actually be produced by a newly recruited motor unit for which no template has been established. For these reasons a detection is confirmed only if the detected waveform is "close" enough to the most likely template. A measure of how close two waveforms are is obtained using signal space representation techniques as described by LeFever and De Luca (1982). For the purpose of the present description, it is sufficient to note that the confirmation of a detection will be automatic only if preset numerical constraints on the value of the squared error are satisfied.

When a detection cannot be confirmed, the decision is transferred to the superposition algorithm described in the next section.

The Superposition Algorithm. The purpose of the superposition algorithm is to resolve an EMG signal waveform, formed by the summation of multiple MUAP waveforms. Only combinations of two waveforms (two templates) are considered, since the computation time is prohibitive for more. Triple or multiple partial superposition of MUAP waveforms can sometimes be solved by repeated application of the decomposition algorithm in various modes. The scheme employed by the superposition decomposition algorithm is similar to the single match criteria. Added to these criteria is a procedure that attempts to fit a second motor unit template to the waveform obtained by subtracting the first template from the EMG signal.

The above approach is implemented as follows. Each template is aligned with the detected highest peaks (in the channel where such a peak occurs) in the "unknown" EMG signal waveform and is subtracted from the signal. (The operation is similar to that performed for a single match.) Then an attempt is made to resolve the remaining waveform by aligning each of the remaining templates with the remainder and subtracting. The template providing the smallest squared error is chosen as the second template. The energy in the remainder after subtraction of the two templates is the squared error used to make the decision. This squared error value is modified using the probability weighting factor corresponding to the first subtracted template.

The superposition algorithm could be particularly useful in clinical applications, because it can determine if polyphasic action potentials are indeed representative of an individual motor unit.

TEST FOR CONSISTENCY AND ACCURACY

As stated in the introduction of this chapter, it is essential to assess the accuracy of any EMG signal decomposition system to validate the results obtained using such a technique. This point cannot be overemphasized. Furthermore, this technique is highly interactive, and during decomposition, many decisions may be made by the operator. Thus, it is also necessary to assess the consistency of the results produced by different operators.

The issue of the consistency is the simplest of the two, and it has been extensively addressed by LeFever et al (1982). Briefly, the following test was performed. Two highly trained operators (each with at least 400 hours of experience in decomposing EMG signals) and a third, less experienced operator (16 hours of EMG signal decomposition) were required to decompose the same EMG signal record independently, which was considered "difficult" (i.e., at the limit of the decomposition technique capabilities according to the two experienced operators). The EMG signal selected contained five MUAPTs which the skilled operators believed had been reliably detected. Both skilled operators were 100% in agreement for the detection of a total of 479 MUAPs from five motor units. The results of the untrained operator decomposition contained a total of 12 discrepancies with respect to the two trained operators. Since the original test reported by LeFever et al and De Luca (1982), the consistency has been tested in a similar fashion on many other occasions. Complete agreement has always been obtained among operators having more than 300 hours of experience with the technique.

The issue of the accuracy is much more complicated. It is impossible to measure the decomposition accuracy in an absolute sense, with real EMG signal, since occurrence times of all the MUAPs and precise definitions of all MUAP waveforms in the EMG signal are unknown *a priori*. So far, this limitation has been circumvented in two ways.

First, the accuracy was tested on synthetically generated EMG signals. For details on the procedure to generate synthetic EMG signals and execution of the test refer to LeFever et al (1982). Briefly, the synthetic EMG was constructed by linearly superimposing eight mathematically generated MUAPs along with Gaussian noise. Prior to the simulation, a real force record was obtained from a muscle contraction, and force thresholds for recruitment were randomly chosen for each "synthetic" motor unit. At any point in time throughout the simulated contraction, the mean firing rate of each motor unit was proportional to the difference between force input to a stochastic event generator and the recruitment threshold. The stochastic event generator used a renewal process to create each firing time. The standard deviation of the zero mean Gaussian noise was 40% of the peak amplitude of the smallest MUAP waveform. A segment of the synthetic EMG signal record used for the test is shown in Figure 4.6. A skilled operator was able to decompose the record with an accuracy of 99.8%, incurring one error in a total of 435 classifications. (The particular error was quite inconsequential in that it occurred as an incorrect classification among two MUAPs which belonged to different motor units but had similar shapes and fired less than 1 ms apart.) This error was immediately obvious when the IPI data were plotted. The location of the error was identified, and that segment of the record was subjected to more rigorous decision criteria which rendered the correct identification. In fact, the IPI plots are routinely used to quickly scan for any obvious errors. This particular record is now used as a benchmark to identify the performance criterion of new operators.

Second, an indirect test of the accuracy of the decomposition technique on a real EMG signal was obtained in the following way. Two quadripolar needle electrodes were inserted in the same muscle (tibialis anterior) about 1 cm apart. The two sets of EMG signals from the two electrodes were recorded simultaneously and decomposed. Some motor units presented motor unit action potential trains in both sets of signals. A comparison of the results from three different contractions with two "common" MUAPTs per contraction showed 100% agreement for a total of 1415 detections of the "common" MUAPs. In this case, an undetected error in the results from the "common" MUAPs detections could occur only if a simultaneous error of the same kind (wrong

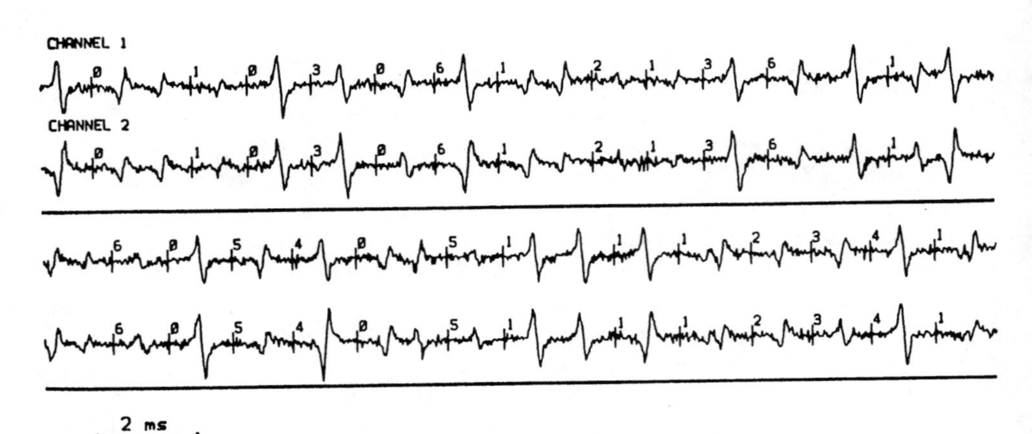

Figure 4.6. An example of a mathematically synthesized and time-compressed EMG signal used to test the accuracy of the decomposition system. As in the previous figure, the *short vertical lines* are skipped interval markers, and the *numbers* represent the milliseconds of time removed.

classification of a MUAP or missed detection) was made in the decomposition of the two records. The chances of such an event are incalculably small. Thus, the consistency of the decomposition data of the same units from two different electrodes provides an indirect measure of the accuracy in real data decomposition.

TIME DOMAIN ANALYSIS OF MOTOR UNIT ACTION POTENTIAL TRAINS

Having described a system which will provide detailed and accurate measures of the characteristics of MUAPTs, it now remains to describe approaches for processing the data. The data reduction techniques that follow may of course be used for any MUAPT data, regardless of how it is obtained. However, the reader is reminded that erroneous data yields erroneous results when analyzed.

Motor Unit Action Potential Characteristics

In clinical environments, the waveform of the MUAP is used to provide what have come to be considered conventional parameters. These are: the amplitude, the time duration, and the number of phases of the MUAP. These parameters are considered to carry information related to the state of the muscle fibers. The decomposition technique described in this chapter renders MUAP waveforms which are not comparable, in terms of amplitude, time duration, and number of phases, to those which are acquired by conventional means. However, because a highly accurate representation of the timing events of the MUAPTs (decomposed from the EMG) is available, it is possible to obtain the conventional waveforms from the fourth channel presented in Figure 4.3, which is bondpassed in a conventional manner. The waveform recovery is realized by trigger averaging the conventional EMG signal with the timing of the motor unit discharges available from the decomposed MUAPTs. Refer to the next section for an explanation of trigger averaging.

The waveforms of a MUAP of one train obtained from the EMG signals that were bandpassed at 1 to 10 kHz are presented in a raster plot in Figure 4.7. This figure is presented for the purpose of indicating that the waveform of the MUAP is not stable during a contraction, even an isometric constant-force contraction, as was the case in this particular example. The plus sign represents situations when a superposition with MUAP of some other train occurred.

Trigger Averaging

The process of trigger averaging is employed to extract a continuously recurring waveform that is enveloped by considerable noise, rendering the quality of the waveform unacceptable at any occurrence. However, if the time at which the waveform occurs is known (as is the case for the decomposed MUAPTs), it may be used to mark the time period in the

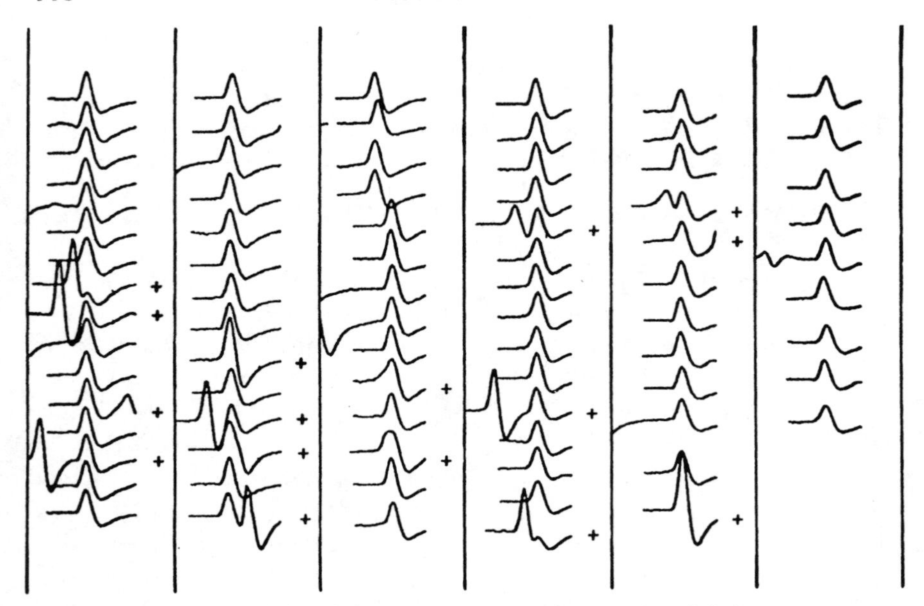

Figure 4.7. An example of the MUAP raster plot. MUAPs (of the same motor unit) are shown as they are detected during the decomposition of the record. MUAPs are displayed sequentially in time from *top* to *bottom* and from *left* to *right*. MUAPs marked with a + sign represent superpositions of the displayed MUAP and of some other MUAP(s) present in the record and simultaneously firing with the detected MUAP.

noisy signal which contains the waveform of interest. This may be done every time the waveform is present in subsequent parts of the signal. The next step consists of removing from the noisy signals all the time periods which have been identified as containing the waveform. Then these time periods are averaged. The noise, being unrelated among the time periods, will cancel out, and the waveform contained in the time periods, being considerably similar, will be enhanced. It is apparent that the greater the number of accurate identifications of the discharge of a motor unit, the more accurate will be the waveform of the recovered MUAP. If the noise in the averaged time periods is independent, then the improvement in the signal-to-noise ratio of the recovered waveform will be proportional to the square root of the number of time periods that are averaged.

The decomposition approach described in this chapter provides the high accuracy required and at the same time provides the means of extracting the waveform of several MUAPs from the same EMG signal.

MUAP Arrival Plots (IPI Bar Plot)

The arrival of MUAPs of the same motor unit is represented as an impulse on a horizontal line which expresses units of time since the beginning of the contraction. An example of such a plot is presented in Figure 4.8. Different horizontal strips correspond to different motor unit action potential trains. The continuous line represents the output force, and it is scaled on the right vertical coordinate in percent of maximal voluntary contraction (MVC). This kind of plot is useful for event timing.

Interpulse Interval vs. Time during the Contraction Plot (IPI DOT Plot)

The left vertical coordinate of each *dot* represents the time (in ms) since the last firing of the same motor unit. An example of such a plot is presented in Figure 4.9 (*top*). Each *horizontal division* indicates the discharges of an individual motor unit; the *horizontal coordinate of each dot* represents the actual time of MUAP arrival. This kind of plot is very useful for identification of errors in the decomposition. In fact, isolated dots out of range (i.e., abnormally long or short IPI) are generally

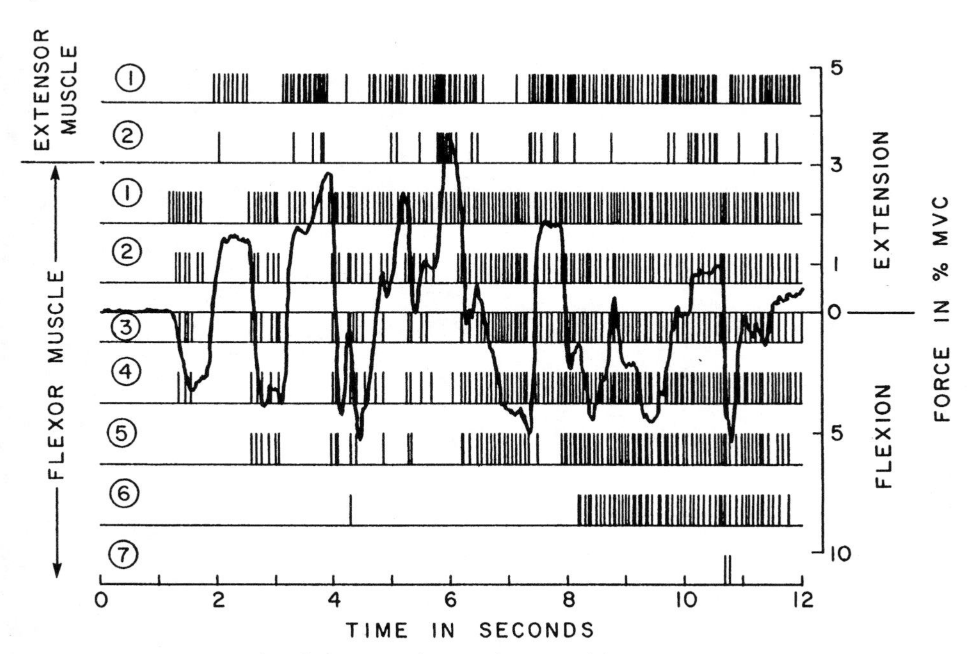

Figure 4.8. Example of IPI BAR plot. Each *vertical bar* represents the arrival of a MUAP at the time (since the beginning of the contraction) indicated on the *horizontal line* at the *bottom* of the *graph*. Each *horizontal strip* presents the activity of a different motor unit. The *continuous line* represents the output of the force transducer scaled on the *right vertical coordinate* in percent of maximal voluntary contraction (MVC). This particular example contains motor unit action potential trains from two antagonist muscles (flexor pollicis longus and extensor pollicis longus) detected during random isometric flexion-extension of the interphalangeal joint of the thumb.

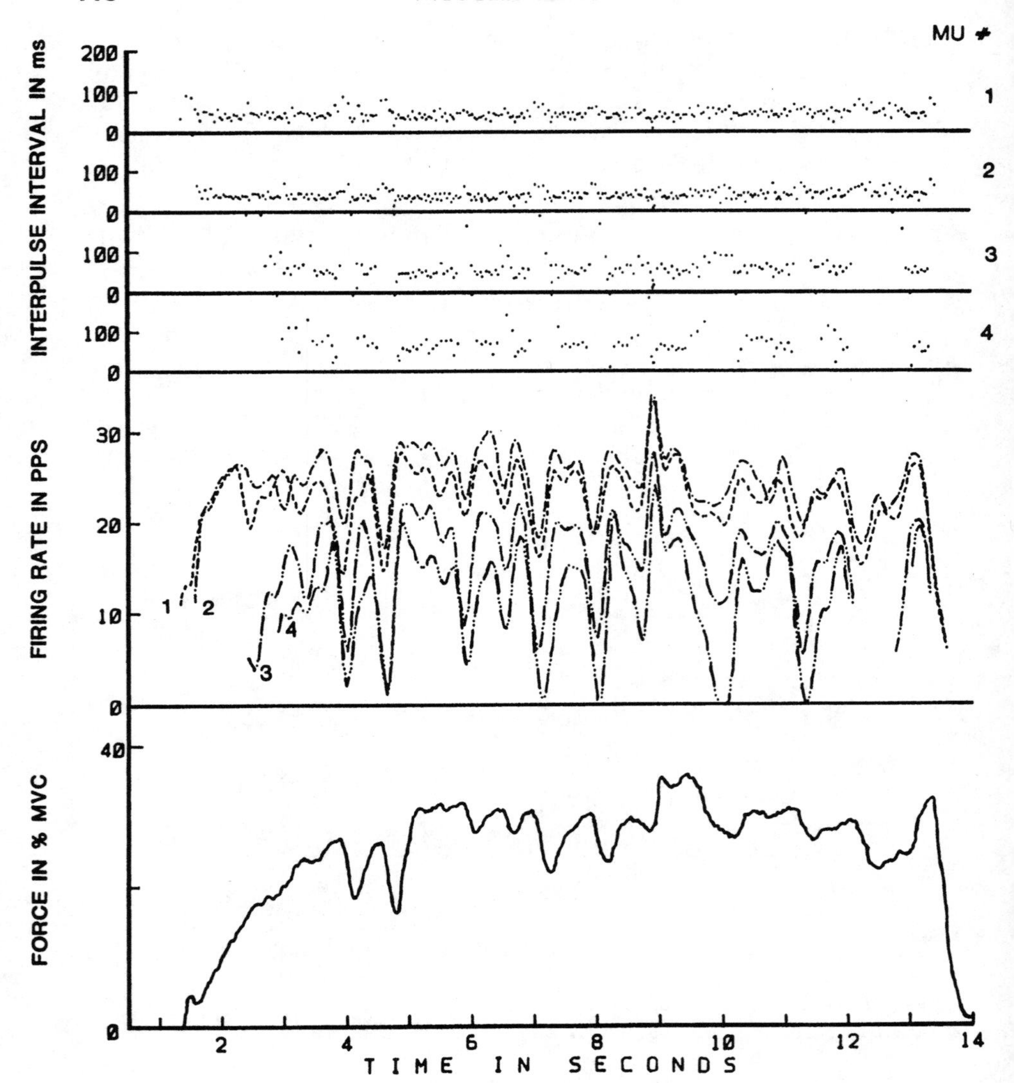

Figure 4.9. Example of IPI DOT plot (*top*) and motor unit firing rate plot (*middle*). In the IPI DOT plot, each dot represents a MUAP arrival at the time indicated on the *horizontal coordinate*. The *left vertical coordinate* is the time since the last firing of the same motor unit (in milliseconds). In the motor unit firing rate plot, the time varying mean firing rate of each detected unit is represented by different *dot-dashed lines*. The firing rate is measured in pulses per second on the *left vertical coordinate*. The *continuous line* (*bottom*) represents the output of the force transducer scaled on the *left vertical coordinate* in percent of maximal voluntary contraction level (*MVC*). The mean firing rates were calculated from the IPI values presented in the IPI DOT plot.

indicative of a missed detection or of a misclassification unless such an event is accompanied by a consistent event in the force record.

Mean Firing Rate Plots

Having available the IPIs of a motor unit, it is possible to calculate the number of occurrences per unit time, that is, the firing rate. It is customary to express this measure as pulses per second. Note that the term *firing rate* is used, rather than firing frequency, which is often found in the literature. The term frequency implies periodicity among the discharges of a motor unit. Such is not the case. The IPIs are semirandom in nature, hence, the concept of *rate* must be used to describe the behavior of the discharge properly. The firing rate value may be obtained by inverting the value of the IPI interval(s). If only one IPI is used, the instantaneous firing rate is obtained. This value will be semirandom in nature. It contains exactly the same information as the IPI value.

A more useful measure of the behavior of the firing rate of a motor unit during a contraction may be obtained by calculating the time varying mean firing rate. This is obtained by averaging the value of the instantaneous firing rate over several consecutive values. Referring back to the descriptions of the averaging process in Chapter 2, the reader is reminded that the average value may be obtained in various ways, each requiring a tradeoff between smoothness and bias of the results. This issue is particularly important in calculating the mean firing rate because it is always preferable to be able to associate the behavior of the mean firing rate of a motor unit at any time with the physiological behavior at that time and with minimal bias on previous behavior. This requirement may be accomplished by estimating the time-varying mean firing rate by convoluting the impulse train presented in the IPI BAR plot (Fig. 4.8) with a noncausal Hanning filter having a symmetric, unit area impulse response. The mathematical expression of this operation may be expressed as follows:

$$h(t) = \frac{1 - \cos(2\pi t/T)}{2}\,; \quad 0 \le t \le T$$

$$h(t) = 0 \text{ elsewhere}$$

where T determines the width of the time window over which the "filtering" is performed. This filter is recommended because it achieves a very sharp attenuation, thus rendering a smooth time-varying mean firing rate with minimal bias. Practical experience has indicated that a T value of 400 to 500 ms provides an acceptable and useful compromise between estimation bias and smoothness. The above operation is equivalent to multiplying each IPI by the filter function and shifting the filter function along the time scale.

An example of the time-varying mean firing rate is presented in Figure 4.9 (*middle*). For each motor unit the firing rate is represented by different dot-dashed lines. Values of the mean firing rate are scaled on the *left vertical coordinate* in pulses per second, and the *horizontal coordinate* represents the time since the beginning of the contraction. This kind of plot is useful for studying relationships among different motor units. In the example shown in Figure 4.9 (lower portion), the continuous line represents the output force, scaled in percent maximal voluntary contraction on the *vertical coordinate*.

Correlation of the Mean Firing Rates

This operation should not be attempted unless the mean firing rate is calculated from IPIs which are identified with at least 99% accuracy, relating to false classification (Shiavi and Negin, 1973). The decomposition system described in this chapter can ensure such occurrence under specific conditions and, hence, makes this analysis procedure possible.

The concept of cross-correlation of two signals, $x(t)$ and $y(t)$, is defined by the mathematical expression:

$$R_{xy}(\tau) = \frac{1}{T} \int_0^T x(t)\, y(t + \tau)\, dt$$

In words, the cross-correlation function is estimated by the following operations:

1. Delay (or shift) the signal $x(t)$ relative to the signal $y(t)$ by a time displacement equal to τ seconds.
2. Multiply the value of $y(t)$ at any instant by the value of $x(t)$ that had occurred τ period of time before.
3. Average the instantaneous value product over the sampling time, T.
4. Repeat the above steps for another value of τ.

This operation provides a measure of the amount of common behavior among the time-varying firing rates of two motor units. In a sense, it is the indication of the common input that drives the motor unit to fire.

Figure 4.10 presents examples of the cross-correlation of motor unit firing rates obtained during constant force contractions. Figure 4.10*A* represents the time-varying mean firing rates of four motor units and the attempted isometric constant-force produced by the muscle. Figure 4.10*B* presents the cross-correlation among the firing rates, and Figure 4.10*C* the cross-correlation of the firing rates and the force output.

Conditional Intensity Functions

This particular operation is useful for measuring the tendency for two motor units to fire at, or nearly at, the same time, i.e., synchronization.

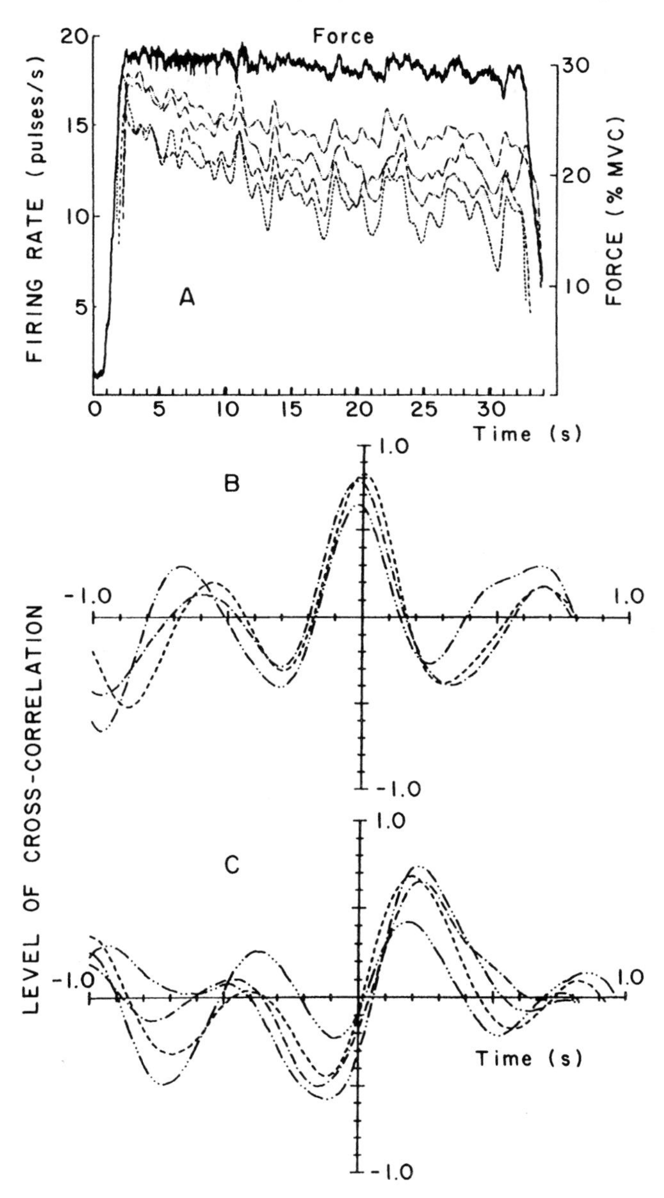

Figure 4.10. (*A*) Firing rate records of four concurrently active motor units (*dashed lines*) are shown superimposed on the force output (*solid line*) recorded during a constant force isometric abduction of the deltoid. The force level is given in percent of maximal voluntary contraction (MVC) at right. (*B*) Functions obtained by cross-correlating between firing rates. (*C*) Cross-correlation between firing rates and force output. Positive shift of peaks in C indicates that firing rate activity leads force output.

Unlike the cross-correlation function, which is used to measure the common behavior of the mean firing rate (obtained over several motor unit discharges), the intensity function measures the common time occurrence of individual discharges. An alternative approach to interpreting the common time occurrence of two motor units is to consider whether the occurrence of the discharge of one motor unit "conditions" the other motor unit to discharge.

The conditional intensity function may be calculated according to the following steps. Consider two motor units, A and B. For each firing of motor unit A (the conditioning motor unit), measure the time difference to the firing of motor unit B (the conditioned motor unit) in both the forward and backward time direction up to a predetermined scanning interval. Accumulate the time values in a histogram. Bin widths of 0.5 ms and scanning intervals of 100 ms are suggested. Divide the total number of discharges of motor unit A within the scanning interval. The units of the histogram are now expressed in pulses per second, similar to the units of the firing rate. The histogram now provides an expression of the *intensity* of firing of the conditioned motor unit B when the conditioning motor unit A fires. This kind of representation has often been referred to in the literature as poststimulus histogram or cross-correlograms, where the conditioning "stimulus" can be the occurrence of any particular event such as tendon tap or the application of electrical stimulus. In this particular case the "stimulus" is the occurrence of a motor unit firing.

A bin width of 0.5 ms may be too small to collect any significant number of firings in a single bin. One possible solution in this case is to group 0.5-ms bins in larger bins. Another approach is to smooth the original conditional intensity plot with a filter having an appropriate impulse response.

The example in Figure 4.11 shows conditional intensity functions of one motor unit with respect to another conditioning unit. The histogram was smoothed with a 4-ms window triangular filter. The two MUAPTs required for this operation were obtained from the flexor pollicis longus muscle during an isometric contraction of the interphalangeal joint of the thumb. The *horizontal axis* represents the time (in milliseconds) since the firing of the conditioning units, and the *vertical axis* represents the conditional intensity in pulses per second. The example may be interpreted in the following way. The conditioned and conditioning motor unit have a tendency to fire together (synchronize) which is manifested by the high conditional intensity values at time zero. The peaks every 75 ms are a result of the periodic nature of the conditioning unit over the computation time interval. If the conditioning motor unit has a mean firing rate of 1/0.075 pulses per second, on the average every 75 ms after (before) the zero time, the conditioning motor unit will fire (fired)

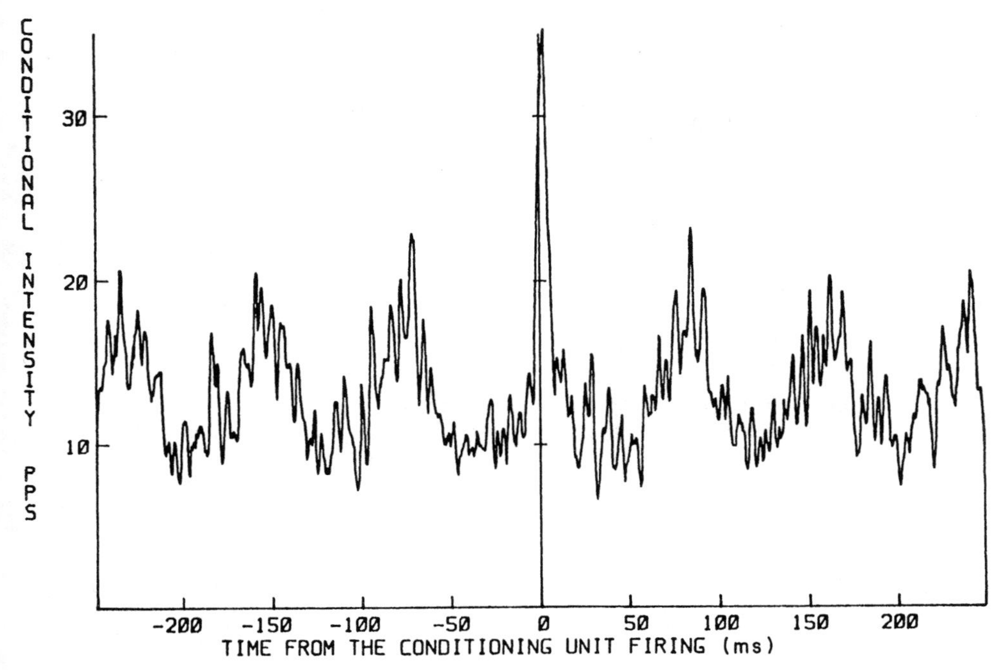

Figure 4.11. Example of a motor unit conditional intensity function. The *horizontal axis* represents the time (in milliseconds) since the firing of the conditioning unit, and the *vertical axis* represents the conditional intensity (in pulses per second). The large peak at time zero indicates that the conditioned and the conditioning motor unit tend to fire simultaneously.

again, and the same tendency to synchronize will produce the periodic peaks in the plots.

SUMMARY

In this chapter, we have described a system for acquiring, processing, and decomposing EMG signals for the purpose of extracting as many MUAPTs as possible with the greatest level of accuracy. This system consists of four main sections.

The first section consists of methodologies for signal acquisition and quality verification. Three channels of EMG signals are acquired using a quadripolar needle electrode designed to enhance discrimination among different MUAPs. An automated experiment control system is devised to free the experimenter from the burden of experiment detailed surveillance and bookkeeping and to allow on-line assessment of the EMG signal quality in terms of decomposition suitability.

The second section consists of methodologies for signal sampling and conditioning. The EMG signal is bandpass filtered (between 1 kHz and

10 kHz), sampled, and compressed by eliminating parts of the signal under a preset threshold level.

The third section consists of a signal decomposition technique where motor unit action potential trains are extracted from the EMG signal using a highly computer-assisted interactive algorithm. The algorithm uses a continuously updated template matching routine and firing statistics to identify MUAPs in the EMG signal. The templates of the MUAPs are continuously updated to enable the algorithm to function even when the shape of a specific MUAP undergoes slow variations.

The fourth section deals with ways of displaying the results. The more frequently used representation formats are:

1. Display of MUAP waveshapes
2. Impulse trains representing motor unit firings
3. IPI plots, where time intervals between successive firings of the same motor unit are plotted vs. time of the muscle contraction
4. Firing rate plots where the estimated time-varying mean firing rate of the detected motor units is plotted vs. time of the muscle contraction
5. Cross-correlation of firing rates which indicate the amount of common drive in the motor units
6. Conditional intensity functions which provide an indication of the amount of synchronization among motor unit discharges

The performance of the system has been tested in terms of:

1. Consistency among results obtained by different operators
2. Accuracy evaluated on synthetic EMG signal
3. Accuracy on real EMG signal by comparing results pertaining to the same MUAPT contained in two EMG signals which were independently and simultaneously detected from two different electrodes

CHAPTER 5

Control Properties of Motor Units

This chapter will deal with those properties of motor units which describe their recruitment and firing behavior during the process of force generation. These properties will be referred to as the *control properties*, because it is through these modes that the central and peripheral nervous systems affect the performance of a muscle or a group of muscles. Motor unit properties such as biochemical structure, twitch response, physical dimensions, etc. may be considered to be properties which specify the identity of the scheme(s) that the nervous system employs to involve (or possibly take advantage of) the different types of muscle fibers. These latter properties will not be addressed directly in this chapter.

Prior to embarking on a detailed description of the control properties of the motor units, it is useful to review the function and potential involvement of peripheral and central mechanisms.

THE PERIPHERAL CONTROL SYSTEM

There are a variety of apparently specialized receptors located in the muscles, tendons, fascia, and skin which provide information to appropriate parts of the central nervous system concerning the state of the force and length characteristics of muscles. A nonreceptor system, the Renshaw system, which resides completely in the anterior horn of the spinal cord will also be described in this section. Although according to the classical anatomical distinction of central and peripheral, the Renshaw system is physically located in the central nervous system. However, as will be seen later in this chapter, some aspects of the Renshaw system are intimately associated with the peripheral feedback mechanisms. Thus, from a control point of view it is advantageous to consider the performance of the Renshaw system in association with the peripheral nervous system. Together, they may be considered to form the output stage of the motor system, or the *peripheral control system*.

An overview of the interactions of the major components of the peripheral control system is presented schematically in Figure 5.1. For completeness, the suprasegmental and supraspinal input are also included in this figure. The representation in Figure 5.1 is purposely arranged to emphasize details of the peripheral control system which are reasonably well understood and accepted. The schematic representation is directed at presenting the behavioral characteristics of the peripheral control system rather than describing anatomical details. Thus, it should not be viewed as a wiring diagram, but rather as a block diagram.

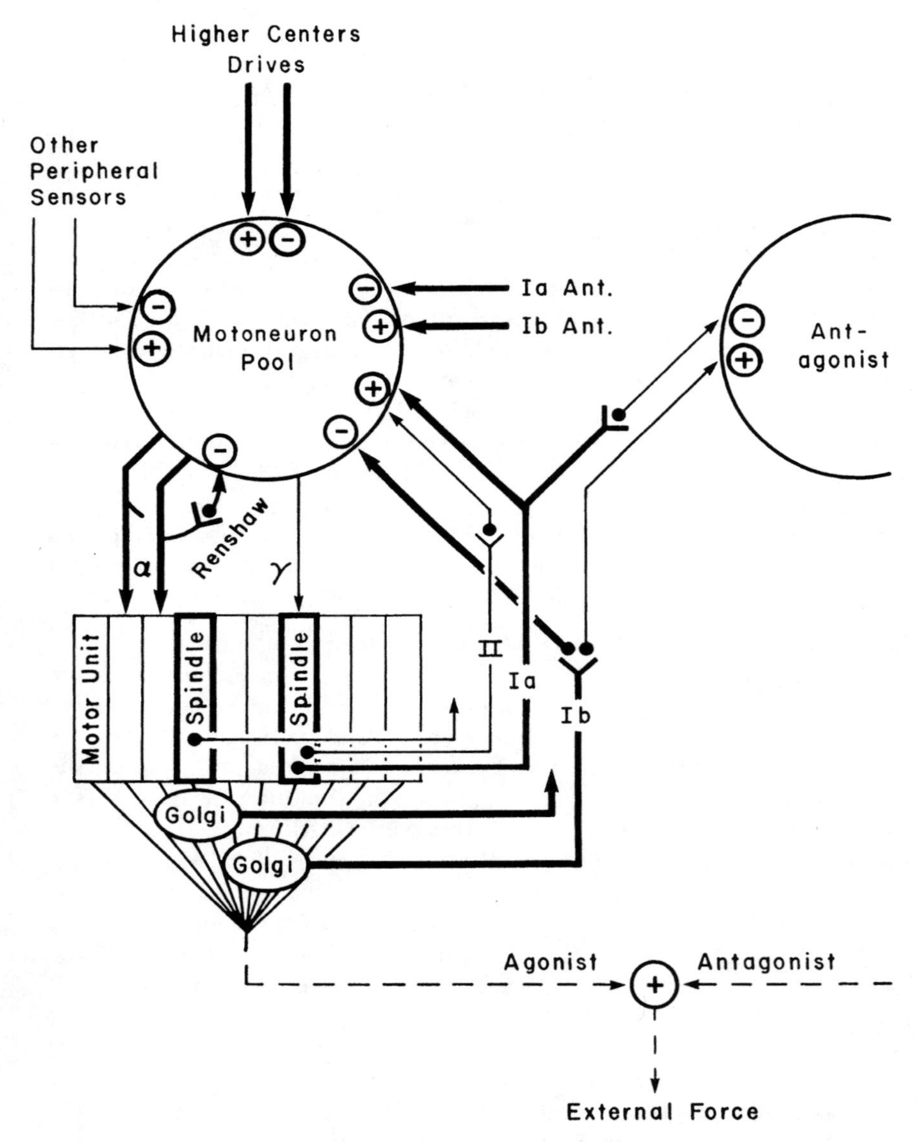

Figure 5.1. A simplified schematic representation of most of the components of the peripheral control system which interact with a muscle. The *arrows* and *signs* indicate excitatory (+) and inhibitory (−) actions on the pool. The *thickness* of the *lines* indicates the dominance of the contributions. The diagram is by no means complete but does contain the more dominant interactions. The *broken lines* represent force interactions. Note that when a spindle is slackened during a muscle contraction, the Ia and II afferent fiber discharge decreases, thus the effect on the motoneuron pool will be disfacilitatory and not excitatory as represented in this diagram.

Note that the concept of the peripheral control system evolves around an entity referred to as a motoneuron pool. This motoneuron pool may be thought of as an aggregate of interacting neurons, located in the anterior horn of the spinal cord, whose joint behavior is associated with the control of a function in either one muscle or a specific group of muscles. The output of the motoneuron pool consists of the (efferent) information transmitted down the α-motoneuron, γ-motoneuron and, possibly, β-motoneuron. The presence of the latter in man remains to be clearly established and, hence, it has not been included in the diagram. The input to the motoneuron pool consists of the (afferent) information from the peripheral receptors, the Renshaw system, and the drive from the higher centers. Note that all the inputs to the motoneuron pool are expressed as either having an excitation (+) or an inhibition (−). An excitatory contribution will increase either the disposition of an α- and γ-motoneuron to begin discharging (recruitment) or to increase its firing rate if it is already active. An inhibitory contribution has the opposite effect.

Let us now proceed to a description of the sensory systems. The forthcoming description is purposely brief. For more specific details the reader should refer to any of the numerous books dealing with mammalian muscle receptors, for example, the chapter Mammalian Muscle Receptors by Hasan and Stuart in *Handbook of the Spinal Cord* (1984).

The Muscle Spindle

The muscle spindle is by far the most studied muscle sensor organ. It is located within the body of the muscle. The spindle consists of a capsule having a fusiform shape, attached at both ends to muscle fibers. It is generally believed to be arranged in parallel with the adjacent muscle fibers. The architectural arrangement is apparently designed to favor the monitoring of muscle length and changes in length. Inside the spindle capsule are located "intrafusal" muscle fibers. These may number from 2 to 25. These fibers have contractile characteristics similar to those of the normal or "extrafusal" muscle fibers and are separable into three categories distinguished on the basis of the arrangement of the nuclei in the middle portion of the intrafusal fiber and, contrary to the depiction in most textbooks, not on the basis of their shape (Matthews, 1972). These fibers are referred to as "bag 1," "bag 2," and "chain" fibers. Bag and chain fibers are distinguishable on the basis of their mechanical properties, with the bag fibers being more dynamic. Whereas bag 1 and bag 2 fibers differ mostly in their content of elastic strands (Banks et al, 1981) and innervation, these intrafusal fibers are innervated by two distinct types of efferent motoneurons: the gamma dynamic, innervating the bag 1 fibers, and the gamma static, innervating the bag 2 and chain fibers. The γ-motoneurons are considerably smaller (2 to 8 μm) than their functionally synonomous α-motoneurons. Wrapped around each of the intrafusal fibers are the endings of two groups of afferent nerve fibers, the larger group Ia and the smaller group II. The Ia afferent fibers connect with a monosynaptic (direct contact) excitatory projection on the motoneuron pool of the same muscle and with a disynaptic (through an interneuron) inhibitory projection on the motoneuron pool of the antagonist muscle(s). It has also been shown that this pathway of "reciprocal inhibition" is accompanied by longer pathways of "reciprocal excitation" via delayed oligosynaptic excitatory and inhibitory Ia projections (Jankowska et al, 1981a and b). The group II afferent fibers also impinge on the motoneuron pool of the same muscle with an excitatory projection but with a disynaptic contact. There is no clear evidence indicating the presence of a group II pathway to the antagonist muscle. A third type of afferent, the β-motoneurons, have been shown to exist in some mammalian subprimate animals. The β-motoneurons innervate both intrafusal and extrafusal fibers. Their presence in human muscles remains speculative and, as such, their role will not be discussed in the context of this chapter.

It is now generally agreed that the Ia and II group fibers modify their discharge rate as the mechanoreceptor endings of these fibers are elongated. *The mechanoreceptor endings may be elongated*, either by stretching the muscle which stretches the spindle capsule and thus the intrafusal fibers, or by contracting the intrafusal fibers via γ fiber excitation. As the mechanoreceptors become elongated, the discharge rate of the receptors increases; as they shorten, the discharge rate decreases. It is also generally agreed that rate of length change of the muscle (fibers) modifies the discharge characteristics of the afferent fibers. Numerous experiments have shown that the Ia afferents respond to length and velocity and the II afferents mainly to length. Thus, the II afferents may be viewed as mainly static sensors and the Ia afferents as both static and dynamic sensors. This distinction is convenient from an analytical point of view, but physiologically the distinction cannot be made so clearly. Several investigators (Houk and Rymer, 1981; Rack, 1981; and others) have further categorized the response of the Ia fibers by demonstrating that they are nonlinear sensors; being much more sensitive to small displacements than to large displacements, thereby keeping the gain of the stretch reflex low when the muscle length changes appreciably. They are more sensitive to lengthening (stretch) than to shortening. Houk et al (1981) have also demonstrated that the dynamic component of the spindle response is proportional to the 0.3 power of velocity. This suggests that spindle Ia afferents are better suited for motion detection (change in velocity) rather than for signaling the precise velocity of the muscle (fibers).

It is a commonly accepted view that the α and γ-systems are coactivated during a muscle contraction, that is, when the motoneuron pool of a muscle is excited, both the extrafusal and the intrafusal fibers contract. It can be reasonably assumed that this coactivation has, as one of its main functions, the task of setting the length of the spindle appropriately with respect to the length of the contracting muscle fibers. This task is equivalent to setting the operating point of the Ia mechanoreceptors, so that their sensitivity remains high over the length variability of the extrafusal muscle fibers. Any length and velocity perturbation applied to the muscle spindles from the adjacent muscle fibers will modify the discharge of the afferent fibers. Note that these perturbations may occur locally within the muscle, even during attempted isometric contractions, because the mechanical disturbances caused by the individual extrafusal muscle fibers may not be (and in all likelihood are not) symbiotic.

The classical approach to conceptualizing the function of the muscle spindle has been to consider it as a servocontroller for compensation of loads applied to the tendon of a muscle. This notion was an outgrowth of the experimental work involving investigations concerning reflexes. In such experiments the whole muscle or limb was commonly perturbed, and the resulting neuroelectric response was monitored.

More recently, a series of studies have addressed the question of the sensitivity of the muscle spindle to the mechanical disturbances of the individual motor unit contractions (Binder et al, 1976; Binder and Stuart, 1980; Cameron et al, 1980 and 1981; McKeon and Burke, 1983). All these investigators recorded the activity of the muscle spindle afferents and the activity of the MUAPTs simultaneously. The time course and magnitude of motor unit contractions were extracted from the force record using spike-triggered averaging techniques. It was found that both group Ia and II afferent activity was strongly coupled to the contractions of some motor units located near the spindle and was more or less indifferent to the contractions of other remotely located motor units. These results support the proposition that muscle receptors (see analogous findings in section on Golgi tendon organ) generate a "sensory partitioning" of the muscles.

Schwestka et al (1981) and Windhorst and Schwestka (1982) showed that the influence exerted by one motor unit on the spindle discharges was more or less strongly affected by the action of other motor units, dependent on the relative timing of their contractions. Windhorst et al (1982) also showed that the muscle spindle is particularly sensitive to "doublet" activations of motor units (successive firing of the same motor unit within 10 ms).

The Golgi Organ

The Golgi organs are located in the relatively stiff aponeuroses extending from the tendon. Thus, these receptors provide almost no information concerning muscle length. They are, instead, sensitive to muscle tension. They are fusiform in shape, approximately 650 μm long and 50 μm in diameter in midsection. They are innervated by the group Ib afferent fibers which, through a disynaptic connection, have an inhibitory effect on the motoneuron pool of the homonymous muscle and a less pronounced but nonetheless measurable excitatory effect on the motoneuron pool of the antagonist muscle(s) (Watt, 1976).

For several decades it was believed that the Golgi organs had a high-force threshold, restricting their involvement to that of a safety valve. Houk and Henneman (1967) showed that discharge of the Golgi organs was highly sensitive to minute increases of the tension applied to the tendon. Subsequently, it has been shown by Binder et al (1977) that the Golgi organs are sensitive to individual motor unit twitches. In fact, the architecture of the muscle fiber insertion into the tendon is structured so that muscle fibers of several (5 to 25) motor units attach to any one Golgi organ. This apparent distribution of motor unit impingement to the Golgi organs serves the purpose of spatially integrating the force emanating from the quasirandomly generated force twitches from the motor units. This arrangement provides the Golgi sensor with the capability of responding to the force contribution of individual motor units, as well as to a more global force contribution from the muscle as a whole.

Behavior of Muscle Spindle and Golgi Organs during a Contraction and a Stretch

The Golgi tendon is essentially a force sensor; therefore, it will respond in a fashion similar to an externally applied tension (during stretch) or an internally applied tension (during a voluntary contraction). The spindle, on the other hand, is sensitive to length and velocity; thus, it will respond differently, depending on whether it is being elongated during a stretch or shortened during a voluntary contraction. The conjoint involvement of the two sensory systems is displayed in a schematic fashion in Figure 5.1.

When an external load applied to a muscle stretches the muscle, probably all (certainly most) of the spindles in a muscle are stretched and respond by providing an excitatory influence on the motoneuron pool of the stretched muscle and an inhibitory influence on the motoneuron pool of the antagonist muscle. The Golgi organs will also be stimulated, and they respond by contributing an inhibitory influence on the motoneuron pool of the agonist muscle and an excitatory influence on the motoneuron pool of the antagonist muscle. In this operating condition, the spindle and the Golgi organ responses conflict. However, if the applied stretch is brisk, that is, if the rate of force applied to provide the perturbation displacement is considerable, a reflex contraction response counteracting the displacement will result. This indicates that under such conditions the Ia afferent fiber stimulation provides the dominant effect, an interpretation which is consistent with the known fact that the Ia fiber endings detect motion changes. This mechanism is the well-known stretch reflex. Thus, viewed from the perspective of external disturbances, the spindle provides a mechanism for load or displacement compensation.

When a muscle is contracted, either under voluntary control or via electrical stimulation, the spindles are slackened as the muscle fibers around it shorten. However, unlike the case of a rapid externally applied stretch, only some of the spindles throughout the muscle will be disturbed at any given force level. (All the spindles will be disturbed only when all the motor units in the muscle are excited.) Thus, the discharge of the spindles will be decreased, providing a reduction in the excitatory influence (which is at times referred to as disfacilitation) on the agonist motoneuron pool and a decrease in the inhibitory influence on the antagonist motoneuron pool. However, the Golgi organs will respond to the increasing tension and increase their discharge appropriately. Thus, they will provide an inhibitory effect on the homonymous muscle and an excitatory effect on the antagonist muscle(s). Hence, during a voluntary contraction, the behavior of the muscle spindle and the Golgi organs is complementary.

Renshaw Cells

Renshaw (1946) discovered that antidromic impulses in motoneuron axons (moving towards the cell body) inhibited neighboring motoneurons, and that such impulses caused discharges of interneurons of the ventral horn, which have since been called Renshaw cells. The Renshaw cells receive collateral branches from the motoneuron axons while the Renshaw cells' axons terminate on the motoneurons themselves. This forms a feedback circuit with "recurrent inhibition" whose significance is not yet fully understood.

It has been reported that Renshaw cells are more strongly excited by collaterals of large motoneurons than collaterals of small ones (Ryall et al, 1972; Pompeiano et al, 1975; Friedman et al, 1981; and others). It also has been shown that Renshaw cells mutually inhibit other Renshaw cells (Ryall, 1970), the γ-motoneurons (Ellaway, 1971), and the interneurons mediating group Ia reciprocal inhibition (Ryall and Piercey, 1971). Renshaw cells may be activated by the discharge of a single motoneuron (Ross et al, 1975), and their discharge rates are nonlinearly related to the motoneuron discharge rate (Ross et al, 1976; Hultborn and Pierrot-Deseilligny, 1979; Cleveland et al, 1981).

It should be noted that the morphological and physiological evidence for the presence of Renshaw cells has been obtained mainly from the lumbarsacral part of the spinal cord of cats. Reports of their existence in man are much scarcer. Pierrot-Deseilligny and Bussel (1975) have provided an elegant demonstration of their activity in the soleus muscle of man.

It is difficult to identify a clear functional role of the Renshaw cells because of their apparent multiple actions. Hultborn et al (1979) have suggested that the supraspinal inputs which converge on Renshaw cells enable the recurrent inhibition to serve as a variable gain regulator at the motoneuronal level. Hultborn et al (1979) also argued that Renshaw cell action on the Ia inhibitory pathway and on the γ-motoneurons is meaningful since all these neurons act together as a functional unit, forming an "output stage" of the motor system.

Other Muscle Receptors

Muscles contain a variety of sensory fibers in addition to those mentioned in the previous sections. Afferent fibers with free nerve endings and having a wide range of diameters (groups II, III, and IV) have been identified. A few Pacinian corpuscles (specific skin sensors sensitive to touch and pressure) are supplied by fibers of group I and II. Joint capsule receptors are also innervated by sensory fibers. Prolonged muscular contractions in conjunction with blood occlusion produce discharges of the unmyelinated C fibers, which also seem to be sensitive to light touch or slight changes in temperature (see Mendell and Henneman, 1980).

Although the actions of most of these receptors are not clearly identified, several withdrawal reflexes which may arise in response to the possibility of injury seem to originate in these types of receptors. It has also been shown (Sabbahi and De Luca, 1981, 1982; and others) that cutaneous afferents may have important inhibitory effects on the α-motoneurons.

BEHAVIOR OF MOTOR UNITS

All the sensory mechanisms, along with the suprasegmental and supraspinal contributions, converge on the soma of the α-motoneurons and contribute to its disposition to discharge. The discharge behavior of the α-motoneurons may be conveniently studied by observing the discharges of the motor units in the muscle.

Interpulse Intervals

When MUAPTs can be properly identified, it is possible to measure the time between adjacent discharges of a motor unit, i.e., the interpulse interval (IPI). The IPI has been observed to be irregular, and can be described as a random variable with characteristic statistical properties (De Luca and Forrest, 1973a).

The most general characterization of the IPI is a histogram, which is a discrete representation of the probability distribution function. The histogram should only be computed for relatively short durations of the MUAPT (less than 10 s). Two common parameters of the probability distribution function, or the histogram, are the mean and the standard deviation. These two parameters have been used to describe the IPIs. Tokyzane and Shimazu (1964) suggested that it might be possible to differentiate between two categories of motor units (tonic and kinetic) by plotting the mean vs. the standard deviation of the IPIs. Their report presented two distinguishably different relationships. Leifer (1969), Person and Kudina (1972), De Luca and Forrest (1973a), and Hannerz (1974) found no such distinction. They, instead, found a continuous range of mean vs. standard deviation relationship.

The shape of the IPI histogram, as reported by various investigators, is not consistent. Buchthal et al (1954b), Leifer (1969), Clamann (1970), and others have reported that the shape has a Gaussian distribution. De Luca and Forrest (1973a and b), Person and Kudina (1972), and others have reported an asymmetric distribution with positive skewness. It is now known that the shape of the IPI histogram will vary as a function of time during a sustained contraction (De Luca and Forrest, 1973a) and as a function of firing rate. Figure 5.2 presents histograms of the IPIs during six consecutive equilength segments of 5 s of a MUAPT detected during a constant force contraction from the deltoid muscle. Note that as the time of contraction progresses, the mean value and standard deviation of the IPI both increase, and the shape of the histogram changes. The MUAP used in Figure 5.2 corresponds to that whose firing rate is displayed as the first recruited motor unit in Figure 5.8. When a motor unit is recruited, the IPIs have a relatively large coefficient of variation. But as the firing rate increases, usually associated with an increase in the force output of the muscle, the coefficient of variation decreases as the motor unit discharges become more regular. This property is demonstrated in Figure 5.3. The *top section* presents the force profile of an isometric contraction of the deltoid muscle superimposed on the IPI values of three MUAPTs which were detected during the contraction. The *bottom section* presents the IPI histograms of four consecutive segments of MUAPT no. 3. The first part of the contraction was maintained at the recruitment level of motor unit no. 3, the second part of the contraction at a higher force level. Note that the IPI values of motor unit no. 3 are highly irregular when it is discharging near its threshold but become considerably more regular as the force output is increased by a small amount (4%). This modification in the IPI behavior is reflected in the histograms.

The rapid decrease of the standard deviation with increasing firing rate after recruitment is found in the data of Hannerz and of Person and Kudina. The latter investigators proposed that after-hyperpolarization may be responsible for reducing the standard deviation and thus increasing the regularity of the intervals. Considerable modifications in

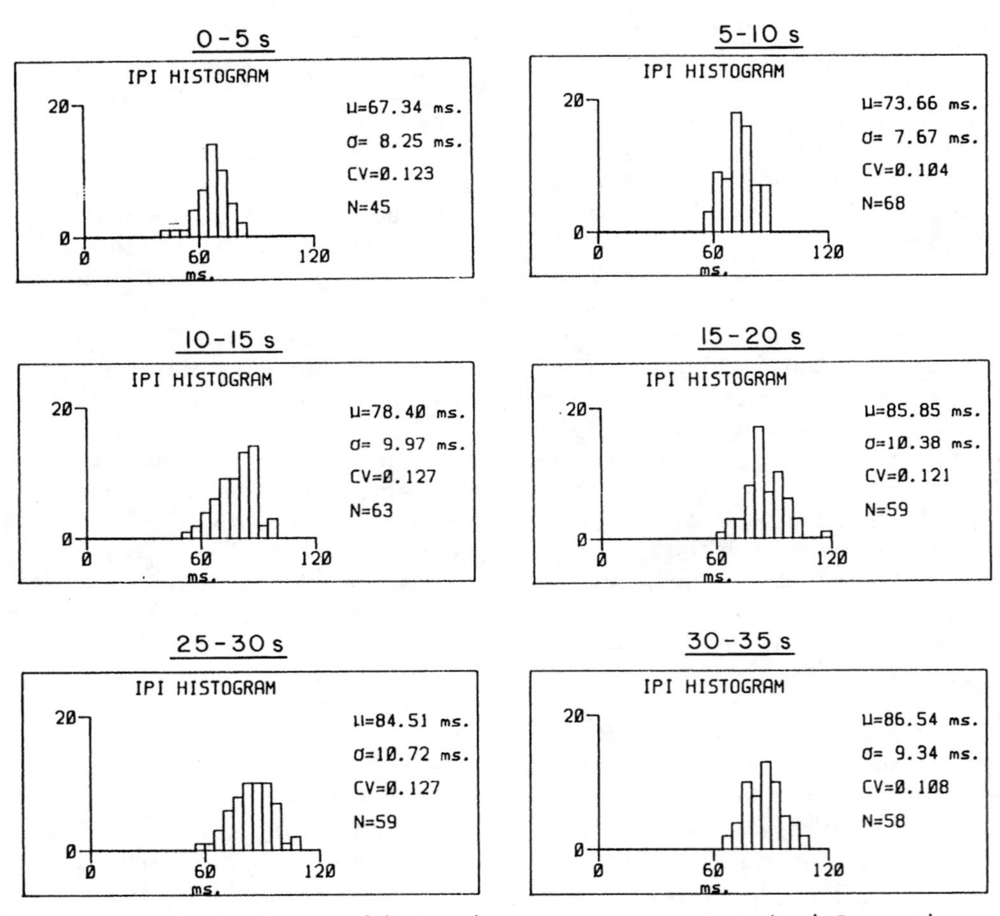

Figure 5.2. Histograms of the IPI of six consecutive segments (each 5 seconds long) of a MUAPT detected from the deltoid muscle during a constant-force isometric contraction at 30% of maximal voluntary contraction. This MUAPT corresponds to the first recruited motor unit in Figure 5.8. Note that as the time of the contraction progresses, the mean, standard deviation and shape of the histogram of the IPI change.

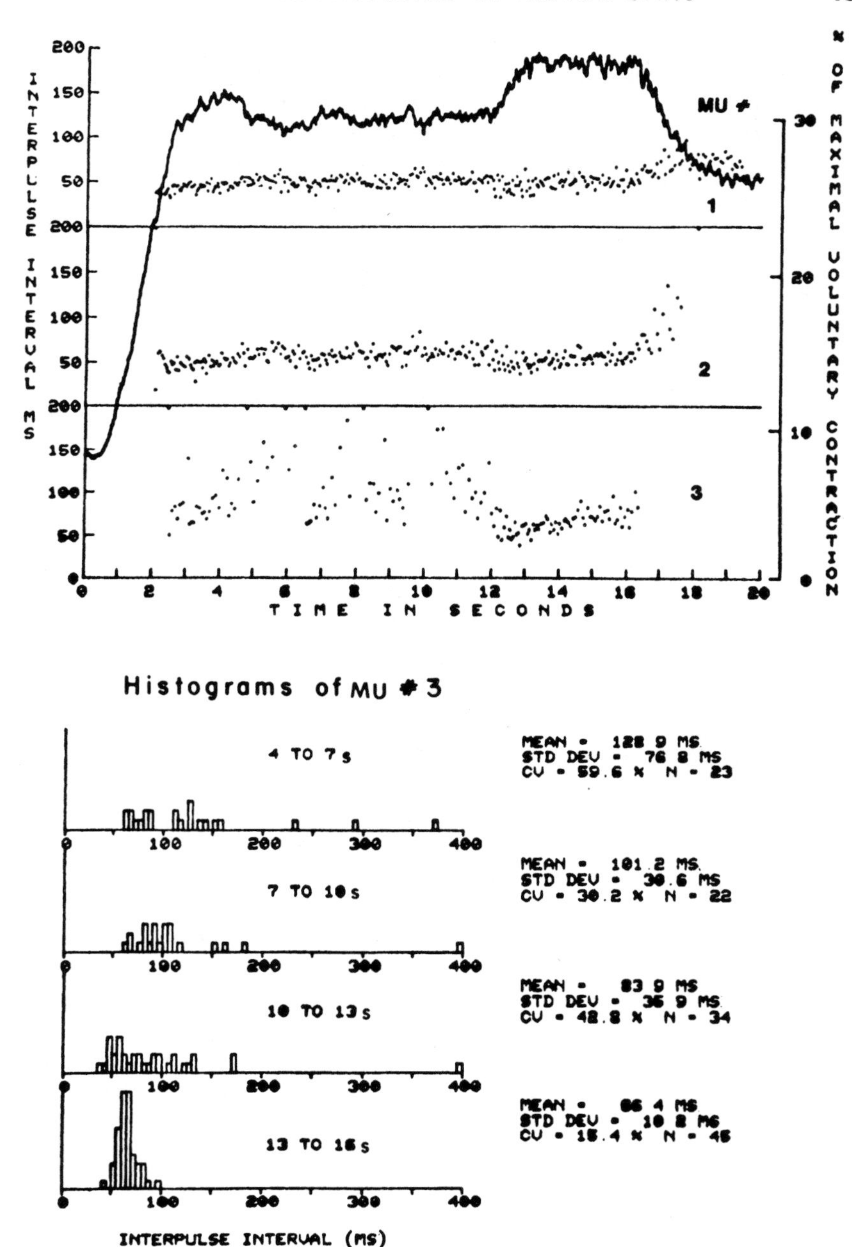

Figure 5.3. (*Top*) The interpulse intervals of three concurrently active motor units during an isometric contraction of the deltoid muscle having the displayed force profile. Note the decreased variability of the interpulse intervals of the third motor unit when the force output increases. (*Bottom*) Histograms of the interpulse intervals of the third motor unit at various time intervals of the contraction. Note the dramatic change in the shape.

the shape of the IPI histograms have also been reported from patients afflicted with supraspinal motor disturbances (Freund et al, 1973).

As the force increases, the rate of discharge of the motor unit increases, and the IPIs become shorter, which is evident in Figure 5.3. This diminution of the IPIs is accentuated during "ballistic" contractions, i.e., those which are performed as fast as possible. During such contractions, IPIs less than 10 ms in duration are present (Desmedt and Godaux, 1978; Bawa and Calancie, 1982).

Gurfinkel' et al (1964) reported several influences on the standard deviation for the IPIs of individual motor units. In patients with disturbances of joint perception, the standard deviation was considerably reduced compared to that of a normal individual, but in patients with cerebellar disturbances, no differences were seen. They also found a tendency for the standard deviation to decrease when normal subjects used surrogate means of control (audio or visual feedback) in addition to proprioception. Although Holonen (1981) reports no discernible difference in the regularity of the IPI between audio and visual feedback, Sato (1963) found that the coefficient of variation for motor units from the dominant hand of right-handed subjects tended to be lower than that of the left hand. Voluntary oscillations were more regular when performed with the right hand. This suggests that a lower coefficient of variation corresponds to greater capability of precision control.

Interdependence. Another statistical parameter of interest for describing the IPIs of a motor unit is their interdependence. The greatest amount of dependence (if any) should occur between adjacent intervals. Dependence may be tested by plotting the values of the adjacent IPIs against each other in the form of a scatter diagram. If the adjacent IPIs are independent and the random process is stationary (time invariant), the points on the scatter diagram will be randomly distributed in a fashion determined by the probability distribution function of the IPIs. In case of dependence, the points on the scatter diagram will have statistically dependent coordinates. An alternative test for dependence is serial correlation. If the average product of the adjacent IPIs is equal to the square of the average of the IPIs, then the serial correlation is zero, and the IPIs are linearly independent. Lesser values indicate a negative serial correlation and the tendency of short IPIs to be followed by long IPIs, and vice versa. If the IPI random process is not stationary, the above tests may indicate dependence, when none exists. Therefore, measurements for IPI dependence must be performed over sufficiently short time periods, to reduce time-varying effects.

Several authors have noted weak, negative correlations between adjacent IPIs of single motor units. Kranz and Baumgartner (1974) found some motor units that exhibited negative serial correlation, some weakly positive, and some with no significant serial correlation in most cases.

Person and Kudina (1971, 1972) found negative serial correlation only for motor units firing at rates above 13 pulses/second (pps). At these firing rates they found a constant small standard deviation (5 ms) and symmetric IPI histograms. They attributed these results to the effect of afterhyperpolarization. De Luca and Forrest (1973a) used a chi-square test on the joint interval histogram for adjacent intervals of MUAPTs detected during constant-force isometric contractions. No dependence of any statistical significance was found.

Few authors report having made calculations from the IPI data of single motor units to test for higher order interval dependence. The 2nd through 10th order serial correlation coefficients computed by Kranz and Baumgartner (1974) were of lesser magnitude than the first order coefficients, and the chi-square test on the 3rd order joint interval histogram computed by De Luca and Forrest (1973a) revealed no dependence.

Synchronization. Synchronization, the tendency for two or more motor units to discharge at a fixed time interval with respect to each other. This includes, but is not limited to, MUAPTs which are phase-locked or entrained. In a mathematical sense, synchronization can be defined as dependence between MUAPTs. A useful technique for observing the synchronous activity among the discharges of pairs of motor units has been described in Chapter 4. It consists of calculating the "intensity function," which is an operation similar to obtaining a cross-correlation function of two discrete variables.

The interest in this property of motor unit discharge has its roots in the observations of Piper (1907, 1908), who noted that on occasions, the surface EMG signal displayed oscillatory (grouped) activity. This occurrence has been accepted as an indication of synchronization of motor unit discharges. Such an interpretation has often been contested and still remains to be proven.

Evidence of the symptoms of synchronization has been reported by several authors. Lippold et al (1957, 1970) found that the MUAPTs from different motor units tended to group at the rate of approximately 9 bursts/second. This grouping became more evident when the muscle became fatigued. Missiuro et al (1962) and others have claimed to observe synchronization by noting the appearance of large periodic oscillations in the EMG signal as the muscle fatigued. Direct evidence was noted by Mori (1973), who observed that motor unit discharges in the soleus muscle synchronized during quiet stance in man. In a later study, Mori and Ishida (1976) demonstrated that the discharge of motor units would indeed become synchronized if the feedback from the muscle spindle in the muscle was sufficiently large.

Kranz and Baumgartner (1974) and Shiavi and Negin (1975) performed a cross-correlation analysis between the MUAPs of simultaneously recorded MUAPTs. They concluded that during nonfatiguing, constant-force, isometric contractions of the first dorsal interosseous, flexor digitorum profundus, extensor digitorum indicis, and tibialis anterior, there was no significant cross-correlation. However, Buchthal and Madsen (1950) and Dietz et al (1976), using the same technique, did find evidence of weak cross-correlation in normal muscles. The amount of the cross-correlation increased in diseased muscles. The degree of cross-correlation also increased as the amplitude of the physiological tremor increased. Milner-Brown et al (1975) reported that it might be possible to accentuate synchronization by exerting large forces for short periods of time.

The phenomenon of motor unit synchronization has not been analyzed and documented as fully as other motor unit properties. Data have been reported which indicates that motor units tend to synchronize when the muscle is fatiguing, during physiological tremor, and in some disease states. However, no detailed description of the behavior of synchronization as a function of measurable parameters such as force and time has been given. This has mainly been due to limitations in the detection and analysis techniques that have been used. Also, a major disadvantage of the studies on synchronization which use indirect measurements from the raw EMG signal is that parameters other than synchronization may cause apparent oscillations in the amplitude of the raw EMG signal.

Firing Rate

Due to the pseudorandom nature of the IPIs, it is useful to measure the discharges of a motor unit in terms of an average firing rate, which is the reciprocal of the average IPI. However, for the firing rate to be meaningful, it should be measured over a representative time interval of 400 to 1000 ms. Measurements made over shorter time intervals may lead to unrepresentatively large firing rate values, and measurements made over larger time intervals may obscure meaningful trends in the firing rate. For details on techniques for calculating the firing rates, refer to Chapter 4.

The procedure of averaging (or filtering) the firing rates over the 400 to 1000 ms time is comparable to the filtering effect which the muscle tissue has on the pulse train from the α-motoneurons which excite it. Solomonow and Scopp (1983) have provided clear evidence that the mechanical response of muscle tissue to the pulse frequency of the electrical stimulation applied to its innervating nerves has the response of a low-pass filter with a 3 dB point in the neighborhood of 1.5 to 2.5 Hz. Thus, the recommended averaging (filtering) interval of 400 to 1000 ms allows the observer to "see" the firing rate in a fashion analogous to the way the muscle tissue "sees" the firing rate.

The first reported study on the firing rates of motor units was that of Adrian and Bronk (1928, 1929), who properly noted that in man, the upper limit of the motor unit firing rate was approximately 50 pps. This was confirmed by other early studies by Smith (1934) and Lindsley (1935).

Firing Rate as a Function of Force. The force dependence of firing rate during an isometric contraction has been studied by many investigators.

In the rectus femoris muscle, Person and Kudina (1972) found that the low threshold motor units began firing at 5 to 11 pps and reached 18 to 21 pps at 45% MVC. They also found that the higher the recruitment threshold of the motor unit, the less the motor unit increased its firing rate with increasing force.

A frequently studied human muscle is the biceps brachii. Clamann (1970) found that the firing rate of motor units recruited at the lowest force levels was 7 to 12 pps. The firing rate increased with increasing isometric force to a maximum of approximately 20 pps. The minimal firing rate of a motor unit increased linearly with the threshold of recruitment. Almost no motor units fired above 20 pps, even near 100% maximal voluntary contraction (MVC), and no recruitment was observed above 75% MVC. Clamann (1970) also found that motor units near the muscle surface had higher thresholds of recruitment than those deep in the muscle.

Leifer (1969), also working with the biceps brachii, found that all motor units fired at approximately 11 to 16 pps throughout the entire range of contraction force. After a motor unit was recruited, its firing rate increased slightly with increasing force and then remained constant at a preferred rate. He found that this preferred rate increased slightly with increasing threshold of recruitment. As the force level decreased, the firing rate decreased to 30 to 40% of the preferred rate before becoming inactive.

Also working with the biceps brachii, Gydikov and Kasarov (1974) found that all motor units had a firing rate of 6 to 10 pps when they were recruited. Minimal recruitment occurred above 60% MVC. For some motor units, the firing rate increased to approximately 13 pps and then remained constant with increasing force, whereas for other motor units, the firing rate increased linearly with force up to 100% MVC. The former were generally recruited at lower force levels than the latter. Based on their data, they proposed the existence of two types of motor units, tonic and kinetic. However, their small sample source (a total of 30 motor units from 15 subjects) limits the significance of their proposal. Clamann (1970) and Leifer (1969) did not describe two types of motor units in the biceps brachii; however, the firing rate characteristics found by these two investigators appear to differ slightly.

Kanosue et al (1979) studied the motor unit firing rate properties of the brachialis muscle, a synergist to the biceps branchii. Their results were in general agreement with those of Leifer (1969) in the biceps brachii. Kanosue et al (1979) provided support for the nonlinear increase of the firing rate as a function of force. However, unlike Gydikov and Kasarov, they did not conclude that two distinctly different categories of motor units are present in brachialis, based on their firing rate characteristics. Kanosue et al (1979) also reported that as the force output of the muscle increased to approximately 75% MVC, the firing rates of some motor units increased dramatically.

Hannerz (1974) and Grimby and Hannerz (1977), working with the tibialis anterior and short toe extensors, reported that the minimal firing rate of motor units recruited below 25% MVC was 7 to 12 pps and the maximal firing rate was 35 pps. For motor units recruited above 75% MVC the minimal firing rate was 25 pps, and the maximal firing rate was 65 pps in the tibialis anterior and 100 pps in the short toe extensors. Thus, both the average firing rate and the initial firing rate at recruitment increased with force. These firing rates are notably higher than those reported by other investigators. This discrepancy may be due to the method used to estimate the firing rate. They also found that all motor units recruited above 80% MVC discharged in bursts with pauses of 1 second or more at constant force levels. In a complementary study on the toe extensor, Borg et al (1978) found that motor units which could be driven continuously at firing rates below 10 pps had axonal conduction velocities between 30 and 40 m/s; those that could be driven only in higher firing rates bursts had higher axonal conduction velocities.

Monster and Chan (1977), working with the extensor digitorum communis in the forearm, demonstrated that the rate of increase of the firing rate increased with the voluntary force output of the muscle for higher threshold motor units. This behavior is consistent with that found by Kanosue et al (1979) in the brachialis muscle. However, it must be noted that the force levels studied by Monster and Chan (1977) was considerably less in terms of percentage MVC than those studied by Kanosue et al (1979).

Tanji and Kato (1973a and b), working with a smaller muscle, the abductor digiti minimi, found that motor units began firing at approximately 8 to 9 pps. They also reported that the earlier recruited motor units had the capability of increasing the firing rate to much higher values than latter recruited motor units during linear force increasing contractions. The firing rate at the beginning of the contraction was noted to increase as the force-rate of the contraction increased.

Milner-Brown et al (1973a) studied the activity of single motor units in the first dorsal interosseous muscle contracting at force levels below

50% MVC. They found that when recruited, motor units began firing at 8.4 ± 1.3 pps and increased their firing rate 1.4 ± 0.6 pps for each 100 g of force output, independent of the force at which each motor unit was recruited. They also found that a change in the force rate affected this result. At slow rates of increasing force (100 g/s), the firing rate had a tendency to reach a plateau, while at faster rates of increasing force (1000 g/s) motor units were recruited at lower force levels but with higher initial firing rates. This difference was not apparent during decreasing voluntary contraction.

Freund et al (1975) also performed an extensive investigation of single motor unit activity in the first dorsal interosseous muscle. They found that all motor units, regardless of their recruitment threshold, began firing at approximately the same rate (6.8 ± 1.4 pps). However, the lower threshold motor units increased their firing rates with increasing force much faster than the higher threshold units. The firing rates increased with force asymptotically to a maximal rate which also depended on recruitment threshold. These maximal rates varied from approximately 25 to 10 pps for low to high threshold motor units, respectively. However, none of the studied motor units were recruited above a force of 700 g. The rate of force increase tested by Freund et al (1975) was slower than that tested by Milner-Brown et al (1973a). This difference might account for some of the observed discrepancy.

Firing Rate as a Function of Muscle. It is now apparent that two distinct behaviors of the firing rate of motor units as a function of force output of the muscle have been reported by several investigators. The behavior is muscle dependent. In smaller muscles, such as those in the hand, the firing rates of motor units reach relatively higher values than the firing rates of motor units in larger limb muscles. Comparable data for highlighting this point has been presented by De Luca et al (1982a), as well as by Kukulka and Clamann (1981). The latter investigators studied the firing rate behavior of motor units in the biceps brachii and the adductor pollicis in the same subjects. De Luca and colleagues investigated the first dorsal interosseous and the deltoid of the same subjects.

De Luca et al (1982a and b), employing the recording and decomposition system described in Chapter 4, were able to study the behavior of up to eight concurrently active motor units. The subjects were requested to generate triangular force-varying isometric contractions up to 80% MVC. The experiments were performed on four categories of subjects: normal sedentary individuals, world class long distance swimmers, world class powerlifters, and world class concert pianists. An example of the results is presented in Figure 5.4, which presents the continual value of the mean firing rate (averaged over 800 ms) and the force produced by

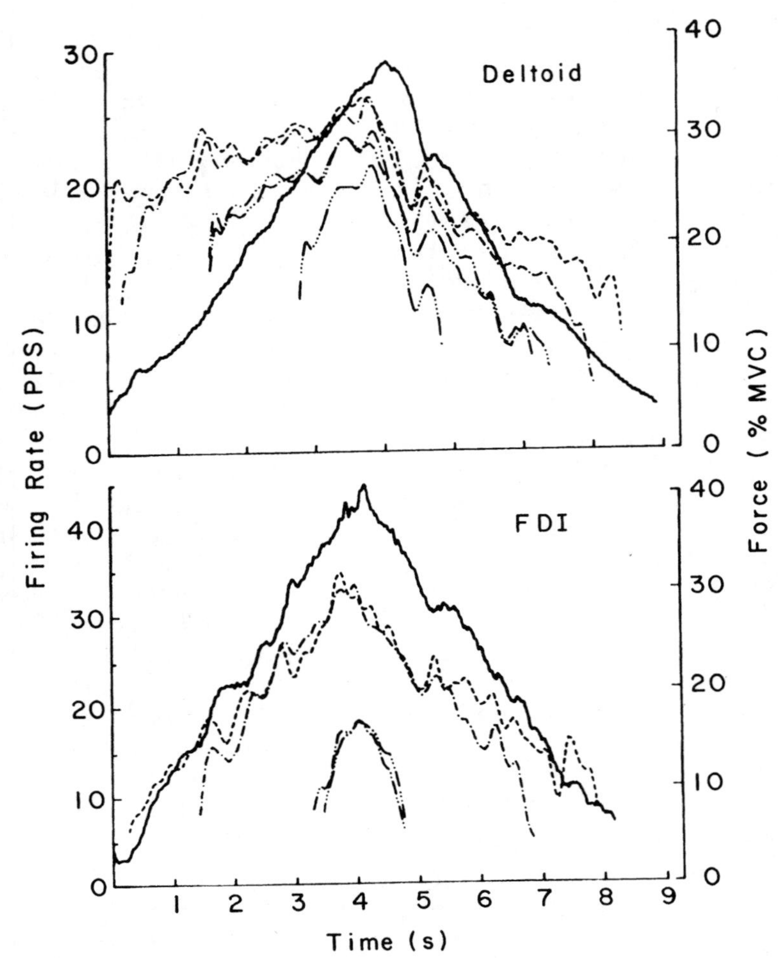

Figure 5.4. Firing-rate records of concurrently active motor units (*dashed lines*) are shown superimposed on the force output (*continuous line*) recorded during triangular force-varying contractions of the deltoid and first dorsal interosseous (FDI). Force levels are given in percent of maximal voluntary contraction (*MVC*) at right. These firing rate patterns are characteristic of those obtained for each muscle at all force rates examined and both peak forces (40 and 80% MVC). Note the presence of separate *vertical scales* for each of the displayed parameters. Firing rate and force values were related through the time axis.

the muscle (about a joint) as functions of the time of the contraction. (The *heavy solid line* represents the force.) Note that in the deltoid muscle the firing rate rises steeply after recruitment, reaches an apparent plateau, and subsequently decreases less rapidly than it increased, although the force rate on the rising and falling phases is somewhat similar. The first dorsal interosseous muscle on the other hand presents a strikingly different behavior during a similar force task. In this case, the firing rates are nearly linearly related to the force and do not display the plateau or asymmetry with force which is evident in the deltoid data. All the subjects clearly demonstrated this distinction in the firing rate behavior.

Another interesting observation that can be made in Figure 5.4 concerns the properties of the firing rate at recruitment and at decruitment. In the deltoid muscle, the recruitment and decruitment firing rates are higher than in the first dorsal interosseous. In the deltoid the decruitment firing rate is lower than the recruitment firing rate. This distinction is not apparent in the first dorsal interosseous. Figure 5.5 presents the results of the grouped data from the separate categories of subjects. The distinction in the behavior of the absolute values of the recruitment and decruitment firing rates is now apparent. Note the relatively minor distinction among subject categories, compared to the distinction between the two muscles. This data provides circumstantial evidence for the adaptation of motoneuronal properties, such as the time course of afterhyperpolarization, during a sustained contraction. It also indicates that the process of adaptation is executed with varying emphasis in different muscles.

The dynamics of the firing rates also differed between the two muscles. Figure 5.6 presents the maximal firing rate values of the motor units at the 40% and 80% MVC level as a function of recruitment force. In the deltoid the firing rate of all the motor units increased only by approximately 16 pps. This increased "swing" in the firing rate range of the first dorsal interosseous muscle is also visible in Figure 5.4.

A comparison of the firing rate properties of the first dorsal interosseous and deltoid muscles is presented in Table 5.1.

This minimal firing rate, or firing rate at recruitment, is generally assumed to be governed by the duration of the motoneuron afterhyperpolarization. Zwaagstra and Kernell (1980) have reported a negative correlation between the size of the motoneuron cell body and the duration of the afterhyperpolarization, indicating that the earlier recruited smaller motoneurons should have a lower initial firing rate. To date, numerous investigations have not revealed a distinct increase in the minimal firing rate as a function of recruitment threshold. (The works of Hannerz (1974) and Grimby and Hannerz (1977) are an exception.) Thus, it is reasonable to conclude that the minimal firing rate is determined by additional factors, possibly the recurrent inhibition of the Renshaw system. This latter suggestion is speculative but nonetheless attractive.

Figure 5.5 and the above table present a clear indication that the firing rate at recruitment is much more variable among motor units in different

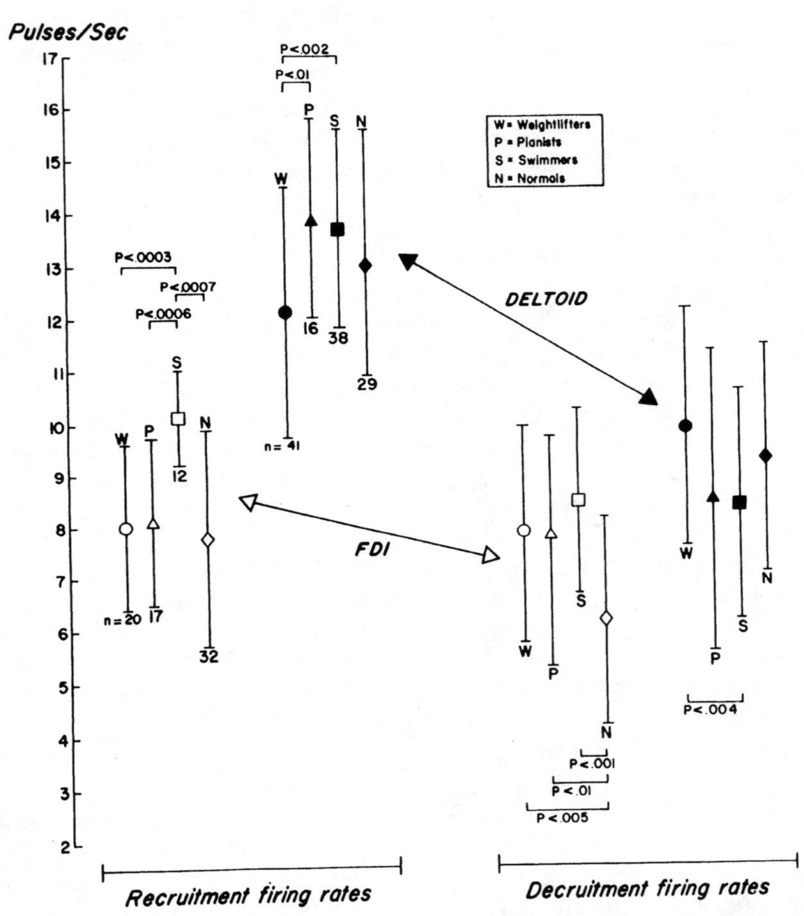

Figure 5.5. The distribution of motor-unit firing rates at initiation (recruitment) and cessation (decruitment) of continuous activity during triangular force-varying contractions reaching 40% MVC. In general, decruitment firing rates are lower than recruitment firing rates, but both of these parameters are greater in the deltoid than in the first dorsal interosseous (FDI). Significant differences between subject groups are indicated by bars showing the upper limits of the *P* values. (From C.J. De Luca et al, © 1982a, *Journal of Physiology*.)

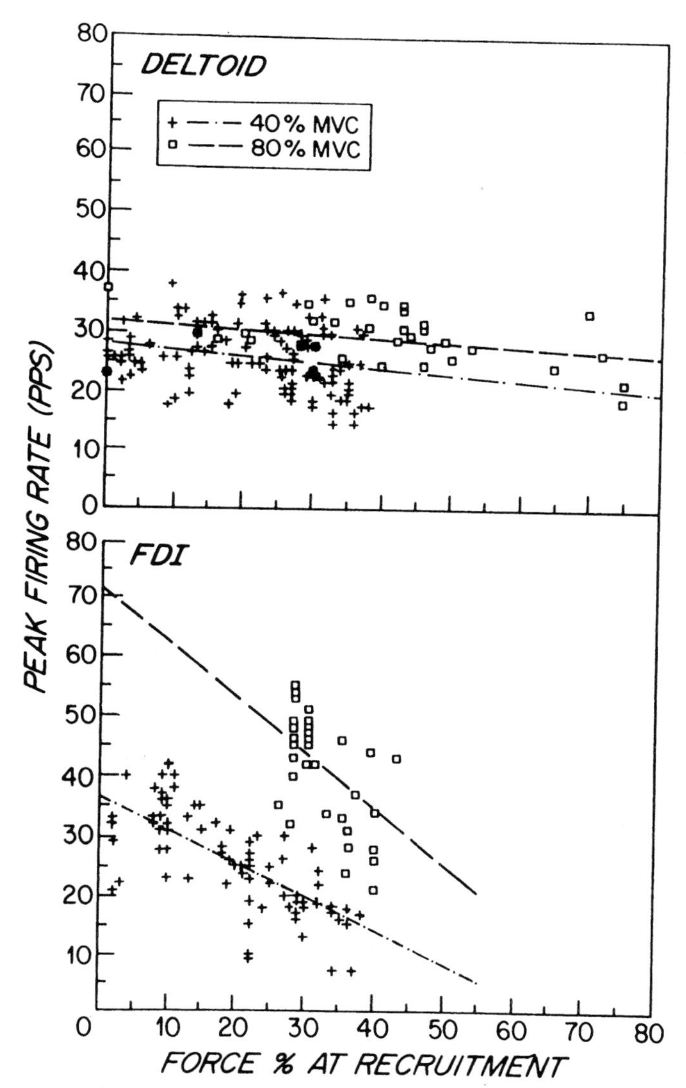

Figure 5.6. Peak firing rates achieved during triangular contractions reaching 40 and 80% MVC as a function of recruitment-force threshold. The computed least-squares linear regressions in the first dorsal interosseous (FDI) show strong negative correlations (linear regression coefficients > 0.56). In the deltoid, the linear regressions had slopes significantly different from zero at 40% MVC ($p < 0.008$), but not at 80% MVC ($p < 0.08$). (From C.J. De Luca et al, © 1982a, *Journal of Physiology*.)

Table 5.1
Motor-Unit Firing Rates in Two Different Muscles[a]

Muscle	Recruitment rate (pulses/s)	Decruitment rate (pulses/s)	Peak rate at 40% MVC (pulses/s)	Peak rate at 80% MVC (pulses/s)
FDI	8.9 ± 2.2[a] (119)	7.3 ± 2.2 (119)	25.3 ± 8.2 (81)	41.4 ± 9.6 (38)
Deltoid	12.9 ± 2.5 (158)	9.1 ± 2.5 (158)	26.3 ± 4.8 (124)	29.4 ± 3.4 (34)

[a] In each case the mean ± SD of an observation is listed with the number of observations (n) in parentheses.

muscles than among motor units within a muscle. It is also apparent that training does not significantly affect the firing rate value at recruitment.

Firing Rate during Strenuous Contractions. Another interesting property of the firing rates of motor units has been observed by Kanosue et al (1979) in the brachialis and by De Luca et al (1982a and b) in the first dorsal interosseous during extremely strenuous isometric contractions. The observations of De Luca et al (1982a) are presented in Figure 5.7. The firing rates of relatively high threshold motor units abruptly double from 30 to 60 pps. Such behavior was only seen occasionally, partly due to the extremely difficult task of accurately identifying motor unit action potential during these strenuous contractions. This phenomenon may also account for the distinction in motor unit behavior noted by Gydikov and Kosarov (1974) in the biceps brachii.

This behavior may represent a means of tapping the force reserves in a muscle. The type of contraction in which this phenomenon is clearly seen (10% MVC/s) is important, since these strenuous contractions were the only type in which the subject reported definite effects of fatigue on performance. As force output slowly increases toward 80% MVC, muscular fatigue causes the actual level of maximal voluntary effort to fall below that possible in a nonfatigued state. During the contraction shown in Figure 5.7, the individual (a normal subject) actually reported the sensation of reaching maximal voluntary contraction at the time of the firing-rate bursts.

Brief firing-rate bursts of 50 pps have been reported in single units during maximal voluntary contractions of the human quadriceps muscles (Warmolts and Engels, 1972) but only in patients with chronic low-grade motor neuropathies. Muscle biopsy revealed that this behavior was only seen in homogeneous fields of fast-twitch fibers (greater than 96%), apparently arising as the result of collateral reinnervation. Because of the difficulties inherent in separating the activities of several rapidly firing motor units, we were unable to determine whether all units displayed this bursting activity or whether large units were activated preferentially. The mechanical contribution of smaller units would pre-

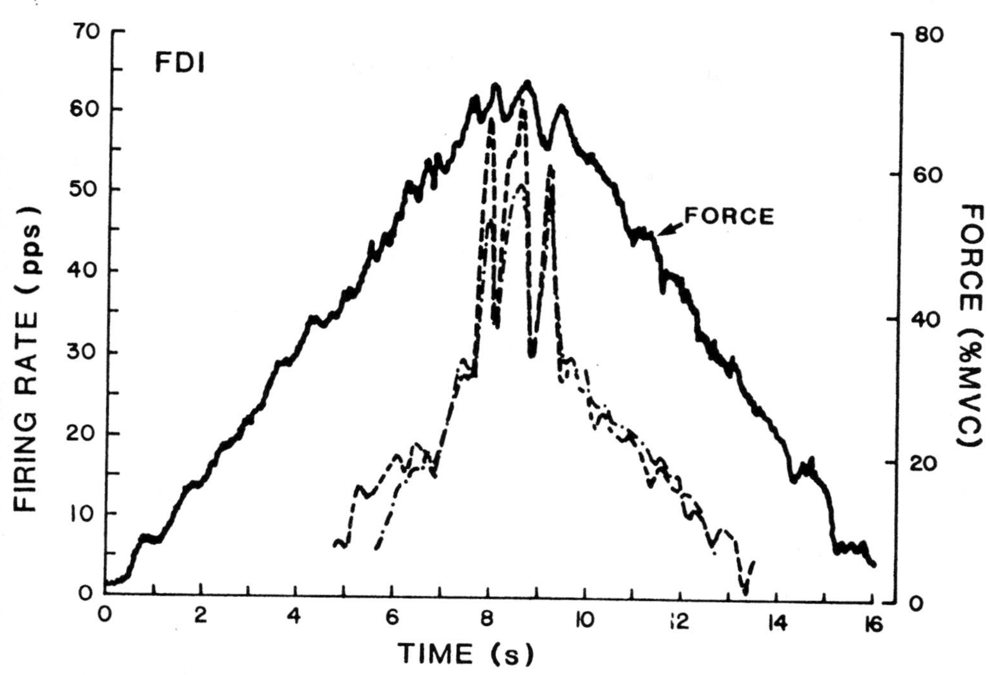

Figure 5.7. Firing-rate activity of two high force-threshold motor units (*dashed lines*) superimposed on the force output (*continuous line*) during a triangular force-varying contraction of the first dorsal interosseous (FDI). The highest rate is achieved by the lower threshold motor unit (recruitment-force thresholds for the two units were 31 and 42% MVC). Note the presence of separate vertical scales for each of the displayed parameters. Firing rate and force values were related through the time axis. (From C.J. De Luca et al, © 1982a, *Journal of Physiology*.)

sumably be small, however, since most are probably fused at the relatively high firing rates observed prior to the bursts (approximately 30 pps).

Although intermittent firing-rate bursts have been observed using single-unit recording, Figure 5.7 is the first evidence of a bursting phenomenon in continuously active motoneurons during nonballistic contractions. Whether these rapidly increasing firing rates result from a sudden increase in synaptic excitation is difficult to determine. However, evidence of this type of nonlinearity is well documented in cat spinal motoneurones subject to increased levels of current injection (Kernell, 1965). Baldisseria et al (1978) have accounted for this effect (the so-called secondary range of motoneuron firing) by a model based on the time course of motoneuronal afterhyperpolarization, indicating that it may well be a passive response to high levels of synaptic current injection. This effect may also account for the extremely rapid firing-rate bursts

(60 to 120 bursts/s) seen by Desmedt and Godaux (1977a and b) during ballistic contractions of the tibialis anterior, where high levels of synaptic current are necessary.

In contrast, no evidence of a firing-rate burst response was seen during voluntary contractions of the deltoid up to 80% MVC. Because of the "tonic" behavior observed in the deltoid, motor-unit firing rates remained much lower than those reached in the first dorsal interosseous. Refer to the previous table for details. If twitch contraction times are comparable for motor units in the two muscles, most fast-twitch deltoid units are probably unfused even at 80% MVC. Interestingly, a recent study of single motor units in the neighboring human brachialis muscle (Kanosue et al, 1979) has demonstrated behavior similar to that seen here in the deltoid but with an increased reliance on rate coding above 70% MVC. De Luca et al (1982a and b) did not investigate the force range between 80 and 100% MVC; however, the presence of unfused motor units indicates that rate coding has a tremendous potential for increasing force output up to (and even beyond) maximal voluntary levels in muscles like the deltoid. If the central nervous system were to increase firing rates or generate firing-rate bursts in the large fast-twitch motor units of the deltoid (such as those seen in the first dorsal interosseous), extraordinary force levels could be achieved for short periods of time. This mechanism may indeed be the explanation for many of the incredible feats performed by humans under high stress conditions and during hypnotic states.

Firing Rates in Abnormal Muscles. Much fewer investigations studying the behavior of the firing rate of abnormal muscles have been performed. However, it appears that interest in this area is increasing as new techniques and equipment develop. Companion reports by Andreassen and A. Rosenfalck (1980) and A. Rosenfalck and Andreassen (1980) have shown that the mean firing rate of motor units in spastic muscles is a reduced firing rate, compared to that of motor units from the corresponding muscle in normal subjects. Kranz (1981) drew similar conclusions in patients with clinically mild lesions of the central nervous system. Conversely, Holonen et al (1981) report that the average firing rate tends to be greater than normal in muscles with myopathic disorders.

Firing Rate as a Function of Time

During sustained contractions of healthy muscles, the firing rate of motor units has a tendency to decrease independently of the force output of the muscle. This behavior appears to be a reflection of motoneuronal adaptation processes and/or a decrease in the excitation to the muscle. This phenomenon was first reported to occur in constant force contractions by Person and Kudina (1972) in the rectus femoris, independently by De Luca and Forrest (1973a) in the deltoid and, more recently, in the first dorsal interosseous by De Luca et al (1982b). An example of this behavior during constant-force contractions is evident in the time progression of the characteristics of the IPI histograms of Figure 5.2 and in the firing rate curves of Figure 5.8. Grimby et al (1981) also observed a decrease in the firing rate. Kranz (1981) observed similar behavior in the first dorsal interosseous, extensor digitorum communis, and flexor digitorum profundus in patients with clinicially mild lesions of the central nervous system.

This phenomenon has also been observed during force-varying isometric contractions. Figure 5.4 and Table 5.1 clearly show that the firing rate at decruitment is less than that at recruitment during a force-varying contraction.

Kernell and Monster (1981) investigated the property by injecting a constant current into the motoneurons of the gastrocnemius muscle of cats. They succeeded in continuously exciting motoneurons which they were able to classify as slow-twitch (fatigue resistant) and fast-twitch (fast-fatiguing). Their results clearly indicated that the firing rates of the fast-twitch (fast-fatiguing) motoneurons decreased, whereas those of the slow-twitch (fatigue-resistant) motoneurons did not alter under similar stimulation of constant-current injection. Thus, it is apparent that the firing rate adaptation is at least in part a motoneuronal property, possibly associated with a modification of the afterhyperpolarization characteristics.

Twitch Potentiation

The decrease in the firing rate during a sustained contraction appears to occur concomitantly with another phenomenon, twitch potentiation, that is, an increase in the twitch tension produced by a motor unit. This phenomenon was first observed in mammalian muscles by von Euler and Swank (1940). Subsequent studies by Stondaert (1964), Nystrom (1968), and Burke et al (1976) have investigated the posttetanic potentiation in the *in vivo* soleus and gastrocnemius muscle of the cat. Substantial potentiation was always evident in the gastrocnemius muscle, in contrast to relatively little potentiation in the soleus muscle. In the cat, the soleus muscle consists almost entirely of slow-twitch fibers, whereas the gastrocnemius contains both slow-twitch and fast-twitch fibers. Thus these results are consistent with and supportive of the observations of the firing rate decrease in fast-twitch fibers made by Kernell and Monster (1982).

Muscles of humans have also been reported to display twitch potentiation. Gurfinkel' and Levik (1976) noted such behavior in the forearm flexor muscles. Vandervoort et al (1983) also demonstrated potentiation of the twitch response of the tibialis anterior and plantar flexor muscles

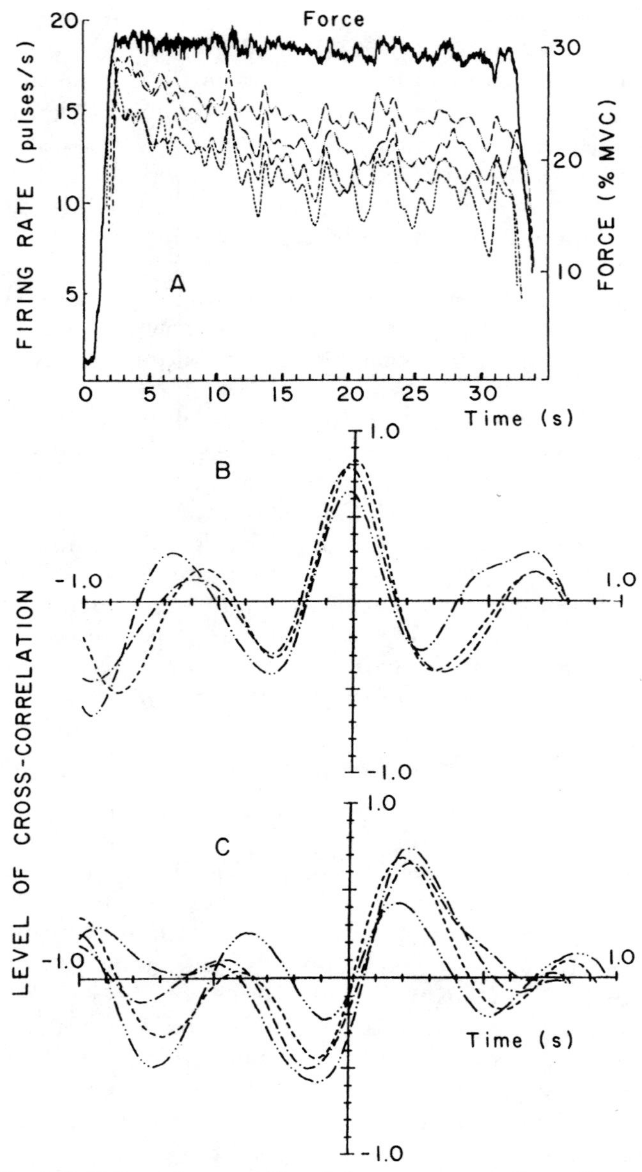

Figure 5.8. (*A*) Firing-rate records of four concurrently active motor units (*dashed lines*) are shown superimposed on the force output (*continuous line*) recorded during a constant-force isometric abduction of the deltoid. The force level is given in percent of maximal voluntary contraction (MVC) at *right*. (*B*) Functions obtained by cross-correlating between firing rates. (*C*) Functions obtained by cross-correlating between firing rates and force output. Positive shift of peaks in *C* indicates that firing-rate activity leads force output. (From C.J. De Luca et al, © 1982b, *Journal of Physiology*.)

of the leg. They noted that the twitch potentiation was always more evident in the tibialis anterior than in the plantar flexors, and that it was only evident in contractions greater than 50% MVC. These observations support the notion that the faster-twitch fibers, which are more abundant in the tibialis anterior and are recruited at relatively higher force levels, are more susceptible to potentiation. They further noted that the potentiation effect was dependent on muscle length and time of contraction. It was greatest after contractions lasting approximately 10 s and in muscles in shortened positions. They speculated that the twitch potentiation was in some fashion related to the normally incomplete activation of the contractile elements in the muscle fibers.

The Concept of the Common Drive

To understand the strategy (or strategies) which the nervous system uses to control motor units for the purpose of generating and modulating the force of a muscle, two central questions arise: (1) Is there a strategy or are there rules which govern the process of motor unit recruitment? (2) Is there a strategy or are there rules which govern the behavior of firing rates of active motor units? The first question has received considerable attention. Notable contributions have been made by Henneman and his colleagues. Details of this question will be addressed in a subsequent section. The second question has not engaged a comparable level of excitement, possibly due to the technical complexity of the experiments necessary to address it.

In order to address the question concerning the behavior of the firing rate properly it is necessary to observe the firing rate as a function of time and force of contraction. The occasional reports in the literature provide the beginning of an indication of the firing rate behavior as a function of force. Several reports (Leifer, 1969; Person and Kudina, 1972; Milner-Brown et al, 1973a; Tanji and Kato, 1973a and b; Monster and Chan, 1977; Monster, 1979; Kanosue et al, 1979) have all demonstrated that the firing rates of active motor units increase proportionally with increasing force output. This implies that increased excitation to the muscle motoneuron pool increases the firing rates of all the active motor units.

This commonality in the behavior of the firing rates was studied in detail by De Luca et al (1982b). They observed the behavior of the firing rates of up to eight concurrently active motor units in the first dorsal interosseous and deltoid muscles during various types of isometric contractions: attempted constant force, linear force increasing, and force reversals. Since that study, we have performed similar investigations on the flexor pollicis longus, extensor pollicis longus, tibialis anterior, extensor carpi ulnaris, and extensor carpi radialis longus.

The studies of De Luca et al (1982a and b) described a unison behavior of the firing rates of motor units, both as a function of time and force.

This property has been termed the *common drive.* Its existence indicates that the nervous system does not control the firing rates of motor units individually. Instead, it acts on the pool of the homonymous motoneurons in a uniform fashion. Thus, a demand for modulation of the force output of a muscle may be represented as a modulation of the excitation and/or inhibition on the motoneuron pool. This is the same concept which comfortably explains the recruitment of motor units according to the size principle.

Figure 5.8*A* provides an example of the behavior of the firing rates of four motor units during an attempted constant-force contraction of the deltoid muscle. The firing rates have been filtered with a 400 ms Hanning window. Note the common behavior of the fluctuations of all the firing rates. This commonality becomes more apparent in Figure 5.8*B*, which presents the cross-correlations of the firing rates. The high correlation values and the lack of any appreciable time shift with respect to each correlation function indicate that the modulations in the firing rates occur essentially simultaneously and in similar amounts in each motor unit. If the firing rates of the motor units are cross-correlated with the force output of the muscle, an appreciably high cross-correlation is also evident (Figure 5.8*C*). The peaks of the cross-correlation functions occur at a time corresponding to the time delays of the force built up after excitation in the muscle fibers. This testifies to the fact that the fluctuations in the force output are causally related to the fluctuations in the firing rates.

The high level of cross-correlation between the firing rates and the force output (Fig. 5.8*C*) points strongly to the fact that a muscle is incapable of generating a pure constant-force contraction under isometric conditions. The fluctuations in force which are ever present during attempted constant-force contractions are a manifestation of the low-frequency oscillations which are inherent in the firing rates of motor units. The dominant frequency of this oscillation is approximately 1.5 Hz. The source of this oscillation has not been identified yet. But, it is interesting to note that the transfer function of the stimulation frequency and mechanical output of a nerve-muscle unit is a low-pass filter having a 3 dB point at approximately 1 to 2 Hz. This observation has been made by several investigators using a variety of paradigms. (Crochetiere et al, 1967; Coggshall and Bekey, 1970; Gottlieb and Agarwall, 1971; Soechting and Roberts, 1975; Solomonow and Scopp, 1983). Therefore, it would be functionally useful to "drive" the muscle near the "critical" frequency of the muscle contractile characteristics. In this fashion, the "drive" to the muscle is continuously poised to affect changes in the force output in the shortest period of time without any overshoot (errors).

Referring back to Figure 5.4, a similar behavior is seen during force-increasing and force-decreasing contractions. In this case, the firing rate fluctuations are superimposed on a "bias" firing rate value. This bias value displays the common and proportional association with force output which has been documented by several investigators, that is, as an increase in the force output of a muscle is required, all the active motor units increase their firing rates proportionally. Given that the initial (or minimal) firing rates of motor units at recruitment are considerably similar, it follows that the higher force-threshold, faster-twitch motor units will always have lower firing rates than the lower force-threshold, slower-twitch counterparts. This arrangement indicates a peculiarity of motor unit control during voluntary contractions, that is, the firing rate behavior is not complimentary to the mechanical properties of the motor units. Higher threshold motor units tend to have shorter contraction times and twitch durations and thus require higher firing rates to produce fused contractions. De Luca et al (1982a) calculated that in some cases, the faster-twitch motor units never achieved a fused contraction during voluntary effort. This behavior provides a basis for the concept that in man the full physical force generation potential of the muscle fibers may not be utilized during voluntary contractions. Conceivably, it may be held in abeyance for occasional dramatic displays of force.

The examples of Figures 5.4 and 5.8 are representative of observations seen in the firing rates of motor units in all the upper and lower limb muscles investigated to date. It has been seen in relatively small and relatively large muscles; in motor units of slow-twitch and fast-twitch fibers.

The *common drive* has also been observed to exist in an agonist-antagonist set of muscles simultaneously. In a recent study involving the flexor pollicis longus and the extensor pollicis longus, the sole controllers of the interphalangeal joint of the thumb, De Luca and colleagues have noted the *common drive* in both muscles. During voluntary stiffening of the interphalangeal joint, the firing rates of motor units in the two muscles were highly correlated with essentially no time shift (see Fig. 5.9 for details). Note that although the force or torque output is approximately zero, the *common drive* remains. This particular example points to the necessity of associating the behavior of the motor unit control to the effect on the motoneuron pool rather than the output of the joint. The same study also reported that during random flexion-extension isometric contractions of the interphalangeal joint, the firing rates of the antagonist motor units were negatively highly cross correlated. This implies the existence of an ordered modulation of the firing rates of motor units in the two muscles; when the firing rate increased in one, it decreased in the other, and vice versa.

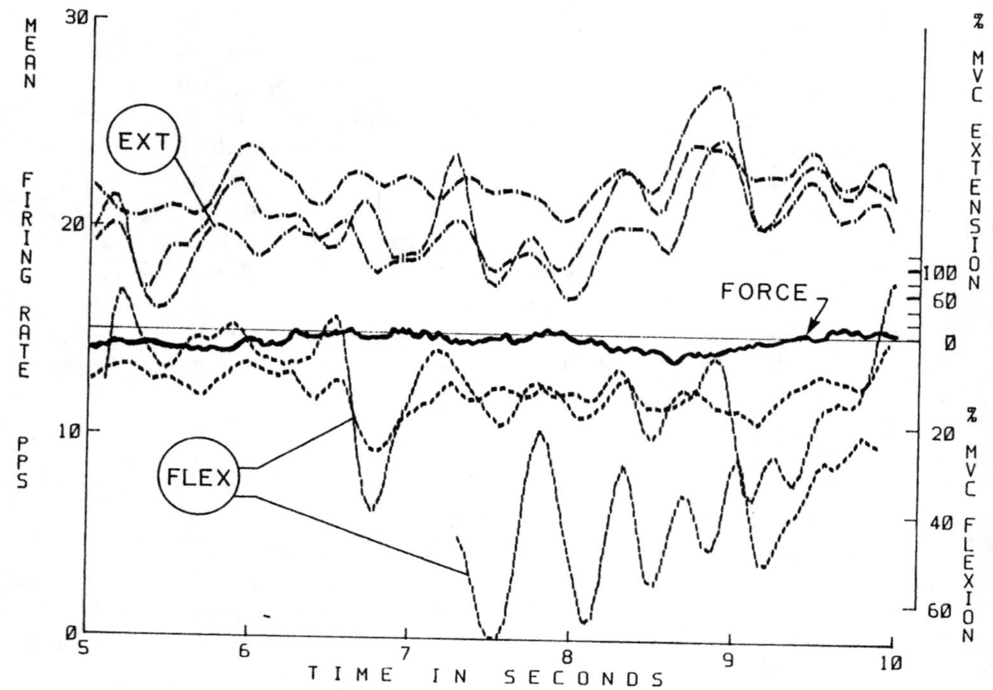

Figure 5.9. Example of motor unit firing rate behavior during thumb phalangeal joint stiffening. FRC line denotes the force (or torque) output from the joint; the *FLEX* lines represent the firing rates of motor units in the flexor pollicis longus; the *EXT* lines represent the firing rates of motor units in the extensor pollicis. These two muscles are the sole controllers of the joint.

These observations of the *common drive* indicate that when two antagonist muscles are activated simultaneously to stiffen a joint, the nervous system views them as one unit and controls them in like fashion. In this case, the homonymous motoneuron pool consists of the motoneuron pools of both muscles. However, when the force output of the joint alternates from flexion to extension, or vice versa, the two pools are controlled reciprocally, with one being inhibited or disfacilitated while the other is excited.

At this point, a cautionary note is in order. The presence of the considerably high level of cross-correlation in the firing rate cannot be interpreted as evidence of motor unit discharge synchronization. It simply means that the average pulses per epoch of time discharged by one motor unit behave similarly to those of all the other active motor units in the same epoch of time. It is, therefore, an indication of the control of motor units over a larger time scale than that which effects the properties of synchronization that relate to individual discharges of motor units.

Firing Rate at Force Reversal. The concept of the *common drive* raises a concern over the control scheme necessary to increase the force output to a precise value and then decrease the force, as would be the case in the execution of an accurate triangular force trajectory required in a skilled task. If the firing rates of all the motor units (slow twitch and fast twitch) are modulated simultaneously, how is an accurate force value generated prior to a force reversal when the contraction times of the different motor units (or muscle fibers) vary from 30 to 150 ms? This question is answered by the data in Figure 5.10. In this particular case the firing rates were filtered with a window of 800 ms in order to emphasize the "bias" firing rate, which is related to the force output. The *shaded area* prior to the force reversal emphasizes the fact that the earlier recruited (slower-twitch, longer contraction time) motor units decrease their firing rates *before* the latter recruited (faster-twitch, shorter contraction time) motor units. The *bottom* of the figure presents the cross-correlation functions of the firing rates and the force, providing a clearer expression of the *lead-time* between the firing rate reversal and the force reversal. This magnificant orchestration of firing rate reversals apparently considers the mechanical properties of the motor units so as to synchronize their contribution to obtain an accurate force output.

The ordered firing rate reversals cannot be explained by differences in axonal conduction velocities. In fact, the conduction velocity gradation is organized in the opposite direction to that required. One explanation for this behavior would be that the nervous system keeps track of the particular mechanical response of each motor unit and delays the firing rate of each motor unit by an appropriate amount. Such an explanation is inconsistent with the common drive, which is in effect during other force generation modalities. In addition, it would require a tremendous amount of processing in the central nervous system. It is indeed highly unlikely in light of other possibilities.

There remain two other possible explanations: a selective sensitivity to a reduction in excitation and/or a selective sensitivity to an increase in inhibition to the motoneuron pool. The possibility of the combined events is particularly attractive since experimental evidence obtained by Clamann et al (1974) suggests that interaction between excitation and inhibition processes might be expressed as simple algebraic values. Lusher et al (1979) have also demonstrated that in anesthetized cats, inhibition apparently proceeds according to the size principle, with the smaller motoneurons being affected first.

The sequence of events might be as follows. As the subject plans or anticipates a force reversal, an increasing inhibitory input is applied to

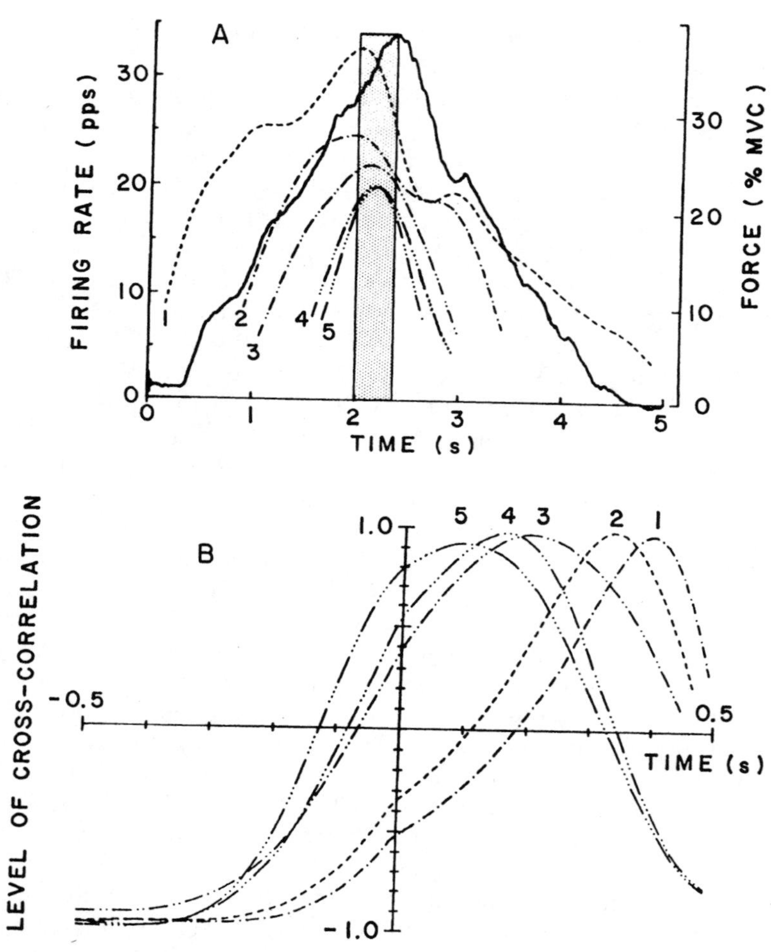

Figure 5.10. (*A*) Firing-rate records of five concurrently active motor units (*dashed lines*) are shown superimposed on the force output (*continuous line*) recorded during a triangular force-varying contraction of the first dorsal interosseous. Width of *shaded area* illustrates the concept of firing-rate reversal lead for the lowest threshold motor unit. (Note the presence of separate *vertical scales* for each of the displayed parameters. Firing rate and force values are related through the time axis.) (*B*) Functions obtained by cross-correlating between the firing-rate and force records for each motor unit shown in *A*. Horizontal positions of peaks are estimates of firing-rate reversal leads. (From C.J. De Luca et al, © 1982b, *Journal of Physiology*.)

the motoneuron pool which competes with the increasing excitatory input in progress. Larger inhibitory postsynaptic potentials are produced in smaller motoneurons, effectively overcoming the excitation and resulting in ordered firing-rate reversals. Either prior to or as the force peak is reached, a reduction in excitatory input augments the firing rate decrease. This simple scheme combines the known electrical responses of motoneurons with the varied mechanical responses of individual motor units to produce sharp force reversals: firing rates of small units with slow-twitch responses are reduced earlier than larger units with fast-twitch responses, effectively synchronizing the mechanical relaxation of the entire motor-unit population.

Thus, the concept of the *common drive* is not violated because the excitation and inhibition act on the motoneuron pool without regard to the individual motoneurons. The specific ordered response is a property of the motoneuron pool architecture and structure.

Recruitment

Ordered Progression. The behavior of the process which controls the recruitment of motor units has received considerable attention. One of the most consistent observations of motor unit behavior reported in the literature concerns the order of recruitment as a function of size. For over two decades Henneman and his colleagues, working with decerebrate cats, have compiled considerable data directed at describing a "size principle." This size principle states that the recruitment order within a motoneuron pool progresses from the smallest to the largest motoneuron. The most convenient *in situ* measure of the motoneuron size in a contracting muscle has been found to be the conduction velocity of the axon. And in fact, Henneman and his colleagues have used this measure (as well as others) to argue their case.

The choice or description of the invariant parameter which describes the recruitment of motoneurons (or motor units) has been questioned by Fleshman et al (1981). In a series of related publications, they have been persistent in arguing that the motor unit type, classified according to the electrical-mechanical properties (or fatigue characteristics), has a dominant involvement in the behavior of the recruitment order. Kernell and Monster (1981), working with anesthetized cats, injected a current in α-motoneurons and confirmed that the axons with lower conduction velocities were consistently more excitable than those with faster conduction velocities. However, among fast-twitch motoneurons of about the same size, as measured by their axonal conduction velocity, the average threshold current was about twice as high for cells innervating fatigue-sensitive muscle fibers than for those supplying more fatigue-resistant ones.

These caveats are not necessarily contrary to the basic concept of the size principle because it is generally accepted that motor unit size and twitch tensions are represented as a continuum in the fiber type classifications. In fact, it is conceivable that the nervous system follows a recruitment order strategy which may reflect the orderly increase of the mechanical contribution of the motor units within a muscle.

Freund et al (1975), working with humans, measured the axonal conduction velocities and found that the slower conduction velocities, and thus the smaller axons, were associated with the lower threshold motor units. Clark et al (1978), working with rhesus monkeys, have also confirmed that motor units recruited at lower force levels have longer contraction times and produce smaller twitch tensions than higher threshold motor units. By averaging the force output of the muscle as each action potential from a single motor unit occurred, Milner-Brown et al (1973b) were able to determine the twitch tension of some motor unit. They found a linear relationship between twitch tension and recruitment force, suggesting that the fractional increment in force ($\Delta F/F$) is constant. Goldberg and Derfler (1977), investigating the masseter muscle, found that motor units with high-recruitment thresholds tend to have larger amplitude MUAPs and twitches with greater peak tension than with motor units recruited at lower force levels.

Experiments directed at studying the recruitment order of motor units during relatively fast force increasing contractions, including ballistic contractions, have been performed by Tanji and Kato (1973a), Budingen and Freund (1976), Desmedt and Godaux (1977b, 1978), and De Luca et al (1982a). All these reports with the exception of Tanji and Kato stated that the recruitment order remained invariant as a function of force output and force rate. Tanji and Kato (1973a), Budingen and Freund (1976), and Desmedt and Godaux (1977b, 1978) all noted that the firing onset of a motor unit occurred earlier as the rate of the force output increased. This phenomenon is particularly evident during ballistic or near-ballistic contractions. This *apparent downshift* in force threshold of recruitment was correctly interpreted by Budingen and Freund (1976), who pointed out that in contrast to firing onset, the peak of the force of a motor unit twitch occurs at approximately the *same* muscle tension, regardless of the force rate.

This latter observation indicates that the time of recruitment of a motor unit must be considered as the time when the motor unit contributes to the force output, not the time at which the action potential is noted. This should serve as a reminder that the action potential is an *artifact* of the contractile process, albeit a useful one for investigatory purposes.

De Luca et al (1982a) verified, in human muscles, that when the force output of a muscle is voluntarily decreased, motor units are *decruited* in the opposite order in which they were recruited. This result had been known to occur in decerebrated cats. The observation implies that disfacilitation of the motoneuron pool obeys a principle of ordered behavior also. The ordered decruitment behavior may be usefully associated with thresholds in the diminishing force. Figure 5.11 presents the

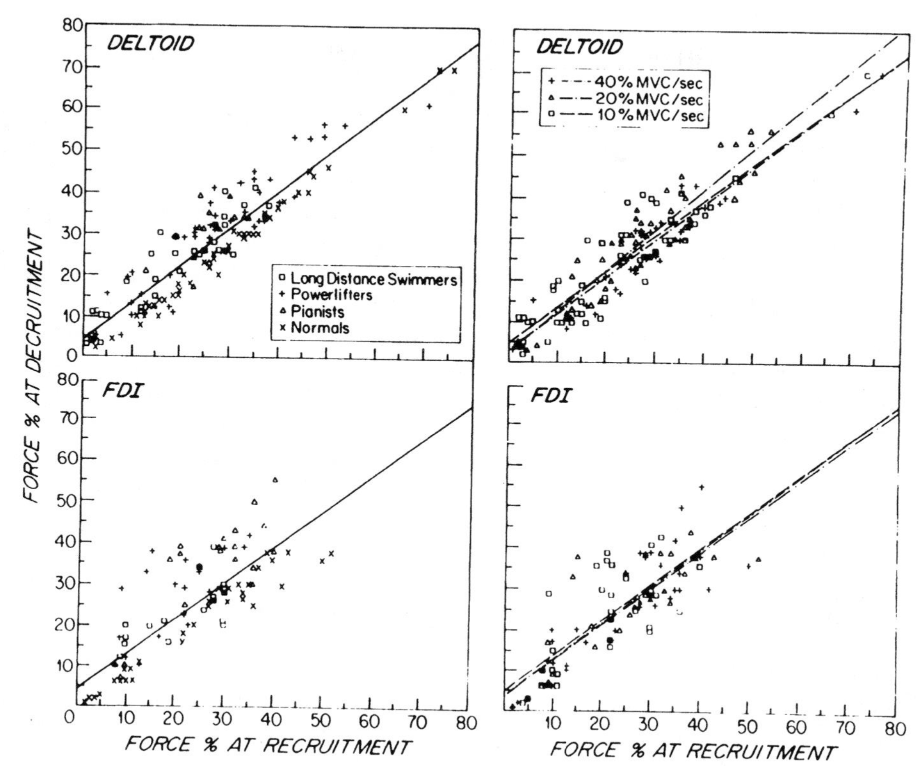

Figure 5.11. Force levels, given as percent maximal voluntary contraction (*MVC*) associated with motor-unit recruitment and decruitment. The linear relationship observed for both muscles (regression coefficients > 0.8; all data grouped) was consistent across all subject groups (*A* and *B*) and was invariant at the three force rates examined (*C* and *D*). Recruitment was only observed up to 52% MVC in the first dorsal interosseous (*FDI*) but was seen as high as 80% MVC in the deltoid. (From C.J. De Luca et al, © 1982a, *Journal of Physiology*.)

force at recruitment plotted against the force at decruitment for motor units which were active during linearly force-increasing and linearly force-decreasing contractions of deltoid and first dorsal interosseous muscle of normal subjects and highly trained athletes. Positive linear correlations were observed in both muscles ($r = 0.94$ for deltoid, $r = 0.83$ for first dorsal interosseous). No significant deviation from this relationship was seen among the four subject groups. Furthermore, Figure 5.11 illustrates invariant behavior across the force rates (nonballistic) of the contractions. This relationship points to a highly ordered recruitment and decruitment scheme which remains invariant with muscle, training, and force rate (nonballistic).

Small upward shifts observed in the regression lines of Figure 5.11 indicate that motor units, in general, have a tendency to cease firing at relatively higher force levels than those at which they began. This effect has also been reported by Milner-Brown et al (1973b). It may simply be an expression of the fact that the force developed by a motor unit lags its discharge.

Modification of Ordered Progression. There have been some reports which have argued that the orderly progression of motor unit recruitment is altered under some conditions.

Tanji and Kato (1973a) reported that the recruitment order is not rigidly fixed among motor units with nearly similar force thresholds. Such observations have also been made in our laboratory. There are at least two possible explanations for such observations. One concerns the instantaneous force-rate at the time of recruitment. If the contractions are not repeated with the identical force rates in the neighborhood of the recruitment thresholds, it is conceivable that the order of two somewhat similarly sensitive motor units may be altered if the rise time of the twitches of the two motor units differs. The other explanation concerns the properties of motoneuron adaptation and/or twitch potentiation. During repeated contractions the twitch responses may be altered so that their mechanical characteristics coincide with the need for an altered recruitment order.

Person (1974) reported that the recruitment order was stable for a given movement task but could be altered when the muscle performed a different movement task. Thomas et al (1978) noted that recruitment order reversal occurred during markedly different orientations in multifunctional muscles such as the abductor pollicis brevis and extensor digitorum; none was seen in the first dorsal interosseous. Desmedt and Godaux (1981), however, did report order reversal in the first dorsal interosseous. Romeny et al (1982), working with the biceps brachii, also reported observing changes in the recruitment order, depending on the function (flexion or supination) performed by the muscle. All three reports stated the reversal was not consistent. Desmedt and Godaux reported a reversal between motor unit pairs only in 11.2% of the motor units which they observed. All three of these investigations used fine-wire electrodes to detect the signals. The reader is referred to the discussion on detection techniques in Chapter 2. Note that a relative movement of 0.1 mm between the detection surfaces of the electrodes and the active fibers may cause considerable modifications in the shapes of the MUAPs, the parameter that is used to determine if a new motor unit is recruited. Considering that the reversal has been noted almost exclusively during contractions of markedly different orientations, any claim for recruitment reversal must first prove that the electrode does not migrate into the territory of other motor units. None of the published studies addresses this question. This necessary query shows that the inquisitiveness of an investigator should be tempered with the technicality of an engineering approach. The reported reversal may in fact occur, but it is yet to be proven.

An unquestionable reversal of the recruitment order of motor units has been artificially induced in the first dorsal interosseous by Stephens et al (1978). They noted that prolonged electrical stimulation of the digital nerves of the index finger induced a reversal of recruitment order. This phenomenon persisted for some time after the stimulation ended. Similar results have been obtained by Mizote (1982) working with the lumbricals of anesthetized cats.

Recruitment as a Function of Muscle. An overview of the available literature reveals that the recruitment scheme varies among muscles. In some muscles all the motor units are recruited at force levels well below maximal while in others, recruitment continues up to maximal. A clear example of this phenomenon may be seen in Figure 5.11, where in the first dorsal interosseous, all the motor units were recruited below 50% MVC, whereas in the deltoid, recruitment persisted up to nearly 80% MVC and may have been present even at higher force values. Milner-Brown et al (1973a) reported similar observations for the first dorsal interosseous. Kukulka and Clamann (1981) reported that in the biceps brachii, recruitment was observed up to 80% MVC, whereas in the adductor pollicis none was observed above 50% MVC. Kanosue et al (1979) found recruitment up to at least 70% MVC in a relatively large muscle, the brachialis.

The accumulation of the above individual pieces of evidence indicates that small muscles, such as those found in the hand, recruit all their motor units below 50% MVC and larger muscles found in the limbs recruit motor units throughout the full range of voluntary force. A possibility may exist that in some muscles such as the soleus and gastrocnemius, not all the motor units are activated during perceived maximal efforts (Belanger and McComas, 1981). Although in a smaller muscle, the adductor pollicis, both Merton (1954) and Bigland-Ritchie (1982) reported that all the motor units do become activated. However, in the latter case it has been correctly pointed out by Belanger and McComas (1981), that the thenar musculature contributing to adduction of the thumb is complex and is not all innervated by one nerve, thus complicating the rationale for comparing the force output during perceived voluntary maximal contractions and tetanic electrical stimulation of muscles via their innervation.

Recruitment as a Function of Time. It is now generally accepted that human muscles contain motor units which have a continuum of mechanical characteristics. The faster-twitch muscle fibers which comprise the motor units that are recruited at higher force-thresholds decrease their mechanical output at a faster rate than earlier recruitment slower-twitch muscle fibers. Thus, a question arises concerning the possibility of motor units being recruited during a sustained contraction.

Several investigators have reported that during a constant-force isometric contraction, new motor units ought to be recruited throughout the duration of the contraction. (Edwards and Lippold, 1956; Vrendenbregt and Rau, 1973; Person, 1974; Maton, 1981; and Kato et al, 1981). In fact we have made observations in our own laboratory which could be construed as supporting this position. However, such an interpretation must be cautioned. All experiments to date which have addressed this question have been restricted to relating the time-dependent behavior of recruitment to the torque measured at the affected joint. This torque is the result of the individual torques of all the agonists minus that of all the antagonist muscles. Thus, it is conceivable that the monitored torque output remains constant while the separate contributions from the agonist and antagonist muscles vary linearly. An extreme example of this complication is presented by Figure 5.9. In this case the joint is stiff, the torque output is approximately zero, and several motor units are active in the agonist and the antagonist.

It is clear that much more work is required on this problem before a generalized statement may be made.

Firing Rate and Recruitment Interaction

Interaction within a Muscle. At the beginning of this chapter a description of the peripheral systems which affect the control properties of motor units was presented. Referring to that text and to Figure 5.1, it may be seen that considerable anatomical and functional coupling exists among the motor units within a muscle. Such an interaction was indeed found by Broman et al (1984) and is displayed in Figure 5.12. In this study it was found that when a motor unit is recruited during slow force increasing (1–2% MVC/s) isometric contraction, it was often observed that previously activated motor units were disfacilitated. This was noted as a *decrease* in the firing rates of previously activated motor units as the firing rate of the newly activated motor unit increased and the force output of the muscle increased. The decrease in the firing rate is accentuated when the new motor unit is recruited with a doublet (first two discharge within 10 ms). The phenomenon has been observed in several muscles (large and small) located in both the upper and lower limb.

This interaction between recruitment and firing rate may be explained by considering the known behavior of the stretch reflex and the Renshaw recurrent inhibition.

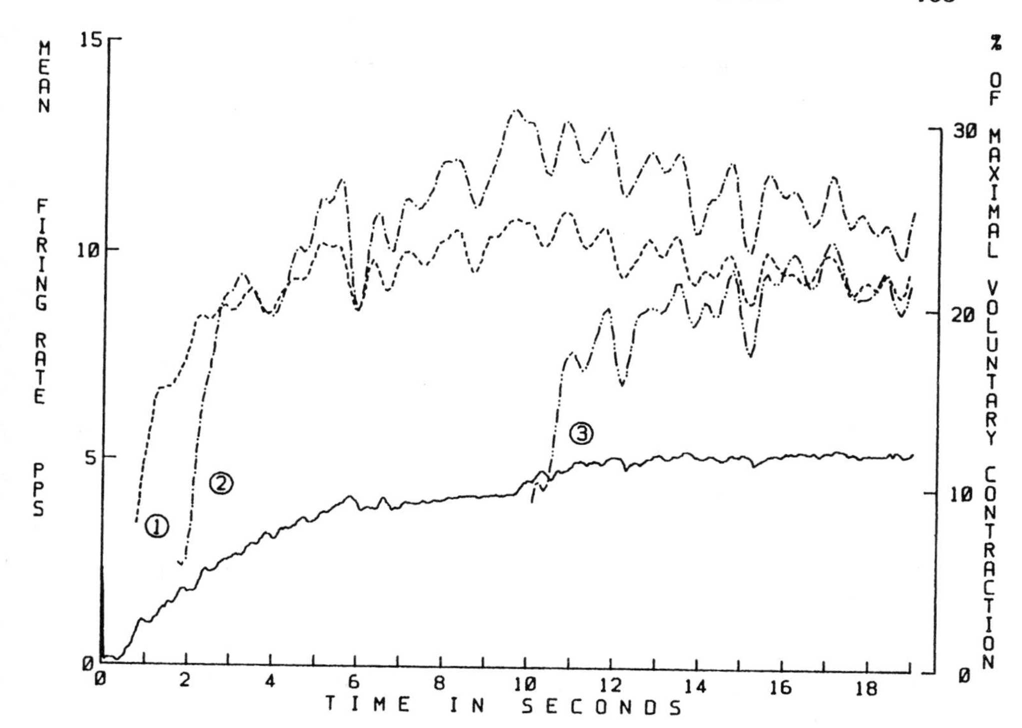

Figure 5.12. Firing rates (*broken lines*) of three concurrently active motor units of the tibialis anterior muscle recorded during an isometric contraction. The force (*solid line*) is presented as a percentage of the maximal voluntary contraction (*right scale*). Note the gradual decrease of the firing rates of the top two tracings as the third motor unit is recruited.

The following sequence of events would explain the phenomenon. As the muscle fibers of a newly recruited motor unit contract, they shorten. If these muscle fibers are located near a spindle, the spindle will slacken, and the discharge of the Ia and II fibers will be reduced, thus decreasing the excitation to the homonymous motoneuron pool. The contracting muscle fibers will also apply tension to the Golgi organs, which will increase the discharge of the Ib fibers, thus producing an increase in the inhibition to the homonymous motoneuron pool. Both effects will disfacilitate the pool and thus decrease the "drive" to the active motoneurons, which is noted as a decrease in the firing rates of the motor units.

The stretch reflex, however, fails to explain two aspects of the interaction: (1) the firing rate increase of the newly recruited motor unit and (2) the slowness of the decrease in the firing rates.

Therefore, the involvement of an additional mechanism, complementing the stretch reflex feedback, is proposed, that is, the Renshaw cell-mediated recurrent inhibition. It has been shown that Renshaw cells can be activated by the discharge of a single motoneuron (Ross et al, 1975) and that Renshaw cells are more strongly excited by collaterals of large motoneurons than by small ones (Ryall et al, 1972; Pompeiano et al, 1975). Consequently, if the Renshaw cell inhibitory action on the α-motoneuron pool is achieved in a size-related fashion (with the small diameter motoneuron being affected more than the large diameter ones), this complementary mechanism could have the desired selective property of preferentially slowing down the motor units which are already active, that is, those having motoneurons with smaller diameter which are recruited earlier and at a lower force level.

The compound effect of the inhibition provided by the Renshaw recurrent inhibition and the stretch reflex inhibition interacting with the common drive excitation on the motoneuron pool are represented schematically in Figure 5.13. In this figure the thickness of the lines expresses the magnitude of the influence.

This interaction between recruitment and firing rate provides an apparently simple strategy for providing smooth force output. Upon recruitment of a new motor unit it may be desirable to produce an increase in muscle force which is less than the minimal incremental contribution of the new motor unit. One way to achieve this goal is to decrease the firing rates of the motor units which are already active, so as to diminish their contribution to the total force output when the new motor unit is recruited. Thus, compensatory decreases of the firing rates of previously activated motor units will enable the muscle to produce a more smooth force output during recruitment. This effect becomes more important as the newly recruited motor units provide an increasingly stronger twitch contribution. Thus, in general, later recruited motor units should have a stronger effect on the firing rates of previously activated motor units, as may be noted in Figure 5.13.

Interaction in Different Muscles. It is apparent that small muscles, such as those in the hand, are controlled by different firing rate-recruitment schemes than larger muscles such as those in the leg or arm. Smaller muscles recruit their motor units within 0 to 50% MVC and rely exclusively on firing rate increase to augment the force output between 50 and 100% MVC. The firing rates of these muscles continuously increase with the force output reaching values as high as 60 pps. Larger muscles recruit motor units at least to 90% MVC, and possibly higher. Their firing rates have a relatively smaller dynamic swing, generally peak at 35 to 40 pps, and tend to demonstrate a plateauing effect.

Smaller muscles rely primarily on firing rate, and larger muscles rely primarily on recruitment to modulate their force. For an example of this phenomenon, refer to Figures 5.6 and 5.11.

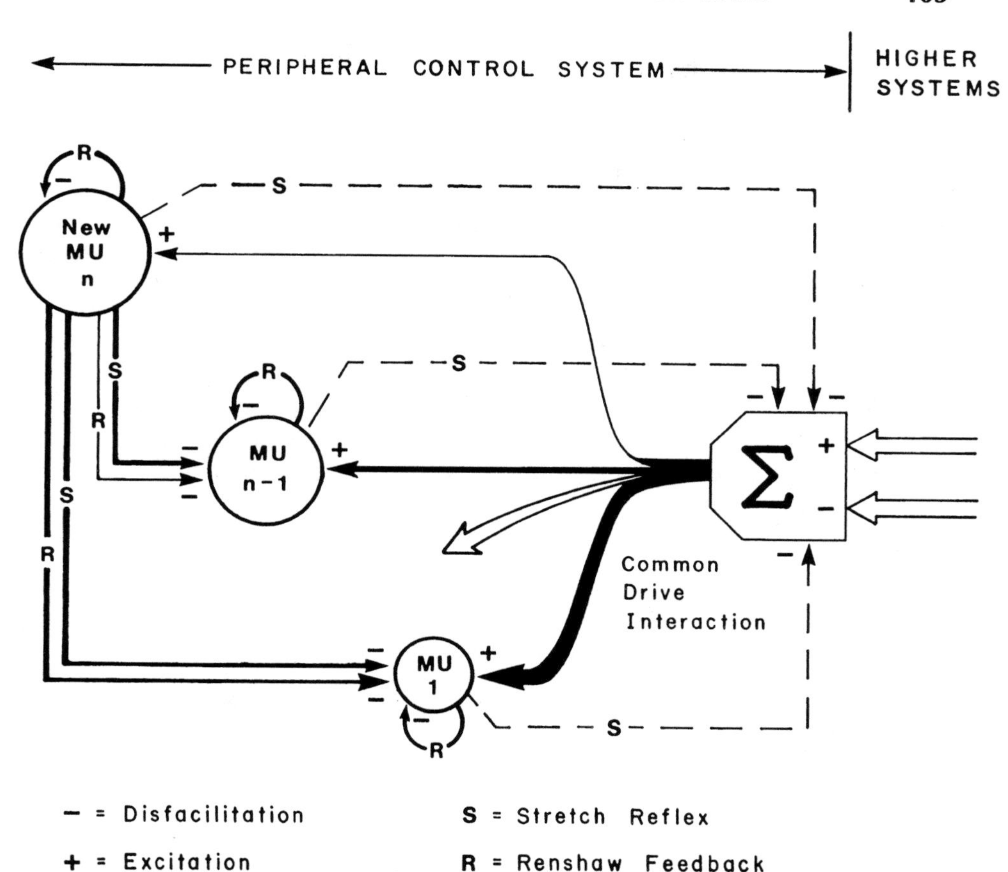

Figure 5.13. Schematic diagram describing the concept of the common drive and phenomenon of recruitment/firing rate interaction during a voluntary contraction. In this representation the excitatory and inhibitory inputs from sources other than the peripheral control system are shown to act on the motoneuron pool as a unit. The increase or decrease in the excitation (+) to each motoneuron has a common origin and is interdependent. The thickness of the line indicates the sensitivity to a change in the state of excitation or inhibition for each motoneuron. The size of the motoneuron (or motor unit) is represented by the size of the circles. Motor unit 1 is the first recruited and motor unit n is the last recruited. The stretch reflex inhibition (−) is represented by the connection S, and the recurrent inhibition by R.

The inhibitory interaction between recruitment and firing rate described above may, in fact, explain the different behavior of the firing rates in muscles with notably different recruitment schemes. A newly recruited motor unit would decrease the firing rate of the motor units which are already active, and the global effect would be to prevent large firing rate increases as long as recruitment occurs. This is consistent with the relatively high increases in firing rate observed above 70% MVC in the brachialis muscle (Kanosue et al, 1979) and above 50% MVC in the first dorsal interosseous (De Luca et al, 1982a) when recruitment is absent or scarce.

The explanation for the need of these contrasting force generation mechanisms may be found by considering the anatomy and function of the muscles. In the human body, smaller muscles are generally involved in performing accurate movements; such movements require small incremental changes in force. In contrast, large muscles are generally involved in either producing large forces or in controlling posture.

Small anatomically confined muscles have relatively few motor units; for example, the first dorsal interosseous contains approximately 120 (Feinstein et al, 1955). When a new motor unit is activated, the average quantal force increase would be 0.8%. If recruitment were the only (or even principal) means by which additional force were developed, small muscles would be incapable of producing a smooth increasing contraction. As force increased, the orderly addition of larger motor units would produce a "staircase" effect in the force output. Yet, generally the function of small muscles is to produce small, accurate movements requiring fine force-gradations. By recruiting its motor units during the first 50% MVC, the average quantum of force augmented by the activation of a new motor unit is one-half the value which would have been increased if the recruitment range extended to 100% MVC. The force above the 50% MVC is generated by the highly dynamic firing rates of motor units in small muscles. As a secondary contribution, the highly dynamic firing rates also assist in smoothing the "staircase" effect.

Large muscles have many more motor units; for example, the biceps brachii contains approximately 770 (Christensen, 1959). Thus, by setting the recruitment to span the full range of force generation, the activation of a new motor unit would provide an average quantal increase of 0.12%. Large muscles generally do not require finer force gradation to accomplish their task. Thus, the firing rates of such muscles do not require continual regulation and do not possess the highly dynamic characteristics seen in smaller muscles.

The functional requirement of the muscle, coupled with the anatomical constraints of it, determine the firing rate-recruitment characteristics which the nervous system engages to achieve the required task. It appears that the nervous system is configured to "balance" the contribution of firing rate control and recruitment control, so as to enhance the smoothness of the force output of the muscle.

SUMMARY

The following description emerges from the information presented in this chapter.

1. The firing rates of motor units are muscle dependent. In small muscles, such as those in the hand, the firing rates begin firing at relatively lower values and reach relatively higher values than those in motor units of larger limb muscles. In larger muscles, the firing rates tend to plateau at 20 to 25 pps, whereas in small muscles the firing rates have a greater dynamic swing, reaching values of 60 pps.
2. During strenuous and high (>70% MVC) level contractions, the firing rate of high threshold motor units may display an abrupt and dramatic increase.
3. In abnormal (dysfunctioned) muscles, the firing rate of motor units appears to behave differently than that of healthy muscles.
4. During sustained contractions, the firing rate of motor units decreases as a function of time. This adaptation of the firing rate is complemented by an increase in the twitch tension of motor units. Data suggest that these two phenomena are more evident in fast-twitch fibers than in slow-twitch fibers.
5. A common drive exists which modulates the firing rates of all motoneurons of a homonymous pool. This indicates that the nervous system does not control the motor units individually.
6. Higher force-threshold motor units consistently have lower firing rates than the lower force-theshold motor units.
7. A muscle cannot generate a pure constant-force contraction in isometric conditions because the firing rates of the motor units are continuously perturbed.
8. Force reversals are accomplished by an ordered progression of firing rate decreases. The earlier recruited motor units decrease their firing rates before the latter recruited motor units.
9. Motor units are recruited and decruited in an orderly progression, possibly according to a size principle. The decruitment occurs in the opposite order of recruitment.
10. Modifications in the ordered progression of recruitment can be induced via electrical stimulation of sensory nerves.
11. The recruitment scheme varies among muscles. In smaller muscles, such as those in the hands, most of the motor units are recruited below 50% MVC, whereas in larger muscles in the limb, recruitment persists up to at least 90%, possibly 100% MVC.
12. The issue of motor unit time-dependent recruitment during constant-force isometric contractions is unsettled.
13. During force increasing contractions, newly recruited motor units have been observed to disfacilitate (decrease the firing rate of) previously activated motor units. This interaction may be explained by invoking the involvement of the stretch reflex and recurrent inhibition. This interaction provides a mechanism which enables the muscle to increase the smoothness of its force output.
14. Smaller muscles rely primarily on firing rate, larger muscles on recruitment to modulate their force. It is conceivable that the nervous system is configured to balance the contribution of the firing rate and recruitment control, so as to enhance the smoothness of the force output of the muscle.

CHAPTER 6

Conscious Control and Training of Motor Units and Biofeedback

Studies of neuromuscular and spinal cord function have been growing increasingly complex in recent years without offering clearer answers to many fundamental problems. Especially confusing and fragmentary are theories on the influence of various cortical and subcortical areas on spinal motor neurons and motor units in man. It was therefore refreshing to be able to develop and advocate a technique that not only proved to be quite simple but also promised to reveal considerable fundamental information. Ironically, the technique was only a modification of ordinary electromyography. This modification consists of regarding EMG signals not for their own intrinsic value but as the direct mirroring of the activity of spinal motor neurons. Thus the group of muscle fibers in a motor unit is considered only as a convenient transducer that reveals the function of the nerve cell.

Perhaps the ultimate irony is that in their classic paper establishing the modern era of electromyography in 1929, Adrian and Bronk suggested that ". . . The electrical responses in the individual muscle fibres should give just as accurate a measure of the nerve fibre frequency as the record made from the nerve itself." Even earlier, Gasser and Newcomer (1921) had shown that "the electromyogram is a fairly accurate copy of the electroneurogram." Perhaps as a reflection of the general turning away from man as an experimental animal in favor of more exotic beasts and preparations, no real use of these early conclusions had been made. In fact, the implications in Gasser and Newcomer's work did not lead to any systematic use of electromyography for studying the behavior of individual spinal motor neurons in any species, even though a MUAP reflects the activity of its spinal motor neuron.

No great progress was made until 1928 and 1929, when Adrian and Bronk published two classic papers on the impulses in single fibers of motor nerves in experimental animals and man. Their method consisted of cutting through all but one of the active fibers of various nerves and recording the action currents from that one fiber. They also succeeded in making records directly from the muscles supplied by such nerves. Somewhat incidentally, Adrian and Bronk introduced the use of concentric needle electrodes with which the activity of muscle fibers in normal human muscles could be recorded. Meanwhile Sherrington (1929) and his colleagues had crystallized their definition of a motor unit as "an individual motor nerve together with the bunch of muscle-fibres it activates." (Universally, later workers have also included in their definition the cell body of the neuron from which the nerve fiber arises.)

Although in subsequent years the concentric needle electrode was seized upon for extensive use, until the Second World War only a handful of papers appeared on the characteristics of action potentials from single motor units in voluntary contraction. In 1934, Olive Smith reported her observations on individual motor unit potentials, their general behavior, and their frequencies. She showed that normally there is no proper or inherent rhythm acting as a limiting factor in the activity of muscle fibers; rather, the muscle fibers in a normal motor unit simply respond to each impulse they receive. Confirming earlier work of Denny-Brown (1929) she set at rest the false hypothesis of Forbes (1922) that the muscle fibers or motor units were fatiguable at the rates they were called upon to reproduce by their nerve impulses.

Forbes had also suggested that normal sustained contraction requires rotation of activity among quickly fatiguing muscle fibers. Smith proved that such a rotation need not occur and that an increase in contraction of a whole muscle involves both an increase in the firing rate of impulses in the individual unit and an accession of new units which are independent in their rhythms. The rates ranged from 5 to 7 pps to 19 to 20 pps, although "highly irregular discharge may occur at threshold both during the onset of a contraction and during the last part of relaxation." Finally, she proved that tonic contraction of motor units in normal mammalian skeletal muscle fiber, the existence of which was widely debated, does not exist. Three generations later, there are people in muscle research still not aware of her definitive studies.

Lindsley (1935), working in the same physiology laboratory as Smith, determined the ultimate range of motor unit rates during normal voluntary contractions. Although others must have been aware of the phenomenon, he seems to have been the first to emphasize that at rest "subjects can relax a muscle so completely that . . . no active units are found. Relaxation sometimes requires conscious effort and in some cases special training."

In none of his subjects was "the complete relaxation of a muscle difficult." Since then, this finding has been confirmed and refined by hundreds of investigators, using much more sophisticated apparatus and techniques than those available in the early 1930s.

Lindsley also reported that individual motor units usually began to respond regularly at rates of 5 to 10 pps during the weakest voluntary contractions possible and some could be fired as slow as 3 pps. The upper limit of firing rates was usually about 20 to 30 pps but occasionally was as high as 50 pps. Earlier, Adrian and Bronk (1928, 1929) had found the same upper limit of about 50 pps for the nerve impulses in single fibers of the phrenic nerve and from the diaphragm of the same preparations.

Gilson and Mills (1940, 1941), recording from single motor units under voluntary control, reported that discrete, slight, and brief voluntary efforts may call upon only a single potential (i.e., a single twitch) of a motor unit being recorded. Twenty years later, Harrison and Mortensen (1962) showed that by means of surface and needle electrodes action potentials of single motor units could be identified and followed during slight voluntary contractions in tibialis anterior. Subjects provided with auditory and visual cues could produce "single, double and quadruple contractions of single motor units . . ." and in one case, ". . . the subject was able to demonstrate predetermined patterns of contraction in four of the six isolated motor units."

Using special indwelling wire electrodes, Basmajian (1963) confirmed these findings and on this basis was able to elaborate techniques for studying the fine control of the spinal motor neurons, especially their training, and the effects of volition. Later, in a series of studies, his group further developed and described a system of testing and of motor unit training. They demonstrated the existence of a very fine conscious control of pathways to single spinal motor neurons (Basmajian et al., 1965). Not only can human subjects fire single neurons with no overflow (or perhaps more correctly, with an active suppression or inhibition of neighbors), but also they can produce deliberate changes in the rate of firing. Most persons can do this if they are provided with aural (and visual) cues from their muscles. Many investigators have documented the qualitative and quantitative aspects (for example: Simard, 1969; Zappalá, 1970; Gray, 1971a, b; Török and Hammond, 1971; Clendenin and Szumski, 1971; Harrison and Koch, 1972; and others, some of whom are cited elsewhere in this chapter).

Following the implantation of fine-wire electrodes and routine testing, a subject needs only to be given general instructions. He is asked to make contractions of the muscle under study while listening to and seeing the MUAPs on the monitors (Fig. 6.1). A period of up to 15 minutes is sufficient to familiarize him with the response of the apparatus to a range of movements and postures.

Subjects are invariably amazed at the responsiveness of the loudspeaker and cathode ray tube to their slightest efforts, and they accept these as a new form of "proprioception" without difficulty. It is not necessary for subjects to have any knowledge of EMG. After getting a general explanation they need only to concentrate their attention. With encouragement and guidance, even the most naive subject is soon able to maintain various levels of activity in a muscle on the sensory basis provided by the monitors. Indeed, most of the procedures he carries out involve such gentle contractions that his only awareness of them is through the apparatus. Following a period of orientation, the subject can be put through a series of tests for many hours.

Several basic tests are employed. Since people show a considerable difference in their responses, adoption of a set routine earlier proved to be impossible. In general, however, they were required to perform a series of tasks. The first is to isolate and maintain the regular firing of a single motor unit (SMU) from among those a person can recruit and display with the technique described. When he has learned to suppress all the neighboring motor units completely, he is asked to put the unit under control through a series of tricks, including speeding up its rate of firing, slowing it down, and turning it "off" and "on" in various set patterns and in response to commands. More elaborate techniques now used are really only controlled versions of the original methods (Basmajian and Samson, 1973). Johnson (1976) has tested and fashioned methods that meet statistical requirements more adequately.

After acquiring good control of the first motor unit, a subject is asked to isolate a second with which he then learns the same tricks, then a

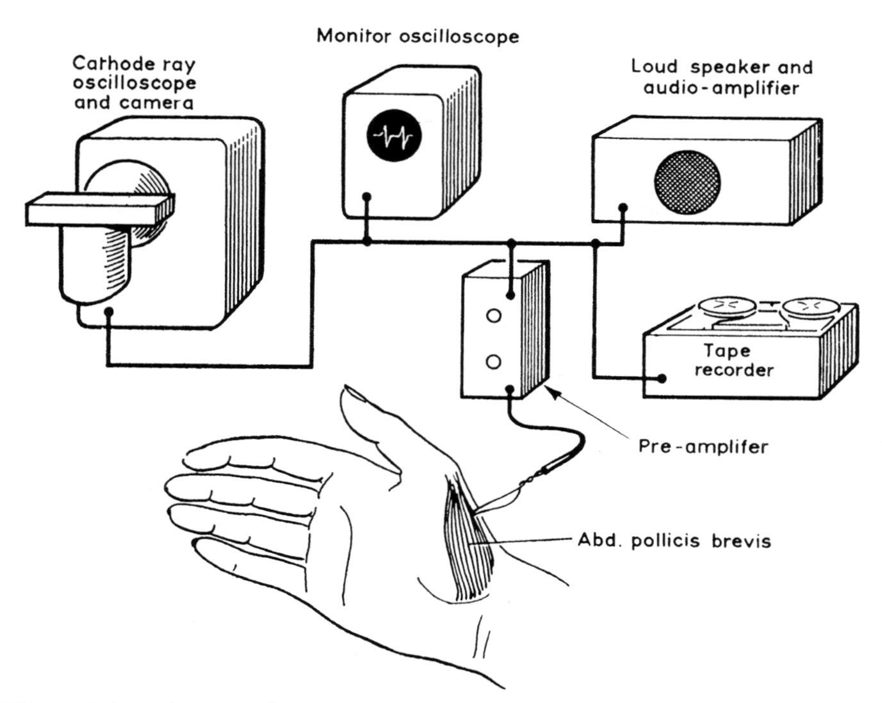

Figure 6.1. Diagram of arrangement of monitors and recording apparatus for motor unit training. (From J.V. Basmajian, ©1963, *New Scientist*.)

third, and so on. His next task is to recruit, unerringly and in isolation, the several units over which he has gained the best control.

Many subjects then can be tested at greater length on any special skills revealed in the earlier part of their testing (for example, either an especially fine control of, or an ability to play tricks with, an SMU). Finally, the best performers can be tested on their ability to maintain the activity of specific MUAPs in the absence of either one or both of the visual and auditory feedbacks, that is the monitors are turned off and the subject must try to maintain or recall a well-learned unit without the artificial "proprioception" provided earlier.

Lloyd and Leibrecht (1971) and Samson (1971) independently showed that the SMU training fulfills the requirements of the learning paradigm. The feedback methodology is not critical; thus, a highly artificial indication of successful training is satisfactory to a considerable degree. Leibrecht et al (1973) went on to show that direct EMG feedback substantially improved initial learning. The nature and amount of learning, including the ability to use proprioceptive cues in controlling an SMU, were not affected; neither was the retention of learning.

Ladd et al (1972) investigated the learning process involved in the fine neuromotor control of SMU training which, of course, also embodies inhibition of motor activity. They employed trained units in five different muscles in 25 subjects. Voluntary inhibition, they found, is a conceptual type of response showing independence of the motor component; it generalizes and transfers positively from one muscle to another. However, the voluntary contractions of an individual unit is a specific perceptual motor type of response; the motor component of the response is essential, and the learned response does not generalize or transfer from one muscle to another. Middaugh (1976) reported that subjects are not relying on peripheral factors in learning, i.e., only limited learning of peripheral sensory information and discrimination occurs. The finding by Vogt (1975) that there is little correlation between self-estimation of success and gross EMG levels of contraction in the forearm supports Middaugh's finding for motor units.

Any skeletal muscle may be selected. The ones most often reported are the abductor pollicis brevis, tibialis anterior, biceps brachii, and the extensors of the forearm. However, it is quite easy to train units in buccinator (Basmajian and Newton, 1973) and in back muscles; Sussman et al (1972) have trained units in the larynx while Gray (1971a) trained them in the sphincter ani!

ABILITY TO ISOLATE MOTOR UNITS

Almost all subjects are able to produce well-isolated contractions of at least one motor unit, turning it off and on without any interference from neighboring units. Only a few people fail completely to perform this basic trick. Analysis of poor and very poor performers reveals no common characteristic that separates them from better performers.

Many people are able to isolate and master one or two units readily; some can isolate and master three units, four units, even six units or more (Fig. 6.2). This last level of control is of the highest order, for the subject must be able to give an instant response to an order to produce contractions of a specified unit without interfering activity of neighbors; he also must be able to turn the unit "off" and "on" at will. The ultimate ability of human subjects was demonstrated by Kato and Tanji (1972a), who found that within 30 minutes their subjects could voluntarily isolate 73% of 286 motor units appearing on the oscilloscope during voluntary contractions.

CONTROL OF FIRING RATES AND SPECIAL RHYTHMS

Once a person has gained control of a spinal motor neuron, it is possible for him to learn to vary its rate of firing. This rate can be deliberately changed in immediate response to a command. The lowest limit of the range of frequencies is zero, i.e., one can start from neuromuscular silence and then give single isolated contractions at regular rates as low as 1/s and at increasingly faster rates. When the more able subjects are asked to produce special repetitive rhythms and imitations of drum beats, almost all are successful (some strikingly so) in producing subtle shades and coloring of internal rhythms. When tape-recorded and replayed, these rhythms provide striking proof of the fineness of the control.

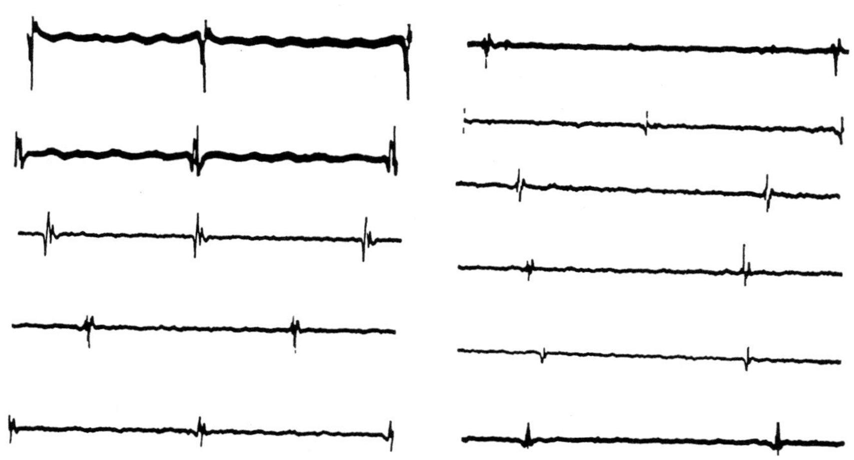

Figure 6.2. Eleven different motor units isolated by a subject in quick succession in his abductor pollicis brevis. (From J.V. Basmajian et al, ©1965, *Journal of New Drugs*.)

RELIANCE ON VISUAL OR AURAL FEEDBACK

Some persons can be trained to gain control of isolated motor units to a level where, with both visual and aural cues shut off, they can recall any one of three favorite units on command and, in any sequence. They can keep such units firing without any conscious awareness other than the assurance (after the fact) that they have succeeded. In spite of considerable introspection, they cannot explain their success except to state they "thought about" a motor unit as they had seen and heard it previously. This type of training probably underlies ordinary motor skills.

VARIABLES WHICH MIGHT AFFECT PERFORMANCE

Tanji and Kato (1971) found that cortical motor potential related to the discharge of an SMU is about the same size as that related to the contraction of whole muscles (e.g., as in key pressing). This led to their obvious conclusion that cerebral mechanisms are involved in an important manner in conscious isolation of individual motor units; they later consolidated these views with more specific tests (Kato and Tanji, 1972b; Tanji and Kato, 1973a,b). However, McLeod and Thysell (1973) did not agree; their studies of evoked EEG potentials revealed no true response in the sensorimotor areas that can be related to single motor unit activity. Intensive research is needed to resolve the question.

No personal characteristics that reveal reasons for the quality of performance have been found (Basmajian et al, 1965). The best performers occur at different ages, among both sexes, and among both the manually skilled and unskilled, the educated and uneducated, and the bright and the dull personalities. Some "nervous" persons do not perform well—but neither do some very calm persons.

Carlsöö and Edfeldt (1963) concluded that: "Proprioception can be assisted greatly by exeroceptive auxiliary stimuli in achieving motor precision." Nevertheless, Wagman et al (1965), using both Basmajian's technique and a technique of recording devised by Pierce and Wagman (1964), emphasize the role of proprioception. They stress their finding that subjects believe that certain positions of a joint must be either held or imagined for success in activating desired motor units in isolation.

Investigations of the various factors which affect motor unit training and control have added interesting features (Simard and Basmajian, 1967; Basmajian and Simard, 1967; Simard, Basmajian and Janda, 1968). They revealed that moving a neighboring joint while a motor unit is firing is a distracting influence but most subjects can keep right on doing it in spite of the distraction. This tends to agree with Wagman and his colleagues who believe that subjects require SMU training before they can fire isolated specific motor units with the limb or joints in varying positions. Their subjects reported that "activation depended on recall of the original position and contraction effort necessary for activation." This apparently is a form of proprioceptive memory and almost certainly is integrated in the spinal cord.

Observations based on trained units in the tibialis anterior of 32 young adults showed that SMU activity under conscious control can be easily maintained despite the distraction produced by voluntary movements elsewhere in the body—head and neck, upper limbs and contralateral limb (Simard and Basmajian, 1967). The control of isolation and the control of the easiest and fastest frequencies of discharge of a single motor unit were not affected by those movements (Fig. 6.3).

Turning to the effect of movements of the same limb, Basmajian's group found that in some persons a motor unit can be trained to remain active in isolation at different positions of a "proximal" (i.e., hip or knee), "crossed" (ankle), and "distal" joint of a limb (Fig. 6.4). This is a step beyond Wagman et al (1965), who observed that a small change in position brings different motor units into action. Consequently they noted the important influence of the sense of position on the motor response. Later investigations by Simard and Basmajian showed that in order to maintain or recall a motor unit at different positions, the subject

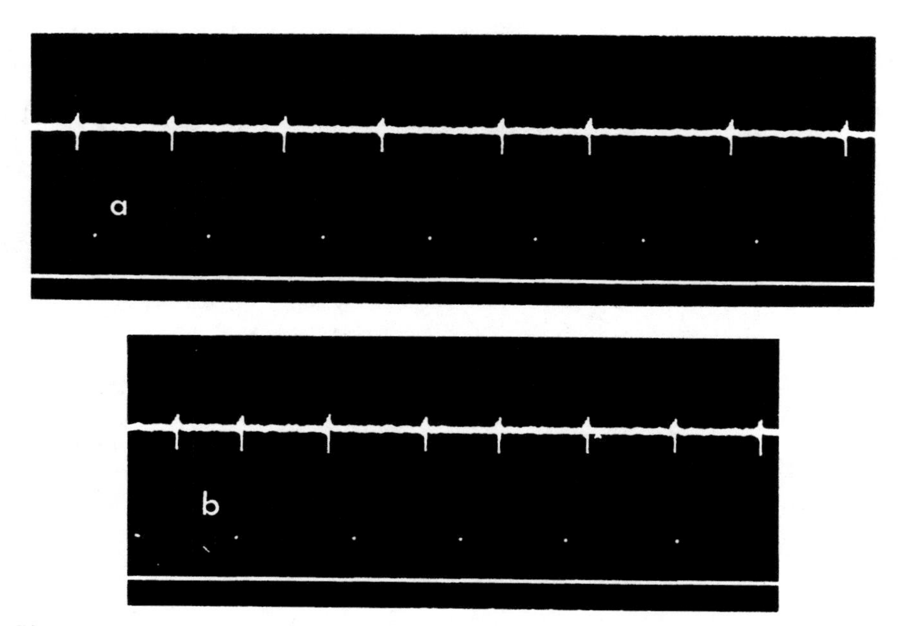

Figure 6.3. Samples of (*a*) the easiest and (*b*) the fastest rate of discharge of a motor unit in the right tibialis anterior during movements of the contralateral limb (time mark: 10 ms intervals). (From J.V. Basmajian and T.G. Simard, ©1967, *American Journal of Physical Medicine*.)

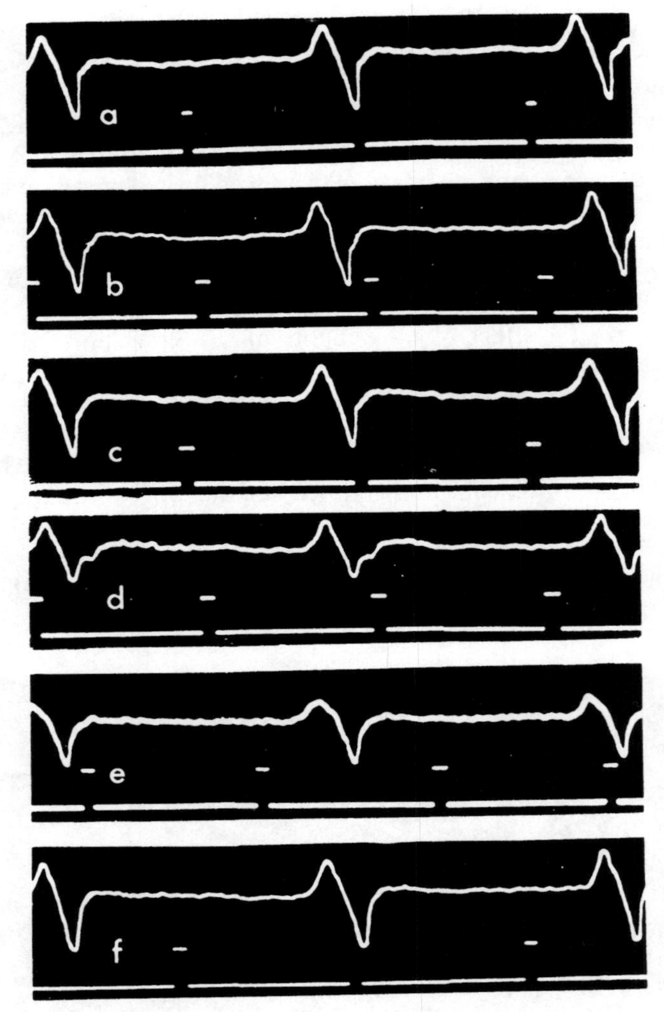

Figure 6.4. Controlled MUAP at different "held" positions of the right lower limb: *a*, neutral; *b*, in lateral rotation at the hip; *c*, in medial rotation at the hip; *d*, dorsiflexion of the ankle; *e*, plantarflexion; *f*, toes extended. (Calibration: 100 μV, 100 ms intervals.) (From J.V. Basmajian and T.G. Simard, ©1967, *American Journal of Physical Medicine*.)

must keep the motor unit active during the performance of the movements and, therefore, preliminary training is undeniably necessary.

The control of the maintenance of activity during "proximal," "crossed," and "distal" joint movements in the same limb has been proved here to be possible, provided that the technique of assistance offered by the trainer is adequate. The control over the discharge of a motor unit during proximal and distal joint movements requires a great concentration on the motor activity. But when one considers the same control during a "crossed" joint movement, there are even greater difficulties for obvious reasons.

The observation that trained motor units can be activated at different positions of a joint is related to the work of Boyd and Roberts (1953). They suggested that there are slowly adaptive end organs of proprioception which are active during movements of a limb. They observed that the common sustained discharge of the end organs in movements lasted for several seconds after attainment of a new position. This might explain why a trained single motor unit's activity can be maintained during movements.

Lloyd and Shurley (1976) studied the hypodynamic effects of sensory isolation on single motor units recorded through wire electrodes in 40 normal subjects. A light panel indicated the trial onset, correct, and incorrect response. Isolation condition was produced by an air-fluidized, ceramic-bead bed in a light and sound attenuating chamber. A relearning session followed the initial session after a 2-week interim rest. Subjects were randomly assigned to the isolation or nonisolation condition for both sessions. The hypodynamic effects of sensory isolation increased the speed of learning to isolate and control an SMU. The results suggested that subjects were better able to attend to the relatively weak proprioceptive information provided by the SMU through the reduction of the amount and/or variety of competing stimuli.

The Level of Activity of Synergistic Muscles

The problem of what happens to the synergistic muscles at the "hold" position or during movements of a limb has been taken into consideration only in a preliminary way. The level of activity appears to be individualistic. Active inhibition of synergists is learned only after training of the motor unit in the prime mover is well established. Basmajian and Simard (1967) clearly demonstrated that as the subject focuses his attention on feedback from an SMU in one muscle (tibialis anterior), the surrounding muscles become progressively relaxed to the point of complete silence when isolation of the SMU is complete. Only such motor units in a limb as are needed to maintain its particular posture are still active. The process of "active inhibition," probably the more dramatic element of motor unit training, is thus achieved.

Influence of Manual Skills

Although the earlier studies failed to reveal any correlation between the abilities of subjects to isolate individual motor units and the variable of athletic or musical ability, a systematic study by Scully and Basmajian (1969) cast some light on the matter. They used the base time required to train motor units in one of the hand muscles as the criterion. Surprisingly, the time required to train most of the manually skilled subjects was *above* the median.

Henderson's (1952) work offers an explanation: the constant repetition of a specific motor skill increases the probability of its correct recurrence by the learning and consolidation of an optimal anticipatory tension. Perhaps this depends on an increase in the background activity of the γ-motoneurons regulating the sensitivity of the muscle spindles used in performing the skill. Wilkins (1964) postulated that the acquisition of a new motor skill leads to the learning of a certain "position memory" for it. If anticipatory tensions and position memory, or both, are learned, spinal mechanisms may be acting temporarily to block the initial learning of a new skill. Perhaps some neuromuscular pathways acquire a habit of responding in certain ways and then that habit must be broken so that a new skill may be learned. The "unstructured" nature of learning a motor unit skill would make this mechanism even more likely (Basmajian, 1972).

Influence of Age and Gender

Although the training of fine control of individual units is complicated when children are involved, it is possible in children even below the age of 6 years (Fruhling et al, 1969). Simard and Ladd (1969) and Simard (1969) have further documented the factors involved.

Zappalá (1970) found only minor gender differences in the ability to isolate SMUs; males showed some superiority. In a different type of experiment, Harrison and Koch (1972) and Petajan and Jarcho (1975) found the opposite, but again the differences were not impressive.

Influence of Competing Electrical Stimulation

Any changes in the action potentials of trained motor units as a result of electrical stimulation of the motor nerve supplying the whole muscle must reflect neurophysiologic changes of the single neuron supplying the motor unit. Therefore, Scully and Basmajian (1969) investigated the influence of causing strong contractions in a muscle to compete with a discrete SMU in it which was being driven consciously. Each of a series of subjects sat with his forearm resting comfortably on a table top. The stimulator cathode was applied to the region of the ulnar nerve above the elbow. The effective stimuli were 0.1-ms square-wave pulses of 70 to 100 V, delivered at a frequency of 90 pulses/min. Because stimuli of this order are not maximal, all axons in the ulnar nerve were not shocked, and slight variation must have existed in axons actually stimulated by each successive shock.

Contrary to expectation, when the massive contraction of a muscle was superimposed on the contraction of only one of its motor units, the regular conscious firing of that SMU was not significantly changed.

These experiments leave little if any doubt that well-trained motor units are not blocked in most persons. Even the coinciding of the MUAP with elements of the electrically induced massive contraction would not abolish the SMU potential.

Influence of Cold

Brief cutaneous applications of ice over the biceps brachii in which an isolated motor unit had been trained elicited facilitation of both background activity and spontaneous activation of the trained SMUs (Clendenin and Szumski, 1971). Wolf, Letbetter and Basmajian (1976) confirmed this finding, using a special electronic cooling device (Wolf and Basmajian, 1973). Seventeen subjects discharging SMUs at a comfortable resting rate (5.2 ± 0.9 pps) tended to get an inhibitory response in the initial minute of cooling. Most subjects (13 of 18) who held SMU discharges to 0.5 pps first got an increase, and then a significant decrease. Apparently the central excitatory state is the mediator of these local motor reactions to cutaneous cooling.

Effects of Handedness and Retesting

When a large number of subjects were studied on two occasions using a different hand each time, Powers (1969) found that they always isolated a unit more quickly in the second hand. Isolation was twice as rapid when the second hand was the preferred (dominant) hand; it was almost five times as rapid when the second hand was the nonpreferred one. The time required to control a previously isolated unit was shortened significantly only when the preferred hand was the second hand. However, in a test-retest situation with much fewer subjects, Harrison and Koch (1972) found no significant improvement from test to retest.

Influence of Disease States

While Basmajian has found that partially paralyzed people can learn SMU controls quite easily, the factor of spasticity introduces considerable difficulty. In clinical studies, one can overcome these difficulties by carefully training the patient to relax spastic muscles. Parkinsonian rigidity seems to be a different matter. Petajan and Jarcho (1975) reported that patients with Parkinson's disease are unable to adjust the firing rate of motor units that initiate contraction from zero to higher rates. Although the frequency modulation is not normal, motor units recruit in an orderly fashion. Levodopa treatment restores normal control of SMUs.

REACTION TIME STUDIES

A number of investigators have used trained SMUs for psychological testing of reaction time (RT). Thus, Sutton and Kimm (1969, 1970) and Kimm and Sutton (1973) have shown stable differences in the RT in triceps and biceps brachii and a slowing of RT following the intake of alcohol. Generally, they concluded that SMU spike RTs were slower than that obtained from the gross EMG signals and lever-press RTs. But Thysell (1969) disagrees, finding them to be comparable and rather like those of Luschei et al (1967). Furthermore, Vanderstoep (1971) questions the finding of inherent differences between muscles when the RT paradigm is used with triceps, biceps, the first dorsal interosseus, and the abductor pollicis brevis. Zernicke and Waterland (1972), on the other hand, were able to show differences between the two heads of biceps brachii. The short head contains motor units that are easier to control than those in the long head. They related this to various morphological and functional requirements of the two heads (e.g., the density of muscle spindles is greater in the short head). The willful fractionization of control between two heads of the same muscle, not entirely unexpected in view of the fineness of willful controls involved in SMU control, once more underlines the discrete nature of controls over the spinal motoneurons.

PRACTICAL APPLICATIONS

Many applications are emerging for the use of motor unit training, e.g., in the control of myoelectric prostheses and orthoses, in neurological studies, and in psychology. The growth of the field of "biofeedback" from this work is the subject of a separate book (Basmajian, 1983). Therefore, only a brief outline is given in the remainder of this chapter as it applies to myoelectric biofeedback only, employing the EMG of a much grosser nature than SMU potentials—although the feedback principles controlling them are common to both.

EMG BIOFEEDBACK

Relaxation. Following confirmation of early studies of the single motor unit principles, Green et al (1969, 1970) rapidly extended biofeedback work into the clinical investigation of the effects of feedback relaxation. They combined this with other forms of electronic feedback and applied the results to a variety of general and local tension states believed to be the cause of pathological physiology. Simultaneously, Gaarder (1971) was exploring practical means to control relaxation in patients with feedback devices.

Hoping to determine whether an ability to produce EMG patterns accurately reflects the ability to achieve specific muscle tensions, Rummel (1974) studied a long series of normal subjects. She was amazed to find no statistically significant correlation. Schwartz et al (1976a, b), on the other hand, revealed patterns of covert activity in facial muscles that could be graded and correlated to states of affective imagery and mood. Earlier, Smith (1973) had found a positive correlation between personality traits of anxiety and EMG activity from the region of the forehead. This finding disagreed with the earlier work of Iris Balshan Goldstein (1962), but it must be remembered that Smith's forehead electrodes often pick up from a wide area (down to the clavicles). Similar findings were reported for the muscles of the jaw by Thomas et al (1973) in explaining temporomandibular joint syndrome. Chapman (1974) showed that EMG activity from the forehead reflected even the fact that the subjects were not alone but were in an audience (i.e., in a social facilitation setting). Biofeedback appears to be superior to verbal feedback in inducing relaxation, at least in the research models used by Kinsman et al (1975) and Coursey (1975), but Alexander (1975) disagreed on the basis of his research.

PSYCHOPHYSIOLOGICAL MECHANISMS

The question continues to arise: Is biofeedback training based on volition or is it operant conditioning? Hefferline and Perera (1963), in their continuing search for the effect of proprioception in behavior, showed that subjects could be conditioned to respond to covert twitches in a thumb muscle (displayed by EMG). After the EMG feedback was eliminated, the response often persisted. By coincidence the muscle used (abductor pollicis brevis) was the same as the one used in the early experiments on SMU training (Basmajian, 1963). Instead of asking the subject to shape the behavior of the EMG signal within the target muscle, Hefferline and Perera conditioned him to press a key using another muscle. Their system was based on the operant conditioning paradigm. Fetz and Finocchio (1971) were able to condition awake monkeys to give bursts of cortical cell activity with and without simultaneous suppression of EMG activity in specifically targeted arm muscles. Operant conditioning methodologies proved sufficient to bring about the correlated response.

In man, Germana (1969) demonstrated quite adequately and not surprisingly that conditioning may be employed in modifying EMG responses; perhaps more importantly, his work has tended to support "cardiac-somatic coupling," with which Obrist and various colleagues have been concerned (see Obrist, 1968). Cohen and Johnson (1971) found a high correlation between heart rate and muscular activity, supporting Obrist's theoretical position. Subtle changes in muscular activity did change heart rate both when subjects were intentionally modifying muscular activity as well as when spontaneous changes were occurring.

Refining his techniques, Cohen (1973) soon after showed a relationship only in subject groups that had a moderately high EMG output from skin electrodes over the muscles of the chin; lower EMG outputs seemed unrelated to heart rate changes; thus the "cardiac-somatic coupling" is not absolute, and mechanisms must exist in the central nervous system for separating cardiac and peripheral motor responses. Other autonomic functions have been linked with covert motor responses; thus, Simpson and Climan (1971) have shown that there is some apparent effect of muscular activity on the pupil size during an "imagery" task in which subjects generated images in response to words.

GENERAL RELAXATION (CONTINUED)

In the 1920s and 1930s, Edmund Jacobson of Chicago became the enthusiastic proponent of a clinical form of EMG monitoring of his patients' progress during relaxation training. Limited by the apparatus available at the time, Jacobson developed methods of electrical measurement of the muscular state of tension and employed his measurements to induce progressive somatic relaxation for a variety of psychoneurotic syndromes (Jacobson, 1929, 1933). Green et al (1969) and Gaarder (1971), using a modification of the SMU training technique, found that EMG biofeedback training would be useful in many states. Mathews and Gelder (1969) studied the effect of relaxation training with phobic patients, showing that the EMG (among other parameters) was altered during relaxation and concluding that relaxation is in some way associated with a controlled decrease in "arousal level" with retention of consciousness. Paul (1969) compared hypnotic suggestion and brief relaxation training, showing the superiority of the latter in reducing subjective tension and distress. Wilson and Wilson (1970), while agreeing that muscle tension could be manipulated by feedback and conditioning, were much less sure of the desirable effects of relaxation. Dixon and Dickel (1967), Jacobs and Felton (1969), Whatmore and Kohli (1968), and Budzynski and Stoyva (1969) also contributed to the literature of EMG biofeedback in relation to general clinical disorders, especially tension headache.

Chronic anxiety is often reflected in overactivity in the general body musculature. Townsend et al (1975) compared treatment of chronic anxiety with EMG biofeedback to treatment with group therapy in a control group. Significant improvements resulted, as they did in a study by Canter et al (1975) in which they compared biofeedback with Jacobsonian progressive relaxation. While the latter was effective, the biofeedback approach proved superior in reducing both muscular tension and chronic anxiety.

Dental specialists are increasingly enthusiastic about the new treatment of the common and distressingly painful jaw pain (temporomandibular joint syndrome) caused by over-active use of muscles that are normally relaxed or only lightly contracted. Myoelectric biofeedback training involves making the patient aware of hyperactivity in the masseter muscle and then training local relaxation of the muscle (Carlsson et al, 1975).

Speech Apparatus and EMG Feedback

As noted in the earlier editions of this book, Hardyck et al (1966) were among the first to modify the lessons of SMU training to applied biofeedback of useful function. Using feedback from surface EMG of the laryngeal muscles during silent reading, they were able to accelerate the reading skills of slow readers. Simultaneously McGuigan and various associates at Hollins College, Virginia were studying the covert oral language behavior as measured by surface EMG of chin muscles (McGuigan, 1966, 1970; McGuigan and Rodier, 1968). Inouye and Shimizu (1970) examined the hypothesis that verbal hallucination is an expression of so-called "inner speech."

The Czech investigators Baštecký et al (1968a, b), using delayed auditory speech feedback and EMG of mimic muscles (primarily mentalis at the chin), found that schizophrenic patients could be differentiated from normal subjects. This area of research, now in its infancy, requires a great deal of investigation. Thus, Sussman et al (1972) have shown that individual units in the laryngeal muscles can be trained. The same group (Hanson et al, 1971; MacNeilage et al, 1972; MacNeilage and Szabo, 1972; and MacNeilage, 1973) have systematically exposed mechanisms of fine control of the laryngeal function which should have far reaching use.

Stuttering has been the special concern of Barry Guitar (1975). He taught stutterers to reduce resting EMG activity in the lips and in the laynx. With myoelectric biofeedback through fine-wire electrodes in the lips and buccinator muscles of the cheek, clarinet players can quickly revise the localized activities in bizarre ways without losing the ability to perform (Basmajian and Newton, 1973). Also trumpet and trombone players have different natural patterns that vary with proficiency and that can be altered with EMG feedback (Basmajian and White, 1973; White and Basmajian, 1973).

TARGETED MUSCLE RETRAINING AND REHABILITATION

Ladd and Simard (1972), building on earlier work on SMU training, trained and studied congenitally malformed children with the aim of using the limited sources of muscle power for myoelectric and other types of artificial limbs and orthoses. Payton and Kelley (1972) explored the factors controlling biceps brachii and deltoid during performance of skilled tasks in a way that lends itself to feedback training.

Practical approaches with practical biofeedback instruments became a reality with commercial equipment being marketed. Booker et al (1969) demonstrated retraining methods for patients with various neuromuscular conditions, and Johnson and Garton (1973) succeeded with hemiplegic patients in retraining functions of the upper and lower limbs where other methods proved inadequate. What has been surprising to many people is the ease with which ordinary patients "take to" the feedback signals and learn to manipulate them by acquiring more precise control over the muscles requiring training or recruitment.

This book is not the place for details of how EMG biofeedback may be used in rehabilitation. The topic has been covered thoroughly in *Biofeedback: Principles and Practice for Clinicians, 2nd Edition, 1983.*

RELATED PSYCHOLOGICAL RESEARCH

Since the valuable start given to it by the Montreal group in the early 1950s (see Malmo et al, 1951; Malmo and Smith, 1955), a sort of electromyographic subculture has existed in the psychological literature.

Following up a previous investigation of limb positioning with kinesthetic cues (Lloyd and Caldwell, 1965), Lloyd (1968) found no statistically significant relationship between position accuracy and the amount of contralateral activity as measured by EMG techniques, but there was no doubt that such activity exists at a low level, especially during passive movement of the ipsilateral limb. Lloyd concluded that a minimal level of activity was required for kinesthetic mediation of accurate limb position. It was this work that led Lloyd and his colleagues to study SMU responses (cited earlier in this chapter).

Wiesendanger et al (1969) measured simple and complex reaction times with the EMG signal of biceps and triceps. While their chief concern was to find differences between normal persons and patients with parkinsonism—there were none in the simple tasks—they showed that the normal reciprocal inhibition of antagonists was modified in different ways, with biceps activity always being present (see the general discussion of agonist-antagonist behavior). Bartoshuk and Kaswick (1966) had shown earlier that general arousal level may not be necessary to produce EMG gradients; instead, selective facilitation may be sufficient.

The influence of environmental and emotional factors on EMG activity is gaining widespread interest. A good example of this type of study is that of Lukas et al (1970), who recorded the effect of sonic booms and noise from subsonic jet flyovers on skeletal muscle tension (in the trapezius muscle) as well as other parameters. The EMG activity increased with sonic booms with lesser effect from the flyover noise.

Phasic changes in muscular and reflex activity during non-REM sleep were demonstrated in man and cats by Pivik and Dement (1970). The suppression of EMG activity from surface electrodes in the submental (chin) area was observed in all subjects during non-REM sleep but occurred with the greatest frequency during sleep stages 2 and 4. The suppressions averaged ¼ minute in duration and exhibited a higher frequency in the 10 minutes prior to the REM period than after. Larson and Foulkes (1969) confirmed that EMG suppression in chin and neck muscles heralds REM sleep onset. The amount of EMG activity during non-REM sleep just prior to being awakened influences the recall frequency of dreams.

Pishkin and Shurley (1968) and Pishkin et al (1968) demonstrated a positive correlation between EMG responses and concept-identification performance which produces cognitive stress. About the same time, Aarons (1968) was exploring possible diurnal variations of myopotentials and word associations related to psychological orientation. Word-association tests revealed qualitative differences among responses before sleep, upon awaking, and at noon. Some differences were related to psychological test variables (kinesthetic orientation, "need for change," and anxiety); the other influences were the time of the tests and, apparently, the intensity of EMG response. EMG levels during sleep correlated highly with electroencephalographic sleep stages.

This brings us back to a group of studies on the effects of stress and anxiety on the EMG, first adequately investigated by Goldstein (1962). Brandt and Fenz (1969) showed a peak of forehead EMG activity in conditions of induced mild stress, suggesting it might reflect inhibitory control. Incidentally, they questioned the specificity of the forehead source as the ideal one for such experiments—and well they might, for with intramuscular wire electrodes the frontalis and corrugator supercilii are silent unless the face shows clear emotive responses (Vitti and Basmajian, 1973). Fridlund et al (1980, 1982) concluded that the general tension factor reflects agitation more than elevated tonic muscle activity.

Searching for a suitable muscle for stress-EMG studies, Yemm (1969a, b) of Bristol, England, concentrated on the masseter—not surprisingly, for he is a dental scientist. He found an increase in masseter EMG activity during the stress of cognitive manual task performances in this postural muscle of the jaw. With patients who have temporomandibular dysfunction, the EMG responses persisted abnormally long (Yemm, 1969c).

The use of muscles active in maintaining human posture has other advocates. Thus, Avni and Chaco (1972) used the EMG activity of supraspinatus muscle (which is described elsewhere in this book in its shoulder-posture role). Reasoning from earlier work (Basmajian, 1961; Basmajian and Bazant, 1969) that drooping of the shoulder should influence supraspinatus activity, they studied a series of depressed patients. While normal controls showed normal antigravity reflex activity, depressed patients all showed significant decrease while they were depressed but recovered the normal pattern on recovery from depression.

NOTES ON TECHNIQUE

The use by Avni and Chaco and by Yemm of postural muscles (noted above) rather than surface EMG of chin and forehead raises the general question of appropriate methods for EMG studies of tension. Unquestionably, some of the techniques employed by investigators naive in electromyography have been less than acceptable. Most EMG activity from the submental region would appear to reflect the frequency of swallowing—which, of course, may be a good criterion of tension. (For swallowing EMGs, see Chapter 19.) As noted before, forehead EMG work also may be questionable, although obvious facial mimicry often represents inner states and so, in a distressed person, may be a satisfactory source of EMG signals.

In the hands of experts good surface EMG is quite adequate for tension studies. Bruno et al (1970) and Kahn (1971) have even demonstrated its usefulness for precise identification of signals. But the factors affecting the reliability of surface EMG signals are many and appear to be ignored by many psychologists; they ought to read and reread the paper by Grossman and Weiner (1966) as well as the details in Chapter 2 in this book.

The foregoing facts mean that to be useful in biofeedback practice, integrated rectified EMG signals from the forehead or frontal region need not come from frontalis muscle. Indeed, a wide source of myopotentials is much to be preferred as a reflection of general nervous tension. But we should admit that (1) wide-source myopotentials are not "frontalis EMG" and (2) the numbers of "microvolts" produced on the meter of a commercial device or any other device simply indicate a microvolt reading at the input of the device. The integrated rectified EMG signal from forehead surface electrodes generally reflects the total or global EMG of all sorts of repeated dynamic muscular activities down to about the first rib—along with some postural activity and nervous tension overactivity. The exact meter readouts can be taken with a grain of salt by the knowledgeable electromyographer at the same time that he is deliberately and wisely using them as (1) a rough indicator of progress in a clinical relaxation training program and (2) a visual placebo in reinforcing the patients' responses. Any higher level of reliance on such inflated numbers is self-deception (Basmajian, 1976).

CHAPTER 7

EMG Signal Amplitude and Force

The surface EMG signal may be conveniently detected with minimal insult to the subject. For this reason it has become very useful in many applications which require an assessment of the muscular effort. The reader is referred to the material in Chapter 2, which addresses the details of this issue.

A considerable controversy exists concerning the description of this relationship. Early theoretical studies (Person and Libkind, 1967; Bernshtein, 1967; Moore, 1967; Libkind, 1968 and 1969) all suggested that, for isometric contractions, the amplitude of the EMG signal should increase as the square root of force generated by the muscle when the motor units are activated independently. These studies were instrumental in generating interest in providing a more structured approach to the interpretation of the EMG signal-force relationship. It is now clear that the assumptions and approximations which were made were simply too generous. The reader is referred to the relationships and associations between the EMG signal, as a function of force and time, with known physiological correlates displayed in Figure 3.9. In those expressions it is apparent that the relationship is complex. In fact, surprisingly few experimental results support the square root relationship. Almost without exception, investigators report either linear relationships or a more than linear increase of the EMG signal with increasing force.

RELATIONSHIP DURING ISOMETRIC CONTRACTIONS

Monotonically Increasing Contractions

Table 7.1 provides a sample of the studies relating to this issue which have been reported in the literature between 1952 and 1979. No attempt has been made to include all the published reports. The contents of the table were designed to represent the wide variety in and disparity among the wealth of studies which have been performed. These investigations are characterized by considerable variability in the muscles examined, the types of contractions performed, and the quantities derived from the raw data to represent the amplitude of the EMG signal.

Beyond the obvious disparities among the reported studies, it is necessary to consider particulars which are specific to the muscle or muscle group which is involved in the force generation process. For example, (*a*) Relatively small muscles, such as those in the hand, and relatively large muscles in the limbs are controlled by different firing rate-recruitment schemes. For additional details on this point, the reader is referred

Table 7.1.
Selection of Previous Investigations of the EMG Signal-Force Relationship

First author	Yr	Muscle	EMG-F relation	Measure of EMG	Contractions		Position	Electrodes			Subjects		
					Type	Angle (from full extension)		Type	Diam (cm)	Sep (cm)	#	Sex	Ages (range or avg)
Inman	1952	Biceps, triceps, pectoral, ant. tibialis	Linear	IEMG	Isom F-vary (ramp)			Surf bip, wire & needle	1.5		11 (Amputees)		
Lippold	1952	Soleus, gastrocnemius	Linear	IEMG	Isom	90° ca. knee	Plantar flexion	Surf bip	0.6		30		
Bigland	1954	Calf, finger extensor	Linear	IEMG	Isom isokin, isot isokin	90° ca. knee	Plantar flexion	Surf bip	0.6			M/F	Young adult
Scherrer	1959	Triceps	Linear	IEMG	Isom (ergonometric reps)			Surf bip, needle	~0.7		5		
Close	1960	Soleus	Linear	MUAP counts	Isom isot & isom isokin	90° ca. knee, angle varied at ankle		Wire			6		Young
Liberson	1962	Biceps	Linear	IEMG	Isom isoton			Surf bip		5.0	10		Adult
Mason	1969	Abd. and ext. digiti minimi	Not regular	Mean amplitude of AP spike	Isom	10° between 4th & 5th fingers	Wrist pronated	Needle			12	M	18–40
Komi	1970	Biceps	(a)Linear (b)Non/incr	IEMG IEMG	Isom isoton, isot isokin	90° ca. elbow 10°–115°	Wrist supinated	Surf bip	1.1	5.0	8 29	M	17–29
Bouisset	1971	Biceps, triceps	Linear	IEMG	Isom isoton	90° ca. elbow		Surf bip			4		
Stephens	1972	FDI	Linear	Smoothed rect. EMG (sre)	Isom F-vary (ramp)			Surf bip	1.0		18		Student

Komi	1973	Biceps, brachialis, brachioradialis	Linear	IEMG	Isot isokin	10°–130° meas. at 80°	Wrist supinated	Wire			10	M	
Vredenbregt	1973	Biceps, brachioradialis	Non/incr	Slope of rect. integrated EMG	Isom isoton	0–140°	Wrist supinated	Surf bip	0.4	4.0	1		
Stevens	1973	Biceps, triceps, brachialis, brachioradialis, coracobrachialis, deltoid	Lin and non/decr	Mean ampl. of APs	Isom isoton	90° ca. elbow	Wrist sup, semisup, pron	Surf bip			47	25M 22F	20–40
Bigland-Ritchie	1974	Quadriceps	Linear	IEMG	Isoton	Varied (on ergonometer)		Surf bip	1.0	15.0	3	2M 1F	19–29
Seyfert	1974	Biceps, vastus	Linear	Mean ampl. of EMG	Isom isoton	90° ca. elbow	Wrist semisupinated	Needle			20	M	20–30
Milner-Brown	1975	FDI	Linear	Mean rect. EMG	Isom			Surf bip	0.9	3.0	6		
Soechting	1975	Biceps, triceps	Linear	IEMG	Isom isoton, isom F-vary (sinusoid)	90° ca. elbow	Wrist supinated	Surf bip	0.9		4		
Komi	1976	Rectus femoris	Non/incr	IEMG	Isom isoton	60° ca. knee		Surf bip	0.4	1.0	12	8M 4F	13–15
Thorstensson	1976	Rectus femoris, vastus lateralis	Non/incr	IEMG	Isom isoton	60° ca. knee		Surf bip	Mini		8	M	24.0
van Hoecke	1978	Biceps, pronator teres	Linear & non/incr	IEMG	Isom isoton	0°–135°	Pron, s-p, sup						
Komi	1978	Rectus femoris	Non/incr	IEMG	Isom isoton			Surf bip	0.4		12	8M 4F	13–15
Moritani	1978	Biceps	Linear	IEMG	Isom	90° ca. elbow	Wrist semisup	Surf bip, surf unip	1.4	5.1	26	M	21.1

Table 7.1—*Continued*

First author	Yr	Muscle	EMG-F relation	Measure of EMG	Contractions		Position	Electrodes			Subjects		
					Type	Angle (from full extension)		Type	Diam (cm)	Sep (cm)	#	Sex	Ages (range or avg)
Hagberg	1979	Biceps	Linear	Rect. & filtered EMG	Isom isoton	90° ca. elbow & shoulder	Wrist semi-sup	surf bip			5	3M 2F	20–34
Lam	1979	Soleus (cat)	Linear	IEMG	Isom, isom isoton			Wire			1		

Notes:

1. Complete references are included in the bibliography.
2. The experiments listed involved nonfatiguing contractions, except for Scherrer (1959).
3. In general, where details are not provided, the information was not explicitly available.
4. Additional specifications which could be included are:
   a. Specific electrode placement
   b. Skin preparation, if any
   c. Type of feedback, if any
   d. Range of force investigated as a percent of MVC
   e. Procedures followed to avoid fatigue
   f. Level of exercise training of subjects
   g. Whether contractions were practiced to achieve some degree of accuracy before actual experiments
   h. Data normalization techniques, if any

Abbreviations:

EDI, first dorsal interosseous
Non/incr, a nonlinear MES-F relation in which the MES increases more than linearly with increasing force
MUAP, motor unit action potential
Isom, isometric
Isot or isoton, isotonic
Isokin, isokinetic
Sup, supinated
Pron, pronated
s-p, semisupinated
Meas, measured
Surf bip, surface bipolar
Surf unip, surface unipolar

to Chapter 5. (*b*) The degree of synergistic action of muscle groups is influenced by the relative spatial orientation of individual muscles. (*c*) Varying amounts of cocontraction among antagonist muscle groups may bias the recorded signals attributed to individual muscles. It is conceivable that any or all of these factors, as well as others, could contribute deterministically to the EMG signal-force relationship.

In addition, variability in detection and data processing techniques may explain some of the inconsistencies which have been reported for specific muscles. The reader is referred to the discussion presented in Chapter 2 which directly addresses the essential technical issues. A review of the reported literature reveals that no consensus has existed on a specific anatomical detection site on the surface muscle from which to detect the signal. All imaginable types of electrodes have been used (including surface, indwelling needle, and wire electrodes) in both the monopolar and bipolar combination. High input impedance amplifiers, employing modern FET-technology eliminate the need to reduce and monitor skin impedance. However, past investigators did not have this benefit and used a variety of methods to prepare the skin. Amplifiers and filter specifications influence the final form of the processed EMG signal, and the specifications which have been used have not been consistent.

There has not been any consistent usage of one parameter measure of the amplitude. This search for a better or different parameter has, to some extent, reflected the lack of universal agreement on which parameter to use, as well as technological developments in electronics which have facilitated the necessary signal processing. Parameters which have been used include the smooth rectified amplitude (Stephens and Taylor, 1972), mean rectified amplitude (Milner-Brown and Stein, 1975), root-mean-squared amplitude (Stulen and De Luca, 1978), and several versions of integrated amplitude (Inman et al, 1952; Komi and Buskirk, 1970; Bouisset and Goubel, 1971; and several others). The distinction and limitations of each of these parameters are discussed and compared in Chapter 3.

The great physiological and technological variability so far described has made comparison and reproducibility of experimental results extremely difficult. Beyond these methodological inconsistencies, a general representation of the system output variables, i.e., the signal amplitude and force level, should contain a formulation which allows for valid comparisons between different muscles and individuals, and for retrials on the same subject. This formulation may be realized by normalizing these variables with respect to some convenient and referable quantities, such as their maximal values in a particular experimental procedure. The absence of normalization often constitutes a deficiency in many reported investigations which have compared or averaged groups of subjects.

In the remainder of this chapter we will consider only the characteristics of EMG signals detected with surface electrodes. It has been shown that the intrasubject variability of the signal is much greater when it is detected with indwelling, rather than surface, electrodes (refer to Fig. 3.12 for an example).

A direct comparison of the relationship between the EMG signal and force ouput has been reported by Lawrence and De Luca (1983). This study investigated the following aspects: (1) Whether the normalized surface-detected EMG signal amplitude vs. normalized force relationship varies in different muscles; (2) whether the relationship is dependent on exercise level; and (3) how much variability exists among the same muscles of different individuals. The data was obtained from the biceps brachii, deltoid, and first dorsal interosseous of accomplished pianists, world-class long distance swimmers, world-class power lifters, and normal subjects during voluntary isometric linearly force-varying contractions.

The signal was detected with two surface electrodes (3 mm in diameter) whose centers were spaced 2 cm apart. The electrodes were oriented parallel to the muscle fibers, and in the case of the deltoid and biceps brachii, the electrode pair was located between the innervation zone and tendinous insertion. The root-mean-squared value of the signal amplitude was used as the variant parameter because, as has been explained in Chapter 3, it is the parameter which more completely reflects the physiological correlates of the motor unit behavior during a muscle contraction.

Figure 7.1 presents the average value of the EMG signal-force data of the three muscles. Each of these three sets of data were obtained from 13 subjects. Each subject performed several contractions. The total number of contractions for each set of data is indicated in the figure. The standard deviation is not indicated because of the visual complexity that would result from the overlap. The standard deviation was generally 25% of the mean value and remained essentially constant over the entire force range. It is apparent that in the case of the first dorsal interosseous, the relationship is quasilinear. A polynomial regression analysis would reveal that a second order (nonlinear) polynomial would only improve the fit by 2 or 3%. Thus, for practical purposes it is safe to consider it as linear. No such claim may be made for the curves of the deltoid and biceps brachii data. They are nonlinear, with signal amplitude increasing more than the force.

Comparable measurements on small and large muscles have been reported (Clamann and Broecker, 1979; Woods and Bigland-Ritchie, 1983). The first group observed the relationship between the smoothened rectified amplitude of the signal from the triceps brachii, biceps

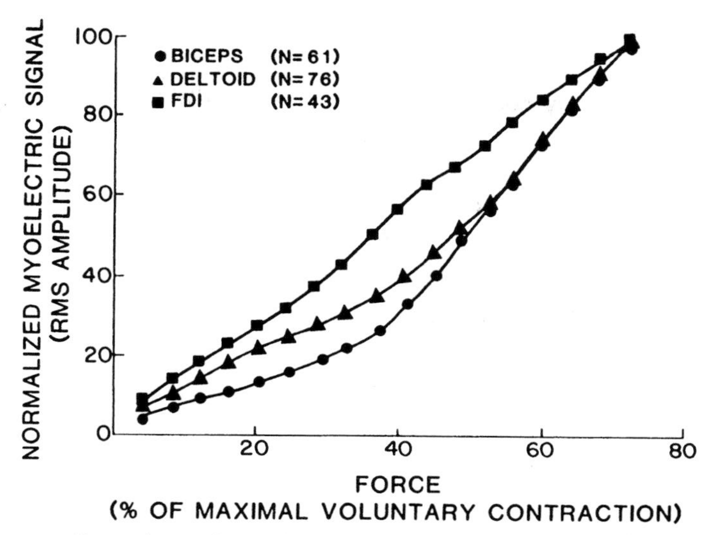

Figure 7.1. Effect of muscle on the EMG signal-force relationship. *N* represents the number of contractions averaged for each muscle. Each set of data was obtained from 13 subjects. (From J.H. Lawrence and C.J. De Luca, © 1983, *Journal of Applied Physiology*.)

brachii, adductor pollicis, and first dorsal interosseous muscles. The second group observed the relationship between the highly smoothened integrated rectified amplitude of the signal from the biceps brachii, triceps brachii, adductor pollicis, and soleus muscles.

Although the methodology was somewhat dissimilar among the three considered studies, one point is strikingly clear. The signal amplitude force output relationship for the small hand muscles, the first dorsal interosseous, and the adductor pollicis was always found to be quasilinear, whereas in the larger muscles, the biceps brachii, triceps brachii, deltoid, and soleus, the relationship was almost unanimously nonlinear. The only exception was the data for the biceps brachii from Clamann and Broecker's study: it presented a quasilinear relationship. This anomaly should serve as a reminder that even among well-executed studies, it is difficult to compare the data because the detected signal is a function of the detection procedure as well as the physiological events.

This distinction in the behavior of the signal-force relationship finds considerable support among the studies listed in the accompanying table. The pattern also appears to be unaltered with rigorous but diverse training regimens, as may be seen in Figure 7.2. These data implicitly indicate that considerable differences in fiber-type constituency do not significantly affect the relationship of the *normalized* signal amplitude and force output. This indirect association is based on the well-docu-

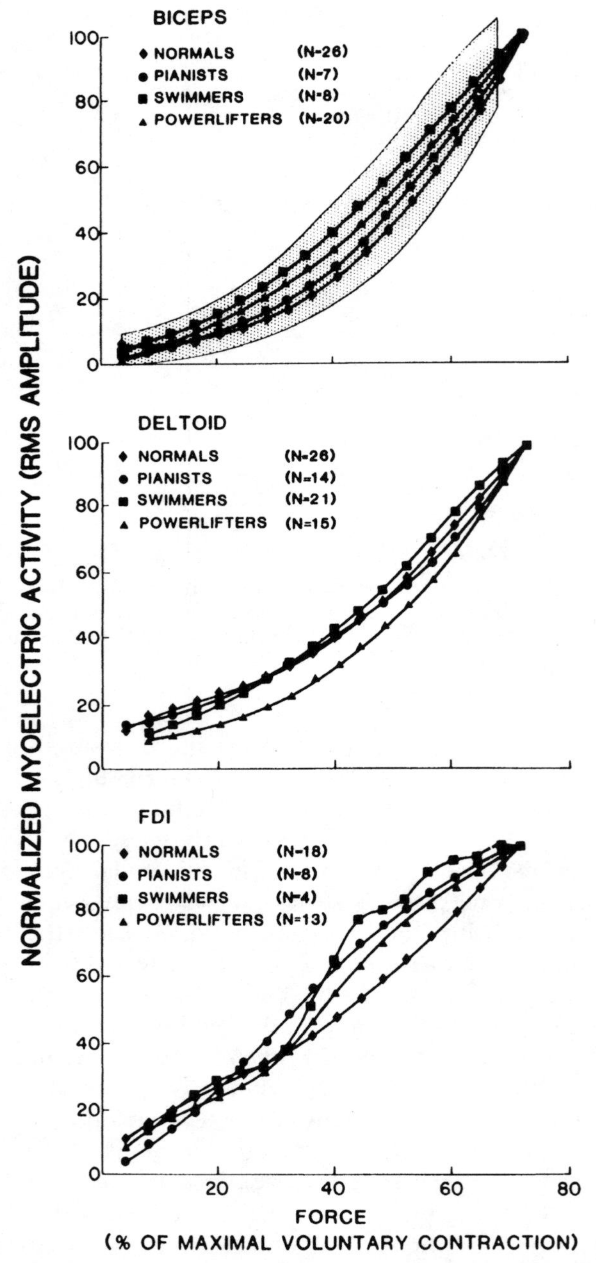

Figure 7.2. Effect of rigorous training regimens on the EMG signal-force relationship in three different muscles. Standard deviations of the raw data are indicated by the *shaded area* in the *top graph*. *N* represents the number of contractions that were averaged to obtain the plotted curves. The data was obtained from 13 subjects. (From J.H. Lawrence and C.J. De Luca, 1983, © *Journal of Applied Physiology*.)

mented fact that elite long distance swimmers have considerably different fiber-type composition in the upper limb musculature than do elite powerlifters (Gollnick et al, 1973).

The distinction in the signal amplitude-force curves remains when the force is generated at different rates ranging from 10% MCV/s to 40% MCV/s (see Fig. 7.3 for details).

A variety of phenomena that may contribute to the muscle-dependent difference in the EMG signal-force relationship can be identified. Some of them are: (*a*) motor unit recruitment and firing rate properties; (*b*) relative location of fast twitch fibers within a muscle and with respect to the detection electrodes; (*c*) cross-talk from signals originating in adjacent muscles; and (*d*) agonist-antagonist muscle interaction.

The agonist-antagonist interaction of simultaneously contracting muscles is an important consideration, especially in isometric contractions, where the joints must be stabilized. In all the studies reported in the literature only the net force or torque resulting from the agonist-antagonist interaction is measured. In many cases this approach provides the correct information with respect to the involvement of the agonist as the muscle of interest. However, in various circumstances involving the need to stiffen the joint, the antagonist(s) may be active. This situation is more likely to occur as the force output increases. In such cases, the net force is customarily assumed to be linear with respect to that of the agonist muscle. However, this relationship may be altered by numerous factors such as joint angle, limb position, and pain sensation. Thus, the signal-force relationship (from the muscle of interest) may be altered.

The electrical cross-talk from adjacent muscles is unquestionably a possible contribution to the behavior of the relationship. Again, its contribution would manifest itself more prominently as the force output of the muscle increases. The presence of cross-talk would be more dominant in smaller muscles, where the electrodes (especially the surface types) are constrained to be located near the adjacent musculature. Cross-talk may also account for the difference in the behavior of the signal-force relationship when the signal is detected with monopolar and bipolar electrodes. Conflicting reports have appeared on occasion (for example, in the work of Moritani and deVries, 1978). As discussed in Chapter 2, these two types of electrodes have considerably different frequency characteristics and detect different amounts of electrical signals from adjacent muscle tissue. The complexity of cross-talk is further compounded by the anisotropy of the muscle tissue itself and the inhomogeneity of the tissues adjacent to the muscle. For this reason, it is often not possible to accurately identify the source of the contaminating physiological signal.

The relative location of slow-twitch and fast-twitch muscle fibers within the muscle is also an important consideration because of the following

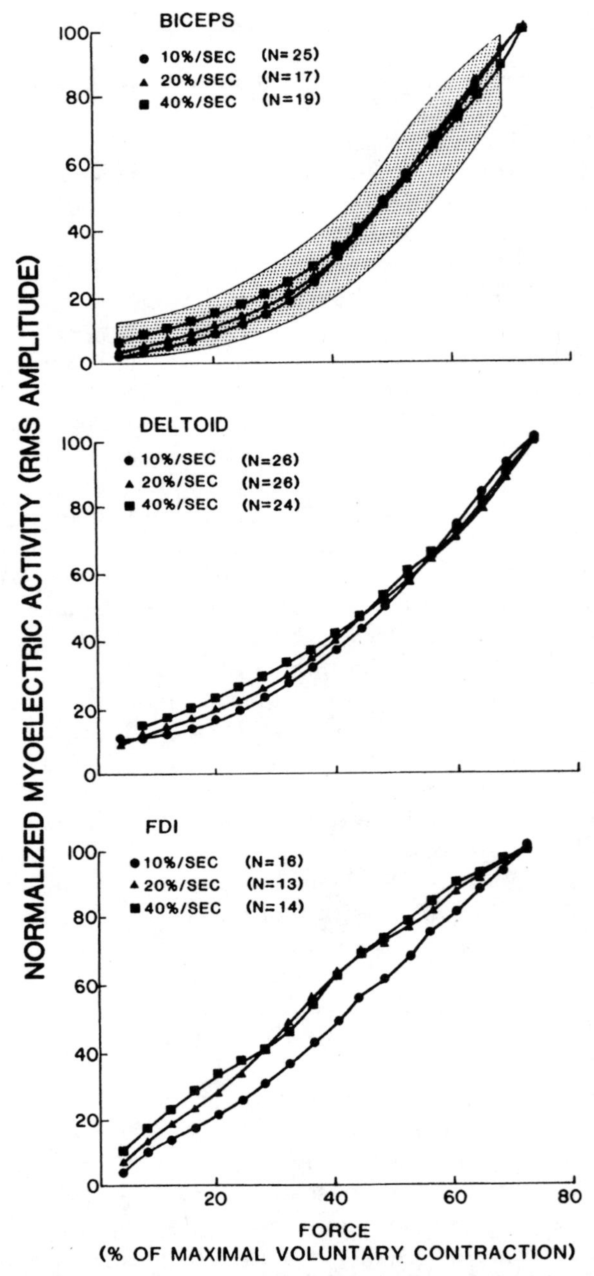

Figure 7.3. Effect of force rate (nonballistic) on the EMG signal-force relationship in three different muscles. The standard deviations of the raw data are indicated by the *shaded area* in the *top graph*. *N* represents the number of contractions that were averaged to obtain the plotted curves. The data was obtained from 13 subjects. (From J.H. Lawrence and C.J. De Luca, 1983, © *Journal of Applied Physiology*.)

reasons. The amplitude of the action potential generated by a single muscle fiber is proportional to the fiber diameter. Fast-twitch fibers, which in the human first dorsal interosseous and biceps brachii muscles are generally larger in diameter (Polgar et al, 1973), have higher amplitude action potentials than slow-twitch fibers. Higher amplitude motor unit action potentials result in a higher amplitude signal. (Refer to discussion in Chapter 3, specifically to Figure 3.11.) However, the amplitude of the motor unit action potential that contributes to the surface signal is a function of the distance between the active fibers and the detection electrodes: the greater this distance, the smaller the amplitude contribution. The larger motor units (containing the larger diameter fast-twitch fibers) are preferentially recruited at high force levels according to the "size principle" (Henneman and Olson, 1965). Therefore, the relative location of the fast-twitch fibers within the muscle and with respect to the recording electrodes determines how the electrical signal from these motor units affects the surface EMG signal.

Although all three causalities described above require consideration when the EMG signal is compared to the force output, it remains difficult to document their intervention quantitatively. The fourth causality mentioned, that is, the motor unit recruitment and firing rate properties, has been documented more concretely. Information described in detail in Chapter 5 provides documentation that the firing rate and recruitment properties of relatively large and small muscles are distinctly different. The equations in Figure 3.11 show that such a distinction has the potential of expressing itself in the value of the parameters that are customarily used to measure the amplitude of the signal.

Oscillating Contractions

An alternative approach to studying the relationship between the EMG signal and the force output of a muscle during isometric contractions is monitoring the signal from the muscle while the force output of the muscle oscillates at a fixed frequency, and repeat the measurement at various frequencies. The modulation of the force output may be induced either by electrical stimulation of the muscle (Buchthal and Schmalbruch, 1970; Crochetière et al, 1967) or by voluntary intention (Gottlieb and Agarwal, 1971; Soechting and Roberts, 1975). The input-ouput relationship (transfer function) between the signal amplitude and the force output as a function of the frequency of the contraction oscillation may be represented as gain and phase functions. (See the discussion at the beginning of Chapter 2 for more details on transfer functions.)

The results of Soechting and Roberts (1975), who detected surface EMG signals, are presented in Figure 7.4. The continuous lines represent a simple transfer function which approximates the data. It is:

$$H(s) = \frac{k}{(s + 5\pi)^2} \qquad \text{where } s = j2\pi f$$

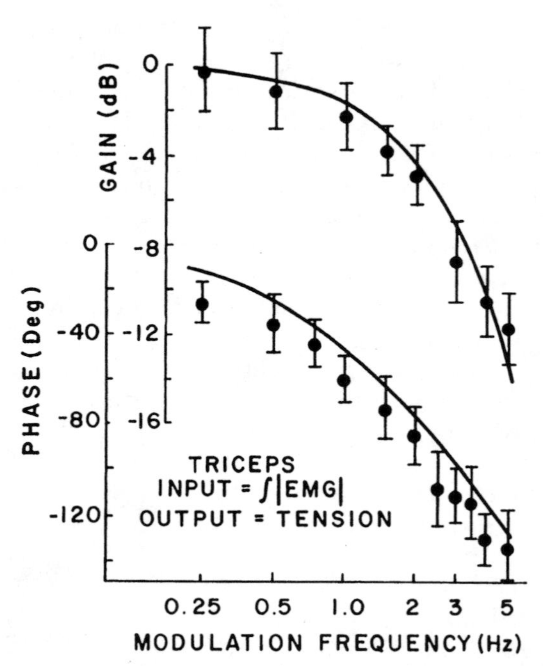

Figure 7.4. Gain and phase relationships between the integrated rectified amplitude of the EMG signal and isometric force in the biceps brachii. (From J.F. Soechting and W.J. Roberts, © 1975, *Journal of Physiology*.)

Although this transfer function does not provide the best fit to the data, it is nonetheless a useful representation of the data because it models the input-output relationship as a simple second order system with a 3 dB point at 2.5 Hz. The data of Figure 7.4 indicate that an increase in the modulation amplitude of the surface EMG signal is required to maintain the same amount of modulation of the output force as the frequency of the modulation increases. In other words, the electrical-mechanical transfer function behaves as a low-pass filter.

RELATIONSHIP DURING ANISOMETRIC CONTRACTIONS

Studies directed at describing the relationship between the signal amplitude and some form of mechanical output of the muscle such as position, velocity, and acceleration are susceptible to additional complications above those associated with isometric contractions. When the contraction is anisometric, that is, one which involves a change in the length of the muscle, the following additional factors must be addressed: (1) the modulation of the EMG signal induced by the relative movement of the electrode with the active fibers; (2) the force-length relationship of muscles; (3) the possible presence of reflex activity; (4) the change in the instantaneous center of rotation of a joint which will effect the moment (force × distance) of the tendon insertion.

A discussion on the complications involving relative movement of the active fibers and electrode is provided near the end of Chapter 2.

The effect of length on the force generation capability is well documented. This nonlinear, nonmonotonic relationship is presented in Figure 7.5. The force produced by the muscle consists of two components: the passive elastic force (curve 1) exerted by the elastic components of a muscle, and the excitation-force response (curve 2). The sum of the two components yields curve 3, which represents the force output of the muscle as a function of its length. Note that the maximal force is generated when the muscle is stretched to approximately 1.2–1.3 times its resting length. This increment of stretch often coincides with the length of the muscle in the relaxed position. It appears that the anatomical architecture of the musculoskeletal system is organized so as to benefit from the force-length characteristics of the neuromuscular system. The instantaneous center of rotation of a joint, that is, the point around which the net torque is zero, is not fixed in most joints. De Luca and Forrest (1973c) demonstrated that in the case of the shoulder joint the instantaneous center of rotation undergoes a considerable excursion as a function of limb abduction.

The involvement of some or possibly all of the above considerations is seen in the findings of Miwa and Matoba (1959). They found that during slow flexion, the biceps brachii is much more active at certain angles of the elbow; it reaches a peak of activity when the elbow is at 160° and falls rapidly to almost nil at 90°, and increases again at maximal flexion. In another study Miwa and Matoba (1963) found similar changes to occur in the muscles of the thigh.

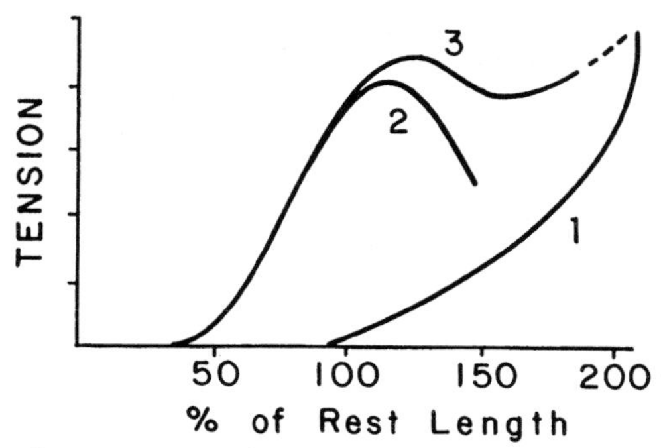

Figure 7.5. Force-length curves for an isolated muscle. Curve 1 is the passive elastic force of a muscle that is stretched. Curve 3 is the total force of a muscle contracting actively at different lengths. Curve 2 is the force developed by the contractile mechanism; it is obtained by subtracting curve 1 from curve 3.

The above discussion should serve as a cautionary warning to resist making facile interpretations concerning the amplitude of the EMG signal when the length of the muscle changes. It is alluring to use the EMG signal amplitude as a parameter to assess the quantitative involvement of the muscle during a functional movement. The literature abounds with reports of such attempts, notably in the area of gait measurements. Admittedly, there are many situations in which the EMG signal does have a useful role in the study of human gait, but in such cases the interpretation of the meaning of the amplitude of the signal should be executed wisely. Appropriate consideration should always be given to the underlying causes and interactions.

SUMMARY

This chapter has described the known behavior of the relationship between the amplitude of the EMG signal and the force output of the muscle. It has been emphasized that the characteristics of the amplitude of the signal can reflect the force output of the muscle, but it may also be affected by the technical details of the detection procedure and by physiological events occurring or originating in muscles not being monitored. Therefore, caution must always be considered when the amplitude of the signal is used as an indication of the muscular mechanical output, especially in contractions that are not isometric.

The relationship between the normalized EMG signal and the normalized force output of the muscle displays the following characteristics.

During a nontonically increasing isometric contraction:

1. There exists a considerable intersubject variation.
2. It is dependent on the muscle; it is quasilinear for the small muscles in the hand and nonlinear (amplitude increasing more than force) for the large muscles of limbs.
3. This distinction in the behavior may possibly reflect the difference in the firing rate and recuitment properties of small and large muscles, as well as other anatomical and electrical considerations.
4. It is independent of training and possibly fiber typing.
5. It is independent of the rate at which the contraction is generated, with the restriction that the contraction be nonballistic and nonfatiguing.

During oscillating isometric contractions the ratio of the force output to the amplitude of the EMG signal decreases as the frequency of the oscillation increases. During anisometric contractions the interpretation of the relationship between EMG signal amplitude and force requires considerable caution because numerous factors other than the force generated by the muscle may affect this relationship.

CHAPTER 8

Muscle Fatigue and Time-Dependent Parameters of the Surface EMG Signal

THE CONCEPT OF FATIGUE

The concept of fatigue as applied to monitoring or measuring the deterioration of a performance of the *human operator* has been ambiguous and often misapplied. Mention the word "fatigue" to a group of health specialists and life scientists, and many diverse and divergent descriptions and explanations will emerge. In man, the issue of fatigue is complex due to the various physiological and psychological phenomena which contribute to it and which demonstrate it. In general most, if not all, currently used methods for measuring fatigue in the active human are, by their very nature, inherently doubly subjective, that is, they rely on the cooperation of the individual performing a prescribed task and the disposition of the observer when assessing the performed task. (It is usually possible for the observer to induce the subject to make an exertion beyond the initial presumption of his capability and/or interest.)

Another source of confusion arises from the fact that far too many health specialists and life scientists appear to have accepted the concept of fatigue as being associated with, demonstrated by, or represented by an event occurring at or over an identifiable period of time. For example, it is common to think of *when* an individual fatigues, or to indicate that an individual is fatigued *when* a particular task cannot be performed or maintained at a specific time. Such a notion of fatigue is inconsistent with that which has been successfully employed by engineers and physical scientists, who have considered the concept of fatigue as a time-dependent process. For example, consider a steel girder that supports the main structure of a bridge. It may well remain in place with no apparent, externally visible structural modification for over 50 years; then suddenly, in one instant a fracture develops, the girder fails, and the bridge collapses. If one, observing from a distance, were to look at the main structure of the bridge for a sign of fatigue, none would be easily noted during the 50-year period. What would be noted in this mode of observation would be the failure point. All the while, however, the crystalline structure of the steel girder was undergoing an alteration caused by chemical and physical processes. In order to monitor the progression of this alteration, specimens of data from within the girder itself or externally observable modifications related to the internal alteration are required.

An analogy in terms of muscle fatigue in the human body would be the task of maintaining a muscle contraction constant for as long as possible. Throughout this task, the involved muscles are continuously fatiguing, but at one instant in time the "failure point" will occur when the desired force output may no longer be maintained and contractile fatigue becomes observable.

Access to biochemical and physiological data within the muscle or the central nervous system could reveal time-dependent changes indicative of a fatigue process, even though the externally observable mechanical performance would not be altered until the failure point. An example of this is presented in Figure 8.1, which demonstrates the force output of the first dorsal interosseous muscle during an attempt at maintaining a 50% maximal voluntary contraction (MVC) constant, along with the median frequency of the power density spectrum of the EMG signal from the same muscle. (It will be explained later that the median frequency, as well as other characteristic frequencies of the power density

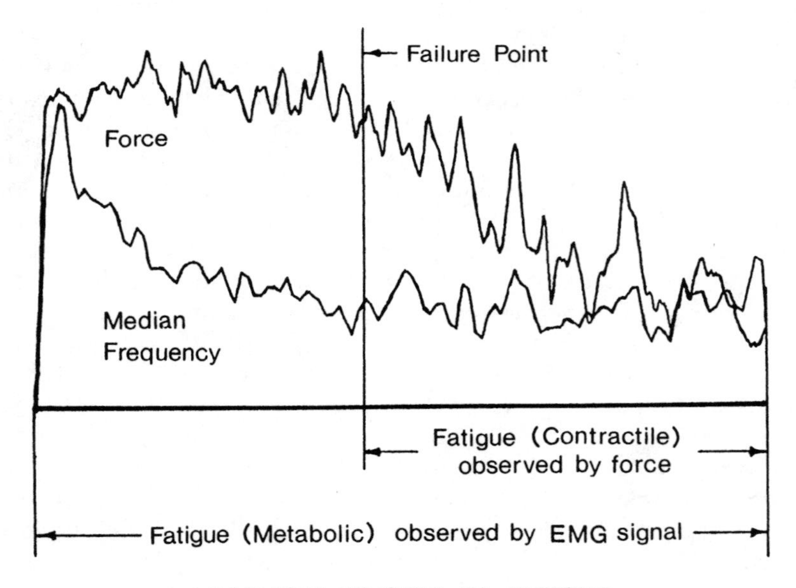

Figure 8.1. Distinction between contractile fatigue and metabolic fatigue. In this case, the force was exerted during an isometric contraction of the first dorsal interosseous muscle. The task consisted of maintaining the force output at 50% of the maximal value for as long as possible, and when this was no longer possible, the subject attempted to produce as much force as possible. The failure point denotes the time when the force output was no longer maintained at the desired average value. The median frequency value was calculated from the power density spectrum of the EMG signal. The time duration of the contraction was 150 s.

spectrum, appear to provide an appropriate representation of biochemical events within the muscle.) Note that even when the force output remains relatively constant, the median frequency is continuously decreasing in value.

This notion of a fatigue process and failure point may well prove useful in describing physiological, biochemical, and mechanical events in the human body.

This chapter will present a review of the published information concerning the relationship of the EMG signal detected on the surface of the skin and fatigue-related events which occur in the muscle. An attempt will be made to explain and unify some of the seemingly disjointed data from different disciplines.

BACKGROUND

Over one-half century ago, Muscio (1921) argued that the then current interpretation of the word fatigue was too general in meaning for scientific use and should be abandoned. This timely advice induced involved professionals to subdivide the concept of fatigue into subsets. This approach was exemplified by Bills (1943), who suggested that fatigue be divided into three major categories. The first was subjective fatigue, characterized by a decline of alertness, mental concentration, motivation, and other psychological factors. The second was objective fatigue, characterized by a decline in work output. The third was physiological fatigue, characterized by changes in physiological processes. These categories have been further subdivided into areas with identifiable origins and symptoms (see Simonson and Weiser, 1976).

One type of physiological fatigue is induced by sustained muscular contractions. It is associated with such external manifestations as the inability to maintain a desired force output, muscular tremor, and localized pain. The effects of this fatigue are localized to the muscle or group of synergistic muscles performing the contraction. This category of fatigue has been termed *localized muscular fatigue* by Chaffin (1973). Although this term originally had its roots in the field of ergonomics, it was subsequently popularized by a research group at Chalmers University of Technology and Sahlgren Hospital in Sweden. However, according to Merton (1954) and various other investigators, even this category of fatigue may have its source peripherally (in the muscle tissue or neuromuscular junction) or centrally (in the brain and spinal cord).

In the study of localized muscular fatigue, analysis of the EMG signal, detected on the surface of the skin over a muscle, has been extensively employed. Since the historic work of Piper (1912), the frequency components of the surface EMG signal have been known to decrease when a contraction is sustained. Cobb and Forbes (1923) noted this shift in frequencies toward the low end with fatigue and also observed a consistent increase in amplitude of the EMG signal recorded with surface electrodes. Many other investigators have also noted an increase in EMG signal amplitude (Knowlton et al, 1951; Scherrer and Bourguignon, 1959; Zhukov and Zakharyants, 1960; Lippold et al, 1960; deVries, 1968; Kadefors et al, 1968; Kuroda et al, 1970; Lloyd, 1971; Vredenbregt and Rau, 1973; Stephens and Usherwood, 1975; Viitasalo and Komi, 1977; Stulen and De Luca, 1978a; Clamann and Broecker, 1979; Maton, 1981; Hagberg, 1981; and others). The frequency shift (towards the lower frequencies) has also been observed often and in a variety of muscles throughout the human body (Kogi and Hakamada, 1962a and b; Sato, 1965; Kadefors et al, 1968; Kwatney et al, 1970; Lindström et al, 1970; Johansson et al, 1970; Chaffin, 1973; Lindström et al, 1974; Viitasalo and Komi, 1977; Lindström et al, 1977; Givens and Teeple, 1978; Komi and Tesch, 1979; Petrofsky and Lind, 1980b; Bigland-Ritchie et al, 1981; Hagberg, 1981; Inbar et al, 1981; Kranz et al, 1981; Palla and Ash, 1981; Stulen and De Luca, 1981; Mills, 1982; Hagberg and Ericson, 1982; De Luca et al, 1983; and others). These two phenomena, which are pictorially represented in Figure 8.2, are in fact related. Lindström et al (1970) and De Luca (1979) explained the interrelationship by noting that during a sustained contraction the low-frequency components of the EMG signal increase and, hence, more EMG signal energy will be transmitted through the low-pass filtering effect of the body tissue. Therefore, the magnitude of the two related phenomena is dependent on many factors, such as force level of contraction, time into the contraction, the type of electrode used to obtain the EMG signal, the thickness of the subcutaneous tissue, and the particular muscle investigated.

A minor digression is necessary at this point. It is commonly observed that the spectral shift is most dramatic near the beginning of a sustained contraction, whereas the amplitude of the EMG signal shows a more pronounced increase near the end of a sustained contraction. Such divergent behavior of these two measurements would seem to indicate that they might have separate origins, were it not for the fact that the firing rates of the motor units decrease, even during constant-force contractions. This decrease in the firing rate is more pronounced near the beginning of the contraction. The decreasing firing rates will decrease the amplitude of the EMG signal and thus offset the increase induced by the frequency shift.

Most of the work in this area has been performed on data obtained during constant-force contractions. Three explanations have been proposed to account for the increase in amplitude and the frequency shift of the EMG signal observed during a sustained, constant-force, isometric contraction. They are: motor unit recruitment; motor unit synchronization, and changes in the conduction velocity of muscle fibers. To this

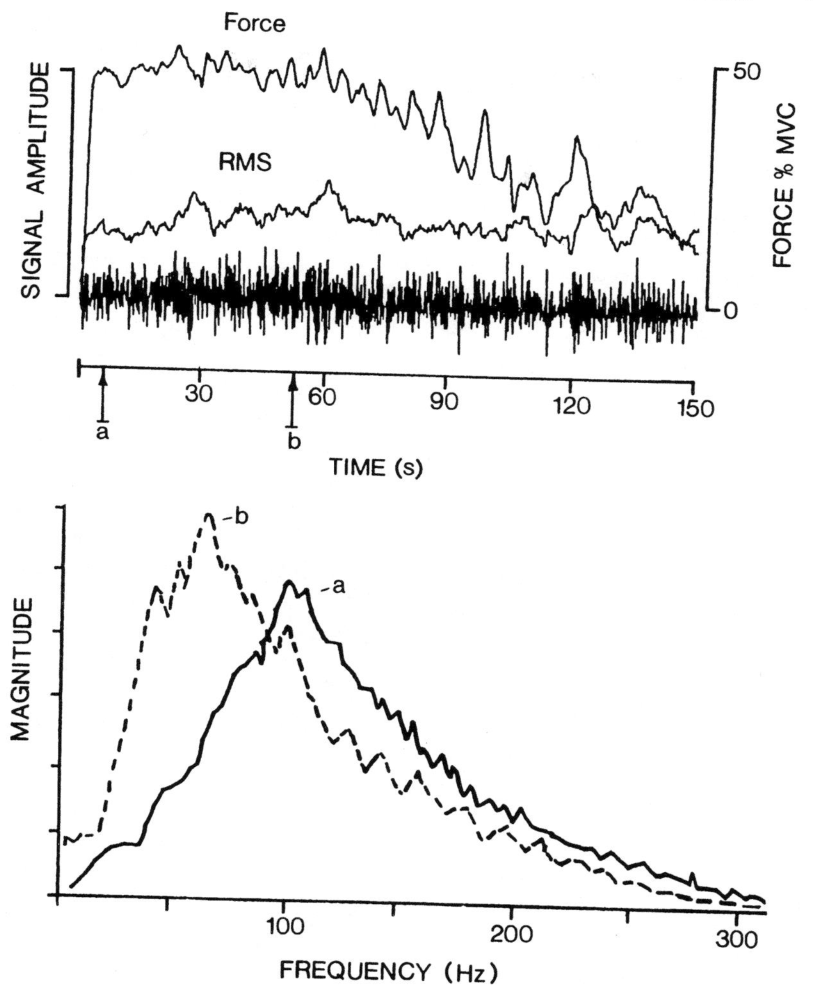

Figure 8.2. (*Top*) EMG signal amplitude and force during an attempted constant-force contraction in the first dorsal interosseous muscle. (*Bottom*) Power density spectra of the EMG signal at the beginning and at the end of the constant-force segment of the contraction.

list should be added the regularity (coefficient of variation) of the motor unit discharge.

Edwards and Lippold (1956), Eason (1960), Vredenbregt and Rau (1973), and Maton (1981) have attributed the increase in amplitude to recruitment of additional motor units. They postulated that as a contraction progressed, additional motor units would be required to maintain the force output constant. Although this is plausible, currently available information does not support this postulate as being a necessary cause of the observed phenomenon. For example, increases in amplitude and frequency shift into lower frequencies have been observed in the first dorsal interosseous muscle during 80% MVC (Stulen, 1980; Merletti et al, 1984), and it has been clearly established (Milner-Brown et al, 1973; De Luca et al, 1982a) that at this contraction level, the first dorsal interosseous muscle does not recruit any motor units. In any case, we have not been able to find one report in the literature that without doubt and ambiguity presented evidence that new motor units are recruited during constant-force contractions of some particular muscle. Arguments for motor unit recruitment during constant-force contractions, such as those presented by Maton (1981), are not convincing without proof that the force output of the muscle under investigation remains invariant. It is important to note that, in all reported experiments in which the EMG signal detected from one muscle has been scrutinized for motor unit recruitment during sustained constant-force contractions, the force output of the muscle has been obtained by monitoring the torque at the joint controlled by the muscle under investigation. Implicit or explicit assumptions have been made that invoke a linear proportionality between these two parameters. However, such is not necessarily the case, because the stiffness of agonist and antagonist muscles may vary without varying the net torque at a joint by increasing the force output of the individual muscle being investigated. Therefore, monitoring the torque of the joint does not provide ensurance that the force of an individual muscle remains invariant.

The above comment should not be misconstrued to argue against the existence of motor unit recruitment during an apparent constant-force contraction. This phenomenon may well exist, but it remains to be proven.

Synchronization, i.e., the tendency for motor units to discharge at or nearly at the same time, has often been cited as the cause of both frequency shift and amplitude increase (Scherrer and Bourguignon, 1959; Missiuro et al, 1962a and b; Person and Mishin, 1964; Lloyd, 1971; Chaffin, 1973; Bigland-Ritchie et al, 1981; and Palla and Ash, 1981; and others). However, synchronization of motor units has been reported to be more evident as the time duration of the contraction progresses (Lippold et al, 1957; Missiuro et al, 1962a; Lippold et al, 1960). Nevertheless, the freqeuncy shift is more pronounced at the beginning of a contraction (see Fig. 8.1). Hence, the behavior of these two phenomena is not complimentary during a sustained contraction, indicating a lack of a powerful association.

Mathematical modeling of the EMG signal has indicated that little, if any frequency shift occurs as the result of motor unit synchronization (Trusgnich et al, 1979; Verroust et al, 1981; Blinowska et al, 1980, Jones and Lago, 1982). Nonetheless, mathematical modeling approaches can only provide limited insight into the modification of the power density spectrum caused by the synchronization of motor unit discharges because of the incomplete knowledge of the detailed behavior of the discharges. In such circumstances, the indications provided by models are highly dependent on the assumptions made about the discharge statistics of the motor units and the shapes of the action potentials. The reports that have accounted for a modifying effect on the power density spectrum are consistent in indicating an increase in the low frequency range of the spectrum, in the range of the firing rate values.

The review presented in the following section will address the effect of the discharge statistics (including firing rate and coefficient of variation of the interpulse intervals) on the power density spectrum. It will be seen that these properties also affect the low frequency components. In fact, the effect of the firing rate, coefficient of variation of the interpulse intervals, and synchronization are inextricably interwoven, and all have the potential of modifying the energy distribution of the lower frequency compartment of the power density spectrum. But, their effect and interaction are complex and are not well understood. However, all current indications strongly suggest that any resulting modification of the spectrum has the potential of being inconsistent in nature. Although they are not suitable candidates for explaining the dramatic frequency shift throughout the whole bandwidth of the EMG signal, their effect on the low-frequency end of the spectrum cannot be disregarded.

The relevance of the third explanation concerning the conduction velocity of the muscle fibers will also become evident in the following section.

FACTORS AFFECTING THE FREQUENCY SHIFT OF THE EMG SIGNAL

A systematic investigation of the behavior of the frequency shift of the EMG signal, requires an analysis of its power density spectrum. This task requires a mathematical model for the spectrum which, in turn, requires a mathematical model for the EMG signal. Such an analysis has been developed in Chapter 3. The equation describing the power density spectrum is repeated here for convenience.

$$S_m(\omega, t, F) = R(\omega, d)\left[\sum_{i=1}^{p(F)} S_{u_i}(\omega, t) + \sum_{\substack{i,j=1 \\ i \neq j}}^{q(F)} S_{u_i u_j}(\omega, t)\right]$$

where

t = time

F = force

$S_{u_i}(\omega, t)$ = the power density spectrum of the MUAPT, $u_i(t)$

$S_{u_iu_j}(\omega, t)$ = The cross-power density spectrum of MUAPTs $u_i(t)$ and $u_j(t)$
$R(\omega, d)$ = the electrode filtering function
p = the total number of MUAPTs that comprise the signal
q = the number of MUAPTs with correlated discharges
d = the distance between the detection surfaces of the bipolar electrode

The power density spectrum of the MUAPT may be expressed as:

$$S_{u_i}(\omega) = S_{\delta_i}(\omega) \cdot |H_i(j\omega)|^2$$

$$S_{u_i}(\omega, t, F) = \frac{\lambda_i(t, F) \cdot \{1 - |M(j\omega, t, F)|^2\}}{1 - 2\{\text{Real}[M(j\omega, t, F)]\} + |M(j\omega, t, F)|^2} \{|H_i(j\omega)|^2\}$$

where

λ_i = the firing rate of the motor unit
$S_{\delta_i}(\omega)$ = The power density function of the impulse train $\delta_i(t)$, which represents the time events of the MUAPs.
$H_i(j\omega)$ = The Fourier transform of the MUAP, $h_i(t)$
$M(j\omega, t, F)$ = the Fourier transform of the probability distribution function $p_n(x, t, F)$ of the IPIs.

These equations emphasize the description of the power density spectrum in terms of the statistical properties of motor unit discharges. Although the waveforms of the MUAPs are represented in $H_i(j\omega)$, any factors which may modify the waveform during a sustained contraction are not represented. One such factor is the conduction velocity of the EMG signal along the muscle fibers. This relationship is expressed in the following alternative representation:

$$S_m(\omega, t, F) = R(\omega, d)\left[\frac{1}{v^2(t, F)} G\left(\frac{\omega d}{2v(t, F)}\right)\right]$$

where

v = the average conduction velocity of active muscle fibers contributing to the EMG signal
G = the shape function which is implicitly dependent on many anatomical, physiological, and experimental factors.

From the above equations and with reference to the related discussion in Chapter 3, it is possible to draw the following observations concerning the behavior of the spectrum:

1. A decrease in the firing rates of the motor units will contribute to a shift of the power density spectrum towards frequencies. This will occur because the probability distribution function (or histogram) will become increasingly skewed. A pictorial explanation of this point is presented in Figure 3.7.

The energy contribution discharge characteristics in the power density spectrum is limited to below 40 Hz.

2. A modification in the discharge characteristics of the motor units may also effect the power density spectrum because the cross-correlation terms will be effected. Factors such as synchronization and regularity (coefficient of variation) of motor unit discharge play a prominent role in determining the behavior of the cross-correlation values. It has been shown (in Chapter 3) that these discharge characteristics have a tendency to increase the energy in the low-frequency part of the spectrum. Their effort is also limited to below 40 Hz.
3. Any modification in the waveform of the MUAPs will be reflected in the Fourier transforms of the waveforms and thus in the power density spectrum. The waveform would have a larger time duration as the conduction velocity decreases during a sustained contraction because the time to traverse the electrode environment would be longer. This would cause a simultaneous increase in the low-frequency components and a decrease in the high-frequency components. Evidence for the increase in the time duration of the MUAP has been provided by Broman (1973), De Luca and Forrest (1973a), Broman (1977), Kranz et al (1981), and Mills (1982).

The first two observations concern control properties of the motor units, which may be either central and/or peripheral in origin. Their involvement in the spectral shift is realistic because it is known that during a sustained contraction, the firing rate of motor units decreases, and other statistical parameters may be altered. (Refer to the discussion in Chapter 5 for details.) The third observation is associated with biochemical and physiological events which occur in the muscle tissue per se. Their involvement in the spectral shift is also realistic and will be discussed in the following section.

FACTORS AFFECTING THE WAVEFORM OF THE MOTOR UNIT ACTION POTENTIAL

It is now apparent that a considerable amount of the frequency shift of the power density spectrum of the EMG signal is caused by a change in the spectral characteristics of the MUAPs which comprise the signal. Such changes may only occur if the waveform of the MUAPs changes. The waveform may be altered by either varying the shape of the waveform or by scaling the waveform by linear operators. In the latter case, the shape of the waveform remains unaltered, but characteristics of the shape are altered. This distinction between the meaning of waveform and shape is important in subsequent discussions. Note that a linear multiplication in either the time scale or amplitude scale does not change the shape. Modifications in the shape are induced by nonlinear transformations. Figure 8.3 presents a schematic diagram incorporating the currently known factors that directly determine or influence the waveform of the MUAPs. One factor, the tissue filtering, determines the actual MUAP shape; the other factor, conduction velocity of the muscle fibers, modifies the characteristics of the waveform.

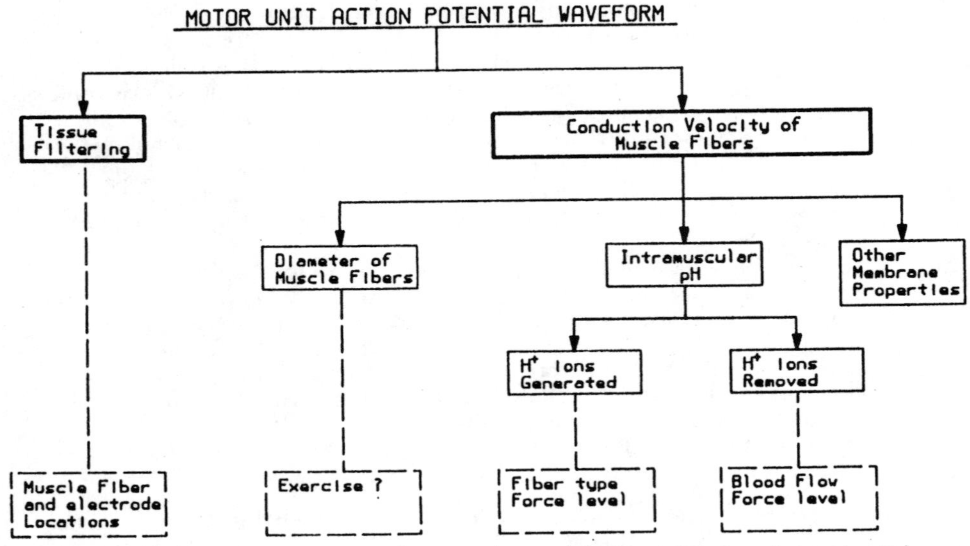

Figure 8.3. Factors affecting the waveform of the motor unit action potentials.

The amount of tissue filtering is determined mainly by three factors: the relative three-dimensional arrangement of the muscle fibers of an active motor unit; the distance between the surface electrode contacts and the active muscle fibers (see Fig. 2.16); and the location of the electrode on the surface of the muscle as a function of the distance between the innervation zone and the tendon of the muscle (Lindström, 1970). Of these two factors, the depth measure of the muscle fiber location is an important concern when additional motor units are recruited or decruited (such as when the force output of the muscle is varied) because their centers of electrical activity are most likely situated in differing locations within the muscle. The distance between the active fiber and the electrode describes the current path between these two points, which is not necessarily the shortest distance, depending on the degree of homogeneity of the tissues.

The conduction velocity of the muscle fibers is monotonically related to the diameter of the muscle fibers and is greatly affected by the intramuscular pH. The relationship between conduction velocity and muscle fiber diameter has been known for many years and is widely accepted. Recent reports have provided conflicting results concerning modifications in the diameter of muscle fibers due to endurance exercise. (For a review of this literature, the reader is referred to Salmons and Henriksson (1981).) The relationship between conduction velocity and pH is less well documented and not as uniformly familiar.

The biochemical-mechanical processes which result in a muscle fiber contraction have as a by-product the formation of lactic acid and pyruvic acid which pass through the muscle fiber membrane into the surrounding interstitial fluid. The amount of hydrogen ions that accumulates inside and outside the muscle fiber membrane will also be dependent on the rate of hydrogen ion removal, either by physical transport or by chemical reaction. The relative and absolute effects of both processes are likely to differ in muscles having considerably different fiber type constituency and tissue consistency. Generally, during repetitive muscle fiber twitches (as would be the case in sustained contractions), the hydrogen ion concentration increases, and the pH decreases. Many reports verifying this point have been published. The earlier work was performed in animals (Ogata, 1960; Beatty et al, 1963). In the past decade, numerous relevant investigations on human muscles in situ have been reported. The reader is referred to the reports of Ahlborg et al, 1972; Hermansen and Osnes, 1972; Sahlin et al, 1975; Tesch and Karlsson, 1977; Sahlin et al, 1978; Tesch et al, 1978; Harris et al, 1981; and Viitasalo and Komi, 1981, among others. Two of these reports by Tesch and his colleagues present data indicating that more lactate is accumulated in muscles that consist mostly of fast-twitch fibers than those that consist mostly of slow-twitch fibers. This observation is consistent with the suggestion that higher activities of glycolytic enzymes, such as LDH and M-LDH, would favor a rapid lactate formation in fast twitch fibers (Sjodin, 1976; Tesch et al, 1978).

It has been postulated that hydrogen ions play a significant role in the generation of action potentials in excitable membranes. They affect the process, possibly by causing physical changes in the arrangement of membrane proteins and/or via the electric field generated by their charge (Bass and Moore, 1973). Experimental evidence for this concept has been provided by Jennische (1982), who demonstrated that the membrane potential decreased as the pH increased in both soleus and gastrocnemius muscles of the cat. Also Tasaki et al (1967) and Orchardson (1978) demonstrated that the membrane excitability decreased when intracellular pH decreased. The conduction velocity is directly related to the membrane excitability. Therefore, introduction of acidic by-products in the membrane environment may be expected to cause a decrease in the membrane conduction velocity. Such behavior in the pH and conduction velocity has been reported in a preliminary report of a study performed on the first dorsal interosseous muscle (Stulen, 1980; De Luca et al, 1983). However, the interaction between pH and conduction and velocity is not firmly established, and additional research in this area would be useful.

It is the net amount of hydrogen ions in the membrane environment that is of importance. Therefore, in addition to the amount of hydrogen ions that are formed during a muscle contraction, it is also necessary to consider any mechanism that removes them from the membrane environment. This point was demonstrated by Mortimer et al (1970), who studied the decrease in conduction velocity in ischemic muscles of cats. In one set of experiments, a muscle was made ischemic by arterial clamping; in another set, a muscle was perfused at normal physiological pressure with nitrogen-bubbled Dextran. In both cases the muscle was not oxygenated. Yet, the decrease in conduction velocity was significantly greater under ischemic conditions than in the dextran perfusion state. Thus, they concluded that the net accumulation of metabolic by-products caused the decrease in conduction velocity.

The removal of acidic by-products will be a function of many factors, the most obvious being the functional capacity of the vascularization in the muscle and the force level of the contraction. As the force output of the muscle increases from zero, the oxygen demand of the muscle increases, requiring an increase in the blood flow. However, the intramuscular pressure also increases, eventually resulting in occlusion of the arterioles and diminution of blood flow in the muscle. Mortimer et al (1971) and Bonde-Petersén et al (1975) employed Xenon-133 clearance techniques to demonstrate that blood flow in the biceps and other muscles of the arm was dependent on the level of contraction. The data in both investigations showed that blood flow peaked at about 25% MVC and decreased to below resting values for contractions performed above 50% MVC. This observation has been qualified by Reis et al (1967), Bonde-Petersén and Robertson (1981), and Petrofsky et al (1981), all of whom demonstrated that muscles consisting mainly of slow-twitch fibers are more heavily dependent on their blood supply for their ability to generate force. Belcastro and Bonen (1975) found that the rate of lactic acid removal was increased during mild self-regulated recovery exercise. It has also been found that sustained exercise will tend to decrease the amount of necessary blood flow into a muscle (Varnauskas et al, 1970; Saito et al, 1980), while apparently increasing the capillary density in the muscle (Andersen and Henriksson, 1977; Ingjer, 1979).

The details of the above discussion are represented in the block diagram of Figure 8.3. In summary, the waveform of the MUAPs, detected by a surface electrode will be a function of the particular muscle that is contracting and the force level of the contraction. These two variables determine the fiber type, number, firing rate, and location of the motor units that are involved, as well as the state of the blood flow. (Some evidence exists that exercise may be a contributing factor by possibly altering the fiber diameter and blood flow.) During sustained fatiguing contractions, two factors will have the greatest effect on the motor unit action potential waveform. In constant-force contractions, in which the number of active motor units is essentially fixed, the dominant factor is the amount of acidic by-products which remain in the muscle fiber membrane environment. During force-varying contractions, the effect of tissue filtering of the newly recruited motor units also plays a prominent role.

PARAMETERS USED TO MEASURE THE FREQUENCY SHIFT

A cautionary note is in order on the technical details associated with the detection and recording techniques used to acquire the EMG signal. (For additional and specific details, refer to the material in Chapter 2.) Special attention should be given to the combined bandwidth of the electrode, amplifier, and recording device in order to acquire a faithful representation of the EMG signal. Ideally, for surface electrodes, the bandwidth (3 dB points) should be 0–500 Hz. However, the DC coupling may cause complications. Therefore, for practical purposes it is recommended that the low-3 dB point be set at 20 Hz. The reader is reminded that the estimate of parameters of the frequency spectrum of the EMG signal discussed in this section will be affected by the recording bandwidth. Wherever possible, in the discussion of this and following sections, it will be assumed that a sufficiently wide bandwidth has been used to acquire the signal.

Several investigators have attempted to use the increase in amplitude of the EMG signal as an empirical measure of localized muscle fatigue (deVries, 1968; Currier, 1969; Lloyd, 1971; Viitasalo and Komi, 1977; Hara, 1980; Petrofsky and Lind, 1980a; Hagberg, 1981; Maton, 1981). These investigators have used either the rectified integrated or the rms value of the EMG signal. Although, as discussed in prior sections, the amplitude is indeed a reflection of the frequency shift (Lindström et al, 1977; De Luca, 1979). Furthermore, the amplitude varies with the type of electrode used to detect the signal, the placement of the electrode, as well as the time-dependent properties of the conducting gels that are commonly used to interface electrodes to the skin. Regardless of these technical limitations, any measure of the total energy content of the EMG signal cannot provide the best representation of the frequency shift. The reader is reminded that the spectrum of the EMG signal is modified by a concurrent increase in the low-frequency components and a decrease in the high-frequency components. These effects tend to offset each other so that the total power of the spectrum will have a reduced sensitivity to any frequency shift in the spectrum.

Some of the earliest attempts at identifying a single parameter of the power density spectrum for representing the frequency shift involved the calculation of the rms value of band-passed EMG signals (Kadefors et al, 1968; Johansson et al, 1970). However, this approach was apparently replaced by the "ratio" parameter, which displayed more dramatic changes that seemed to be related to the frequency shift. The ratio parameter is the ratio of the rms of the low-frequency components to the rms of the high-frequency components (or the inverse). The separation point between the high- and low-frequency regions may be any convenient characteristic frequency, such as the mean or median of the spectrum, chosen at the initiation of the contraction. Alternately, it has been calculated by taking the ratio of the rms values of the signal passed through two band-pass filters, one located in the low-frequency end of the spectrum, the other in the high-frequency end. The ratio parameter is convenient to monitor and has been used by several investigators (Gross et al, 1979; Muller et al, 1978; Bellemare and Grassino, 1979 and 1982; Schweitzer et al, 1979; Bigland-Ritchie et al, 1981).

The main attraction of the ratio parameter is that it presents dramatic changes in value of the frequency shift. However, Stulen and De Luca (1981) have shown through mathematical calculation that the ratio parameter has several drawbacks which make it inferior to other special parameters. This point has been demonstrated empirically by Schweitzer et al (1979) and Hary et al (1982). The ratio parameter is sensitive to the shape of the EMG signal spectrum. This is a hindrance, because in some muscles the frequency components of the EMG signal may vary during force-varying contractions, due to the recruitment of motor units that have significantly different action potential shapes. Furthermore, this parameter is dependent on the initial value of the characteristic frequency chosen to divide the spectrum or on the center frequency and bandwidths of the band-pass filters chosen. For the case where the median frequency is used as the partitioning frequency, a statistical analysis has shown that the ratio parameter estimate is biased and has a covariance approximately 30% greater than the covariance of the estimate of the median frequency. Also, the ratio parameter is not linearly related to the conduction velocity of the muscle fibers (Stulen and De Luca, 1981). In fact, this nonlinear relationship explains the misinterpretation of Bigland-Ritchie et al (1981), who inappropriately arrived at the correct conclusion that the decrease in the conduction velocity does not provide sufficient cause for the change in the frequency spectrum of the EMG signal. This interpretation was subsequently remedied by an elegant study from the same laboratory (Bellemare and Grassino, 1982).

Other parameters have also been proposed but have not been widely used. De Luca and Berenberg (1975) proposed a polar representation consisting of a plot of the low-frequency rms value versus the high-frequency rms value. In such a representation, the plot remains on a line of slope 1 when no frequency shift is present and curves continuously towards the low-frequency rms axis as the shift progresses, thereby describing a characteristic pattern. Sadoyama and Miyano (1981), noting that the frequency shift appeared to be a quasiexponential function of time during a sustained contraction, derived an expression which they termed "measure H." This expression consisted of the integral of the spectral function multiplied by the log of the spectral function. This parameter has the advantage of providing a more linear representation of the frequency shift as a function of time. Hagg (1981), Inbar et al (1981), and Masuda et al (1982) have proposed yet another parameter, the number of zero crossings of the EMG signal. This parameter has the advantage of being relatively simple to implement in hardware. However, the number of zero crossings may be severely affected by noise, a deficit whose impact may be reduced by introducing a nonzero threshold to the signal. A more important disadvantage of this approach is introduced by the fact that the number of zero crossings is approximately linearly dependent on the force of the contraction during relatively low efforts (Lindström and Petersén, 1981).

Since, according to the equation relating the conduction velocity to the power density spectra, all the frequencies are scaled by the same factor, a frequency shift may be observed by tracking any characteristic frequency. Three have been used by various investigators. They are: the median frequency; the mean frequency; and the mode frequency. The median frequency is the frequency at which the power density spectrum is divided into two regions with equal power; the mean frequency is the average frequency; and the mode frequency is the frequency of the peak of the spectrum. All three are (in a mathematical sense) linearly related to the conduction velocity of the muscle fibers (Stulen and De Luca, 1981).

The mean frequency has been used by Herberts et al (1969), Lindström et al (1977), Lindström and Magnusson (1977), Broman and Kadefors (1979), Hagberg (1979), Komi and Tesch (1979), Lynne-Davies et al (1979), Ortengren et al (1979), Hagberg (1981), Hagberg and Ericson (1982), and Ladd et al (1982). The median frequency has been used by Stulen and De Luca (1978b, 1979), Sabbahi et al (1979), Petrofsky (1980), Petrofsky and Lind (1980a), Inbar et al (1981), Palla and Ash (1981), Stulen and De Luca (1982), De Luca et al (1983), van Boxtel et al (1983), Sadoyama et al (1983), Kranz et al (1983), and Merletti et al (1984).

Of these characteristic frequencies, the mode frequency is the least useful, although superficially it might appear to be a useful parameter, because even for relatively poor signal-to-noise ratios it is always theoretically possible to obtain the best estimate. However, this is not the case, because the EMG signal is a stochastic signal which does not have a smooth and sharply defined region near the peak value of its spectrum; hence, the variance of the spectrum would strongly influence the estimation accuracy of the mode. This point has been confirmed empirically by Schweitzer et al (1979). They found that the coefficient of variation for the estimate of the mode frequency was five times greater than that of the mean frequency for EMG signal obtained from the human diaphragm.

Recently, Stulen and De Luca (1981) have shown that the median frequency provides a reliable, consistent, and relatively unbiased estimate of a parameter of the spectrum that is related to the muscle fiber conduction velocity. In general, the estimate of both the median and mean frequencies provides an acceptably good representation of the frequency shift. Both are superior to other parameters (Stulen and De Luca, 1981; Hary et al, 1982). However, both have relative advantages and disadvantages, depending on the quality of the EMG signal, the shape of the spectrum, and other related factors. These two frequency parameters offer the additional advantage that the calculation of their estimate may be implemented in analog circuitry, allowing them to be obtained on-line and in real-time (Broman and Kadefors, 1979; Stulen and De Luca, 1978b, 1982).

BEHAVIOR OF CHARACTERISTIC FREQUENCIES

It is now apparent that according to available information the median and mean frequencies are the preferred characteristic frequencies for monitoring the frequency shift. The median frequency (Stulen and De Luca, 1981) and the mean frequency (Lindström, 1970) have both been mathematically demonstrated to be linearly related to the average conduction velocity of the muscle fibers. Sadoyama et al (1983) have provided experimental verification for a linear relationship. Kranz et al (1983) have suggested that the change in conduction velocity may, in fact, account for nearly all the spectral shift seen in the signal.

Both these characteristic frequencies have been shown to decrease as a function of time during a sustained contraction (Fig. 8.1). During sustained constant-force contractions, the rate of decrease has been found to be either quasilinear (Petrofsky, 1980; Petrofsky and Lind 1980a and b; Inbar et al, 1981; Mills, 1982) or quasiexponential (Lindström et al, 1977; Stulen, 1980; Hagberg, 1981; Stulen and De Luca, 1982; De Luca et al, 1983). This apparent discrepancy may easily be attributed to either different muscles that may have been used to perform the measurements or to different processing schemes employed to calculate the characteristic frequencies. It should be noted that an exponential response may appear to be quasilinear if the time constant of the processing scheme or device is long relative to the time constant of the event being monitored. The rate of decrease is a function of the contraction force; the higher the force, the greater the rate (see Fig. 8.4).

The mean or median frequencies may decrease by more than 50% in value from the beginning to the end of a sustained isometric constant-force contraction. However, the amount of decrease appears to be

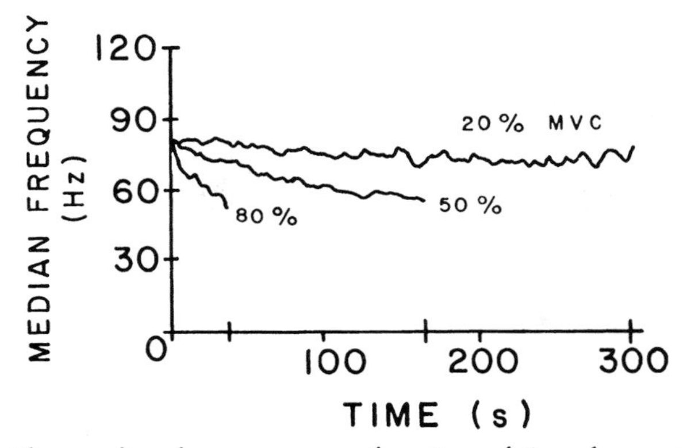

Figure 8.4. The median frequency as a function of time for contractions performed at 20, 50, and 80% of the maximal voluntary contraction level in the first dorsal interosseous muscle.

dependent on the muscles being investigated; some reveal much less dramatic decreases. Stulen (1980) found that in the first dorsal interosseous muscle and the deltoid muscle the greatest decrease in the value of the median frequency occurred at 50% MVC. This observation is consistent with that of Clamann and Broecker (1979), who found that the amplitude of the EMG signal demonstrated the greatest increase during sustained 50% MVC contractions and with that of Tesch and Karlsson (1977), who found that maximal lactate concentrations were found in muscles which contracted isometrically at 50% MVC to exhaustion.

After termination of a sustained contraction, the median and mean frequencies monitored in a muscle have been observed to recover (increase towards their initial value) within 4–5 minutes (Sabbahi et al, 1979; Stulen, 1980; Petrofsky and Lind, 1980a; Mills, 1982; Merletti et al, 1984). This behavior is consistent with that of the conduction velocity observed by Broman (1973), and with the time required to remove lactic acid after cessation of exercise that induces localized muscle fatigue (Harris et al, 1981). Other reports, such as that of Sahlin et al (1978), described much longer lactate removal times after exercises which cause a systemic exhaustion, such as exercising on an ergonometer until total exhaustion. The distinction between these two types of tasks is important. In the latter case, most of the skeletal muscles in the body would be involved. Hence, a relatively high concentration of lactate would be present in the systemic blood stream. Whereas in the latter case, only a localized group of muscles would be primarily involved in the prescribed exercise, less lactate would be present in the blood stream, and the time required for absorption would be less. The latter case is more indicative of localized muscular fatigue and may more accurately be associated with the characteristic frequency measurements made from one muscle.

The absolute value of the mean and median frequencies has been found to vary inconsistently as a function of force within any one muscle (Petrofsky and Lind, 1980a and b; Stulen, 1980; Palla and Ash, 1981). In another study, Hagberg and Ericson (1982) found that the mean frequency increased with force at relatively low levels of contraction; at levels exceeding 25% MVC, the value of the mean frequency became independent of force output. These results are consistent with the explanation presented in a previous section, that is, the relative position of the newly recruited motor units with respect to the recording electrode and the diameter of the muscle fibers of the newly recruited motor units determine the modification of the EMG signal frequency spectrum caused by the additional MUAPTs present in the signal.

Changes in median and mean frequencies have also been shown to be affected by blood occlusion within the muscle. Recent studies by Hara (1980), Mills (1982), as well as our own work (Merletti et al, 1984) have all indicated that the frequency shift of the EMG signal is more pronounced when the blood in the contracting muscle is occluded by external compression. These results are consistent with the fact that when the blood is occluded, acidic by-products accumulate in the environment of the muscle fiber membrane and decrease the conduction velocity of the muscle fibers. The median frequency has also been shown to be affected by the muscle temperature. Petrofsky and Lind (1980a) found it to increase as the muscle temperature was increased from 10 to 40°C. Merletti et al (1984) found it to decrease linearly with decreasing muscle temperature during cooling. This latter observation is consistent with the known fact that the conduction velocity of the muscle fibers is proportionately related to temperature.

All the reported observations of the mean and median frequencies are consistent with the series of events displayed in the block diagram form in Figure 8.3. The only published work which argues against this thesis, with the weight of properly interpreted data, is that of Naeije and Zorn (1982). These investigators simultaneously measured the mean frequency of the EMG signal and the average conduction velocity of the muscle fibers contributing to the EMG signal from the biceps brachii of eight subjects. In four of the subjects, the decrease in the mean frequency and conduction velocity were linearly correlated, and in the rest the mean frequency decrease was not accompanied by a decrease in the conduction velocity. This apparent dichotomy might be resolved by considering the technique used to measure the conduction velocity. They used a cross-correlation technique to measure the difference in the time of arrival of the EMG signal at two locations along the direction of the muscle fibers away from the innervation zone. This technique is not consistently reliable, especially when the cross-correlation value is less than 0.7. A possible explanation may be provided by the anisotropy of the muscle, fatty, and skin tissues. The impedance of the path between a muscle fiber source and a location on the surface of the skin may vary with the location. Therefore, the arrival time of the signal to any point on the skin will be a function of the path taken by the signal to reach that point as well as the conduction velocity of the muscle fibers. Therefore, measurements of this kind require caution in their interpretation. Sadoyama et al (1983) repeated essentially the same experiment, using the same muscle and the same cross-correlation technique. They were careful to resolve signal analysis conflicts, and their data showed a consistent simultaneous decrease in both the mean frequency and conduction velocity.

APPLICATIONS OF QUANTITATIVE MEASURES OF LOCALIZED MUSCULAR FATIGUE

The technique of monitoring the frequency shift of the EMG signal for the purpose of measuring localized muscular fatigue has several advantages: it is noninvasive; it may be performed on muscle in situ; it may be performed in real-time; and it provides information relating to events which occur inside the muscle. In addition to that, Lindström and Petersén (1981) have recently shown that the decrease of the mean frequency is directly related to increased subjective sensation of perceived exertion during a sustained muscle contraction performed at moderate levels, although it is possible that the perceived sensation is not causally related to events occurring in the muscle per se. Many applications of this approach for measuring muscle fatigue are envisioned; some are only concepts and still require experimental verification, whereas others have already been put into practice. The use of the EMG signal frequency shift, however, is far from routine. Some of these applications are seen below.

Athletic Training. The effects of athletic training and exercise on muscle fiber metabolism and architecture are currently an issue of considerable discussion. Numerous studies have been reported, with a variety of conflicting results. For a review of these details, the reader is referred to an article by Salmons and Henriksson (1981). As discussed in previous sections, these modifications within the muscle are theoretically accompanied by corresponding observable changes in the frequency spectrum of the EMG signal.

Industrial Applications. The use of the frequency shift as an indicator of localized muscular fatigue has been applied in the field of ergonomics (Broman et al, 1973; Kadefors et al, 1976; Petersén et al, 1976). Herberts et al (1979) and Hagberg (1979, 1981a) have studied the effect of elevated arm positions on localized muscular fatigue in the shoulder muscles, which is experienced by workers in a variety of work environments. Each study concluded that the change in the EMG signal power density spectrum is useful for measuring the progression of fatigue as a function of arm position. In fact, Herberts et al (1979) suggest a preferred position to minimize the "sensation" of fatigue.

The technique may conceivably be used to distinguish between psychological fatigue derived from boredom and physiological fatigue derived from sustained effort in a work station. It is conceivable that this technique may prove useful in designing work stations in which individuals may comfortably and productively interact with their tasks.

Physical Therapy. In rehabilitation programs involving muscle reeducation and exercise, it is often necessary to assess the effectiveness of a prescribed physical therapy program. Manual muscle tests are currently the primary procedure for determining muscular strength and the progression or regression of strength. Yet, these tests are subjective, and their accuracy depends on the training, skill, and experience of the clinician performing the examination (Kendall et al, 1971). In a relatively recent report, Edwards and Hyde (1977) stated that there are no quantitative methods for measuring muscle function in clinical use today for the diagnosis and management of patients complaining of weakness.

During a physical therapy session, it might be possible to assess the response of the impaired muscle(s) to treatment by measuring the frequency shift. If a characteristic frequency obtained from the impaired muscle decreases, then the muscle is indeed being exercised and is undergoing a fatigue process. If, on the other hand, the characteristic frequency does not change, it may indicate that the muscle is not being adequately exercised and/or the unaffected synergists are generating most of the force. When a muscle or a group of muscles is weakened, there is a tendency for subtle shifts in the pattern of muscle activity to occur to enable the synergistic muscles to generate the required force. This is known as "muscle substitution," and it denies the impaired muscle the intended exercise. Muscle substitution is difficult to detect by current manual testing, which depends greatly on the experience of the clinician. With the frequency shift technique, muscle substitution might be observed by noting a modification in the behavior of a characteristic frequency. For example, if a characteristic frequency obtained from a muscle decreases and then abruptly levels off or begins to increase without a decrease in force output, it may indicate that other muscles are now generating most of the force, allowing the impaired muscle to relax in a relative sense. This information alone would make the frequency shift technique a useful aid for the physical therapist.

The effectiveness of a prescribed treatment program could be determined by changes in the behavior of the characteristic frequency obtained during a series of treatments. Possibly, both the time constant of decay and/or the percentage decrease in the characteristic frequency obtained from a sustained contraction should increase with the number of trial sessions. If the muscle is severely atrophied at the start of the therapy, then the initial value of the characteristic frequency may also change significantly over several of the initial test sessions.

Diagnosis and Prognosis of Neuromuscular Disorders. The effectiveness of the use of the EMG signal in the assessment of neuropathic and myopathic disorders has often been investigated. The level of these investigations has included: single fiber electromyography promoted by Stålberg and his coworkers (Stålberg et al, 1975; Schwartz et al, 1976); so-called quantitative electromyography (Buchthal et al, 1957) based on the temporal characteristics of the MUAPs and other factors observable with signals obtained with needle electrodes; macroelectromyography based on information obtained by a needle electrode with considerably large pickup area (Stålberg, 1980); and also frequency analysis of the EMG signal (Larsson, 1975; Kopec and Hausman-Petrusewicz, 1966).

Larsson (1975) studied neuropathies induced by lesions of the peripheral motoneurons. His results suggest that the spectrum of the EMG signal is shifted into lower frequencies in neuropathies with a clinical history of at least 6 months. Since a characteristic frequency is sensitive to the "average" shape of the MUAPs, changes in the characteristic frequency may be useful in following the development of the disorder. The opposite effect on the shape of the MUAPs is characteristic of myopathies, that is, they are generally shorter in duration than normal and are more often polyphasic (Kugelberg, 1947). Each of these factors leads to a shift of the spectrum into higher than normal frequencies, which has been confirmed by Kopec and Hausman-Petrusewicz (1966).

For a measure of the characteristic frequency to be useful in the diagnosis or prognosis of either type of disorder, the frequency shift must be measurably greater than that which is normally expected due to the stochastic nature of the signal. However, if the disease is known to be present or is suspected, then the change in the characteristic frequency obtained from subsequent examinations may be useful in monitoring the progression or regression of the disorder.

Other Clinical Applications. One interesting application of the EMG signal frequency shift has been in the assessment of diaphragm fatigue. In this particular case, contractile fatigue is inconvenient to measure and, moreover, provides a rather late assessment of the functional capability of the diaphragm. Researchers in Canada (Bellemare and Grassino, 1979; Gross et al, 1979; and Solomon et al, 1979), as well as researchers in the USA (Lynne-Davies et al, 1979; Schweitzer et al, 1979) have investigated this possible application. In particular, Solomon et al (1979) have indeed shown that metabolic fatigue, as indicated by the frequency shift of the EMG signal, occurs before contractile fatigue. Hence, they proposed to use a measure of this shift to set the resistance to breathing for safely exercising the diaphragm in quadriplegic patients.

Ladd et al (1982) suggested that a measure of the mean frequency could be used to monitor the progression of peripheral nerve regeneration. This possibility is based on the concept that different muscle fiber populations and/or different motor unit architectures would emerge throughout the regenerative process. Such a concept has merit; however, serious consideration should be given to the accompanying modifications of the vascular network in the reinnervating muscle, because it may influence the quantity of acidic by-products which are retained in the muscle.

One area that has not yet been explored for possible consideration, but which from a logical perspective holds promise, involves peripheral vascular diseases. The diminished capacity of the vascular network to remove acidic by-products from contracting muscles in limbs decreases the pH within the muscles (O'Donnell, 1975); hence, the characteristic frequency would reflect this decrease by a corresponding decrease in its value.

Basic Research. It is self evident that the use of the frequency shift in this area of investigation holds a myriad of possibilities and applications. The process(es) of localized muscular fatigue are numerous and apparently complex, requiring a host of techniques for its study and analysis. The frequency shift is a prime candidate because of its noninvasive nature and its direct and indirect relationship(s) to physiological, anatomical, and biochemical events and modifications within the muscle. To date, very little work of this nature has been reported in the literature. Preliminary reports by Sabbahi et al (1979) and Merletti et al (1984) provide some data indicating that the median frequency of the EMG signal is affected by a decrease in temperature and ischemic conditions in the muscle. Preliminary reports by Rosenthal et al (1981) and Hakkinen and Komi (1983) have indicated that there might be a relationship between the fiber type composition of a muscle and the value of the median and mean frequencies. If this initial observation is substantiated, the technique could provide a noninvasive alternative to muscle biopsy.

CHAPTER 9

Muscle Interactions

In this chapter, we shall consider several concepts and studies concerned with the involvement and interaction of muscular action. The concepts deal with matters of fundamental importance.

Before proceeding it will be useful to clarify some points which will be encountered in the text. A muscle is a contractile organ; it can only generate tension and, hence, can only pull. Thus, at least two opposing muscles are required to control the simplest possible joint functionally. The *agonist* muscle is the muscle which initiates a desired contraction; thus, it is the prime mover muscle. The *antagonist* muscle is any muscle which actively provides a negative contribution to a particular function during a contraction. A *synergist* muscle actively provides an additive contribution to a particular function during a contraction. Thus, the classification of any given muscle in these categories is entirely dependent on the attempted movement or task.

COACTIVATION AND RECIPROCAL INHIBITION

Antagonist Muscles

Every joint is acted upon by muscles which generate forces in opposite directions. Therefore, it is possible to control separately both the torque and the stiffness (force/length) of the joint. The net torque which may be monitored at the joint is the difference of the torques of the agonist and antagonist sets. The stiffness is the sum of the individual stiffness of the agonist and antagonist sets. Thus, the value of these variables will range from high torque and low stiffness to zero torque and high stiffness, depending on the relative activation of the agonist and antagonist sets. If both agonists and antagonists are activated simultaneously (coactivation), the stiffness of the joint will be high, the net torque low. If the antagonists are relaxed (reciprocal inhibition), the net torque will be high.

The description and explanation of the relative involvement of these two modalities of motor control has captured the interest of many investigators during the past century.

Four mechanisms have been found to govern the contractions of an agonist-antagonist muscle set. These are: centrally mediated reciprocal inhibition; centrally mediated coactivation; peripherally mediated reciprocal inhibition; and peripherally mediated coactivation.

Some of the earliest and surely influential work on this subject was performed by Sherrington (1906, 1909). He demonstrated the existence of *centrally mediated reciprocal inhibition* when he applied electrical stimuli to specific areas of the motor cortex of a cat and noted that some muscles contracted while their antagonists relaxed. Sherrington also observed that in decerebrate cats, one stimulus applied peripherally elicited opposite reactions in an agonist-antagonist muscle set. This led him to postulate the concept of *peripherally mediated reciprocal inhibition*. (Some of the muscle sensors and neural pathways which are involved in these segmental reflexes have been described in Chapter 5; see Fig. 5.1 for details.) Although Sherrington elaborated on the concepts of reciprocal inhibition, he did recognize that agonist-antagonist sets could be volitionally excited to contract at the same time. This *centrally mediated coactivation* is a realizable and, at times, used control strategy.

A relatively recent finding in this area has been the existence of *peripherally mediated coactivation*. This mechanism was first reported by Fényes et al (1960). They observed a reflex response in both agonist and antagonist muscles in response to a tendon tap in premature babies and very young normal children. Less powerful demonstrations of this phenomenon have also been reported in normal adult individuals (Rao, 1965; Agarwal et al, 1970; Gurfinkel' and Pal'tsev, 1972 and 1973; Kudina, 1980). The phenomenon appears to be pronounced in spastic patients, be they children (Kenney and Heaberlin, 1962; Feldkamp et al, 1976; Myklebust et al, 1981) or adults (Gottlieb et al, 1982). It is important to note that the current understanding of this phenomenon is very limited and somewhat insecure. For example, Rao (1965) admits that although peripherally mediated coactivation may be elicited in response to a tendon tap, peripherally mediated reciprocal inhibition is, nonetheless, unquestionably involved in the execution of normal voluntary actions. Gottlieb et al (1982) reported that the neural pathway involved could be short latency oligosynaptic or even possibly monosynaptic. The insecurity is further reflected in the appellation of the phenomenon Fényes et al (1960) referred to as a "co-reflex." Gottlieb et al (1982) called it a "reciprocal exictation." We prefer the term peripherally mediated *coactivation*. This appellation established a parallelism between coactivation schemes of central and peripheral origin.

Sherrington's dual observation has led to an active debate concerning central versus peripheral influences in agonist-antagonist muscle control. Lasheley (1951) pointed out that there are examples of motor tasks involving movements that are too rapid to be controlled by proprioceptive afferent cues from the moving limb. In such cases a central program would be required to generate the pattern of excitation and inhibition needed to produce these movements.

The involvement of higher centers in affecting reciprocal inhibition has been reasserted. For example, Cheney et al (1982) recorded from a single motor cortex cell in awake monkeys during wrist flexion-extension movements and used spike-triggered averaging techniques on the EMG signal to test the effect of each cell output on both agonist and antagonist forearm muscles. They concluded that the synaptic terminations of some motor cortex cells with flexor and extensor spinal motoneurons are reciprocally organized. Along the same line of investigation, Terzuolo et al (1973a and b) demonstrated that in ballistically initiated movements performed by monkeys, the agonist-antagonist reciprocity is eliminated after cerebellar ablation. These results have been confirmed by studies of human patients with cerebellar lesions (Hallett et al, 1975b). These studies suggest that separate systems control cocontraction and reciprocal inhibition in antagonist muscles. They also support the idea, originally suggested by Tilney and Pike (1925), that the cerebellum plays an important role in switching from reciprocal inhibition to cocontraction.

Subsequent studies (Hufschmidt and Hufschmidt, 1954; Barnett and Harding, 1955; Basmajian, 1957 and 1958; Bierman and Ralston, 1965; Gottlieb et al, 1970; Patton and Mortensen, 1970; Terzuolo et al, 1973a and b; Terzuolo and Viviani, 1974; Simoyama and Tanaka, 1974; Hallett et al, 1975a; deSousa et al, 1975; Morin et al, 1976; Jacobs, 1976; Hallett and Marsden, 1979; Brown and Cooke, 1981a and b; Ghez and Martin, 1982; and others) have all shown that, during rapid movements, the activity of both agonist and antagonist muscles displayed a triphasic pattern: an initial burst of agonist activity with the antagonist silent (limb acceleration), followed by a reduction of agonist activity with burst of activity in the antagonist (limb deceleration), and a subsequent resumption of agonist and antagonist activity. This sequence of activity may be seen in Figure 9.1, where the rectus femoris is the agonist and biceps femoris is the antagonist. The generally accepted explanation for this triphasic sequency is that the nervous system avoids damage which might result from the explosive force being generated by the agonist muscle during ballistic contractions.

Most of these studies on ballistic movements indicate that if the subject is instructed to precontract the antagonist muscle against a fixed load and to activate the agonist as quickly as possible, the earliest sign of the rapid voluntary movement is not the activation of the agonist with a rapid burst, but rather the inhibition of the antagonist muscle. This has been referred to as the silent period.

Hallett et al (1975a) suggested that in ballistic movements the triphasic putter is originated by purely "preprogrammed" signals with little influence from the periphery. Conversely, Angel (1977) did not support the notion of complete preprogramming of the agonist-antagonist triphasic pattern of activation, arguing that the contraction is affected by feedback signals from the periphery. Pursuing the same point, Jacobs et al. (1980) analyzed the latencies between cessation of agonist (triceps) and initiation of antagonist (biceps) EMG signal activity during rapid, voluntarily, and involuntarily terminated movements with variable peripheral loads.

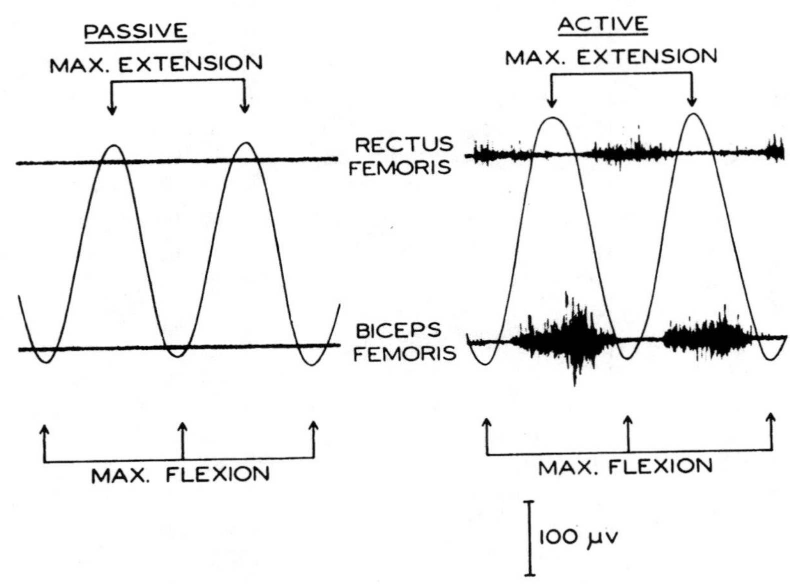

Figure 9.1. EMG of rectus femoris and biceps femoris during passive and active flexion of the knee. (From W. Bierman and H.J. Ralston, © 1965).

From the measured latencies, they concluded that both spinal and supraspinal control mechanisms are necessary in regulating agonist-antagonist functions. Chez and Martin (1982) come to a similar conclusion. They studied the EMG activity associated with rapid limb movements in the cat and concluded that, in the triphasic activity pattern, the antagonist burst and the subsequent agonist burst represent responses to muscle stretch whose amplitude is modulated by descending commands.

Waters and Strick (1981) studied the EMG activity in human subjects required to track a visual target, with ballistic movements of a hand manipulandum. They concluded that antagonist muscle activity during ballistic movements may be infuenced by the subject's movement strategy. Our work is consistent with this observation. We have noted that during an isometric constant-force contraction in which the antagonist muscles are not active, the strategy used to stop the contraction differs, depending on the requirements of the individual. If the contraction may be terminated with no time constraints, the agonist simply relaxes, and the contraction ends. If the contraction is to be terminated faster than the agonist can relax, then the antagonist becomes active to achieve the task. This behavior may be seen in Figure 9.2.

Lagasse (1979) suggested that separate motor systems control speed and force, and that, while the interaction between agonist and antagonist muscle is fundamental for maximal speed of joint movement, it is

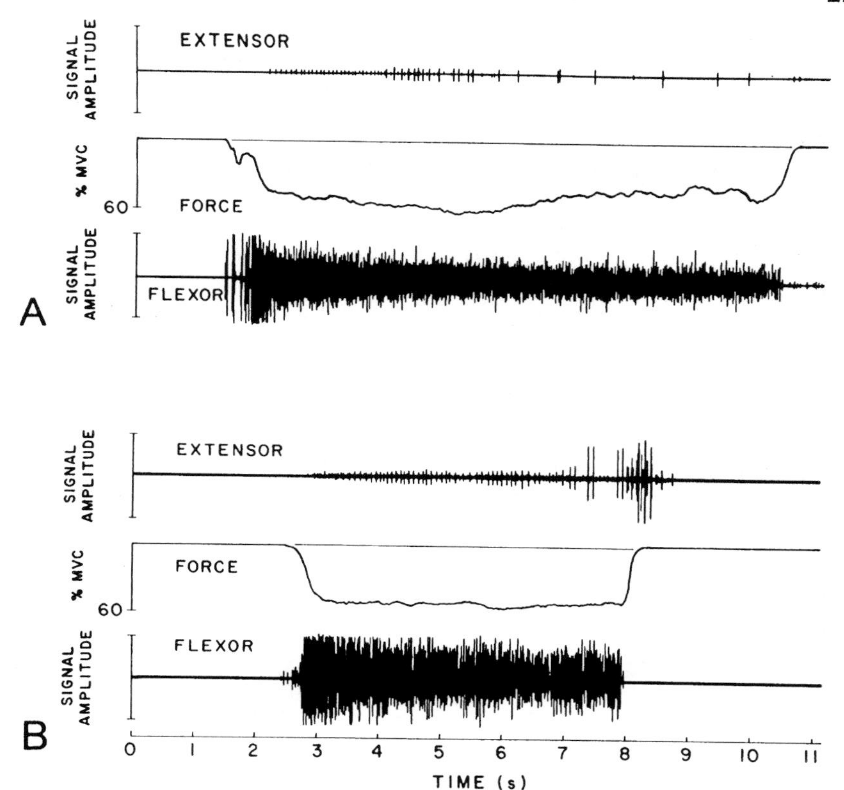

Figure 9.2. Two examples (*A* and *B*) of the raw EMG signals from the extensor pollicis longus (*upper traces*) and from the flexor pollicis longus (*lower traces*) during an attempted constant force isometric contraction at 60% of maximal voluntary contraction (*MVC*) in flexion. The *central traces* represent the net output force measured at the thumb distal phalanx, and the downward direction indicated flexion. Note that as the force decrease at the end of the contraction becomes more rapid, the antagonist (extensor) contracts in example *B*.

relatively unimportant in static force generation. This point of view is consistent with the results of numerous investigations of a group of French coworkers (Goubel and Bouisset, 1967; Bouisset and Goubel, 1967; Lestienne and Bouisset, 1968; Goubel et al, 1968; Pertuzon and Lestienne, 1968; Lestienne and Goubel, 1969; Bertoz and Metral, 1970). In summary, they found a pattern of responses in which low unsustained activity occurs in antagonists at low speeds of voluntary flexion and extension of the elbow; at middle speeds, there were successive activities in the agonist and antagonist, including common electrical silence; at high speed of flexion and extension, there was partial overlapping of phasic activities in agonist and antagonist. They focused their attention not on the speed per se but on the tension in the agonist.

It is generally accepted that slow movements are under peripheral influence at least in the process of learning new skills (see Desmedt, 1978a). However, Polit and Bizzi (1979) showed that trained deafferented (dorsal root severed) monkeys can perform arm target-acquisition movements in the presence of external mechanical disturbances and in the absence of afferent feedback. This was thought to be possible because a given level of coactivation of agonist and antagonist muscles defines an equilibrium position for the joint. Modifications in this level of activation will change the equilibrium position. The work of Day et al (1981) conflicted with this interpretation; their results showed that target acquisition with the thumb distal phalanx in a deafferented patient could not be properly achieved in the presence of external mechanical disturbances. In a later study, Bizzi et al (1982) argued that posture control could indeed be achieved by presetting the level of coactivation of antagonist muscles; however, during movements there is also active control of the trajectory, in addition to control of final position. Thus, some control algorithm other than simply presetting agonist-antagonist activation is used during movement.

In primates, the isometric clasping functions of the hand are good examples of cocontraction of antagonist muscles. Extensive EMG-signal studies of the forearm and the hand during both power grip and precision grip have shown that almost all the extrinsic (finger and hand muscles located in the forearm) and intrinsic (finger muscles located in the hand) muscles are active when large forces are exerted during gripping (Basmajian, 1978; Long et al, 1970; Rash and Burke, 1974).

Synergist Muscles

It is a self-evident fact that the contraction of any one muscle may be accompanied by the synergistic activity of a companion muscle acting on the same joint.

Such activity is particularly evident in mechanically complicated joints surrounded by small muscles. An example of this type of synergistic behavior is found in the work of Weathersby (1966), who reported considerable activity in certain forearm flexors during ordinary movements of the thumb.

What is less apparent, but nonetheless instinctively obvious, is the often noted synergistic activity in other muscles to stabilize adjacent joints. Gellhorn (1947) demonstrated the role of far-removed synergists in movements of the wrist. While flexor carpi radialis was activated in very slight flexion of the wrist, triceps brachii became active with the increasing effort in the prime movers (the extensors of the wrist remaining relaxed meanwhile). Only with very strong static flexion of the wrist would activity—and that only occasionally—appear in the antagonists.

Using as a model the act of prehension of the hand, Livingston et al

(1951) of Paris demonstrated the plasticity of synergists during voluntary movements. Thus, the interplay of activity of the flexors of the fingers and of the thumb with those of the forearm was shown during normal activity to vary significantly, depending on the information of peripheral origin, e.g., position of joints, angle at which the synergists act, the nature of objects grasped, etc.

Missiuro and Kozlowski (1961) illustrated the ultimate plasticity of synergists. In a study of rabbit muscle transplanted to the place of its "antagonist," they found the transplant took on the function of the anatomical and functional antagonist. Obviously the nervous system is able to adapt readily to such changes.

Effect of Training

It has been shown that the control strategy for activating the individual motor units in a given muscle does not alter with training (De Luca et al, 1982a, b). Thus, it may be concluded that the interaction of muscles *must* be altered with training; otherwise, individuals would not be able to improve their performance with practice.

One of the earliest studies on this topic was performed by the Russian physiolgoist Person (1958), who analyzed the EMG signal of the biceps and triceps brachii while subjects were being trained in certain types of work (e.g., chopping and filing). Before the training, the rhythmical flexion and extension of the elbow were effected by exuberant, apparently wasteful activity of the antagonist which is overcome by the greater activity of the agonist. With training, there is a progressive inhibition of the antagonist during the movements of flexion and extension until, with advanced training, the inhibition becomes complete. Kozmyan (1965) found that the latency of antagonist inhibition and agonist excitation varied most frequently during movements responding to nonrhythmic stimulation. With rhythmic repetitive movements, the latencies as well as dissociation of reciprocal inhibition diminished. Thus, inhibition of the antagonist muscles was to be expected in rhythmic activity with any element of supraspinal control or learning. Bratanova (1966) of Sofia found essentially the same thing with rhythmic activity of biceps and triceps brachii. In the "training" stages, coactivation was common apparently as the result of excitation radiation, but later it was extinguished. Kamon and Gormley (1968) and Hobart and his colleagues (see, for example, Hobart et al, 1975) also confirm these views in their studies of changes that occur in motor functions during the acquisition of novel throwing tasks. Similar findings have been reported by Lloyd and Voor (1973) for the acquisition of a special skill during competition.

The diminution of antagonist activity as a function of training comfortably explains the results of Ciriello (1982) and Kaman (1983). Ciriello noted that subjects who were requested to practice strenuous flexion-extensions of the knee daily demonstrated a continual rise in the torque output during the first 6 days, followed by a subsequent plateau. Biopsies of the vastus lateralis revealed no statistically significant alteration in fiber area or type. Kamen studied the task of producing maximal voluntary force at a joint in isometric conditions. He also found that the *net* force measured at the joint increased daily for 6 days. It is surprising how often one finds that modification in the interaction of the agonist-antagonist musculature explains the results of numerous biomechanical studies reported in the literature.

It can be argued that the most widespread training regimen is one that involves daily activities of living. Thus, the basic process of physically interacting with our environment should provide the nervous system with information to modify the reciprocal inhibition control scheme. Such a progression is found in the performance of aging infants and children.

Among others, Janda and Stara (1965) demonstrated that in children a high incidence of grouped responses in a predictable pattern exists, even in muscles that are far removed from those which produce a required movement. As children mature this overactivity disappears and is generally absent in normal adults. It reappears in adults under psychological stress, but people can be trained to inhibit it to varying degrees. Gatev (1967), working independently, also found that as infants mature the excessive cocontraction typical of childhood diminishes progressively. This appears again to be the result of learned supraspinal control eliminating "undersirable" or "useless" coactivation. A group of Moscow investigators, led by Yusevich (see, for example, Okhnyanskaya et al, 1974), attribute normal motor hyperactivity in infants and children to synkinesis or synergies of suprasegmental origin, pointing out the fact that they normally disappear by the time a person is adult.

It is the confirmation of these observations which provides the basis for exercises and training regimens that constitute the rehabilitation programs administered to patients afflicted with various forms of spasticity. In such patients, the normal inhibitory pattern often is found to be dysfunctioned. (For an example, see Fig. 9.3.)

Protective Muscular Responses

A number of reflex phenomena have been demonstrated with EMG. Carlsöö and Johansson (1962) showed that when subjects fall to the ground on the outstretched hand all the muscles which surround the elbow joint are "strongly activated some tenths of seconds before the hand touches the surface." Consequently the musculature is prepared to protect the joint. This is partly a conditioned reflex and partly an unconditioned reflex arising from tonic neck and labyrinthine reactions.

Independently Watt and Jones (1966) found rather similar results in

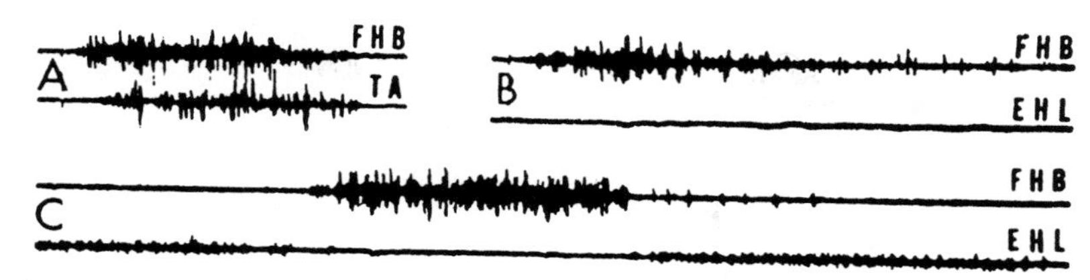

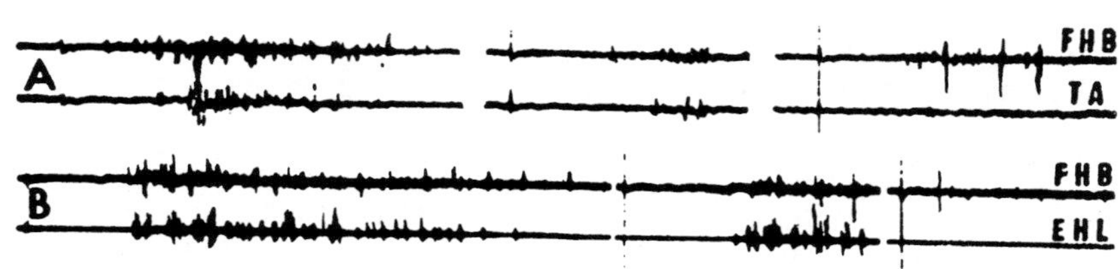

Figure 9.3. Normal flexor response (*upper set of traces*) compared with the abnormal extensor response (*lower set*). EMGs from flexor hallucis brevis and extensor hallucis longus. (Composite of segments of two illustrations from W.M. Landau and M.H. Clare, © 1959, *Brain*.)

the lower limb. EMG activity in gastrocnemius began 80 ms before and 40 ms after the landing impact of the foot. They suggested that there is "a preprogramed open-loop sequence of neuromuscular activity virtually unaided by myotatic feedback." Myotatic reflexes were found to play no significant role in the decceleration, for they came much too late.

Jones and Watt (1971) later showed by EMG that the human gastrocnemius has a stretch reflex (which they called the functional stretch reflex) elicited by a sharply applied and maintained dorsiflection at the ankle; it occurs after a 120-ms delay. Gravitational forces elicit a "body jerk" in muscles of the trunk and lower limbs whenever external force is removed suddenly (Denslow and Gutensohn, 1967). O'Connell (1971) has demonstrated the effect of sensory deprivation on the ability of the human being to achieve the erect posture when he is dropped vertically some moderate (but to him unknown) distance.

SUMMARY OF THIS SECTION

In this section we attempt to identify specific conditions of motor control which may be associated with the control strategies discussed thus far.

Although reciprocal inhibition may be demonstrated as either central and/or peripheral in origin in specific animal preparations, no such distinction can yet be made during the execution of purposeful voluntary contractions. Therefore, the two phenomena will be presented conjunctly as reciprocal inhibition. It appears to be present in:

1. Rhythmical motor process, such as mastication, locomotion, respiration, where the activity of the agonist-antagonist muscles set alternates. (This is most likely peripherally mediated.)
2. High velocity (near ballistic) limb movements. In these cases, a triphasic pattern is noted. First, the agonist contracts to accelerate the limb; then the agonist becomes inactive while the antagonist muscle contracts to decelerate the limb; and finally, if the joint requires further restraint the agonist and antagonist cocontract.
3. Contractions where an external resistance prevents displacement of the joint by the agonist (except in isometric prehension).

Coactivation, be it central and/or peripherally mediated, appears to be associated with the performance of tasks which require assurance that they be realized effectively. In a sense it's an *insurance* mechanism. It appears to be present in:

1. Isometric prehension.
2. Contractions during situations when a joint is required to be stiff, such as when balance is insecure or when large forces are expected to traverse a joint.
3. Contractions where the force or torque at the joint must decrease faster than the relaxation rate of the agonist muscles.
4. Unskilled movements. It has been found to diminish as a function of training.
5. The earlier stages of our lives. It is generally more prominent in infants and children.
6. Patients afflicted with spasticity.

TWO-JOINT MUSCLES

A two-joint muscle is one that not only crosses two joints but is also known to have an important action on both. The best examples are found in the thigh, crossing the hip and knee joints—rectus femoris, the hamstrings, gracilis and sartorius —but the anatomist is soon reminded that such important muscles as gastrocnemius, biceps brachii, and the long head of triceps also cross two joints. Moreover, the tendons of many muscles of the forearm and leg cross an even larger number of joints. Generally, however, there is little confusion about the significant actions and functions of this last group, and the unsolved problems are centered more on the functions of the simple two-joint muscles. For lack of exact knowledge about these functions, they are usually dealt with superficially and, at best, theoretically.

Markee and his associates (1955), basing their conclusions on dissections of human cadavers and nerve-stimulation studies in dogs, stated that two-joint muscles of the human thigh can act at one end without influencing the other end. This is an astonishing concept that appeared to run directly against the logical understanding of muscular action. The explanation they offered is that the middle of the muscle bellies may be moored in various ways and the pull on each end can then be from the middle. If this were true, the functional implications would be extremely interesting and important, and the clinical applications would be obvious.

We tested the above thesis. A series of normal male volunteers were examined electromyographically, using a row of 3 to 5 needle electrodes in each muscle examined (Basmajian, 1957a). In the analysis of the activity in the proximal, middle, and distal parts of the muscles we consider significant movements of the hip (proximal) joint and the knee (distal) joint, disregarding whether these are flexion or extension. In none of the muscles tested was there greater activity in the proximal part of the muscle when the proximal or hip joint was acted upon (Figs. 9.4 and 9.5). Only one muscle in one subject, a semitendinosus, showed greater activity in the distal part when the distal joint (knee) was acted upon (the formula being middle > distal > proximal). However, this particular muscle showed the same formula with movements and postures of the hip, thus indicating that it has a relatively constant pattern, regardless of the joint moved (Basmajian, 1975a). Miwa et al (1963) confirmed these findings in a similar study.

The thesis put forward by Markee and his colleagues appeared at first to be attractive and important, but the electromyographic results showed that it is completely untenable in the case of normal human two-joint

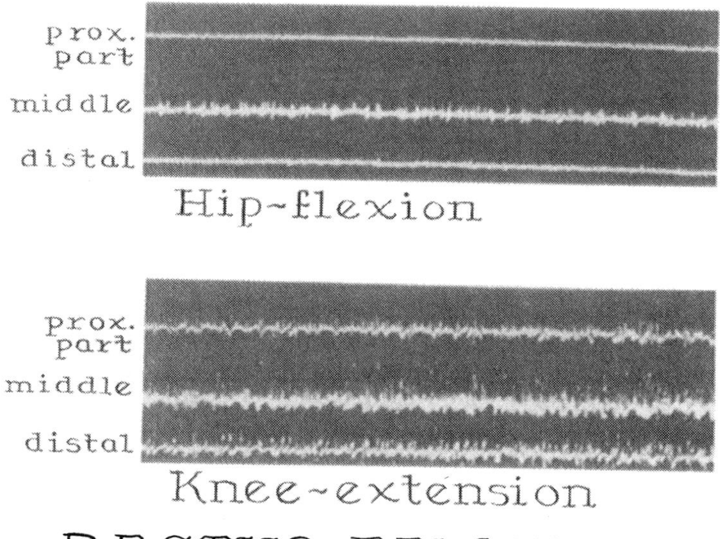

Figure 9.4. EMGs of three parts of one subject's rectus femoris during hip flexion and knee extension (with other joint "relaxed"). (From J.V. Basmajian, © 1957, *Anatomical Record*.)

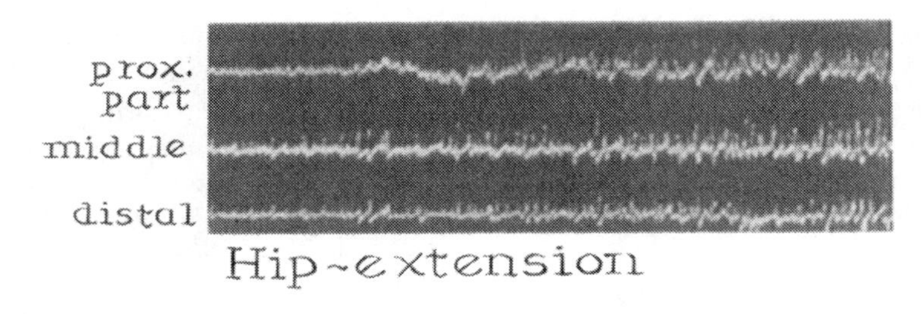

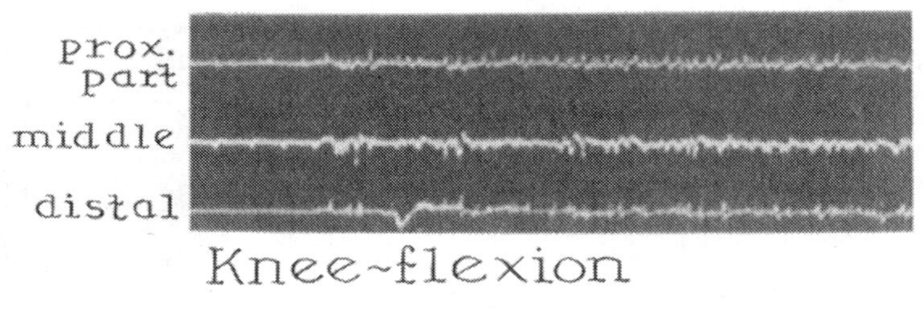

SEMITENDINOSUS

Figure 9.5. EMGs of three parts of one subject's semitendinosus during hip extension and knee flexion (with other joint "relaxed"). (From J.V. Basmajian, © 1957, *Anatomical Record.*)

muscles. In fact, the evidence is overwhelmingly in favor of the orthodox view. These muscles pull directly from one end to the other simply because all parts of the muscle belly contract together, the greatest activity being at the middle of the belly. What has been said above is not true for muscle bellies in parallel or parallel heads of large muscles. For example, in another study we have demonstrated that the two heads of biceps brachii may act relatively independently (Basmajian and Latif, 1957).

To clarify the unresolved problem of how reciprocal innervation appears in a two-joint muscle that is being elongated (stretched) at one end and simultaneously shortened (activated) at the other, we studied the rectus femoris and medial hamstrings (Fujiwara and Basmajian, 1975). In 10 healthy volunteer adults, we inserted bipolar wire electrodes into rectus femoris and the medial hamstrings, as well as the iliopsoas, vastus medialis, and gluteus maximus. For iliopsoas, the wire electrodes were situated in fleshy fibers near the hip joint, and for the other four muscles, the electrodes were in the middle of their bellies.

Each subject lay in a supine position, and the leg was suspended in a sling with a balancing weight so that the hip and the knee were held passively flexed at right angles. The subject was ordered to maintain his limb in this position against new forces exerted by adding 5-kg weights.

These forces were directed to make: (1) the hip or the knee extend or flex as a monoarticular motion; (2) the hip and knee extend simultaneously; (3) the hip extend and the knee flex; (4) the hip flex and the knee extend; and (5) the hip and knee both flex as a biarticular motion. Thus, the reactive contraction included both isometric and isotonic elements.

Monoarticular motions of the hip or the knee (Fig. 9.6). Iliopsoas was active in hip flexion and knee extension. Rectus femoris was active in hip flexion and in knee extension, the activity being rather prominent in knee extension. Vastus medialis was active not only in knee extension but also in hip extension. Gluteus maximus was active in hip extension. More activity was observed from the medial hamstrings in knee flexion than in hip extension.

Biarticular motions (Fig. 9.7). Iliopsoas showed activity in simultaneous hip and knee flexion, and hip flexion with knee extension. Vastus

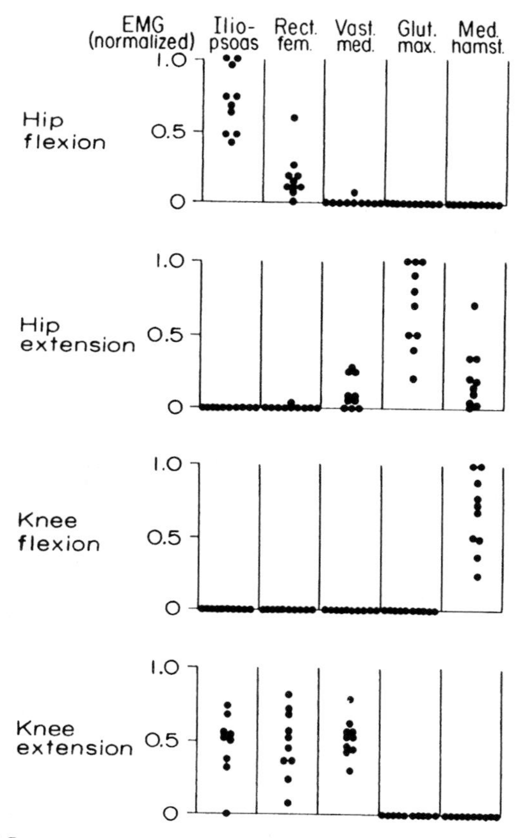

Figure 9.6. EMGs in monoarticular motions (10 subjects). (From M. Fujiwara and J.V. Basmajian, © 1975, *American Journal of Physical Medicine*.)

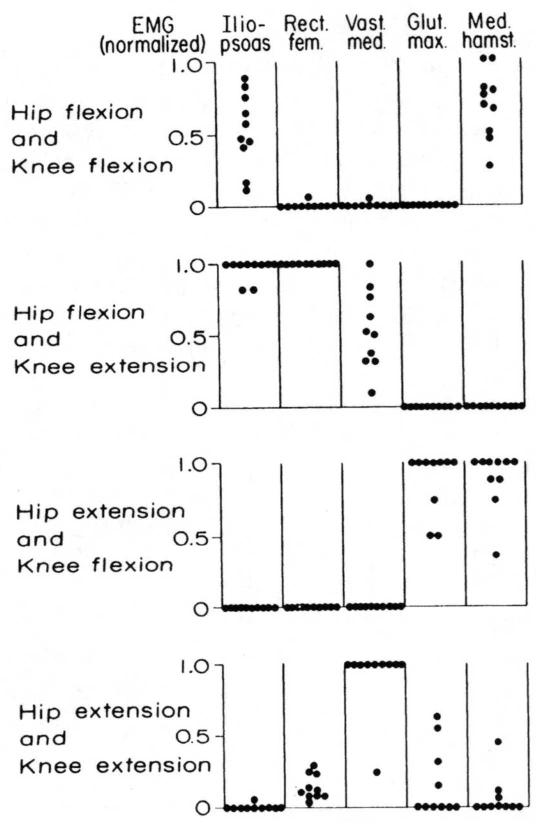

Figure 9.7. EMGs in biarticular motion (10 subjects). (From M. Fujiwara and J.V. Basmajian, © 1975, *American Journal of Physical Medicine*.)

medialis showed activity only in the motions that included knee extension and gluteus maximus in hip extension with knee flexion. During hip flexion with knee extension, we found maximal activity from rectus femoris in all subjects. Activity of rectus femoris was less in knee extension with hip flexion than in knee extension without hip motion, and no activity was observed in hip flexion with knee flexion. The other two-joint muscles, the medial hamstrings, showed maximal activity during hip extension with knee flexion; however, less activity occurred during knee flexion combined with hip flexion, and no activity in hip extension combined with knee extension.

The activity of two-joint muscles more or less influences both joints. If one joint is stabilized by other muscles, the actual kinetic effect would be limited on the second joint. Monoarticular motions involving two-joint muscles are a result of a coordination with other muscles. Alone, a two-joint muscle cannot work as a one-joint muscle.

In countercurrent movements (Fig. 9.8), rectus femoris and the medial hamstrings show maximum activity. However, in concurrent movements (Fig. 9.9), the activity of the two-joint muscle is not the simple sum of the activity of two monoarticular motions. Two examples: (1) rectus femoris shows activity in hip flexion (Fig. 9.8) but not in hip flexion accompanied by knee flexion (Fig. 9.9), and (2) rectus femoris works as a knee extensor, but when hip extension accompanies knee extension, almost all rectus activity disappears (Fig. 9.9).

In example (1), the activity of rectus femoris muscle is completely depressed to allow knee flexion. In example (2), the activity of rectus femoris is depressed for hip extension, and the vastus muscles compensate

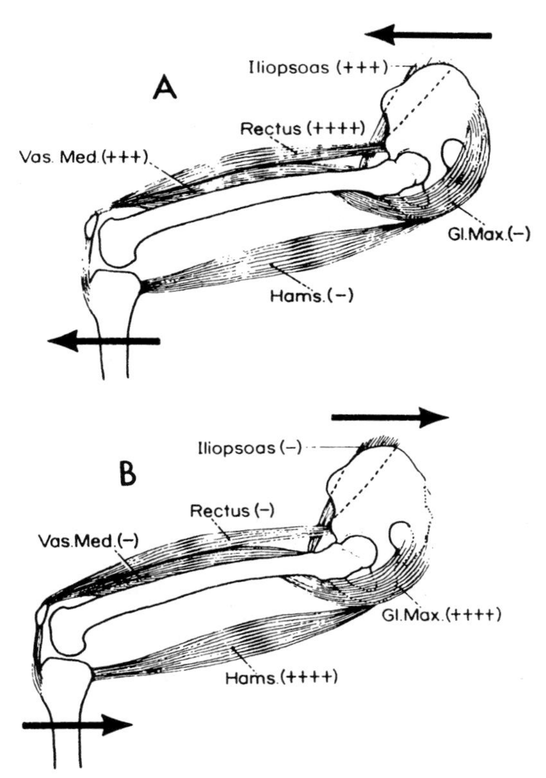

Figure 9.8. (*A*) Hip flexion and knee extension. Biarticular motion—countercurrent movement (modified after Frost). (*B*) Hip extension and knee flexion. Biarticular motion—countercurrent movement (modified after Frost). (From M. Fujiwara and J.V. Basmajian, © 1975, *American Journal of Physical Medicine*.)

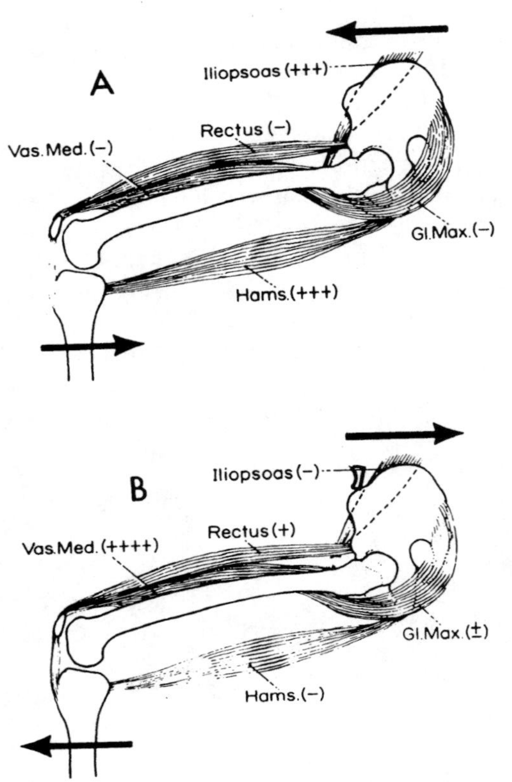

Figure 9.9. (*A*) Hip flexion and knee flexion. Biarticular motion—concurrent movement (modified after Frost). (*B*) Hip extension and knee extension. Biarticular motion—concurrent movement (modified after Frost). (From M. Fujiwara and J.V. Basmajian, © 1975, *American Journal of Physical Medicine*.)

for it to maintain or produce knee extension (Fig. 9.9). Depressed activity of rectus femoris in concurrent movements is in response to a mechanical demand to flex the knee or to extend the hip. From the neurological standpoint, this may be called an antagonistic inhibition of reciprocal innervation. Contraction of rectus femoris influences knee extension more than hip flexion. Like the rectus femoris, the medial hamstrings are also depressed in some concurrent movements, and they are knee flexors more than hip extensors (Fig. 9.9) (Fujiwara and Basmajian, 1975).

To summarize: The effect of contraction of two-joint muscles is never limited to one joint; whenever a two-joint muscle participates in a monoarticular motion its role shifts in close coordination with the other muscles. In biarticular concurrent motion the activity of the rectus femoris and the medial hamstrings is inhibited when they are antagonists, especially where motion of the knee is concerned.

Related conclusions were drawn from a different approach by Yamashita (1975) of Kyoto University. He was concerned most with mechanisms that generate and transmit resultant leg extension forces by maximal isometric contractions in two directions while the hip and knee are kept at 90°. Single-joint muscles were most active when they crossed the joint where they limit extension force. Two-joint muscles were only moderately active then, even though the forces applied were great.

It may be argued that, in spite of the normal findings, occasions may arise when the proximal or distal part of a two-joint muscle does act independently. Such occasions must be very rare indeed. In fact, in upper motor neuron disease the reverse is seen. In such cases the patient employs mass response of many neighboring, and often unrelated muscles. It is difficult to imagine his employing one isolated muscle, let alone the proximal or distal half of one.

Carlsöö and Molbech (1966) attempted to answer the underlying principle of two-joint muscles by comparing a number of them in the thigh both in free movements and during bicycling. In bicycling, during one revolution the effect of a muscle contraction can change from a flexing to an extending effect in the case of hamstrings and gastrocnemius. With the latter, during the knee extension of pedaling there is paradoxical activity, and Carlsöö and Molbech feel that this is part of a steered movement in a closed kinematic chain.

Finally, we must admit that under artificial experimental conditions, proximal and distal parts of two-joint muscles can be made to contract independently with relatively isolated effects. To accept this observation there is no need to invoke species differences, and probably it would be reproducible in a human "preparation" if such were available.

MUSCLES SPARED WHEN LIGAMENTS SUFFICE

Hardly any informed person would doubt that when gravity acts on the upper limb, and certainly when the limb carries a heavy load, muscles are the chief agents in preventing the distraction of the joints. Yet, as a result of many studies, we have concluded that this is a false belief. The original study (Basmajian and Bazant, 1959) involved the shoulder joint. Subsequent studies extended our observations to include the elbow region.

Essentially, the fundamental conclusions can be made that ligaments play a much greater part in supporting loads than is generally thought and, in most situations where traction is exerted across a joint, muscles play only a secondary role. A review of experiments on the foot (discussed on p 349) adds further confirmation to the idea that normally ligaments, and not muscles, maintain the integrity of joints.

Our broader studies in the shoulder and elbow region will be described in greater detail below (p 273), and the full details have been published previously (Basmajian and Latif, 1957; Basmajian and Bazant, 1959; Basmajian, 1961). This electromyographic investigation dealt with supraspinatus, infraspinatus, deltoid, biceps, and triceps muscles in a series of normal persons. In the case of the deltoid, the needle in the anterior fibers was 5 cm below the lateral end of the clavicle; that in the middle fibers was 5 cm below the lateral border of the acromion; and that in the posterior fibers was about 7.5 cm below the spine of the scapula. The electrodes in the supraspinatus and the infraspinatus were placed in or near the middle of their bellies. The electrodes in the biceps were placed in the middle of the muscle whereas those in the triceps were placed in the middle of its long head.

The subject was seated upright with his arm hanging in the relaxed neutral position (the forearm midway between pronation and supination). Two types of load were added to the subject's arm. The first of these was a load of about 7 kg (lead weights held in the hand to the limit of individual endurance which proved to be a variable factor). The other load, less precise but more effective, was a sudden heavy sustained downward pull by one of the observers on the subject's hanging arm. In five persons a longitudinal pull was applied to the arm which had been abducted to the horizontal plane and completely supported by another observer so that no abduction activity was required of the subject's muscles.

For studies of the elbow, electrodes were inserted in both heads of biceps, in brachialis, and in brachioradialis of 24 adults. In addition, the pronator teres muscles of eight other subjects were studied later, and our findings were published (Basmajian and Travill, 1961).

In all instances, electromyographs were made with the subject seated upright and the upper limb hanging straight downwards in a comfortable position. Thus considerable numbers of biceps muscles were studied, some with heavy and moderate loads and others with light loads. In making the electromyograms of the pronator teres, only a strong downward pull was used as the added load since experience had already shown the ineffectiveness of lesser loads.

Contrary to expectation, the vertically running muscles that cross the shoulder joint and the elbow joint are not active to prevent distraction of these joints by gravity (Figs. 9.10 and 9.11). Much more surprising is the fact that they do not spring into action when light, moderate, or even heavy loads are added unless the subject voluntarily decides to flex his shoulder or his elbow and thus to support the weight in bent positions of these joints. Quite often he may do this intermittently or, when

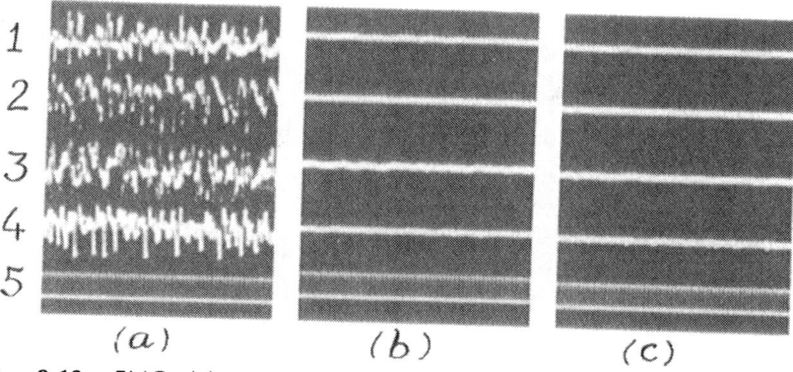

Figure 9.10. EMGs (*a*) During abduction; (*b*) unloaded arm hanging; and (*c*) heavy downward pull applied to arm. Lines, *1*, *2*, and *3*: anterior middle and posterior fibers of deltoid; line *4*: supraspinatus; line *5*: time marker. Intervals, 10 ms (From J.V. Basmajian and C.R. Dutta, © 1961, *Anatomical Record*).

uninstructed, from the very onset. But it must be clear that such muscular action is a voluntary action and not a reflex one.

Carlsöö and Guharay (1968) confirmed our findings of muscular inactivity in the heavily loaded shoulder and elbow joints—the muscles being biceps, triceps, brachialis, and brachioradialis. In addition, they found that the temperature fell in these muscles, apparently because of a lower oxygen demand.

Even while the muscles were quiescent, our subjects rapidly felt local fatigue. What, then, is the sensation of fatigue in the heavily loaded limb? Normally, it would be thought of as "muscular fatigue," but we see now that this is incorrect. The sensation of "fatigue" that is experienced probably originates from the painful feeling of tension in the articular capsule and ligaments, not from overworked muscles. In fact, as we have seen, the muscles need not be working at all. Cain (1973) confirmed that feelings of fatigue in static contraction arise in part from structures outside the muscles.

An analogous situation occurs in the foot where we found, some years

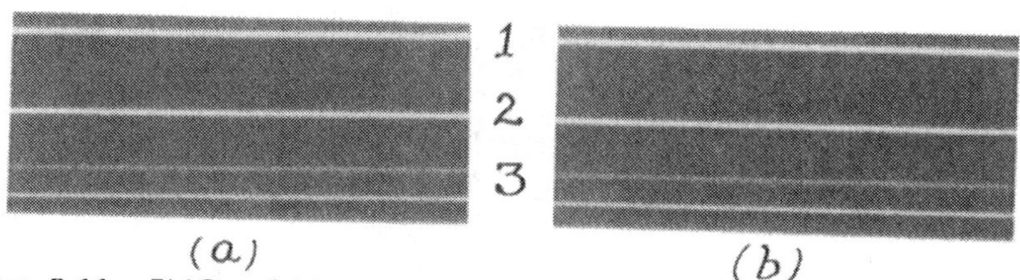

Figure 9.11. EMGs of biceps (line *1*) and triceps (line *2*). (*a*) Hanging arm (unloaded); (*b*) Heavily loaded. No change. Line *3*, time marker. Intervals 10 ms. (From J.V. Basmajian and C.R. Dutta, © 1961, *Anatomical Record*.)

ago that the muscles that are usually supposed to support the arches continuously were generally inactive in standing at rest (see p 258). Independently Hicks (1954) argued by deduction that the plantar aponeurosis and plantar ligaments were the chief weight-bearers in this position. It would seem, then, that in the normal foot the fatigue of standing is not a muscular phenomenon.

The dual conclusion that articular ligaments suffice to prevent the downward distraction of joints in the upper limb and that the sensation of fatigue is chiefly a form of pain in the ligaments appears to be of fundamental importance. It not only runs counter to "common sense," but it is of practical interest, for example, in explaining why dislocations by traction on normal limbs are rare. It should be noted especially that the capsule on the superior part of the shoulder joint including the corachohumeral ligament is extremely tight only when the arm hangs directly downward and the scapula is in its normal position. The special mechanism that includes this ligament, together with the supraspinatus muscle and the normal slope of the glenoid cavity, will be described elsewhere (p 273). When the shoulder joint is abducted or flexed, however, the capsule is extremely loose, and the shoulder joint depends for its integrity on the well-known "rotator-cuff" muscles.

All the experiments reported above finally led to one that has in turn led to new ideas. Elkus and Basmajian (1973) found that healthy subjects suspended by their hands from a trapeze can hold on for less than 3 minutes, even when their fists are kept closed by a special gauntlet. Severe discomfort in the hands and forearms and uncontrollable slipping of the fingers (when no gauntlet was used) were the main causes for failure. Action potentials from a large number of muscles were unremarkable and all evidence pointed to a significant ligamentous force rather than muscles preventing articular distraction. Similar EMG studies by Tuttle and Basmajian (1973) on apes (gorilla, chimpanzee, and orangutan) confirm these findings. Furthermore, Brantner and Basmajian (1975) found a clear-cut training effect in a remarkably increased endurance time of normal young human subjects; this was best explained by psychological adaptation.

Quite independent from us, Stener, Andersson, and Petersén of Göteborg, Sweden, were arriving at similar conclusions in regard to ligament sparing from a different type of experiment in cats and man. When Andersson and Stener (1959) greatly increased the tension in the medial ligament of the knee of the cat in specially designed experiments, no reflex muscular contractions appeared in the muscles of the thigh, as would have been expected if the usual hypothesis of "ligamento-muscular protective reflexes" were valid (Fig. 9.12). Furthermore, they showed convincingly that the absence of reflex motor effects was *not* due to

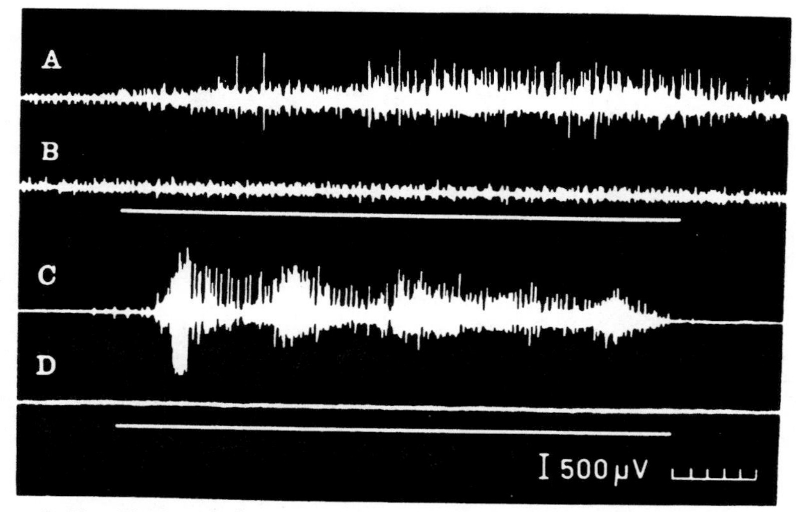

Figure 9.12. EMGs of (decerebrate) cat from: *A* and *B*, vastus medialis; *C* and *D*, semitendinosus. *A* and *C* show reflex responses to stimuli other than stretch, to compare with the lack of response in *B* and *D* when the tendons were stretched by transverse loading (duration shown by *straight horizontal white lines*). (From J.G. Andersson and B. Stener, © 1959, *Acta Physiologica Scandinavica*.)

absence of afferent discharges which were well detected from the articular nerves.

Petersén and Stener (1959) carried the above experiments forward to human subjects, again using the medial ligaments of the knee. Their results were a complete vindication of the conclusions made in the animal experiments described previously. In addition their work suggests that if injured ligaments are pulled until pain results, muscles *do* show reflex contraction; if the torn ligament is then anesthetized, they do not.

Following almost the same line of reasoning, deAndrade et al (1965) distended human knees with nonirritating plasma (which emphasized the pressure phenomenon, as opposed to pain). There was a definite and even marked inhibition of quadriceps contraction with a depression of motor unit activity. This is undoubtedly a reflex inhibition and helps to explain further the muscle weakness, atrophy, and deformity that follow knee injury and disease. Czipott and Herpai (1971) reported that atrophy of the quadriceps that develops in connection with meniscal and ligamentous injuries is of neurogenic origin with disuse atrophy only secondary. However, Carlsöö and Norstrand (1968) could find no qualitative difference in muscle coordination in most of the thigh muscles examined by EMG when they compared patients with ruptured cruciate ligaments and normal subjects.

Freeman and Wyke (1966, 1967) obtained a definite and chronic drop in reflex postural tonus in cats by cuting the sensory nerve supply of the knee joint capsule. The mechanoreceptors in the joint are involved in reflex muscular activity to maintain posture in quadrupeds; undoubtedly the same mechanisms occur in man as well. Surely all the above observations are of great importance in orthopedics and surgery of joint injury; they deserve wide attention.

CHAPTER 10

Miscellaneous Neural Influences

MUSCLE TONE AND RELAXATION

Most neurophysiologists now agree that electromyography shows conclusively the complete relaxation of normal human striated muscle at rest (Clemmesen, 1951; Basmajian, 1952; Ralston and Libet, 1953). In other words, by relaxing a muscle, a normal human being can abolish neuromuscular activity in it. This does not mean that there is no "tone" (or "tonus") in skeletal muscle, as some enthusiasts have claimed. It does mean, however, that the definition of "tone" should include both the passive stiffness of muscular (and fibrous) tissues and the active (although not continuous) contraction of muscle in response to the reaction of the nervous system to stimuli. Thus, at complete rest, a muscle has not lost its tone, even though there is no neuromuscular activity in it (Basmajian, 1957b).

In the clinical appreciation of tone, the more important of the above two elements is the reactivity of the nervous system. One can hardly palpate a normal limb without causing such a reaction. Therefore, the clinician soon learns to evaluate the level of "tone", and it may seem of little consequence to him that the muscle he is feeling is, in fact, capable of complete neuromuscular inactivity. In spite of this, he would be surprised to learn that an experienced subject can simulate hypotonia or even atonia of lower motor neuron disease and successfully deceive—if only for a brief period—the most astute physician.

During the course of various electromyographic studies on spastic patients and spastic rabbits, we were impressed with the relative ease with which most spastic muscles also can be completely (although only temporarily) relaxed (Fig. 10.1). Magoun and Rhines (1947) and Hoefer (1952) and others have also noted and commented on this, and it has been demonstrated by Kenney and Heaberlin (1962) in spastic children lying quietly, and by Holt (1966) also. The speed with which voluntary relaxation can occur is quite impressive: Miyashita et al (1972) found the mean values for relaxation reaction time in the biceps brachii of normal healthy adults to be the same as those for contraction reaction time.

In thousands of electromyograms on normal human muscles there has been complete and almost instantaneous relaxation when the subject has been ordered to relax. However, a small number of normal subjects do have great difficulty in relaxing quickly.

"Complete rest" requires some qualification. A normal person does not completely relax all his muscles at once. Reacting to multiple inter-

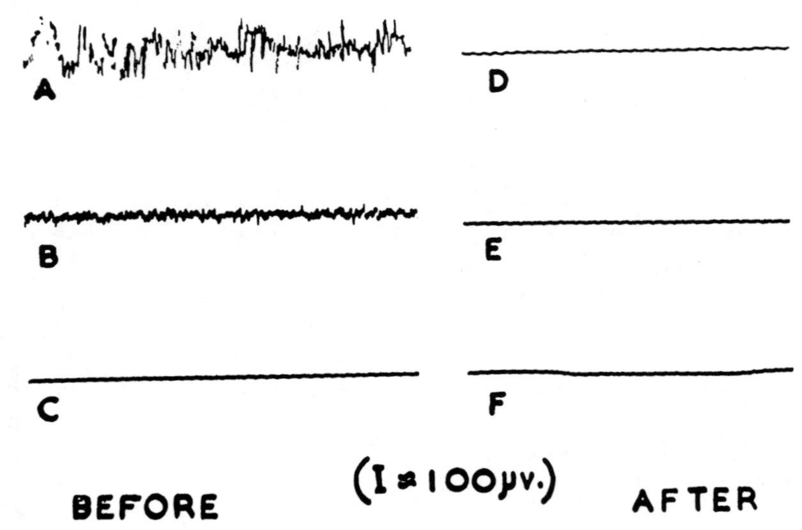

Figure 10.1. Tracings *A, B,* and *C* showing varying degrees of EMG activity at rest in the spastic quadriceps of three different patients with severe spasticity. Many subjects can be quieted down to the "EMG silence" of *C* without drugs. (Tracings *D, E,* and *F* show the results of intravenous chlorpromazine in the same three patients.) (From J.V. Basmajian and A. Szatmari, © 1955a, *A.M.A. Archives Neurology.*)

oceptive and exteroceptive stimuli, various groups of muscles show rising and falling amounts of activity. The idea that some individuals might show overall higher levels of tension than others is a familiar one. Iris Balshan Goldstein (1962) showed that 16 muscle groups measured at rest were related to a general muscular tension. Goldstein (1965) showed later that hysterics and certain other neurotics have very little increase in overall muscular activity over that of normal persons.

These findings do not necessarily imply that all muscle groups in a person tense and relax in unitary fashion. Fridlund et al (1980) found little evidence for a within-subject "general tension factor" in frontalis EMG biofeedback training. Moreover, Fridlund et al (1982) attempted a partial replication of Balshan (1962) and found that the "general tension" found by her occurred only when data from many subjects were taken together. In within-subject analysis, the muscle activity was largely unrelated. Thus, the "general tension factor" was taken by Fridlund et al (1982) to reflect agitation more than elevated tonic muscle activity.

Smith (1973) also disagreed with Goldstein's conclusions on the basis of widely separated electrodes on the forehead ("frontalis") which pick up EMG from most of the head and neck and hence reflect tension reasonably well. On the other hand, Alexander (1975) tended to agree with the earliest findings; his research revealed poor generalization between the forehead muscles and the whole body. Berger and Hadley (1975) found that the EMG activity of observers taken from their arms and lips rose or fell according to whether they watched others ("models") arm-wrestle or stutter. A group in Los Angeles (deVries et al, 1977) found that the flexors of the forearm are "probably best related to the factor of 'general resting muscular tension'." Schloon et al (1976) found that the newborns showed most of their resting activity in the region of the chin and neck.

During sleep, tonus of most head and neck muscles falls off but, according to Jacobson et al (1964) "trunk and limb muscles exhibit stable levels of tonic activity throughout the night." Tauber et al (1977) have recently confirmed this statement in an elegant study.

Leavitt and Beasley (1964), in accepting the absence of EMG potentials at rest as true relaxation, wondered whether such EMG silence could be maintained while the muscles being recorded were passively stretched. Quite surprisingly, they obtained absolute silence in both flexors and extensors of almost all subjects whose knee joints were flexed and extended passively, regardless of whether this was done slowly or rapidly. Although they actually were more interested in reciprocal inhibition of antagonists during active movements, Bierman and Ralston (1965) reported similar findings. Obviously, subjects could relax consciously during a state when normal myotatic reflexes would be expected to occur.

Muscular "tone" is a useful concept if we keep in mind that at rest a muscle relaxes rapidly and completely. This has now been common knowledge among neurophysiologists for more than a decade. If one keeps one's hands off a resting normal muscle, it shows no more neuromuscular activity than one with its nerve cut. In fact, it shows *less* because the fibers of denervated muscle engage in many fine random contractions invisible through the skin but detected by electromyography as "fibrillation potentials." The muscles in lower motor neuron denervation actually exhibit very fine invisible contractions while normal resting muscles exhibit complete neuromuscular silence. (These fibrillations are not to be confused with fasciculation, the coarse contractions of motor units visible through the skin and also often called fibrillations by older neurologists.)

Where did the false concept of continuous neuromuscular activity during rest originate? Chiefly it seems to be a widespread misinterpretation of Sherrington's *postural tonus*. There is no denying that any muscle that is helping to hold the subject upright shows various degrees of activity. On the other hand, our group, among a number of others, has shown that not all the muscles of the leg need to be active in the upright position. That is, the human upright position allows many of the limb muscles to relax completely. Yet these muscles will respond immediately to any change endangering the loss of balance.

Spasticity

In another set of experiments we found that the spastic limb muscles of most human beings with lesions of the central nervous system could be relaxed completely (Fig. 10.1). In the remaining spastic subjects the activity could be materially reduced but, under the experimental conditions, complete relaxation was not obtained—probably due to the presence of considerable environmental stimuli (e.g., a well-lighted room, the activity of the investigators, apprehensiveness). In all spastic subjects, very slight stimuli (even conversation) causes immediate electromyographic activity. Similarly, the muscles of spastic rabbits can be completely relaxed, as we noted above, but much more careful effort is required on the part of the handler. As with spastic human subjects, the slightest stimuli causes marked activity which takes some considerable soothing to abolish.

Shimazu et al (1962) confirmed our conclusion with the spastic and rigid muscles of patients with parkinsonism. They found no spike discharges in resting muscles, but there was a greatly exaggerated stretch reflex.

The findings described for spastic human beings and rabbits at rest therefore are not surprising. While the increased tone is simply an overactive reflex contraction, the limbs can be relaxed completely, albeit with greater "effort" than that required by normal subjects.

An incidental, perhaps inevitable, result of our electromyographic studies of human spasticity and the effects of chlorpromazine (Basmajian and Szatmari, 1955a and b) seems to have been the general adoption of electromyography for studying the effect of the newer relaxant drugs and other types of therapy (Brennan, 1959). In view of what has been already discussed, it hardly seems necessary to warn against a naive acceptance of "electromyographic" evidence if the only criterion used is the amount of the activity "at rest." Unfortunately some workers have already published such data. In our later studies of the effects on spasticity of new drugs, we used reflex responses to controlled constant stimuli (see, for example, Basmajian and Super, 1973).

EFFECTS OF CROSS EXERCISE

The hypothesis that there is a transfer of activity to the contralateral limb during prescribed exercise on one side has been frequently postulated, but now it is being seriously questioned. Probably it is invalid, except in very special circumstances. Any productive inquiry directed at documenting the concept of overflow must address the issue of stability. That is, if a contraction in one limb is sufficiently powerful, will it require the activation of musculature in the opposite limb to stabilize the trunk which is being subject to force disturbances.

Gregg et al (1957) found that overflow to the unexercised, contralateral muscles did not occur during simple nonresistive exercises or during isometric contractions of one biceps brachii. As the exercise stress increased, however, there was some "overflow" to the opposite triceps and, after even greater stress, to the biceps. Increasing fatigue played an important role in the "overflow" but was reversible. After a rest of 2 minutes, "overflow" would at first be absent. Moore (1975) found overflow activity to be between 10 and 20% of the maximal intensity of activity in the exercised limb.

Samilson and Morris (1964) confirm the finding that in normal man activity of one upper limb is not accompanied by activity in the contralateral resting limb. However, in spastic children, there is such a spread. On the other hand, Podivinsky (1964) of Bratislava, Czechoslovakia, finds that a slight motor irradiation occurs from the strong contraction of finger flexors to the related muscles of the opposite limb ("crossed motor irradiation"). This, perhaps, is related to the findings of Hellebrandt and her colleagues regarding indirect learning, i.e., the improvement of strength in one limb by exercising the opposite limb (Hellebrandt and Waterland, 1962a and b). Its practical significance in ordinary life is unknown and appears to have been exaggerated since the days of Scripture et al (1984). We have shown that at the finest levels of control in motor unit training, the role of cross-training is not significant (Basmajian and Simard, 1966).

Panin et al (1961) seem to have delivered a serious blow to the concept of "cross exercise." In their extensive study they found that the spread of activity was minimal to insignificant. Insignificant potentials of low amplitude and frequency appeared in all nonexercised muscles in a widespread distribution in all four limbs. They appeared most in areas required for postural stabilization of the subject's body. Even then the amount of activity was slight so as not to constitute exercise effect.

Our own studies on quadriceps and those of Sills and Olsen (1958) largely confirm the conclusions of Gregg and his colleagues. We found in our studies of spastic patients, however, that an exuberant overflow occurs to the opposite limb. Walshe (1923) and, more recently, Hopf et al (1974) and Soto et al (1974) have written about a similar phenomenon in hemiplegia. We must conclude that "cross education" is, at best, of dubious value in *normal* subjects.

ELECTROMYOGRAPHY OF THE FETUS AND NEWBORN

Until the 1950s no reliable information was available in regard to the earliest muscle potentials during fetal development. The characteristics of the earliest potentials are of great interest both in embryology and in the related fields of electromyography and neurology. Particularly important is the relationship of the time and innervation to the time of appearance of the earliest potentials. Our laboratory (then at Queen's University) developed a program employing special techniques with which to examine a series of living vertebrate fetuses.

The details of our methods and the detailed results have been published (Lewis and Basmajian, 1959; Ranney and Basmajian, 1960). A large number of rabbit fetuses and a limited number of goat fetuses were studied.

In the rabbits (which have a gestation period of 32 days) all fetuses of age 18 days or more showed electromyographic potentials. At 17 days, only some of the fetuses had EMG activity. This is the earliest fetal age in which we first observed visible movements. At 16 days, only one of 16 fetuses exhibited true muscle potentials. None of the younger fetuses did so.

The EMG potentials of these fetuses ranged from as low as 13 μV to as high as 250 μV. Their durations were too long to classify them with the short-duration small potentials which are known as fibrillation potentials and are diagnostic of denervation in postnatal life. Therefore, it has not been possible to state that the potential we found indicated a lack of innervation of the fetal muscles.

A few authors have described early visible movements in mammalian fetuses and embryos, but these have not been intrinsic or spontaneous; rather, they have been in response to prodding or electrical stimulation. Straus and Weddell (1940) found contractions of the forelimb in fetal rats stimulated electrically in the latter half of the 15th day (of a 21-day gestational period). This is in general agreement with the finding of Windle et al (1935). Furthermore, Windle (1940) states that the site of earliest activity is in the lower cervical region; it was nearby, in the shoulder region, that we picked up our earliest muscle potentials. Apparently, the onset of spontaneous activity requires a further degree of maturation beyond the stage at which muscles respond to an external stimulus.

Boëthius and Knutsson (1970) found in fetal chicks that an increase in membrane potentials of individual muscle fibers coincides with the transformation of the majority of the cells from myotubes to myocytes.

Marinacci (1959), in a report on the EMG of prematurely born infants, concluded that at the 6th month of intrauterine life (relatively much older than our rabbit fetuses) about 20% of the muscle fibers have still to be innervated. At the time of birth, 5% apparently have not yet received their nerve supply. At the end of the 4th postnatal month, practically all the muscle fibers have been innervated. The delayed innervation is largely in the lower extremities, especially in the intrinsic muscles of the feet.

Eng (1976) revealed the occurrence of spontaneous potentials—quite similar, if not exactly the same as, fibrillations—in almost half of 19 normal but premature infants. These disappeared in subsequent months. Normal newborns also had similar findings in a third of cases studied extensively and intensively.

Prior to innervation, primitive muscle fibers theoretically should possess an inherent tendency to spontaneous fibrillation and related electrical activity. Marinacci found that muscle fibers do fibrillate in premature infants corresponding to a stage in intrauterine life when, he believes, they might not all be innervated. Our own extended studies of fetuses in goats and rabbits (Ranney and Basmajian, 1960) failed to reveal any of the signs of spontaneous preinnervation potentials. Botelho and Steinberg (1965) confirmed this in the canine fetus. We can offer no real explanation for this discrepancy; perhaps the finding of spontaneous fibrillation potentials in premature infants by Marinacci and by Eng does not necessarily prove their existence in normal fetuses in utero. Finally, there may be a species difference. In fetal sheep, Änggård and Ottoson (1963) found that skeletal muscles could be made to contract at the 50th day, considerably before the time at which myelination (and therefore normal functioning) of the axons occurs.

During prenatal development the fetal sheep shows marked changes in neuromuscular function as studied by means of motor nerve stimulation. Änggård and Ottoson found marked changes from the early stage (50th day) to more advanced stages when spontaneous movements are common (100th day). Although not strictly an EMG study, this work bears upon our present concern. It clearly indicates that, in mammalian, embryos, speed of conduction of motor axons is related to the amount of progressive myelination. However, the excitable properties of the axons is independent. A key factor in neuromuscular function is the clear establishment of motor endings on the muscle fibers. In fetal sheep this occurs after about 50 days.

In normal newborn babies, Schulte and Schwenzel (1965) of Göttingen found irregular spontaneous bursts of normal motor unit activity in upper and lower limb muscles. These were sometimes reciprocal and sometimes strictly alternating between antagonistic muscles. In the upper limb the flexors were preferred. With hypertonic newborns, there was often constant tonic activity in certain muscle groups. This was widespread in the more severe cases.

CHAPTER 11

Posture

Electromyographic studies on postural muscles were begun by various investigators soon after the modern form of the technique was introduced near the end of World War II. Previously, it is true, rather crude attempts had been made to investigate the function of muscle groups by picking up and recording the electrical discharges that accompany the contraction of muscle. However, these attempts were thrown into the shade by the rise of modern electronics. Today, we must admit that, following the epoch-making stimulation studies of Duchenne in the 19th century, nothing very useful had been contributed to our knowledge of human posture by primarily electrical techniques until the past two or three decades.

It will be seen, therefore, that, narrowly defined, the subject to be discussed does not have a long history. Nonetheless, it cannot be divorced from an historical background because the object of inquiry, i.e., posture, has been the concern of anatomists, biologists, and orthopedic surgeons for many years. Therefore, this discussion will not be confined strictly to EMG; neither will it be confined to studies on posture in its extremely limited sense.

The definition of posture can be altered for the sake of argument according to how broad or how narrow one wishes to make it. In the narrowest sense, posture may be considered to be the upright, well-balanced stance of the human subject in a "normal" position. In this sense, the EMG of posture would deal with the maintenance of the erect subject's position against the force of gravity. The present account will, of necessity, emphasize this aspect of posture, but a broader, more generous, and more palatable definition would not exclude the multiplicity of normal (and abnormal) standing, sitting, and reclining positions that human beings assume in their constant battle against the force of gravity. In the final analysis, the intrinsic mechanisms of the body that counteract gravity make up the essence of the study of posture. One of these, the muscular mechanism, shall be our chief concern.

In a series of investigations concerning the organization of motor programming in man, Gelfand et al (1966) stressed the concept that the nervous system should be concerned not only with the muscles involved in the intended movement but also with those involved in maintaining posture so as to minimize the disturbance of balance. The existence of postural activity prior to and during a movement was established in man by Belenki et al (1967). Since then, this neural organization has been investigated in greater detail by Coulmance et al (1979) and Bouisset and Zattara (1981). The latter group found that anticipatory movements are present in the lower limbs, hips, and trunk before the onset of voluntary movements in the arms. They further noted that the postural changes induced by these anticipatory movements were specific to forthcoming intentional movement and therefore must be of central origin.

Posture of the entire body may be considered as a unit, and because such considerations are rather artificial, they often lead to a facile neglect of the posture in those parts of the body which do not intercept the main line of gravity for the trunk and lower limbs. As a result, posture in the upper limbs, both while hanging freely downwards and in various other positions, too often gets ignored. In still another direction, the posture of the mandible may also be ignored by the general anatomist, but it certainly is emphasized by the orthodontist, for whom it assumes a considerable practical importance.

The problems of static posture, then, revolve around the truism that the balance or equilibrium of the human body or its articulated parts depends on a fine neutralization of the forces of gravity by counter-forces. These counter-forces may be supplied most simply both internally and externally by a supporting horizontal surface or series of horizontal surfaces that are inert. The "easiest" posture in which a human being can achieve equilibrium with gravity is the recumbent one. We should not lose sight of the fact that this is our normal posture for the first year or so of our lives and for about half of our lives thereafter. When we lie down, we bring the center of gravity of the entire body, as well as any or all of its parts, closest to a supporting antigravitational surface.

Lundervold (1951) of Stockholm demonstrated by EMG that healthy persons who do not tense their muscles can sit comfortably and relax in many positions, and can even work in many different manners without pronounced increase in muscular activity. Nervous subjects do not relax completely in more than a few positions, and they cannot change their individual optimal working positions without a markedly increased exertion of muscle power.

Returning to the support of the erect body, we find that in the 19th century many laborious studies, some fruitful and others not, were performed to determine the line of gravity and the center of gravity of the whole human subject. The simplest estimate and one that is most easily appreciated is that of von Meyer (1868), who found that the weight center is situated at the level of the second sacral vertebra. (On the surface of the body this vertebra is at the level of the posterior superior iliac spines.) In the coronal plane the exact point lies 5 cm or less behind the line joining the hip joints and, of course, it is in the midline.

It will be seen that to maintain an equilibrium in the standing posture with the least expenditure of internal energy, a vertical line dropped from the center of gravity should fall downward through an inert supporting column of bones. This is the ideal, and it is surprising how closely the human supporting mechanism approaches it, if only intermittently.

The idealized normal erect posture is one in which the line of gravity drops in the midline between the following bilateral points: (1) the mastoid processes; (2) a point just in front of the shoulder joints; (3) the hip joints (or just behind); (4) a point just in front of the center of the knee joints; and (5) a point just in front of the ankle joints (Fig. 11.1). Muscular activity is called upon to approximate this posture or, if the body is pulled out of the line of gravity, to bring it back into line. Using a force platform, Stribley et al (1974) and Murray et al (1975) found a large area of stability over which weight is or can be shifted and maintained. The pressure fluctuates incessantly around a mean point. It

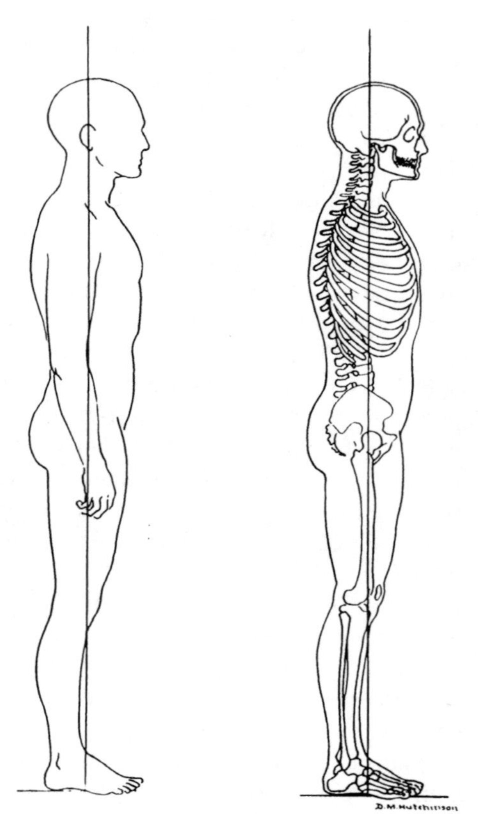

Figure 11.1. Line of gravity in total erect man (see text).

appears to become more stable in children between the ages of 4 and 8 years (Shambes, 1976).

Most people do not appreciate that, among mammals, man has the most economical of antigravity mechanisms once the upright posture is attained. The expenditure of muscular energy for what seems to the student of phylogeny to be a most awkward position is actually extremely economical. Most comparative anatomists certainly seem to be ignorant of this fact. A quadruped that is required to maintain the multiple joints of its limbs in a state of partial flexion by means of muscular activity demonstrates a much more wasteful antigravity machinery. An exception to this seems to be the elephant, whose limbs serve as static columns to maintain an enormous weight. On the other hand, the specialization of the elephant's weight-bearing limbs is so great that it cannot produce a true jump for even short distances. Relative to its size, the muscles of its limbs are quite puny compared with man's. The reason for this disproportion is that, unlike the elephant, man is constantly challenging gravity by his continued wide range of postures, and great power is required only to achieve them. Thus we find that man's so-called antigravity muscles are not so much to maintain normal standing and sitting postures as they are to produce the powerful movements necessary for the major changes from lying to sitting to standing. Therefore it is wrong to equate the antigravity muscles of man with those of the common domestic animals which stand on flexed joints.

In man, the column of bones that carries the weight to the ground constitutes a series of links. Ideally, these links should be so stacked that the line of gravity passes directly through the center of each joint between them. But even in man this ideal is only closely approached and is never completely reached—and then only momentarily. As Steindler (1955) showed, a completely passive equilibrium is impossible because the centers of gravity of the links and the movement centers of the joints between them cannot be all brought to coincide perfectly with the common line of gravity. In spite of this, Steindler and many others have greatly exaggerated the amount of effort required to maintain the upright posture. The fatigue of standing is emphatically not due to muscular fatigue and, generally, the muscular activity in standing is slight or moderate. Sometimes it is only intermittent. On the other hand, the posture of quadrupeds, which is maintained by muscles acting on a series of flexed joints, is highly dependent on continuous support by active muscular contraction. Of course, the same is true for the human being in any but the fully erect standing posture.

Dudley Morton (1952) anticipated much of what has been recently proved by EMG. Unfortunately, he incorrectly ascribed the fatigue of prolonged standing to a continuous activity of the muscles. This error is surprising because his calculations are otherwise quite valid. What he and others have ignored is that walking is usually less fatiguing than standing. Although extreme exertion can produce muscular fatigue, most fatigue in the lower limbs caused by standing is more intimately associated with the inadequacies of the venous and arterial circulation and with the direct pressures and tensions upon inert structures.

As Carlsöö (1961) has emphasized, certain muscle groups can be called "prime postural muscles." Among these are the neck muscles, sacrospinalis, hamstrings, and soleus. Carlsöö also points out that such postural muscles are among the most powerful. However, one must note that this is not an absolute case.

Carlsöö found that during stooping most persons failed to use a well balanced position, placing "too large a part of the load on the anterior part of the foot." They then powerfully engaged the soleus, gastrocnemius, flexor hallucis longus and peroneus. "Others placed too much of the load on the heels, so that the tibialis anterior and the peroneus muscles, which do not well tolerate continuous loading, were strongly activated."

Carlsöö considers the shifting from foot to foot in ordinary standing as a relief mechanism. "By assuming asymmetric working postures, and using the right and left leg alternately as the main support, the leg muscles are therefore periodically unloaded and relaxed." One should add that the relief to the inert structures is perhaps even more significant (p 239).

POSTURE OF LOWER LIMB

Leg

The function of the large muscles of the leg in relationship to posture has been studied by a number of investigators and quite early made the main subject of a book by Joseph (1960). Not infrequently, different conclusions resulted from different techniques. For example, Joseph and Nightingale (1952, 1956) concluded from their study with surface electrodes that the soleus of all persons and the gastrocnemius of many show well-marked activity when the subject is standing at ease; meanwhile, they claimed, the tibialis anterior is "silent." Their explanation, which agrees with the conclusion of Åkerblom (1948), is that the line of gravity is found to fall in front of the knee joint and ankle joint, necessitating activity in gastrocnemius. On the other hand, Basmajian and Bentzon (1954) showed with needle electrodes that there is actually a wide range of findings for each of these muscles, although, indeed, the posterior calf muscles are generally much more active than the tibialis anterior (Fig. 11.2*B*). Furthermore there is frequently a periodicity in the activity, and this is apparently related to an almost imperceptible forward-and-backward swaying of the body (Basmajian and Bentzon, 1954). Perio-

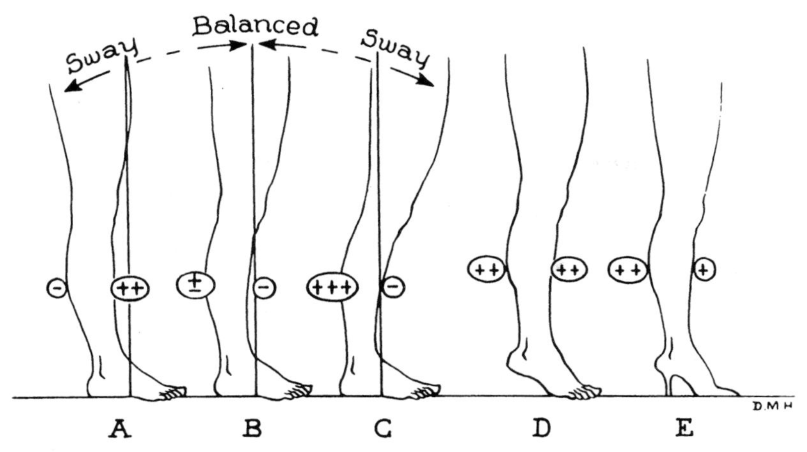

Figure 11.2. Diagram of EMG activity in anterior and posterior muscles of the leg under differing conditions.

dicity was first noted in this regard by Floyd and Silver (1950), and it has been commented on by Portnoy and Morin (1956) and others. Granit's (1960) statement that, in general, soleus is tonic while gastrocnemius is phasic may explain some of the discrepancies in the findings for the leg reported above. Carlsöö (1964) found activity regularly in soleus during quiet, symmetric standing but never in tibialis anterior. With a heavy load held either in front of the thighs or carried on the back, activity becomes pronounced in soleus, apparently to counteract the forward leaning of the body. Tibialis anterior remains completely inactive.

As would be expected, any deliberate leaning forwards or backwards of a standing subject produces compensatory activity in the muscles to prevent the occurrence of a complete imbalance (Fig. 11.2*A–C*). A very finely regulated mechanism is in control, and the slightest shift is reacted to through the nervous system by reflex postural adjustments; sometimes the motor responses are so fine that they can only be detected electromyographically.

Houtz and Fischer (1961) showed that these muscles respond to many influences, such as postural changes, the shifting of body weight, and the resisting of external forces applied to the upper part of the trunk. More recently, detailed studies have confirmed their findings (Okada, 1972, 1975).

In women who wear high heels, there appears to be a modification of the muscular response. Both Basmajian and Bentzon (1954) and Joseph and Nightingale (1956) found that the wearing of high heels increases the activity of the calf muscles in individual subjects, apparently due to a shifting forwards of the center of gravity (Fig. 11.2*E*). One would have expected that in women wearing high heels the well-known compensatory spinal lordosis would be a sufficient adjustment. Apparently it is not.

Thigh

The muscles of the thigh obey the same rules as those of the leg. By and large, the activity during normal, relaxed standing is usually slight. Indeed, it may be absent in most of the muscles for varying periods of time. The reports of Weddell et al (1944), Åkerblom (1948), Arienti (1948), Wheatley and Jahnke (1951), Floyd and Silver (1951), Joseph and Nightingale (1954), Basmajian and Bentzon (1954), Portnoy and Morin (1956), Oota (1956a), Joseph and Williams (1957), and Jonsson and Steen (1966) agree in principle. These overlapping and detailed studies include most of the large muscles of the gluteal region and thigh and no purpose would be served in recapitulating the details here. The main generalization to be extracted from all this is that the activity in these muscles is surprisingly slight during relaxed standing. The effect of swaying, mentioned above for leg muscles, appears in some hip muscles also (Jonsson and Synnerstad, 1967). Gluteus medius an tensor fasciae latae show such bursts but not the gluteus maximus, which remains silent.

When subjects carried a load either held in front of the thighs or strapped to the back, Carlsöö (1964) found that quadriceps remains completely inactive. Meanwhile, the ischiocrural muscles (hamstrings) show individual variations—from very active to completely inactive—apparently depending on the degree of flexion of the hip and on whether or not the line of gravity had been shifted anterior to the hip joint.

Foot

The postural function of the muscles of the foot in relation to the normal support and the abnormal flattening of the arches has always posed a question of some fundamental interest. Using needle electrodes, Basmajian and Bentzon (1954) showed that the intrinsic muscles are generally quiescent during normal standing but become extremely active when a person rises on tip-toes and during the take-off stage of walking (Fig. 11.3). This was confirmed in general by Sheffield et al (1956), but it appears that others have not paid sufficient attention to this fundamental consideration in the posture of the foot.

In the past, the peroneal and tibial muscles have often been considered to play an important role in maintaining the longitudinal arches of the foot in standing. This theory seems to have been discredited by the findings of Basmajian and Bentzon and by the indirect contributory evidence of other investigators. During standing these muscles of the leg are generally quiescent. Furthermore, they remain inactive even when a subject suddenly lowers himself to a normal standing position from an elevated seated position. However, if in the standing posture the foot is

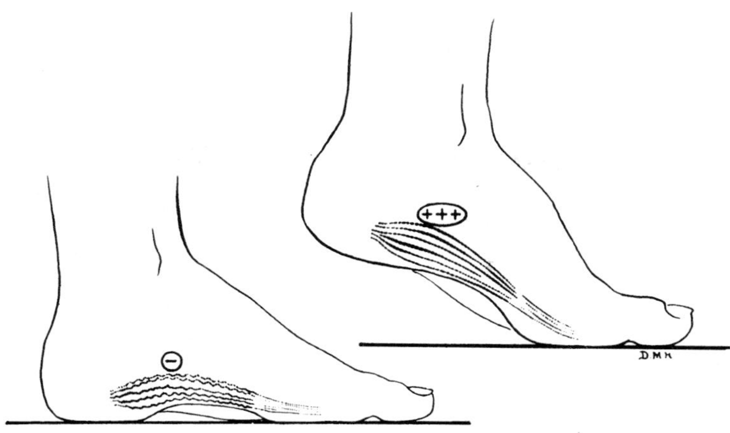

Figure 11.3. Diagram of EMG activity in intrinsic foot muscles (see text).

obviously inverted by tibialis anterior, activity is quite intense. During locomotion, peroneal and tibial muscles show marked activity (Fig. 11.2).

Apparently, the first line of defense against flat feet is a ligamentous one (Basmajian and Stecko, 1963). This is considered further on p 349. But the added stresses of walking require special mechanisms (Basmajian, 1955a and 1960). Independently, a similar view has been advanced by Dudley Morton (1952) who, on the basis of various calculations, predicted essentially the same thing. He showed that static strains upon the ligaments of the arch to sustain its elevated position are low in intensity and fall well within the capabilities of the ligaments. His calculations showed that only acute, heavy but transient forces (such as in the takeoff phase of walking) required the dynamic action of muscle. Meanwhile, further confirmation was provided by Hicks (1951 and 1954), who has demonstrated the importance of the plantar aponeurosis.

Hip and Knee

The hip and knee each have at least one muscle with a special postural function that can be demonstrated electromyographically. Experiments (Basmajian, 1958b) have shown that iliopsoas remains constantly active in the erect posture (in contrast to the large thigh muscles). It would appear that iliopsoas functions as a vital ligament to prevent hyperextension of the hip joint while standing (see p 313). (The supraspinatus has a similar activity at the shoulder joint; see below.)

At the knee, Barnett and Richardson (1953) have shown a constant activity in the popliteus in the crouching or "knee-bent" posture. This apparently is related to a stabilizing postural function to help the posterior cruciate ligament prevent an anterior dislocation of the femur (Fig.

11.4). There is no similar popliteal activity in the erect posture when dislocation is not threatening the joint.

POSTURE OF TRUNK, NECK, AND HEAD

Spine

While standing erect, most human subjects require very slight activity and sometimes some intermittent reflex activity of the intrinsic muscles of the back according to Allen (1948), Floyd and Silver (1951, 1955), Portnoy and Morin (1956), and Joseph (1960). These authors showed that during forward flexion there is marked activity until flexion is extreme, at which time the ligamentous structures assume the load and the muscles become silent (Fig. 11.5). Floyd and Silver (1955) proved (with both surface and needle electrodes) that in the extreme-flexed position of the back, the erector spinae remained relaxed in the initial stages of heavy weightlifting. This observation appears to confirm strongly the dangers of the vertebral ligaments and joints of lifting "with the back" rather than with the muscles of the lower limb (see p 354).

Asmussen (1960) first concluded from an EMG study that continuous activity of the back muscles during standing is the rule because "the line

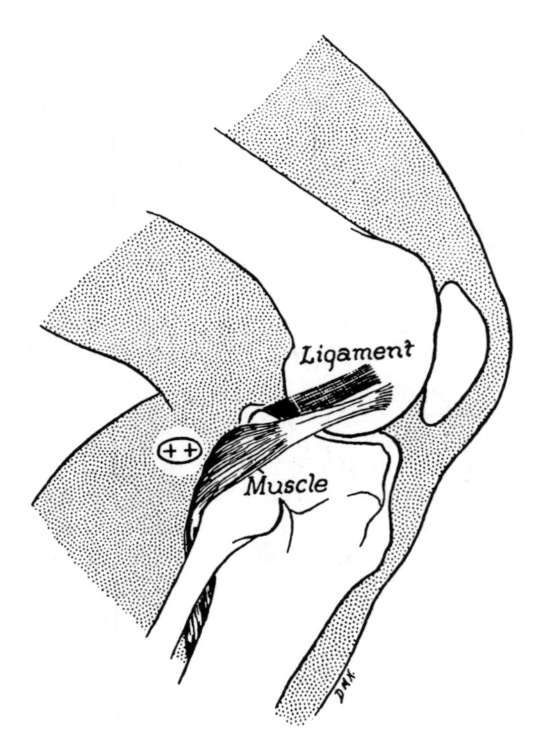

Figure 11.4. Popliteus shows EMG activity in the knee-bent stance.

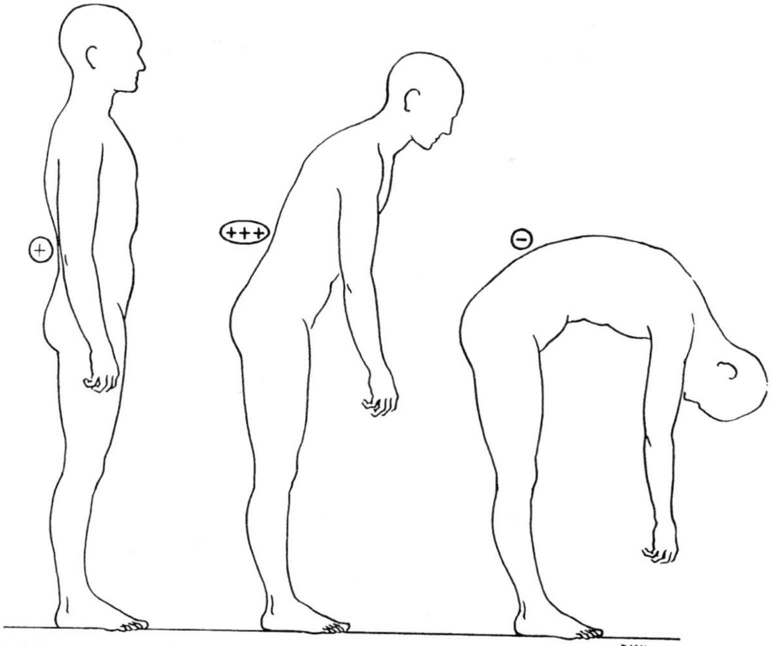

Figure 11.5. Diagram of activity in erector spinae during forward bending (see text).

of gravity passes in front of the spinal column." The last part of this conclusion is contrary to previous opinions and requires confirming. Later, Asmussen and Klausen (1962) modified the earlier extreme view to conclude that "the force of gravity is counteracted by one set of muscles only, most often the back muscles, but in 20 to 25% of the cases the abdominal muscles. The line of gravity passes very close to the axis of movement of vertebra L4 and does not intersect with the curves of the spine as often postulated." Carlsöö (1964) regularly found activity in sacrospinalis in the symmetric, rest position.

Later, Klausen (1965) investigated that effect of changes in the curve of the spine, the line of gravity in relation to vertebra L4 and ankle joints, and the activity of the muscles of the trunk. He concluded that the short, deep intrinsic muscles of the back must play an important role in stabilizing the individual intervertebral joints. The studies of Donisch and Basmajian (1972) and Wolf and Basmajian (1980) support this view (see p 361). The long intrinsic muscles and the abdominal muscles stabilize the spine as a whole. An increased pull of gravity always is counteracted by increased activity in one set of muscles only, i.e., either in the back or anterior abdominal wall.

Placing a load high on the back automatically causes the trunk to lean slightly forward. The increased pull of gravity is counteracted by an increased activity in the lower back muscles. A load placed low on the back reduces the activity of the back mucles. This last finding was duplicated by the independent work of Carlsöö (1964). However, Carlsöö found increased activity in sacrospinalis with a load held in front of the thighs. Thus, the position of the load—either back or front—either aids the muscles or reflexly calls upon their activity to prevent forward imbalance. Nachemson (1966) claims the vertebral part of psoas major helps to maintain the posture of the lumbar vertebrae.

Abdomen

During relaxed standing, only slight activity has been recorded in the abdominal muscles by Floyd and Silver (1950), Campbell and Green (1955), and Ono (1958). The first investigators appear to have proved that the activity is greatest in the internal oblique to provide the protection that the muscles afford to the inguinal canal in the upright posture (see p 395). Carlsöö (1964) reported that carrying a load on the back always increases activity in rectus abdominis, but carrying one in front of the thighs left rectus abdominis completely silent.

Thorax

Jones et al (1953) were the first to suggest that the intercostal muscles play a part in posture or, at least, in the maintenance of certain flexed postures and adjustments of position. Certainly worthy of further study is the interesting proposal advanced by Jones and Pauly (1957) that the intercostals have as their chief function the maintenance of a proper distance between the ribs while the rib cage is actively elevated by the neck muscles during inspiration (see p 409). Credence in their theory is strengthened by the independent work of Campbell (1955), who has recorded activity in the scalenes and sternomastoid muscles during quiet respiration, and by that of Koepke et al (1958). Surprisingly, these seemingly fundamental problems have not been attacked with any vigor by other investigators (see also p 425).

Mandible

Moyers (1950), Carlsöö (1952), and MacDougall and Andrew (1953) claimed that the muscles of the jaw which act against gravity in maintainng posture are all those that raise the mandible. Latif (1957), however, while working in Basmajian's laboratory, demonstrated that temporalis alone is responsible for keeping the teeth in apposition and is constantly active in the upright posture. During the resting mouth-closed position, there is strikingly greater activity in the posterior fibers, which run almost horizontally backwards, than there is in the anterior fibers, which run vertically. Temporalis is discussed further in Chapter 20.

POSTURE OF UPPER LIMB

Posture in the upper limb is chiefly a matter of maintaining the integrity of the series of joints in the hanging position. However, in the recumbent posture, if the upper limb is raised to a vertical position, many of the factors which normally govern the posture of the erect whole body come into play for the first time in the upper limb. In the standing position, on the other hand, the hanging limb poses different problems in posture because the force of gravity produces tensions rather than compressions. These tensions are easily carried by the bones which are rigid, but the logical question is often asked: What prevents dislocation of the series of joints? As we have seen already, the most frequent answer—that it is muscular action—is not correct.

Shoulder

Not surprisingly, low grade postural activity occurs in the upper fibers of trapezius in supporting the shoulder girdle. Basmajian often made this observation incidental to other studies in normal subjects over many years (unpublished studies). Minimal postural activity in the serratus anterior has also been described by Catton and Gray (1951) and confirmed by many scattered observations during clinical EMG.

At the shoulder joint, the main muscular activity in resisting downward dislocation occurs in supraspinatus (and to a slight extent in the posterior, horizontal-running fibers of deltoid) (Basmajian and Bazant, 1959). The bulk of the deltoid, and the biceps and triceps show no activity in spite of their vertical direction. Surprisingly, this is true even when heavy weights are suspended from the arm. The function of supraspinatus is apparently associated with what Basmajian (1961) described as a *locking mechanism* dependent upon the slope of the glenoid fossa. The horizontal pull of the muscle, along with an extreme tightening of the superior part of the capsule only when the arm hangs vertically, prevents downward subluxation of the humeral head (see p 239).

Elbow

At the elbow joint, without an added load there is no activity in the muscles, suggesting that the ligaments carry the weight (Basmajian and Latif, 1957; Basmajian and Travill, 1961). The addition of a small or moderate load does not produce any activity in biceps, triceps, brachioradialis, or pronator teres. This is also true for the main elbow-crossing muscles when a human or gorilla subject is hanging from a trapeze (Elkus and Basmajian, 1973; Tuttle and Basmajian, 1973; Tuttle, Velte, and Basmajian, 1983) (see p 242). Perhaps it is superfluous to note that in flexed positions of the elbow the maintenance of the flexed posture is shared by the brachialis and biceps muscles. However, brachioradialis shows little if any activity in maintaining flexed postures, even against added loads, because it is theoretically an example of a "shunt" muscle, as first postulated by MacConaill (1946).

Wrist

At the wrist and hand, a minimum of activity is required to overcome the ordinary force of gravity. Repeated EMG investigations have revealed a general silence of the forearm and hand while hanging at rest. The "postural" activity of the wrist flexors and extensors which accompanies the making of a fist or the grasping of a handle is perhaps more properly a "synergistic" function and is possibly outside the limits of our subject. In this regard, the only significant postural EMG of this region was reported by Dempster and Finerty (1947). They loaded the hand of the horizontally held forearm and recorded the activity in various muscles that cross the wrist. When a muscle was in a superior position and was working to support the load against gravity, its activity was three to four times as great as when it was below and maximally aided by gravity, i.e., when it was serving a stabilizing or synergistic function only.

RECUMBENT POSTURE

In conclusion, let us return to the recumbent posture. Here, in this pleasantest of postures, the force of gravity is counteracted by mechanisms that are entirely passive. Repeated EMG studies by many investigators have demonstrated beyond the shadow of a doubt that resting muscles exhibit no neuromuscular activity (Basmajian, 1955a). It is time, then, that all anatomists and physiologists firmly alter their teaching. Contrary to widespread belief, there is no random activity of motor units in a resting muscle to provide what is often hazily called *muscular tone* (see section on tone in Chapter 10).

SITTING POSTURE

In view of the confusing welter of theories regarding "ideal" sitting postures, the paucity of scientific EMG experiments is surprising. Åkerblom (1948), who first published a monograph on the subject in Sweden, remains almost alone in the field (see also Åkerblom, 1969). His investigations showed good relaxation of the spinal muscles when good support is given to the lumbar and thoracic regions, or when good lumbar support is given, and even when the subject is allowed to slump or sink into a ventriflexed position. Knutsson et al (1966) duplicated Åkerblom's studies. Rosemayer (1971), working in Munich and in South Africa with Professor Lewer Allen, has extended this type of work to motorcar driving postures; related work was done in Naples under Carlo Serra (1968), with only preliminary results reported. The work on back muscles of seated subjects by Donisch and Basmajian (1972) is reported on p 361.

CHAPTER 12

Upper Limb

Many scattered EMG studies of the upper limb followed upon the landmark work in California of Inman et al reported in 1944. This chapter will bring together most of the available information in an organized form and deal with the actions of groups of muscles topographically. Especially in the 1970s we saw a rapid growth of studies applying electromyography in gymnastics (e.g., Landa, 1974), in athletics (e.g., Kemei et al, 1971; Okamoto and Kumamoto, 1971; Toyoshima et al, 1971; Yamashita et al, 1971, 1972; Yoshi et al, 1974; Matsushita et al, 1974; Yoshizawa et al, 1976; Okamoto, 1976; Okamoto et al, 1976; Goto et al, 1976; Tokuyama et al, 1976, 1977; Kazai et al, 1976), and in ergonomics (e.g., Jonsson and Jonsson, 1975, 1976; Jonsson and Hagberg, 1974; Carlsöö and Mayr, 1974; Avon and Schmitt, 1975; Hagberg, 1981; Bjelle et al, 1981). The wrist, hand, and fingers will be dealt with in the next chapter.

A series of papers on EMG studies of the forelimbs of various types of nonhuman primates have also been published with the prime purpose of unravelling the evolution of hominoid and human locomotion on the ground (Tuttle and Basmajian, 1974, 1975, 1977, 1978; Stern et al, 1977, 1980 a and b; Susman and Stern, 1979, 1980; Tuttle et al, 1983).

TRAPEZIUS

Following the classical study of Inman et al, whose chief concern was with the dynamics of the shoulder (see below), Yamshon and Bierman (1948) and Wiedenbauer and Mortensen (1952) studied various parts of the trapezius during voluntary movements in a series of normal adults. The trapezius was found to be considerably active during elevation or retraction of the shoulder and during flexion or abduction of the upper extremity through a range of 180°. During scapular elevation the greatest activity was recorded, as would be expected, from the upper parts of the muscle; during retraction, from the middle and lower parts; and during flexion, from the lower half. The greatest activity in trapezius appears during abduction of the limb and chiefly in the lower two-thirds of the muscle. These findings were confirmation of the California studies and have been confirmed in detail by Thom (1965) of Heidelberg, and in general in Basmajian's laboratory. No one can find confirmation for Duchenne's belief that trapezius is a respiratory muscle, and this idea should be dropped.

In his study of static loading, Bearn (1961b) discovered that the upper fibers of trapezius, contrary to the universal teaching, "play no active part in the support of the shoulder girdle in the relaxed upright posture." This was confirmed by Fernandez-Ballesteros et al (1964). Some of Bearn's subjects initially showed a low level of activity in this part of the trapezius; but upon their being instructed to relax, the activity stopped entirely. As he notes, this observation is surprising. Indeed, the upper part of the muscle shows through the skin in thin people and appears to be under some tension even when no weight is borne by the limb.

When a load of 10 lbs (about 4.5 kg) was held in the hand, fully three-quarters of Bearn's subjects were able to relax the trapezius either immediately or within 2 minutes. The remainder showed very little activity compared with the result of slight shrugging movements. With a 25-lb (11.4 kg) load, a third of the subjects could support the weight without EMG activity in trapezius. Bearn cautioned against the interpretation that this is a desirable way to carry loads: moreover, he ascribed various abnormalities to the habitual depression of the clavicle.

Trapezius muscle is not just a postural or supporting muscle. Its role in the adjustments of the scapula during elevation of the upper limb is vital. In fact its activity is essential both in raising the arm (see Scapular Rotation, p 267) and in preventing downward dislocation of the humerus (p 273).

SUBCLAVIUS

Only one specific study has been done with needle electrodes in this small and relatively hidden muscle. Reis et al (1979) placed needle electrodes in the subclavius of 12 normal persons and recorded the EMG during 31 different movements and postures. They concluded that the muscle's main function is on stability of the sternoclavicular joint, supplementing the ligaments of the joint.

PECTORAL MUSCLES

Inman et al (1944) were the first to examine pectoralis major electromyographically. In abduction of the arm, no part of this muscle is active. In forward flexion, the clavicular head is the active part, reaching its maximum activity at 115° of flexion; the sternocostal head remains inactive. With the exception of the studies by Ravaglia (1958) in Italy, Scheving and Pauly (1959) in Chicago, and Okamoto et al (1967) of Kyoto, the pectoral muscles seem to have been otherwise ignored until de Sousa et al (1969) and Jonsson et al (1972) published their works. Ravaglia had been concerned with their alleged accessory functions in respiration. He demonstrated the presence of moderate activity in them during forced inspiration but none with quiet breathing. Scheving and Pauly confirmed many of the findings of Inman et al and further confirmed the standard teaching regarding the important activity of the sternocostal head in adduction. However, they found that medial rotation must be against resistance for the pectoralis major to be called into action. With this, de Sousa et al (1969) disagreed emphatically, pointing out that the clavicular head is active almost always during either free or resisted medial rotations of the humerus. But they agreed that the sternocostal head remained relaxed, except when adduction occurred; this is true regardless of the starting position of the limb (even behind the back) according to Jonsson et al (1972). Shevlin et al (1969) tested and confirmed standard textbook views of the functional differences between the two parts of the pectoralis major during isometric effort in different positions of the arm.

Following removal of the pectoral muscles during radical mastectomy, Flint et al (1970) found a surprising lack of functional disturbance in ordinary activities. Surrounding muscles compensate, except for stabilization of the shoulder joint in movements where the humerus is forced upwards and laterally.

SERRATUS ANTERIOR

Though serratus anterior must be considered further under "Scapular Rotation," we should note the work of Catton and Gray (1951), who proved beyond question that this is not an accessory respiratory muscle. Their EMGs failed to demonstrate any activity in serratus anterior during voluntary deep breathing, during breathing that was obstructed by forcing the subject to breathe through a narrow tube, and even during coughing. The final blow to the concept of this being an accessory respiratory muscle was struck by Jefferson et al (1960), who demonstrated that during respiration action potentials were generally absent in the nerve to serratus anterior.

SCAPULAR ROTATION

Just as scapular rotation is a distinct and important function, well-known since Duchenne, so the muscles which produce the movement are a distinct functional group. The studies of Inman et al first drew special attention to them. As they showed, the upper part of the trapezius, the levator scapulae, and the upper digits of serratus anterior constitute a unit whose main activities are in concert: they passively support the scapula (slight continuous activity), elevate it (increasing activity) and act as the upper component of a force couple that rotates the scapula (Fig. 12.1).

The lower part of trapezius and lower half or more of the serratus anterior constitute the lower component of the scapular rotatory force couple; they were found to act with increasing vigor throughout elevation of the arm. The lower part of trapezius is the more active component of the lower force couple during abduction; but in flexion it is less active than serratus anterior, apparently because the scapula must be pulled forward during flexion.

The middle fibers of trapezius are most active in abduction especially

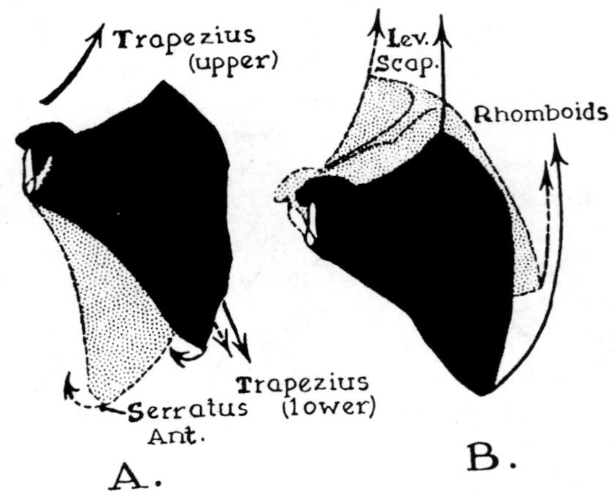

Figure 12.1. Scapular rotation and force couples. (*A*) Glenoid "up". (*B*) Glenoid "down." (From J.V. Basmajian, © 1955, Williams & Wilkins, Baltimore.)

as the arm reaches the horizontal plane (90°). In forward flexion, the activity of the middle fibers of trapezius decreases during the early range but builds up toward the end. In general, then, the middle trapezius serves to fix the scapula but must relax to allow the scapula to slide forward during the early part of flexion.

The rhomboid muscles (major and minor) imitate the middle trapezius, being most active in abduction and least during early flexion. During most movements of the head and trunk, they are relatively inactive (de Freitas et al, 1979; de Freitas, 1980), but in free movements of the arm (shoulder abduction, adduction, flexion, extension and hyperextension), de Freitas and Vitti (1981a and b) found middle trapezius and rhomboideus major consistently active at the moderate to marked levels.

MOVEMENTS AND MUSCLES OF THE GLENOHUMERAL JOINT

The chief muscles that act upon the shoulder (glenohumeral) joint are the deltoid, the pectorales (discussed above), the latissimus dorsi, teres major, and the four rotator cuff muscles—subscapularis, supraspinatus, infraspinatus and teres minor. The considerable interest these muscles have aroused among electromyographers is not surprising, for the movements and protection of the shoulder joint are of paramount importance. Both heads of biceps brachii play a modest role in shoulder function. Below are the composite results derived from the work of various authors.

Abduction

The activity in the deltoid increases progressively and becomes greatest between 90° and 180° of elevation (Fig. 12.2). The activity of supra-

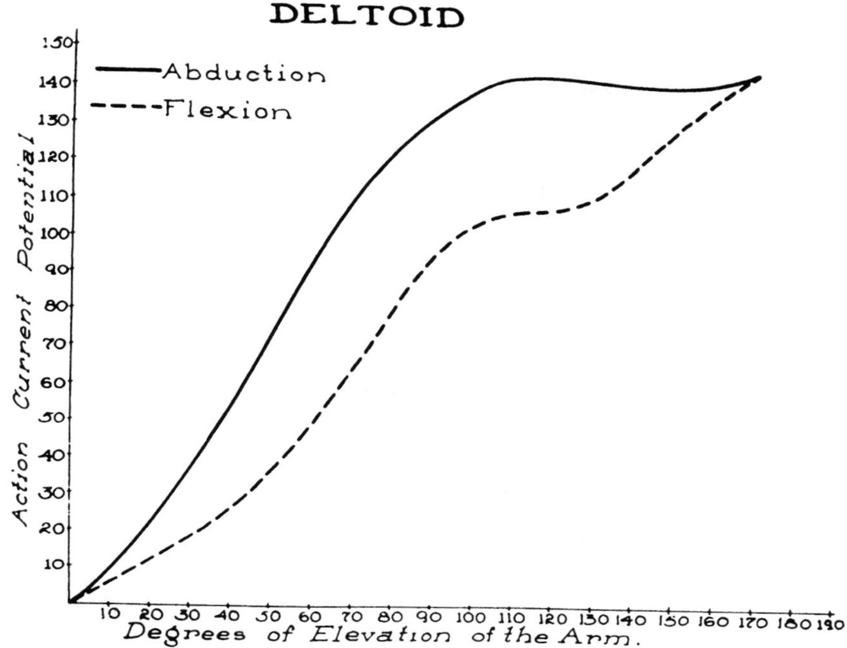

Figure 12.2. Relation of EMG activity in deltoid (in arbitrary units) to degree of elevation of arm. The "action current potential" was in reality the smoothened voltage of the EMG signal; for a more complete explanation, see p 95. (From V.T. Inman et al, © *Journal of Bone and Joint Surgery*.)

spinatus increases progressively, too (Inman et al). Thus, it is not simply an initiator of abduction, as was formerly taught. Many clinical studies by Basmajian's group have conclusively substantiated these statements, as have those of Wertheimer and Ferraz (1958) and Comtet and Auffray (1970). Quite surprising but rational is the discovery by Van Linge and Mulder (1963) that complete experimental paralysis of supraspinatus in man simply reduces the force of abduction and power of endurance. They concluded that in abduction, supraspinatus plays only a quantitative, not a specialized role. No part of pectoralis major is active during abduction. The role of biceps brachii in abduction seems to be confined to a contribution in maintaining this position while the arm is laterally rotated and the forearm is supine. When the arm is medially rotated and the forearm prone, biceps does not contribute to abduction (Basmajian and Latif, 1957).

Flexion

The clavicular head of pectoralis major, along with the anterior fibers of deltoid, are the chief flexors (Shevlin et al, 1969; Okamoto et al,

1967). Okamoto (1968) showed that there are substantial variations in local EMG activity, depending on slight changes in resistances applied in varying ways. Both heads of biceps brachii are active in flexion of the shoulder joint, the long head being the more active (Basmajian and Latif, 1957).

Depressors of Humerus

Subscapularis, infraspinatus, and teres minor were shown by Inman et al to form a functional group which acts as the second or inferior group of the force couple during abduction of the humerus. They act continuously during both abduction and flexion. In abduction, activity in infraspinatus and teres minor rises linearly while activity in subscapularis reaches a peak or plateau beyond the 90° angle and then falls off.

Adduction

Pectoralis major and latissimus dorsi produce adduction. The posterior fibers of deltoid are also very active, perhaps to resist the medial rotation that the main adductors would produce if unresisted (Scheving and Pauly, 1959). While the teres major acts only when there is resistance to the movement (Broome and Basmajian, 1971a), Jonsson et al (1972) report that when the arm is behind the coronal plane, adduction recruits teres major activity without added load.

Teres Major and Latissimus Dorsi

Teres major never exhibits activity during motion "but plays a peculiar role in that it only comes into action when it is necessary to maintain a static position, it reaches its maximum activity at about 90 degrees." These statements of Inman and his colleagues have been neither challenged nor confirmed until recently. Kamon (1966) finds that teres major is very active in movements of the free arm during gymnastics on the pommeled horse. On the other hand, even during the vigorous activity of shot-putting, the teres major remains relatively quiet (Hermann, 1962). De Sousa et al (1969) reported that teres major was indeed active during active motion of the arm.

The Broome and Basmajian study (1971a) was undertaken to clarify the disagreements because of the obvious importance of this large muscle. The closely related latissimus dorsi was studied simultaneously as a reference in 10 normal adult subjects. Bipolar wire (75 μm) electrodes were inserted into the mass of the teres major 3 cm lateral and 3 cm superior to the inferior angle of the scapula, and into the latissimus dorsi muscle 4 cm below the inferior angle of the scapula. Subjects were directed through a series of motions of the shoulder and arm: (1) standing at rest with the arms hanging at the side; (2) medial rotation; (3) lateral rotation; (4) abduction; (5) adduction; (6) flexion; and (7) extension. These were first done without resistance and then against a resistance force that either allowed or prevented completion of the arc of the motion.

The teres major had no electrical activity in motions without resistance. But against active resistance, it consistently showed electrical activity during medial rotation, adduction, and extension in both the static and the dynamic exercises. The latissimus dorsi had similar activity during both static tension and resisted motion; without resistance, in five of the seven acceptable subjects latissimus dorsi had activity during medial rotation, adduction, and extension.

The more precise and sensitive techniques used in the study by Broome and Basmajian appear to resolve the controversy over the functions of teres major. The key to the solution seems to be whether a movement or an attempted movement is always active. If added resistance is lacking, free movement of the shoulder joint in all its directions does not recruit the teres major, although it usually recruits its close relation, latissimus dorsi; when the arm is behind the back, teres major activity does appear "without resistance" during (hyper)extension and in adduction (Jonsson et al, 1972).

As far as latissimus dorsi alone is concerned, there is little doubt of the powerful extensor activity of this muscle, and no one has found reason for doubting it during EMG studies. This is not the case for medial rotation. While an early study by Scheving and Pauly (1959) suggested that the muscle was the essential medial rotator, later investigations by de Sousa et al (1969) suggested that this view was fallacious and came from the concomitant adductor effect. Thus, while de Sousa et al clearly deny any rotatory function for latissimus dorsi, Broome and Basmajian confirmed earlier views.

Jonsson et al (1972) also confirmed the common opinion that latissimus dorsi and the sternocostal part of pectoralis major depress the humerus.

The only additional study of latissimus dorsi is that of Ito et al (1976) of Tokyo. With surface electrodes over six parts of the muscle in 10 healthy men they found that the upper and lower parts have functional differences during varying movements. The upper part was markedly active during hyperextension; the lower, during lateral tilting of the pelvis.

Deltoid

Scheving and Pauly (1959) found that the three parts of deltoid are active in all movements of the arm, as did Yamshon and Bierman in an earlier and less sophisticated study (1949). In flexion and medial rotation, the anterior part is more active than the posterior; in extension and lateral rotation, the posterior is the more active; and in abduction the middle part is the most active. Scheving and Pauly believed that, although one part of deltoid may act as the prime mover, the other parts contract to stabilize the joint in the glenoid cavity. They further recommended the inclusion of deltoid with the four rotator cuff muscles as stabilizers of the joint. However, later work (reported below) does not endorse their recommendations.

Wertheimer and Ferraz (1958) and Shevlin et al (1969) found that the anterior part of deltoid shows its principle action in forward flexion of the shoulder joint but also participates in elevation and (slightly) in abduction of the arm. Deltoid does not participate in medial rotation. The intermediate portion acts strongly in abduction and elevation of the arm and also participates slightly in flexion and extension. The posterior part has its principal action in extension, but the action is inconstant and slight in abduction and elevation of the arm. Its participation in lateral rotation is minimal, being practically absent.

If the humerus is flexed into the horizontal plane (in reference to the anatomical position), it may then be swung in backward arcs called *horizontal extension-abduction* and may be produced by the appropriate muscles of abduction and extension of the shoulder joint (principally, middle and posterior deltoid, infraspinatus, and teres minor). The opposite movement—*horizontal flexion-adduction*—is produced mostly by the anterior deltoid and pectoralis major (Rasch and Burke, 1978). These movements are of special interest to sports kinesiologists.

In sports, Hermann (1962), in an EMG study of shot-putting, found that ideally the anterior deltoid is active during the entire maneuver. The greatest contracting occurs during the thrust phase of the shoulder and arm but before the shot is released from the hand. The middle fibers come into play—and into very strong action—after the thrust is initiated. The posterior fibers play no important role until the moment just as the shot leaves the hand. Other applications of EMG analysis of the shoulder region in sports situations were reported by Kamon (1966), Kamon and Gormley (1968), Vitti and Bankoff (1979), and Hinson (1969).

Biceps Brachii

The finding of slight action of the biceps during flexion of the shoulder joint and *nil* activity during abduction with the arm medially rotated (Basmajian and Latif, 1957) confirmed the accepted teaching. More recently, Furlani (1976) agreed, but he added some useful details. During flexion *with resistance* (and with the elbow straight), both heads of biceps are always active. During abduction, resistance again recruits activity in both heads; the opposite movement, adduction against resistance, recruits activity in half of the short heads and nothing in the long head. The long head plays no role in rotatory movements of the shoulder, but the short head occasionally acts during medial rotation.

PREVENTION OF DOWNWARD DISLOCATION OF THE HUMERUS

The part played by various muscles during movements of the shoulder joint has been the subject of investigation and argument for more than a century. Even though most of the important questions on movements in the shoulder area have been answered, little reliable information had been available regarding the role of such muscles in *maintaining joint stability*. In particular, the mechanism preventing downward dislocation or subluxation of the shoulder joint has not been adequately explained—indeed, it has been largely ignored. Cotton in 1921 and Fairbank in 1948, considering the matter in connection with fractures of the humeral neck, both assigned the greatest importance to the vertically running scapulohumeral muscles, for example, deltoid and biceps. During some incidental studies of the region, Basmajian's group were surprised to find the exact opposite. Therefore, to clarify the part played by the muscles and the capsule of the joint in preventing downward dislocation of the vertical or adducted humerus, the following two types of systematic investigation were performed (Basmajian and Bazant, 1959): (1) an EMG study of the deltoid, supraspinatus, infraspinatus, biceps, and triceps of a series of young men, using multiple concentric-needle electrodes; and (2) a study of gross dissections of the shoulder joint.

The findings did not support the hypothesis advanced by Cotton and endorsed by Fairbank. In fact, they completely disagreed. It was apparent from the electromyographic results that the deltoid (the muscle one would expect to be especially active in preventing downward dislocation of the humerus) is inactive even with heavy pulls. Other muscles running vertically from the scapula to the humerus, particularly the biceps and the long head of triceps, were conspicuously inactive as well. Therefore, there now seems to be little, if any, reason to doubt that downward dislocation is prevented by the superior part of the capsule, along with the supraspinatus (and to a lesser extent the posterior fibers of the deltoid). Strangely enough, these structures run in a horizontal and not in a vertical direction (Fig. 12.3). Bearn (1961) confirmed the findings (in considerable detail) quite independently and using loads of 25 lbs.

The mechanism by which these horizontally placed structures succeed is dependent upon a well-known, but previously unexplained fact, namely, the obliquity of the glenoid fossa. When the scapula is examined in its correct orientation, invariably it is found to face somewhat upward in addition to forward and laterally (Fig. 12.4). This slope of the glenoid fossa—particularly its lower part—plays an important role in preventing downward dislocation or subluxation. As the head of the humerus is pulled downward, it is of necessity forced laterally because of the slope of the glenoid fossa (Fig. 12.5). If this lateral movement could be stopped,

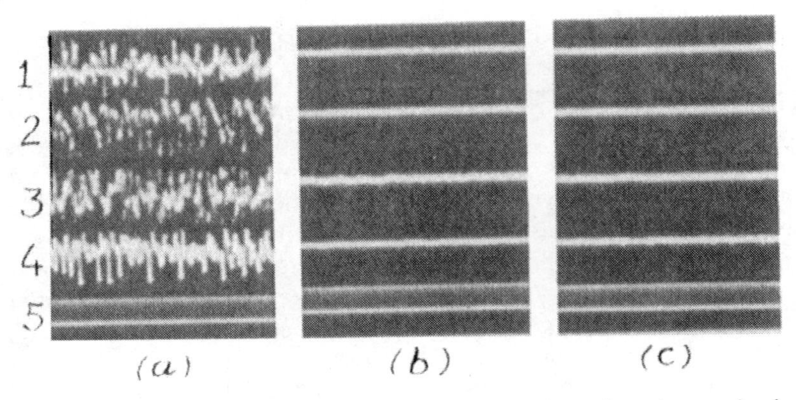

Figure 12.3. EMGs (a) during abduction; *b*, unloaded arm hanging; and *c*, heavy downward pull applied to arm. Lines *1*, *2*, and *3*: Anterior, middle, and posterior fibers of deltoid; line *4*: supraspinatus; line *5*: time marker, 10-ms intervals. (From J.V. Basmajian, © 1961, *Canadian Journal of Surgery*.)

the result would be a stopping of the downward movement. The superior part of the capsule of the joint and the supraspinatus (as well as the posterior fibers of the deltoid) are so placed that they can—indeed they must—tighten to prevent the downward dislocation (Fig. 12.6). Simple

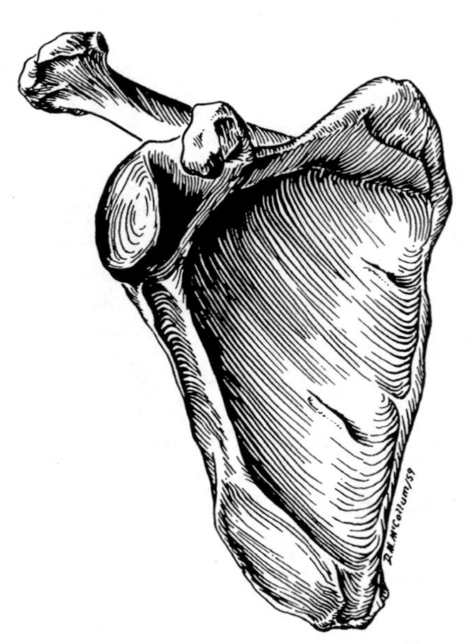

Figure 12.4. Correct orientation of the right scapula viewed directly from in front at eye level. (From J.V. Basmajian and F.J. Bazant, © 1959, *Journal of Bone and Joint Surgery*.)

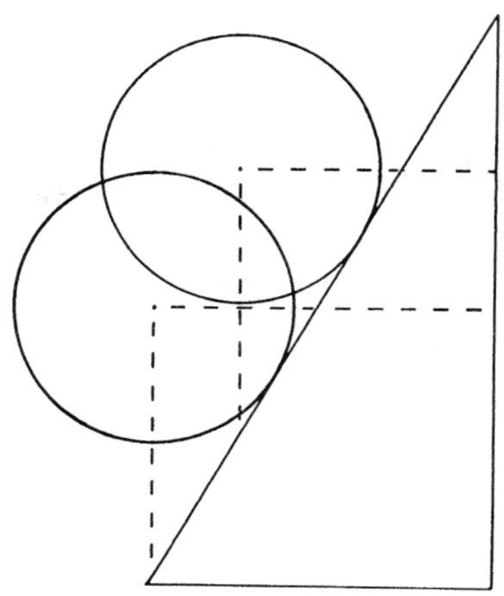

Figure 12.5. Diagram to illustrate locking mechanism at shoulder. The farther down the slope the ball slides, the farther laterally it is displaced. If the lateral displacement can be prevented, the ball cannot move downward. (From J.V. Basmajian and F.J. Bazant, © 1959, *Journal of Bone and Joint Surgery*.)

as this explanation may seem, it is dependent on our findings that (1) the vertically placed muscles definitely remain relaxed while (2) the supraspinatus (and posterior deltoid) become quite active and (3) the superior part of the capsule becomes taut. The very rareness of downward dislocation of the normal shoulder joint confirms the effectiveness of this locking mechanism.

The ordinary effect of gravity on the unloaded arm is counteracted in many persons by the superior part of the capsule. In this area the coracohumeral ligament forms a real thickening, and apparently this is an important function of the ligament. With moderate or heavy loads, the supraspinatus is called upon to reinforce the horizontal tension of the capsule; in some persons it is required even without a load. The posterior fibers of the deltoid, imitating in general the direction of the supraspinatus, must act in the same way but to a lesser extent.

The well-known subluxation that occurs in the humeral head of many patients after a stroke now can be explained and alleviated. The drooping shoulder region causes relative abduction of the humerus, unlocking the natural locking mechanism. Chaco and Wolf (1971) have gone further to demonstrate that in flaccid hemiplegia, actual supraspinatus weakness plays an important part in the subluxation. More recently Peat and Grahame (1977) described an excellent method for EMG evaluation of the shoulder region in hemiplegic patients.

The locking mechanism cannot operate when there is abduction of the humerus. As a result, the head can be easily subluxated when it is in the abducted position in the cadaver. Fairbank convincingly demon-

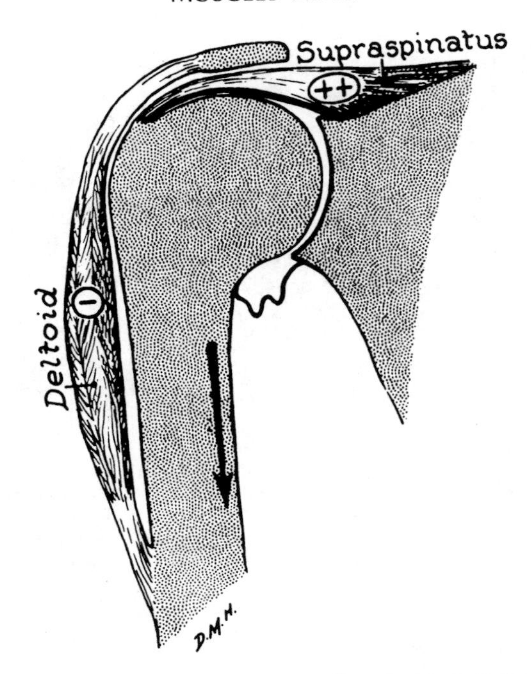

Figure 12.6. Diagram contrasting presence of moderate activity in supraspinatus with none in deltoid during loading in the direction of *arrow* (see text).

strated that subluxation can also be produced in anesthetized normal men, but his roentgenograms of the subluxated shoulders show in each case a real degree of abduction of the humerus in relation to the position of the scapula. It would appear that in these unconscious subjects the subluxation was preceded by a drawing downward of the glenoid fossa (that is, relative abduction of the humerus and thus a neutralization of the locking mechanism). The muscles that prevent dislocation of the joint in its unstable position are the "rotator cuff" muscles—supraspinatus, infraspinatus, teres minor, and subscapularis (and perhaps teres major and other muscles spanning the joint).

ELBOW FLEXORS

Although they can be felt quite easily, the biceps brachii, the brachialis, and the brachioradialis have not been fully understood as far as their integrated functions are concerned. Furthermore, both Duchenne and Beevor introduced and perpetuated errors that require correction. And again, although it is obvious that these muscles are primarily concerned with flexion, a variety of theories have obscured the role played by each during flexion and other movements of the elbow. It is probably fair to say that the few formal studies of these muscles (both with and without objective techniques) have been either too sharply circumscribed in approach or too highly generalized and deductive.

For these reasons, a detailed EMG study was made of both heads of the biceps, the brachialis, and the brachioradialis in a long series of young adults (Basmajian and Latif, 1957). A careful consideration of the time sequence of activity indicated there is a completely random selection in the sequence of appearance and disappearance of activity in these muscles. For example, during slow flexion with the forearm supine, all the muscles that showed any activity began this activity simultaneously in about half of the subjects. However, in only about one-quarter of them did the activity end simultaneously. In a small number in whom the activity did not begin simultaneously, it did, however, end simultaneously.

Any of the muscles that was to show activity during a movement functioned first or last in an unpredictable fashion, i.e., there was no set pattern. In the same way, the activity ceased in the muscles in an unpredictable order. Morever, the muscle that was to show the greatest activity in individual subjects only occasionally began first and ended last.

These results provided convincing evidence that in the movements produced by the biceps, the brachialis, and the brachioradialis there was a fine interplay between them; this was to be expected. What is more striking, however, was the wide range of response from any one muscle in the series. Thus, although a general trend may be described, there was rarely any unanimity of action. For example, the brachialis was generally markedly active during quick flexion of the supine forearm, but in one subject it was completely inactive.

These findings reemphasize the general biological principle that there is a range of response in any phenomenon. It would seem that anatomists and clinicians have taken too little heed of this wide range of individual pattern of activity in something even so simple as elbow-flexion.

In the Basmajian/Latif study, the long head of the biceps showed more activity than the short head in the majority of the subjects during slow flexion of the forearm, during supination of the forearm against resistance, and during flexion of the shoulder joint (although there was little difference between the activity of the two heads during isometric contraction and during extension of the elbow). Sullivan et al (1950) reported similar findings in a more limited but fine experimental series with surface electrodes during flexion only.

The biceps is generally active during flexion of the supine forearm under all conditions and during flexion of the semiprone forearm when a load (of about 1 kg) is lifted (Fig. 12.7). However, with the forearm

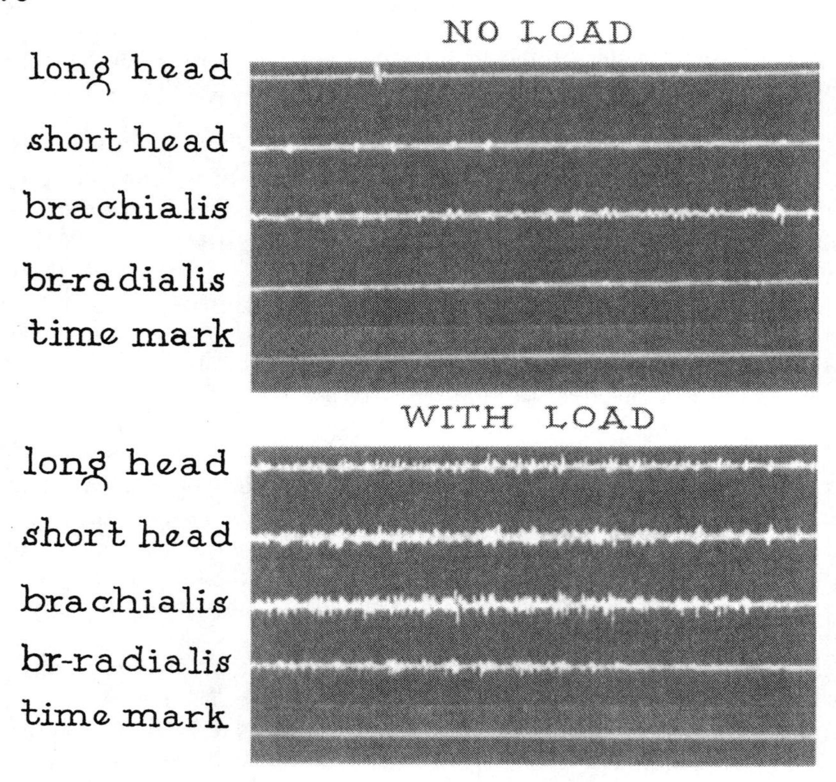

Figure 12.7. EMGs of the two heads of biceps, the brachialis, and the brachioradialis during slow flexion of semiprone forearm.

prone, in the majority of instances biceps plays little if any role in flexion, in maintenance of elbow flexion, and in antagonistic action during extension, even with the load. Beevor (1903, 1904) stated that if the forearm is in supination the biceps acts during flexion when there is a resistance of as little as 120 g, but that in a position of complete pronation it does not act until the resistance is at least 2 kg. The results of the EMG study support Beevor's observations in regard to flexion.

Bankov and Jørgensen (1969) convincingly showed that both in isometric and in dynamic contractions the maximum strength in the elbow flexors is smaller with the pronated, as compared with the supinated, forearm. The integrated rectified EMG signals of biceps were considerably reduced in the pronated forearm. Confirmatory studies were done by Settineri and Rodriguez (1974).

The biceps is usually described as a supinator of the forearm. In the Basmajian/Latif study, no activity in the muscle was demonstrated in the majority of the subjects during supination of the extended forearm through the whole range of movement, except when resistance to supination was given. However, activity was observed in all the subjects when supination was strongly resisted. It follows that, generally, the biceps is not a supinator of the extended forearm unless supination is resisted.

It is necessary to explain why the biceps does not ordinarily supinate the extended forearm. It appears that because of the tendency of the biceps to flex the forearm, it is reflexly inhibited. Thus, the extended position of the forearm is maintained while the supinator does the supinating. On the other hand, when supination is resisted, the biceps comes into strong action, and we have noted that usually the previously extended forearm is partly flexed as well during supination against resistance. For actions of biceps at the shoulder, see p 272.

The brachialis has been generally and erroneously considered by anatomists to be a muscle of speed rather than one of power because of its short leverage. Basmajian and Latif found it to be a flexor of the supine, semiprone, and prone forearm in slow or quick flexion, with or without an added weight. McGregor (1950) has correctly described it as a "flexor *par excellence* of the elbow joint." Apparently brachialis is called upon to flex the forearm in all positions because the line of its pull does not change with pronation or supination.

Maintenance of specific flexed postures of the elbow, i.e., isometric contraction, and the movement of slow extension when the flexors must act as antigravity muscles both generally bring the brachialis into activity in all positions of the forearm. This is not the case with the other two flexor muscles. Thus, the brachialis may also be designated the "workhorse" among the flexor muscles of the elbow.

A short burst of activity is generally seen in all the muscles during quick extension. This activity can hardly be considered antagonistic in the usual sense. Rather, it may provide a protective function for the joint. Biceps is particularly active during quick extension with an added load.

In the past, the brachioradialis has been described as a flexor, acting to its best advantage in the semiprone position of the forearm. Basmajian and Latif found in most subjects that the brachioradialis does not play any appreciable role during maintenance of elbow flexion and during slow flexion and extension when the movement is carried out without a weight. When a weight is lifted during flexion, the brachioradialis is generally moderately active in the semiprone or prone position of the forearm and is slightly active in the supine position. There is no comparable increase in activity with the addition of weight during maintenance of flexion and during slow extension. Also, in most instances the brachioradialis was quite active in all three positions of the forearm during quick flexion and extension. It follows that the muscle is a reserve for occasions when speedy movement is required and when weight is to be lifted, especially in the semiprone and the prone positions. In the latter position, the biceps usually does not come into prominent action. Furthermore, the activity of the brachioradialis in speedy movements is related to its function as a shunt muscle.

The brachioradialis has been described since Duchenne's day as a supinator of the prone forearm and a pronator of the supine forearm acting to the semiprone position in both cases. The Basmajian/Latif study showed that it neither supinates nor pronates the extended forearm unless these movements are performed against resistance. Here, at most, brachioradialis acts only as an accessory muscle, coming into action when strength is required to supinate or to pronate the forearm. More probably, it acts only as a synergist.

All of the observations strongly suggest that the biceps, the brachialis, and the brachioradialis differ in their flexor activity in the three positions of the forearm (prone, semiprone, and supine). However, all three muscles act maximally when a weight is lifted during flexion of the semiprone forearm. The semiprone position of the forearm has been described as the natural position of the forearm, the position of rest and the position of greatest advantage for most functions of the upper limb (Basmajian, 1955a).

Pronator Teres and Elbow Flexion

Basmajian and Travill (1961) showed that pronator teres contributes to elbow flexion only when resistance is offered to the movement. It shows no activity during unresisted flexion, whether the forearm is prone, semiprone, or supine (see Fig. 12.10 on p 284.)

Before leaving the flexor muscles of the elbow, we might note that Wells and Morehouse (1950) believed that biceps and triceps (and latissimus dorsi) act as cocontractors in exerting a pull such as on an aircraft control stick. They found that the extent of the "contribution" each muscle makes is altered when the position of the arm is changed (see p 277). When the arm is pulling in an extended position, the biceps dominates the action, but in the flexed or intermediate position, triceps is brought strongly into action. These findings can hardly be valid evidence of cocontraction because of many complicating factors in the set-up. However, Wells and Morehouse did show that, as far as muscular dynamics is concerned, "the best arm position of a pilot seated in a conventional upright position and operating a control stick is one which is intermediate between flexion and extension."

TRICEPS BRACHII AND ANCONEUS

Travill (1962) found that the long head of triceps is surprisingly quiescent during active extension of the elbow, regardless of the position of either the subject or his limb. The medial head, however, is always active and appears to be the prime extensor of the elbow; meanwhile, the lateral head shows some activity as well. Against resistance, the lateral and long heads are recruited. Therefore, we might compare the medial head of triceps to the brachialis, which we noted above to be the workhorse of the elbow flexors; it is the workhorse of the extensors. The lateral and long heads are reserves for extension, just as the two heads of biceps are reserves for flexion. Isometric contractions against immovable resistances at 60°, 90°, and 120° produce a rising amount of myoelectric activity (Currier, 1972). There is also a considerable amount of decrease during learning acquired with repetition (Payton, 1974).

Travill confirmed Duchenne's view that, of the two superficial heads, the long head is the less powerful during extension. This is probably due to the lack of fixation of the scapular origin and the necessity of adducting the shoulder with the forearm either flexed or extended. Too strong a contribution from the long head would tend to give extension during adduction of the arm.

In a series of three papers, investigators at the Pitié-Salpêtrière in Paris have further clarified the functions of the three heads of triceps and the related anconeus (Le Bozec et al, 1980a and b; Maton et al, 1980; Le Bozec and Maton, 1982). During different levels of static work at a right angle, the integrated EMG of the three heads of triceps increases quadratically with torque with somewhat similar (but not identical) increases in anconeus. During voluntary extension at different velocities and inertias, the muscular activity increases linearly with the external work but the activity in the three heads of triceps has different chronologies, depending on shifting conditions. Anconeus is particularly active with slow movements. During rapid flexion, the braking or antagonist effects are shared by muscular responses in the two muscles; they are much reduced during slow flexion. (Anconeus is discussed further at the end of the chapter.)

PRONATION AND SUPINATION OF FOREARM

Pronators

Until recently no accurate authoritative information existed on the relative functional roles of the two pronator muscles, although their gross anatomy is adequately described in the standard textbooks. Investigation of the pronator teres and the pronator quadratus in a series of volunteers revealed that the few remarks on function that are presently available in books are largely misleading (Basmajian and Travill, 1961).

In each subject three groups of records were made from needle electrodes (Fig. 12.8). In the first group, the elbow was kept in the extended, fully supported position on the table top (Fig. 12.9, α). In the second group, the elbow was flexed to a right angle, with the forearm

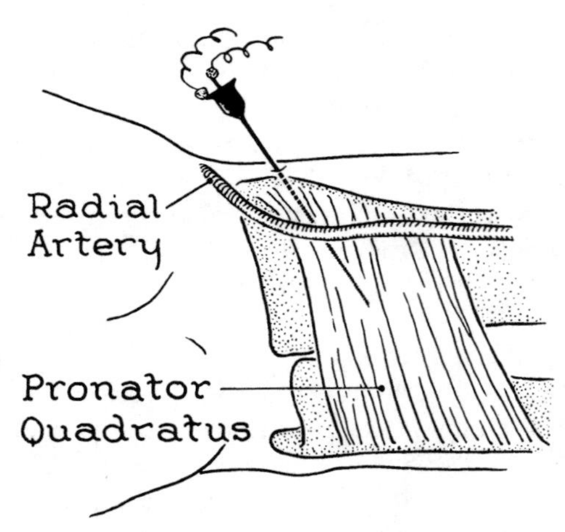

Figure 12.8. Diagram of electrode placement in pronator quadratus. (From J.V. Basmajian and A. Travill, © 1961, *Anatomical Record.*)

vertical and the arm and elbow supported (Fig. 12.9, β). In the third group, the elbow was flexed to an acute angle while it was still fully supported (Fig. 12.9, γ). In each of these three groups of tests, records from the two muscles were made during the following movements and positions: (1) slow pronation from the comfortable supine position to the fully prone position; (2) fast pronation through the above range; (3) "hold" in the fully prone position; (4) slow supination through the whole

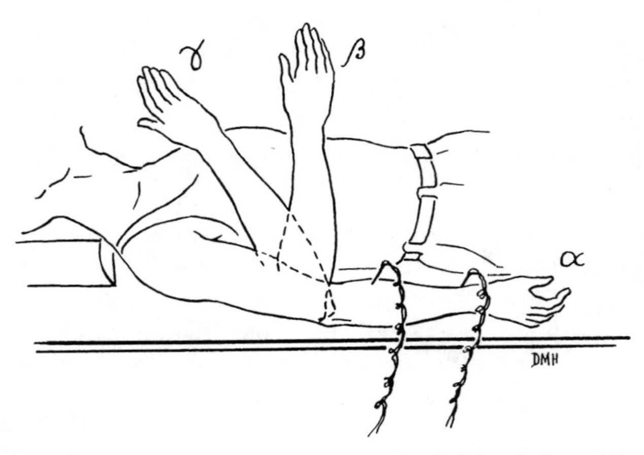

Figure 12.9. Diagram of primary positions of limb during three groups of tests (see text). (From J.V. Basmajian and A. Travill, © 1961, *Anatomical Record.*)

range to full (forced) supination; and (5) fast supination through the above range.

Both pronator quadratus and pronator teres are active during pronation, the consistent prime pronating muscle being the pronator quadratus (Fig. 12.10). This is true, irrespective of the positions of the forearm in space or the angulation of the elbow joint. In general, the pronator teres is called in as a reinforcing pronator whenever the action of pronation is rapid. Similar reinforcement occurs during pronation against resistance.

Whether pronation is fast or slow, the activity in the pronator quadratus is markedly greater than that in the pronator teres. This observation conflicts with the opinions offered in a number of the internationally recognized textbooks (Steindler, 1955; Johnston et al, 1958; Hamilton and Appleton, 1956; Lockhart et al, 1959). Some writers sit squarely on the fence and suggest that both muscles pronate without preference. Only two books suggested before 1960 that the pronator quadratus is the main pronator (Hollinshead, 1958; Basmajian, 1960).

A number of authors likewise expressed the view that the pronator teres displays its greatest activity either during midflexion of the elbow (Steindler, 1955; Lockhart, 1951) or during full extension (Hamilton and Appleton, 1956; Hollinshead, 1958). However, regardless of whether the pronating action is carried out swiftly or slowly, the angle of the elbow joint has no bearing on the amount of activity of the pronator teres.

During slow supination there is no activity whatsoever in either of the pronators—although some have suggested that the deeper layer of the pronator quadratus acts as a supinator. De Sousa et al (1957, 1958), using a different approach, have independently arrived at the same conclusion.

During fast supination, there is negligible activity in the pronators. This is rather surprising in view of earlier work, mentioned above, on the EMG activity of the biceps and triceps during flexion and extension of the elbow (Barnett and Harding, 1955; Basmajian and Latif, 1957). In those muscles, a sharp burst of antagonistic activity occurs during fast movements, this activity being thought to be the manifestation of a protective stretch reflex.

De Sousa and his colleagues (1957, 1958, 1961) in São Paulo, Brazil, showed that pronator quadratus is a pronator only. They agree with Basmajian and Travill that it participates in normal pronation, but there is some difficulty in reconciling their finding that there is little early activity in pronator quadratus during the early stages of pronation. Their explanation of the difficulty, which indeed seems to be a valid one, is that at the beginning of unresisted pronation the natural elastic recoil from complete supination is quite enough. Basmajian/Travill "prona-

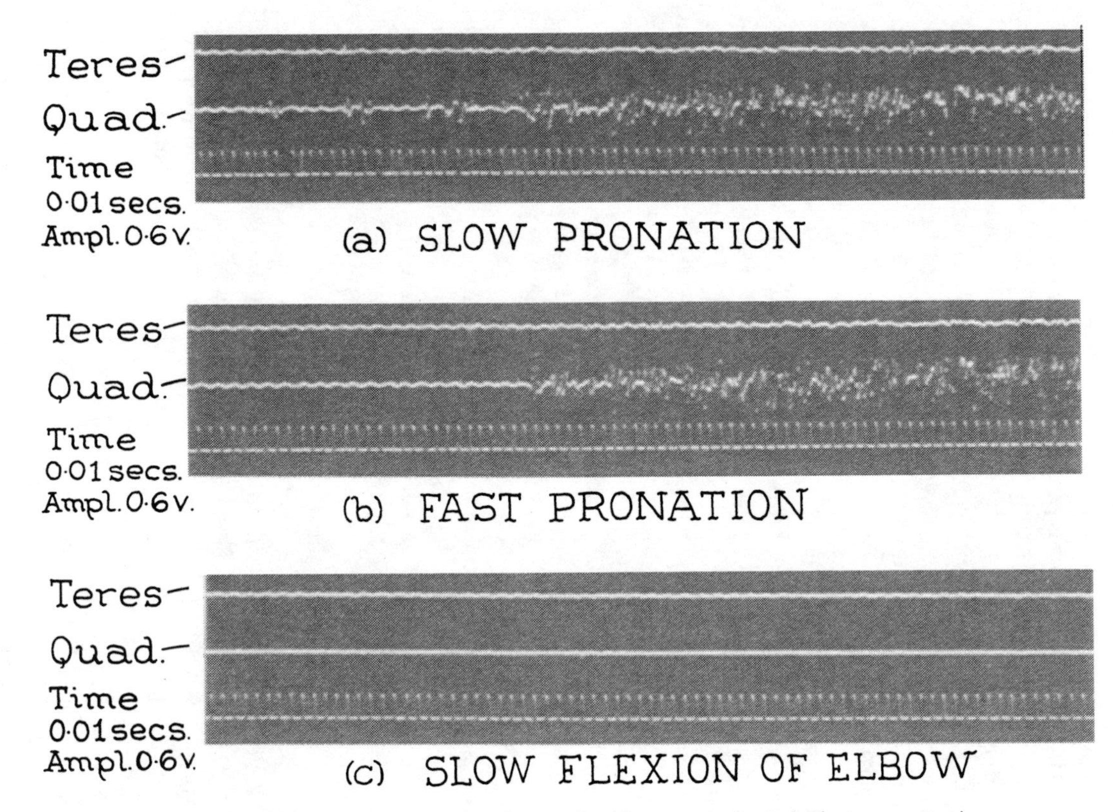

Figure 12.10. Typical EMG recordings of *a* (slow pronation), *b* (fast pronation), and c (slow flexion of elbow). (From J.V. Basmajian and A. Travill, © 1961, *Anatomical Record*.)

tions" were begun from the comfortably supine, not the fully supine, position. In any case, pronator teres acted no earlier than quadratus in their series. Moreover, the flexor carpi radialis, brachioradialis and extensor carpi ulnaris were shown to have no pronating function.

Supinators

Supination of the forearm in man is undeniably of fundamental importance and yet, all too often, the parts played by the chief supinator muscles are either ignored or taken for granted. Throughout the textbooks, there is no thread of consistency, and the truth appears to be so tangled with hopeful guesses it cannot be recognized. For example, only in recent years did electromyography confirm Beevor's strong insistence in 1903 that brachioradialis (known for years as "supinator longus") is not a true supinator (see p 278).

Therefore, an electromyographic study of the supinator and the biceps brachii was justified (Travill and Basmajian, 1961). This study was complementary to and, in part, overlapped by that on the pronator muscles outlined above. A needle electrode was inserted into the middle of each of the following four muscles: the supinator; the biceps brachii; the pronator quadratus; and the pronator teres, the pronators being tested simultaneously as controls only.

Two series of recordings were made from each subject: the first series with the elbow extended, the second with the elbow flexed to 90°. With each of these two positions recordings were made during: (1) the movement of supination from full position; (2) the "hold" position of maximum supination; and, finally, (3) the return movement of pronation to the original comfortably supine position. Recordings were made during slow movements, fast movements, and forceful movements against the resistance offered by the grip of an observer.

Slow unresisted supination, whatever the position of the forearm, is brought about by the independent action of the supinator (Fig. 12.11). Similarly, fast supination in the extended position requires only the supinator, but fast unresisted supination with the elbow flexed is assisted by the action of the biceps. All movements of forceful supination against resistance require the cooperation of the biceps in varying degrees.

This last-mentioned cooperative activity of the supinator and the biceps during resisted supination, especially when the elbow is flexed, had never been seriously questioned in the past. Primacy of the supinator during the unresisted movement has not, however, received such universal acceptance, even though it was first suggested by Duchenne and broadly hinted at by Bierman and Yamshon (1948). For example, Steindler (1955) and Gardner et al (1960) emphasize only the power of the biceps against resistance; this power is undeniable.

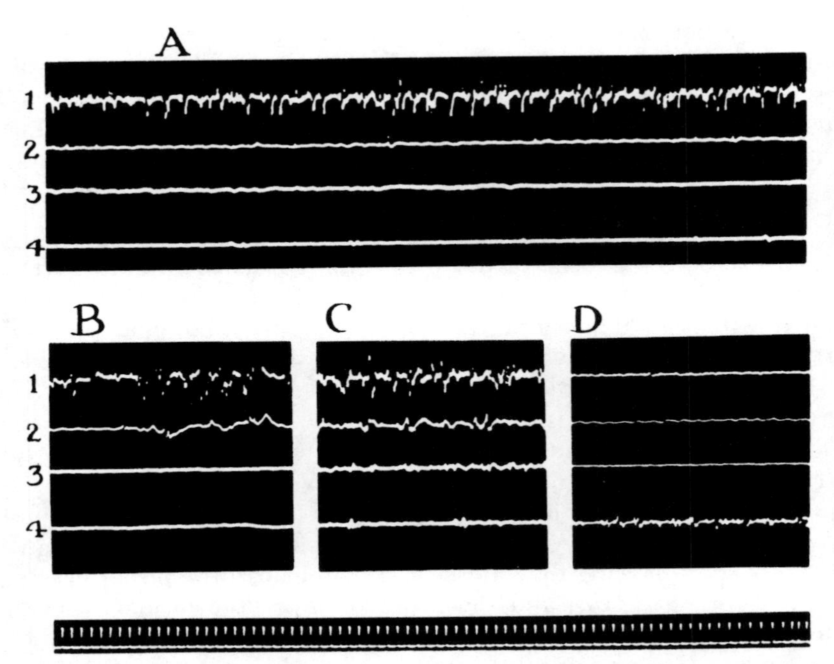

Figure 12.11. EMG recordings during *A* (slow supination), *B* (fast supination), *C* (forceful supination against resistance), and *D* (pronation of the forearm). Channel *1*, supinator; *2*, biceps brachii; *3*, pronator teres; *4*, pronator quadratus. Time marker, 10-ms intervals. (From A. Travill and J.V. Basmajian, © 1961, *Anatomical Record*.)

During supination, the action of supinator is augmented by that of the biceps. This was similar to the earlier findings for pronation, where pronator quadratus is augmented by the pronator teres when required (Basmajian and Travill, 1961).

Both the supinator muscles are completely relaxed during pronation (slow, fast, or resisted). This is again similar to the findings for pronation, wherein complete relaxation of the pronator quadratus and the pronator teres during supination is the rule.

The "hold" or static position of supination depends on activity in supinator for the maintenance of the supine posture. Against added resistance, however, the biceps always becomes active. The movement of supination is initiated and mostly maintained, by the supinator; it is only assisted by the biceps as needed to overcome added resistance.

Anconeus

This small muscle has been the subject of controversy since Duchenne suggested a special role for it, that of abduction of the ulna during pronation of the forearm. On the other hand, one never sees cases of localized paralysis of anconeus from which its role as an abductor could be confirmed. Furthermore, pronation is equally efficient with or without abduction. However, some simple observations reveal that the usual way in which most persons carry out pronation includes the slight abduction of the ulna. Thus the hand can be "turned over" without its shifting away from its original position. The anconeus is in the ideal position to perform this secondary movement.

Under the direction of Professor de Sousa of São Paulo, Brazil, Da Hora (1959) of the University of Recife reinvestigated anconeus morphologically and electromyographically. From the latter study he concluded that anconeus was always active during extension of the elbow. What is surprising, however, is his finding that it is active in both pronation and supination, whether these movements are resisted or unresisted. He also reported activity during flexion of the elbow, especially against resistance.

Ray et al (1951) reported that anconeus was very active during the whole of pronation. However, Travill (1962) concluded from his EMG study of this muscle that it is only active when resistance is offered to the movement—but it is quite as active during resisted supination. All authors confirm the classical view that anconeus is most active during extension of the elbow.

Because bipolar wire electrodes and other refinements in EMG might solve the enigma and because orthopedic surgeons continued to ask questions about the anconeus, a formal electromyographic study of the anconeus in 10 subjects, as well as their related muscles for comparison, was undertaken (Basmajian and Griffin, 1972). Bipolar wire electrodes were inserted into the middle of the bellies of the anconeus, supinator, triceps (medial head), and pronator teres. An electrogoniometer monitored and recorded elbow angle, which was held constant at a series of angles (Fig. 12.12) by a special holding device.

The electromyographic results for supination, pronation, and extension in anconeus alone are summarized in Figure 12.12 (*left*). It shows that extension recruits somewhat more than moderate activity while pronation and supination elicit moderate activity in the anconeus through most of the range of elbow motion.

Comparing the anconeus with the medial head of triceps shows similarity in activity levels with some mild variations in detail (Fig. 12.12, *right*).

During supination the (expected) marked activity in supinator and relative quiet of the pronator teres are accompanied by moderate activity in the anconeus and slight to moderate activity in the triceps. During pronation, moderate activity in the anconeus is accompanied by slight to

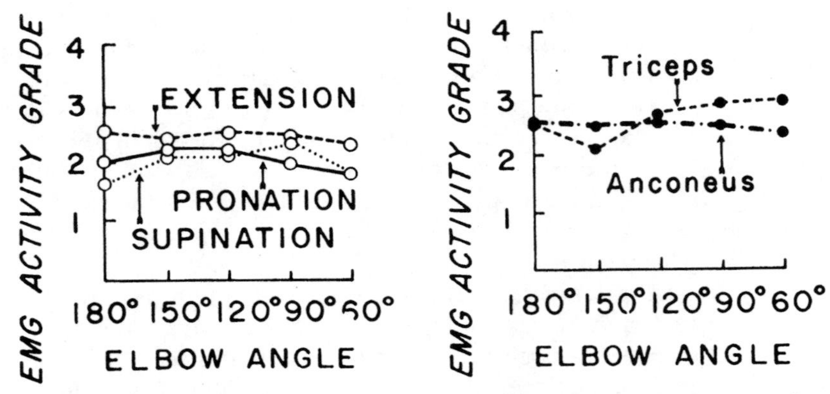

Figure 12.12. (*Left*) Graph of electromyographic activity in anconeus during supination, pronation, and extension of elbow. (*Right*) Graph comparing electromyographic activity in anconeus and triceps (medial head) during extension of elbow. (From J.V. Basmajian and W.R. Griffin, © 1972, *Journal of Bone and Joint Surgery*.)

moderate activity in the supinator and triceps (see Fig. 12.13). Gleason et al (1983) reported in a "Letter to the Editor" that pronation around the axis joining the radial head to the middle finger recruits more activity, suggesting an ulnar-abduction function.

A definite function for the anconeus is still not proved, but its moderate activity during both pronation and supination (especially at their extremes), also described by Da Hora and by Travill, gives little support to Duchenne's proposed function of ulnar deviation during pronation, except perhaps in the above type of pronation. Rather, it suggests some form of joint stabilization. This synergy is apparently shared by the

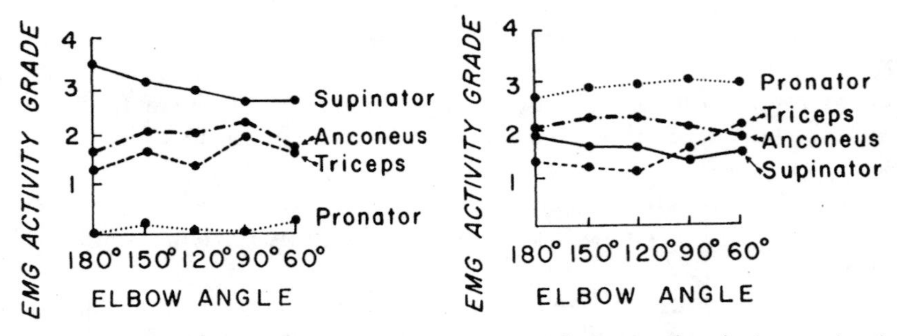

Figure 12.13. (*Left*) Graph comparing activity in four muscles during supination of the elbow at various set angles of the elbow. (*Right*) Graph comparing activity in four muscles during pronation of the elbow at various set angles of the elbow. (From J.V. Basmajian and W.R. Griffin, © 1972, *Journal of Bone and Joint Surgery*.)

closely related medial head of triceps and perhaps the supinator (Basmajian and Griffin, 1972).

FOREARM-HAND SYNERGIES

All forearm muscles contribute to hand function, including the supinators and the pronators; they contract to various degrees during finger movements (see p 307).

CHAPTER 13

Wrist, Hand, and Fingers

The two radial extensors of the wrist were the early special subject of electromyographic examination by Tournay and Paillard (1953). They showed that during pure extension of the wrist the extensor carpi radialis brevis is much more active than the longus, whether the movement is slow or fast. Actually, except with fast extension, the longus was essentially inactive. However, the roles of the two muscles are completely reversed during prehension or fist-making; now the longus is very active as a synergist. This appears to be an extremely important observation. The two muscles are both quite active during abduction of the wrist, as one would guess from their positions.

Bäckdahl and Carlsöö (1961) found that during extension of the wrist there is a reciprocal innervation between extensors and flexors. Extensores carpi radiales (longus and brevis) and extensor carpi ulnaris, as well as the extensor digitorum, work synchronously; none seems to be the prime mover. This was confirmed by McFarland et al (1962). During forced extreme flexion of the wrist, there is a reactive cocontraction of the extensor carpi ulnaris, apparently to stabilize the wrist joint; this does not occur with the extensor digitorum and extensores carpi radiales.

During flexion of the wrist, the flexores carpi radialis, ulnaris, and digitorum superficialis act synchronously, according to Bäckdahl and Carlsöö—none is the prime mover. Flexor digitorum profundus plays no role. Two possible muscles in the antagonist position (the radial extensors of the wrist and the extensors of the fingers) are passive, even in extreme flexion of the wrist, but the extensor carpi ulnaris shows marked activity as an antagonist.

In abduction and adduction, the appropriate flexors and extensors act reciprocally as one might expect, with the antagonist muscles relaxing. Extensor digitorum contracts during abduction (radial abduction), but Bäckdahl and Carlsöö found that this contraction is not limited to the radial part of the muscle. Apparently this last activity has a synergistic function. McFarland et al also found activity in extensor digitorum during extremes of abduction and adduction of the wrist; moreover, the flexor digitorum superficialis was active too. This group of investigators also emphasizes the uniform occurrence of antagonist activity in the flexors when the wrist is extended and the metacarpophalangeal joints is hyperextended.

FINGERS

Increasing scientific attention to the movement of the fingers has appeared. Person and Roshtchina (1958) of Moscow were concerned with the nervous mechanisms that enable a person to perform isolated movements of a single finger. They studied the common flexors and extensors of the fingers during rhythmic flexions and extensions. When all the fingers are moving simultaneously, the activity of the "antagonist" muscles conforms to the principle of reciprocal inhibition (p 223). When only the little finger or ring finger is moving while the others are extended, the extensor is active during both extension and flexion. When one finger is moved while the others are kept bent, the flexor is active during both movements.

If a single finger moves, the "antagonist" must remain active to immobilize the other fingers. However, if the other fingers are held immobile by an observer, there is no activity in the antagonist muscle.

Backhouse and Catton (1954) studied the lumbricals of the hand and proved conclusively that they are only important in extension of the interphalangeal joints, reinforcing the action of the extensor digitorum and interossei. They agree in large measure with the Australian anatomist Sunderland (1945), who suggested that the importance of lumbrical-interosseus extension at the interphalangeal joints is in the prevention of hyperextension of the proximal phalanx by the extensor digitorum. This preventive action allows a more efficient pull on the dorsal expansion which extends the interphalangeal joints (Figs. 13.1–13.3).

Metacarpophalangeal flexion is performed by a lumbrical only when the interphalangeal joints are extended. Backhouse and Catton concluded that a lumbrical has no effect on rotation or radial deviation of its finger during opposition with the thumb (as first suggested by Braithwaite et al, 1948, in their classic morphological study).

At Washington University in St. Louis, Lake (1954, 1957) made a simultaneous study of the extensor digitorum (communis), flexor digitorum superficialis (or sublimis), and the second and third dorsal interossei. She found that the extensor digitorum begins or increases its activity with the inception of interphalangeal (IP) joint extension, regardless of the position of the metacarpophalangeal (MP) joints. During extension or hyperextension of the MP joint, extensor digitorum alone was active.

Flexor digitorum superficialis is active during flexion of the middle phalanx (proximal IP joint), and it is active in flexion of the MP joint, provided that the next distal joint is stabilized. Surprisingly, the superficialis is active during rapid, forceful IP extension, regardless of the position of the MP joint.

The interossei in Lake's research were found to be markedly active from the very onset of flexion of the MP joint, even with moderate effort. IP joint position was of no consequence in this. The interossei also showed activity before the onset of visible extension in either the proximal or distal IP joint. In the case of extension of the proximal joint,

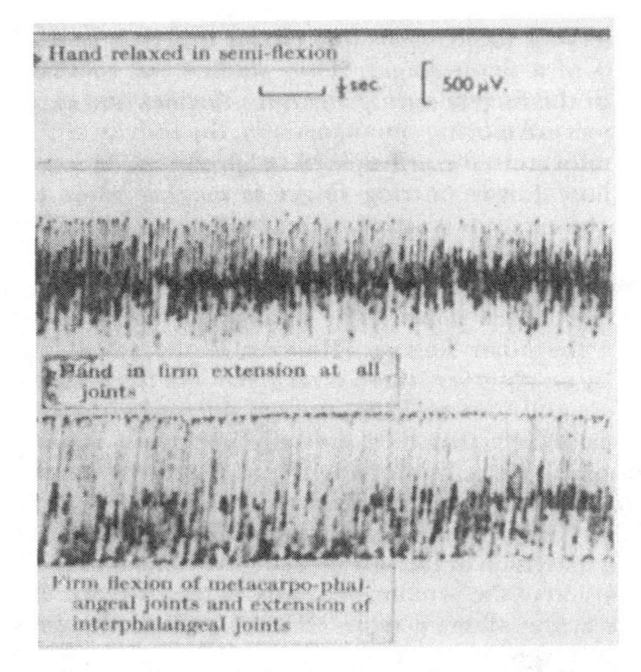

Figure 13.1. EMG recordings of a lumbrical muscle in several positions of the hand. (From K.M. Backhouse and W.T. Catton, © 1954, *Journal of Anatomy*.)

the distal joint had to be extended simultaneously, but the position of the MP joint was not important.

Meanwhile, Brown et al (1960) and Long et al (1960, 1961, 1970) at Case Western Reserve University in Cleveland completed a comprehensive and ingenious study of the hand musculature, the results of which add to our understanding of its kinesiology. Using multiple indwelling pliable wire electrodes, this group concluded that the interossei of the hand act as MP flexors only when their other action of IP extension does not conflict. Therefore they act best and strongest when combined MP-flexion–IP-extension is performed. During all IP extension, the intrinsic muscles of the hand contract, regardless of MP posture.

They concluded that the long tendons of the fingers provide the gross motion of opening and closing of the fist at all the joints simultaneously. However, the intrinsic muscles perform their major function during any departure from this simple total opening or closing movement. Thus, they are the primary IP extensors while the MP joints are flexing.

Long and Brown (1962, 1964) confirmed and expanded on their preliminary findings for the lumbricals. In general, these findings confirm those of Backhouse and Catton (see above). The lumbricals are

Figure 13.2. EMG recordings for second lumbrical muscle while middle finger is in different positions. (From K.M. Backhouse and W.T. Catton, © 1954, *Journal of Anatomy*.)

silent during total flexion of the entire finger but are very active whenever the proximal or distal IP joints are extended actively or are held extended while the MP joint is flexed actively. The lumbricals can be kept very quiet during MP movements in any direction by keeping the IP joints fully flexed.

To summarize, Long and Brown conclude that the interossei and the lumbrical of one finger do not form a functional unit—they act discretely. The interossei participate in IP extension only when the MP joint is either flexing or held flexed. The lumbrical always takes part in

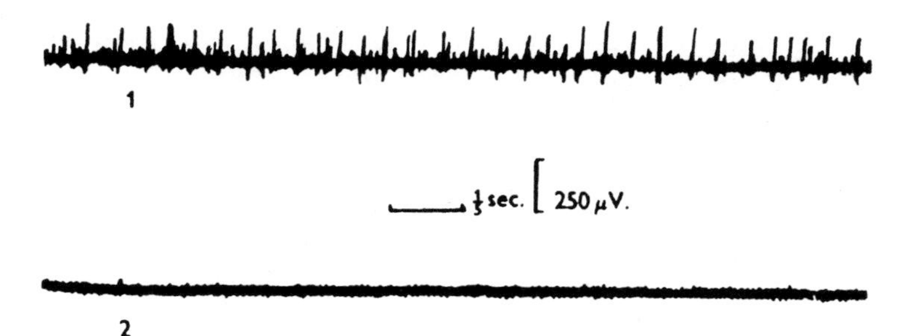

Figure 13.3. EMG recordings of lumbrical. *1*, relaxed in extension with ulnar deviation; *2*, radial deviation against resistance. (From K.M. Backhouse and W.T. Catton, © 1954, *Journal of Anatomy*.)

interphalangeal extension. It shows a crescendo of activity throughout the movement, reaching a peak at full extension. This suggests that its function may include prevention of hyperextension at the MP joint. Neither the interossei nor lumbrical of the middle finger acts during closing of the full hand, suggesting that in this total movement they are not synergists.

The activity of the long extensors and flexors occurs in special sequences. Extensor digitorum acts during MP extension—in both the movement and the "hold" position. But it is also active in many flexion movements of that joint, apparently acting as a brake. The flexor profundus is the most consistently active flexor of the finger. Joined by the flexor superficialis, the profundus may act as a flexor of the wrist joint also. The superficialis has its maximal action when the hand is being closed or held closed without flexion of the distal IP joint.

Power Grip and Precision Handling

The important paper by Long et al (1970) is the bedrock of all recent EMG confirmation and clarification of the fine functioning of the human hand. One can do no better than to quote or paraphrase their main conclusions and to recommend consulting the original to readers who want further details.

POWER GRIP

In power grip the *extrinsics* provide the major gripping force. All of the extrinsics are involved in power gripping and are used in proportion to the desired force to be used against the external force.

The major *intrinsic muscles* of power grip are the interossei, used as phalangeal rotators and metacarpophalangeal flexors. The lumbricales, with exception of the fourth, are not significantly used in power grip. The thenar muscles are used in all forms of power grip except hook grip.

Ohtsuki (1981) investigated the electromyography and biomechanics of the decrease in grip strength induced by simultaneous bilateral exertion. Voluntary maximum isometric grip strength was measured under the conditions of simultaneous bilateral and separate unilateral exertion on 16 male and 8 female subjects. Strengths of each finger and surface electromyograms of finger flexors in the forearm were recorded concurrently with grip strength recording. The following results were obtained: (1) grip strength and integrated rectified EMG signals of both arms were significantly reduced by simultaneous bilateral exertion; (2) the decrease ratio in strength was 5–14%; (3) there was an obvious linear relationship between grip strength and integrated rectified EMG signal both for original values and for decrease ratios; (4) parallelism between strength and electrical activity of muscle suggests that neurophysiological mechanism subserve this strength decrease; (5) the middle finger occupied the largest share of grip strength and the ring, index, and little finger followed in decreasing order; and (6) the proportion occupied by each finger in decrease ratio of grip strength was largest in the case of the middle finger, followed by other fingers in the same order as strength share, especially for male subjects.

PRECISION HANDLING

In precision handling, specific extrinsic muscles provide gross motion and compressive forces.

In rotation motions the interossei are important in imposing the necessary rotational forces on the object to be rotated; the motion of the metacarpophalangeal joint that provides this rotation is abduction or adduction, not rotation of the first phalanx. The lumbricales are interphalangeal joint extensors as in the unloaded hand, and additionally are first phalangeal abductor-adductors and rotators.

In translation motions towards the palm, the interossei provide intrinsic compression and rotation forces for most efficient finger positioning; the lumbricalis is not active. Moving away from the palm the handled object is driven by interossei and lumbricales to provide intrinsic compression and metacarpophalangeal-joint flexion and interphalangeal-joint extension.

The thenar muscles in precision handling act as a triad of flexor pollicis brevis, opponens pollicis, and abductor pollicis brevis to provide adduction across the palm, medial rotation of the first metacarpal, and maintenance of web space depth.

The adductor pollicis is used in specific situations when force is required to adduct the first metacarpal towards the second.

PINCH

In pinch, compression is provided primarily by the extrinsic muscles. Phalangeal rotational position is adjusted by the interossei and perhaps also by the lumbricales. Compression is assisted by the metarcapophalangeal joint flexion force of the interossei and flexor pollicis brevis and by the adducting force of the adductor pollicis. The opponens assists through rotational positioning of the first metacarpal (Long et al, 1970).

Brandell (1970) used the manipulation of a ruler, pencil, and paper to determine the fine function of the index musculature. He found that the first dorsal interosseus muscle was the most consistently active when the ruler was carried with the three joints of the index finger flexed, but the flexor digitorum superficialis and extensor digitorum muscles appeared to play an important role in the maintenance of the grip when the ruler was carried with the interphalangeal joints of the index finger straightened.

Of the five subjects Brandell tested, one had practiced the procedure over a period of several weeks before the combined records were made, while the performance of the remaining four was entirely unrehearsed. The methods used by the trained and untrained subjects to grasp and release the ruler involved contrasting uses of the index finger and its musculature. The straight, or straightening, index finger, which was correlated with flexor digitorum superficialis activity, performed an unexpectedly important role in the basic phases of ruler manipulation.

FOREARM-HAND SYNERGIES

In a comprehensive multichannel study with fine wire electrodes, Sano et al (1977) of Tokyo revealed the integrated role of forearm and finger muscles during movements of fingers in different postures. Some forearm muscles participate as synergists early in voluntary finger motions, others only late in the movement. During movements of the index finger with various forearm positions the supinator muscle appeared to take part in stabilization of the proximal radioulnar joint and the pronator quadratus in that of the distal radioulnar joint (especially when the forearm is pronated).

Sano et al found anconeus muscle active in all movements of the index, apparently for stablization of the humeroulnar joint. Flexor carpi radialis, known to be a synergist during finger extension, also is active during extension. Both in wrist extension and in index finger movements, extensor carpi radialis brevis is more active than the longus. These muscles are apparently as active as the extensor digitorum itself, according to Sano et al. They emphasized the complex activity of all the forearm synergists during unrehearsed normal movements of the digits.

Deformities of Fingers

In the rheumatoid hand with its developing deformities, Backhouse (1968) and Swezey and Fiegenberg (1971) showed how electromyography of intrinsic hand muscles reveals the role of inappropriate muscle actions.

Thenar and Hypothenar Muscles

The EMG of the thenar and hypothenar muscles was slow in coming because of the close packing of these small muscles. The French pioneers Tournay and Fessard (1948), as well as Weathersby (1957), made useful but brief preliminary reports; this was followed by a longer report by Weathersby et al. Otherwise, no substantial work appeared until Forrest and Basmajian (1965) published a systematic study with wire electrodes; this study will be heavily drawn upon in the subsequent paragraphs.

The report of Tournay and Fessard was concerned with general phenomena rather than with particular actions of the thumb muscles. Weathersby found with surface and needle electrodes that each of the thenar muscles was involved to some extent in most of the movements of the thumb. The abductor pollicis brevis contracts strongly during opposition and flexion of the thumb, as well as in abduction. The oppenens shows strong activity in abduction and flexion of the metacarpal, as it does in opposition. The flexor pollicis brevis shows considerable activity in opposition as well as flexion, and in adduction. The adductor pollicis (which is, of course, not properly a thenar muscle) is active in adduction and opposition and, to a slight extent, in flexion of the thumb.

It is interesting to note that Sala (1959) of Pavia, Italy, used EMG combined with nerve stimulation to determine the innervation of the muscles of the thenar eminence. He found that one of every four flexor brevis muscles was exclusively supplied by the ulnar nerve, while in almost all the remainder both ulnar and median nerves shared in the supply. These findings are quite contrary to classical teaching. Opponens pollicis was supplied exclusively by the median nerve in two-thirds of cases only, most of the remaining one-third having a double supply. He reported frequent bilateral asymmetry; this is disturbing, to say the least, when one considers the practical difficulties introduced into clinical examinations.

The fundamental study of the thenar and hypothenar muscles (Forrest and Basmajian, 1965) included 25 young adult subjects. A preliminary detailed study in the dissecting room provided a thorough knowledge of the anatomy, relationship, and landmarks of the muscles concerned and provided an opportunity to practice and perfect the placement of electrodes. Their location in the middle of a muscle belly gives the truest picture of the general activity of that muscle. In the case of the flexor pollicis brevis, the electrode was inserted into superficial fibers corresponding to the portion of the muscle described by Jones (1942) as the superficial or external head; this arises from the flexor retinaculum and trapezium, passes along the radial side of the tendon of the flexor pollicis longus, and inserts into the radial sesamoid bone of the metacarpophalangeal joint and the base of the first phalanx of the thumb. In a later study, Forrest and Khan (1968) showed clear differences between this superficial head and the deep head insofar as specific fine movements are concerned.

All movements began with the subject's hand in the rest position, in which, of course, there were no action potentials. The subject then moved the hand into a series of prescribed positions. Movement was performed slowly; each position was held for several seconds; and then the hand was returned slowly to the rest position. Positions of opposition were held either softly (with thumb and finger just touching) or firmly (with just enough pressure to resist the withdrawal of a sheet of paper from between the thumb and finger). Objects (a cup, glass, or dowel) were held firmly, that is, securely, but with much less than maximum strength and effort. Although recordings were made during the entire movement, only those from the active positions are considered here, viz:

1. Of the thumb:

A. Extension (movement away from the radial side of the palm and index finger in the plane of the palm).

B. Abduction (movement away from the radial side of the palm and index finger in a plane 90° to that of the palm).

C. Flexion (flexion of the interphalangeal, metacarpophalangeal, and carpometacarpal joints of the thumb in a plane parallel to that of the palm so as to scrape the ulnar side of the thumb lightly across the palm).

2. Of the little finger:

A. Extension (full extension of all the joints of the little finger).

B. Abduction (movement away from the ringer finger in the plane of the palm).

C. Flexion (90° flexion of the little finger at the metacarpophalangeal joint with both interphalangeal joints almost fully extended).

3. Eight positions of opposition in which the thumb was held *softly* opposed

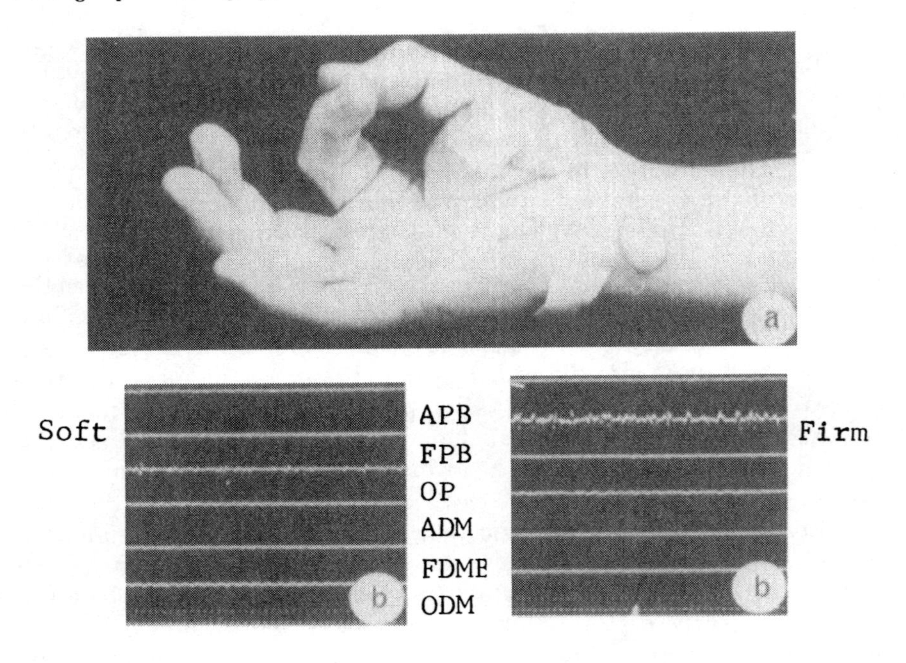

Figure 13.4. (*a*) Opposition of the thumb to the side of the index finger (position one). (*b*) Electromyographic recordings during soft and firm opposition, respectively. (From W.J. Forrest and J.V. Basmajian, © 1965, *Journal of Bone and Joint Surgery*.)

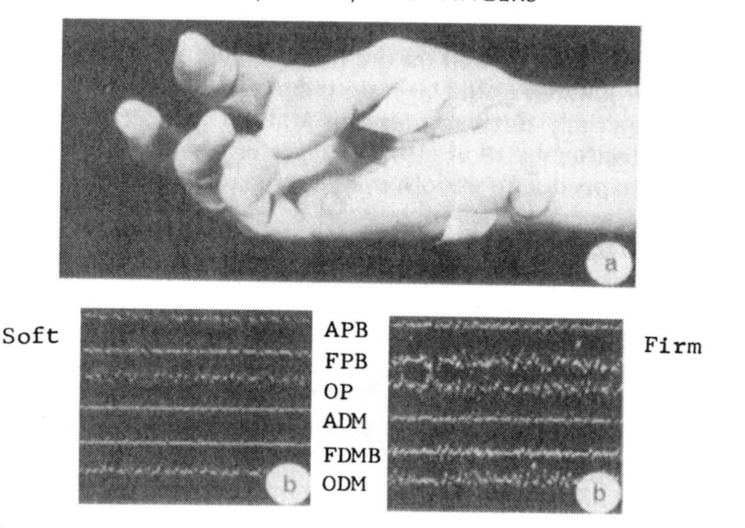

Figure 13.5. (*a*) Opposition of the thumb to the little finger, tip-to-tip (position 8. (*b*) Electromyographic recordings during soft and firm opposition, respectively. (From W.J. Forrest and J.V. Basmajian, © 1965, *Journal of Bone and Joint Surgery*.)

to each finger in two ways—with the pad of the thumb to the lateral side of the bent finger near its tip, and with the thumb and finger tip-to-tip, roughly forming the shape of the letter O (this series began with position 1, opposition to the side of the index finger—as in Figure 13.4*a*—and position 2, opposition to the tip of the same finger, and the proceeded in a similar fashion to the long, ring, and little fingers, ending with position eight, tip-to-tip opposition to the little finger—as in Figure 13.5*a*.

4. *The same eight positions* with *firm* opposition.
5. *Clasping* firmly a wooden dowel 1 inch (2.5 cm) in diameter.
6. *Holding*, in turn, 2 inches (5 cm) above the table, first a glass of water and then a cup of water by the handle while the subjects sat with elbow unsupported and flexed to 90°.

POSTURES OF THUMB

During extension, only the opponens pollicis and abductor pollicis brevis showed appreciable activity, which was moderate on the average. During abduction, the same two muscles showed marked activity on the average whereas the activity of the flexor pollicis brevis was slight. During flexion, the mean activity of the flexor pollicis brevis was moderate to marked, but the opponens pollicis was only slightly active, and the abductor pollicis brevis was essentially inactive.

The occurrence of equal levels of activity in both the abductor pollicis brevis and the opponens pollicis during extension and abduction of the thumb cannot be rationalized on the basis of their insertions. These are such that these muscles would be expected to move the thumb in opposite directions, especially during extension and, to a lesser extent, during abduction. Weathersby et al (1963) suggested that stabilization of the part in order to produce a smooth even motion was a possible explanation for the significant activities of muscles in situations such as this. This would seem to be a valid explanation.

Not all thenar muscles were active during extension and flexion of the thumb. Only three subjects showed more than slight activity in the flexor pollicis brevis during extension, the mean activity being nil-to-negligible. During flexion, the abductor pollicis brevis exhibited negligible activity; the opponens pollicis, slight activity on the average; and the flexor pollicis brevis, moderate-to-marked activity. Indeed, in the position of flexion, 10 of the 25 subjects had *nil* or negligible activity in both the opponens and abductor while the flexor was significantly active.

In one other position there was coincident activity and inactivity in the thenar muscles. During firm pinch between the thumb and side of the flexed index finger (position 1), only negligible activity was recorded from the abductor pollicis brevis. Yet the opponens pollicis and, in particular, the flexor pollicis brevis were significantly active.

POSTURES OF LITTLE FINGER

During extension of the little finger, all three hypothenar muscles were rather inactive on the average, but in many subjects the activity in one or more of the three muscles was negligible or *nil*. During abduction, although the abductor digiti minimi fulfilled the function indicated by its name and was the dominant muscle (with a mean of moderate-to-marked activity), the two other hypothenar muscles were also significantly active. During flexion, moderate-to-marked activity occurred in all three hypothenar muscles.

The abductor digiti minimi was very active during flexion of the little finger at the metacarpophalangeal joint. (The participation of this muscle in this position of the finger is obvious also by palpation). Part of the explanation for this activity depends on the muscle's insertion into the ulnar side of the base of the proximal phalanx. The abductor digiti minimi was also significantly active when the thumb was held opposed to either the ring or little finger. Some of this activity is possibly associated with the small degree of flexion at the fifth metacarpophalangeal joint that is required when the thumb and little finger are opposed. Yet, such flexion is obviously not required during opposition of the thumb and ring finger. Some of the activity of the abductor digiti minimi, then, may be to provide stability, and simple abduction of the little finger may be the least important function of the abductor of this finger.

POSITIONS OF OPPOSITION

During soft opposition of the thumb to the side and tip of each finger (positions 1 through 8), gradual increases in activity were recorded from all six muscles, starting at position 1 in the case of the thenar muscles and beginning at position 5 in the case of the hypothenar muscles (Figs. 13.5*b* and 13.6). The opponens was the most active of the thenar muscles; the flexor was the least active (Fig. 13.4*b*). The opponens digiti minimi was the most active hypothenar muscle. All the thenar muscles were more active than the hypothenar muscles.

When opposition was firm (Fig. 13.7) the flexor pollicis brevis replaced the opponens pollicis as the dominant muscle, particularly in positions 1 to 4 (index and long fingers) (Fig. 13.4*b*). In positions 5 to 8, the activity of the opponens pollicis approached and then equalled that of the flexor

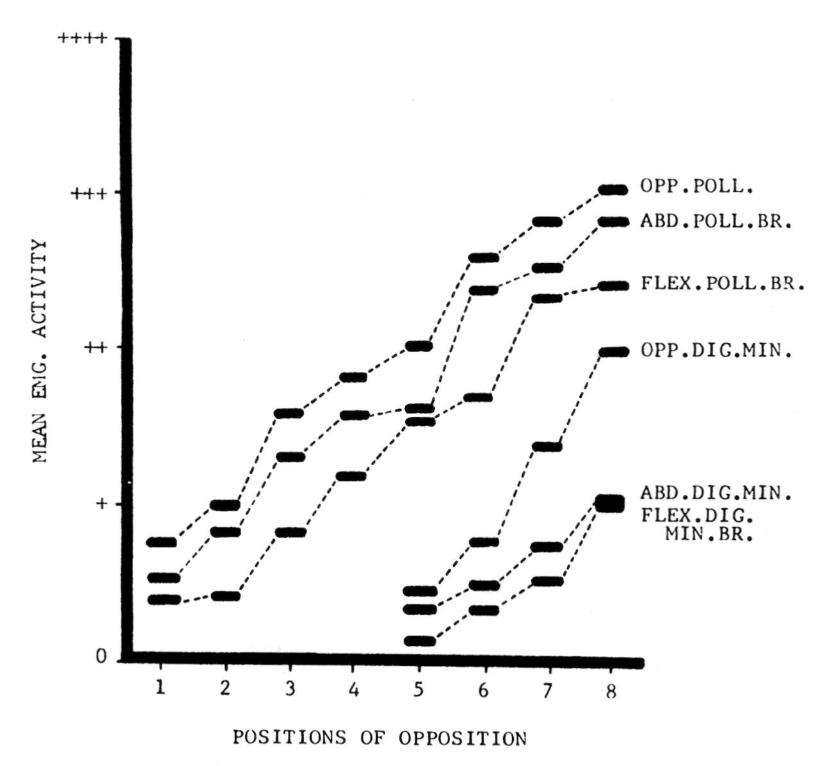

Figure 13.6. Mean electromyographic activities during soft opposition of the thumb to side and tip of each finger, beginning with the side of the index finger (position 1) and ending with the tip of the little finger (position 8). (From W.J. Forrest and J.V. Basmajian, © 1965, *Journal of Bone and Joint Surgery*.)

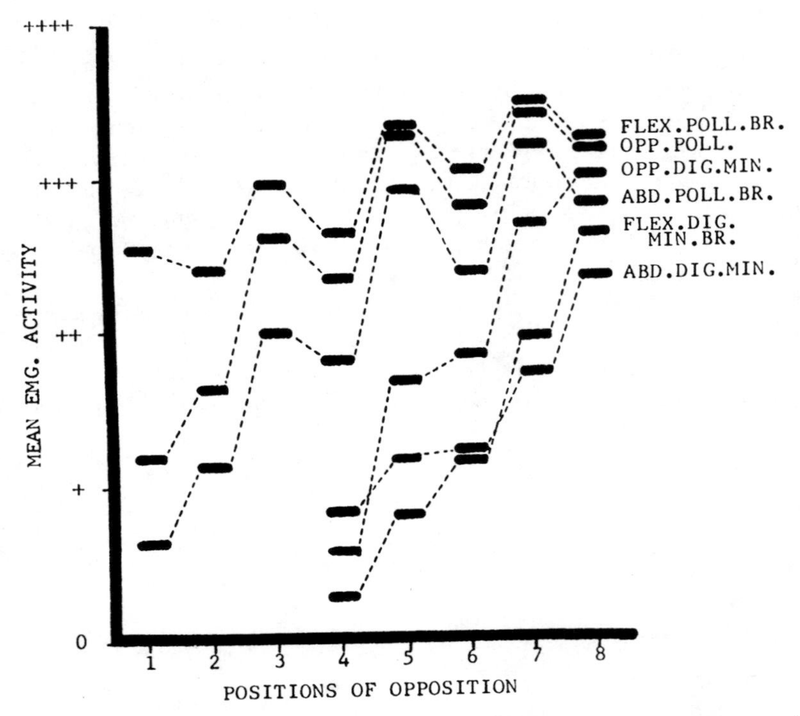

Figure 13.7. Mean electromyographic activities during firm opposition of the thumb to the side and tip of each finger, beginning with the side of the index finger (position 1) and ending with the tip of the little finger (position 8). (From W.J. Forrest and J.V. Basmajian, © 1965, *Journal of Bone and Joint Surgery*.)

pollicis brevis. The abductor pollicis brevis was the least active of the thenar muscles (Fig. 13.5*b*).

The steady increase in thenar muscle activity from positions 1 through 8 seen during soft opposition was not observed during firm opposition. Instead, higher levels of activity were usually recorded during firm opposition of the thumb to the side of the finger, as compared with the activity during tip-to-tip opposition with the same finger.

The hypothenar muscles showed a steady increase in activity during firm opposition, beginning at position 4 of the thumb. The opponens digiti minimi was again the most active muscle, and its mean actvity was even slightly greater than that of the abductor pollicis brevis in position 8 (Fig. 13.5).

When the thumb was opposed firmly to the index and long fingers, the flexor pollicis brevis replaced the opponens pollicis as the most active of the six muscles (Fig. 13.4*b*). The opponens, however, approached and then equalled the flexor in its activity during firm opposition to the ring and little fingers (positons 5 to 8). Firm pinch between the thumb and the index and long fingers is a grip position of day-to-day importance.

Little (1960) attributes to the flexor pollicis brevis and to the mechanically advantageous position of the adductor pollicis the great power of the thumb, which enables this digit to balance the combined power of the fingers. Ignoring the contribution of nonthenar muscles such as the adductor (which is expanded on below) our findings emphasize the importance of the short flexor in thumb power. Weathersby et al (1963) also noted an increase in flexor activity as the subject pressed lightly with the thumb and index finger in a position corresponding to our position 2.

One may observe that the greater the medial rotation of the first metacarpal, the greater is the tendency of the head of the fifth metacarpal to be drawn in an anterolateral (volar-radial) direction. The opponens digiti minimi is mainly responsible for this movement of the fifth metacarpal, and its action is almost reflexive in nature. The more active the opponens pollicis is in medially rotating the first metacarpal, the more

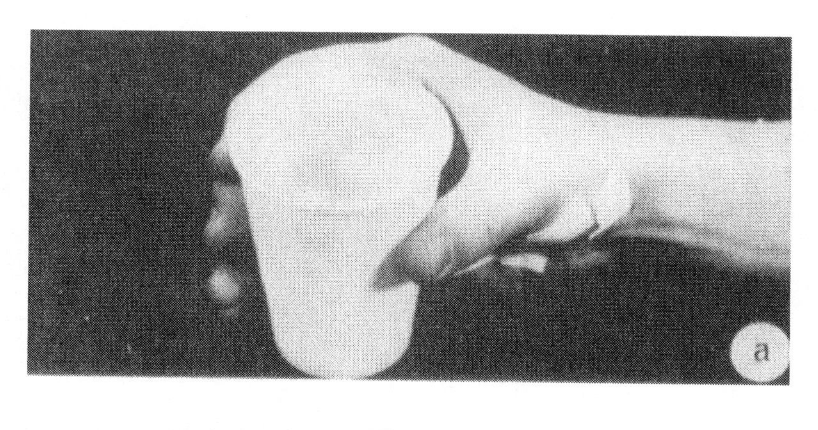

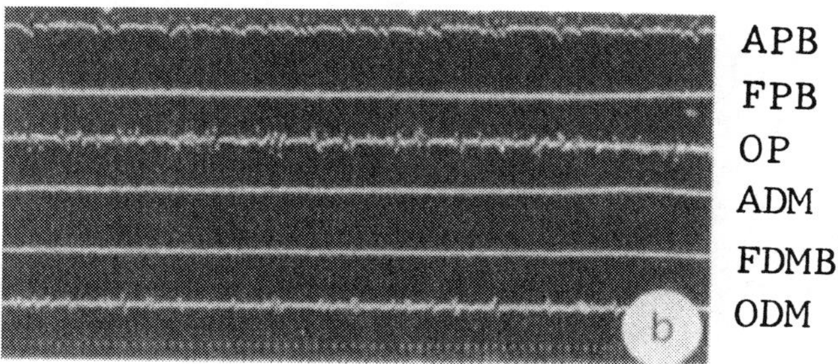

Figure 13.8. (*a*) Holding a glass of water. (*b*) Electromyographic recording. (From W.J. Forrest and J.V. Basmajian, © 1965, *Journal of Bone and Joint Surgery.*)

active the opponens digiti minimi becomes. But the opponens pollicis is always the more active muscle. It is possible that beyond a certain degree of medial rotation of the first metacarpal, the two opponens muscles begin to act in unison to form the transverse metacarpal arch mentioned by Littler (1960) and by others. Indeed, this might be expected when one views the two opponens muscles, with the flexor retinaculum between, linking up the first and fifth metacarpal bones.

POSITIONS OF GRIP

The important role of the flexor pollicis brevis in firm grasp is illustrated in the positions of firmly clasping a dowel and of holding a cup of water (Fig. 13.8). Although the flexor pollicis brevis was the most active muscle while the dowel was grasped firmly, this was not the case when the glass of water was also held firmly. Both the opponens pollicis and the abductor pollicis brevis were then more active (Fig. 13.8). This finding, and other preliminary work tht we have done has led to a tentative conclusion that the more the thumb is abducted (as in holding the glass), the less the flexor brevis contributes to a firm grip. The activity of this muscle, which provides firmness of grip when only a small degree

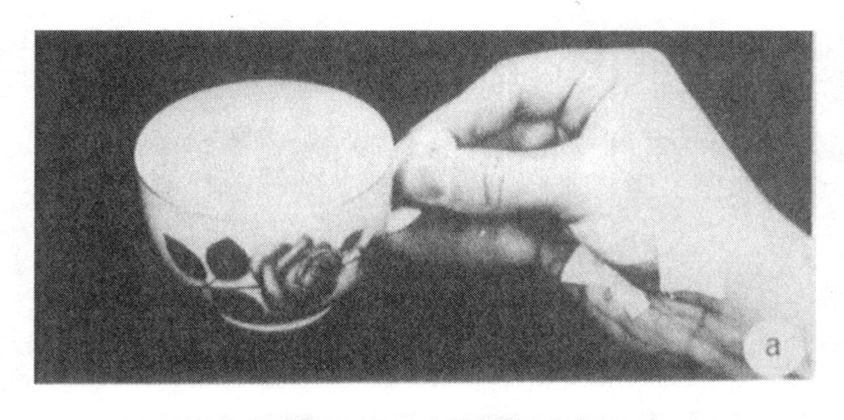

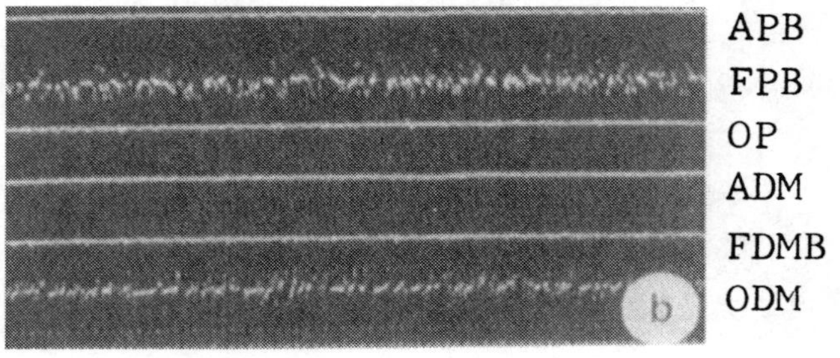

Figure 13.9. (*a*) Holding a cup of water. (*b*) Electromyographic recording. (From W.J. Forrest and J.V. Basmajian, © 1965, *Journal of Bone and Joint Surgery*.)

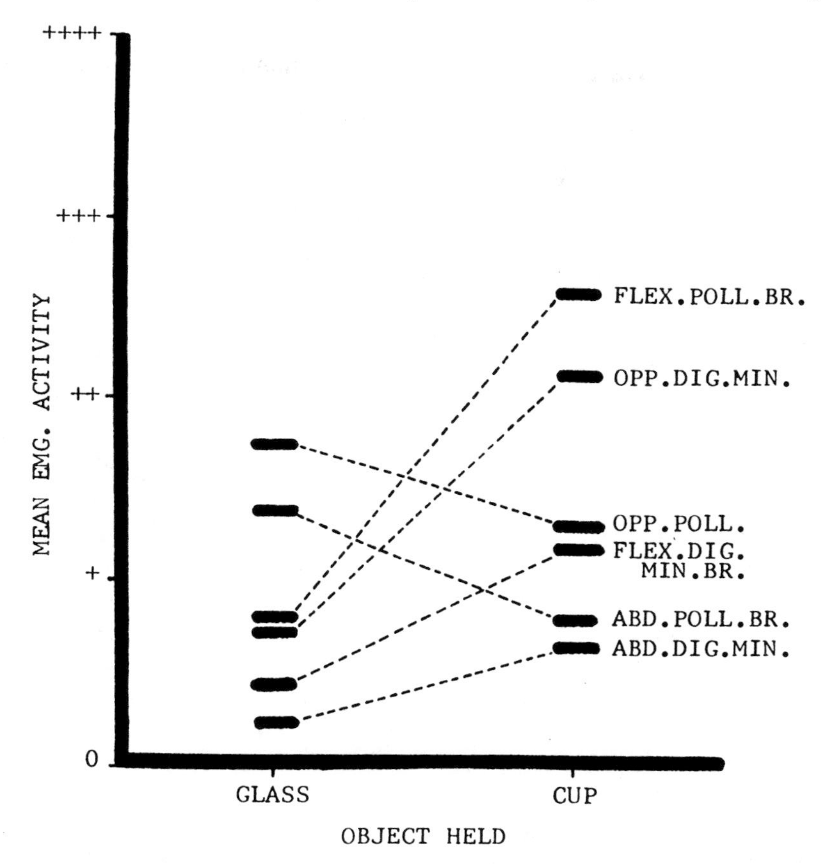

Figure 13.10. Holding a glass and cup of water: mean EMG activities compared. (From W.J. Forrest and J.V. Basmajian, © 1965, *Journal of Bone and Joint Surgery*.)

of abduction exists (as in holding the cup), is replaced by that of the opponens when a large amount of abduction is present (Figs. 13.9 and 13.10). In the absence of significant flexor activity, this activity of the opponens, coupled with that of the abductor, provides the power of a firm grip.

In summary, not all thenar muscles are active in all thumb positions, but all hypothenar muscles are active in three basic postures of the little finger. Two somewhat different patterns of activity occur when the thumb is first softly and then firmly opposed to each of the fingers in a sequence that begins at the index and ends at the little finger. The flexor pollicis brevis is dominant in firm grip, particularly in grip between the thumb and two radial fingers, but a large degree of abduction of the thumb might possibly be a limiting factor in the activity of this muscle.

The two opponens muscles seem to act as a unit in opposition of the thumb to both the ring and little fingers. Certain activity in some of the six muscles and inexplicable on a morphological basis probably serves to provide stability.

Flexor Pollicis Brevis vs. Adductor Pollicis

Forrest and Khan (1968) and Khan (1969) of Queen's University performed many elegant experiments to carefully discriminate between the various fine functions of the superficial and deep heads of the flexor pollicis brevis and of the transverse and oblique heads of the adductor pollicis. These "four" muscles radiate from their packed insertions like a fan. Then Johnson and Forrest (1970) elaborated the interplay of functions among the abductors and flexors of the thumb, including both intrinsic and extrinsic muscles.

Patterns of activity emerged for each part of the flexor brevis and adductor—which is not a thenar muscle (Forrest and Khan, 1968). As might be expected there were marked differences between the role of the transverse head of the adductor and the superficial head of the flexor during most functions. Most of the time, but not always, the two heads of each muscle act in concert. Surprising to those who hold that the deep head of the flexor belongs either with or to the oblique adductor was the finding that they often acted apart; this justifies the view that they are clearly different muscles belonging to flexion or to adduction, respectively. Forrest and Khan showed that in very fine functions each part of the two muscles is a functional entity with precise influence on the stabilization of the thumb.

Khan's (1969) detailed findings include the following.

The deep flexor is more flexor than adductor in its direction of fibers and observed activity, making its present name appropriate. Similarly, the oblique adductor is more adductor than flexor in its direction of fibers and observed activity; its present name is also appropriate.

Adductors are more active than flexors during combined adduction-lateral axial rotation of the thumb. Conversely, flexors are more active than adductors during combined abduction-flexion-medial axial rotation of the thumb.

The introduction of moderate resistance to a series of positions of simple pinch between the thumb and first two fingers resulted in no appreciable increase of activity in all muscles except the superficial flexor. This may be evidence, although inconclusive in this study, that the prime function of some of the intrinsic muscles of the thumb is to move and position it, not to provide it with strength.

Flexors are quite active during opposition of the thumb to the little finger and medial side of the hand, but significant adductor activity in this situation does not occur unless pressure is exerted by the thumb.

A small decrease in the activity of each of the four muscles apparently occurs as the thumb exerts moderately firm pressure against a series of objects requiring gradually increasing abduction of the thumb. A future study should deal with the role of the thumb and its muscles during grips of the hand.

Intrinsics vs. Extrinsics

Johnson and Forrest (1970) studied the intrinsic-extrinsic muscle interplay of paired muscles in a series of 20 subjects. A comparison of the first pair, the abductor pollicis brevis and longus, revealed that the brevis was usually more active than the longus; this was true in simple abduction of the thumb and in movements of the thumb having abduction as one component, with or without a load applied. In thumb extension, however, the longus was more active than the brevis when no load was applied; with a load, activities were equal. In comparing the second pair, the flexor pollicis brevis (superficial head) and longus, the brevis was usually more active than the longus in various postures of the hand requiring flexion-medial rotation of the thumb. In full flexion of the thumb (where medial rotation is minimal), the longus exhibited greater activity than the brevis. An interesting finding with respect to the two flexors was that they flexed the metacarpophalangeal joint of the thumb either independently or together, depending upon the position of the interphalangeal joint or the load applied.

The detailed findings reported by Johnson (1970) in his thesis include the following most significant points.

FLEXOR POLLICIS LONGUS VS. BREVIS

(1) The position of the distal phalanx in movements of the thumb apparently determines, to some extent, the interrelationships of these two muscles with respect to their action on the metacarpophalangeal joint of the thumb. As the interphalangeal joint is increasingly flexed, the flexor pollicis longus becomes the prime mover of the metacarpophalangeal joint. This is particularly evident when the thumb is flexed, and when its tip is placed near the bases of the two radial fingers. When the thumb is positioned near the distal ends of fingers, the short flexor appears to act as the prime mover of the proximal phalanx at the metacarpophalangeal joint.

(2) Most daily activities involving the thumb require that it be positioned near the tips of the fingers. The relatively high levels of activity seen in the short flexor as the thumb assumes these positions (no load applied) indicates that its primary role may be to position the thumb with the assistance of other thenar muscles and adductor pollicis. Throughout these movements, the long flexor remains relatively quiet and is probably involved only minimally in these positioning activities.

(3) The long flexor appears to provide much of the force necessary in overcoming moderate loads applied to the thumb while its tip is positioned near the tips of fingers. This is true regardless of the position of the distal phalanx. With a load applied to the thumb during movements such as opposition to the tips of fingers, activity in the long flexor increases sharply—more so than in the short flexor. This conclusion is further reinforced by a morphological characteristic of this muscle. Some of its tendinous fibers insert onto the midportion (palmar surface) of the distal phalanx. As the distal phalanx is flexed, this factor apparently brings about an increase in the muscle's line of application of force to the center of rotation of the joint.

ABDUCTOR POLLICIS LONGUS VS. BREVIS

The brevis is more active than the longus in all movements of abduction or movements with abduction as a component without or with a load applied. This includes those postures of the thumb described as opposition and lateral pinch.

FLEXOR POLLICIS LONGUS VS. EXTENSOR LONGUS

During adductory movements of the thumb, the intrinsic muscles seem to require the assistance of the flexor pollicis longus and the extensor pollicis longus when a load is applied. In the process of balancing the other muscle's tendency to flex or extend the thumb, the resultant action of the two muscles is to adduct the thumb.

Functions of Individual Muscles

According to Johnson (1970) the following clear functions are apparent.

ABDUCTOR POLLICIS LONGUS

In addition to its role as a primary mover in abduction of the carpometacarpal joint of the thumb, this muscle appears also to bring about some flexion of the joint. This conclusion was previously based on non-electromyographic evidence only, and is now supported by his study.

EXTENSOR POLLICIS LONGUS AND EXTENSOR POLLICIS BREVIS

These muscles are the prime movers in extension of the thumb. Perhaps their most valuable contributions to the function of the thumb lies in their ability to assist the abductor muscles in repositioning the thumb from positions of opposition or, by continued action, to spread the thumb out widely in order that the hand may grasp large objects. These muscles evidently act also to stabilize the interphalangeal and metacarpophalangeal joints of the thumb during some movements of opposition, perhaps enabling the thumb to perform fine, smooth, and precise movements.

ABDUCTOR POLLICIS BREVIS

Johnson's (1970) conclusions confirm findings made by previous authors using EMG in their investigations. (1) The abductor pollicis brevis, through its insertion into the extensor expansion of the thumb, assisted the extensor pollicis longus in extending the distal phalanx—particularly when Johnson applied a load. (2) Through its insertion into the proximal phalanx, it assists the flexor pollicis brevis and opponens pollicis in flexion-medial rotation of the thumb.

Conclusion

This section on the muscles of the hand must be ended by admitting the need for much systematic work on this most important functional region. Finally, a careful review of Duchenne's classical experiments is time well spent for those who wish to enlarge further upon the now available EMG findings reported above. There they will discover that much of our current understanding of the interossei and long muscles of the fingers actually springs from ancient times. Indeed, they will learn that it was Galen, in the second century A.D., who first described the actions of the lumbricals and interossei.

CHAPTER 14

Lower Limb

Because of its importance in posture and locomotion and because of its accessibility and large size, the lower limb has been the subject of electromyography from the earliest days of this science. The quality of the research done in this region of the body has been spotty, and many unwarranted conclusions have been made and—quite fortunately—have been ignored by textbook writers. Part of the trouble stems from poor technique that in turn arises from inexperience. Novices seem to be especially prone to doing their earliest studies on the lower limb. Furthermore, some of them seem to have become completely overcome by their initial effort and seem to have stopped publishing completely. One can only hope that this was occasioned by remorse.

The muscles of the limb will be discussed from above downwards. Reference has already been made to the postural functions of these muscles in Chapter 11. Therefore, some repetition is unavoidable and, indeed, desirable. Locomotion is treated as a special chapter (p 367) but will be referred to wherever necessary in this chapter also.

HIP REGION

The muscles of this region which have been studied by various investigators and which will now be considered are: iliopsoas, the gluteal muscles and tensor fasciae latae. Other muscles that cross the hip joint (adductors, hamstrings, rectus femoris, and sartorius and gracilis) will be considered with the muscles of the thigh (p 319).

Iliopsoas

Orthopedic surgery of the hip joint focussed attention on the muscles of the region. Interest in the functions of iliopsoas in particular was first renewed by the novel surgical procedure introduced by Mustard of Toronto (1952) and modified by Sharrard (1964) in which the insertion of iliopsoas is transplanted to the greater trochanter to substitute for paralyzed abductor muscles. Surgeons found that the resulting restoration of stability to the pelvis greatly outweighs the reduction of flexor power. The muscle remains alive in most cases (Broome and Basmajian, 1971b), and the remaining flexor muscles are quite capable of providing any needed flexion for ordinary functions (Mustard, 1958).

When one begins to search the usual source books for precise information about the actions and functions of iliopsoas, the only point that is agreed upon by all is that the muscle is obviously a flexor of the hip and probably has some influence on the lumbar vertebrae. There is a confusing disagreement about the other influences produced by the muscle. Last (1954) ascribed medial rotation of the hip mainly to the "powerful pull of ilio-psoas." The American *Gray's Anatomy* edited by Goss (1959) stated that it rotates the hip medially while the British *Gray's* edited by Johnston and Whillis (1954) was more cautious with "it produces a slight degree of medial rotation . . . ." Lockhart (1951), in *Cunningham's Textbook*, agreed with this. Woodburne (1957) agreed with Steendijk (1948) that iliopsoas rotates medially when the limb is extended and laterally when it is flexed. To complete the spectrum of opinion, at least a dozen major reference works (see Steendijk, 1948) stated that iliopsoas is a lateral rotator.

In this running controversy that is now more than a century old, almost everyone lost sight of the principle that a muscle so close to a joint must have an important postural or stabilizing function.

There also was some disagreement in the EMG information of this muscle, partly resulting from different techniques. Using surface electrodes over this deep and almost inaccessible muscle, Joseph and Williams (1957) concluded erroneously that iliopsoas is inactive in standing subjects. This was confirmed with wire electrodes for iliacus by LaBan et al (1966), although an investigation of iliacus with long needle electrodes (Basmajian, 1958b) indicated a continuous slight to moderate activity during relaxed standing in four persons (Fig. 14.1*A*). Using bipolar wire electrodes, Basmajian and Greenlaw (1968) found that the psoas major also shows some slight to moderate activity during relaxed standing in many normal young adults. The bipolar, fine-wire electrodes (p 29) were introduced from behind directly into the psoas in the midlumbar region. Similar findings have been made by Keagy et al (1966) of Northwestern University in five subjects and in several normal subjects by Close (1964). Although there seems to be a wide division of opinion here, the truth may be that there is very little real difference. What disagreement there is probably arises from differences both in technique and in the stance of subjects. In any case, the activity is not very marked.

ILIACUS

As one would expect, action potentials are recorded during flexion of the hip in almost any posture of the whole subject and in almost the whole range. The amount of activity varies directly with the effort or resistance. We find marked activity in iliacus throughout flexion of the hip during "sit-up in the supine position" (Greenlaw and Basmajian, 1968, unpublished). However, LaBan et al found there was little or no activity in iliacus during the first 30° of hip flexion. But, during "sit-up" from the "hook-lying" position, considerable activity occurred during the entire movement. Flint (1965) reported considerable variation in the

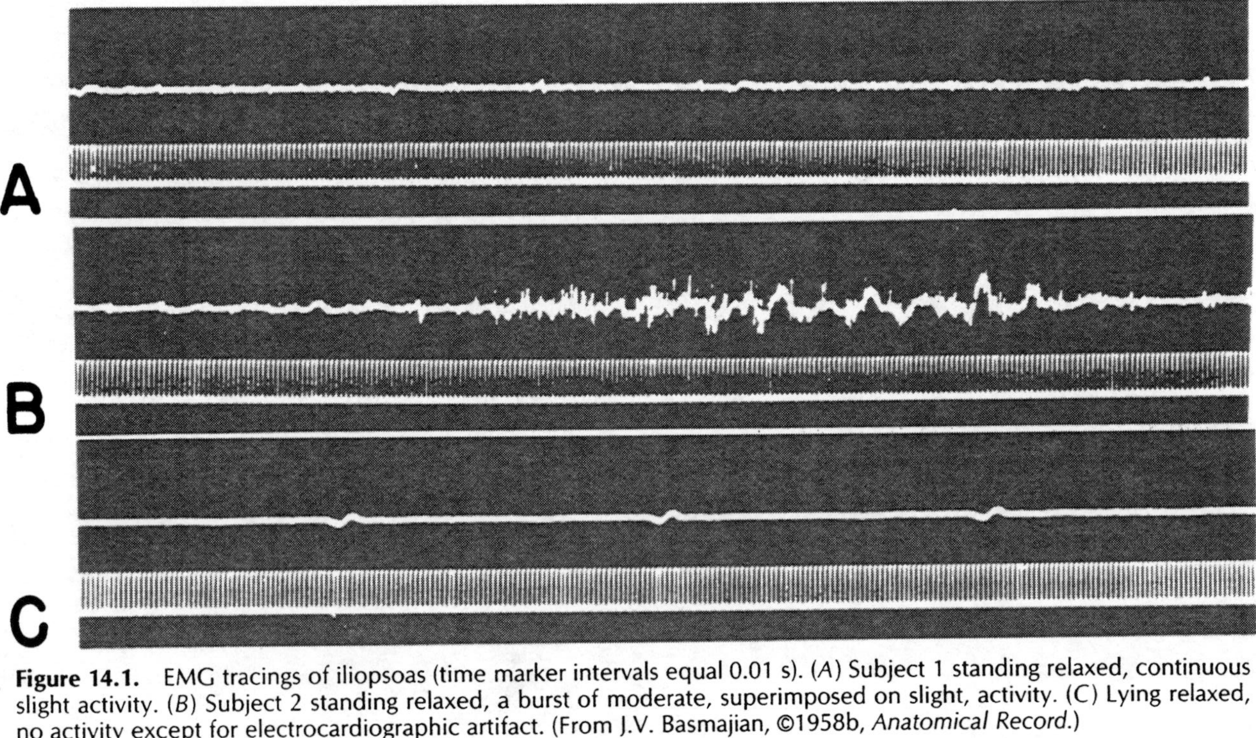

Figure 14.1. EMG tracings of iliopsoas (time marker intervals equal 0.01 s). (*A*) Subject 1 standing relaxed, continuous slight activity. (*B*) Subject 2 standing relaxed, a burst of moderate, superimposed on slight, activity. (*C*) Lying relaxed, no activity except for electrocardiographic artifact. (From J.V. Basmajian, ©1958b, *Anatomical Record*.)

styles of doing sit-ups but generally got little activity through surface electrodes in the first 45° of flexion.

Sometimes the iliacus muscles show intermittent short bursts of marked activity at irregular, short intervals during quiet standing; these apparently occur with invisible changes of position of either the limb or the trunk (Fig. 14.1*B*).

When subjects lie down or sit at ease, and during extension and adduction of the joint, there is no activity (Fig. 14.1*C*). Both medial and lateral rotation of the hip joint may produce some slight activity, whether the joint is passively or actively held in any of the extended, semiflexed, or flexed positions (Fig. 14.2), but usually there is clearly more on lateral rotation. LaBan et al confirm these findings in general.

PSOAS MAJOR

The direct recordings from psoas are strikingly similar to those from iliacus. Thus there was a slight activity during relaxed standing, strong activity during flexion in all of many trial postures, slight-to-moderate activity in abduction and in lateral rotation (depending on the degree of accompanying flexion), none during most medial rotations, and little during most other conditions involving the thigh. Tönnis (1966a and b) agrees with these rotation findings but, although he got similar results from abduction, he discounts his results, feeling the iliopsoas "cannot" be an abductor but rather must be an "antagonist to the glutei"—a tortuous and unnecessary conclusion. The only lumbar movement which consistently recruits psoas is a deliberate increase in lumbar lordosis while standing erect. (The details of this study by Greenlaw and Basmajian unfortunately remain unpublished.) Robert Close (1964) emphasized the abductor function of iliopsoas, which he found to be quite active during extreme abduction.

Nachemson's (1966) study of the vertebral part of psoas with coaxial needle electrodes dealt with vertebral effects. He concluded that psoas has a significant role in maintaining upright postures.

Iliopsoas—both its parts—appears to be an active postural or stabilizing muscle of the hip joint as well as a flexor (Fig. 14.3). The controversy as to whether it is a medial or a lateral rotator should be abandoned because, in fact, it is only a weak lateral rotator. Indeed, reviewing the work of Duchenne, one would find that he did not disagree with this conclusion, although his view has often been misrepresented.

A number of clinical studies have appeared which reflect on normal functioning and may be of special interest to some readers (Sutherland et al, 1969; Baumann, 1969, 1971, 1972; Baumann and Behr, 1969; Stotz and Heimstädt, 1970).

The Glutei and Tensor Fasciae Latae

The gluteus maximus has usually been considered separately, even where many muscles have been studied simultaneously, while the gluteus

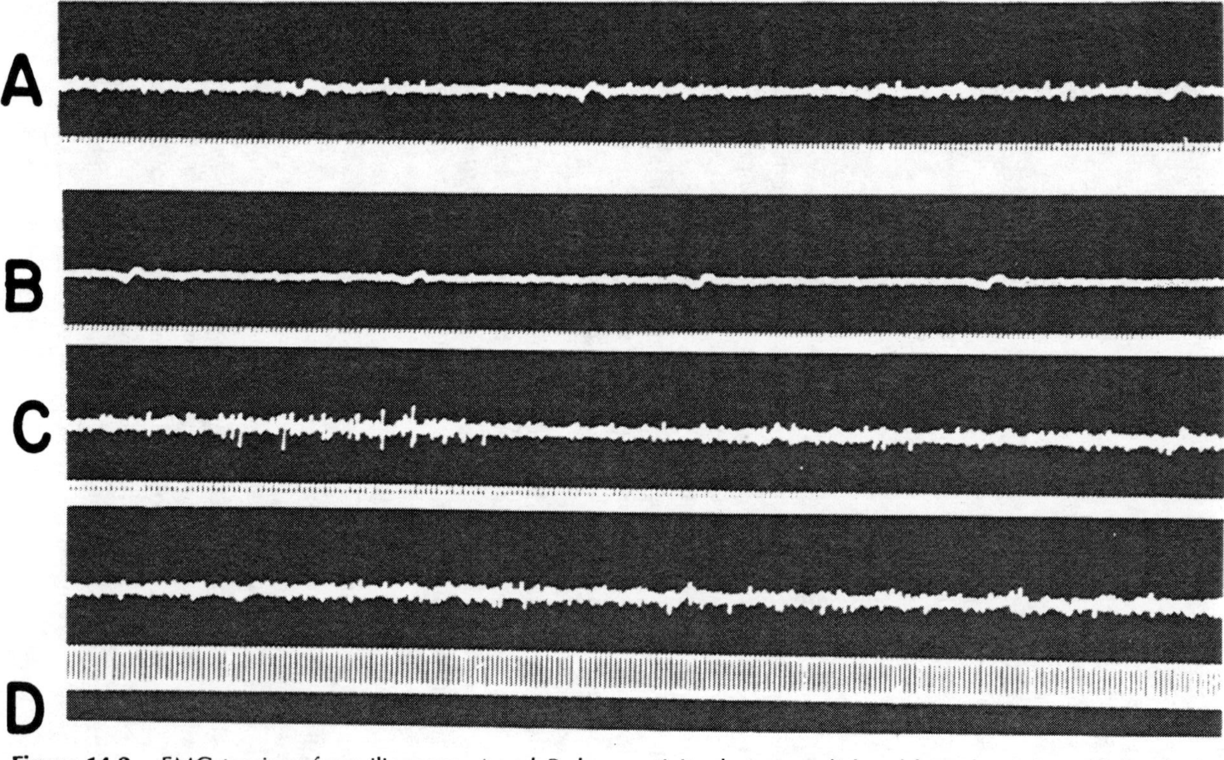

Figure 14.2. EMG tracings from iliopsoas. *A* and *B* show activity during medial and lateral rotations of the thigh, respectively, in one subject. (*C* and *D*) Medial and lateral rotation in another subject. (From J.V. Basmajian, ©1958b, *Anatomical Record*.)

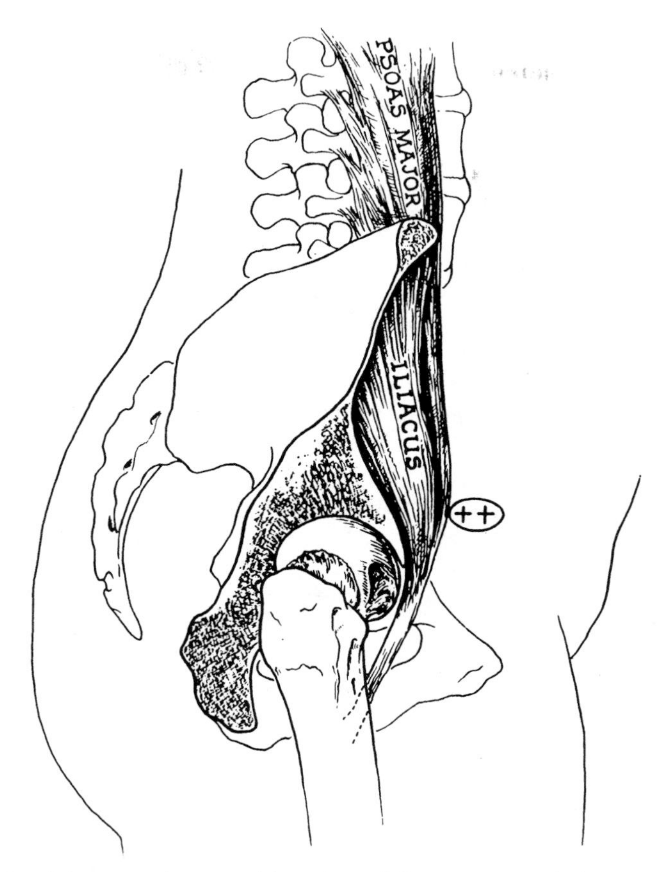

Figure 14.3. Diagram of postural function of iliopsoas. It is slight-to-moderately active continuously during standing.

medius and gluteus minimus have usually been considered together because of their close association. Tensor fasciae latae is, of course, closely associated with the glutei.

GLUTEUS MAXIMUS

Wheatley and Jahnke (1951), Karlsson and Jonsson (1965), and Greenlaw and Basmajian concluded that the gluteus maximus was active only when heavy or moderate efforts were made in the movements classically ascribed to this muscle (Greenlaw, 1973). It was active during extension of the thigh at the hip joint, lateral rotation, abduction against heavy resistance with the thigh flexed to 90°, and adduction against resistance that holds the thigh abducted. Lateral rotation (but not the opposite) also produced activity in gluteus maximus. While the whole muscle is engaged in extension and lateral rotation, only its upper part is abducent. As an abductor, gluteus maximus is a reserve source of power. Furlani et al (1974) are in general agreement with these findings.

The studies of Joseph and Williams (1957) show that the gluteus maximus is not an important postural muscle, even during forward swaying. In bending forwards it exhibited moderate activity. When straightening up from the toe-touching position it showed considerable activity throughout the movement. Tests of this muscle with wire electrodes tend to confirm all the above findings which were made with surface electrodes (Greenlaw and Basmajian, in Greenlaw, 1973). They also confirm the report of Inman (1947) that the gluteus maximus is not a postural abductor muscle when the subject is standing on one foot (as are the medius and minimus). However, Karlsson and Jonsson found that when the center of gravity of the whole body is grossly shifted, activity of gluteus maximus occurs. In positions where one leg sustains most of the weight, the ipsilateral muscle is active in its upper or "abducent" part; apparently this is to prevent a drooping of the opposite side. They also found that, during standing, rotation of the trunk activates the muscle that is contralateral to the direction of rotation (i.e., corresponding to lateral rotation of the thigh). Forward bending at the hip joint and trunk recruits gluteus maximus, apparently to fix the pelvis. One of the chief values of the work of Karlsson and Jonsson is their showing the *range* of responses from their subjects—who showed considerable normal variation.

Duchenne's observation that complete paralysis of gluteus maximus in no way disturbs relaxed walking has often been noted, but we should emphasize that this does *not* mean that normal walking does not recruit activity in the muscle. In fact, with fine wire electrodes in the upper, middle and lower parts of gluteus maximus in a long series of subjects Greenlaw and Basmajian found two considerable bursts in specific sections of the walking cycle (Greenlaw, 1973). This is enlarged upon under the section on gait on p 374. Finally, Houtz and Fischer (1959) have found it to be unimportant in bicycle pedaling.

Hominid Evolution

Our extensive EMG study of apes, led to the conclusion that no major factual impediment exists in its functions to theorizing that the human gluteus maximus evolved from an ape-like condition (Tuttle et al, 1975). In the gorilla, the upper and middle parts act prominently as a hip extensor and lateral rotator. The lower part of the muscle is active as a hip extensor.

GLUTEUS MEDIUS AND MINIMUS

Though Inman's demonstration (1947) of marked EMG activity in the abductors when the subject stands on one foot is hardly surprising, the reader is referred to his valuable 1949 paper in which he described various other factors as well (Fig. 14.4). Similarly, the finding of Joseph and Williams (1957) that gluteus medius and minimus are quiescent during relaxed standing is to be expected. Detailed studies with wire electrodes (Greenlaw and Basmajian, in Greenlaw, 1973) have also confirmed the usual teaching about these abductors—emphasizing their importance in preventing the Trendelenburg sign, during abduction of the thigh and in medial rotation (anterior fibers). This last action is the only controversial one and seems now to be confirmed by the locomotion studies with wire electrodes by Greenlaw and Basmajian (see p 379); these show triphasic activity for gluteus medius and biphasic activity for gluteus minimus during each cycle of walking (Greenlaw, 1973). Houtz and Fischer (1959), from their EMG studies, concluded that the activity in all the glutei was minimal in bicycle pedaling (Fig. 14.5).

During elevation (flexion) of the thigh in the erect posture, Goto et al (1974) found that the anterior part of the gluteus medius was also active in the initial stage only. With resistance it increased.

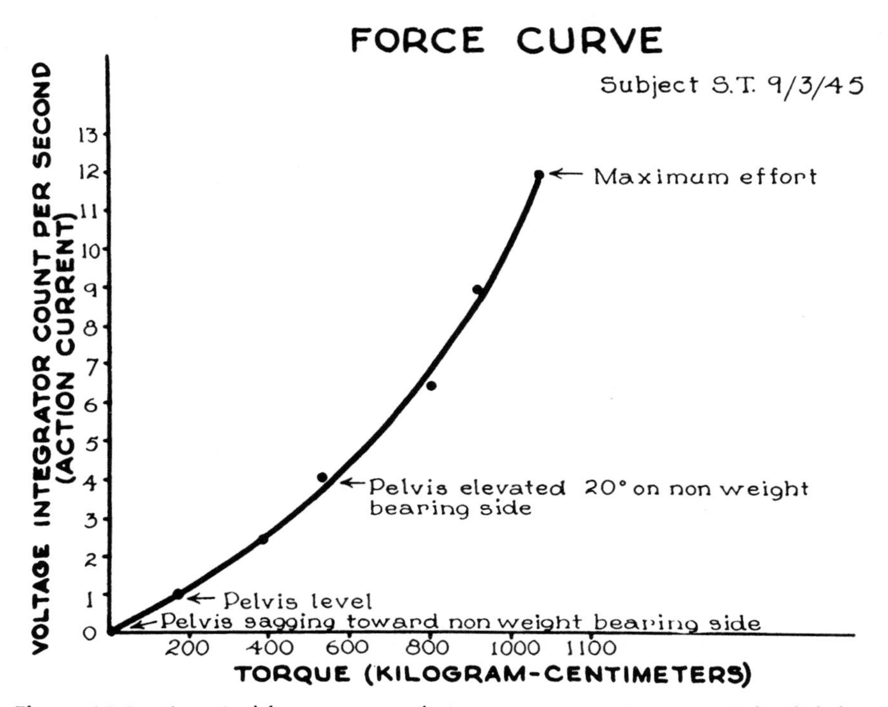

Figure 14.4. A typical force curve, relating torque to action potentials of abductor muscles of the hip. (From V.T. Inman, ©1947, *Journal of Bone and Joint Surgery*.)

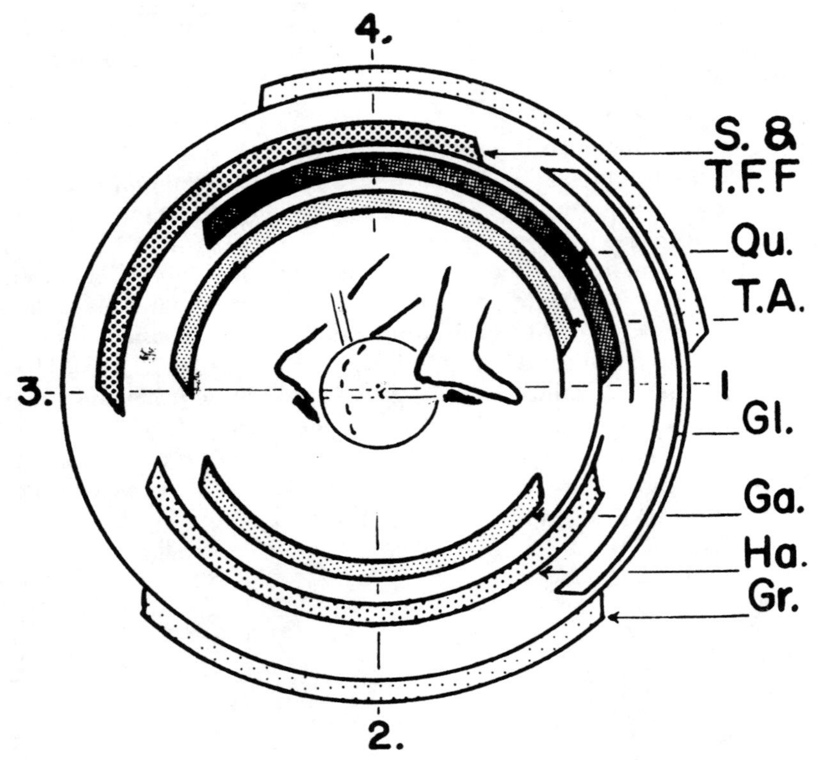

Figure 14.5. Diagrammatic summary of EMG activity in lower limb during one cycle of bicycling. Greatest activity is indicated by *shaded areas,* but where cycle is completed with a single line, it means that slight activity continues. *Gr.,* gracilis; *S. & T. F. F.,* sartorius and tensor fasciae latae (femoris); *Qu.,* quadriceps; *T.A.,* tibialis anterior; *Gl.,* gluteus maximus and medius; *Ga.,* gastrocnemius; *Ha.,* hamstrings. (From S.J. Houtz and F.J. Fischer, ©1959, *Journal of Bone and Joint Surgery.*)

TENSOR FASCIAE LATAE

Wheatley and Jahnke (1951), Carlsöö and Fohlin (1969), Goto et al (1974), and Carvalho et al (1972) found action potentials in this muscle during flexion, medial rotation, and abduction of the hip joint. It was a medial hip rotator in all positions. Duchenne clearly stated that the power of tensor fasciae latae as a rotator (in response to faradic stimulation is weak, and this appears to be correct. Carlsöö and Fohlin discount the rotary influence of tensor fasciae latae at the knee, finding no activity. During walking, Greenlaw and Basmajian found the muscle active biphasically during each cycle (see p 380). Unlike the glutei, tensor fasciae latae are active during bicycling, showing their greatest activity as the hip is being flexed (Houtz and Fischer, 1959).

THIGH MUSCLES

The groups and single muscles to be considered now are the adductors (longus, brevis, magnus, gracilis, and pectineus), the hamstrings (semimembranosus, semitendinosus, and biceps femoris), sartorius, rectus femoris, the vasti (medialis, lateralis, and intermedius), and the popliteus. Some of these cross the hip joint only (adductors); others cross both hip and knee (hamstrings, rectus, gracilis, and sartorius); and still others cross the knee only (vasti and popliteus).

Adductors of the Hip Joint

In the first edition of this book, it was necessary to admit that "a surprising hiatus appears in our knowledge of the adductors. Forming an enormous mass on the medial side of the upper thigh, they must have considerable importance. In spite of this, their exact function is usually a matter of guesswork."

Since these words were written, in addition to the extensive studies of Greenlaw and Basmajian (1968, unpublished except in Greenlaw's 1973 thesis), Janda and Véle (1963) and Janda and Stará (1965) have helped to correct the situation. They studied the role of the adductors in children and adults during flexion and extension of both the hip and the knee, with and without resistance. (Care was taken to avoid rotation.)

In almost every child the adductors were activated during flexion or extension of the knee, and they were very active against resistance. Most adults showed activity during flexion of the knee, but only a minority were active during extension. With resistance, almost all adult subjects showed great activity.

During movements of the hip the role of the adductors was localized to their upper parts. During flexion against resistance, all the children and half the adults showed activity. During resisted extension, all were active.

Janda and Stará suggest that this labile response of the adductors is related to a postural response. They believe that these muscles are facilitated through reflexes of the gait pattern rather than being called upon as prime movers. With this view one can readily agree. Spruit's (1965) theoretical analysis of the adductors adds conviction to the opinion.

De Sousa and Vitti (1966) studied the adductores longus and magnus (upper and lower parts) during movements of the hip joint. During free adduction the longus is always active while magnus is almost always silent unless resistance is offered. Both muscles are active during medial rotation but not during lateral rotation of the hip, settling a classic argument that usually leaned in the other direction. The upper fibers of the adductor magnus showed the greatest activity.

During flexion of the thigh, de Sousa and Vitti found the main activity occurring in the adductor longus while the magnus is often completely silent. The results of Goto et al (1974) tend to agree. While standing in a relaxed natural posture, both muscles are inactive. However, weak activity sometimes appears when standing on one foot. Okamoto et al (1966) found similar results.

Using fine-wire electrodes, Greenlaw and Basmajian examined a long series of normal adult subjects during both free test movements and various postures and locomotions. Electrodes were in adductor longus, adductor brevis, and the upper and lower parts of adductor magnus. Even when standing on one foot, the adductors on that side remain silent. Medial rotation (but not lateral) recruits them all (except perhaps for the vertical part of the magnus). During walking, the adductors showed different types of phasic activity. There is marked difference between the two parts of the adductor magnus; the upper (truly adductor) part is active almost through the whole cycle. Adductor brevis and longus show triphasic activity with the main peak at toe-off (Greenlaw, 1973).

Gracilis, which belongs "officially" to the adductor group, will be considered separately below.

Hamstrings

The three hamstrings, biceps femoris, semimembranosus, and semitendinosus, act on both hip joint and knee joint. Various studies, including those of Basmajian, have shown that the first of these is active in ordinary extension of the hip joint (in contrast to gluteus maximus, which acts only against resistance), and in flexion and lateral rotation of the tibia at the knee. Wheatley and Jahnke (1951) have shown biceps is active also in lateral rotation of the extended hip and in adduction against resistance of the abducted hip. Furlani et al (1973, 1977) found with a needle electrode in each head of the biceps femoris of normal men that both heads were active in less than half of cases during lateral rotation of the knee. The muscle is an obvious flexor of the knee joint. During the standing-at-rest position and standing on one foot, both heads fell silent, as they did with adduction of the thigh (Furlani et al, 1973, 1977).

The semimembranosus and semitendinosus are quite active in extension and adduction against resistance of the abducted hip, and in flexion and medial rotation of the tibia at the knee joint. With medial hip rotation, recruitment is slight (Greenlaw and Basmajian, 1968, unpublished; Greenlaw, 1973).

Joseph and his colleagues at Guy's Hospital Medical School (1954 et seq) demonstrated the much greater stabilizing function of the hamstrings, as compared with gluteus maximus, but emphasized their quiescence in ordinary standing. Portnoy and Morin (1956) agreed with them, as do Greenlaw and Basmajian; even standing on one foot does not recruit much activity in hamstrings.

In flexion at the hip and in leaning forwards, the hamstrings are much more active as supports against gravity. Arienti (1948a and b) of Milan, Italy, showed in early EMG studies performed while the subjects were walking on a treadmill that the hamstrings come into action at different stages of walking. It is not possible to forecast the exact phase of activity in a muscle during walking by only examining it while the limb is put through artificial tests of prime movers. For example, if semitendinosus and semimembranosus are examined while the two are producing a deliberate test movement, such as flexion of the knee, they are found to act synchronously. On the other hand, the semitendinosus has a triphasic pattern during each walking cycle while semimembranosus is biphasic (see p 380).

Arienti believed that, although both heads of biceps femoris act synchronously during a free-moving test of flexion, the short head acts during the swing phase of walking while the long head acts as a stabilizer when the foot is on the ground. Greenlaw and Basmajian found that an increase of walking speed changes a biphasic pattern to a triphasic one in the long head (p 380).

Hirschberg and Nathanson of New York University made similar studies, reported in 1952. The hip muscles, quadriceps, and hamstrings showed specific individual patterns, being active (in general) during the transition from the swing phase to the stance phase (Fig. 14.6). Only gluteus medius continued to contract beyond the middle of the stance phase; the others stop contracting within the first third of the stance phase. At the transition from the stance to swing, the adductor muscles (and sometimes the hamstrings) contract, according to Hirschberg and Nathanson.

There is now no doubt that the hamstrings do not by regional contraction act only on one joint. The studies of these muscles described in Chapter 9 (p 232) showed that the entire muscle contracts, regardless of whether the upper or lower joint is moved. Which joint is to move as a consequence depends on the immobilization of the other joint by other agencies.

Rectus Femoris

Undoubtedly, the rectus femoris is a flexor of the hip and extensor of the knee joint, and electromyography can only contribute information of secondary importance, such as the timing of its activity in various movements. Because flexion of the hip is closely associated with extension of the knee, one is not surprised to find a single muscle performing these two movements. As with the hamstrings, Basmajian's studies on two-joint muscles demonstrated that normally the whole muscle contracts even with isolated movements of only one of the two joints. With truly isolated movements of single joints, Carlsöö and Fohlin (1969) revealed that *no*

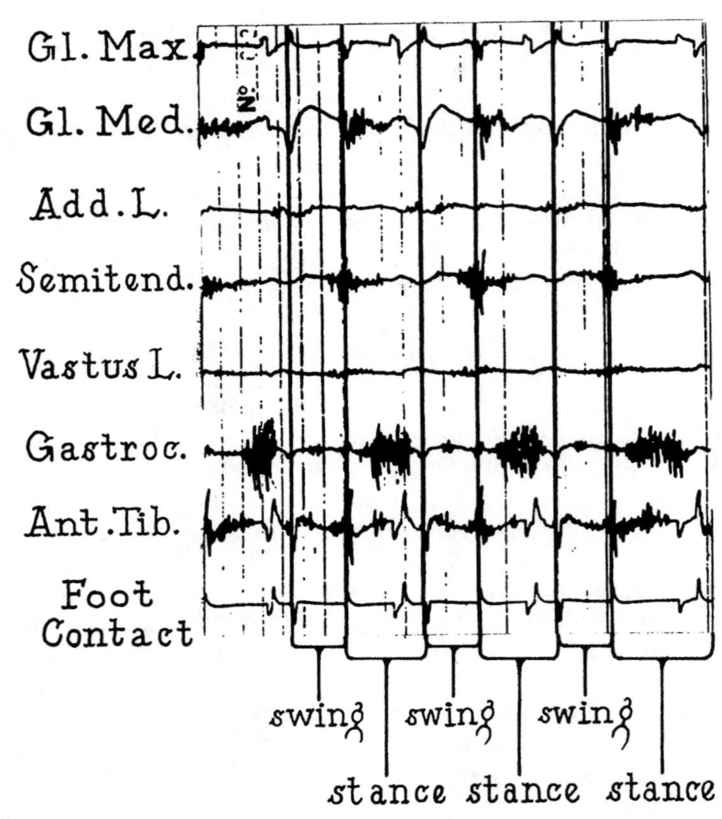

Figure 14.6. EMG activity of normal gait. (From G.G. Hirschberg and M. Nathanson, ©1952, *Archives of Physical Medicine*.)

activity occurs in rectus femoris during knee extension. Okamoto (1968) found that the hip joint must be stabilized for full function to occur in rectus during knee extension.

Greenlaw and Basmajian found rectus femoris negligibly active when standing on one foot; it is a lateral rotator but not a medial rotator. During moderate walking it shows slight biphasic activity (Greenlaw, 1973).

Rectus femoris also aids in abduction of the thigh (Wheatley and Jahnke, 1951) and apparently is relaxed in ordinary standing. The studies of Joseph and Nightingale (1954), Joseph and Williams (1957), Portnoy and Morin (1956), Floyd and Silver (1950), and Basmajian's observations all agree on this. Houtz and Fischer (1959) showed a marked activity in rectus, together with the vasti, during the thrusting motion of bicycle-pedaling but not during the flexion of the hip joint.

Gracilis and Sartorius

Although belonging to the adductor mass, the gracilis crosses and therefore acts on both hip and knee. It is active in flexion of the hip with the knee extended but is inactive if the knee is allowed to flex simultaneously (Wheatley and Jahnke, 1951). It adducts the hip joint (and therefore it rightfully belongs to its parent group, the adductors), and it rotates the femur medially (Jonsson and Steen, 1966; Greenlaw and Basmajian, 1968, unpublished; Greenlaw, 1973). Jonsson and Steen found that during flexion of the hip joint, gracilis is most active during the first part of flexion, both in free "basic movements" and during walking and cycling. In walking on a horizontal level and on a staircase, its activity occurs during the swing phase. At the knee it is a flexor and medial rotator of the tibia, although in medial rotation its activity appears to be slight, according to Jonsson and Steen. Both groups found it is insignificant in maintaining the standing posture. In bicycle pedaling it is not very active (Houtz and Fischer, 1959), but during walking, gracilis, like adductor longus, shows prolonged activity through most of each walking cycle (Greenlaw, 1973).

Sartorius is active during flexion of the thigh, regardless of whether the knee is straight or bent and during flexion of the knee joint or medial rotation of the tibia (Wheatley and Jahnke, 1951). Both sartorius and gracilis may play a role in the fine postural adjustments of the hip and knee, although Joseph (1960) and Greenlaw and Basmajian found no activity in sartorius during relaxed standing. In the bicycling experiments of Houtz and Fischer (1959), sartorius showed its maximal activity during the thrusting phase of pedaling, as would be expected. Only one real peak of activity occurs during walking—it comes a short time after toe-off (Johnson et al, 1972; Greenlaw, 1973). During level walking, there is some activity throughout the swing phase rising to about the middle of the swing phase (Johnson et al, 1972). During the remainder of the gait, there was only *nil*-to-slight activity. The same pattern occurred in descending steps. However, in ascending steps, the sartorius appeared to be more active immediately before heel strike (Carvalho et al, 1972).

Flexion of the hip evoked the greatest activity of the sartorius muscle when only one joint was put through any movement. With combined movements of two joints, the greatest activity was recorded when hip flexion was accompanied by maximum knee flexion, although knee extension after the hip was flexed did show increased activity also (Johnson et al, 1972).

Only when it was resisted did abduction of the hip produce slight-to-moderate activity. Medial rotation of the hip recruited little or no activity. There was only slight activity during lateral rotation of the hip while supine, and there was slight to moderate activity while sitting.

As expected, flexion of the knee was generally accompanied by more activity than any other motion confined to the knee, being most marked against resistance in a sitting position. Extension of the knee was accompanied by slight activity in most subjects, but in three it showed more activity than in flexion of the knee (Johnson et al, 1972).

Variations in the insertion of the sartorius are well known and may account for this. Sartorius is the most anterior muscle of the "pes anserinus" muscles, and it is easy to see how the pull could occasionally be at, or anterior to, the knee axis; this condition would be enhanced when the knee is already in extension.

The sartorius may play a stabilizing role in strong knee extension as previously proposed (Houtz and Fisher, 1959). The role of the sartorius in walking appears to be that of a regulator. The swing phase of a normal gait is a low-energy phase. Once initiated, the weight of the leg swings forward as a pendulum, but its course is regulated by several muscles of the thigh and leg, one of which is the sartorius. Slight activity in this muscle occurs during the entire swing phase, being at its peak (moderate activity) at about the middle of the swing phase. The hip is laterally rotating during the entire swing phase due to pelvic rotation in conjunction with the leg's forward controlled momentum. More hip and knee flexion is needed to clear the toe at the middle of the swing phase, perhaps explaining the rise in activity in the middle of the swing phase.

To clear the foot while ascending stairs requires more flexion of the thigh as well as dorsiflexion of the foot and appears to be accompanied by more lateral rotation of the thigh at the end of the swing phase. This would explain the increase in its activity during the latter part of this swing phase.

Sequential Motor Patterns

Two-joint muscles play an important integrated function in both skilled and distorted activities of the hip and knee joints. These interactions are discussed on p 232.

Vasti

The vasti, of course, are powerful extensors of the knee joint. The experience of the Basmajian group since 1949 agrees with that of many other investigators that the vasti are, however, generally quiescent during relaxed standing (Åkerblom, 1948; Kelton and Wright, 1949; Floyd and Silver, 1950; Portnoy and Morin, 1956; Joseph et al, 1954 et seq). Joseph and Nightingale (1956) found that when women wear high heels, activity appears in the vasti in a substantial proportion of subjects. Arienti (see p 321) has demonstrated that during walking on a treadmill, the three vasti and rectus femoris do not act simultaneously but have a phasic pattern.

Why are there four heads for a muscle that has most of its insertion on the restricted edges of the patella? Do the various parts have individual functions? Although Lieb and Perry (1968) answered this question in part by ingenious morphological and biomechanical studies of amputated limbs, electromyography offers an added dimension to their findings by revealing the actual timing of activity and the interplay of function in the four parts.

Earlier EMG studies by Pocock (1963) and Close (1964), limited in technique, offer provocative but useful ideas that required confirmation, expansion, and quantification. An incentive for gaining new and precise information arises from the growing interest in myoelectric assistive devices for the physically handicapped (including programmed muscle stimulators for recruiting of lower limb muscles in proper sequence). For these reasons, a detailed study of the interplay of activity in the three vasti and the rectus femoris was made in a group of normal adults (Basmajian et al, 1972). Bipolar fine-wire electrodes injected in the middle of the belly of each head of quadriceps provided excellent EMG response with no "crosstalk."

TIMING OF ONSET AND CESSATION

Figures 14.7 to 14.9 summarize the tabulated results of the pooled data. In summary, the most striking features are: (1) the variation between the different heads; (2) the considerable standard deviation arising from individual differences; and (3) the late onset and early cessation of activity in rectus femoris during the squatting maneuvers.

GRADING OF ACTIVITY

Figure 14.10 summarizes the findings. The most revealing findings are the similarities between muscles in the pooled data with the considerable variation in individuals at certain times (the most marked variation occurring at the end of straightening up from the squatting posture). There is no question that weighted extension recruits the greatest activity. Vastus medialis does appear to increase its activity more rapidly toward the end of unweighted extension, but it is not greatly different from the other heads during the act of squatting.

It is now becoming widely recognized that vastus medialis acts through the whole range of extension, not just at its terminal phase. Also it is in this terminal phase that the muscle is supposedly the most active because it completes extension. However, another very real function of vastus medialis at the end of extension is its prevention of lateral dislocation of the patella. The special direction and insertion of the lowest fibers of the muscle point directly at this being the real role of that lowest part of the muscle which bulges so prominently here.

Most clinicians and functional anatomists believe that an unstable knee is due to an inability to produce the final 15° of extension. In turn, many clinicians continue to believe that this failure is primarily due to weakness of the medial head of the quadriceps. It is difficult to deny that the final

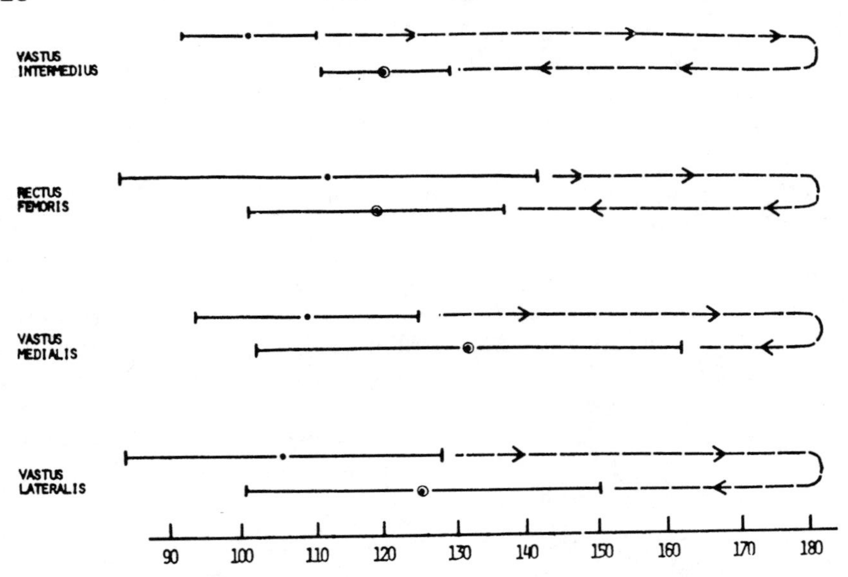

Figure 14.7. Diagrammatic representation of onset and cessation of activity in the four heads of quadriceps during unweighted extension of the knee from 90° to 180° and flexion back to 90°. *Subject seated; no added load.* Mean and standard deviation of the angle at which activity begins starts each line; the *broken arrows* indicate the direction of movement to full extension and back to the mean and standard deviation of the angle at which the activity ceases. (From J.V. Basmajian et al, 1972, *Anatomical Record.*)

15° of extension is important. Indeed, the aim of quadriceps retraining is to enable a patient to extend his knee fully and to maintain that extension. Whether the fundamental part is vastus medialis itself is open to serious doubt. Lieb and Perry (1968, 1971) concluded that early atrophy of the vastus medialis prominence, coupled with the loss of terminal extension, is simply indicative of a *general* quadriceps weakness. Earlier Hallén and Lindahl (1967) reported essentially the same findings. Both groups established the general role of vastus medialis rather than its widely touted terminal-extensor function. Hallén and Lindahl also stressed the importance of pain inhibition and adhesions in the limitation of extension in its last 10°.

The visible prominence of the vastus medialis is really related to the marked obliquity of the distal fibers of the muscle, its lowness of insertion, and the thinness of fascial covering in that area. The extensor lag accompanying knee extension is a function of great loss in the mechanical advantage of the whole muscle during the final 15° of the extensor range because a 60% increase of force is needed to complete extension

(Lieb and Perry, 1968). Thus, the only selective function attributable to vastus medialis is patellar alignment. One cannot deny the importance of the latter, but it has no special part to play as a prime mover in the final extensor movement insofar as mechanical advantage is concerned; this appears to be confirmed by the findings of Basmajian et al and more recently in a study of gait by Adler et al (1983). A surprising burst of activity during the squatting test as subjects pass from 150° to 135° remains unexplained.

The above detailed studies supplemented those of Pocock (1963), Close (1964), Lesage and Le Bars (1970) and Lieb and Perry (1971), and offers quantification in all four heads (including vastus intermedius). True, this does not greatly alter our understanding of the vastus intermedius which seems to have no discrete role. The question of "Why four heads?" is still unanswered, except in a negative way: the muscles act in concert to achieve a common end. Minor differences (such as the burst of activity in vastus medialis during squatting) may alone explain the structural discreteness. The relatively short period of significant grades

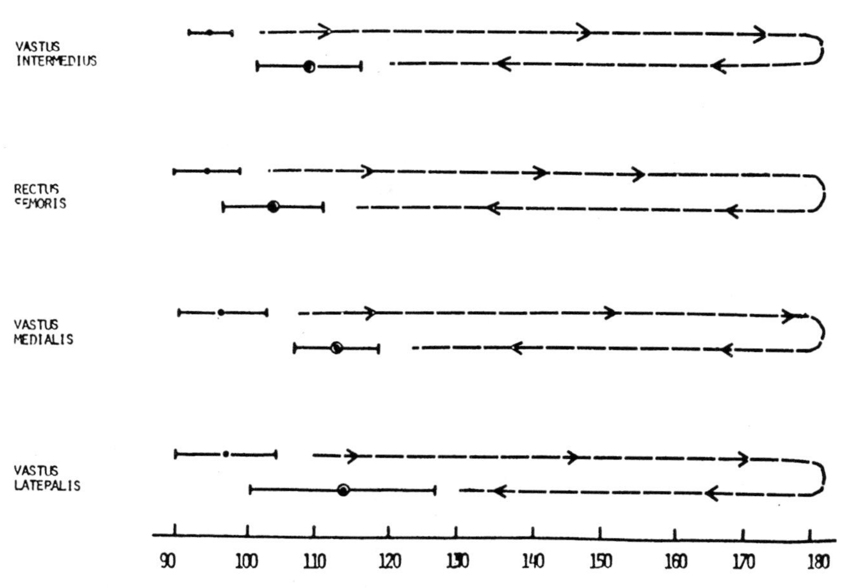

Figure 14.8. Diagrammatic representation of onset and cessation of activity in the four heads of quadriceps during weighted extension of the knee from 90° to 180° and flexion back to 90°. *Subject seated; 30 lbs (13.6 kg) added load.* Mean and standard deviation of the angle at which activity begins starts each line; the *broken arrows* indicate the direction of movement to full extension and back to the mean and standard deviation of the angle at which the activity ceases. (From J.V. Basmajian et al, ©1972, *Anatomical Record.*)

of activity in the rectus femoris (compared with the vasti), also noted by Pocock, may also have some importance which is not now apparent. Any judgments of the timing of the muscles based upon biomechanical calculation appear to be especially fallible where the quadriceps femoris is concerned.

Pocock's finding that quadriceps activity is less during the down phase of squatting is contradicted by the findings which show a parallelism between the up and down stages. Some difference in technique may account for his conclusion.

An important additional finding in the study by Basmajian et al is the variation from subject to subject. This is particularly marked at the conclusion of standing up from the squatting exercise. Individual variations in responses of muscles must be taken into account in the design of myoelectric devices. Rideau and Duval (1978), Elorante and Komi (1981), and Stratford (1982) enlarged on these findings; they believe that the slope of the integrated EMG-force relationship is linear with the speed of contraction and with the position of the knee joint, but more convincing evidence is needed for a final decision. In nonhuman primates

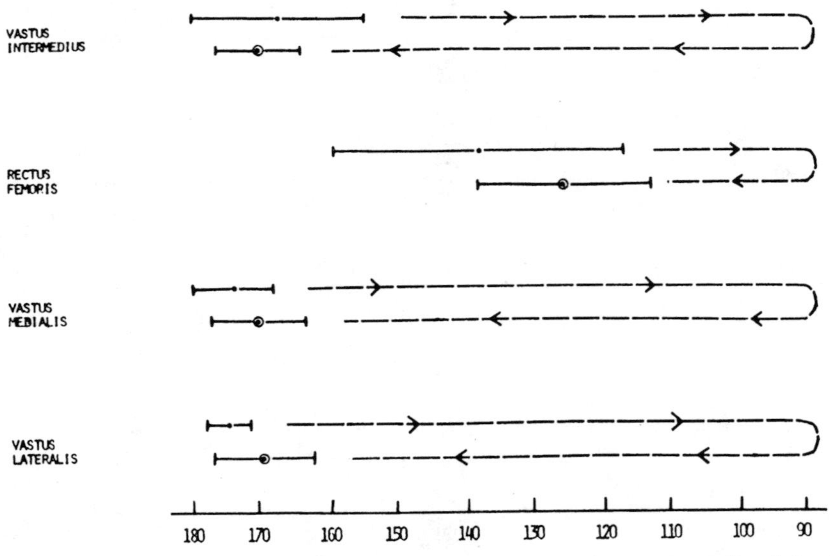

Figure 14.9. Diagrammatic representation of onset and cessation of activity in the four heads of quadriceps during squatting to 90° from the fully erect posture and back up again. *Subject standing; no added load.* Mean and standard deviation of the angle at which activity begins start each line; the *broken arrows* indicate the direction of movement to 90° flexion (squatting) and back to the mean and standard deviation of the angle at which the activity ceases. (From J.V. Basmajian et al, ©1972, *Anatomical Record.*)

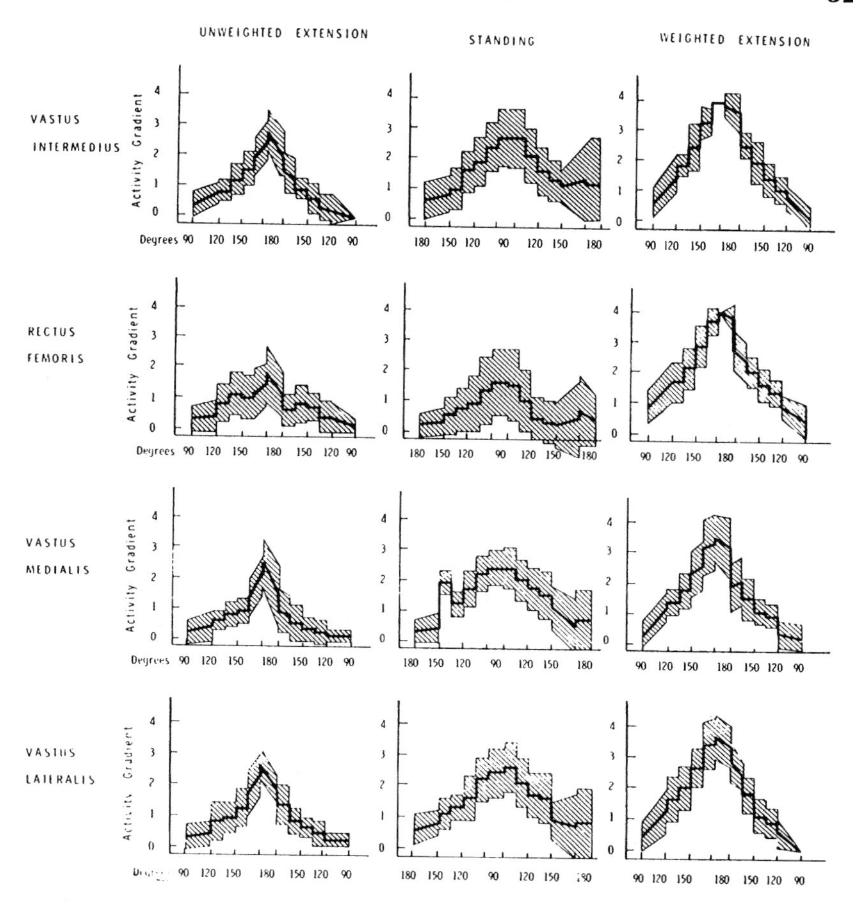

Figure 14.10. Composite diagram of changing EMG activity rated 0 to 4 during the movements of the knee reflecting mean and standard deviations at epochs of the movements shown on the horizontal axis. *Heavy lines* indicate the means; *shaded areas* the standard deviations. (From J.V. Basmajian et al, ©1972, *Anatomical Record.*)

(Jouffroy et al, 1979; Jungers et al, 1980), the results of EMG studies tend to agree with human studies. During bicycle pedaling, considerable quadriceps activity was obvious in the study by Mohr et al (1981).

Duarte Cintra and Furlani (1981) compared slow and fast movements both free and resisted in the four heads of quadriceps recorded via wire electrodes. In general they confirmed that the vasti acted simultaneously, with medialis and intermedius "more active." Rectus femoris acted late in (a) hip flexion with passive motion at the knee; (b) in bending backward at the hip; (c) in squatting; and (d) in the sitting-down movement. In standing up, rectus femoris acts briefly at the start only.

The study by Murphey et al (1971) of Dallas, Texas, has special interest for they were concerned with the activity of the quadriceps during standing in subjects who habitually show *hyperextension* and *flexion* of the knee when standing up. In all instances, flexion subjects showed more activity than hyperextension subjects.

Ravaglia's (1957) earlier investigation of quadriceps femoris is also of interest. In a series of normal subjects he recorded *via* surface and needle electrodes from the vastus medialis, vastus lateralis and rectus femoris simultaneously. During the movement of rising from the sitting to the standing position and vice versa, the activity in the three heads was not synchronized and equal. The vastus medialis was retarded and was not as active as the other two. In erect standing the activity in the three heads fell rapidly. The vastus medialis was more active than the other two muscles examined. Ravaglia demonstrated conclusively that the three heads acted in different ways in various phases of movement.

Wheatley and Jahnke (1951) found a greater activity in vastus medialis when the knee was held in extension with hip joint flexed or the knee joint (tibia) laterally rotated. On the other hand, the vastus lateralis was more active in extension of the knee when the hip was flexed or the knee joint (tibia) medially rotated. During resisted extension of the knee, the various parts of quadriceps came into action at different phases of the movement (Fig. 14.11).

In an unpublished 1957 study by J. V. Basmajian, W. E. K. Brown, and Rita Harland of electromyographic examinations of quadriceps femoris, the results were quite interesting. Subjects included 11 young women in whom simultaneous recordings were made from the vastus medialis, vastus lateralis, and rectus femoris using skin electrodes. The aim was to evaluate a number of standard procedures sporadically used in rehabilitation work, ostensibly to help strengthen the quadriceps. For example, associated movements of the toes have been advocated to

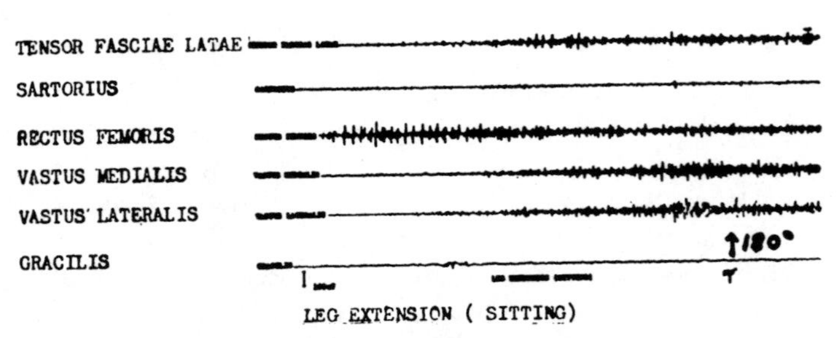

Figure 14.11. EMG records of thigh muscles during extension of the knee from 90° to 180° (subject seated, limb hanging). (From M.D. Wheatley and W.D. Jahnke, 1951, *Archives of Physical Medicine*.)

augment the activity of the quadriceps. More than half of the subjects showed no such augmentation. In those in whom augmentation of quadriceps did occur, the actually effective toe movement was flexion in some and extension in others.

Associated foot and ankle movements were somewhat more effective than toe movements in causing augmentation of quadriceps activity; this was true in most (but not all) of the subjects. However, there was no clear-cut difference among the effects of any of the following: dorsiflexion, plantar flexion, inversion, and eversion. All seemed to augment in some subjects, while one or another of the movements proved to be the most effective in others.

Medial or lateral rotation of the hip joint performed simultaneously with contraction of quadriceps had essentially no effect on the EMG activity of quadriceps (except for some slight augmentation in one subject). Not only did simultaneous hip flexion fail to augment the amount of activity in quadriceps in most subjects, but it even decreased it in some.

The most effective technique for maximal motor unit activity was having the subject actively perform extension of the knee against resistance—not in a static position but during motion. Nonetheless, in many subjects, static contractions were just as effective—or even more effective—and therefore cannot be categorically condemned. The greatest activity during motion occurred in the last half (i.e., 90°) of extension. With static contractions, the position of the knee most effective for showing maximal activity in quadriceps was almost always the fully extended one.

In all subjects, concentric actions caused more activity than eccentric ones, i.e., activity of the muscle during its primary action or shortening was considerably more than the activity while the muscle was acting as an antagonist or being forced to lengthen as it acted (negative work).

Flexion of the trunk or isolated contraction of the opposite quadriceps had little if any effect on the quadriceps under examination, suggesting that such techniques are practically useless for rehabilitation work. Simultaneous bilateral contraction of the quadriceps did not augment the activity in the one under study in some subjects while it did so in others. Finally, having the subject "push down" at the hip and knee regions (in an effort to hyperextend the knee joint actively) did not increase the activity of quadriceps but actually diminished it in 3 of the 11 subjects. Instructing a patient to "push down your knee" is therefore not an acceptable procedure for producing maximal quadriceps activity.

The lesson to be learned by physiotherapists and rehabilitation specialists from these findings is one of healthy skepticism for many of the dogmatic teachings that bear on the best methods of evoking maximal quadriceps activity. Some of the findings appear to confirm the dogmas;

others flatly contradict them; and still others show that different subjects react in different ways.

Popliteus

In a pioneer piece of EMG research with needle electrodes on this small, deeply set muscle lying behind the knee, Barnett and Richardson (1953) confirmed the classical teaching (often denied) that it is a medial rotator of the tibia. Its activity at the start of flexion of the knee is related to the unlocking of the knee joint. When a person stands in the semi-crouched knee-bent position, continuous motor unit activity of the muscle was demonstrated (see Fig. 11.4 on p 260). When the knee is bent, the weight of the body tends to slide the femur downward and forward on the slope of the tibia. It seems that the continuous marked activity of popliteus aids the posterior cruciate ligament in preventing forward dislocation. Reis and Carvalho (1973) emphasized their findings based on equilibrium functions for the control of torsional forces.

Using improved wire electrode techniques, Basmajian and Lovejoy (1971) delineated the functions of popliteus further in 20 normal persons. The data were recorded and stored on magnetic tape and were analyzed after analogue-to-digital conversion by a LAB-8 computer. The digital information was normalized by making the greatest activity for each subject equal to one; lesser degrees of activity were then compared with this normalized maximum activity.

STATIC EXTENSION

Each subject was tested in the sitting position. Data were recorded with the leg and foot maintained by the subject in the neutral position, then in full medial rotation, and in lateral rotation, beginning with the knee joint at 90°, and then at approximately 120, 135, 160, 175, and 180° (full extension) (Fig. 14.12). The EMG activity in neutral and lateral rotation remained approximately the same in all degrees of extension (Fig. 14.13). However, in medial rotation of the tibia on the femur there was marked increase in activity. The activity was greatest through the first 40° of extension and decreased as full extension was reached.

STATIC FLEXION

Each subject was tested in the prone position, and data were recorded at the same degrees of flexion and rotation as previously (Fig. 14.14). Again, the electromyographic activity remained essentially constant in neutral and lateral rotation of the tibia. There was a marked increase in activity through the first 20° of flexion with the leg medially rotated, and it gradually decreased as 90° of flexion was reached.

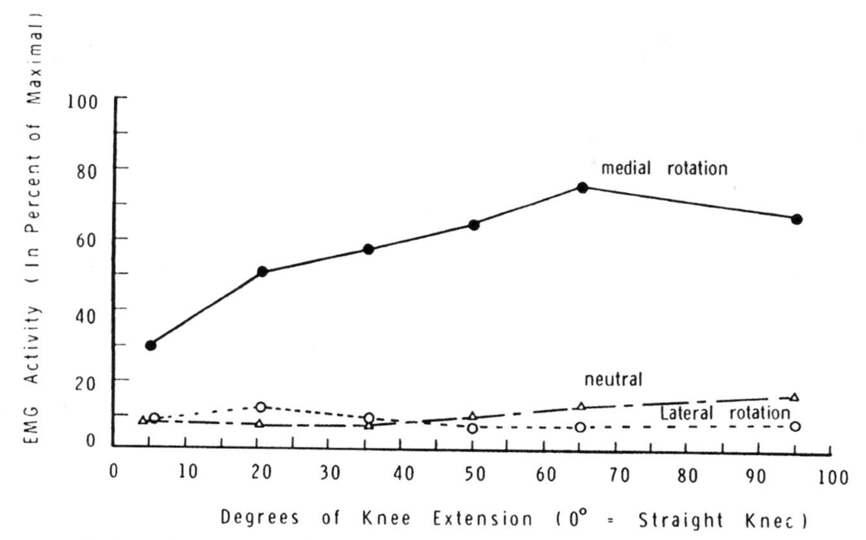

Figure 14.12. Composite of mean EMG results in popliteus with subject seated and unsupported knee held at different angles. (From J.V. Basmajian and J.F. Lovejoy, ©1971, *Journal of Bone and Joint Surgery*.)

MOVEMENTS OF KNEE

The EMG activity was recorded beginning with the knees locked, then with the subject flexing to 150°. This was repeated with the right shoulder and trunk rotated forward, then with the left shoulder and

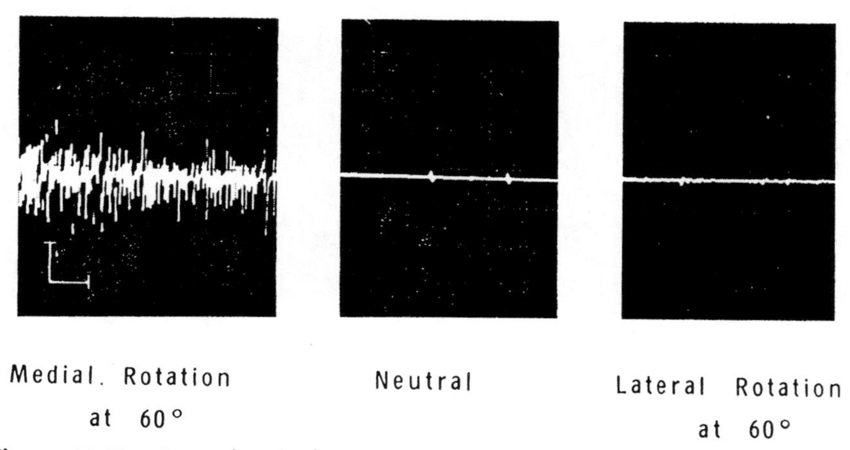

Figure 14.13. Example of EMG activity in popliteus during rotations of tibia. (From J.V. Basmajian and J.F. Lovejoy, ©1971, *Journal of Bone and Joint Surgery*.)

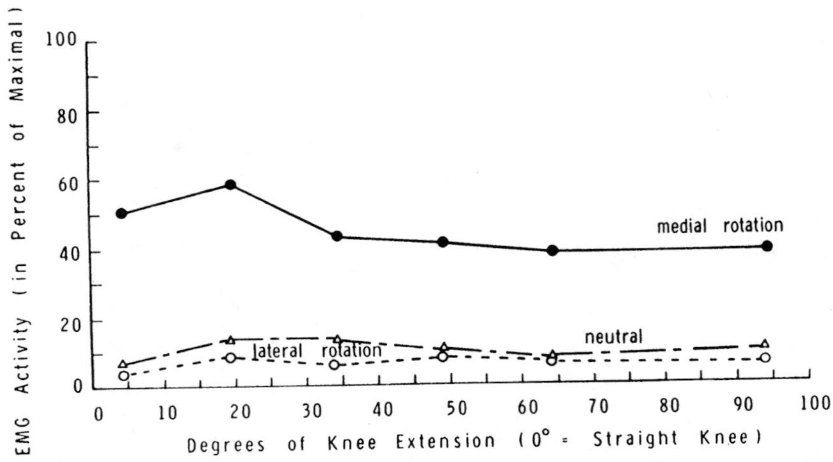

Figure 14.14. Composite of mean EMG results in popliteus with subject lying prone and knee held at different angles. (From J.V. Basmajian and J.F. Lovejoy, ©1971, *Journal of Bone and Joint Surgery*.)

trunk rotated forward. In each instance recordings were made with the feet in three positions: neutral, medial rotation, and lateral rotation.

The activity in the right popliteus muscle remained constant with the right shoulder rotated anteriorly and the feet at neutral, in medial rotation, or in lateral rotation. With the left shoulder rotated anteriorly and the feet laterally, the activity increased with knee flexion. With the feet at neutral or in medial rotation, the activity was greatest at the initiation of flexion. The activity was persistently elevated above that obtained with the right shoulder rotated and with the feet in the same positions.

Gait was analyzed in all subjects using foot switches on the heel, toe, and lateral margin of the shoe. The subject was tested on a treadmill at 1.6 and 3.2 km/hour with normal gait, toe-in, and toe-out gaits. In general the activity was greatest at heel-strike and from foot-flat to toe-off, that is, during most of the weight-bearing phase of gait.

LEG MUSCLES

An early EMG study with needle electrodes of three muscles of the leg (and three intrinsic muscles of the foot) was reported by Basmajian and Bentzon (1954). The muscles of the leg were the tibialis anterior, the peroneus longus, and the lateral head of gastrocnemius. Most of the subjects were tested in different postures and during a variety of movements. Often, other muscles were examined as well. The first part of this section concerns only the findings in the muscles named during the "relaxed standing at ease" position. In this regard, of special interest was testing the validity of the popular theory that the peroneus longus, the tibialis anterior, and the intrinsic muscles of the foot maintain the arches of the foot in ordinary standing. Special attention was also paid to the possible influence of certain variable factors, such as the sex of the subject, the type of feet, and the wearing of high heels by women.

For each muscle, records were made while the subject was reclining (as a control), immediately upon assuming the standing barefooted posture, and after 2 minutes of standing. Considerable testing showed that the initial activity invariably found on changing a position falls off rapidly to the resting level and that the 2-minute interval is adequate. Moreover, many of the subjects were tested at longer intervals (up to 15 minutes) with no change in the results.

The "relaxed standing at ease" posture is the comfortable well-balanced stance with the feet several inches (about 8 cm) apart and bearing equal weight, and with the hands clasped loosely behind the back. The relative position of the feet would show minor individual variations. However, the slight changes that were necessary for some subjects to conform to the standard position made no difference in their comfort or in the findings.

In the women, additional similar records were made from the muscles of the leg while high heels were being worn. The heels were all 2½ inches (about 6 cm) high, except in one case in which they were 3 inches (7.5 cm).

The lateral head of gastrocnemius was active in the majority of subjects. In the women, the group showing continuous activity was definitely smaller when the subjects were barefooted but, when high heels were worn, was almost the same as in the males. A quarter of both men and women showed no electromyographic activity in this muscle when the subject stood barefooted. Analysis of individual cases revealed that only one of the women continued to show no activity with high heels.

Almost half of the men and a quarter of the women showed no EMG activity in tibialis anterior when standing barefooted (Fig. 14.15). But another quarter of both the men and the women showed pronounced activity which could be abolished by leaning forward. Each of the women in this latter group exhibited at least the same degree of activity on standing in high heels. One woman showed moderate activity while standing with high heels but only slight activity while standing barefooted.

Only 1 of 16 men and 2 of 16 women showed continuous activity in peroneus longus while standing barefooted. An additional man and 5 women showed intermittent activity. Half of the men and a third of the

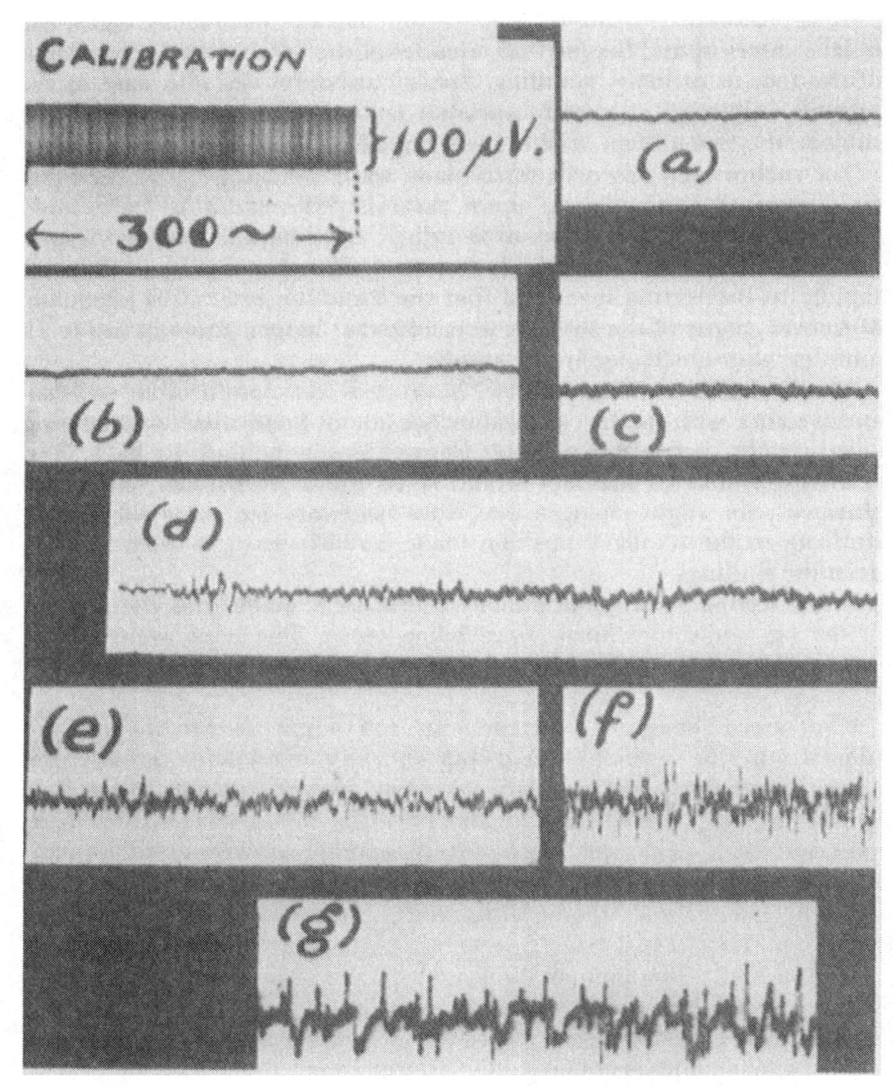

Figure 14.15. Representative EMGs from leg muscles during standing: *a* and *b*, *nil* to negligible activity; *c*, slight; *d*, intermittent bursts; *e*, moderate continuous; *f* and *g*, marked continuous. (From J.V. Basmajian and J.W. Bentzon, ©1954, *Surgery, Gynecology, and Obstetrics*.)

women showed inactivity when barefooted. With high heels, half of the women showed continuous marked activity and none showed inactivity in peroneus longus.

There was no consistent significant relationship between the types of feet and types of activity in any muscle.

It now seems to be beyond controversy that the tibialis anterior and the peroneus longus (and, as we shall see, the intrinsic muscle of the foot) play no important active role in the normal static support of the long arches of the foot. These muscles are completely inactive electromyographically in many normal individuals while standing. Smith (1954) came to the same conclusion after examining six subjects who all showed EMG quiescence in the anterior crural muscles during standing. The same situation holds for both peroneus longus and peroneus brevis (Jonsson and Rundgren, 1971), and for tibialis anterior, the peronei, flexor and extensor digitorum longus, and abductor digiti quinti (Suzuki, 1956).

Although this investigation of predominantly normal feet showed no obvious relationship between types of feet and the types of activity in the muscles concerned during standing, no attempt is made to suggest that the muscles play no role in the abnormal flat foot. Furthermore, one cannot dismiss the role of these muscles in the maintenance of the arch during locomotion. Indeed, the intrinsic muscles of the foot are always very active when one rises on the toes to even the slightest degree (see below).

In so far as the muscles of the leg are concerned, the results showed a biological range of activity and were not in accord with some of the absolute findings of Smith, Joseph and Nightingale, and Åkerblom. We find that in the relaxed standing at ease position, there is activity in some individuals in the tibialis anterior and peroneus longus and that this can be abolished easily by unusual and varying stances or by the removal of weight from the limb. The disagreement probably stems largely from our use of the more sensitive needle electrodes rather than simple skin electrodes (Fig. 14.16).

Joseph and Nightingale were concerned with soleus, gastrocnemius, and tibialis anterior. In their subjects they found continued activity in every soleus and in some gastrocnemii, but none in tibialis anterior (Fig. 14.17). Since their original paper they soon added to their series with no change in their conclusions (Joseph, 1960). Granit's (1960) statement that, in general, mammalian soleus is tonic and gastrocnemius phasic may bear on this matter. However, Smith (1954) found that what postural activity there was in human legs during standing was intermittent and confined to gastrocnemius, not soleus. Levy (1963) found that in man soleus produces greater reflex contraction than gastrocnemius during the ankle jerk. He suggested that a greater density of muscle spindles in soleus (now generally accepted) accounts for this. One might expect a greater sensitivity to stretch in soleus. Herman and Bragin (1967) further elaborated on the differences between gastrocnemius and soleus in man. In general, the former is more sensitive to conditions of length, strength, and rate of contraction; soleus plays a more constant role and is more active at ankle dorsiflexion and in minimal contractions. Campbell et al (1973) in Biggs' laboratory at the Baylor College of Dentistry were even more specific. Recently they showed with fine-wire electrodes that the stability of the foot support affects the activity of even selected parts of these two muscles. The medial part of soleus has distinct functions: it is both a strong mover of the foot on the leg and a stabilizer of the leg on the foot; however, the lateral part gives little power to moving the ankle and is largely a stabilizer—especially when the platform is unstable. The two heads of gastrocnemius are quiet until motion is required at the ankle, thus acting as an auxiliary plantar flexor. The main dynamic and static flexor, the medial part of the soleus, has never before had such a lime-light exposure.

Closely related to the soleus and gastrocnemius muscles is the much less obvious *plantaris* (p 345). A study of 12 normal subjects with fine-wire electrodes revealed that its primary function is plantar flexion and inversion (Iida and Basmajian, 1973). Only in a loading situation does plantaris assist in knee flexion. Its function is not always the same as that of the closely related lateral head of gastrocnemius, which always has a slight activity in eversion.

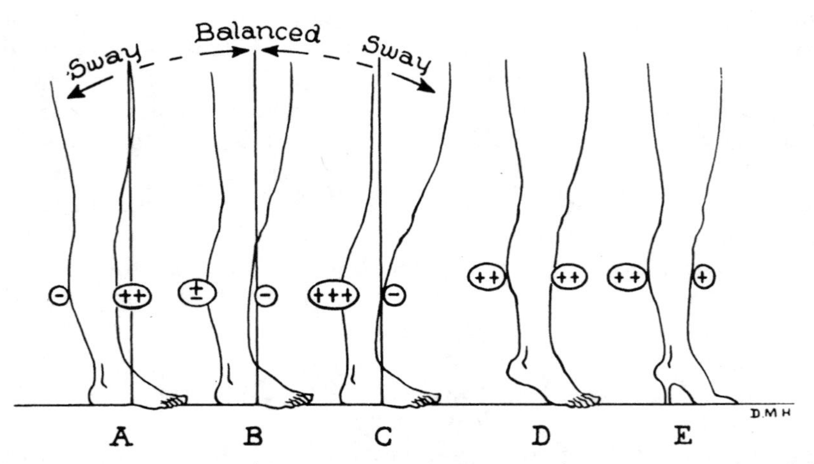

Figure 14.16. Diagram of EMG activity in anterior and posterior muscles of the leg under differing conditions.

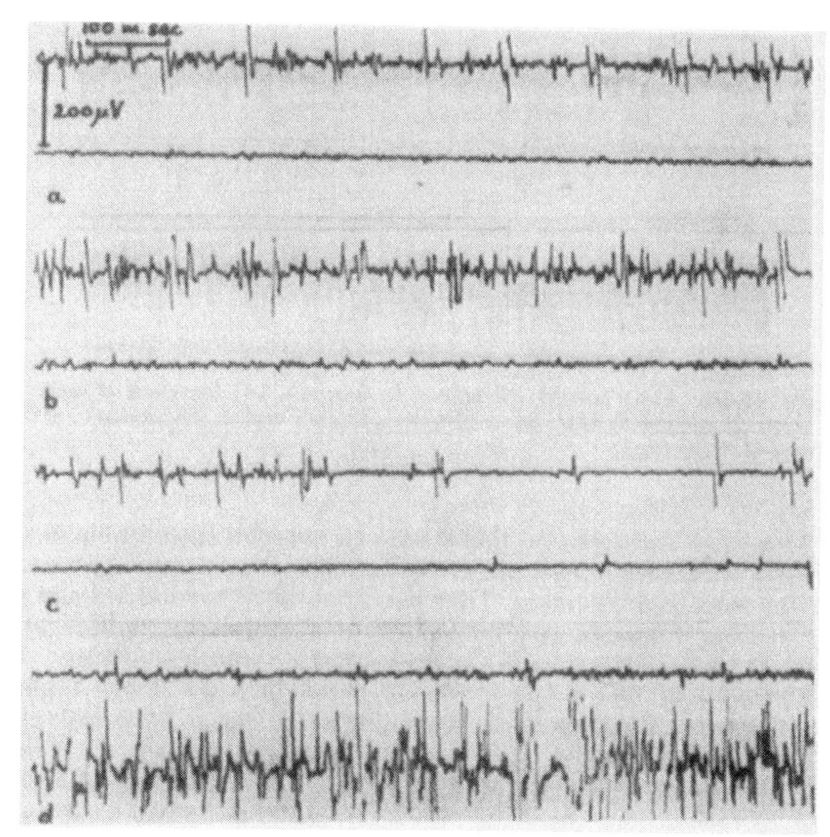

Figure 14.17. EMG records from soleus (*upper trace*) and tibialis anterior (*lower trace*. (*a*) Standing at ease. *b*) After swaying forwards. (*c*) While swaying backwards. (*d*) After swaying backwards. (From J. Joseph, ©1960, *Charles C Thomas*.)

The increased activity in all three of the muscles of the leg when high heels are worn may seem to be due to the element of instability introduced by the posture (Fig. 14.16). But it will be noted that gastrocnemius and peroneus longus are involved and that tibialis anterior is not affected to any great extent. In another series of experiments peroneus longus was found to be markedly active in plantar flexion of the ankle. It appears, then, that the wearing of high heels shifts the center of gravity forward to a new position in many women, with a resultant increase in the activity of gastrocnemius and peroneus longus.

Joseph and Nightingale (1956) confirmed the finding that the wearing of high heels caused an increase in activity of the calf muscles (specifically, soleus) in most women (Fig. 14.18). They investigated the line of gravity

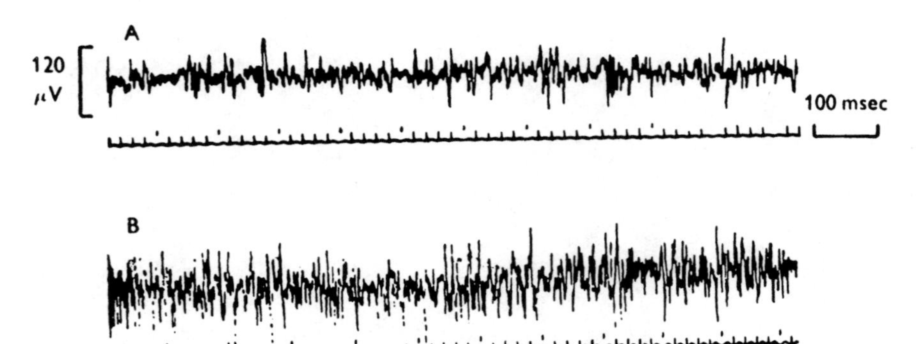

Figure 14.18. EMG records of soleus in women. (*A*) Standing at ease. (*B*) Standing at ease with high heels. (From J. Joseph and A. Nightingale, ©1956, *Journal of Physiology*.)

in 11 women and reported that it bore no constant relationship in spite of their reluctance to agree to the occurrence of intermittent activity of the leg muscles in standing. They also established beyond a doubt that the swaying forwards and backwards of a subject by as little as 5° produces reflex activity of the posterior and anterior leg muscles.

Portnoy and Morin (1956) tended to confirm the above findings, reporting that 5 gastrocnemii showed intermittent activity while 9 (of 16) showed continuous activity during relaxed standing. Naponiello (1957) reported similar intermittent activity in tibialis anterior with or without high heels, as did Floyd and Silver (1950), who first drew special attention to the intermittent, unconscious, back-and-forth swaying which causes it.

Ferraz et al (1958) have concluded from a study of seven subjects that the peroneus longus and peroneus brevis act intermittently as postural muscles, becoming very active in leaning forward and silent when leaning backward. Their activity is pronounced during the propulsive phase of normal walking, and the activity of the two muscles is synchronous.

ARCH SUPPORT OF FOOT

The mechanism of arch support in the foot remains controversial despite years of investigation. According to one theory, the arches are maintained by the contraction of muscles; according to a second, by the strength of passive tissues; and according to a third, by the combination of both muscles and passive structures.

A century ago, Duchenne stated that by faradization of the peroneus longus in flat-footed children he was always able to produce the progressive formation of a normal plantar arch. Keith, in 1929, concluded that muscles are all-important in the support of the arch and that ligaments come into play only after the muscles have "failed." Morton (1935) disagreed. He concluded that the structural stability of the foot is not dependent on muscles. He claimed that appreciable muscle exertion is needed only when the center of gravity of the body moves beyond the margins of structural stability, whereas only a slight controlling action by the muscles is required when the center remains between those margins. In 1952, Morton and Fuller further showed that static strains upon the foot are relatively low in intensity, falling well within the capabilities of the ligaments. Their calculations showed that only acute, heavy, but transient forces, such as in the takeoff phase of walking, require the dynamic action of muscles.

Thus the controversy continued and was kept alive by others, including Kaplan and Kaplan (1935) and Lake (1937). In 1941, R.L. Jones, using the method of palpation in the living and direct observation in cadavers, concluded that not more than 15 to 20% of the total tension stress on the foot is borne by the posterior tibial and peroneal muscles. Much the greater part of this stress is borne by the plantar ligaments of the foot, but the short plantar muscles, being in an advantageous position, also contribute to the support.

After World War II, Harris and Beath (1948) concluded from their extensive survey in the Canadian Army that both passive supporting structures and muscles are responsible for a normal arch. They frankly favored the role of the passive structures but admitted the readiness of the muscles to assume a role in arch support. Wood Jones (1949) agreed that maintenance of the normal arched form of the foot results from the dual control exerted by the passive elasticity of the ligaments and the active contractility of muscles. He concluded that the plantar aponeurosis and plantar tarsal ligaments hold the anterior and posterior pillars of the arch together and that the actively contracting intrinsic muscles between the aponeurosis and tarsal ligaments also play an important part.

From a general EMG study of the leg and foot with needle electrodes, the tibialis anterior, the peroneus longus, and the intrinsic muscles of the foot were found to play no important role in the normal static support of the long arches of the foot (Basmajian and Bentzon, 1954). As noted before, many if not most of these muscles showed inactivity during standing in a relaxed position. This was confirmed in general terms by Smith (1954), who used skin electrodes. Many standard textbooks, however, still overemphasize the part played by muscles in the support of the arches of the foot. For example, *Gray's Anatomy* (Johnston et al, 1958) stated that the tibialis posterior is the most important factor when the foot is bearing weight, and that the peroneus longus, tibialis anterior, flexor hallucis longus, abductor hallucis, and flexor digitorum brevis also contribute to the support.

In an attempt to settle the controversy, a special study was performed of the muscular support of the loaded arch (Basmajian and Stecko, 1963). Because earlier studies ran the risk of confusing the muscle activity required for postural adjustment with that for the support of the arches, the subjects were seated, and the leg and foot were loaded artificially in a special apparatus (Fig. 14.19) that provides graded loads of up to 400 lbs (182 kg).

Six muscles of particular interest were chosen for this EMG study. They were: the tibialis anterior (since by its insertion it would appear to raise the summit of the medial longitudinal arch); the tibialis posterior and the peroneus longus (since by acting together, these two might provide a sling support); and the flexor hallucis longus, abductor hallucis, and flexor digitorum brevis (since all three are in a position to act as longitudinal bowstrings). These muscles were studied with indwelling fine-wire electrodes, and simultaneous recordings were made with high-gain amplifiers.

The special load-applicator consisted of a lever made from an oak beam, fixed at one end to a heavy frame by a hinge to provide the fulcrum (Fig. 14.19). The bent knee of the subject could be placed under the beam, and by use of leverage, loads of 100, 200, and 400 lbs (about 45.5, 90, and 182 kg) could be applied through the vertical leg to the foot. These loads were chosen because 100 lbs approximates or exceeds the normal load on each foot in the upright bipedal stance; 200 lbs approximates or exceeds the load on the arch in upright unipedal stance; and the 400-lb load is the maximum that can be applied without extreme discomfort at the knee. The system provides a convenient method of loading the arches while eliminating any postural effect that muscles might have on the leg and foot. To test the influence of various positions of the foot and ankle, the foot was supported on a specially constructed adjustable platform (Fig. 14.19).

In these experiments, all six of the muscles, which are often considered to be important contributors to arch support, did not react to loads that actually surpassed those normally applied to the static plantigrade foot. One hundred pounds elicited little if any contraction. With loads of 200 lbs applied to one foot, a small number of the muscles showed some activity, but this was exceptional and varied with the muscle and the posture of the foot. The peroneus longus was dramatically quiescent except, perhaps, when 400 lbs was applied to the inverted foot. With this load, however, a substantial number of the other muscles also came into play (Fig. 14.20).

An analysis of the forces on the arch by the method of Steindler reveals that 400 lbs does not exceed the normal forces imposed on the arch in the takeoff position of walking. Earliest study (Basmajian and Bentzon, 1954) showed a great deal of activity in the tibialis anterior, peroneus longus, and intrinsic muscles of the foot when the subjects stood on

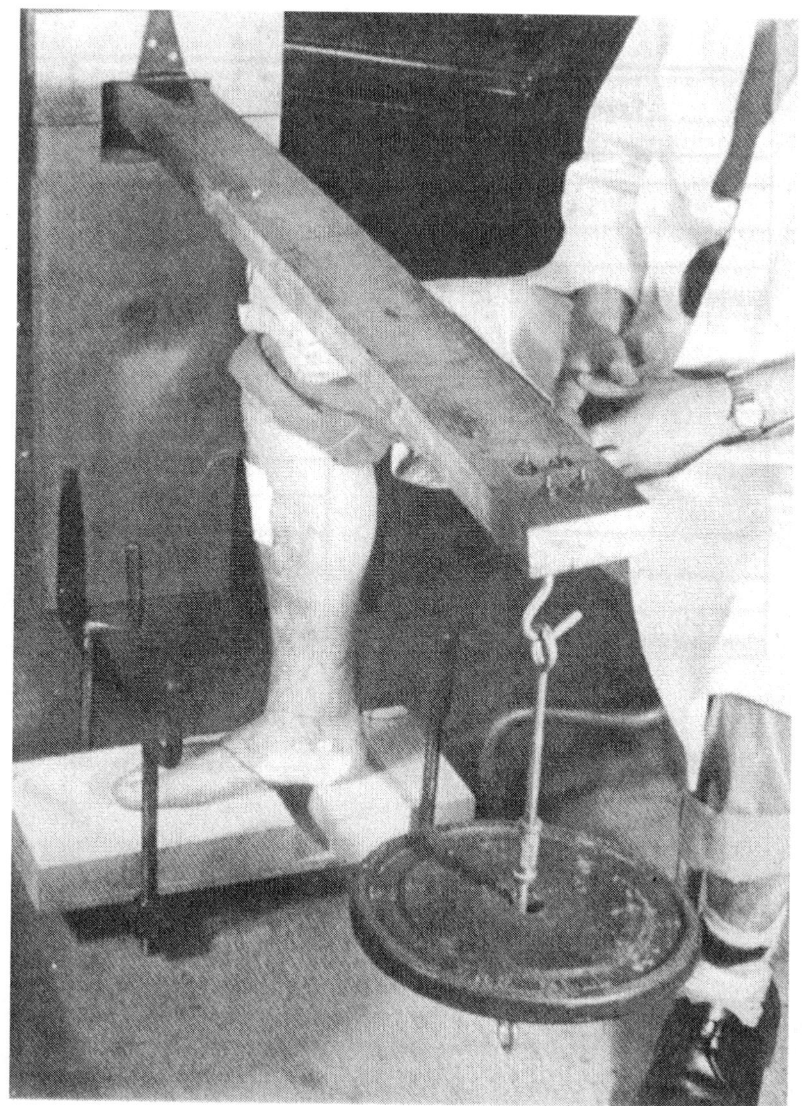

Figure 14.19. Arrangement of a subject with the load applied by leverage and the foot on the adjustable platform (here set in the horizontal position). (From J.V. Basmajian and G. Stecko, ©1963, *Journal of Bone and Joint Surgery.*)

tiptoe. From these earlier findings and those of the newer study, one may conclude that in the standing-at-ease posture muscle activity is not required, and the muscles are inactive; however, in positions in which excessive stresses are applied, as in the takeoff phase of gait, the muscles

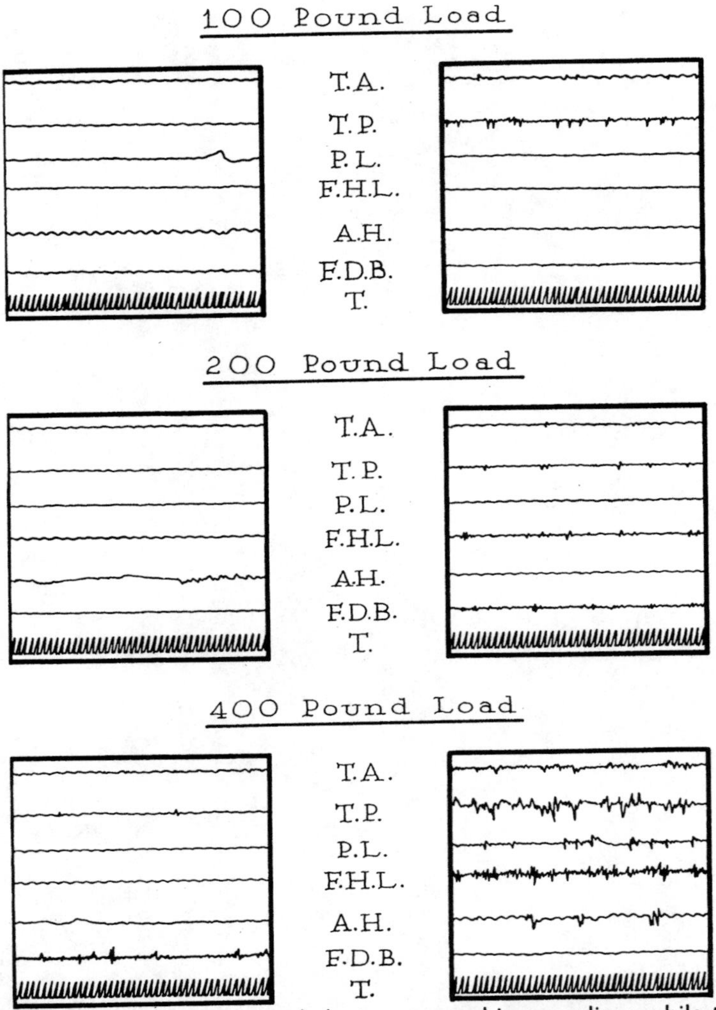

100 Pound Load

T.A.
T.P.
P.L.
F.H.L.
A.H.
F.D.B.
T.

200 Pound Load

T.A.
T.P.
P.L.
F.H.L.
A.H.
F.D.B.
T.

400 Pound Load

T.A.
T.P.
P.L.
F.H.L.
A.H.
F.D.B.
T.

Figure 14.20. Sample tracings of electromyographic recordings while the foot is on the horizontal platform to show range of activity from minimum (*left side*) to maximum (*right side*) activity with loads of 100, 200, and 400 lbs. T.A., T.P., P.L., F.H.L., A.H., and F.D.B. are abbreviations for tibialis anterior and posterior, peroneus longus, flexor hallucis longus, abductor hallucis and flexor digitorum brevis. T. is the time signal with a frequency of 100 cycles per second and an amplitude of 200 μV. (From J.V. Basmajian and G. Stecko, ©1963, *Journal of Bone and Joint Surgery*.)

do react. Without any question, the first line of defense is provided by the passive structures. During activity the muscles would appear to contribute to the normal maintenance of the longitudinal arches.

This brings us back to the conclusion of Harris and Beath that the normal foot is supported both by passive factors (bones and ligaments) and by active factors (muscles) and that these factors are reciprocal. They stated that, in the average or strong foot, most of the support is provided by passive factors, with little load being supported by the muscles. Much greater stability and strength for the foot can be provided by the passive support of a well-designed skeleton than from the active support provided by muscles. However, the concluded that the muscles always play some part in maintaining balance and in supporting the load. This latter opinion is shown by the experiments of Basmajian and Stecko and by those of Miyoshi (1966) to be incorrect for the static foot.

Harris and Beath stated further that the strong foot is one in which the tarsal bones are so articulated with each other that the weight of the body is borne without appreciable movement between them. With a strong foot, the muscles are used to maintain balance, to adjust the foot to uneven ground and, of course, to propel the body in walking and running. The weak foot is one in which the tarsal bones are so shaped and are so disposed that they are unstable and shift in position when weight is superimposed. Only by increasing the support provided by the muscles can the normal shape of the foot be maintained and the body weight supported.

There is a limit to the contribution that muscles can make. They cannot function unremittingly, nor can they provide the powerful support furnished by the skeleton. The first line of defense of the arches is ligamentous. The muscles form a dynamic reserve, called upon reflexly by excessive loads including the takeoff phase in walking.

Undoubtedly, a study of a series of subjects with flat feet, using the techniques of the present research, would help to clarify the situation. Thus Gray and Basmajian (1968) found that more than half of the flat-footed subjects had activity in the muscles tested (tibialis anterior, tibialis posterior, peroneus longus and soleus). They concluded that the use of these muscles as a dynamic reserve in attempting to maintain the arch was real. They greatly expanded this study using fine-wire electrodes in tibialis anterior, tibialis posterior, flexor hallucis longus, peroneus longus, abductor hallucis, and flexor digitorum brevis. Their findings lend further support to the theory that in the *imbalanced* foot musclear activity does occur, apparently reflexly. Gray (1969) elaborated further on these findings, showing that flat feet always are accompanied by activity in tibialis anterior, tibialis posterior, and peroneus longus. The interesting finding of Novozamsky and Buchberger (1970) lends strong support to all that is written in the previous pages: 44 persons went on a 100-km hike; those who had normal feet maintained their arches, but those who had fallen arches before had a further drop.

Plantaris

Clinically this small and sometimes absent muscle attracts the attention of surgeons after achilles tendon rupture, when the plantaris tendon usually remains intact. Even with complete achilles tendon ruptures, patients can plantar flex to some degree. Therefore, the precise action of normal plantaris in 12 persons during voluntary movements of the knee and ankle and everyday activities of standing, walking, and stair-climbing was studied by Iida and Basmajian (1975). To provide a reference, the lateral head of gastrocnemius was recorded simultaneously.

The plantaris is very active when plantar-flexion occurs in full knee extension. As the angle of the knee decreases, the amplitude of activity falls progressively, apparently because the muscle becomes shorter and loses its mechanical advantage. Also, plantaris is a two-joint muscle (p 232).

There are two types of inversion. In "inversion 1," there is no change of the ankle angle; in "inversion 2" the foot is inverted while plantar flexing the ankle. In inversion 1, plantaris activity is slight and unaffected by knee angle; the foot is adducted and twisted in such a way that the medial border is raised and there is little or no plantar flexion; however, the plantaris may contract slightly to assist the twisting movement. In inversion 2, the plantaris muscle is strongly active, although somewhat less than in plantar flexion, because plantar flexion always accompanies the medial rotation of the pendant foot so that the sole is directed inward.

In eversion, the plantaris muscle is slightly active only in full knee extension (in some subjects only) because knee flexion shortens the muscle. In flexion of the knee in the prone position without resistance there is no activity until resistance is offered—when there is moderate activity. The moderate activity in some subjects during flexion of the knee in the standing position is related.

The moderate activity in climbing upstairs for all subjects—and in level walking and knee flexion in standing in some—suggests that the plantaris muscle assists the function of the knee in loading situations, although perhaps this is not a primary action of plantaris muscle. Slight activity in plantaris in some subjects while leaning forward occurs because the mechanical axis falls in front of the geometric center of the supporting bone. As the heel loses contact with the floor, strong contraction of the plantar flexors, including plantaris, quickly checks the forward movement of the trunk, and the subject attains a new stable position on the balls of the foot.

Silence of plantaris is pronounced in level walking; knee flexion in standing; leaning backward, left and right; standing on forward and backward slopes; descending stairs; and knee flexion in the prone position without resistance. Apparently all these movements need neither plantar flexion nor inversion 2, nor strong flexion of the knee joint, all of them being the chief actions of plantaris.

Difference of function between the plantaris and the lateral gastrocnemius emerges in eversion and in leaning right. The origin of plantaris muscle on the lateral femoral condyle is higher and thus closer to the axis of the femur than is the origin of lateral gastrocnemius, and the insertion is more central on the posterior aspect of the calcaneus. The momental line of plantaris is more vertical and parallel to the leg axis than that of the lateral gastrocnemius, in which the line is oblique from lateral to medial. Thus, in leaning right, when the body weight is shifted right, the mechanical axis falls lateral to the geometric center of supporting bone, and in order to hold the balanced standing position, the right foot must be everted, resulting in moderate activity in the lateral gastrocnemius muscle.

Free Movements of Ankle

The studies of Basmajian et al over many years and those of O'Connell (1958) confirmed the classical teaching in regard to the importance of tibialis anterior in dorsiflexion of the ankle with the assistance of extensor digitorum longus and extensor hallucis (Fig. 14.21). The peronei are inactive during dorsiflexion but active during plantar flexion. The activity of the peronei seems to be transmitted chiefly to the transverse tarsal joint and not the ankle joint. Walmsley (1977) showed that the peroneus longus and brevis muscles act synchronously during the stance phase. Peak activity was at full-foot, both on the level and on an incline.

O'Connell and Basmajian's group have each found independently that tibialis anterior is not very active in producing inversion *unless dorsiflexion*

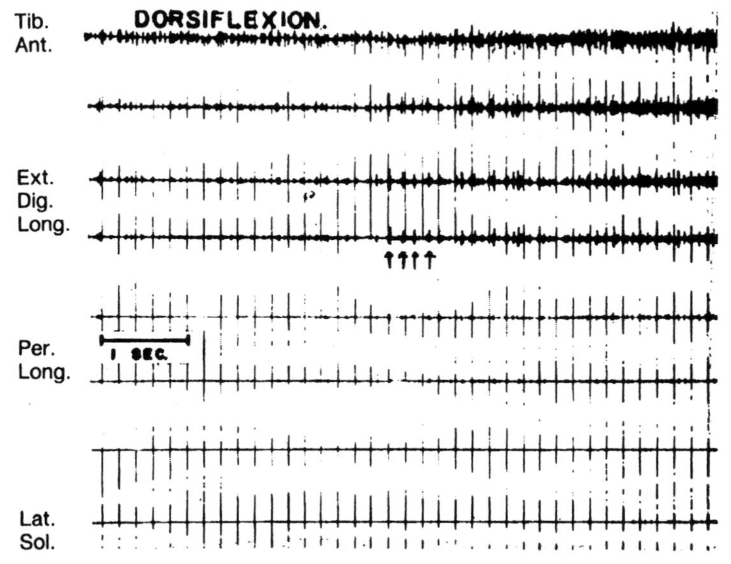

Figure 14.21. EMG records of various leg muscles during dorsiflexion. (From A.L. O'Connell, ©1958, *American Journal of Physical Medicine.*)

occurs simultaneously. Scattered experiments on the tibialis posterior indicate that it is a powerful invertor only when the ankle is simultaneously plantar flexed.

O'Connell proved the lack of a consistent pattern of activity in the two main peronei during eversion, sometimes the one showing activity first and sometimes the other. Moreover, the lateral part of the soleus appeared to become active during eversion while the medial part was active during inversion. This suggests a bipartite behavior of that muscle which requires further investigation. Arienti (1948) has also suggested patterns of activity of different parts of the triceps surae (i.e., gastrocnemius and soleus) during walking on a treadmill rather than unanimity of action. (In the upper limb, similar differences occur in the heads of biceps brachii (Basmajian and Latif, 1957).)

O'Connell and Mortensen (1957) reported on the activity of various muscles of the leg with the limb elevated. The action of tibialis anterior was variable during inversion (which, as seen above, must vary with the concurrent dorsiflexion or plantar flexion of the foot). It was strongly active during forced eversion, a finding which remains unexplained and unconfirmed. One would question this result, except for the knowledge of the integrity of the observers.

Houtz and Walsh (1959) compared the activity of the soleus and gastrocnemius in walking and in rising on tiptoes. During the "stance phase" of walking, the activity was less in these muscles than in rising on tiptoes. In other words, they seem to be stabilizers during walking. Apparently, rising up on the toes is not a normal part of the ordinary gait; this conclusion threatens the hallowed "push-off" concept of walking and might well be correct. Sheffield et al (1956) showed in a simultaneous study of leg muscles during walking that the dorsiflexors (tibialis anterior and the two long extensors of the toes) act in unison during the swing phase, obviously to provide adequate clearance of the ground. Early in the stance phase there is a greater burst of activity in them, apparently to stabilize the foot on the ground. Sheffield's group appears to have shown that the plantar flexors are stabilizers during the stance phase. They also noted a paradoxical activity in soleus (with none in gastrocnemius) during the swing phase.

Houtz and Fischer (1959) found a surprising amount of activity in tibialis anterior during the pedaling of a bicycle. This must be due to the stabilizing function since the foot is already forced into dorsiflexion by the pedal itself. Gastrocnemius showed considerably less activity than tibialis anterior and, predictably, its occurrence was in the exactly opposite phase of pedaling (see Fig. 14.5 on p 318).

In an elaborate early study of walking, Hirschberg and Nathanson (1952) described the patterns of activity in many of the lower limb muscles. In individuals, a consistent pattern of activity was found. Only one group, the calf muscles, started to contract in the middle of the stance phase, and these were the most active muscles. During the swing phase only the anterior tibial group contracted strongly. Frequently, a burst of activity occurred in gastrocnemius in the middle of the swing phase (see Fig. 14.6 on p 322). In a brief article, Richter (1966) reported similar findings.

Intrinsic Foot Muscles

Since the first edition of this book a number of new EMG studies have been added to the first report by Basmajian and Bentzon (1954), who used needle electrodes in the abductor hallucis, flexor digitorum brevis, and abductor digiti minimi in 12 men and 2 women. Generally, there was very little activity in these muscles while standing. This was confirmed by Mann and Inman (1964). Almost all the abductors of the great toe and the short flexors of the toes showed electromyographic silence. A quarter of the abductors of the little toe showed no activity while more than half showed negligible activity. In several cases there was marked activity in abductor hallucis, which was found to be due to "digging in" of the great toe. This activity was more or less abolished immediately when the subject straightened his toe.

When the subjects rose on tiptoes, there was a marked activity in the intrinsic muscles. This was confirmed in a later study by Oota (1956b). In the takeoff stage of walking there was a similar marked activity which was confirmed by Sheffield et al (1956) (see also p 258).

There was no consistent significant relationship between the types of feet and the types of activity in any muscle. However, Mann and Inman found that the pronated foot requires greater intrinsic muscle activity than does the normal foot "to stabilize the transverse tarsal and subtalar joints."

As suggested before, it now seems to be beyond controversy that tibialis anterior, peroneus longus, and the intrinsic muscles of the foot play no important active role in the normal static support of the long arches of the foot. The results show that these muscles are completely inactive electromyographically in many normal individuals while standing. Furthermore, when a subject suddenly lowers himself rapidly from a raised seated position to a direct relaxed standing position, there is little or no appearance of activity. However, voluntary and visible efforts to increase the arch of the foot are accompanied by marked activity.

Mann and Inman (1964) in the excellent detailed study showed that the intrinsic muscles of the foot act as a group in many movements, especially the abductors of the great and little toes, the flexor hallucis, and the flexor digitorum brevis. During walking on the level, they become active at or about the 35% mark of the whole walking cycle (Fig. 14.22). (But activity is earlier with flat-footed subjects.) Activity always

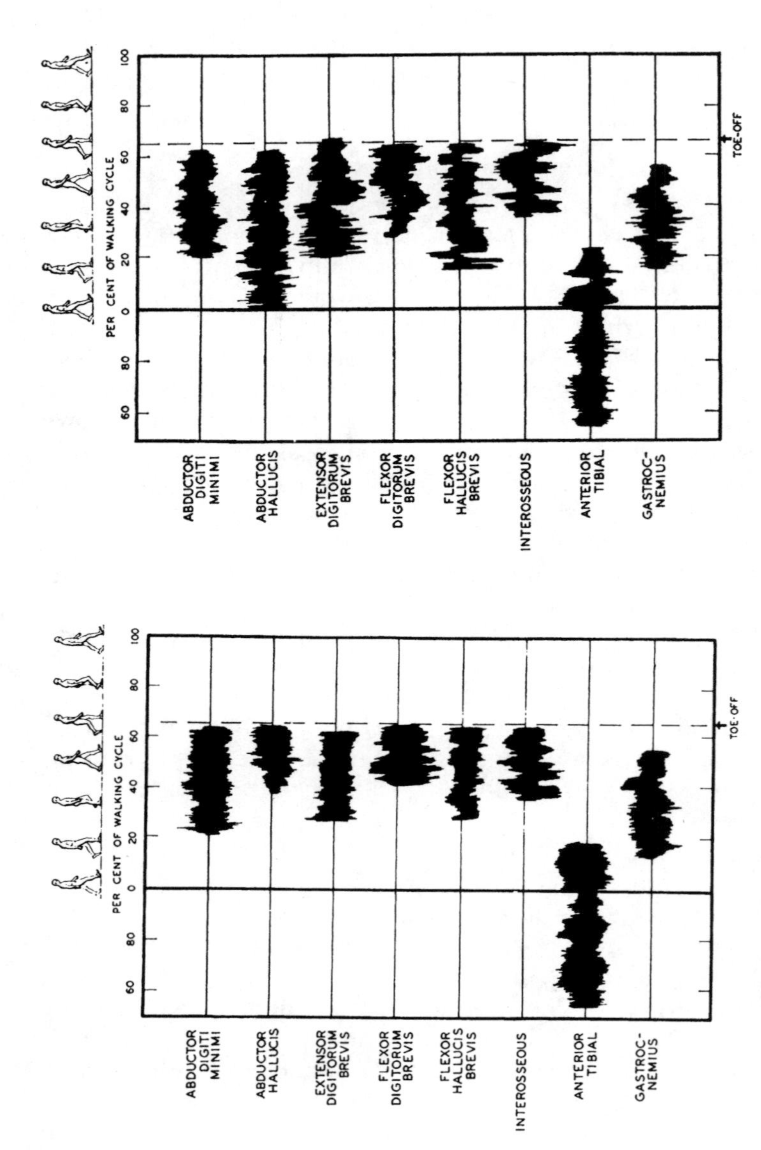

Figure 14.22. Cycle of EMG activity during walking in muscles of the leg and foot (composite of superimposed multiple records). (*Left*) Normal. (*Right*) Flat-footed subjects. (From R. Mann and V.T. Inman, ©1964, *Journal of Bone and Joint Surgery*.)

ceases just before toe-off. When walking on a downslope, the start of activity again is advanced, often occurring from the onset of the cycle (as the heel strikes). Mann and Inman relate the activity in intrinsic muscles to the progressive supination at the subtalar joint. They believe that an important role of the intrinsic muscles is the stabilization of the foot during propulsion, acting mainly at the subtalar and transverse tarsal joints. The pronated foot requires greater intrinsic muscle activity than does the normal foot. Reeser et al (1983) emphasize the combined function of the three toe flexors in resisting toe extension during the stance phase of locomotion; the abductors "affect the mediolateral distribution of pressure by positioning the forefoot" both in posture and locomotion.

Mann and Inman agreed with Basmajian and Bentzon (1954) that activity is not needed to support the arches of the fully loaded foot at rest. To fully investigate this, experiments were done with loading of the static foot while completely removing the factor of posture (Basmajian and Stecko, 1963); as noted above (p 341), they were in complete accord, as were Reeser et al (1983) more recently.

Extensors of Toes

The activities of the extensor brevis of the little and great toes were compared by Carvalho et al (1967) with needle electrodes in 20 persons. In half the cases the muscles did not act entirely synchronously. During walking they did not show a clear pattern. In different subjects the recruitment pattern showed different behavior. The extensor digitorum brevis was often silent during gait. Carvalho and Vitti (1965) also described the EMG of the long extensor of the great toe, which they found to be silent in ordinary stance but becoming active when the subjects swayed backwards and also during dorsiflexion of the ankle.

Movements of the Hallux

The great toe is not just a useless appendix. In human locomotion it is the last contact with the ground at takeoff. However, conscious control of its functions is usually primitive. Furthermore, hallux valgus (with associated bunions) is a major surgical problem blamed on muscle imbalance. The opposite—hallux varus—also has been the object of surgical curiosity and intervention (Thompson, 1960). By implication or direct statements, authorities cite the roles of contracture, imbalance, or predominance of activity in the adductor hallucis muscle over those in the abductor, or vice versa. But no clear experimental proof of muscle imbalance exists. Therefore, an EMG investigation was made of 25 normal persons and 10 patients with "idiopathic" hallux valgus to test the hypothesis of muscle imbalance (Iida and Basmajian, 1974). The muscles that insert at the MP joint of the great toe—abductor hallucis, adductor hallucis (both transverse and oblique heads), and flexor hallucis brevis were chosen. Although other muscles also act on the great toe, for simplification the inquiry was limited to muscles acting on the MP joint of the great toe. The experimental results indicate that the abductor force in normal subjects balances the adductor force at the MP joint. However, in hallux valgus, while the adductor force is markedly decreased, the abductor force actually falls to *nil* and so a weak adductor force becomes operative.

In hallux valgus the activity of (1) adductor hallucis is markedly decreased in adduction; (2) flexor hallucis brevis is slightly decreased only in loading situations; (3) abductor hallucis has completely *nil* activity during abduction; but (4) adductor hallucis and abductor hallucis muscle show their strongest activity in flexion.

The changed patterns of activity can be explained as secondary to the following mechanical changes: with bony malalignment, the distance between origin and insertion of the intrinsic muscles is increased, so the insertion of adductor hallucis moves toward the medioplantar side and the tendons of flexor hallucis brevis are displaced laterally; now, they will bowstring in the loading situation. As a result, the adductor hallucis and flexor hallucis brevis are stretched, with the adductor losing its force markedly and the flexor slightly. The tendon of the abductor hallucis moves plantarwards in relation to the metatarsal head, and so it completely loses its abductor force, instead gaining flexor force. Thus, the imbalance is in the direction of adduction. But there also occurs a rotation of the proximal phalanx into an inverted position; this is due to imbalance caused by the changed function of the abductor muscle, which is dragged into the orientation of a flexor. Now the long axis of the first ray is shortened by the pull of the adductor force—further enhancing the valgus tendency.

Morphology of Hallux Muscles

In an unpublished morphological study, J.W. Kerr and J.V. Basmajian made the following observations on 22 adult feet dissected with particular emphasis on the insertion of the abductor hallucis muscle.

A great variation in the mode of insertion of the abductor hallucis tendon was revealed. In only one of the specimens did the tendon lie on the medial border of the foot and insert into the medial side of the base of the proximal phalanx in such a fashion as to be an obvious abductor.

At the other end of the scale, in several specimens the abductor hallucis and medial head of the flexor brevis had a common insertion into the base of the medial sesamoid, with the abductor tendon lying on the plantar aspect of the foot as an obvious flexor.

Between these two extremes, in some specimens the abductor tendon was on the plantar aspect of the foot and, passing over the medial portion of the seasmoid without attachment, it inserted into the base of the proximal phalanx. In a quarter of the cases, a slip of insertion was given off the lateral side of the abductor to the medial sesamoid before its insertion on the phalanx. In another quarter, there was a common slip of insertion with the medial head of the short flexor into the sesamoid.

It was concluded that in about one-fifth of specimens the abductor hallucis was so placed as to be capable of true abduction. In most, the abductor must have acted at the metatarsophalangeal joint to flex the great toe. The abductor hallucis was always closely attached to the capsule of the metatarsophalangeal joint as it crossed it.

Great variation in the attachment of the medial head of the flexor brevis to the abductor tendon was found. In every case, there was an attachment between the two muscles proximal to the sesamoid. The insertion of the medial head of the short flexor was always into the ventral surface of the abductor hallucis tendon.

The Plantar Reflex

Landau and Clare (1959) analyzed the normal and abnormal plantar reflex (Babinski sign) electromyographically. The flexor response shows variable patterns of muscle contraction, while the abnormal extensor response shows both hyperexcitability and stereotypy. The unique feature of the abnormal extensor response is the recruitment of anterior crural muscles—extensor hallucis longus, tibialis anterior, and extensor digitorum longus. Then there is an actual mechanical competition between the flexors and extensors of the great toe, and it is the latter that triumph. If, perchance, the extensors are weak or denervated, flexion occurs when the Babinski extensor sign would be expected. Thus Landau and Clare concluded that the extensor reflex is really just a hyperactive flexor response with radiation to the extensors which, proving stronger as a rule, produce extension of the great toe. However, caution must be exercised in accepting this conclusion, for Grimby (1963b) has shown that the plantar response is a complex phenomenon and not a simple reflex. It depends to a great deal on the exact area of skin stimulated (Grimby, 1963a; Engberg, 1964). While not disagreeing with this point of Grimby's, van Gijn (1975, 1976) more emphatically agrees with Landau and Clare and greatly stresses the powerful EMG reaction of extensor hallucis longus during the Babinski response.

CHAPTER 15

Back

The muscle masses filling the space between transverse and spinous processes on both sides of the vertebral column are known as the deep, oblique, or transversospinal muscle groups. Commonly they comprise (from superficial to deep): the semispinales, the multifidi, and the rotatores muscles. Generally these are not considered to be part of the erector spinae. The superficial muscles of the transversospinal tract are accepted as the semispinales in the neck and thoracic region. Multifidi exist deep to the semispinales but are superficial in the lumbar area, where there is no semispinalis. The next layer, the rotatores, are 11 pairs of small muscles of the thorax deep to the multifidi, best seen in the thoracic region.

Although others have studied the intrinsic muscles of the back, the pioneers in this field were Floyd (a physiologist) and Silver (an anatomist) of Middlesex Hospital Medical School in London. During the 1950s these two investigators broadened our understanding of the erector spinae, and this chapter continues to lean heavily on their reports. Further detailed studies are still called for because the regional or local differences in the structure and function of the many muscles in this group must be explained. Furthermore, clinical conditions exist—including low back pain and scoliosis—that demand clarification.

The tentative effort of Riddle and Roaf (1955) was a step in the right direction. However, their conclusion that deep rotator muscle paralysis is the cause of idiopathic scoliosis cannot be substantiated by their published findings, although it may well be the truth (Basmajian, 1955b). Żuk (1960, 1962a, 1962b) of Warsaw demonstrated increased muscular activity on the convex side of the scoliotic curve. He believed it to be a secondary reaction of the body in an attempt to compensate for the curvature, the cause of which he blamed on "muscle imbalance." His series of patients examined electrically numbered some 250.

Hoogmartens and Basmajian (1976), in a 1972 research effort together in Atlanta, employed computer analysis of vibration-induced electromyography with 24 members of scoliosis families who had some measurable degree of scoliosis. Their hypothesis was that a concave-sided hypersensitivity of muscle spindles is responsible for idiopathic scoliosis. Since spindles are sensitive to vibration, they induced a response bilaterally in the back muscles. Of 23 thoracic curves, 15 showed vibration hypersensitivity of the spindle system on the concave side and 5 on the convex side. These results have been replicated and expanded by Trontelj et al (1977).

Horn (1969) and Redford et al (1969) and Alexander and Season (1978) also have made valuable contributions to the electromyography of scoliosis, while various authors have published clinical papers that reflect some light on spinal function (Pauly and Steele, 1966; Yamaji and Misu, 1968; Thomas, 1969).

Turning to the EMG studies of the normal back, one must note the early work of Lewer Allen (1948) in South Africa. He reported a study with the emphasis on erector spinae, and his chief conclusion was that the erector spinae is active during forward flexion of the vertebral column. Therefore, one main function is to control "paying out," which, indeed, is as important as the function of extension (Fig. 15.1).

In very rapid flexion, little or no activity is required or in fact appears. As the slowly flexing trunk is lowered, the activity in erector spinae increases apace and then decreases to quiescence when full flexion is reached. If an attempt is made then to force flexion further, silence continues to prevail in the erector. In full flexion, then, the weight of the torso is borne by the posterior ligaments and fasciae—the posterior common ligament, the ligamentum flavum, the interspinous ligaments, and the thick dorsal aponeurosis. The erector spinae again comes into action when the trunk is raised once more to the erect position.

In standing erect, Lewer Allen believed that activity in the erector spinae is not required, except for forced extension. He seems to have concluded in this admittedly early and not completely elegant study that no activity in the erector spinae results during extension, except with added resistance. This is a dubious conclusion.

Soon after, Floyd and Silver (1950) confirmed the main finding of Lewer Allen. They pointed out that Fick (1911), without the benefits of EMG, had hypothesized the complete relaxation of the erector spinae in full flexion of the spine. Their experiments were, on the whole, more elegant than any others done on the back in the 1950s. With multiple surface electrodes on the skin over the muscle at the levels of T_{10} and T_{12} and L_2 and L_4 and the added use of needle electrodes for confirmation, they were able to show the activity of multifidus as well as erector spinae (Fig. 15.2).

Their findings showed that in the initial stages of flexion of the trunk in bending forward, the movement is controlled by the intrinsic muscles of the back. They agreed with Lewer Allen that the ligaments take over and were quite sufficient in the fully flexed position.

They also showed that the position of full flexion while seated (usually considered by school teachers as a "bad" posture) is maintained comfortably for long periods and that during this time erector spinae remains relaxed. Quite correctly, they warned against jumping to conclusions, pointing out that "certain people experience backache if they sit in the fully flexed position for a sufficient time—e.g., patients sitting in bed with only the thoracic part of the vertebral column supported, motor-

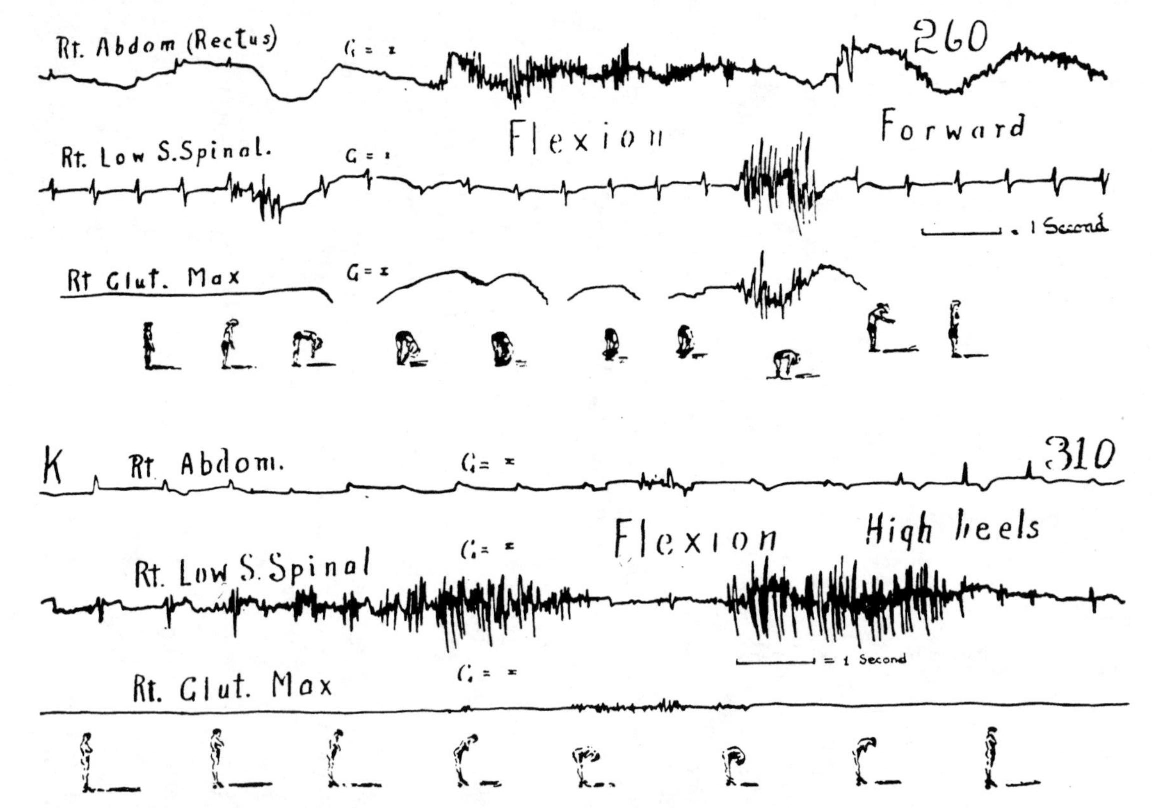

Figure 15.1. EMG records of rectus abdominis, erector spinae (sacrospinalis), and gluteus maximus in two different subjects, one wearing high heels (*lower tracings*). Synchronized drawings to show phase of forward flexion of trunk. (Composite of parts of two illustrations from Allen, 1948; photograph retouched to improve engraving.)

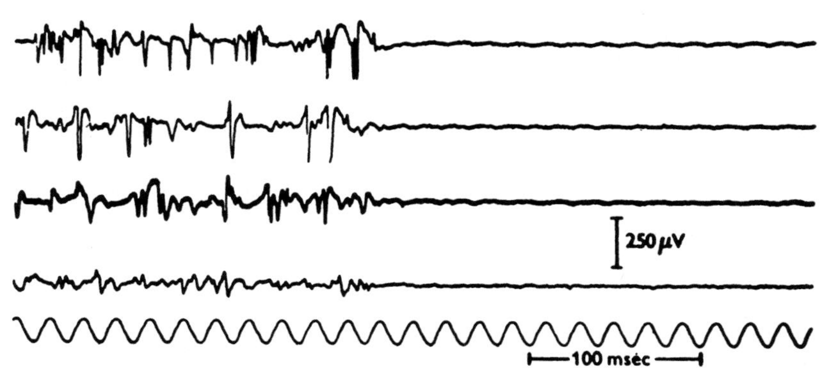

Figure 15.2. EMG records of flexion-relaxation of erectores spinae. Recorded with needle electrodes at depths of 1, 2, 3, and 4 cm, at the level of L_3 vertebra. Similar EMG signals were obtained at all depths. (From W.F. Floyd and P.H.S. Silver, © 1955, *Journal of Physiology*.)

car drivers, etc." Floyd and Silver suggested, perhaps too cautiously in the light of our present knowledge, that a reflex inhibitory mechanism explains the complete relaxation of erector spinae in full flexion. Finally, they suggested that this relaxation of the muscle and the dependence on the ligaments, including the intervertebral disc, had implicit dangers, including injuries to the disc.

In a later extensive series of investigations, Floyd and Silver (1955) examined the function of the erector spinae in certain postures and movements and during weight-lifting. They used both surface and confirmatory needle electrodes for the thoracolumbar parts of the erector spinae. Posture was recorded by photography, direct measurements, and radiography.

Most subjects standing in a relaxed erect posture showed a "low level of discharge" in the erector spinae. Small adjustments of the position of the head, shoulders, or hands could be made which would abolish the activity of the muscle, i.e., an equilibrium or balance could be achieved.

From the easy upright posture, Floyd and Silver found that extension (hyperextension) of the trunk is initiated, as a rule, by a short burst of activity. Their findings during flexion of the trunk were described before (see p 260 and Fig. 11.5).

While standing upright, flexion of the trunk to one side is accompanied by activity of the erector spinae of the opposite side, i.e., the muscle is not a prime mover but an "antagonist." However, if the back is already arched in extension (hyperextension), not even this sort of activity occurs.

Floyd and Silver state that erectores spinae contract (apparently vigorously) during coughing and straining. This occurs even in the midst of their normal silence, whether the subject is erect or "full-flexed." The clinical implications of this last observation have not been, from our point of view, adequately explored by orthopedic specialists.

With the subject standing, the activity in erector spinae ceases earlier during forward bending than it does when he is seated. In some patients they found complete relaxation in the sitting but not the standing posture.

Finally, Floyd and Silver reported that the erector spinae remained relaxed during the initial movement of lifting weights of up to 56 lbs (28.5 kg). They proved that it is movement at the hip joint that accounts for the earliest phase of apparent extension of the trunk. However, the ligaments of the back were required to carry the added weight without help from the adjacent muscles (Fig. 15.3).

The studies on the erector spinae of Åkerblom (1948), Portnoy and Morin (1956), and Joseph and McColl (reported by Joseph in his book *Man's Posture*, 1960), are especially concerned with posture and are discussed in that chapter on p 260. In essence they agree with Floyd and Silver.

In 1958, Friedebold of Berlin reported a study on the mode of action of erector spinae in a series of women who carried out a series of movements and postures of the trunk. In addition to confirming in general the earlier studies, this report enlarged upon the activity during lateral flexion. Most impressive is the recording of activity from both right and left erectores during bending to either side, although there seems to be a pattern of cooperative activity and not a simple simultaneous antagonism (Fig. 15.4).

A new dimension in the EMG of intrinsic muscles of the back was added by Morris et al (1962) and Waters and Morris (1972) of San Francisco. They investigated the activity of different layers and parts of the spinal musculature—iliocostalis in the thoracic and lumbar parts, longissimus and rotatores in the region of the 9th and 10th ribs, and multifidus abreast of the 5th lumbar spine.

During the performance of various trunk movements, muscles showed patterns of activity that clearly showed two functions—sometimes they initiate movement and at other times they stabilize the trunk. Almost all the movements recruit all the muscles of the back in a variety of patterns, although the predominance of certain muscles is also obvious.

In compound movements, when subjects are not trying to relax, there is constantly more activity than when the movement is carried out deliberately and with conscious effort to avoid unnecessary activity of muscles. Complete relaxation and lower levels of contraction are the "ideal" rather than the rule for normal bending movements. Morris et al found that muscles that might be expected to return the spine to the vertical position often remain quiet; they suggest that such factors as ligaments and passive muscle elasticity play an important role.

During easy standing, *longissimus* is slightly to moderately active; it can be relaxed by gentle ("relaxed") extension of the spine. During forced

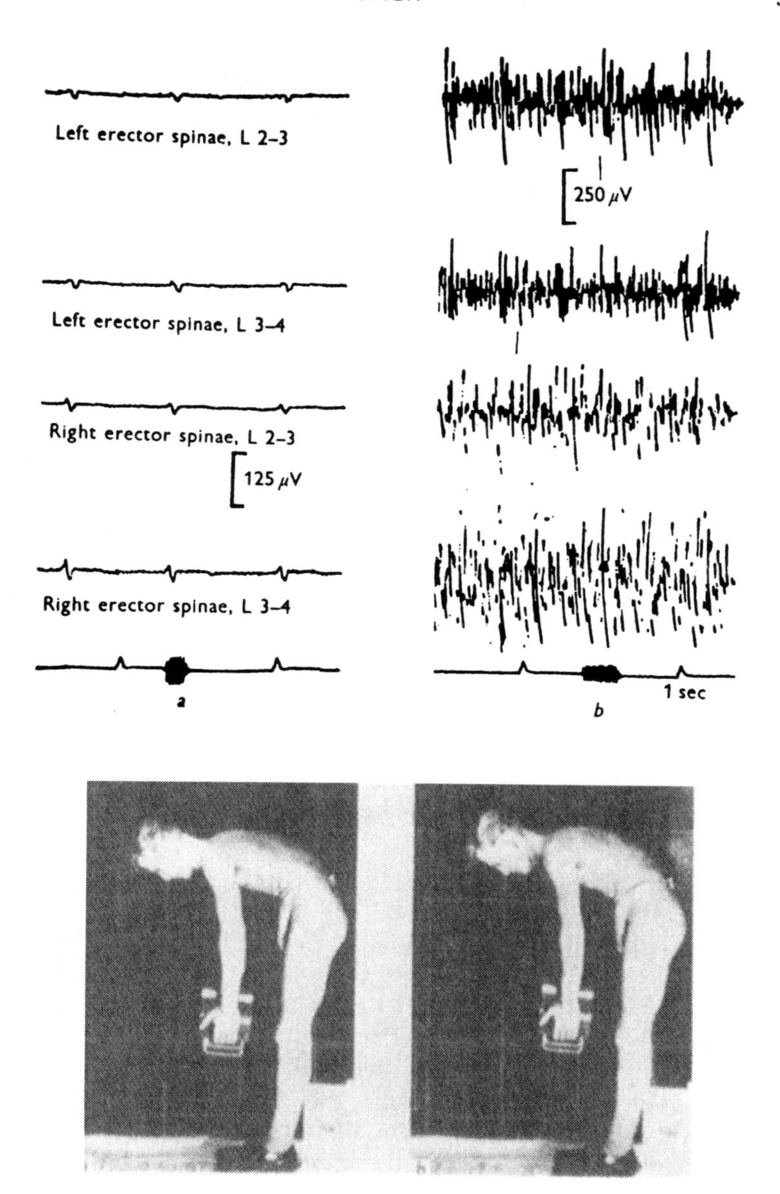

Figure 15.3. EMG records of erectores spinae during weight-lifting. Left and right sides at levels L_2 to L_3 and L_3 to L_4, with corresponding photographs. The subject lifted a 28-lb weight from the ground without activity in erectores spinae until the trunk reached a position intermediate between those shown in photographs. (From W.F. Floyd and P.H.S. Silver, © 1955, *Journal of Physiology*.)

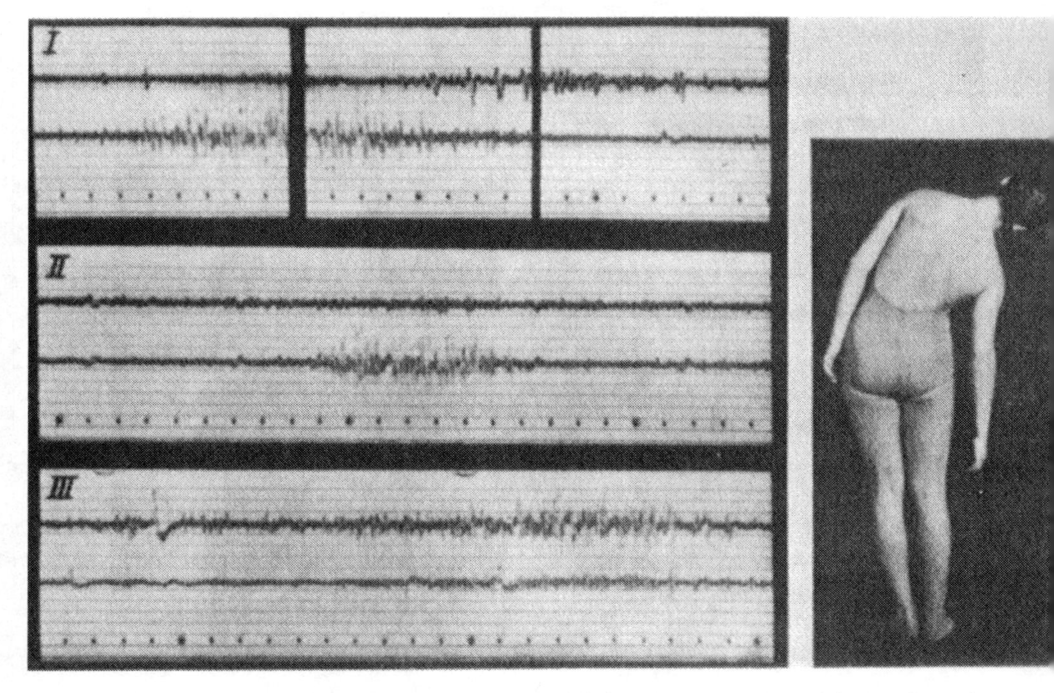

Figure 15.4. EMG records of right and left erectores spinae during lateral bending. *I*, upper lumbar region; *II*, middle lumbar region; *III*, corresponding points at both levels reciprocally. (Surface electrodes used, but are not shown.) (From G. Friedebold, © 1958, *Zeitschrift fuer Orthopaedie und ihre Grenzgebiete*.)

full extension, flexion, lateral flexion and rotation in different positions of the trunk, it is almost always prominently active.

A position of complete silence is easily found for *iliocostalis* in the erect position, but with slight forward swaying activity is instantly recruited. Forward flexion and rotation in the flexed position bring out its strongest contractions, but it is also fairly active in most movements of the spine.

Multifidus and *rotatores* have rather similar but not identical activity. With movements in the sagittal plane, they are active as they also are in contralateral rotary movements. Yet, like all the other muscles, these too relax almost completely during full flexion, leaving the trunk practically hanging on its ligaments.

Pauly (1966) conducted a systematic exploration of the intrinsic muscle of the spinal column during various exercises widely advocated for physical fitness. At the same time he clearly confirmed the earlier studies of the back and revealed new aspects of the normal functioning of these muscles. Unlike some earlier workers, he finds that semispinalis capitis and cervicis apparently help to support the head by continuous activity during upright posture.

In almost all vigorous exercises performed from the orthograde position, Pauly finds that the most active muscle is spinalis; next in order is longissimus, and least active is iliocostalis lumborum. Nevertheless all three muscles and the main mass of erector spinae act powerfully during strong arching of the back in the prone posture. During push-ups, there is considerable individual variation but, typically, the lower back muscles remain relaxed.

Simple side-bending exercises of the trunk do not recruit erector spinae as long as there is no concomitant backward or forward bending. This clearly refutes earlier opinions whose authors had ignored movements in the ventrodorsal plane that do involve erector spinae. Much of this work has been confirmed by Jonsson (1970), and the technique has been adopted for ergonomics by Tichauer (1971).

DEEP MUSCLES

Donisch and Basmajian (1972) tried to clarify the following question: what can one find out about the deep muscles close to the vertebral column by examining them bilaterally at different levels? Although it might have been interesting to duplicate the experimental conditions of the previous investigators, present-day regulations, improved equipment, and a different population of subjects ruled against a simple replication.

The deep layers of the transversospinal back muscles were studied in 25 healthy human subjects. Bipolar wire electrodes were inserted bilaterally at the level of the sixth thoracic and the third lumbar spinous processes. Activity was registered simultaneously in sitting and standing, and during movements while in these positions. The same muscle group displayed different patterns of activity in the thoracic, as compared to the lumbar level. Variations in the pattern of activity during forward flexion, extension, and axial rotation suggest that the transversospinal muscles adjust the motion between individual vertebrae. The experimental evidence confirms the anatomical hypothesis that the multifidi are stabilizers rather than prime movers of the whole vertebral column.

Because of the complexity of the intervertebral joint mechanism and the impossibility of efficiently preventing motion between individual vertebrae, actual free movements, rather than resisted ones, were studied—even at risk of electrode displacement. While it was the purpose of this investigation to learn about the functions of these muscles during postures and movement, the interpretation of electrical activity presented some difficulties. Did a muscle showing activity produce the movement, prevent the movement, or was it contracting isometrically? Therefore emphasis was placed on the occurrence of electrical silence, knowing that the muscle tested was not taking part in the movement under observation. Decreasing or increasing activity during a movement also seemed to be functionally more important than unchanging activity.

Andersson et al (1977), in their efforts to relate paravertebral muscle function to disc pressures, found that the amplitude of the EMG signal and pressure increased both with angles of forward flexion and with increasing static loads in flexion. During asymmetric loading, pressure values and myoelectric activity increased, being greater on the contralateral side of the lumbar region and ipsilateral side of the thoracic region. The disc pressure, intra-abdominal pressure, and semi-integrated rectified EMG signal were higher throughout when the trunk was loaded in rotation, rather than in lateral flexion.

Changes in the Lumbar Curvatures During Sitting and Standing

In most subjects of the Donisch-Basmajian series, the lumbar muscles were inactive during relaxed sitting but showed some activity in straight sitting and in the standing posture. This finding is in agreement with the results of most other workers, including Eastman and Kamon (1976) and Andersson et al (1975). This last very productive group in Goteborg, Sweden, found that disc pressure and myoelectric activity change together (as noted above). When the back of a seated subject is supported, levels of both pressure and EMG signal fall. They also confirmed that intramuscular wire electrodes are superior to surface electrodes in the study of intrinsic back muscles (Andersson et al, 1974, 1977).

Forward Flexion in Sitting and Standing Positions

As noted before, Floyd and Silver (1951) first showed that the back muscles became electromyographically inactive at a critical point during extreme flexion. Morris et al (1962) found that flexion-relaxation can occur, but they felt that in normal bending movements the back muscles remained frequently active. Donisch and Basmajian found spontaneous electrical silence of the lumbar muscles in extreme flexion in most subjects, but only half of them showed spontaneous inactivity of their thoracic muscles in both seated and standing postures.

During the Valsalva maneuver with increased intrathoracic and abdominal pressure while holding a sandbag of 11.25 kg, all thoracic and a number of lumbar muscles showed activity instead of electrical silence. This might be explained partly by the fact that most subjects were no longer in extreme flexion and might have reached the "critical point" of activity.

Extension from the Flexed to the Upright Posture

While inactivity of the back muscles during the last stage of flexion can be explained in that the muscles are no longer needed and ligaments are holding the vertebral column, there is no explanation of why these muscles do not always become active immediately when extension is begun. Instead, there frequently are short "bursts" of activity that occur (especially in the lumbar region) when the movement of extension is half completed. It therefore appears that in most persons the lumbar transversospinal muscles do not initiate extension from the fully flexed position.

Lifting weights with different mechanical advantages seemed to indicate that in most instances more energy is used in the lumbar and thoracic back muscles when the object lifted cannot be brought close to the line of gravity of the subject. Pauly (1966) and other investigators also noticed increased activity of the back muscles when the center of gravity was shifted forward.

Kumar's study (1980) clarified some of the physiological responses to weight lifting. A weight of 10 kg was lifted by 11 normal male volunteers (mean age 34.2 years) from ground to knee, hip, and shoulder levels in the sagittal, lateral, and oblique planes. During these lifting maneuvers, intra-abdominal pressure was measured by telemetry, and the activity of erector spinae and external obliques were recorded by electromyography. The values obtained for peak and sustained intra-abdominal pressure and the averaged electromyographic activities of erectores spinae and external obliques were subjected to analysis of variance and correlation analysis. A significant difference between the responses in these three planes was found: the sagittal plane activities evoke least response. Intra-abdominal pressures, erector spinae activity, and external oblique activity were highly significantly correlated in each of the three planes.

Axial Trunk Rotation

In the Donisch-Basmajian experiment less than half of the examined subjects showed this expected activity of the transversospinal muscles of the thoracic region, whereas more than half of our subjects showed the expected activity in the lumbar region. This finding is somewhat surprising considering that most of the actual rotatory movement occurs in the thoracic region. Paradoxical activity of the deep muscles was found in five subjects at the thoracic and in three persons at the lumbar level. Morris et al (1962) also occasionally found activity of the rotatores and multifidi during ipsilateral rotation. These workers further stated that little activity was seen when their subjects returned to the original position.

All subjects showed bilateral activity in the thoracic level (Donisch and Basmajian); in more than half this activity did not appear to be related to the direction of rotatory movement (Fig. 15.5). In the lumbar region the muscular activity seemed more often to support the theory of rotatory function. On the other hand, the position of articular facets in relation to the direction of muscle pull casts doubt on the anatomical feasibility of such a function.

Perhaps the designation of specific function is almost impossible in the back, where we have a complex arrangement of muscle bundles acting on a multitude of equally complex joints. Those who insist on finding prime movers, antagonists, and synergists in the genuine musculature of

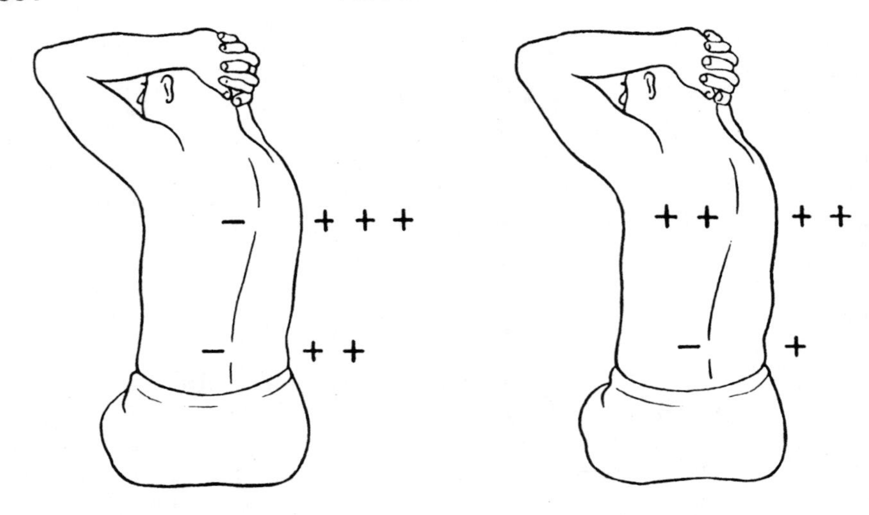

Figure 15.5. EMG activity of deep back muscles during rotation. (From E.W. Donisch and J.V.Basmajian, 1972, *American Journal of Anatomy.*)

the back will be always disappointed. For example, in their wire-electrode study of the intrinsic muscles, the Tokyo orthopedics team of Iida et al (1976) were unable to provide validation for reeducation exercises widely advocated and used in France (Piette, 1974). Rather than abandon the exercises, they suggested that the transversospinal muscles are stabilizers and that this function is important.

Standing

Joseph (1960) found continuous activity of the back muscles in the lower thoracic region during standing. He concluded that the activity of these muscles depended on their relation to the line of gravity. The segments of the vertebral column located further posterior to the line of gravity had the tendency to fall forward, a movement that was counteracted by the back muscles. Pauly's (1966) results were similar. In his experience the spinalis muscle showed more activity during the orthograde position than the erector spinae muscles at lower and more lateral levels. Donisch and Basmajian found this greater activity of the thoracic transversospinal muscles not only in the orthograde posture but also frequently during other postures. The thoracic muscles showed a greater tendency to remain active, while the lumbar muscles acted with "bursts" of electrical potentials.

Both Joseph (1963) and Pauly (1966) stated that some muscles apparently contracted unnecessarily. These contractions were more often seen in women and untrained men. We found the same tendency in our subjects.

Asymmetry

There are some differences in activity of the transversospinal muscles at the same levels. This asymmetrical activity occurred during quiet sitting and standing but was also noted with movements in the sagittal plane. Jonsson (1970, 1973) explained these differences of electrical activity by asymmetrical posture.

Wolf and Basmajian (1980) and Wolf et al. (1979) assembled and analyzed quantified data correlating normal back movements with the EMG activity in 121 adult subjects who reported no history of low back discomfort. EMG records were obtained from vertical pairs of surface electrodes placed bilaterally 3 cm from the midline at the L_{3-4} and L_{4-5} levels. Recordings were made of a range of possible movements while standing and sitting (with the pelvis stabilized).

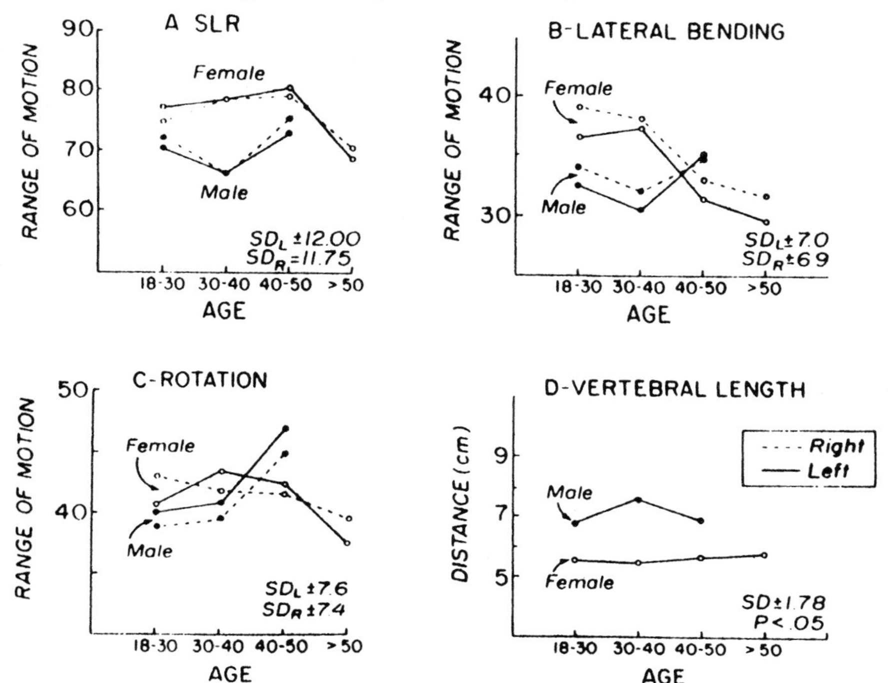

Figure 15.6. Mobility measurements for the first 66 (of the total 121) normal subjects. Mean values are expressed for (*A*: straight leg raising (SLR) in the supine position (*A*), Lateral bending (*B*), and trunk rotation in the standing position (*C*), change in vertebral length between cervical 7 and sacral 1 during maximal trunk flexion in the standing position (*D*). Changes in vertebral length were signifiantly greater for males. SD_R, standard deviation (*right side*); SD_L, standard deviation (*left side*).

The range for bilateral movements and trunk flexion is shown in Figure 15.6. Differences in left- and right-sided mobility are minimal and, as expected, men showed a significantly greater excursion in vertebral separation during complete trunk flexion (Fig. 15.6*D*).

Quantified EMG activity was averaged by gender for each age group at each of the four electrode pair combinations. Significantly greater activity occurred during extension from the flexed trunk position than vice versa for each electrode pairing. For rotational movements in the standing or sitting postures, greater activity was seen during rotation contralateral to the location of a unilaterally placed electrode pair. The magnitude of this activity level was not significantly greater for male or female subjects.

During stooping or squatting movements, males demonstrated significantly greater activity than females for recordings at all electrode placements, except the lower bilateral pair.

All results in relation to the mechanical advantage, center and line of gravity, and the possible axis of movement confirm the idea that the transversospinal muscles act as dynamic ligaments (MacConaill and Basmajian, (1969). These adjust small movements between individual vertebrae, while movements of the vertebral column probably are performed by muscles with better leverage and mechanical advantage (for details, see Donisch and Basmajian, 1972).

CONCLUSIONS

This chapter will be ended by emphasizing what must have become apparent to the informed reader. Electromyography has a great deal of practical value in this area and, aside from some general—but important—observations recorded above, much remains to be learned by this technique, especially about the fine functioning of various areas and depths of the intrinsic muscles of the back.

CHAPTER 16

Human Locomotion

Of all the electromyography done in recent years, studies on locomotion give the greatest promise of practical application. Yet, the sad fact is that such application is slow in coming. At least one reason for this delay is a lack of synthesis of the various findings. These remain relatively isolated and therefore meaningless to those who might use the information. This chapter will be devoted to an attempt at such a synthesis. Taken with the foregoing chapters on posture, limbs, and back, it gathers in one place all that appears to be significant on the use of EMG in studies of normal human locomotion. Locomotion and gait are not synonymous, but relatively little EMG research has gone beyond simple walking. Yet, the title of this chapter remains what it is in the hope that it will act as a goad for new investigations, especially clinical ones (e.g., Burdet et al, 1979; Berger et al, 1982).

The neurophysiology of locomotor automatism of terrestrial animals is extensively considered in a review by Shik and Orlovsky (1976). In general, many of the most important mechanisms—cortical, subthalamic, midbrain, spinal, and peripheral afferent—are still confusing or obscure. Cohen and Gans (1975) revealed both constancies and variations in the limb-muscle EMG activity in rats; however, temporal relationships between the onset of activity in specific muscles and the gait cycle are remarkably constant. In view of some similarities to human gait among the apes, readers may wish to refer to our papers—Ishida et al (1974), Tuttle and Basmajian (1976), and Tuttle et al (1975).

LOWER LIMB IN GAIT

Although movements of the trunk and upper limb play a role in normal walking (and will be discussed at the end of the chapter), the activity of the lower limbs evokes the greatest interest. In study after study, Basmajian and his colleagues have noted that walking elicits very slight activity in the thigh and leg muscles, compared to voluntary free movements. This has been remarked on by others also (e.g., Koczocik-Przedpelska et al, 1966).

A number of laboratories have devoted many years of study to the movements of the joints and accompanying muscular activity. Prominent among them was the Biomechanics Laboratory in the University of California at San Francisco, which conducted a long series of studies, first reported in 1944. The early investigations there dealt with the problems of amputees, and their work remains unique. Normal body mechanics and gait have occupied a greater part of the Laboratory's time. Particularly useful, but outside the scope of the present review, are other California studies on the energy cost of various types of gait (reviewed by Ralston (1964) and added to by Delhez et al, (1969)). The EMG studies of Liberson and the biomechanical studies of Elftman are also especially important and will be drawn upon below.

In any study of gait, electromyography by itself would be of limited value. If quantitative evaluations are to be made, they must be supplemented with other biomechanical techniques. Photographic methods, particularly high-frequency cinematography, have been used since the classic studies of Muybridge (1887). Marey (1885) in France, Braune and Fischer (1895) in Germany, and Bernstein (1935, 1940) in Russia all greatly improved the techniques of photographic cyclograms. Liberson (1936) was the first to combine these methods with electronic accelerometers while Schwartz et al (1936) introduced the recording of contacts of various parts of the foot with the ground (with an instrument called an electrobasograph). This later led to the use of walking on force-plates in which multiple, electronic, force, and displacement transducers are incorporated. The latter technique has been used intensively in several centers, but its expensiveness in equipment and time has proved forbidding for most laboratories. Nevertheless, Carlsöö (1962, 1973) and Carlsöö and Skoglund (1969) designed a fairly simple apparatus, as did the group at Purdue University (Ismail et al, 1965). Magora and his team (1970) at Hadassah University Hospital in Jerusalem and Grundy et al (1975) in Manchester also developed excellent apparatus.

A number of investigators have explored the use of telemetering for gait studies (Joseph, 1968; Rainaut, 1971). Rainaut has also described a workable pneumographic foot switch. Brandell et al (1968) developed a miniaturized portable EMG tape recorder device which is carried by running subjects and free of all external connections. Winter and Quanbury (1975) and Dubo et al (1975) described an excellent working technique developed by a large team of workers in Winnipeg, Manitoba, led until the mid-70's by David A. Winter, now of the University of Waterloo in Ontario. Much of their work employed multifactorial computer analysis (see below). In Italy, the work of telemetering by Casarin et al (1974) and Pedotti (1977) is worthy of note.

Accelerometers appear to offer additional useful and tidy results when combined with multichannel EMG techniques. While early experience tended to confirm this, difficulties in calibration and jiggling of the transducers on the subjects led away from these devices. Liberson and his colleagues (1962, 1965) of Chicago (now in Miami) and Smidt's group at Iowa University (Smidt et al, 1971; Smidt, 1974) have the longest and most fruitful experience with accelerographs applied to gait. In vivo recording of tendon strain during walking has been reported in sheep by Kear and Smith (1975); some day it may be applicable to man.

The results of the San Francisco studies led Saunders et al (1953) to define the six major determinants of human gait as: (1) pelvic rotation; (2) pelvic tilt; (3) knee flexion; (4) hip flexion; (5) knee and ankle interaction; and (6) lateral pelvic displacement. Actually, the phenomenon of walking is much more complex; yet, these are the components that provide the unifying principles. Locomotion is "the translation of the center of gravity through space along a path requiring the least expenditure of energy." Pathological gait "may be viewed as an attempt to preserve as low a level of energy consumption as possible by exagerations of the motions at the unaffected levels." When a person loses one determinant from the above six, compensation is reasonably effective. Loss of the determinant at the knee proves the most costly, according to Saunders et al. Loss of two determinants makes effective compensation impossible; the cost in terms of energy consumption triples and apparently discourages the patient to the point of his admitting defeat.

MULTIFACTORIAL ANALYSIS WITH COMPUTER

The use of computer analysis for gait has gained easy acceptance as improved techniques of pattern recognition developed in many research centers; its use with indwelling wire electrodes was first reported from the National Research Council of Canada in Ottawa (Milner et al, 1971); Kasvand et al, 1971). Those studies were designed to provide some insight into the precise timing of vastus lateralis, the long head of biceps femoris (the lateral hamstring), tibialis anterior, and lateral gastrocnemius in the course of normal walking. There were two sets of experimental conditions: (1) level walking at various speeds in the range of 2.2 to 7.5 ft/s (0.67 to 2.28 m/s); and (2) level walking at a speed nominally fixed at 4.5 (1.37 m/s) but with pace periods set by metronome beats in the range of 0.2 to 0.8 beats/s. Are average EMG values dependent, as suggested by Eberhart et al (1954), upon walking speed and pace frequency?

EMG signals from indwelling electrodes and footswitch data, together with other control signals (including footswitches) to facilitate automatic processing of the data, were collected on five channels of a seven-channel FM tape recorder. The tape was input to the A/D converters of an SDS 920 computer via signal conditioners. A series of computer programs enabled the automatic processing of the data, which were first digitized. Next, the footswitch timing characteristics were analyzed, and averages were obtained. Using these analyzed data, the EMG data were subdivided, and a mean wave form for each muscle at each step was computed, as well as various data output in typewritten and graphic form. From the processed data average EMG characterstics were plotted as functions of speed and pace periods; phasic activities for these various conditions were plotted for six normal adult male subjects.

For details, see our lengthy study (Milner et al, 1971), with its many illustrations. Only the highlights are given here. The average EMG activity in each of the several muscles showed clearly that there is a strong dependence in each instance of the EMG values upon speed. Perhaps the most interesting result was exhibition of a minimal value of activity in tibialis anterior. Since EMG amplitude might be interpreted as a measure of muscular energy, it was concluded that a minimal energy condition can be found for most subjects in the range of 3 to 5 ft/s (0.9 to 1.5 m/s). A generalization about comfortable, optimal walking speeds seems probable when the EMG-speed characteristics of the other muscle groups are examined in this speed range. The most comfortable walking speed lies close to 3 ft/s (0.9 m/s) to minimize "energy consumption" and permit a reasonable propulsion speed. The "energy cost" would appear to increase considerably as speeds in excess of 6 ft/s (1.8 m/s) are attained, at least under experimental conditions.

When a subject is permitted to walk without the imposition of a pace frequency constraint, he selects a walking pace for the set speed in such a manner as to allow a minimum of muscular activity. During the course of such a walk, random variations in EMG activity seem to be more marked. This gives an indication of some measure of ongoing adaptive control. The technique of averaging records was designed primarily to eliminate the differences between successive steps and to determine a meaningful average value (Milner et al, 1971).

More recent studies have polished up those findings and extended them (Bajd et al, 1974; Richards and Knutsson, 1974; Wait et al, 1974; Winter et al, 1974; Cheng et al, 1974; Cappozzo et al, 1975; van der Straaten et al, 1975; Kasvand et al, 1976; Hershler and Milner, 1976; Takebe and Basmajian, 1976). Pearson (1976) drew special attention to studies of neural mechanisms in his semipopular article in *Scientific American*.

PHASES IN WALKING CYCLE

Traditionally, human gait is composed of two phases: (1) stance, beginning when the heel strikes the ground; and (2) swing, beginning with toeing-off. The fairly precise division of the whole cycle into 10% segments is illustrated in Figure 16.1 Carlsöö (1966) showed that the initiation of walking from a stance posture consists of the body losing its balance as a result of cessation of activity in postural muscles (including erector spinae and certain thigh and leg muscles). The various torques of the body weight displace the line of gravity, first laterally and dorsally,

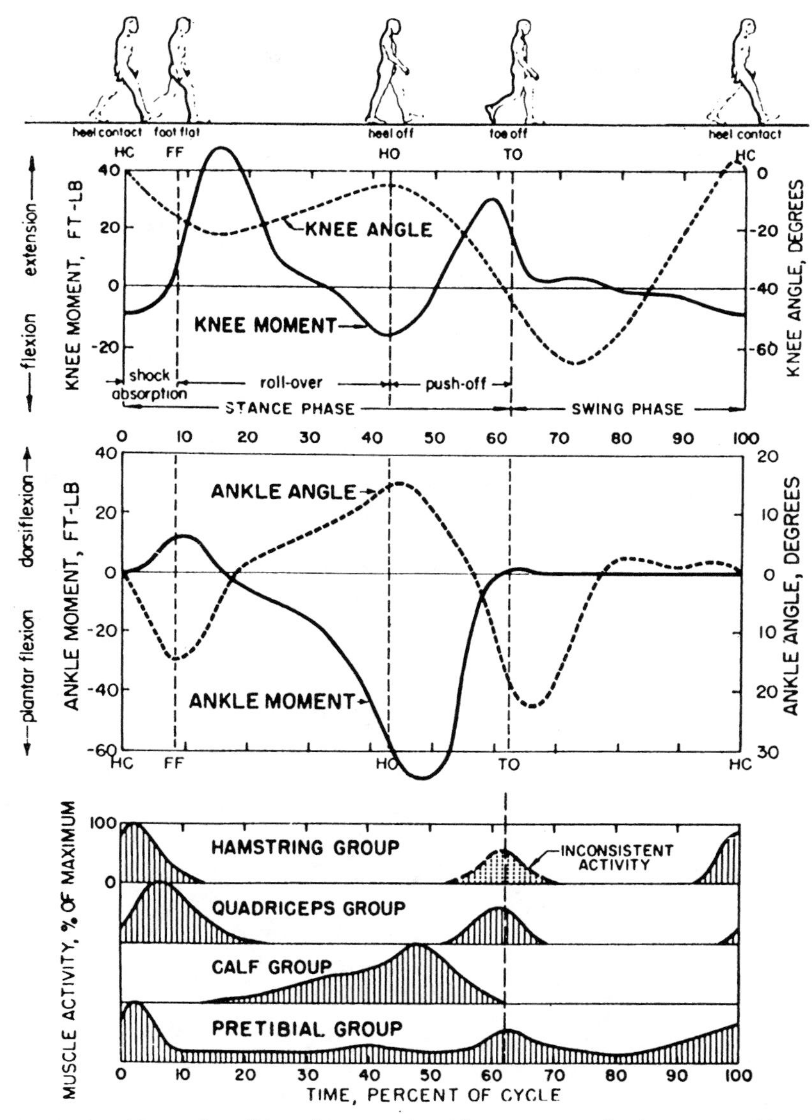

Figure 16.1. Normal walking: knee and ankle moments (in foot-pounds) compared with muscular activity during one cycle of walking (right heel to right heel contact) on level ground. (From C.W. Radcliffe, ©1962, *Artificial Limbs*).

then ventrally, to a position in which the propulsive muscles are able to contribute to and complete the first step.

Radcliffe's (1962) diagram (Fig. 16.1) illustrates the interaction between the knee and ankle joints and the phasic action of the major muscle groups recorded electromyographically. (The terms "knee moment" and "ankle moment" refer to the action of muscles about the knee or ankle which tend to change the angle of these joints towards either flexion or extension.)

In Figure 16.1, one should not miss the following features:

1. As the heel strikes the ground the hamstrings and pretibial muscles reach their peak of activity.
2. Thereafter, the quadriceps increases in activity as the torso is carried forward over the limb, apparently in maintaining knee stability.
3. At heel-off the calf group of muscles build up to a crescendo of activity which ceases with the toe-off. Before and during toe-off, the quadriceps and sometimes the hamstrings reach another (but smaller) peak of activity.
4. The pretibial muscles maintain some activity all through the cycle, rising to a peak at heel-contact and a smaller peak at toe-off. This has been often confirmed (e.g., by Dubo et al, 1976).

The complex phasing of these normal actions led the San Francisco investigators to studies of amputees and prostheses (Radcliffe, 1962). Similar work on the muscle functions in lower amputees was pursued in Poland under the direction of J. Tomaszewska (1964) of Poznan, and M. Weiss (1959, 1966) of Warsaw. There is little doubt that all such work will continue to stimulate improvements in prosthetic appliances of both the conventional and myoelectrically controlled types.

In a sober review, Elftman (1966), a pioneer in the multifactorial approach for the study of gait, gave the apt warning that EMG requires the addition of other criteria for the monitoring of tension in nonisometric contractions. Studies of the calf and foot during walking have in the past two decades employed such adjuncts with notable success. Eberhart et al (1954) showed that the function of the calf muscles during walking is limited to the push-off. They lift the body against gravity on the forepart of the foot.

Radcliffe (1962) postulated that during the stance phase the stabilizing function of the ankle plantar flexors at the knee is most important. This was confirmed by Sutherland (1966) by means of combined EMG techniques and motion pictures of gait. The period of activity in the calf muscles and of knee extension and dorsiflexion of the foot corresponded. Only at the end of plantar flexion of the ankle did plantar flexion of the foot occur. A bizarre finding was that knee extension occurred after quadriceps activity had ceased. This is related to the fact that full extension of the knee never occurs during walking in the way that it does in standing (Murray et al, 1964).

Sutherland believes that knee extension in the stance phase is brought about by the force of the plantar flexors of the ankle resisting the dorsiflexion of the ankle; this dorsiflexion is in turn the resultant of extrinsic forces—kinetic forces, gravity, and the reaction of the floor. Because the resultant of extrinsic forces proves to be greater, increased dorsiflexion of the foot continues until heel-off begins. The restraining function of the ankle plantar flexors in decelerating forward rotation of the tibia on the talus proves to be the key to their stabilizing action.

Using similar techniques, Houtz and Fischer (1961) had earlier produced evidence that a movement of the torso and hip region that shifts their position over the feet initiates the movements of each foot during walking. Movements initiated in the trunk lead automatically to changes in the position of the leg and foot. Houtz and Walsh (1959), by showing that soleus functions to stabilize and adjust the tibia on the talus, gave additional evidence for the view that movements of the ankle during walking occur as a reaction to muscular forces far removed from the foot (p 348).

Liberson (1965b), combining the techniques of motion picture photography, accelerograms, electrogoniograms, myograms, and EMGs, has reported the following correlation of activity (Figs. 16.2–16.4):

1. Contraction of the triceps surae is followed by that of gluteus maximus on the opposite side.
2. Contraction of iliopsoas occurs simultaneously with that of gluteus maximus of the opposite side.
3. Dorsiflexion of the foot begins at the time of maximum acceleration of the lower leg.
4. Extension of the knee begins at the time of maximum velocity of the leg.
5. Contraction of the triceps surae corresponds to the first hump of the vertical accelerogram.
6. Contraction of the gluteus maximus on the opposite side corresponds to the second hump of the vertical accelerogram.
7. In many cases, two-joint muscles show an increase of tension without EMG potentials because they act as simple ligaments during the contraction of the antagonists.

Two comprehensive EMG studies of gait have been completed on muscles in the region of the leg and foot by Gray and Basmajian (1968) (described below), and in the region of the hip joint by Basmajian and Greenlaw (1968) and Greenlaw (1973), reported elsewhere in this volume and in this chapter. Recently, Adler et al (1983) demonstrated with finality that the oblique head of quadriceps acts in concert with its long head and the other vasti, both in slow and fast walking. How long will it take for textbook writers and orthopedic surgeons to accept this convincing study?

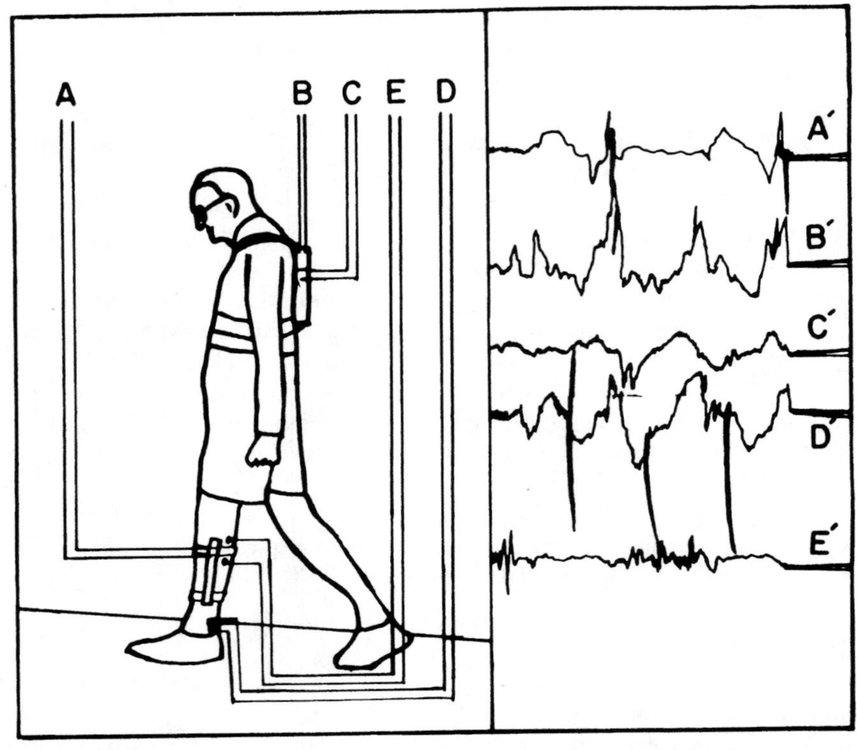

Figure 16.2. Diagram of typical multifactorial gait-recording, showing on the *left* a motion picture frame and on the *right* oscillograms, the terminal points of which correspond to the instant the picture of the walking subject was taken. (*A*) Angular accelerometer on left leg. (*B* and *C*) Vertical and horizontal accelerometers, respectively. (*D*) Lissner strain gauge tensiometer on the left gastrocnemius muscle; (*E*) Electrodes in the left gastrocnemius muscle . (*A'*) Accelerogram of the left leg. (*B'*) Vertical accelerogram. (*C'*) Horizontal accelerogram. (*D'*) Tensiogram from left gastrocnemius muscle. (*E'*) Electromyogram from left gastrocnemius muscle. Note that EMG activity precedes the major tensiogram deflection and the relationship of the latter to the accelerograms. (From W.T. Liberson, ©1965b, *Archives of Physical Medicine*.)

LEG AND FOOT MUSCLES

Tibialis Anterior

This muscle has been a favored object of attention, and so it is commonly accepted that peak EMG activity occurs in it at heel-strike of the stance phase. Movies show the foot to be inverted and dorsiflexed at this time (Gray and Basmajian, 1968).

Notwithstanding the above, there has been no general agreement as to the function of tibialis anterior at heel-strike. Without offering direct evidence, some suggest only that it counteracts forces applied to the heel

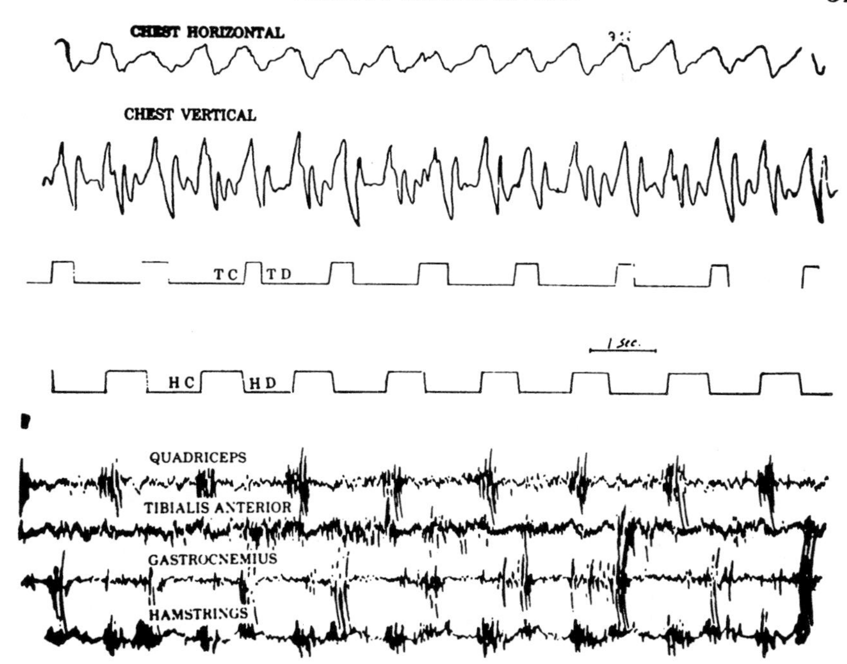

Figure 16.3. Series of multifactorial gait-tracings from a normal subject when walking. Listed from top down: *horizontal* and *vertical* accelerograms; right toe switch (TC, toe contact and TD, toe off); right heel switch (HC, heel strike and HD, heel off); and four electromyograms from muscles in the right lower extremity. (From W.T. Liberson, ©1965b, *Archives of Physical Medicine.*)

by the ground, while others propose that the tibialis anterior decelerates the foot at heel-strike and lowers it to the ground by gradual lengthening (eccentric contraction). Perhaps the clinical condition known as "drop-foot" due to paralysis of the tibialis anterior forces this conclusion.

During the more central movements of the stance phase (full-foot, midstance and heel-off) modern techniques reveal no tibialis anterior activity in "normal" subjects. The flatfooted subjects of Gray and Basmajian (1968) and those of Battye and Joseph (1966) are like "normals," except for extended activity into full-foot. Curiously, the movies of flatfooted subjects show the foot staying inverted during full-foot, maintaining inversion in order to distribute the body weight along its lateral border.

A peak of EMG activity that occurs at toe-off of the stance phase is apparently related to dorsiflexion of the ankle, presumably to permit the toes to clear the floor.

Although earlier workers believed that there is a slight fall in the activity of tibialis anterior at midswing, there is, in fact, a period of

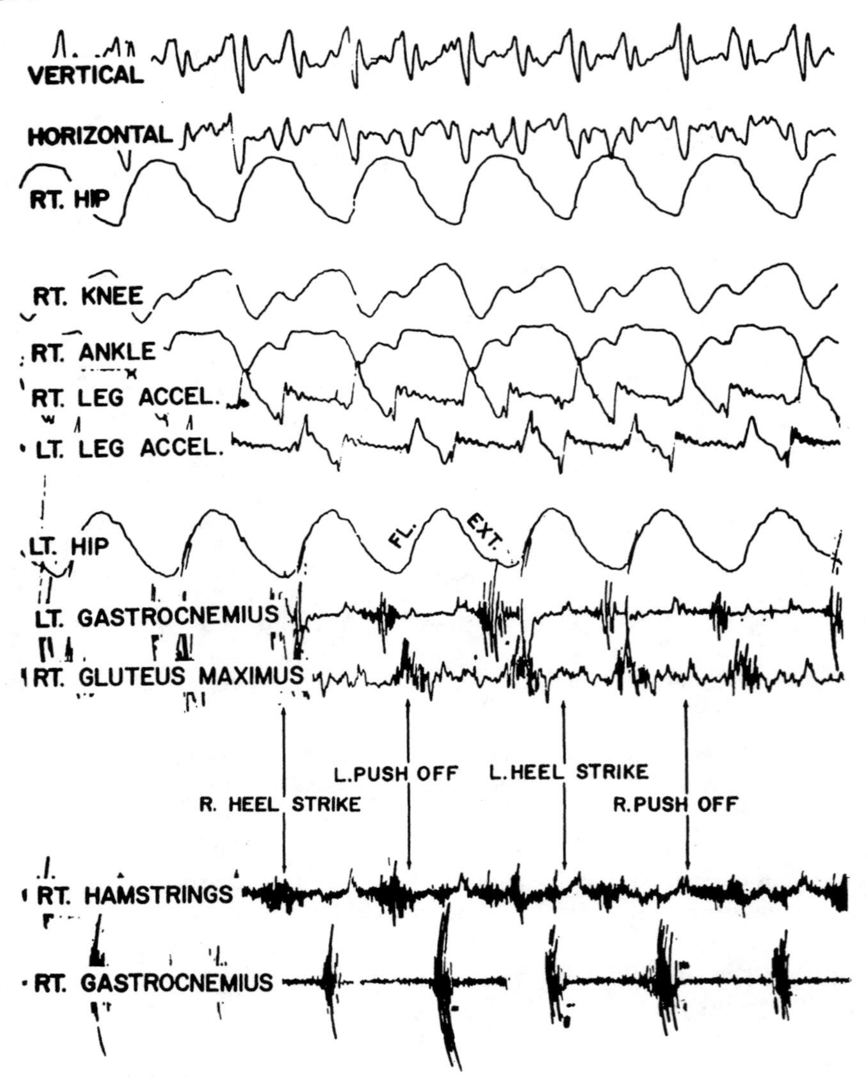

Figure 16.4. Series of multifactorial gait-tracings from normal subject during walking. Listed from top down: vertical and horizontal accelerograms; goniograms from right hip, knee, and ankle; angular accelerograms from right and left legs; left hip goniogram; and electromyograms from left gastrocnemius, right gluteus maximus, right hamstrings, and right gastrocnemius muscles. (From W.T. Liberson, ©1965b, *Archives of Physical Medicine.*)

electrical *silence* at midswing. The explanation emerges from our movies which show the foot everting at the end of "acceleration" and remaining everted through midswing. This allows for adequate clearance, while the inactivity of the invertor fits the concept of reciprocal inhibition of antagonists. Apparently, the brief period of electrical silence of tibialis anterior is essential.

The peak of activity at toe-off tapers to a slight-to-moderate level of activity during acceleration of the swing phase. Conversely, prior activity in deceleration of the swing phase builds up to a peak of activity at heel-strike. Thus, the pattern of activity of tibialis anterior is biphasic. Apparently, tibialis anterior is in part responsible for dorsiflexion during acceleration and for inversion of the foot during deceleration of the swing phase.

The pattern of activity of tibialis anterior suggests that it does not lend itself to direct support of the arches during walking. At heel-strike, when the muscle shows its greatest activity, the pressure of body weight is negligible. Conversely, during maximum weight-bearing at midstance when all the body weight is balanced on one foot, the tibialis anterior is silent. When the activity resumes at toe-off, the weight-bearing of the involved foot is minimal.

Tibialis Posterior

During ordinary walking tibialis posterior shows activity at midstance of the stance phase. The movies show the foot remaining inverted throughout full-foot and turning to a neutral position (between inversion and eversion) just before midstance. First, the fourth and fifth metatarsal heads make contact; then, as the foot everts increasingly toward neutral, more of the ball of the foot makes contact at midstance until the tibialis posterior acts. It is an invertor in nonweight-bearing movements of the foot, but its role at "midstance" appears to be a restraining one to prevent the foot from everting past the neutral position.

R.L. Jones (1941, 1945) showed in human cadaveric preparations that the tibialis posterior distributes body weight among the heads of the metatarsals. In living subjects he showed that a lateral torque on the tibia results in an increase or shift of body weight onto all but the first metatarsal head; a medial torque has the opposite effect. He concluded that by inverting the instep of the foot the tibialis posterior increases the proportion of body weight borne by the lateral side of the foot. Sutherland (1966) concluded that the plantar flexors, including the tibialis posterior, have a restraining function, to control or decelerate medial rotation of the leg and thigh observed at midstance; by controlling the eversion of the foot at midstance, the tibialis posterior provides an appropriate placement of the foot. In the flatfooted subjects of the Gray/Basmajian study, the EMG activity of tibialis posterior in the early stance phase is consistent with the maintenance of an inverted position during full-foot. By maintaining inversion the foot is supported in order to keep the body weight on the lateral border of the sole.

The foot must be inverted to accomplish lateral weight-bearing in the early "moments" of the stance phase. This of course is because the middle part of the medial border of the foot does not bear body weight in "normal" subjects; the lateral border with its strong plantar ligaments is well-equipped to bear the stresses of body weight in walking (Napier, 1957).

Although tibialis posterior is often considered to be a plantar flexor of the ankle, during level walking with an accustomed foot position it shows *nil* activity at heel-off (when plantar flexion of the ankle takes place to raise the heel). (This is not to deny that tibialis posterior may be a plantar flexor of the ankle when more powerful contractions are needed.)

Flexor Hallucis Longus

At midstance, when the entire body weight is concentrated on one foot, flexor hallucis longus shows its greatest activity. Flexing the big toe apparently positions and stabilizes it during midstance. During heel-off, our movies show the big toes hyperextended. Although Napier (1957) felt that the flexor hallucis longus helps maintain overall balance and prevents instability induced by excessive extension of the big toe, the EMG observations support this only for the flatfooted subjects and then only weakly. There is a slight activity during heel-off which may be related to preventing overextension and so giving a better balance. In contrast, the "normal" subjects show negligible activity. Consequently, one may conclude that the flexor hallucis longus is not needed in most "normal" subjects to play this role (Gray and Basmajian, 1968).

Peroneus Longus

The pattern of activity of the peroneus longus in the Gray/Basmajian study confirms the findings of many who have suggested that the peroneus longus helps to stabilize the leg and foot during midstance. The movies and electromyograms show how the peroneus longus and tibialis posterior, working in concert, control the shift from inversion during full-foot to neutral at midstance. Thus, the opinion of R.L. Jones is again confirmed: from static studies, he inferred that peroneus longus is related to eversion of the foot at midstance during level walking. Sutherland further concluded that peroneus longus, like tibialis posterior, is involved in controlling rotary movements at the ankle and foot. Our movies showed that eversion of the foot and medial rotation of the lower limb occur together. One may conclude that the peroneus longus is in part responsible for returning the foot to, and maintaining it in, a neutral position at midstance. As noted before, Walmsley (1977) found peroneus brevis to act synchronously with the peroneus longus during ordinary walking.

Throughout most of the stance phase, peroneus longus is generally more active in flatfooted subjects than in "normal" subjects. This appears to be a compensatory mechanism called forth by faulty architecture.

During heel-off, the movies showed some inversion while peroneus longus, an evertor, is active, and the invertors are relaxed. Duchenne (1867) first suggested that the inversion is caused by triceps surae. The activity in peroneus longus apparently affords stability by preventing excessive inversion, thus maintaining appropriate contact with the ground.

In flatfooted subjects, the interplay of activity between peroneus longus and tibialis posterior appears to play a special role in stabilizing the foot during midstance and heel-off. At midstance the tibialis posterior is notably more active, but at heel-off the emphasis shifts to peroneus longus.

Abductor Hallucis and Flexor Digitorum Brevis

These two muscles become active at midstance and continue through to toe-off in "normal" subjects; in flatfooted subjects most show activity from heel-strike to toe-off. Perhaps the flexor digitorum brevis and abductor try to grip the ground since they are flexors of the toes. Although others are not opposed to this idea, they believe that the muscles are also in an ideal location to help support the arches. The findings of Gray and Basmajian (using wire electrodes) tend to confirm this opinion only for flatfooted subjects because they showed higher mean levels of muscular activity.

Triceps Surae

Dubo et al (1976) found a single high peak of activity during the "push-off" of the gait cycle. At heel contact, mean muscle activity was a fifth of maximum, and this continued thus through foot flat to midstance. As the foot enters push-off the gastrocnemius sharply increases its activity. Following this there is a rapid drop that continues as the swing starts and reaches zero at midswing. Then as the transition to stance begins, activity once more increases.

HIP AND THIGH MUSCLES

Greenlaw and Basmajian's locomotion studies of the muscles that cross the hip joint have appeared extensively only in the form of a Ph.D. thesis by Greenlaw (1973) as well as being given at meetings (Basmajian and Greenlaw, 1968). The essence of their findings (from wire electrodes) is given below.

Gluteus Medius and Minimus

In the *anterior fibers of gluteus medius* there is moderate activity at heel-contact that persists through to midstance. There is also a brief burst at toe-off and another just before heel-contact. The posterior fibers are rather (but not exactly) similar. *Gluteus minimus* has only a biphasic response (at heel-contact to 40% and at midswing).

Tensor Fasciae Latae

Its pattern is biphasic with a peak during early stance through midswing and another short smaller peak during toe-off.

Gluteus Maximus

Upper, middle, and lower parts were all tested simultaneously. The *upper part* shows a clearly biphasic pattern with a small peak at heel-strike and one near the end of swing phase. The *middle part* is more triphasic with an additional high peak just before to just after toe-off. The *lower part* is biphasic, rather like the upper fibers. See also the work of Lyons et al (1983) summarized under the section Ascending and Descending Stairs, which follows.

Hamstrings

Semitendinosus has a triphasic pattern with an initial low peak at heel-contact, a second peak of 50% of the cycle, and a small third one just before the end (90%) of the cycle. *Semimembranosus* is biphasic, lacking the peak in the middle. *Biceps femoris* is also biphasic, but more crisply so.

Sartorius and Rectus Femoris

Rectus femoris is generally biphasic and triphasic, depending on cadence. Sartorius really shows only one peak, immediately during toe-off.

Iliopsoas

Iliacus acts continuously through the walking cycle with some rises and falls. The highest rise is during the swing phase, but there is another in midstance.

Psoas is triphasic (except with slow cadence); the main peaks correspond to those of iliacus, with a third peak at 50% of the cycle.

Adductors

Adductor magnus is really two muscles. The upper (horizontal) part is active nearly continuously; it reaches *nil* only at midswing. The lower (vertical) part acts like a biphasic hamstring.

Adductor brevis varies with the speed of walking. At moderate speed it is biphasic—at 40% and 90% of the cycle. *Adductor longus* and *gracilis* also have a main peak of activity at toe-off and additional peaks at late-stance and early-swing phases.

Other Gait Studies

In a brief preliminary report, Joseph (1965) described his findings of telemetered EMG signals from a number of muscles used in gait. Generally, his findings were similar to those reported in the 1940's and 1950s by the San Francisco group. In the swing phase, the hamstrings were inactive (even though knee flexion occurred) and the tibialis anterior was also inactive—but only for a brief period. In the supporting phase, activity occurred early in the calf muscles, hamstrings, and gluteus maximus but ceased toward the end. Two periods of activity were found in the sacrospinalis (at level L3): one during the swinging and one during the supporting phase.

Later, Battye and Joseph (1966) reported details of a study which, because of its clarity, will be heavily drawn upon below, supplemented by the findings of Greenlaw and Basmajian.

In general, Battye and Joseph found more similarities than differences in the walking patterns of 14 persons (8 men and 6 women). They also emphasized the importance of the inertial forces as factors in producing certain movements.

Soleus begins to contract before it lifts the heel from the ground; it stops before the great toe leaves the ground. Apparently these are supportive rather than propulsive functions.

Quadriceps femoris contracts as extension of the knee is being completed, not during the earlier part of extension when the action is probably a passive swing. Quadriceps continues to act during the early part of the supporting phase (when the knee is flexed and the center of gravity falls behind it). Quadriceps activity occurs at the end of the supporting phase to fix the knee in extension, probably counteracting the tendency toward flexion imposed by gastrocnemius. Similar findings have been reported (for vastus lateralis) by Dubo et al (1976).

The *hamstrings* contract at the end of flexion and during the early extension of the thigh, apparently to prevent flexion of the thigh before the heel is on the ground and to assist the movement of the body over the supporting limb. In some persons, the hamstrings also contract a second time in the cycle during the end of the supporting phase; this may prevent hip flexion.

Gluteus medius and *gluteus minimus* are active at the time that one would predict, i.e., during the supporting phase; however, we found with wire electrodes that gluteus medius is actually triphasic while minimus is biphasic during each walking cycle (Greenlaw and Basmajian, in Greenlaw, 1973).

Gluteus maximus shows activity at the end of the swing and at the beginning of the supporting phase. This is contrary to the general belief that its activity is not needed for ordinary walking (Battye and Joseph, 1966; Greenlaw, 1973). Perhaps gluteus maximus contracts to prevent or to control flexion at the hip joint.

The same six female subjects that provided the above Battye/Joseph findings were also studied while they wore high heels (Joseph, 1968). There were comparatively few differences from the findings previously noted. These differences were: tibialis anterior contracted less strongly but more continuously; soleus was more active; quadriceps femoris contracted (either continuously or intermittently) during the stance phase; and gluteus maximus contracted during swing phase when walking with high heels.

The gait pattern of women also interested Finley et al (1969), who compared elderly women with young ones. Rectus femoris, tibialis anterior, peroneus longus, and gastrocnemius were recorded through surface electrodes along with a multijoint electrogoniometer complex. The elderly women showed greater activity in tibialis anterior, peroneus longus, and gastrocnemius. Except for rectus femoris, the muscles of the elderly women showed considerably greater activity during foot-flat. The other differences do not appear to be functionally significant.

Comparing changes in integrated rectified EMG signal of the quadriceps with those of the calf muscles under increased degrees of stress (increased slope and/or speed), Brandell (1977) reported that the vasti responded relatively more vigorously to the stress. The peak activity for the calf occurs when the extending activity of the knee stretches the gastrocnemius across the back of the knee joint, thereby helping the calf muscles as a group to lift the heel (plantar flexion), the "most essential actions for producing the push-off and thrust in the normal walking cycle."

A pilot study of backward walking in one subject offers some suggestions of changes in EMG activity and gait patterns (Kramer and Reid, 1981). Nashner (1980) showed that minor perturbation of human gait produced EMG changes in the muscles that far exceeded the amount of activity needed for gait. Human gait on a flat surface may be usefully described as similar to bicycle riding at a steady comfortable rate and described by Basmajian (1976) as "the human bicycle" (Fig. 16.5).

ASCENDING AND DESCENDING STAIRS

Although a number of investigators have included stair climbing in their work, Joseph and Watson (1967) give the best early report. They found that during both ascent and descent, each limb has a support and swing phase. In walking up, the body is raised by contraction of soleus, quadriceps femoris, hamstrings, and gluteus maximus (along with gluteus medius activity for hip stability). During the swing phase, tibialis anterior and the hamstrings are active, as would be expected.

Walking downstairs recruits the same muscles in lowering the body, with the exception of gluteus maximus. Tibialis anterior inverts the foot at the beginning of the support phase and dorsiflexes it during the swing. The hamstrings are also active during the swing.

Both erector spinae muscles contract twice in each step, apparently in controlling bending which occurs in walking both up and down stairs.

In a more recent and scientifically elegant paper concentrating on the hip, Lyons et al (1983) investigated the timing and relative intensity of EMG activity of hip abductor and extensor muscles during free and fast

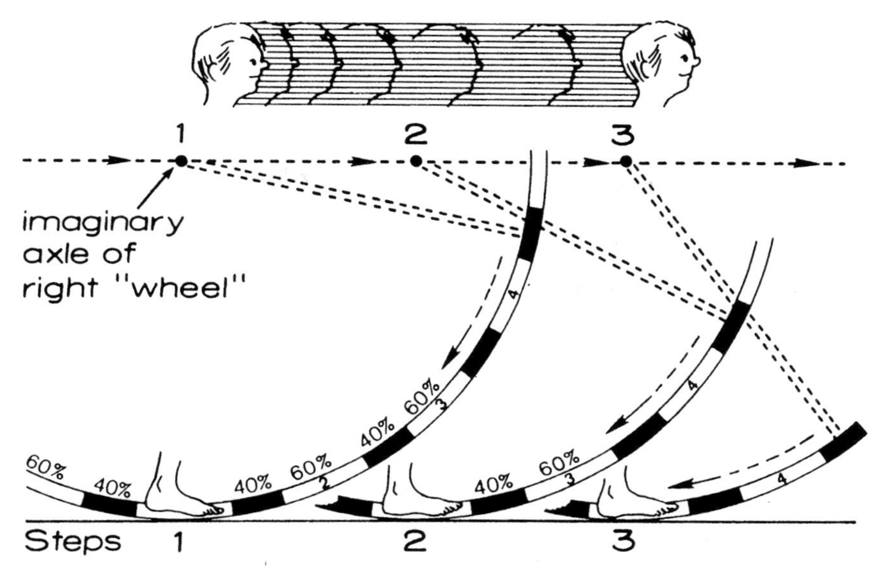

Figure 16.5. The human bicycle.

velocity walking and during ascent and descent of stairs. Eleven healthy subjects were tested using wire electrodes. Data were quantified by normalizing all activity during gait with EMG activity occurring during a sustained maximum isometric effort resisted either manually or with a dynamometer. The results indicated that the hip extensor muscles had different phasic patterns and moments of peak activity. During level walking, the semimembranosus and long head of the biceps femoris muscles displayed the greatest swing phase activity (beginning in mid-swing). The adductor magnus muscle followed with its onset in terminal swing. Both this muscle and the gluteus maximus were the principal hip extensors active during the loading response. For ascending stairs, the lower portion of the gluteus maximus muscle proved to be the main hip extensor during the loading response and midstance. The upper portion of the gluteus maximus muscle functioned more like the gluteus medius muscle than its own lower portion during both level and stair walking.

Walking up and down stairs by athletes has been shown by Townsend et al (1978) to affect the EMG activity of certain leg muscles, but there is no predictable pattern correlated with athletic training.

GAIT OF CHILDREN

The most systematic EMG studies of children's gait have been carried out by several investigators in Japan who continue with this valuable area of investigation. Okamoto and Kumamoto (1972) made long-term serial multichannel EMG recordings (combined with foot contact switches) in two infants learning to walk and in 30 subjects (ranging from infant to adult). The muscles investigated were tibialis anterior, gastrocnemius, vastus lateralis, rectus femoris, biceps femoris, gluteus maximus, sacrospinalis, and deltoid.

Okamoto and Kumamoto found a varying time in development at which these muscles acquire their adult patterns. Tibialis anterior acted in an adult fashion within 1 month after infants learned to walk, but the others were delayed to the middle of 2 years of age (vastus lateralis) or even later: thus, in gastrocnemius, biceps femoris, gluteus maximus, and sacrospinalis adult patterns appeared at the end of 2 years of age, and in the deltoid at 4 years of age.

During the 10 days after learning to walk (about a dozen steps), there is a continuous discharge in tibialis anterior during the stance phase, but it drops off, and by day 15, it has become adult in its pattern. Between the 50th and 85th days after learning to walk, the reciprocal discharge pattern of tibialis anterior and gastrocnemius is clearly established. Tsurumi (1969) independently found that in children, EMG activity of gastrocnemius appears during the push-off at 6 years of age. Although his findings agreed with those of Okamoto and Kumamoto for gastrocnemius, he found that tibialis anterior does not assume the adult role until the 5th year. Differences in technique could account for the discrepancy.

While the general adult mode of gait and its EMG patterns are recognizable from 2 years of age on, the child must reach its 7th birthday before the walking pattern truly becomes adult in form and function (Okamoto and Kumamoto, 1972; Okamoto, 1973; Kazai et al, 1976). Similar results have been reported by Marciniak (1975) of Poznan, Poland.

RUNNING

Increasingly more thorough papers have appeared on EMG patterns during running. Hoshikawa et al (1973) revealed a nonlinear correlation between speed and muscular activity. Elliott and Blanksby (1976) found a high reproducibility of results from one day to the next when surface electrodes and automated analysis techniques are used.

TRUNK MUSCLES DURING GAIT

Erector Spinae

Erector spinae shows two periods of activity, according to Battye and Joseph (as noted before by other investigators). They occur "at intervals of half a stride when the limb is fully flexed and fully extended at the hip at the beginning and end of the supporting phase." Battye and Joseph's explanation is that the bilateral activity of the erectores spinae prevents falling forward of the body and also rotation and lateral flexion of the trunk.

More recently, and using intramuscular wire electrodes, Waters and Morris (1972) reported striking periodic electrical activity in the trunk muscles of all subjects during walking at two different speeds (4.39 and 5.29 km/hour). It occurred faithfully with each cycle and was very constant in relationship to parts of the cycle from subject to subject. The muscles included paravertebral (erector spinae) muscles (iliocostalis in thoracic and lumbar parts, longissimus, multifidus, and rotatores) as well as quadratus lumborum, obliqui externus and internus, and rectus abdominis (Figs. 16.6 and 16.7).

Abdominal Muscles

In addition to the work of Waters and Morris noted above, Sheffield (1962) performed a study of abdominal muscles which differs surprisingly: the abdominal muscles were reported as inactive in level walking which the wire-electrode investigations of Water and Morris negate. Very little investigation of gait has been done otherwise in this part of the body.

SWINGING OF THE ARM DURING GAIT

Fernandez-Ballesteros et al (1964) of Copenhagen recorded the activity in the arms of normal subjects during walking. Buchthal and Fernandez-Ballesteros (1965) applied the same techniques to a study of patients with Parkinson's disease. The first EMG abnormality to appear in the early stages of the disease is a nonrhythmical pattern with random or wrongly timed activity. Later, swinging movements disappear, which becomes quite obvious clinically. In fact, it is characteristic of Parkinson's disease.

In normal persons, the posterior and middle parts of deltoid begin to show activity slightly before the arm starts its backward swing, and this activity continues throughout the backward swing. The upper part of latissimus dorsi and the teres major acts from the onset of the backward swing until the arm reaches the line of the body. Similar results were reported by Hogue (1969), who also showed that the activity of scapular muscles and the muscles of the arm and forearm are silent. The following summary of the work of Fernandez-Ballesteros et al does not agree with Hogue in some details, however.

During forward swing of the arm, activity is confined to some of the medial rotators (subscapularis, upper part of latissimus dorsi and teres major); the main flexors are strikingly silent.

In half of their subjects, they found activity in both forward and backward swing in the rhomboids and infraspinatus. This was most marked in persons who walk with a stoop.

Apart from brief silent periods in the extreme positions of swing, trapezius is active in both phases to maintain elevation of the shoulder. Similar activity occurs in supraspinatus; this obviously is related to the prevention of downward dislocation (p 273). However, Fernandez-Bal-

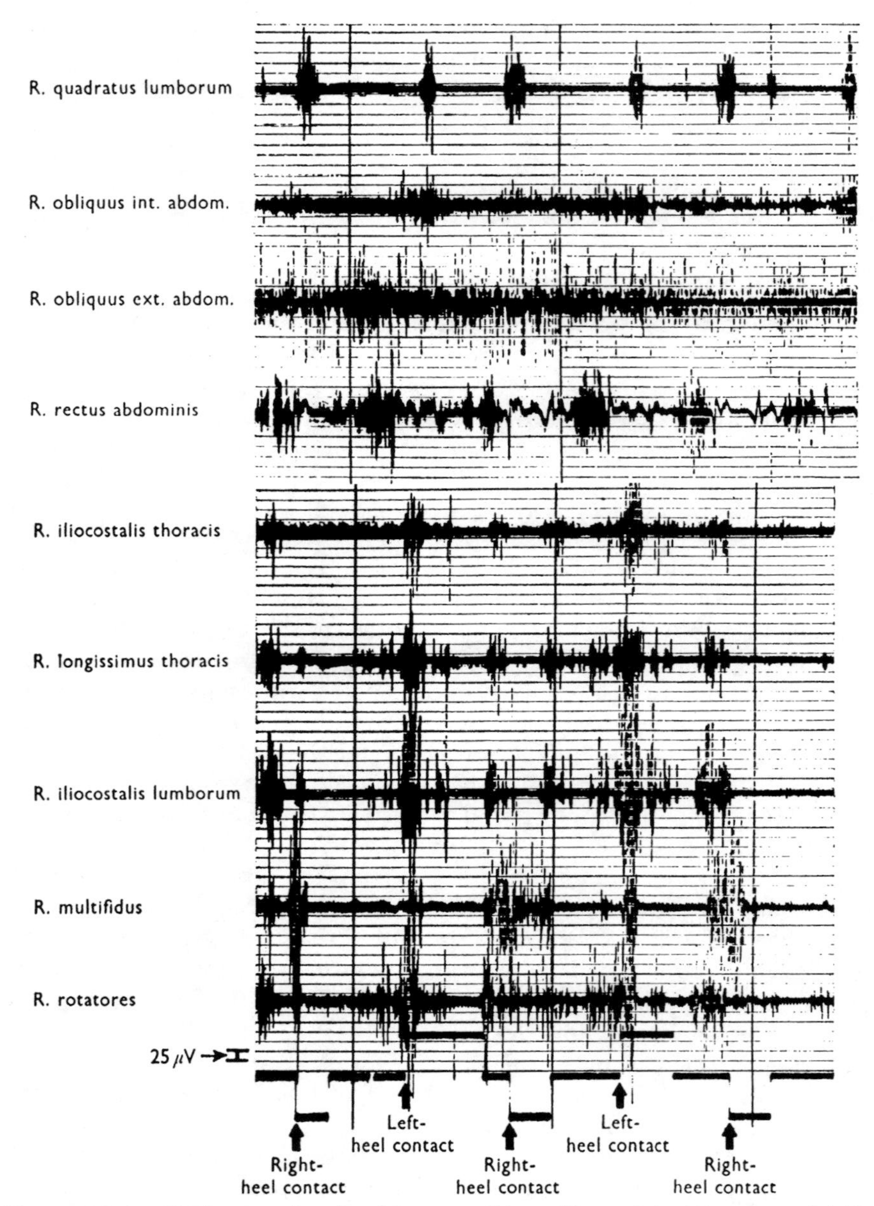

Figure 16.6. EMG records of subjects walking. (From R.L. Waters and J.M. Morris, 1972, ©*Journal of Anatomy*.)

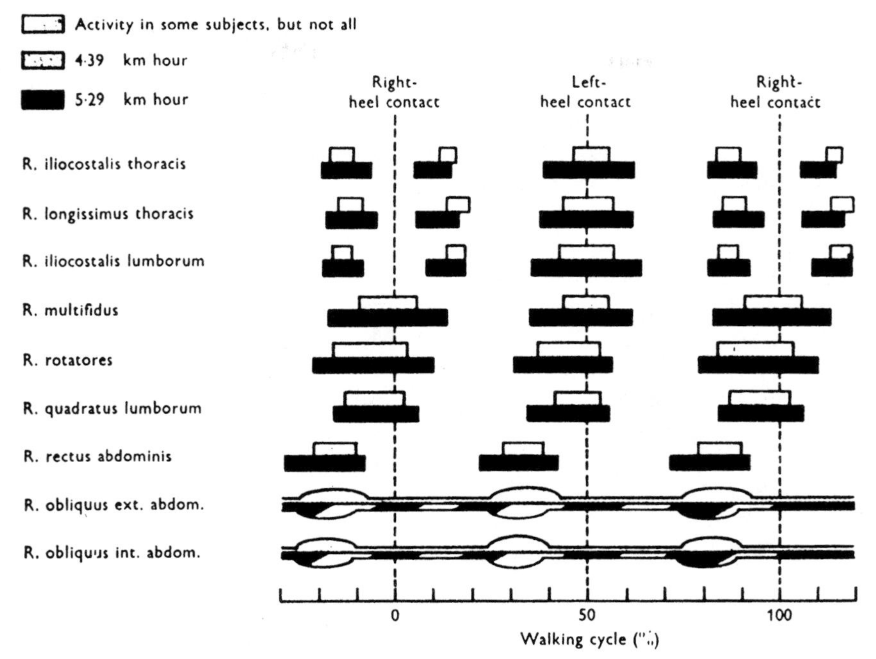

Figure 16.7. Average durations of EMG activity in trunk muscles (10 subjects) during walking. (From R.L. Waters and J.M. Morris, 1972, ©*Journal of Anatomy*.)

lesteros et al ascribe it (unconvincingly) to abduction of the shoulder to allow the arm to bypass the trunk. Trapezius falls silent immediately after a subject comes to a halt, contrary to earlier opinions, but corroborating Bearn's (1961b) finding.

PRIMATE EVOLUTION: EMG STUDIES

EMG studies on the flexor muscles in the forearm of a gorilla suggest that future comparative morphological studies on the wrists of African apes may reveal special bony features related to certain close-packed positions imperative to knuckle-walking (Tuttle and Basmajian, 1972). These features may then be employed to discern evidence of knuckle-walking heritage in the wrists of other extant hominoids and to trace the history of knuckle-walking in available fossils.

The fact that the flexor digitorum profundus muscle, which constitutes approximately 44% of total forearm musculature in the gorilla, is relatively inactive during many knuckle-walking behaviors indicates that special close-packed positioning mechanisms may be operant in the metacarpophalangeal joints of digits II to V. But these mechanisms probably are not exclusive of muscle activity since the flexor digitorum superficialis and perhaps also the lumbrical and interosseous muscles may participate severally in knuckle-walking episodes.

The relative inactivity of the extensor carpi ulnaris muscle during knuckle-walking is probably related to the fact that the same basic posture of the wrist is maintained in the swing and stance phases of most slow and moderately paced progressions. During swing phase, when activity of the wrist extensors might be anticipated, elbow flexion elevates the hand clear of the floor, and shoulder movements are probably chiefly responsible for its placement anteriorly.

Tuttle and Basmajian, assisted by others, conducted a series of extensive EMG studies in gorilla, orang and chimpanzee from 1970 to 1975 to determine the function of both forelimb and hindlimb muscles in posture and locomotion. These studies, carried out at the Yerkes Regional Primate Research Center of Emory University, have been reported in a series of papers (Tuttle and Basmajian, 1972 et seq; Tuttle et al, 1975).

CHAPTER 17

Anterior Abdominal Wall and Perineum

Since 1948 a considerable number of papers have appeared which deal with the actions of either specific abdominal muscles, e.g., rectus abdominis, or in connection with specific functions, e.g., posture of the vertebral column or breathing. A few have dealt with the abdominal wall in a more general way, and this method of approach will be our first concern. Clinical examinations with needle electrodes have confirmed almost all of the systematic observations now to be noted.

ABDOMINAL WALL IN GENERAL

Floyd and Silver (1950) were the first to make an extensive EMG study of the abdominal musculature in normal people. With a grid of paired multiple electrodes on the anterior abdominal wall (Fig. 17.1), they recorded simultaneously from various parts of the rectus abdominis, the external oblique, and the internal oblique where it is not covered by external oblique, i.e., in the triangular area bounded by the lateral edge of the rectus sheath, the inguinal ligament, and the line joining the anterior superior iliac spine to the umbilicus. Here in this triangle, the external oblique is represented only by its aponeurosis and, quite fortu-

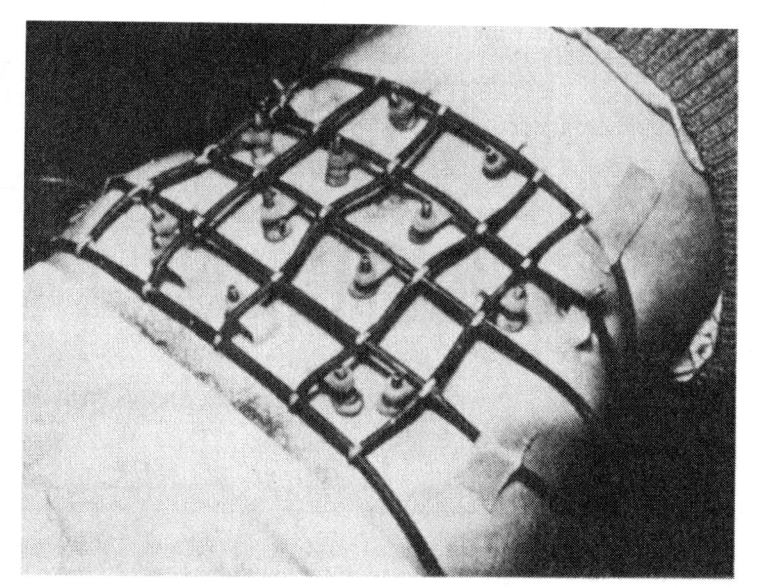

Figure 17.1. Photograph from Floyd and Silver (1950) to show their electrode grid for recording of abdominal EMGs.

nately, it is this very part of internal oblique which is of greatest interest. (They did not try to study the transversus abdominis because of its depth, admitting the possibility of a pickup from it through the electrodes for internal oblique.).

Floyd and Silver frequently found some difference between the right and left sides of the abdominal musculature, even when electrodes were carefully matched. They ascribed this to a basic asymmetry in function and found individual variations in the amount of difference.

With the subject lying supine and resting, slight activity was found in some nervous subjects, but none was found with relaxed comfortable persons (Fig. 17.2). Even in nervous subjects the activity could be reduced or abolished by proper positioning, e.g., by propping up the upper part of the trunk. Lowering the trunk caused an exaggeration of activity.

With the "head raising" movement used commonly as an exercise for strenghtening abdominal muscles, the recti were powerfully active while external oblique and the part of internal oblique that was studied were

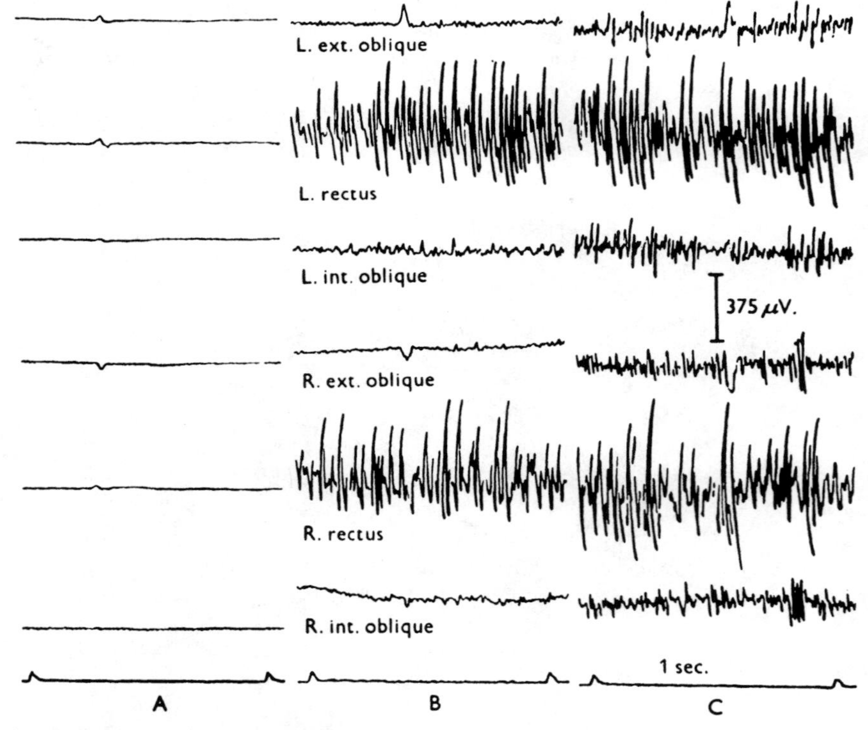

Figure 17.2. EMG records of abdominal muscles. (*A*) Supine. (*B*) Raising head. (C) Greater effort in head raising. (From W.F. Floyd and P.H.S. Silver, ©1950, *Journal of Anatomy*.)

only slightly active at first. Even with increased effort they become only moderately active (Fig. 17.2). These findings were confirmed later by Campbell (1952) using needle electrodes, and by Ono (1958) of Hirosaki University. One might conclude from their finding that only the rectus is benefited maximally by the head raising exercise, in spite of the exercise being advocated to increase the general "tone" of the abdominal wall. In contrast to the head-raising exercise, the bilateral leg-raising exercise brought all the abdominal musculature into activity to steady the pelvis. One-sided leg-raising was much less effective, calling upon activity predominantly on the same side of the abdomen.

In the relaxed standing position, all the electrodes except those over the lower part of the internal oblique picked up no activity. Internal oblique apparently is on constant guard over the inguinal region.

When the subjects (whether recumbent or erect) were made to strain or to "bear down" with the breath held, the external obliques and the internal obliques (lower parts) contracted to a degree that was direcly related to the effort, but rectus abdominis, in contrast, was very quiet (Fig. 17.3). This was later confirmed by Ono (1958). Surely it is surprising that physical therapists have not seized upon these findings for application in the strengthening of weak or stretched obliques. Perhaps they are not dramatic enough!

Floyd and Silver found no inspiratory or expiratory activity in the abdominal muscles during quiet breathing, a finding that was later enlarged upon by Campbell (1952) (see p 395) and by Ono (1958), Carman et al (1971), de Sousa and Furlani (1974), and Rideau et al (1975). With forced expiration, with coughing, and with singing, the pattern was similar to that in straining, i.e., marked activity in the obliques and none in the rectus.

Most investigators quite correctly emphasize the importance of the rectus sheath in protecting the abdominal area occupied by the rectus during all these physiological functions which are *not* accompanied by contraction of the rectus. Therefore, they point out, repairs of the sheath and maintenance of its integrity during surgical closures is vital. However, most surgeons erroneously think abdominal hernia is actively prevented by the activity of rectus abdominis in the region it covers. Floyd and Silver have proved conclusively that the apparent hardening of the recti on straining is usually only a passive bulging of the muscles and their sheaths. One can only hope that this knowledge, available now for several decades, will soon reach the practicing surgeon. But de Sousa and Furlani (1974), while they agreed in general with all previous workers about the rectus abdominis, disagreed violently with the lack of activity reported by others in the recti *during coughing*. In every one of 20 subjects studied with needle electrodes (instead of the usual surface

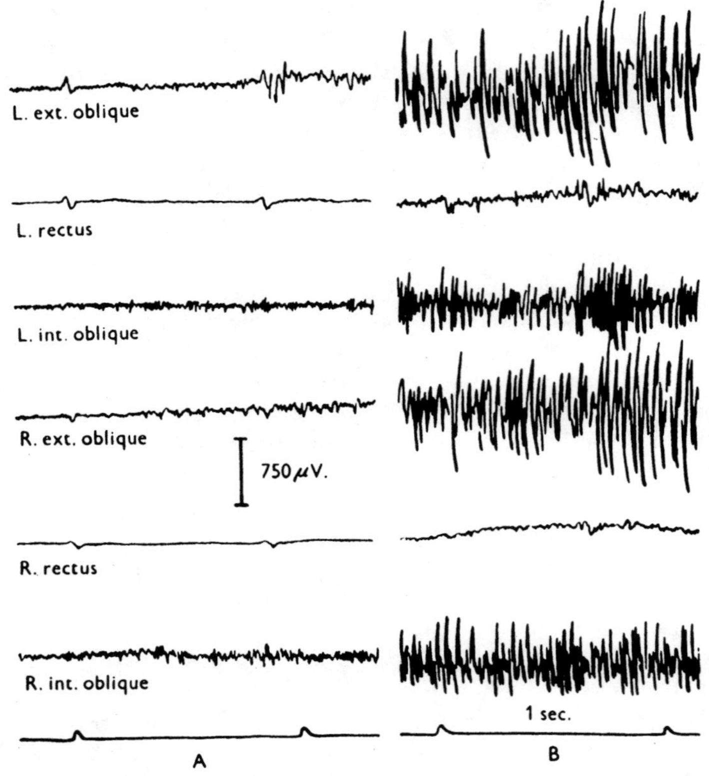

Figure 17.3. Abdominal EMGs of "straining." (*A*) During start of straining. (*B*) Four seconds later (maximal effort). (From W.F. Floyd and P.H.S. Silver, ©1950, *Journal of Anatomy*.)

electrodes), they found exuberant activity in the recti bilaterally during a cough. So there is a difference, apparently, between the reactions of the recti to the increase of intra-abdominal pressure from "bearing down" and the sharp, short increase of coughing.

Walters and Partridge in 1957 (then at the University of Illinois) reported on the EMG of the abdominal muscles during exercise and later added further information (Partridge and Walters, 1959). They found that in movements of the trunk performed without resistance in either the sitting or the standing posture, the obliques and recti remained quiescent. However, lateral bending of the trunk does produce activity in the more posterolateral fibers of external oblique (a fact that is also mentioned by Campbell, 1952). Inclining the trunk backwards gives activity in all the muscles, but forward bending was, as Floyd and Silver also found, unaccompanied by activity. They also confirmed, in general, the findings of Floyd and Silver concerning forced expiration and coughing, i.e., the recti remain relaxed while the obliques contract. During forced trunk-twisting exercises the internal obliques of the side to which the twisting occurred were greatly active while both external obliques showed some slight activity, and the recti showed none (unless the subject violently flexed the trunk simultaneously).

Because Partridge and Walters were concerned about exercises in the bedridden patient, they also studied the abdominal muscles during exercises in the supine position (Fig. 17.4). They found that all portions of the external oblique and rectus abdominis were activated best by "a lateral bend of the trunk, pelvic tilt, straight trunk curl, and trunk curl executed with rotation." They state, in rather naive jargon, that "rotation of the pelvis on the thorax (hip roll) and the reverse curl are excellent activators of the internal oblique in this (supine) position."

Concerned more with athletic training, Flint and Gudgell (1965) put a series of subjects through vigorous exercises while recording EMGs from the rectus abdominis and external oblique. Their most effective exercises for bringing out the greatest activity were: the "V-sit," "basket hang" on the horizontal bar, "side-lying trunk raise," backward leaning, and "curl-up." (Detailed definitions of these and other technical but explicit terms are given in their article). Less effective were "chin-up," "pull-up," "pelvic tilt" in the supine position, isometric contraction of the abdominal wall, "low bicycle with tilt," vertical jumping, and straight leg-raising in the supine position. Least effective, to the point of being useless, were "full-waist circling" and vertical reach in the standing position, controlled leg extension in the supine position, and "hip-roll" while lying on the mat.

Flint (1965a and b) showed further that the upper and lower parts of the rectus abdominis vary in response to different movements. Most of the activity in the recti during trunk flexion from the supine position occurs during the first half of the movement. Trunk raising elicits more activity than trunk lowering. Maximal abdominal muscle activity occurs during hook-lying (knees bent) unsupported slow and fast sit-ups (Godfrey et al, 1977).

POSTURAL ROLE OF ABDOMINAL MUSCLES

Lewer Allen was the first to prove conclusively with electromyography that the rectus abdominis does not draw a resistant spine forward but that gravity does it. Only in full flexion does rectus show activity, apparently in an effort to force the trunk further downwards against resistance. In hyperextension (at the other end of the range of motion) the rectus abdominis shows activity while being stretched; this apparently steadies the torso. Floyd and Silver (1950), Ono (1958), and Partridge and Walters (1959) have confirmed these findings which were predicted many years ago by Duchenne and others. Campbell (1952 et seq) and

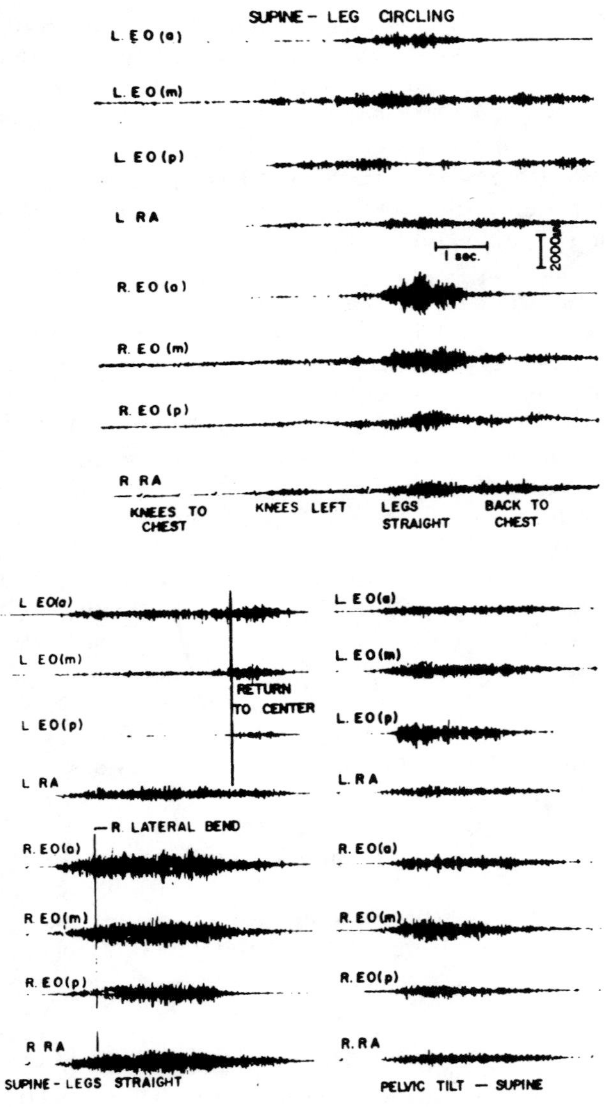

Figure 17.4. EMG records of abdominal muscles during various activities while subject supine. (From M.J. Partridge and C.E. Walters, ©1959, *Physical Therapy Review*.)

Campbell and Green (1953 et seq), using needle electrodes, found some activity in quiet standing which might be missed by surface electrodes, but this activity was never very marked.

Morris et al (1961), Waters and Morris (1970), and Bearn (1961a) have stressed the importance of the abdominal muscles in the developing of positive pressure in the abdomen. This is an important adjunct to the vertebral column in stabilizing the trunk.

As was mentioned above, in ordinary standing the only muscle to show important continuous activity was the internal oblique, but this activity was not related to maintenance of the general posture; it will be now considered below.

CONTROL OF INGUINAL CANAL

In the investigations of the abdominal wall posture, we must consider in particular the lower part or region of threatening hernia. Here the inguinal canal tunnels through the muscular layers of the abdominal wall and so provides a weak spot. Through this area, excessive intra-abdominal pressures (particularly while the person is standing) may force a hernia. Since, in the male, the opening transmits the ductus (or vas) deferens, it must be protected without, however, causing complete occlusion. This delicate, but dynamic, function is performed by the internal oblique and transversus abdominis—in particular by their lowest fibers which arise from the inguinal ligament. These fibers arch over the inguinal canal and insert medially on the pubic bone. One would imagine that they must be in constant contraction during standing. The work of Floyd and Silver (described above) gives ample evidence to prove this long-held opinion of anatomists and surgeons. Furthermore, one would imagine that, regardless of a person's position, straining and coughing would require increased activity in the muscular protection of the canal. Indeed, the evidence now is overwhelmingly favorable to this view (Fig. 17.3).

RESPIRATORY ROLE OF ABDOMINAL MUSCLES

Following upon the work of Floyd and Silver, their graduate students at Middlesex Hospital Medical School expanded that part which dealt with respiration. Campbell, joined in parts of his research by Green, performed both extensive and intensive studies, combining EMG of the abdominal muscles and direct and accessory respiratory muscles with various other techniques, such as spirometry. The respiratory muscles will be considered in the next chapter, but here we should consider the findings for the abdominal musculature only.

Campbell's first paper (1952) confirmed and underlined the work of the earlier electromyographers who found that there was no activity in the external oblique and rectus abdominis of supine normal subjects breathing quietly. The new dimension added by Campbell was his use of needle electrodes. He showed that with maximal voluntary expiration these muscles contracted as they also did towards the end of maximal voluntary inspiration (Fig. 17.5). Yet they did not contract under the

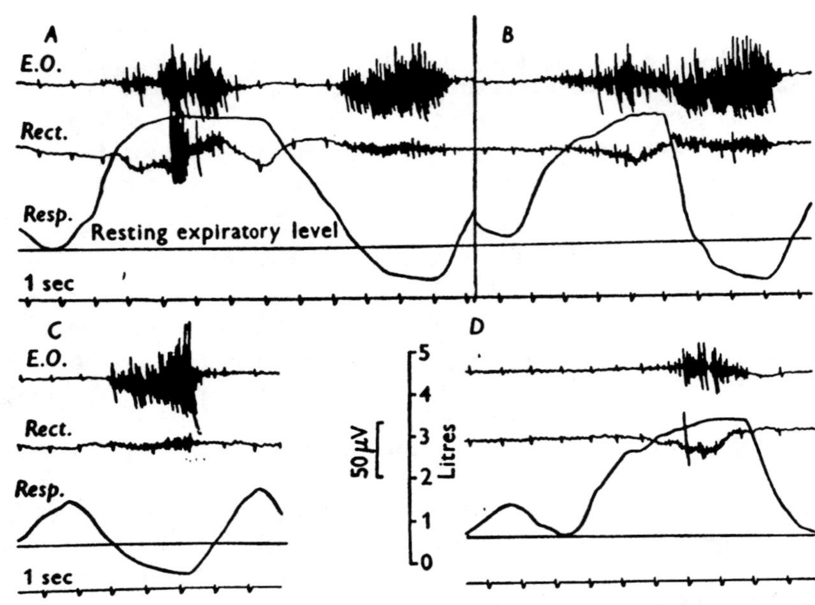

Figure 17.5. EMG records of external oblique and rectus abdominis and spirometry of maximal inspiration and expiration. *Resp* is the spirometry trace (inspiration upward). (*A*) Inspiration held for 4 s and released, then followed by maximal inspiration. (*B*) Forced expiration within a second of attaining full inspiration. (*C*) Maximal expiration followed by normal inspiration. (*D*) Maximal inspiration followed by relaxation to the resting respiratory level. (Note: superimposed ECG spikes should be ignored.) (From E.J.M. Campbell, ©1952, *Journal of Physiology*.)

latter condition when the breathing was increased by imposing asphyxia. In contrast, the activity in maximal expiration was increased further when the volume of breathing was increased by asphyxia.

Campbell reported that the great activity during expiration with hyperpnea appeared first towards the end; it was never prominent at the beginning. He concluded that abdominal contraction was a factor in limiting voluntary inspiration, but in the presence of very rapid deep breathing due to asphyxia it was inhibited. It would seem, then, that contractions of the abdominal muscles to aid expiration only occur in severe cases of greatly increased pulmonary ventilation under stress. In any case, they do not initiate the expiratory phase, but rather they help to complete it quickly. Campbell showed that a pulmonary ventilation of more than 40 liters/minute was required for the abdominal muscles to play their accessory respiratory role (Fig. 17.6).

Campbell and Green (1955) showed that the early findings were essentially true both in the supine and in the erect posture, but that normally in the erect posture there is some continuous activity in the

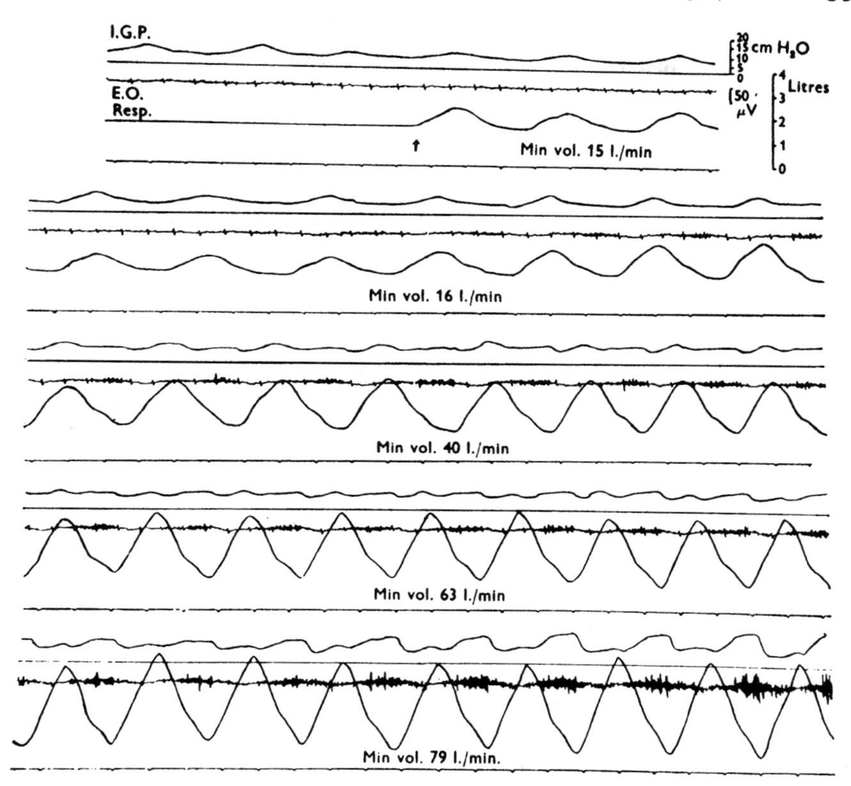

Figure 17.6. Simultaneous EMG records (*E.O.*), spirometry (*Resp.*), and recording of intra-abdominal (intragastric) pressure (*I.G.P.*) showing effects of progressively increasing pulmonary ventilation on the abdominal muscles. Continuous recording (From E.J.M. Campbell and J.H. Green, ©1955, *Journal of Physiology.*)

muscles. This activity can be abolished by certain postures and is not particularly related to respiration.

In order to summarize the present knowledge of the role of abdominal muscles in respiration, it seems wise to present below the essence of the paper by Campbell (1955c):

1. The abdominal muscles are the most important and are the only indisputable muscles of expiration in man. The obliques and transversus are much more important than the rectus abdominis.
2. Vigorous contraction occurs in all voluntary expiratory maneuvers (such as coughing, straining, vomiting, etc.).
3. The abdominal muscles (almost exclusively the obliques) contract at the end of maximum inspiration to help limit its depth but, in normal persons, they do not contract in hyperpneic asphyxia, apparently being inhibited by central mechanisms.

Hyperpnea with a ventilation rate of greater than 40 liters of air per minute calls upon activity of these muscles at the end—and only at the end—of expiration.

The finding by Fink (1960) of phasic expiratory activity in patients while being anesthetized does not (as Fink himself believed) invalidate Campbell's conclusion for normal subjects. Fink, however, described a practical use of the integrated rectified EMG signal of the abdominal wall muscles. He advocated it for monitoring the relaxation of the abdomen during operations in which the muscle-relaxant succinylcholine is used. The activity occurring during expiration under these circumstances is real, and the EMG signal provides an excellent tool for determining relaxation. Bishop (1964) showed that continuous positive pressure breathing initiates expiratory activity of the abdominal muscles in cats. Youmans et al (1974) furthe reviewed the control of "involuntary" or tonic responses of abdominal muscles, demonstrating in their studies that the tonus is under neural inhibitory centers which react to many influences, including intragastric and intracardiac pressure, respiration, etc.

Recently, Martin and De Troyer (1982) studied the behavior of the abdominal muscles during inspiratory mechanical jogging. Abdominal muscle activity was studied in eight normal subjects, while seated and supine, during inspiratory resistive and elastic mechanical loading. EMG signals from the external oblique and rectus abdominis muscle were recorded using unipolar needle electrodes, and changes in anteroposterior dimensions of the abdomen were measured using linearized magnetometers. Elastic loading evoked abdominal muscle activity during expiration, especially in the rectus abdominis, in seven seated subjects, while resistive loading did so in five. In four runs, activity was also detectable throughout inspiration. Abdominal muscle activity was associated with a reduction in abdominal dimensions throughout the respiratory cycle. When the cycle subjects were supine, loads did not evoke abdominal muscle activity. Martin and Troyer concluded that abdominal muscle recruitment during inspiratory mechanical loading may facilitate inspiration by increasing diaphragmatic length.

PERINEUM

The muscles of the human perineal region that have been investigated are the external anal sphincter, the striated sphincter of the male urethra, and the striated muscles of the pelvic floor and urethra in normal women and patients with "Manchester repairs" for prolapse. These studies will be considered in that order, following which a brief account of reflex control of micturition will be given.

Sphincter Ani Externus

Aside from Beck's early EMG study (1930) of the anal sphincter almost exclusively in dogs, only Floyd and Walls (1953) have reported on the EMG of this important muscle. Beck's results are now more provocative than useful but those of Floyd and Walls are extremely interesting and practical. Some of them have been confirmed by Basmajian's group during a study of the urethral sphincter (see below).

Floyd and Walls found that the anal sphincter is in a state of tonic contraction (Fig. 17.7). The degree of this tone varies with posture and the subject's alertness, falling to a very low level during sleep. Presumably, the internal sphincter is the main agency for keeping the rectum closed during sleep. They found that the subjects can voluntarily produce an outburst of activity in the anal sphincter. J.V. Basmajian and W.B. Spring (1954, unpublished study) found in a series of men that the contraction of sphincter ani externus is not isolated but is accompanied by general contraction of the perineal muscles, especially the sphincter urethrae. Since these muscles are of common origin from the cloacal musculature, these findings are not surprising.

With increased intra-abdominal pressure produced by straining, speaking, coughing, laughing, or weight-lifting, Floyd and Walls found increased sphincteric activity related in amount to the degree of pressure. This has been confirmed by Cardus et al (1963) and Scott et al (1964). However, actual efforts to defecate were usually (but not always) accompanied by relaxation of the sphincter ani.

Duthie and Watts (1965) found that electrical activity in the striated external sphincter, although greatly reduced, persisted even under general anesthesia. In response to rectal distension, the sphincter showed an increased activity as the maximal rate of diminution in pressure occurred.

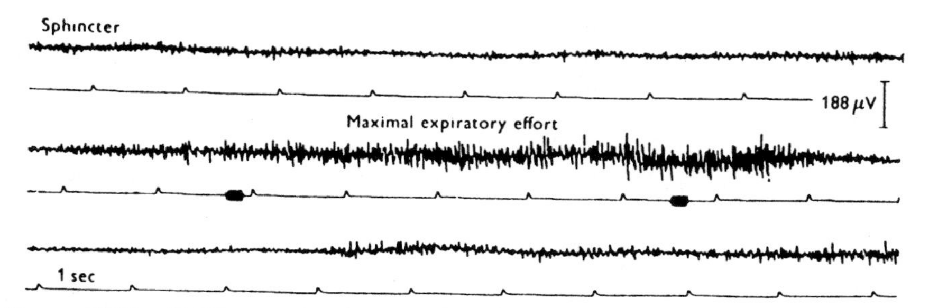

Figure 17.7. EMG records of external anal sphincter, showing resting tone and (between signal marks) the increase in tone during a maximal expiratory effort. The record is in three serial horizontal strips and reads continuously from left to right. (From W.F. Floyd and E.W. Walls, ©1953, *Journal of Physiology*.)

Thus relaxation in the anal canal is independent of the action of this sphincter, which contributes to pressure only when a bolus is present.

Porter (1960) also has shown that the external sphincter and the puborectalis show continuous activity at rest, heightened activity with effort and coughing, and inhibition with defecation and micturition. A critical volume brings about the desire for a bowel movement with sphincteric inhibition. After evacuation, tonic activity is restored in the external anal sphincter and puborectalis at the same time as the internal (smooth-muscle) sphincter returns to its normal contracted state (Ihre, 1974).

A growing clinical interest in EMG of the external anal sphincter has led to the appearance of special papers on the subject. Some of these reflect normal function and may be of interest to readers (e.g., Archibald and Goldsmith, 1967; Haskell and Rovner, 1967; Bailey et al, 1970; Jesel et al, 1970; Ruskin, 1970; Waylonis and Aseff, 1973; Chantraine, 1974; Ihre, 1974; Blank and Magora, 1975; Schuster, 1975; Kiviat et al, 1975; Lane, 1975; Frenckner and Euler, 1975; Shafik, 1975; Kiesswetter, 1976; van Gool et al, 1976; Whitehead et al, 1981).

Striated Male Sphincter Urethrae

An early report (Basmajian and Spring, 1955) dealing with this almost inaccessible muscle, after years of virtual eclipse, has been referred to so often in recent years that it can be repeated here to advantage. Very fine, self-retaining wire electrodes were inserted through the perineum into, or very near, the sphincter urethrae in six men during cytoscopic examination. Action potetials of the muscles were obtained and recorded on Stanley Cox 6-channel electromyograph. When the bladder is empty, there are a few occasional, small potentials in the sphincter urethrae with long periods of inactivity. As the bladder is filled slowly through the cystoscope, the action potentials increase in number. There is a continuous low level of activity in the striated muscle surrounding the membranous urethra as long as the bladder contains fluid. When the subject is instructed to micturate after the cystocope is removed, the potentials disapper as micturition begins and remains absent during the whole period of micturition and remain absent if the bladder is empty (Fig. 17.8). Sudden voluntary stopping of micturition before the bladder is empty is accompanied by a marked outburst of potentials, the frequency of which then falls off rapidly to "the resting level."

Nesbit and Lapides (1959) in a study of male patients concluded that the striated sphincter is necessary for sudden interruption of micturition and for maintenance of continence when the vesical neck is incapacitated. Lapides is emphatic in his belief that micturition may be initiated and terminated consciously by voluntary effort without the use of any striated muscles and that urinary continence is normally maintained by the

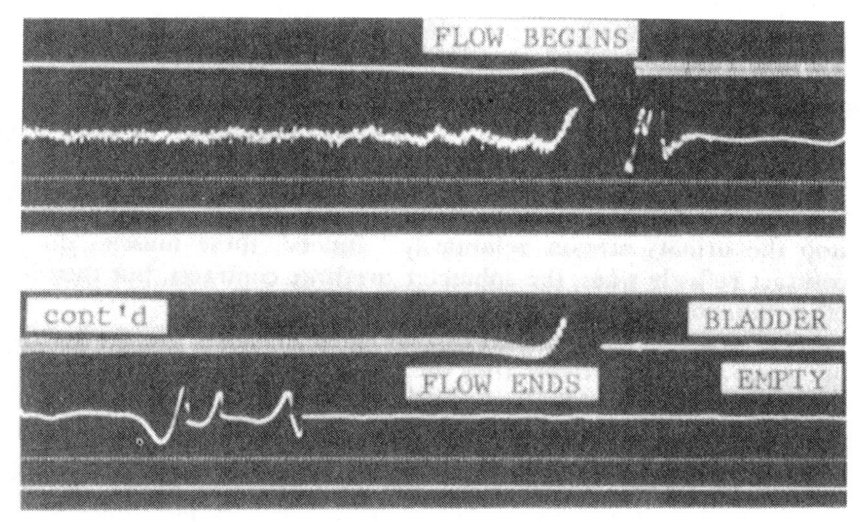

Figure 17.8. EMG records of male striated sphincter urethrae with partly filled bladder (continuous activity) during micturition (relaxation of sphincter) and with bladder empty (relaxation continues). *Uppermost trace* is a recording of urine flow (*flow begins* to *flow ends*). *Middle trace* is EMG. and *lowest trace* is 10-ms time marker.

internal vesical sphincter, not by striated muscles. (See Lapides et al, 1957, 1960.) On the other hand, Susset et al (1965) clearly showed that under conditions of stress, such as coughing or straining, the external sphincter is necessary and very active. A full bladder even results in minimal opening of the bladder neck on coughing, throwing added responsibility on the striated muscles. Susset et al also proved the importance of striated muscle spasticity in patients with upper motor neuron lesions; in such cases the external sphincter takes over the function and, being spastic, obstructs micturition.

Meyer Emanuel's (1965) review of control of the bladder outlet appears to rationalize the findings of many types of investigation. In men, the urinary sphincter system consists of a tubular extension of the bladder containing elastic tissue with a collar of striated muscle at the urogenital diaphragm; in women this system accounts for the whole length of the urethra. Incontinence occurs only if both ends of the sphincteric system become damaged. The striated muscle is important for interrupting micturition, but it does not maintain sustained contraction indefinitely.

Petersén and Franksson of Stockholm (1955) reported the electromyography of the male striated urethral sphincter and the bulbocavernosus muscle in 10 and 11 patients, respectively. When their patients were asked to contract their muscles to stop micturition, there was a sudden burst of activity in the sphincter urethrae, as in the Basmajian-Spring findings, and similar activity in bulbocavernosus (Fig. 17.9). This agrees, in general, with Basmajian's early opinion mentioned above, namely, that the cloaca-derived musculature contracts simultaneously and indiscriminately when one or the other muscle is suddenly contracted. It also accounts for Mueller's (1958) erroneous belief that levatores ani and the pubococcygeus are the "primary muscles used to stop the urinary stream voluntarily." Indeed, these muscles do also contract reflexly when the sphincter urethrae contracts, but they have other functions to perform.

Scott et al (1964) have found such a close association in the activities of cloacal musculature that they use EMG of sphincter ani externus for the routine indication of activity in the striated urethral sphincter. Study of their results reveals that they are indistinguishable from direct recordings from the urethral muscles.

To elaborate this point further, the electromyography and the morphology of the external anal sphincter and external urethral sphincter were studied in rabbits (J.V. Basmajian and J. R. Asunción, 1964, unpublished; J.V. Basmajian, S. McKay, and R. Hons, 1964, unpublished). The outstanding finding by both techniques was the inseparable nature of the two sphincters. The morphology is intricately linked; bundles of one sphincter intermingle in a complex fashion with bundles of the other. In rabbits (as in some other mammals) the structure is quite primitively cloacal in nature when compared with that in human beings. Recently, Gallai et al (1983) convincingly showed that cloaca-derived muscles (particularly the female bulbocavernosus, which they recorded during labor) have synchronous "and simultaneous" activity.

Frankkson and Petersén (1955) have also reported EMG studies of patients with various disorders in micturition, convincingly demonstrating neurogenic sphincteral disturbances in some. Such studies must, of

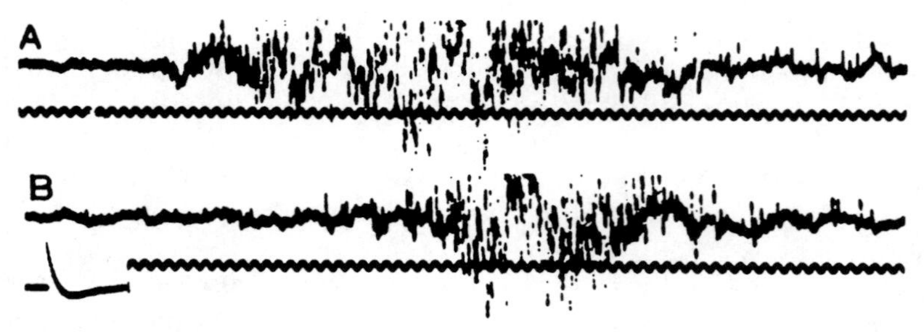

Figure 17.9. EMG records from bulbocavernosus. (*A*) In voluntary contraction. (*B*) In coughing. Time in milliseconds. Calibration (*lower left corner*): 100 μV. (From I. Petersén and C. Franksson, ©1955, *British Journal of Urology*.)

necessity, be expanded by urologists. Work along this line has been reported by Hutch and Elliott (1968).

Giovine (1959) of Milan has also presented some EMG results from the striated sphincter. In addition, he reported histological findings from which he concluded that the muscle is made up of red fibers to a large extent. The significance of red and white fibers, not being firmly established, must await further study. Meanwhile, Vereecken et al (1975) have demonstrated that fatigue ensues rapidly during sustained voluntary contraction (even for 1 minute); thus the intensity of contraction waxes and wanes, rather than being maintained steadily. Fatigue in the levator ani is less pronounced. Women have a high voluntary control of the activity in those pelvic floor muscles (Vereecken et al, 1975).

Pelvic Floor and Urethra in Women

The pelvic floor or pelvic diaphragm is mostly muscular, is very important in parturition, and is generally misunderstood. EMG one might suppose, would have been invaluable in clearing up misunderstanding, but only one group has actually studied the musculature of the female pelvic floor—Petersén et al (1955). This is the same Stockholm group mentioned before with the addition of a gynecologist, who then assumed senior authorship of a second paper on abnormalities of the pelvic floor (Danielsson et al, 1956).

In the first study, Petersén et al, using needle electrodes and without general anesthesia, explored: (1) the pubococcygeus, which is the medial or most important part of the levator ani and (2) the urethral sphincter. The electrodes were inserted through the vaginal wall in 24 normal women (about half of whom had borne children). They concluded that some subjects were able to relax the sphincter urethrae completely, while others were unable to relax it. However, none could relax the pubococcygeal part of levator ani, even though they were in the "lithotomy position."

Diminution or complete cessation of activity in the sphincter urethrae at micturition (or attempted micturition) agrees with the Basmajian/Spring findings in men (Fig. 17.10). Furthermore, their finding that voluntary efforts to contract the one muscle automatically recruits the contraction of the others agrees with our impression, already noted above, that individual contraction in the perineum is difficult if not impossible. Petersén et al (1962) proved in women that voluntary complete relaxation of the external sphincter was possible, even with a partially filled bladder. Voluntary interruption of micturition results in a rapid closing of the striated external sphincter; only afterwards does the posterior urethra empty relatively slowly in a proximal direction. Considerable variations were found in normal persons.

Petersén et al (1955) reported regional differences in the activity of

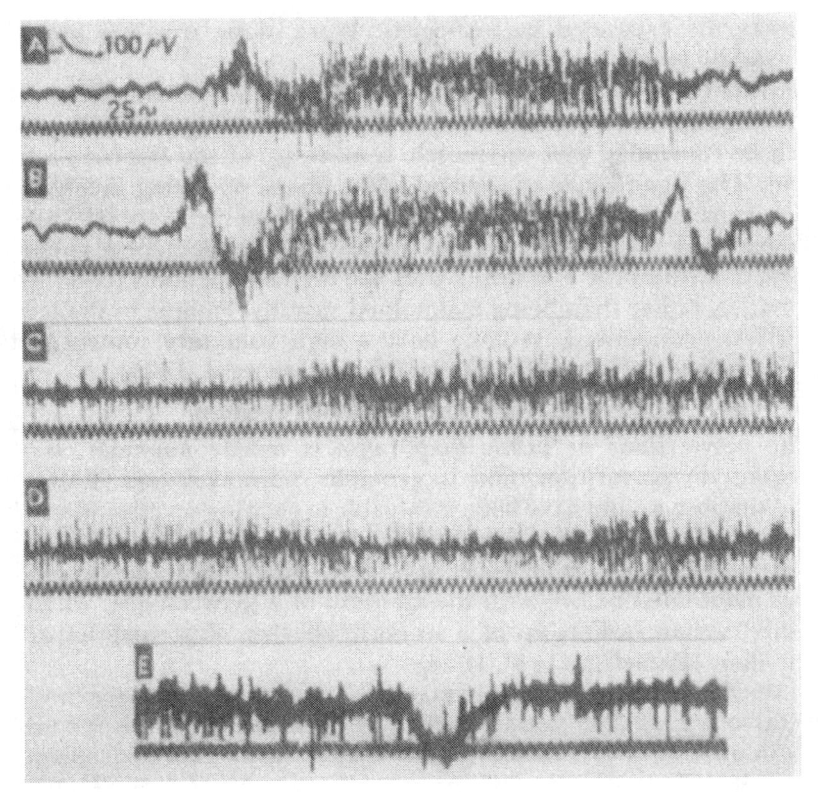

Figure 17.10. EMG records from normal female sphincter urethrae during voluntary contraction (*A*) and a cough (*B*) from pubococcygeus during voluntary contraction (*C*), a cough (*D*), and an attempt to micturate (*E*). (From I Petersén et al., ©1955, *Acta Obstetricia et Gynecologica Scandinavica*.)

the sphincter, and they stated that the pattern was more or less related to whether the subject had borne children. Nulliparous subjects showed little difference in the response of the whole circumference of the sphincter while the multiparous subjects had much less or even no activity in the dorsal part of the sphincter. This dorsal part, related as closely as it is to the vagina, might have been destroyed by laceration in the course of childbirth, according to these investigators. Pubococcygeus, however, showed no difference attributable to parity. As noted earlier, women have a high voluntary control of levator ani (Vereecken et al, 1975).

Danielsson et al performed EMG explorations with similar techniques on women who had recovered from the Manchester reparative operation for prolapse but who were still complaining of urinary stress incontinence. In all of these women there were no EMG signals obtainable from the dorsal part of the urethra, suggesting again that the part of the sphincter urethrae next to the vagina had been torn by parturition (Fig. 17.11). Lesions of the pudendal nerves, either at childbirth or at the time of the Manchester operation, may have also played a role.

In the 1980s, the enormous increase of interest in urinary incontinence has led to many new clinical studies, including EMG studies (e.g., Barrett and Wein, 1981; Zinner and Sterling, 1981).

Spinal Reflex Activity from the Bladder

Bors and Blinn (1957) of Long Beach, CA presented convincing evidence that demonstrates the importance of the bladder mucosa in influencing the striated musculature of the pelvic floor. Thus, the sphincter ani and the sphincter urethrae contract in response to mucosal stimuli of the bladder wall. The Basmajian/Spring finding that filling of the bladder with more than a few cubic centimeters of water starts up activity in the male sphincter urethrae is in agreement with this. The further findings of Bors and Blinn (1957, 1959) and of Pierce et al (1960) demonstrate that all these reflexes may be grouped simply under one generic term, "the bulbocavernosus reflex"; this reflex may be elicited from the stimulation of any genitourinary mucosal surface, including the glans penis. (See also Nakaarai, 1968; Aranda and Jara, 1969; Allert, 1969; Ertekin and Reel, 1976; but see the contrary and less convincing view of Rattner et al 1958.)

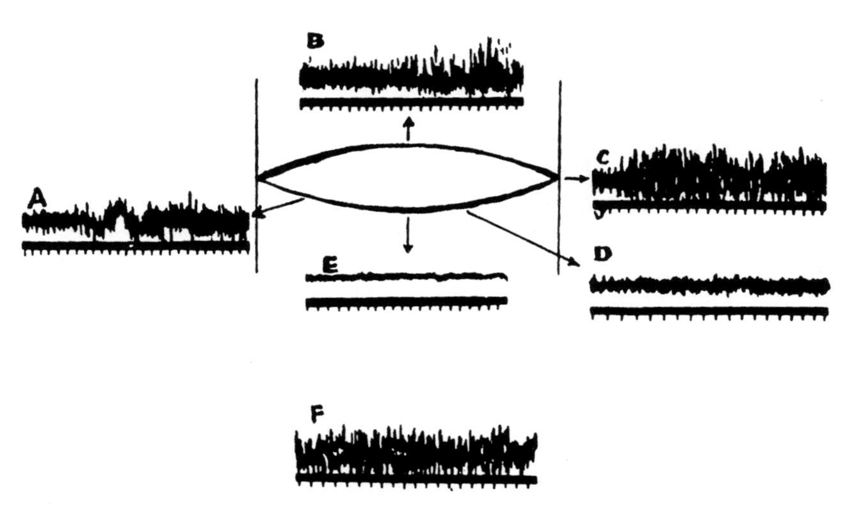

Figure 17.11. EMG records of abnormal sphincter urethrae of woman who had borne three children and had had a Manchester operation. (*B*) Ventral part of sphincter. (*E*) Dorsal part (nearest vagina). (*A, C,* and *D*) From sides; (*F*) from normal pubococcygeus. (From C-O. Danielsson et al, ©1956, *Acta Obstetricia Gynecologica Scandinavica*.)

Detrusor Muscle of Bladder

Having developed a valid technique with wire electrodes for recording EMG activity from smooth muscles of the bladder wall, La Joie et al (1976) proved that the spike potentials were of local origin. Simultaneous EMG recordings were made from the detrusor muscle and the overyling part of the rectus abdominis, showing disparity in their actions. The normal detrusor muscle is electrically silent when the bladder is empty; it shows increasing activity with filling. If the filling is interrupted, accommodation occurs. When the bladder is completely full, EMG activity remains high; it gets even higher at the initiation of voiding. Then, progressively, EMG activity falls and reaches zero as the bladder empties.

Relationship of Abdominal and Perineal Muscles

Bors and Blinn (1965) investigated the activity of the rectus abdominis in relation to micturition and to contractions of perineal muscles. Before or at the onset of micturition, whether *on desire* or *on volition*, the EMG activity of rectus abdominis usually remained unchanged; also at the normal cessation of micturition there was generally little or no reaction in most subjects. However, when either micturition was suddenly interrupted or the external anal sphincter was consciously contracted, activity in rectus abdominis usually increased. There would seem to be a related contraction of abdominal and perineal muscles, but the exact relationship is still obscure. Bors and Blinn's view is "that phasic contractions of the pelvic floor cast their shadow upon the abdominal muscles."

EMG of Ejaculation and of Penile Muscles

In a preliminary study, Kollberg et al (1962) reported the train of events recorded electromyographically from the striated external sphincter urethrae and adjacent striated muscles. Some seconds before ejaculation occurs there is a lively contraction in muscles of the urogenital diaphragm. The cause or effect of this remains obscure. It may play some role in penile engorgement just before ejaculation. On ejaculation, the sphincter (and probably its neighbors) contract rhythmically for 15 to 20 times in about 25 s. These contractions, which must have some part in propelling the semen, appear as salvoes of action potentials alternating with quiet intervals. Some reciprocity is also noted, i.e., activity in the muscles alternates.

The preliminary study was enlarged and reported by Petersén and Stener (1970) and Kadefors and Petersén (1970). The rhythmic EMG activity in the striated urethral sphincter is split up into two parts. The first is characterized by a comparatively high-frequency content. The later part has fewer high-frequency components. Perhaps, these enterprising Swedish investigators will also succeed in recording the state of contraction of the smooth muscle at the neck of the bladder which is so often believed to prevent reflux of sperm into the bladder.

Hart and Kitchell (1966), using special bipolar needles, recorded the EMG activity from the penile muscles of dogs—ischiourethralis, bulbocavernosus, and ischiocavernosus. Different patterns of reflex activity were obtained on light stimulation of different parts of the penis. Rubbing behind the bulbus glandis elicited tonic contractions of the ischiourethralis, rhythmic contractions of the other two muscles, and rapid penile detumescence. This reflex would seem to be related to normal responses in ejaculation. Penile tumescence was then obtained by application of pressure behind the bulbus glandis and by rubbing the urethral process; the tumescence was accompanied by the same pattern of activity in the three muscles. Stimulation of the corona glandis resulted in tonic contraction of the bulbo- and ischocavernosus muscles, again accompanied by rapid detumescence, but this reflex is not as amenable to explanation.

CHAPTER 18

Muscles of Respiration

The muscles usually considered to be the primary muscles of respiration are the diaphragm and the intercostals, but the following also have been implicated as either primary or accessory in respiratory function: the scalenes and the sternomastoid in the neck, the musculature of the shoulder region including the pectoral muscles and serratus anterior, and the anterior abdominal muscles. These "accessory" muscles are discussed, but only briefly at the end of this chapter, and reference is also made in appropriate places elsewhere to the broader aspects of each group of muscles.

Since the earliest recorded medical history respiration and its mechanical production have been the subject of inquiry. Both before and after Galen, theories waxed and waned. Galen in the second century was perhaps the first to direct attention to the action of the intercostals, although he did not belittle the role of the diaphragm in breathing. Furthermore, he was aware of the two layers of intercostals—external and internal. His assignment of inspiration to the former and expiration to the latter still reaches down to the present day, causing renewed controversy. Such illustrious names as Willis, Hamburger, and Magendie continue to appear in any history of the respiratory function, but a review of their work here would serve no particular use.

Toward the end of the 19th century, Martin and Hartwell (1879), from crude experiments in anesthetized cats and dogs, made conclusions that were not much different than those of Galen. The history of our knowledge of diaphragmatic function unfolded until the turn of the century. Newer techniques and medical advancement have increased our knowledge of the function of the muscles of respiration but have not lessened the arguments.

In the past decade, EMG has provided a tool which promises to remove much of the uncertainty. Already there has been definite progress which will be reported in some detail in this chapter. The names of Jones, Pauly and Beargie, then of Chicago; Campbell, first in England and now at McMaster University in Canada; and Koepke, Smith, Murphy, Rae, and Dickinson of Ann Arbor, Michigan all stand out prominently in any review of this subject. Contributions have also been made by others; these will be mentioned in the appropriate places. Studies on the diaphragm have been particularly voluminous.

It is convenient first to consider the movements of the ribs and the muscles that produce these movements.

COSTAL RESPIRATION

Jones et al (1953) were the first to make a substantial EMG contribution to the knowledge of costal respiration. With surface electrodes over the upper four internal and external intercostals, the scalene muscles in the neck, and the abdominal muscles, they put their subjects through various tests.

The usual concept of normal quiet breathing is that the scalenes anchor or fix the first rib while the external intercostals elevate the remaining ribs towards the first—this in spite of radiographs showing no approximation of the ribs. Jones and his colleagues showed that both sets of intercostals in man were slightly active *constantly* during quiet breathing and showed no *rhythmic* increase and decrease. In contrast with this, the scalenes did show a rhythmic increase during inspiration (Fig. 18.1). Their role in quiet breathing was confirmed with needle electrodes by Raper et al (1966).

With forced inspiration, the scalenes, sternomastoid, and internal and external intercostals showed marked activity (Jones et al, 1953; Raper et al, 1966). In contrast, with forced expiration, the scalenes were quiescent while the intercostals were still active. Attention was focused anew on the scalenes as fundamental muscles of inspiration.

In 1968 and 1969, with Dr. Mica Fruhling at Queen's University, J.V. Basmajian inserted fine-wire electrodes into the scalenus anterior of eight young men and recorded their EMGs in quiet and deep breathing (Figs. 18.2–18.4). Alas, the premature death of Dr. Fruhling soon after the analysis of the data (compounded by Basmajian's move to Atlanta) left these results in limbo until this edition. They clearly support the hypothesis that the scalene muscles are respiratory muscles. We publish them now as a memorial to Dr. Mica Fruhling.

Jones et al were led to the conclusion that the function of the intercostal muscles in respiration is to supply the tension necessary "to keep the ribs at a constant distance from each other while the chest is expanded from above and contracted from below." They insisted that passive membranes between the ribs instead of muscles would be inadequate because they would be sucked in and blown out during respiration. Therefore, they would not provide the constant fine control of the rib positions. Moreover, the intercostals were shown to function in flexion of the trunk (as in sitting up from the supine position) and, probably for the first time, they were suggested as being postural muscles.

This group of investigators emphasized the belief that the main role in human respiration is performed by the diaphragm, while the intercostals were necessary for markedly increasing the intrathoracic pressure. Thus they were agreeing with Hoover (1922), who showed that a person with paralyzed intercostals had a sharp reduction of sucking and blowing

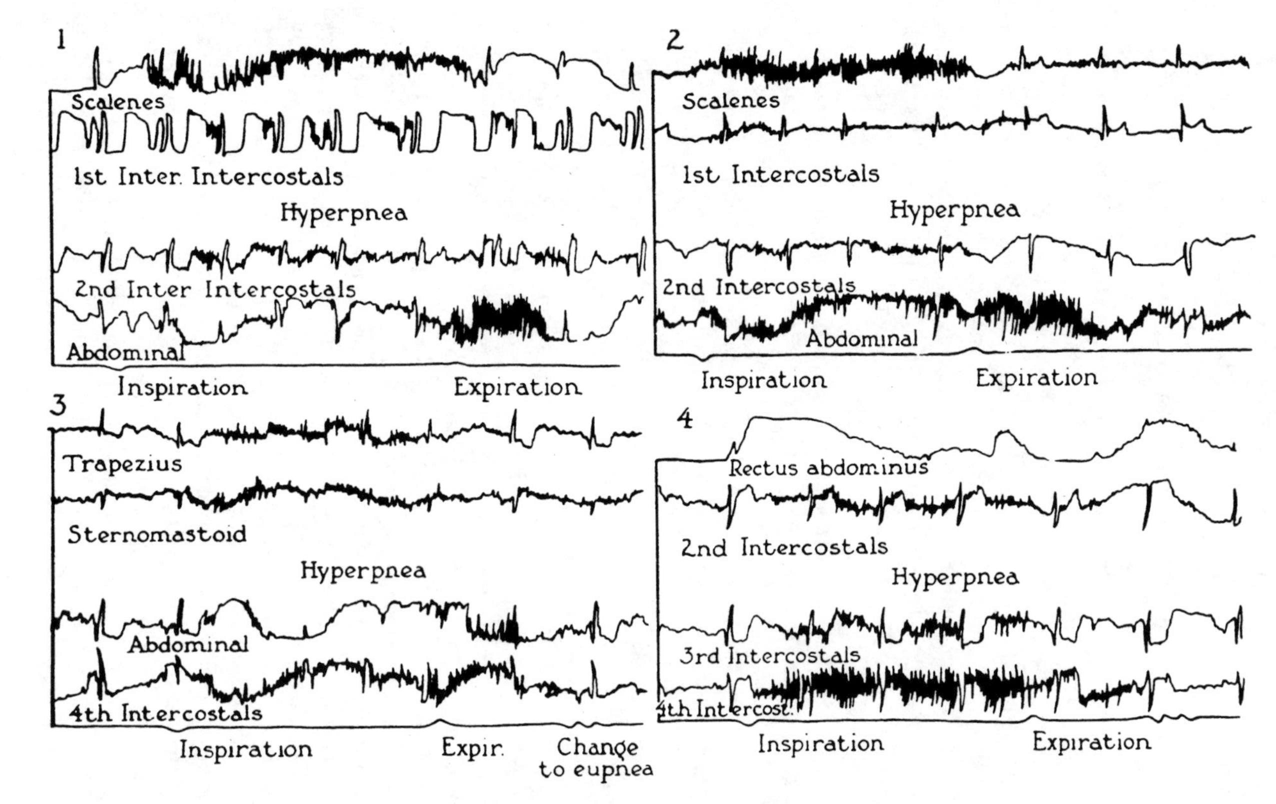

Figure 18.1. Four EMGs of scalenes and intercostals and other muscles during four different respiratory cycles. (From D.S. Jones et al., ©1953, *Anatomical Record.*)

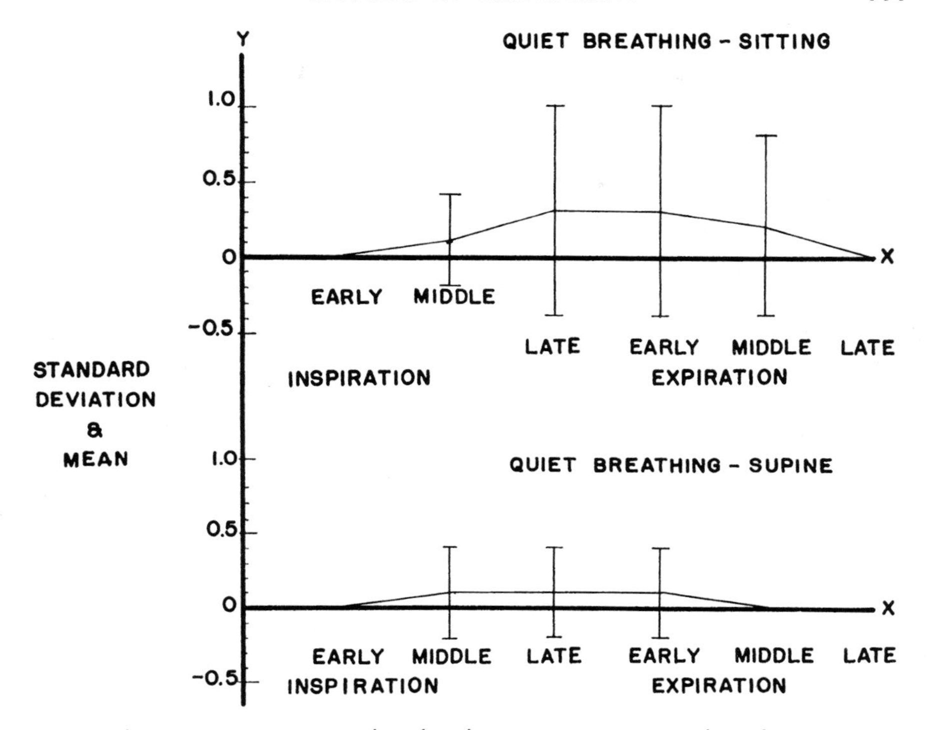

Figure 18.2. EMG results of scalenus anterior—quiet breathing.

power with comparatively much less embarrassment of quiet respiration. In a later paper, Jones and Pauly (1957) state that perhaps the intercostal muscles are "used in nonrespiratory activity more than in ordinary respiration." (See also Pauly, 1957).

Campbell (1955a) did not agree completely with these American workers. Using interspaces (6th, 7th, and 8th) lower than those that they had used for recording, he concluded that in quiet breathing the intercostals contract during inspiration and completely relax during expiration. He went further in his disagreement by stating that considerable hyperpnea (forced breathing) still did not recruit intercostals in expiration (Fig. 18.5) (see also Campbell's 1958 monograph, *Respiratory Muscles*, for further details).

It appears that the findings of both groups may be correct—the apparent disagreement arising from the use of different interspaces for the two studies. Koepke et al (1958) showed electrical activity in the 1st, 2nd, and 3rd interspaces in quiet breathing but none in the 4th, 5th, and 7th. The inherent differences that have been revealed have not yet been explained, but in their rationalization may lie the final solution of

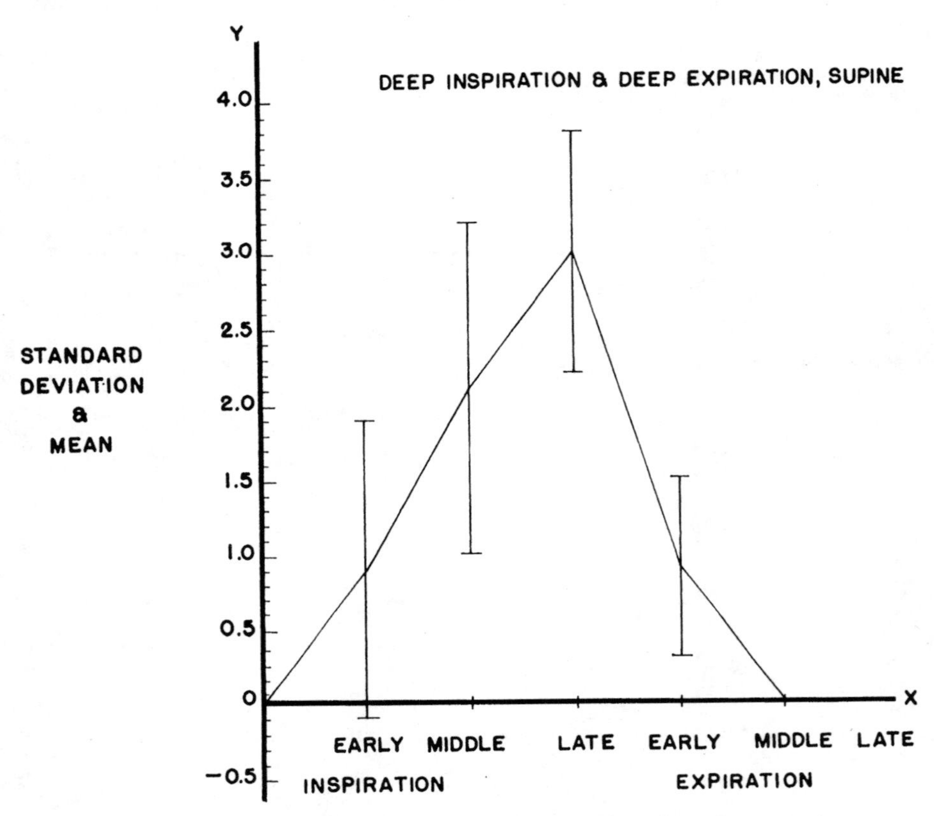

Figure 18.3. EMG results of scalenus anterior—deep breathing, supine.

the role of the intercostals in breathing. In spite of this indecision, the following cannot be ignored: (1) the vital role of the scalenes as ordinary respiratory muscles; (2) the possible role of the intercostals as primarily psotural muscles; (3) the marked differences in activity which are found between different interspaces; and (4) the primary role of the diaphragm in respiration.

In deeper breathing, Koepke et al (1958) showed that the lower intercostals that were so noticeably silent in quiet breathing became progressively recruited in descending order until even the lowest became active on very deep inspiration. This pattern of recruitment is further discussed below (p 421). Hirschberg's (1957) studies on patients with partial respiratory paralysis tended to confirm the findings for normal patients. Generally, the intercostals are recruited during inspiration but not during expiration. Morosova and Shik (1957) of Moscow, from an electromyographic study of respiratory muscles, concluded that in patients with various respiratory deficiencies extreme augmentation of the electrical activity occurs. This suggests that there is some form of compensatory stimulation *via* the respiratory center of the central nervous system.

A clear direct relationship has been shown by Viljanen (1967) of Helsinki between the electrical and the mechanical recording of human intercostal muscles during voluntary inspiration.

Draper et al (1957) and then Taylor (1960) reawakened the Galenic teaching that the external and internal intercostals had different functions. Although many other workers could find no evidence for this with standard needle electrodes, Taylor succeeded in a long series of human subjects by probing with very fine needle electrodes. He demonstrated that there are two functionally distinct layers of intercostal muscle

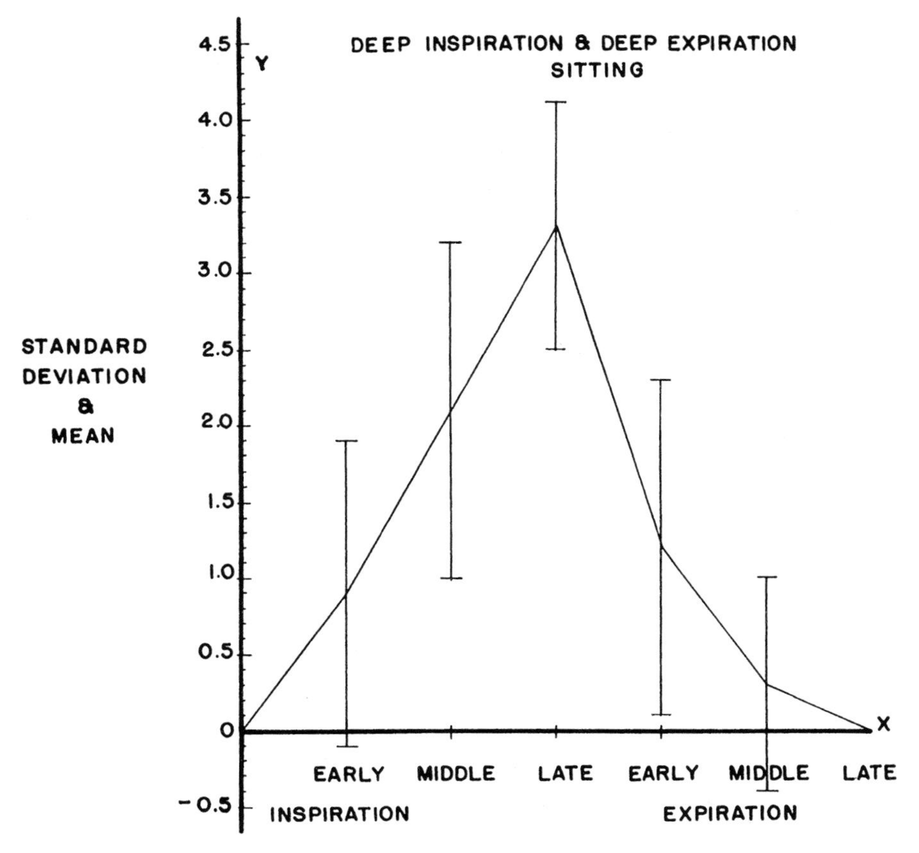

Figure 18.4. EMG results of scalenus anterior—deep breathing, seated.

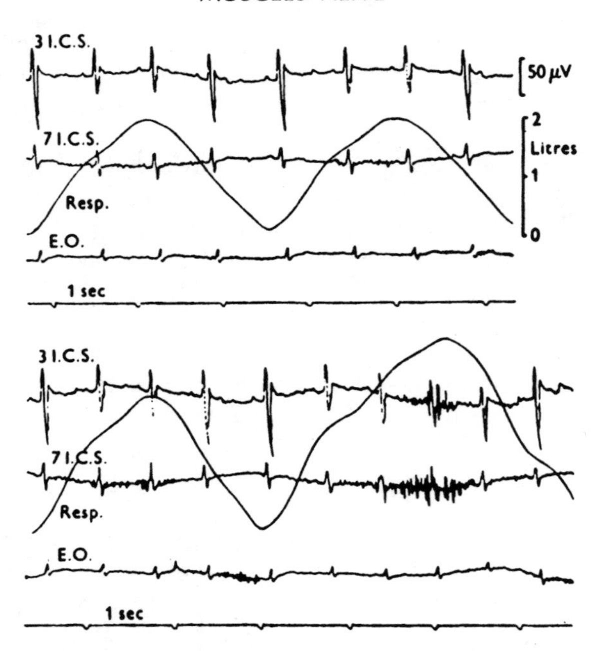

Figure 18.5. EMG records of intercostal muscles (3rd and 7th intercostal spaces (I.C.S.)—3 *I.C.S.* and 7 *I.C.S.*) and of external oblique of the abdomen (E.O.) along with spirometry (*Resp.*). (From E.J.M. Campbell, ©1955a, *Journal of Physiology*.)

everywhere *except* anteriorly in the interchondral region and posteriorly in the areas medial to the costal angles.

Where there are two functional layers, the superficial one (external intercostal) acts only during inspiration, and the deeper (internal intercostal) acts during expiration. Thus, Taylor seems to have proved in man what has been common experience in experimental animals for centuries.

Taylor found that in quiet breathing what EMG activity that occurs is limited to the parasternal region during inspiration—here there is only one functional layer, and it is exclusively internal intercostal. This explains why other workers, who usually study internal intercostal in this very place because here it is not covered by the external intercostal, have generally considered internal intercostal to be inspiratory. Indeed, this part of it *is* inspiratory, according to Taylor. Yet, to confuse the issue, Nieporent (1956) found this same part to be active in expiration, and Boyd (1968, 1969) found and emphasized that in rabbits both external and internal layers acted only in inspiration. (See also the discussion of work of Hirschberg et al on p 426).

During expiration, in quiet breathing, Taylor found EMG activity limited to the lower lateral part of the thoracic cage and coming from the internal intercostals. Apparently, he found no external intercostal activity whatsoever in quiet breathing.

His finding of intercostal expiratory activity even in quiet breathing, if true, shakes the old theory that expiration is entirely passive. Probing deeper, Taylor found that the transversus thoracis is purely expiratory in function—including its parts known as sternocostalis, intercostales intimi, and subcostalis.

More vigorous respiratory effort brings the layers of muscular activity into reciprocal action all over the chest wall. Now the external layer is entirely inspiratory and the internal layer expiratory. Taylor suggests that these two layers exert opposite rotational forces on individual ribs around their long axes. At this point, obscurity creeps into the interpretation that one might put on the role of such possible rotation; no purpose is served in speculating further here.

EMG of Intercostals during Phonation

Since the early and unsubstantiated work of Stetson (1933), no great progress was made in this type of study until the 1960s. Hoshiko (1960, 1962) investigated the sequence of activity during phonation from the intercostals and rectus abdominis. Whatever the speech material and rates of utterance may be, the internal intercostals are always recruited first, followed by the rectus abdominis and then the external intercostals. Action potentials disappear at the end of phonation in the external intercostals.

Hoshiko believes (confirmed verbally in 1981 to J.V.B.) that the internal and external intercostals cooperate in releasing a syllable, i.e., initiating the simple pulse associated with syllabication. Contrary to Stetson's teaching, the external intercostals are not involved in terminating the syllable movement.

Extending this type of work to *connected speech* in native English-speakers, Munro and Adams (1971, 1973) found similar EMG patterns in the internal intercostal muscles for the utterances of rhymes, fabricated sentences with rhythm similar to rhymes, and prose passages. Thus, a preliminary burst of activity before each phrase was followed after an interval by increasing activity in the utterance of the phrase. Activity increased further through each phrase of the sentence. However, phasic activity was *not* apparent in the external intercostals. The localized burst of activity in the internal layer seems to contribute to the increased intrathoracic pressure necessary to overcome the resistance of the glottis and later—during the utterance—it may help to maintain an adequate intrathoracic pressure despite diminution of lung volume.

DIAPHRAGM

As indicated above, the main respiratory muscle, at least in man, is the diaphragm. Most EMG studies of this vital muscle—except the most recent ones—have been indirect. Admittedly, studies of the diaphragm in experimental animals under anethesia are not rare, but, being conducted under highly unphysiological conditions, they cannot give definitive answers to a vital problem. For example, an EMG study by Di Benedetto et al. (1959), which appears to be as technically acceptable as most such studies, used dogs that were actually hepatectomized, in addition to their bellies being left wide open during the tests. To be fair to these workers, they were not attempting to categorize the actions of the diaphragm in detail. However, many of the statements on diaphragmatic function in the textbooks are based on exactly this type of observation, with or without EMG.

The function of the diaphragm has been of longstanding interest. Recognition of its importance dates back to Hippocrates, but he attributed no movement to the diaphragm. During the past 100 years, various comprehensive studies (see Boyd and Basmajian, 1963) showed that the diaphragm—at least when stimulated electrically—expands the base of the thorax by moving the ribs upward and outward. Other studies combining fluoroscopy and spirometry showed that a change from the erect to the supine position causes marked alteration in the pattern of diaphragmatic movement in man.

J.C. Briscoe (1920) and G. Briscoe (1920) postulated that the diaphragm is a tripartite organ consisting of the right and left costal parts and the crura forming one part. This view was supported by the fact that the costal and the crural parts develop from different muscular sheets.

Research on anesthetized rabbits by Wachholder and McKinley (1929) showed that the diaphragm was almost continuously active during quiet breathing with only brief relaxation during expiration. Campbell's (1958) attempts to confirm these findings in normal human subjects with surface electrodes met with disappointing results because of the intervening mass of bone, cartilage, and lung. Studies of diaphragmatic activity in many species, ascribed to Chennells by Campbell (1958), apparently showed that there is no activity during expiration. In human subjects Nieporent (1956) and Draper et al (1957) found that diaphragmatic activity could be obtained with needle electrodes during inspiration, but there was none during expiration. However, other workers, also using needle electrodes, showed that diaphragmatic activity in man could be recorded during both inspiration and expiration (Koepke et al, 1958; Murphy et al, 1959; Petit et al, 1960). Bahoric and Chernick (1975) showed with EMG techniques that in fetal sheep diaphragmatic activity is present although minimal.

Over the years, then, little if any advance had been made in our understanding of the total function of the diaphragm. This fact led to a detailed investigation in which recordings were made simultaneously from indwelling multiple electrodes (up to 16) along with spirometric tracings in conscious rabbits. The correlated results provided the first complete account of the activity of the whole diaphragm in quiet breathing (Boyd and Basmajian, 1963).

A series of 25 adult male rabbits had multiple clip electrodes implanted in their diaphragms at open operation. Following postoperative recovery, EMG and spirometric data were collected under normal physiological conditions. A color motion picture demonstrates the techniques (Basmajian and Boyd, 1960).

The rabbit's diaphragm, like man's, is divided clearly into eight left and eight right muscular slips or digits. Each side of the diaphragm has one sternal slip, six costal slips, and a lumbar slip from the aponeurotic arches and crus. The muscular slips may be numbered anteroposteriorly: sternal slip, 1; costal slips, 2 to 7; and lumbar slip, including fibers from the crus, 8.

The question as to whether the diaphragm is functionally a single muscle, or two functional halves, or a tripartite organ with a lumbar or crural portion and two costal portions was answered conclusively: all portions of the diaphragm contract simultaneously. Almost certainly, the whole diaphragm normally functions as a unit. This is true also in the cat, as shown by Sant'Ambrogio et al (1963) and Grassino et al (1976). Most likely it is true in all mammals.

PHASES OF RESPIRATION

One of our main findings was the quiet respiration includes not only the two simple opposite phases of inspiration and expiration but also a static phase before each. These we have called preinspiration and preexpiration. In duration, they are much shorter than the air-moving phases; nonetheless, the static phases make up an appreciable part of the respiratory cycle. They would appear to correspond to the well-known inspiratory and expiratory "pauses" in man.

Preinspiration

This static phase, which occurs prior to inspiratory air movements, lasts from 20 to 120 ms. During preinspiration, the activity is rarely greater than 1 +, or slight.

Other workers noted activity occurring in the diaphragm before the onset of active inspiration without recognizing a static phase. With needle electrodes, Koepke et al (1958) determined from spirometric tracings that the onset of contraction in human subjects occurred as much as ¼ second before the onset of inspiration. This finding was confirmed by Taylor (1960); potentials began immediately before inspiratory airflow, increased rapidly to a maximum, and died away in the first half of the expiratory phase. Similar findings were reported by Petit et al (1960), using esophageal electrodes in man.

Inspiration

Inspiration is an increase in the volume of the thorax with an actual inward flow of air. The diaphragm contracts and increases the cavity in a caudal direction and perhaps in the ventrodorsal and lateral directions as well. The motor units of the diaphragm increase their rate of firing at the onset of inspiration, and as this phase proceeds, new units are added or "recruited" so that the inspiratory effect gains force as it proceeds (Bergström and Kertula, 1961; Lourenço et al, 1966). As inspiration continues, the individual motor units in the diaphragm accelerate in rate, resulting in a progressive increment in the strength of contraction of each unit. With an increase in the number of active units and by augmentation in the firing rate of each unit, inspiration reaches its peak. In all but an insignificant number of records, activity was recorded throughout the entire inspiratory phase. Indeed, activity was continuous throughout both preinspiration and inspiration. However, the level of activity fluctuates over these phases (Boyd and Basmajian, 1963).

The entire inspiratory phase in rabbits last from 300 to 550 ms (and rarely longer). This inspiratory phase may be divided for convenience into four quarters. Of course, the length of these varies with the length of each phase, which in turn depends on the respiratory rate.

The peak of the inspiratory motor unit activity terminates some 30 to 40 ms before the end of inspiration. In most instances, the peak is followed by a rapid decline—and in some instances by a very sharp drop—to the baseline, where varying degrees of activity continue to the end of the fourth quarter of the inspiratory phase. A similar finding was reported by Koepke et al, indicating that the diaphragmatic voltage pattern in man consistently reached its greatest amplitude at or slightly before the end of inspiration; then it tapered to its stopping point in what was considered to be expiration in man, but EMG silence following inspiration occurred in 44% of the recordings in rabbits.

At the onset of inspiration, almost all of our recordings (98%) showed a carryover of preinspiratory activity into the first quarter of inspiration. Then, in all of the recordings, activity was present throughout the second and third quarters of inspiration. During the first quarter of inspiration, in general, the diaphragm is only slightly active. The motor units of the diaphragm do not begin to increase their firing rate during quiet breathing until the second quarter of the inspiratory phase is reached; most of the recordings (74%) show moderate activity by that time. One might postulate that the intercostals play the main role in the first quarter.

The third quarter of inspiration forms the peak for inspiratory activity during quiet normal respiration with 76% of our recordings showing marked activity. Nonetheless, in a substantial number (12%), the activity remains only slight during the third quarter.

The activity in the third quarter of inspiration, whatever its character, is carried into the fourth quarter. In 76% of the recordings the great activity recorded during the third quarter continued into the fourth quarter and stopped some 20 to 40 ms before the termination of the inspiration. However, the great activity at the peak of the inspiration does not continue throughout the entire fourth quarter. The fourth quarter thus can be divided into two unequal parts, the line of demarcation being the terminal point at which the muscular activity ceases. Thus, during the first part of the last quarter, 76% were markedly active, and the remainder were moderately or slightly active (12% each).

Great activity, when it did occur at the end of inspiration (7%), was never carried into preexpiration (Boyd and Basmajian, 1963).

Preexpiration

The static phase of preexpiration usually lasts from 20 to 50 ms (and somewhat longer in some cases). EMG activity is slight or absent (44% and 56%, respectively) in this phase.

Preexpiration is a regular precursor of expiration. Studies on the human diaphragm by Murphy et al during passive expiration seemed to show that electrical activity occurred even though each subject had been instructed to relax after taking in a breath. They found that an increase in the duration of activity during expiration was directly related to the depth of the preceding inspiration. These findings could not be confirmed by Boyd and Basmajian, for most of the recordings were taken during quiet respiration and, of course, they were in nonhuman subjects.

Expiration

Expiration, the last phase in the respiratory cycle, consists of a decrease in the volume of the thorax with air moving outward. One way that this might be accomplished actively is the contraction of the abdominal muscles forcing the diaphragm up into the thorax. During quiet breathing, however, expiration is generally regarded as passive. Nonetheless, activity in the diaphragm is recorded during expiration. In almost every instance, the (slight) activity that Boyd and Basmajian recorded during expiration lasted for a longer period of time than did the greater activity that occurred in inspiration. Previous to those findings, Murphy et al reported that the activity continued through as much as 98% of expiration; Agostoni et al (1960) found activity persisting into the early part of expiration in human diaphragms.

Many varied opinions exist on the activity of the diaphragm during expiration. The literature includes only one study specifically in the rabbit (Wachholder and McKinley, 1929). This report indicates that the diaphragm is almost continuously active during quiet breathing with only a very brief period of nonactivity during expiration. In contrast, using needle electrodes, Nieporent (1956), Campbell (1958), and Draper et al (1957) found no diaphragmatic discharge during the expiratory phase in man or in animals.

Murphy et al (1959) found no activity originating in the human diaphragm during forced expiration; they reported that activity occasionally carried over from deep inspiration but always stopped abruptly at the onset of a forced expiratory effort. In patients with transverse myelitis, irrespective of the size of the preceding inspiration, diaphragmatic voltages were not found during a forced expiratory effort. During passive expiration, activity always continued from inspiration into expiration, and this pattern was like that in normal subjects.

During quiet breathing in man (Petit et al, 1960) activity started in the diaphragm at the beginning of inspiratory flow, increased in intensity through inspiration, and persisted after the onset of expiration but with decreasing intensity. A similar pattern occurred during increased ventilation, with the difference that the activity started before the beginning of inspiration. Some workers tend to agree that there is a carry-over of inspiratory activity into the expiratory phase, but none indicates either how far into the expiratory phase this activity extends or the degree of activity carried into expiration.

In most of the Boyd/Basmajian recordings (63%), some activity occurred throughout the entire expiratory phase. Complete absence occurred in only 14%. Almost always the diaphragm is active during the last quarter of expiration. One possible reason why activity carries into or continues throughout expiration is that activity in the diaphragm during passive expiration is a braking action to oppose the normal elastic recoil of the lungs (Murphy et al, 1959). In effect, it is not a true expiratory effort. Agostoni and Torri (1962) hold like views from like findings in man. However, they implicate reflexes to balance the antagonist activity of the abdominal wall muscles. Delhez et al (1963 et seq) discount the importance and even question the occurrence of activity at the end of forced expiration.

The anatomical structure of the diaphragm supports this view because all of its fibers are arranged in radiating fasciculi inserting into the central tendon. Shortening of the fibers can only cause a flattening of the dome and so actually resist the production of an expiratory force. Furthermore, no electrical activity could be recorded during forced expiration in human subjects.

According to Campbell, the rate of airflow at the onset of expiration does not rise rapidly to a maximum (as would be expected if the muscles of inspiration relaxed immediately), indicating that during the early part of expiration the muscles of inspiration decrease their force of contraction only gradually. He further states that there is a persistence of intercostal activity during the early part of expiration. Although no direct reference to the diaphragm is made, he reports that measurements of the work of breathing based on pressure-volume diagrams suggest that the muscles of inspiration may exert considerable force in opposing the elastic recoil of the lungs during expiration. One may conclude, then, that the slight activity recorded from the diaphragm during expiration is a braking action to oppose the normal elastic recoil of the lungs.

Koepke et al (1958) at Ann Arbor, MI, also made a concerted attack on problems of the respiratory muscles, including the diaphragm. They used needle electrodes in the diaphragm in several human subjects. These electrodes were inserted through the 11th intercostal space into the muscular digit of the diaphragm that arises from the 12th rib. We must keep in mind the limitations imposed by this localization of pickup (in a muscle which, in the findings on rabbits, may act somewhat differently in its various parts). With ordinary inspiration, the diaphragm became active before any of the intercostals and before the flow of air by as much as a quarter of a second. During quiet breathing the diaphragm never failed to act in inspiration, although some of the intercostals were only recruited with deeper breathing (see above).

The same group, with Murphy as senior author (1959), reported on the EMG activity during expiration. They insisted that the diaphragm always shows electrical activity as a carry-over into the early stage of passive expiration (Fig. 18.6). The diaphragm consistently showed its greatest activity at or slightly before the end of inspiration; this tapered

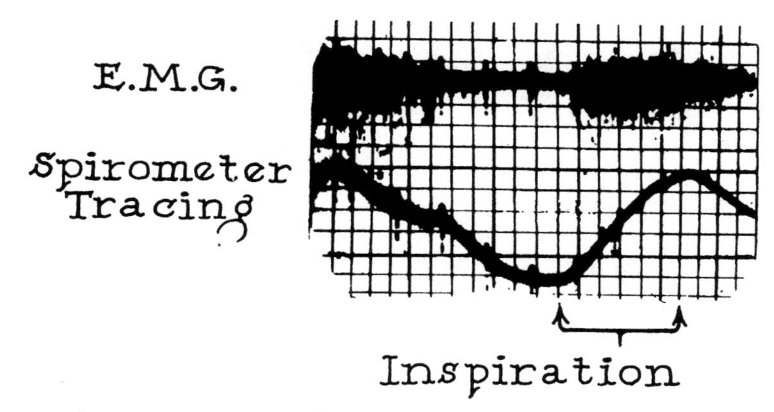

Figure 18.6. EMG signal from diaphragm. (From A.J. Murphy et al, ©1959, *Archives of Physical Medicine*.)

off to silence in early expiration. The duration of activity in the diaphragm during expiration was directly related to the depth of the previous inspiration. In some instances, the activity continued through as much as 98% of expiration. None of the intercostal muscles showed the same degree of carry-over of activity as did the diaphragm. During quiet breathing, *only* the diaphragm showed such expiratory activity.

During induced hyperinflation, Martin et al. (1980) found that the volume of hyperinflation is influenced by persistent activity of the respiratory muscles in expiration. This is due to the inspiratory intercostal and "accessory" muscles (e.g., scalenes) rather than the diaphragm, which is almost silent in ordinary expiration.

Murphy et al suggest that the activity of the diaphragm during passive expiration may be a braking action "to oppose the normal elastic forces of the lungs rather than the exertion of a true expiratory effort." Observing that the fibers of the diaphragm radiate from the central tendon, they too believe that activity of the muscle can only flatten and lower the dome. Furthermore, no activity in the diaphragm was seen during forced expiration, which appears to confirm their thesis.

A large and vigorous group of investigators in Liège, Belgium (Petit et al, 1960) devised a novel technique for diaphragmatic EMG in conscious man. The electrical activity was detected by means of electrodes passed down the esophagus to the level of the diaphragmatic esophageal hiatus. In four normal persons they found the activity to be synchronous with the respiratory variations of intra-abdominal and intrathoracic pressures. Potentials occurred from the onset of inspiration and increased in intensity. They continued into expiration for a varying length of time with decreasing intensity. During increased ventilation, potentials began immediately before inspiration rather than just at its onset. For further details of the work of the Liege group, see Delhez and Petit (1966), Delhez et al (1968), and Petit et al (1968).

Admittedly, the intraesophageal electrodes pick up potentials only from the crural fibers of the diaphragm, but this does not invalidate the results. Although in the future investigators may show some variations between various parts of the diaphragm, no major differences are likely to be revealed. In any case, the approach of Petit's group to diaphragmatic EMG is a promising one.

Similar (and successful) techniques were used by Miranda and Lourenco (1968), Hixon et al (1969), and Önal et al (1979) for technical investigations beyond our present concerns. The recent demonstration by Trelease et al (1982) of the efficacy of the fine-wire electrode technique inserted chronically into the crura of cats, promises a major advance in research on the diaphragm. Somewhat related to the transesophageal EMG technique for the diaphragm are the esoteric EMG studies of the esophagus itself, although they present a quandary as to where to fit these interesting studies in this book. Interested readers should consult Coman and Car (1970) for a description of swallowing as it affects esophageal reflex contractions produced by stimulation of the vagus and superior laryngeal nerves and Tokita et al (1970) for a study of normal and abnormal functions.

BREATH-HOLDING AND SNORING

Agostoni (1963) has reported a curious diaphragmatic activity during human breath-holding. After an initial silence, a marked discharge occurs and repeats at progressively higher rates until the breaking point. These are ineffectual respiratory efforts that cause a fall of intrathoracic pressure apparently because the glottis is clamped shut by effort. Agostoni et al (1960) found a rather similar fluctuation during the very brief episode of either coughing or laughing. Readers interested in intercostal EMG during snoring should consult the work of Laugaresi et al (1975).

QUADRATUS LUMBORUM AS AN ACCESSORY

In a later study using the earlier Boyd/Basmajian techniques, Boyd et al (1965) showed that quadratus lumborum acted simultaneously with the diaphragam to stabilize the rib, which might otherwise be elevated. Thus quadratus acts in concert with the diaphragm as a respiratory muscle; furthermore, its activity coincides with the diaphragm to produce the normal braking action during expiration noted on p 420.

DIAPHRAGM-INTERCOSTAL INTERRELATIONSHIP

Although many generalized statements have been made in the literature (both of respiration and of EMG) about the interplay of diaphragmatic and intercostal activity, only the work of the above-mentioned Ann Arbor group (Koepke et al, 1958; Murphy et al, 1959) and the rabbit studies appear to cast any direct light on their relationship (Boyd and Basmajian, 1963; Boyd, 1969).

In human quiet respiration, the diaphragm became active a ¼ second before the onset of inspiratory air flow as measured by spirometry. Furthermore, this occurred before any activity in any intercostal muscle. The intercostals that were recruited earliest were the first topographical pair (i.e., the first intercostal muscles). Then, the recruitement proceeded progressively to include lower and lower intercostals. With quiet breathing the diaphragm was always active in inspiration, the 1st intercostals usually were, and the 2nd intercostals occasionally were. All the others were inactive. During inspirations deep enough to call upon all the intercostals, the onset of activity advanced progressively in successively lower intercostals.

During natural passive expiration (as distinct from forced expiration) these workers found that a carry-over of activity was also present in the intercostals. Among them, the lowest intercostals were more important than the highest, the duration decreasing sequentially from the 11th pair upwards.

Although no activity was found in the diaphragm during forced expiration, intercostal activity was almost always present. The likelihood of such activity was related to the volume of air in the lungs at the end of inspiration. When intercostal activity was present during forced expiration, the recruitment again was progressively upwards from the lowest intercostals to the highest.

Murphy and his colleagues postulated that when the intercostals are recruited sequentially from below upwards during forced expiration, the lower portion of the thorax becomes relatively smaller than the upper to produce a desirable pressure gradient within the chest to empty the lungs. This remains pure hypothesis and is not particularly convincing.

During active REM sleep in the supine position, a sharp drop occurs in intercostal muscle activity and rib cage movements compared with quiet sleep and the awake upright posture—which are about the same (Tusiewicz et al, 1977).

OTHER RESPIRATORY EMG STUDIES

Fink et al (1960) used the integrated rectified EMG signals from patients during general anesthesia to study the threshold of the respiratory center. Although their study revealed nothing about the normal functioning of the diaphragm, it did show the feasibility of human diaphragmatic EMGs. They inserted unipolar needle electrodes into the diaphragm through the 8th, 9th, or 10th intercostal space and successfully obtained recordings. Their main findings relating to the onset of apnea revealed—as indicated by cessation of diaphragmatic potentials—that it occurs when the average alveolar carbon dioxide tension or pressure falls to 38 mm Hg. Diaphragmatic activity reappears as the CO_2 tension rises above this value, the stimulus acting through the CO_2-sensitive respiratory center.

The studies of Björk and Wåhlin (1960) on the effects of the muscle-relaxant drug succinylcholine upon the cat diaphragm would be of interest to a very few readers and are mentioned here for completeness only. The effect of the drug is to disturb the synchronization of parts of the motor unit and to cause a progressively decay of motor unit potentials down to individual fiber potentials. This effect is peripheral rather than central.

Di Benedetto et al (1959) used EMG combined with phrenic nerve stimulation to investigate the innervation of the diaphragm in dogs. They found that, contrary to widely held opinion, the muscle mass to the right of the esophageal hiatus was commonly innervated by both phrenic nerves, i.e., the right crus was bilaterally innervated in about one third of cases. However, they found no instance in which left crus was bilaterally innervated.

Jefferson et al (1949), who also worked with dogs, found complete paralysis of the left hemidiaphragm with left phrenicotomy. This (1) confirms the above findings and (2) confirms the teaching (which has been often challeged without adequate evidence) that the only innervation of the diaphragm is the phrenic nerve.

Sant'Ambrogio and Widdicombe (1965) have studied respiratory reflexes from the diaphragm and intercostals in rabbits to assess their strength. They used individual MUAPs as a direct and quantitative measure of activity before and after vagotomy. There is little doubt from their results that proprioception plays an important part in driving the respiratory muscles. Guttmann and Silver (1965), approaching the problem quite differently, i.e., by studying the reflex activity in the intercostals of tetraplegics, confirm the role of stretch reflexes in respiratory muscular activity.

Youmans and his group at the University of Wisconsin studied the "abdominal compression reaction" by means of EMG records from the diaphragm and intercostal muscles of anesthetized dogs (Briggs et al, 1960). This reaction is initiated by procedures which cause a decrease in central blood volume, and it consists of a steady state of activity of abdominal muscles rhythmically interrupted by breathing. In some instances,the intercostal muscles show no action currents while in others they show a burst of activity during inspiration and again during the abdominal compression reaction. The steady-state contraction of the external oblique abdominal muscles commonly begins while expiration is in progress and reaches a maximum after completion of expiration. When a strong abdominal compression reaction is present, the initial phase of inspiration is the movement of the diaphragm caudally, related to sudden inhibition of the abdominal compression reaction and a corresponding decrease in intra-abdominal pressure. The diaphragm begins to move caudally because of less pressure on the abdominal side (and not because of motor activity), and it continues to move as a result of its contraction, according to Youmans and his colleagues (1963).

Delhez and various colleagues in Liège, Belgium, have made an important series of contributions to the literature of respiration and diaphragmatic function over several years (1963 et seq). The amount of activity in the diaphragm in human subjects was found to comply proportionately with the ventilation up to 50 or 60 liters/minute. Above that, EMG activity rises more rapidly than the ventilation, apparently to counteract elastic forces and antagonistic activity (Delhez, 1964; Delhez et al, 1964a–d). Their insistence that strong diaphragmatic activity occurs during forced expiration has created considerable attention and controversy without satisfactory resolution as yet. Their views, expanded upon in a long review written by Petit et al (1965), are interesting, but fuller consideration is out of place here. Also interesting to some readers will be the EMG work of Kelsen et al (1976) on the neural responses of respiratory muscles during elastic loading and that of Sharp et al (1976) and Lopata et al (1976) in evaluation of respiratory regulation.

Motor Units in Diaphragm

Even in the highly developed human diaphragm, the muscle is quite thin, and likely there is a great lateral spread of fibers belonging to one motor unit. To test this hypothesis, Krnjević and Miledi (1958), at the Australian National Univesity in Canberra, investigated the distribution of single motor units in the rat diaphragm electromyographically using a "phrenic-hemidiaphragm" preparation. They found the fibers of one motor unit irregularly scattered over an area of several millimeters. They also found that the motor units were considerably intermingled.

ACCESSORY RESPIRATORY MUSCLES

The muscles which are usually considered to be accessory respiratory muscles are the following: the muscles of the vocal cords (discussed on p 442); sternomastoid and scalenes in the neck (pp 466 and 467); abdominal muscles (p 395); pectoral muscles (p 266); serratus anterior (p 267); and trapezius (p 265). As pointed out in the appropriate sections of this book, the scalenes should be considered primary respiratory muscles; the abdominal muscles are certainly accessory respiratory muscles; and, generally, the upper limb muscles (including serratus anterior) take no part in quiet or even labored respiration—except under highly abnormal conditions, diseases, and postures (Grønbaek and Skouby, 1960).

Nieporent (1956) found that there was no activity in pectoralis major during quiet breathing, but during maximal inspiration some slight to moderate activity appeared. During dyspnea, it functions primarily in inspiration as an accessory muscle. Campbell (1954, 1958) found some activity only during very deep inspirations in trapezius (upper part), latissimus dorsi, pectoralis major and minor, the serratus anterior. On the other hand, Tokizane et al (1954) reported activity with ordinary breathing in some of these muscles but, being unconfirmed by later studies, their findings are subject to serious doubt.

In patients with complete diaphragmatic paralysis, Hirschberg et al (1962) found that most could breath quietly by using their accessory muscles. Surprisingly, the intercostals (whose viability could be proved by other tests) were the least active during quiet breathing; the abdominal muscles were the most active. Accessory muscles in the neck (although they were partially paralyzed in the patients with poliomyelitis) were also quite active. Of course, readers of p 406 would expect this. Guttmann

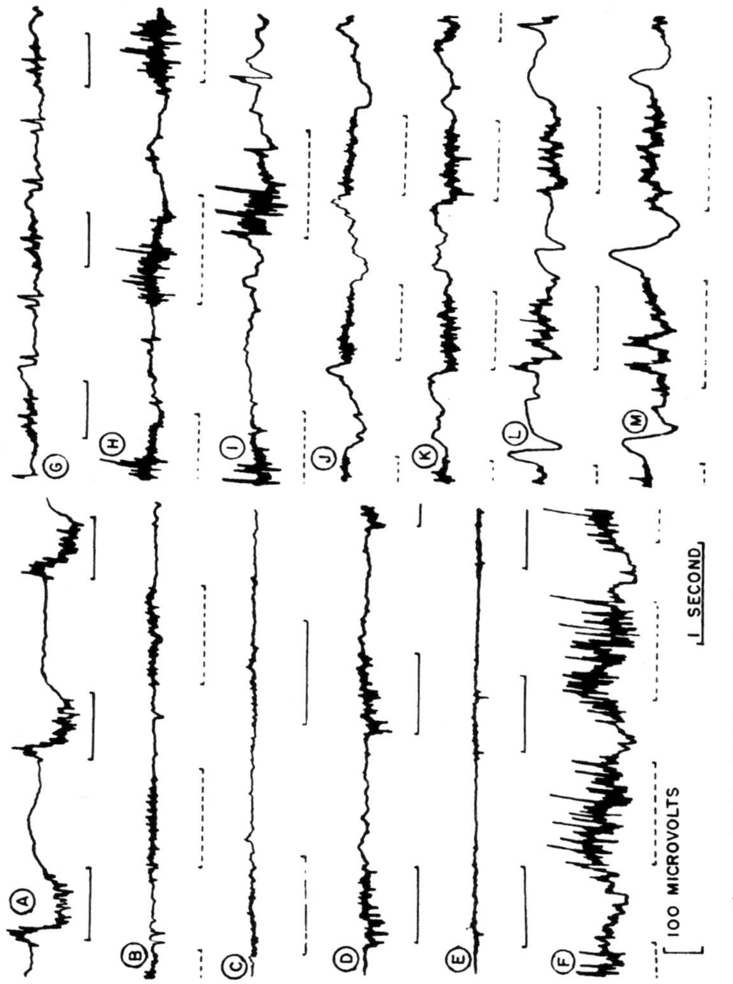

Figure 18.7. EMG records of various canine muscles during respiration. *Solid marker line*, inspiration; *dashed line*, expiration. (*A*) Dilator naris. (*B*) Mylohyoid. (*C*) Sternohyoid. (*D*) Sternothyroid. (*E*) Hyothyroid. (*E*) Scalenus anterior. (*G*) Scalenus posterior. (*H*) External intercostal. (*I*) Internal intercostal. (*J*) Rectus abdominis. (*K*) External oblique. (*L*) Internal oblique. (*M*) Transversus abdominis. (From T. Ogawa et al, ©1960, *American Journal of Physiology*.)

and Silver's (1965) finding that the intercostals of tetraplegics resumed reflex activity is of some importance. This developed in response to stretch reflexes initiated by the diaphragm or the "accessory" muscles of the neck (mainly sternomastoid).

Ogawa et al (1960) at Michael Reese Hospital in Chicago found in dogs that the following muscles had no respiratory function: digastric; masseter; levator nasolabialis; scutularis; cervicoauricularis; splenius; brachiocephalicus; trapezius; rhomboideus; supraspinalis; infraspinalis; deltoid; semispinalis; serratus anterior; pectoralis superficialis and profundus; serratus posterior superior and inferior; and psoas. In contrast, they consistently found respiratory activity in the following canine muscles: nostril; intrinsic laryngeal; scalenus anterior; intercostals; rectus abdominis; external and internal obliques; and transversus abdominis (Fig. 18.7).

CHAPTER 19

Mouth, Pharynx, and Larynx

Until 1958, electromyography of the mouth and pharynx was virtually an unexplored frontier, and even now the tongue—although it is an obvious and accessible muscular mass—has not been adequately explored. Indeed, only one systematic study had been done on the functions of the intrinsic or extrinsic muscles of the tongue before the second edition of this book appeared in 1967. Furthermore, the easily accessible musculature of the floor of the mouth (i.e., mylohyoid, etc.) also has been badly neglected until later. On the other hand, the palate and the pharyngeal constrictors have now been well explored by several groups.

Insofar as the palate and pharynx are concerned, this chapter will deal chiefly with studies employing intramuscular electrodes, supplemented by those of clinical electromyographers. Also to be discussed at the end are the reports on the EMG of the larynx.

TONGUE AND MOUTH

Bole (1965), the first to use inserted wire electrodes for the tongue, performed definitive studies of the actions of genioglossus muscle. He found that the right and left muscles act together with approximately similar EMG response during many general movements of the tongue—even lateral shifts. The greatest activity appeared when the tongue met resistance. Surprisingly, little activity accompanied protrusion of the tongue unless it was against the back of the incisor teeth.

Bole showed that there may be several different patterns of glossal movement during swallowing. The duration of activity in genioglossus ranged from 1.42 to 2.74 s and appeared in two or more bursts. Generally the bursts were at the onset of swallowing and after the substance had left the tongue. Bole's pioneer work on the human tongue, because it gave great promise of revealing the true functions of this vital organ, was partially followed up by Basmajian and his colleagues and by scattered investigators elsewhere. Thus, in an investigation of the activity of the paired genioglossus and geniohyoid muscles of 26 human subjects during deglutition, a general pattern of muscular activity was revealed involving an initial buildup, gradual summation, and tapering of electrical potentials during swallowing of both saliva and water (Cunningham and Basmajian, 1969). There is an observable difference in the pattern of swallowing of individuals within a group and among the individual swallows of a single subject (see Fig. 19.1).

There are longer periods of electrical activity during a saliva swallow

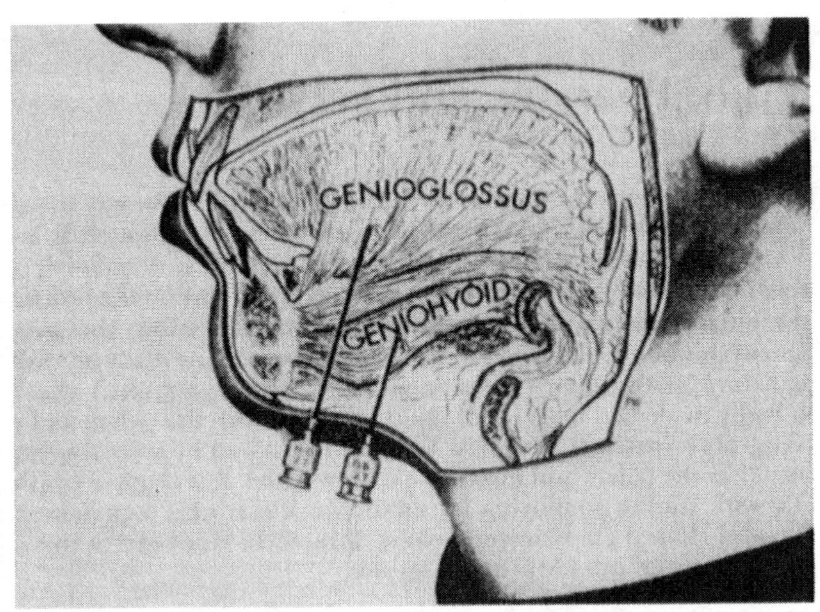

Figure 19.1. Schematic drawing of genioglossus and geniohyoid muscles showing the position of the needle cannulae before their withdrawal. Small hooks at end of fine wires will hold them in place. (From D.P. Cunningham and J.V. Basmajian, ©1969, *Anatomical Record*.)

than during a water swallow. The type of bolus also seems to affect the pattern of activity in the individual muscles as well as the length of time that they are working. The geniohyoid muscles do not appear to begin their activity with the genioglossus muscles but rather lag behind, and they do not appear to be active for as long. Both pairs of muscles appear to remain active during and after the time that the bolus has passed the area of the laryngopharynx. A period of electrical silence occurs prior to the characteristic burst of activity associated with a swallow. This appears to be the result of an active inhibition (Cunningham and Basmajian, 1969).

The "tongue-safety" function of the genioglossus muscle has been the special interest of Sauerland in Galveston, Texas (Sauerland and Mitchell, 1975; Sauerland and Harper, 1976). More active during inspiration both during working and sleeping hours, the muscle prevents obstruction of the airway. The glossopharyngeal part of the superior pharyngeal constrictor ("myloglossus muscle") acts as an antagonist (Jüde et al, 1975). Of special interest to dentists is the finding of Lowe and Johnston (1979), whose computer-based system to compare activities of genioglossus, masseter, and orbicularis oris muscles revealed that postural tongue activity could play an important role in the development of the anterior dentition.

Muscles in the floor of the mouth related to the tongue were studied by Basmajian and various colleagues. The geniohyoid, anterior belly of digastric, mylohyoid, and genioglossus muscles of 20 human subjects were studied electromyographically to determine the temporal relationships of their activities during the act of swallowing. Although the firing order of the four muscles varied within the same subject, the best estimate of the "true" firing sequence was established for each of the 18 subjects who provided statistically significant data. However, no definite universal pattern could be established for the four muscles because there was great intersubject variability in both the duration and the sequence of activity. Therefore, at least with respect to these four muscles, each individual has his own swallowing pattern, but different people may swallow quite differently (Hrycyshyn and Basmajian, 1972; Vitti and Basmajian et al, 1975).

The type of bolus (saliva *vs.* water) may influence the duration of the muscles' activity. On the other hand, posture (semireclined *vs.* sitting) did not seem to have any influence. There was no evidence to indicate that posture and/or the type of bolus are correlated with the sequence of muscular activity.

The anterior belly of the digastric muscle was not active in one-quarter of the swallows studied. When active during deglutition, all muscles had a general EMG pattern of one to many summations of activity separated by relatively quiet periods before and after each swallow (Hrycyshyn and Basmajian, 1972).

The mylohyoid muscle has also been studied by Lehr et al (1971) to determine its activity relative to isolated movements of the tongue and mandible and during various functions involving multiple parts of the oral apparatus. Data were obtained from 20 subjects. Using bipolar fine-wire electrodes, the anterior fibers were found to be more active than the posterolateral fibers in a majority of activities performed. Tongue movements produced slightly more activity in the posterolateral fibers; the anterior fibers were more active during mandible movements. During mastication, deglutition, sucking, and blowing, both the anterior and posterolateral fibers were markedly active. Other EMG work on mylohyoid and geniohyoid, among other muscles, has been reported by Miller (1972a and b) and by Car (1970), who were concerned with the swallowing reflex induced by peripheral nerve and brainstem stimulation.

The lack of muscle spindles in mylohyoid led Kleppe et al (1982) to compare single motor unit controls in it with abductor pollicis brevis. The latter proved to be a much easier muscle to train, reinforcing the hypothesis that spindles are important in fine control.

Other studies of the tongue have been concerned with speech production and disturbances and use rather global pick-ups from electrodes applied to the tongue's surface (MacNeilage et al, 1967; Shankweiler et al, 1968; Huntington et al, 1968). Fine-wire electrodes have been used in such studies only sparingly (Harris, 1971; Borden and Harris, 1973; Raphael, 1975; Raphael and Bell-Berti, 1975). As techniques improve—and fine-wire electrodes simply must gain acceptance—these studies should flourish.

Buccinator

Buccinator EMG research is increasing. This facial muscle, which has sunk deep and become the lining of the mucosa of the cheeks, has many roles to play. Its part in facial expression will be described with facial muscles on p 461. Blanton et al (1970) made an analysis of the buccinator muscle using indwelling, fine-wire electrodes. Data were obtained from 22 subjects with normal dentition during various oral activities. The buccinator muscle was found to be markedly and consistently active during swallowing, blowing, sucking, masticating, and various lip and mandibular movements. Its activity during mandibular movements was not believed to be for the sake of direct propulsion but rather as an "expression of an effort to perform the movement." The studies of de Sousa and Vitti (1965) and by Basmajian's group are related chiefly to mobilizing the face and will be described on p 461.

PALATE

Cho-luh Li and Arne Lundervold (1958), while working in Montreal, had the opportunity of examining electromyographically a series of normal human palates as controls for a broader study of cleft palates. They were able to record separately from the tensor palati (which becomes aponeurotic as it turns around the pterygoid hamulus into the palate) and the various fleshy muscles in the palate itself. Examinations of the soft palate in a series of rabbits (Basmajian and Dutta, 1961a) were compared with examinations in a series of normal human volunteers by Basmajian and Dutta (1961b).

[Extensive investigations by Doty and Bosma (1956) on dogs, cats, and monkeys were concerned chiefly with swallowing. These investigators did not report on the soft palate as such (but see below). Broadbent and Swinyard in 1959 reported on their EMG studies following human cleft palatal repairs in which they had used a "dynamic pharyngeal flap." These studies are mentioned here for the sake of completeness.]

The experiments on the palate of rabbits were of two types. In the first type, rabbits under general anesthesia were tested for reflex swallowing with a bipolar needle electrode inserted into the part of the palate that contains the bulk of levator palati. In the second type of experiment, EMGs were recorded by two different techniques in a series of conscious rabbits made to swallow "normally" by placing water on the tongue. The experiments on human subjects were on conscious normal volunteers in whom recordings were made with special long, fine, bipolar electrodes inserted under direct vision into the levator palati.

In conscious human subjects, always found, as did Li and Lundervold, a burst of activity upon inserting the electrode (Fig. 19.2). This lasted for several seconds, but it never lasted more than 10 s in either set of investigations. Then complete relaxation of the levator palati and tensor palati ensued and continued for as long as the subject remained at rest. This was confirmed by Fritzell (1963) for both muscles by direct recordings. When subjects sucked water through a straw, the levator palati became slightly active and remained thus as long as the water was held in the mouth (Fig. 19.2) (Basmajian and Dutta, 1961b).

During swallowing, potentials came as a burst lasting about 1/3 second and were followed by complete relaxation (Fig. 19.2). Li and Lundervold comment only on the normal appearance of the potentials obtained during voluntary swallowing, but one gets the impression that their findings were similar.

The results of the experiments on the soft palate of conscious rabbits were similar to those on the palates of human beings. In contrast, the most striking finding in anesthetized rabbits was that reflex swallowing (caused by prodding the pharyngeal mucosa) showed very little activity in the palate (see Fig. 19.5 on p 438).

Palatal Activity During Speech

Fritzell (1963) found that activity started simultaneously in tensor and levator palati just before speech begins. This "acoustically silent" period

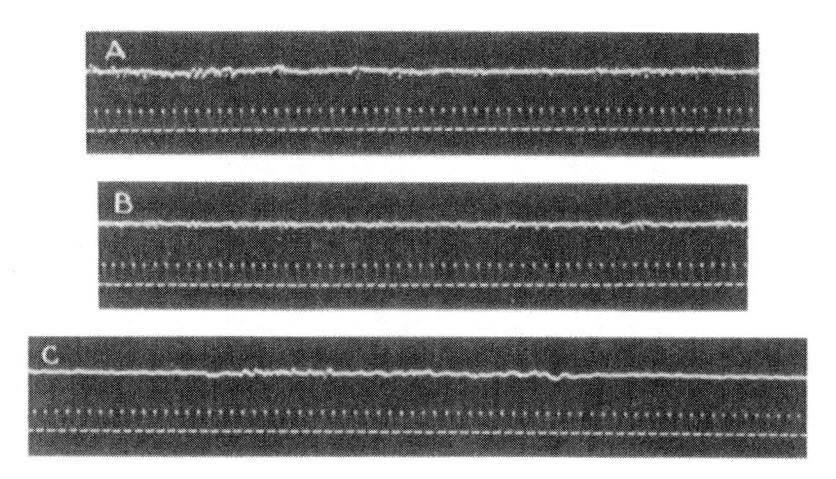

Figure 19.2. EMG records of human palate. (*A*) "Insertion" potentials. (*B*) Activity during sucking water through a straw. (*C*) Swallowing. Calibration signal: 10-ms intervals; amplitude, ca. 500 μV.

of palatal activity varied in different subjects. The potentials diminish and usually disappear before an utterance is finished.

When words beginning with a nasal sound are spoken, action potentials precede the microphone signal. But when nasal sounds appear within a word the potentials disappear or diminish only to return when oral sounds are made. "The production of nasal is regularly announced in the EMGs before the sound appears in the microphone record," according to Fritzell.

The hypothesis of a very simple organization of muscular activity (for example, two discrete levels) underlying a more complex pattern of velar positioning was tested by Lubker (1968). The data obtained in his study were not consistent with this hypothesis. Palatal EMG activity during speech appeared to vary in a relatively continuous manner and was positively correlated with velopharyngeal positioning.

A majority of Lubker's subjects demonstrated little or no velar movement during production of the sustained, detached nasal consonant. A relatively marked positive correlation was obtained between tongue and palatal positioning during speech production, thus suggesting that a high tongue position is associated with a high palatal position. The Moll-Shriner hypothesis predicts such a relationship and suggests that it is due to anatomical interconnections between tongue and palate. An alternative is supported by the data obtained in Lubker's research: greater palatal elevation may accompany vowels with high tongue position simply because such elevation is needed to prevent the vowel from being detected as nasal in quality. The latter explanation may be more consistent with the high correlation observed between palatal position and EMG activity. Further EMG studies of palatoglossus muscle by Bell-Berti and Hirose (1973) suggest that during speech this muscle is associated more with glossal rather than palatal movements. Ushijima and Hirose (1974) and Bell-Berti (1974, 1976) have greatly expanded knowledge of palatal muscles during speech using wire electrodes.

Lubker joined Fritzell and Lindqvist in Sweden to expand this work with fine-wire electrodes (Lubker et al, 1970). This work suggests that palatoglossus does not contract maximally during speech, in contrast to its vigorous action in swallowing and nasal breathing. Later Lubker and May (1973), using bipolar fine-wire electrodes, found that palatoglossus functions in a variable fashion during speech, assisting to move the tongue or reacting to tongue movements when the levator palati is contracting, and lowering the palate when the levator is inactive. Fritzell and Kotby (1976) in turn compared the activity of the levator palati with that of thyroarytenoideus. They are clearly different. During the rest position, the levator palati remains silent but thyroarytenoideus (as part of its respiratory function) is constantly active. Although both are intimately involved in speech, their latencies before utterances differ significantly, depending on the characteristics of the sounds produced. Lubker (1975) further enlarged upon studies of the palatal muscles with fine-wire electrodes. Of special interest were his findings of a temporal organization of speech, with considerable controlled variations among the various muscles involved.

Clinical application of palatal EMG is promising. Thus, Chaco and Yules (1969) proposed that a candidate for tonsilloadenoidectomy should have EMG of the soft palate because of possible velopharyngeal incompetence. Šurina and Jágr (1969) have made similar recommendations for cleft palate patients.

Palatal Activity During Sleep

Anch et al (1981), under the guidance of Professor E.K. Sauerland in Galveston, have provided us with the most lucid EMG findings from tensor palati during sleep. They were particularly interested in oropharyngeal patency in the Pickwickian syndrome and the pathogenesis of occlusive sleep apnea (OSA), whose pathomechanics are obscure (Sauerland et al, 1981a and b). They reported finding a striking and consistent reduction in EMG activity of the tensor with the onset of OSA. However, there was no association between the rises and falls of activity in tensor with rises and falls of activity in the genioglossus muscle both during sleep and awake periods.

PHARYNX

In view of the relative inaccessibility of the striated muscles of the pharynx, electromyography of the sphincters has not been widely attempted. Nonetheless, a considerable number of publications that make reference to muscular action have appeared on the process of swallowing. In most of this material, no actual objective recordings of the sequence of events inthe involved muscles have been reported. The superior constrictor's activity during respiratory phases has been investigated by Hairston and Sauerland (1981) with wire electrodes inserted into 10 subjects. During normal breathing, the muscle is silent in inspiration and shows considerable activity during expiration.

The resurgence of practical interest in the mechanism of swallowing and in the reparative surgery of the pharynx has made a bold approach to pharyngeal electromyography a necessity. Doty and Bosma (1956) performed EMG studies of reflex swallowing in the areas of the mouth, pharynx, and larynx of anesthetized monkeys, cats, and dogs. Broadbent and Swinyard (1959) then made similar studies during operations on anesthetized human patients with cleft palate. Car (1970) and Miller (1972a and b) are other workers who have made real contributions to the field of swallowing reflexes, using indwelling electrodes in individual oral and pharyngeal muscles of experimental animals.

Shipp et al (1970), in a study of patients who had had laryngectomies, found two types of swallowing patterns: the first, called "Type I," was observed when EMG sampling was conducted early in surgery before any major structures were altered or divided. This pattern consisted of a single large burst of activity from the inferior constrictor with a coincident inhibition of the preswallowing low level activity from the cricopharyngeus. The second pattern, labeled "Type II," was found in EMG sampling later in surgery and in all subsequent EMG procedures on laryngectomized subjects up to 3 years postoperatively.

The typical Type II muscle activity pattern showed both muscles simultaneously initiating activity during the first burst and the quiet segment, whereas for the second burst the cricopharyngeus followed the inferior constrictor by 60 to 180 ms. The second burst appears to be analogous to the pattern found in Type I or "normal swallow."

Type II swallowing patterns were similar in the awake and anesthetized conditions, although duration measures differed.

From the start of studies of the pharynx, it seemed necessary not only to examine reflex swallowing under anesthesia, but also swallowing under conditions that are as normal as possible. Therefore, various procedures were devised towards that end, and extensive dissections were performed to clarify the anatomy of the region. The first report (Dutta and Basmajian, 1960) dealt with anatomical studies in rabbits. It was followed by a report on the rabbit's pharyngeal and palatal muscles (Basmajian and Dutta, 1961a). The third report dealt with EMG studies on conscious, normal human beings (Basmajian and Dutta, 1961b).

In about half the series of rabbits, direct recordings with concentric needle electrodes were made of the electrical potentials of the soft palate and individual constrictors during reflex swallowing under anesthesia.

In the other half of the series, operative exposure was followed immediately by implanting into the three constrictors special indwelling, flexible, wire electrodes to be used for postoperative recordings. EMG testing was usually done after recovery from the operation—a delay of several days being the rule.

In some of the rabbits in which implants were used, simultaneous records were made also from the fleshy part of the soft palate *via* a concentric needle electrode. Active voluntary swallowing was induced by running some water from an "eye-dropper" onto the tongue of the rabbit.

In the studies of conscious normal volunteers, the electrodes were passed under direct vision through the open mouth, using a laryngeal mirror and clinical headset, when necessary. A special type of bipolar

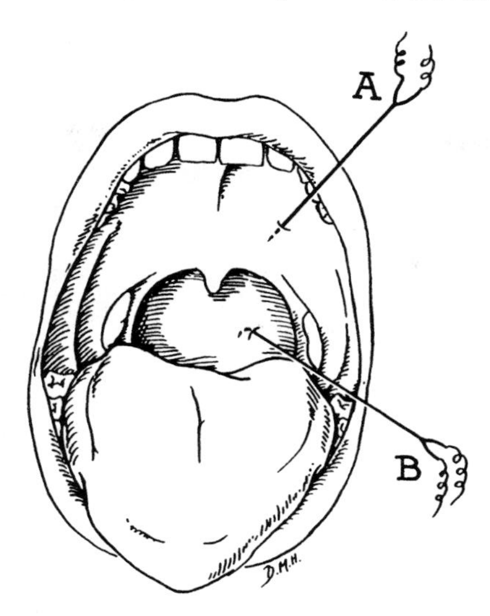

Figure 19.3. Diagram of special bipolar electrodes. (*A*) In soft palate. (*B*) In pharyngeal wall.

electrode was designed for use in these experiments. It consists of two fine surgical stainless-steel wires about 10 cm long, glued together, yet insulated from each other, by lacquer. This type of electrode has the advantage of strength, lightness, and flexibility (Fig. 19.3 and 19.4).

Except during obvious swallowing, in all the experiments on human beings, conscious rabbits, and anesthetized rabbits, there is little or no activity in any of the constrictors. That is, there is no resting tonus.

During swallowing each constrictor of the rabbit contracts for about ¼ second, whether it is part of reflex or of conscious swallowing (Fig. 19.5). The contraction of the superior constrictor begins simultaneously with that in the soft palate; that in the middle constrictor is delayed by

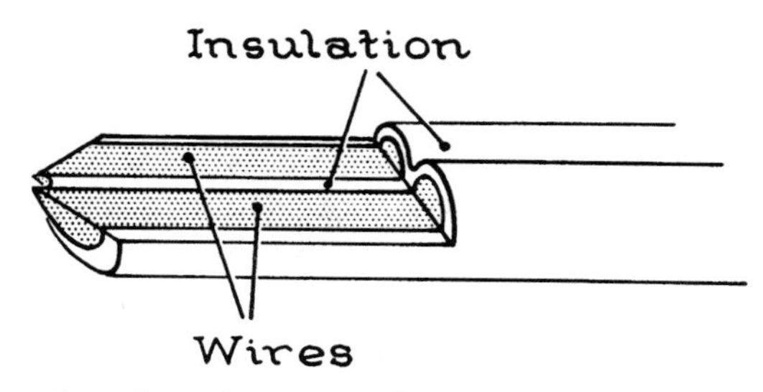

Figure 19.4. Greatly enlarged cutaway diagram of tip of special electrodes (cf. Fig. 19.3).

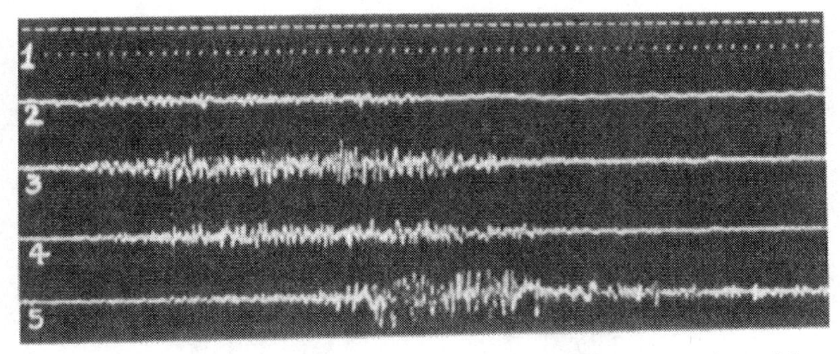

Figure 19.5. Simultaneous EMG records during swallowing in the rabbit. (*1*) Calibration: 10-ms intervals; amplitude, ca. 300 μV. (*2*) Soft palate. (*3*) Superior constrictor. (*4*) Middle constrictor. (*5*) Inferior constrictor. (Tracings slightly retouched to improve engraving.)

about 25 ms; and that in the inferior constrictor, by about 75 ms. The entire duration of activity in the constrictors lasts about ⅓ second. These figures must be accepted as broad generalizations of the detailed results (Basmajian and Dutta, 1961a).

The amplitude of activity in the three constrictors of the rabbit gradually increases and becomes maximal just beyond the middle of its total period of activity. Then the amplitude falls to the base-line due to rapid relaxation. It is important to note that the base-line of *nil* activity is maintained, except during swallowing, i.e., to repeat, there is no tonic activity of the palate and pharyngeal constrictor muscles. They act in an "all or none" fashion.

In the human volunteers, any activity picked up from these muscles following the insertion of the electrode apparently is a reflex "tightening up"; it can be abolished by the subject's voluntary relaxation. When the subject is relaxed and resting between swallows, the pharyngeal constrictors are inactive, as is the levator palati. During the sucking of water through a straw, all three constrictors remain silent, while—as might be expected—the levator palati is active. With each swallow, the duration of activity is very close to ½ second in each of the pharyngeal constrictors (Fig. 19.6).

Because the recordings in the human muscles were not simultaneous (as in the study on rabbits) the exact sequence of activity in man is not proved. However, there is no reason to suppose that it is different from that in rabbits.

Doty and Bosma (1956) described the EMG pattern of activity during swallowing in the muscles of the mouth, pharynx and larynx of monkeys, cats and dogs. They found a definite pattern or sequence of activity, i.e., with minor variations, the "schedule of excitation and inhibition among

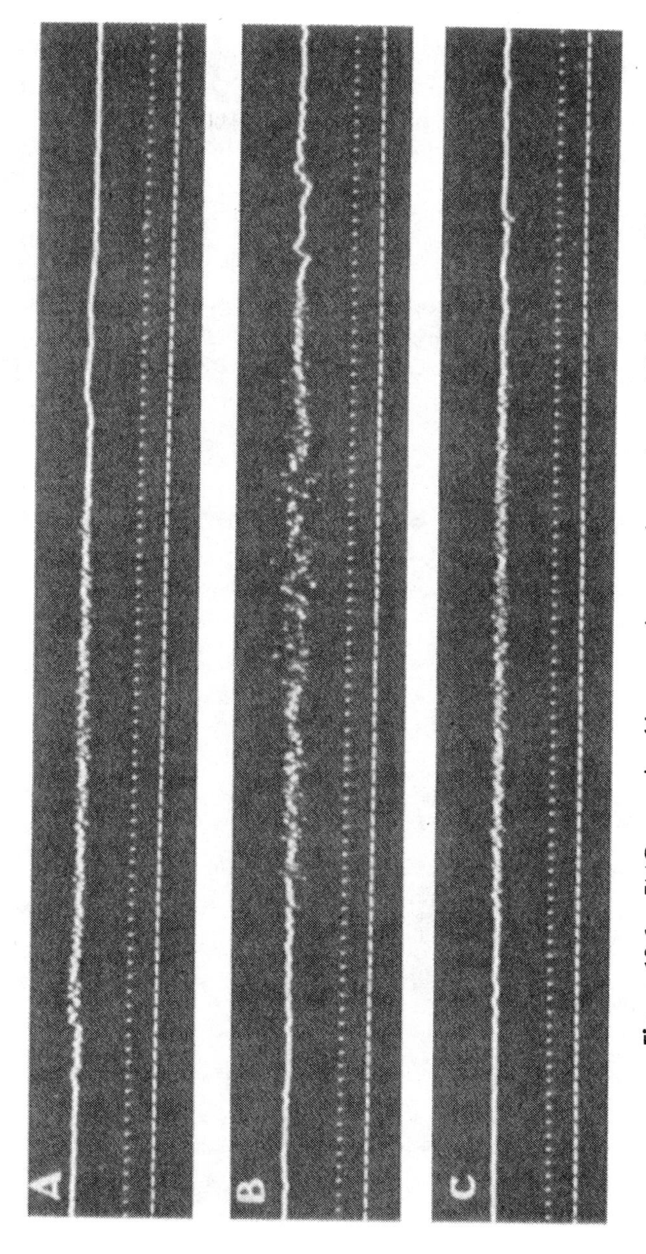

Figure 19.6. EMG records of human pharyngeal constrictors. (*A*) Superior. (*B*) Middle. (*C*,) Inferior.

the participating muscles was highly constant." No difference was found in reflex swallows evoked by various means, including stimulation of the superior laryngeal nerve, stimulation with a cotton swab, or the rapid squirting of water into the pharynx. A leading "complex," consisting of the superior constrictor, palatopharyngeus, posterior intrinsic muscles of the tongue, and various muscles attached to the hyoid bone, becomes active for 1/4 to 1/2 second to initiate the act. The middle constrictor, inhibited at first (as in our experiments), follows. The inferior constrictor is the last to become active, being deferred until the leading complex is nearly over (Fig. 19.7).

Rather similar findings in dogs have been reported by Kawasaki et al (1964) whose studies of air pressure gradients are particularly valuable.

Cricopharyngeus muscle has been the special concern of Levitt et al and

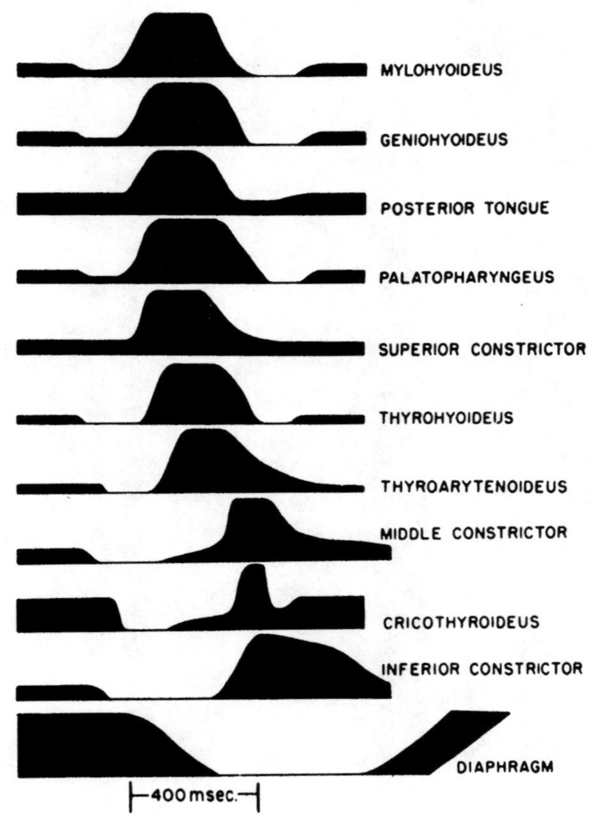

Figure 19.7. Doty and Bosma's (1956) schematic summary of EMG activity during swallowing in dogs.

Ogura (1965) of St. Louis, who have recently made the greatest contribution to its understanding. In dogs they found that these sphincteric fibers at the junction of pharynx and esophagus relax much of the time, contrary to the widely held view that they are tonically contracting. Bursts of activity do occur but usually in response to external factors such as breath-holding, straining, or stimulation of the hypopharynx. Murakami et al (1972) found EMG evidence of double-innervation (somatic and vagal) to the cricopharyngeus.

The striated muscle of the upper esophagus has been investigated by Arlazoroff et al (1972). There was steady "high grade" EMG activity in all subjects (detected with surface electrodes attached to a small inflated balloon); the activity increased further during the Valsalva maneuver.

LARYNX AND VOCAL CORD

As Professor Georges Portmann of Bordeaux and Paris pointed out in his Semon Lecture (published in 1957), the larynx has two functions. The first comprises respiration and the protection of the pulmonary apparatus, these being inseparable. The other is phonation or the production of sound. Surprisingly, to the present time only provocative results have been obtained by EMG for this vital function. Serra (1964) has given a thorough review of the various neuromuscular studies of this area.

Portmann and his colleagues first made some important contributions through their experiments on patients who had had operations that left the glottis exposed postoperatively. These investigators were able to insert needle electrodes directly into different parts of the vocal folds and the thyroarytenoid muscles.

During each expiration, the thyroarytenoid becomes very active, especially in the middle of expiration. At the beginning of inspiration, activity stops and does not reappear until expiration begins. Spontaneous and involuntary activity during expiration is quite independent of every effort of phonation.

Portmann's belief that during phonation the EMG oscillations have the same frequency as the sound emitted—strong support for the theory advanced by Husson (1950)—has been rejected widely (Rubin, 1960; Spoor and Van Dishoeck, 1960; Kirikae et al, 1962; Milojevic and Hast, 1964; Dedo and Ogura, 1965; and still others). According to Floyd et al (1957), closure of the glottis is caused by tonic coordinated action of the sphincteric muscles in which thyroarytenoid plays its part. They have not succeeded in reproducing in dogs a higher frequency of vibration of the cords by stimulation of the laryngeal nerves with graded electrical frequencies. On the other hand, there is evidence being produced by independent workers which shows that complete tetanic fusion does not occur in laryngeal muscles until stimulation frequencies as high as 400

Hz are used (Mårtensson and Skogland, 1964). Also Louis-Sylvestre and MacLeod (1968) have advanced the provocative idea that the human vocal muscle mechanism may be similar to that of asynchronous insect muscle. Thus, an electrodynamic servo-system, linked to vibration frequency, allows instantaneous modifications of the mechanical impedance sensed by the muscle, which is kept in its natural environment. These investigators believe vibration frequency follows without latency the changes in impedance parameters; they find the sign and amplitude of variations to be identical in both insect flight muscle and human vocal muscle.

Green and Neil (1955) noted that impulses in the recurrent laryngeal nerve of cats coincide with inspiration and EMG activity in the posterior cricoarytenoid. They found activity in the abductor muscles of the cords during inspiration alternating with activity in the adductors during expiration.

Under the general direction of Fritz Buchthal of the Institute of Neurophysiology at Copenhagen, Faaborg-Andersen completed a monumental study on the normal and abnormal muscular control of the vocal cords and published it as a monograph in 1957. In a long series of normal persons, he inserted needle electrodes under direct vision into the cricothyroid, vocalis, arytenoideus, thyroarytenoid, and posterior cricoarytenoid muscles, two at a time, and recorded the subject's phonation and EMG data simultaneously on a tape recorder.

At rest, there was a slight activity in all the adductors even if the breath was held quietly. This then is "postural activity," according to Faaborg-Andersen. It increased from the resting level during inspiration but was unchanged during expiration. At "rest," the abductors were very active by contrast, but during the inspiratory phase this activity decreased somewhat, although it was uninfluenced by expiration.

With phonation, an increase in electrical activity was found in all the adductor muscles. The change began 0.35 to 0.55 s before the audible sound. Unlike Portman, Faaborg-Andersen found that single unit potentials fell below the range of 20 to 30/s in basic frequency. This is much like ordinary skeletal muscles.

During phonation with increasing pitch there was no corresponding increase in the electrical activity of the adductor muscles. However, with increasing pitch the increase in the total electrical activity was marked, provided that the increase in pitch was in the same register, i.e., within the same octave. If the increase in pitch was accompanied by a shift in register, the change was only slight.

Thus Faaborg-Andersen does not support the theory that the frequency of vibration of the vocal cords during phonation of a tone changes directly with the frequency of nerve impulses and contractions of the muscle fibers. Faaborg-Andersen upholds the orthodox view that the vibrations of the vocal cord are passive and independent of the frequency of motor unit contractions.

With phonation of different vowels there was no apparent electromyographic change in Faaborg-Andersen's series (Fig. 19.8). With whispered voice or "silent speech," there was activity in the adductor muscles, but it was far less than with ordinary voice.

Buchthal and Faaborg-Andersen (1964) found that the average time between the onset of the increase in electrical activity in cricothyroid muscle and the onset of phonation is between $^1/_{10}$ and $^1/_5$ second. The interval may be considerably shorter for some sounds.

During a cough and during swallowing there was a considerable increase in the electromyographic potentials in all the adductors just before the onset of audible sound; conversely, the abductors relaxed during the cough.

The earlier morphological and EMG studies have been summarized in an article by Gomez Oliveros (1969). In recent years laryngeal electromyography has gained a much wider following, partly because of Shipp's introduction of improved techniques employing wire electrode methodologies (Shipp, 1968; Shipp et al, 1968; Shipp et al, 1970). Figure 19.9 illustrates his technique of inserting electrodes into the posterior cricoarytenoideus. For other muscles, the direct anterior route through the skin also may be used in conscious human subjects (Hirose and Gay, 1972, 1973; Kotby, 1975; Dejonckere, 1975). Dedo and Hall (1969), in experiments with dogs, have used wires and needle electrodes, finding them both reliable. Payne et al (1980) have described a new, miniature, bipolar, surface electrode which can be swallowed down to a position behind the posterior cricoarytenoideus and produces excellent results.

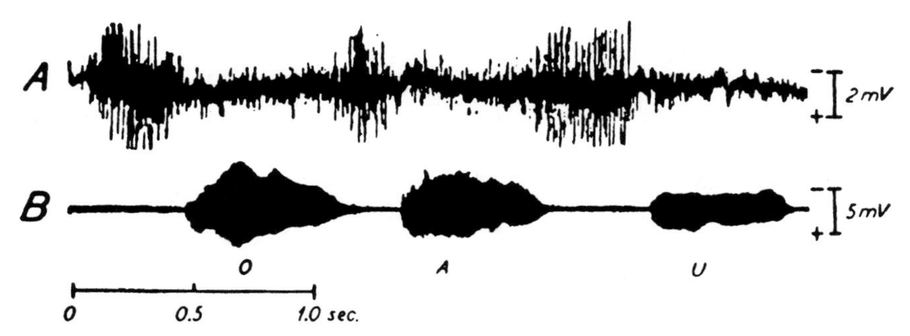

Figure 19.8. EMG records in thyroarytenoid (*tracing A*) and microphone recordings (*tracing B*) during phonation of vowels o—a—u at frequency of 200 Hz. (From K.L. Faaborg-Andersen, ©1957, *Acta Physiologica Scandinavica*.)

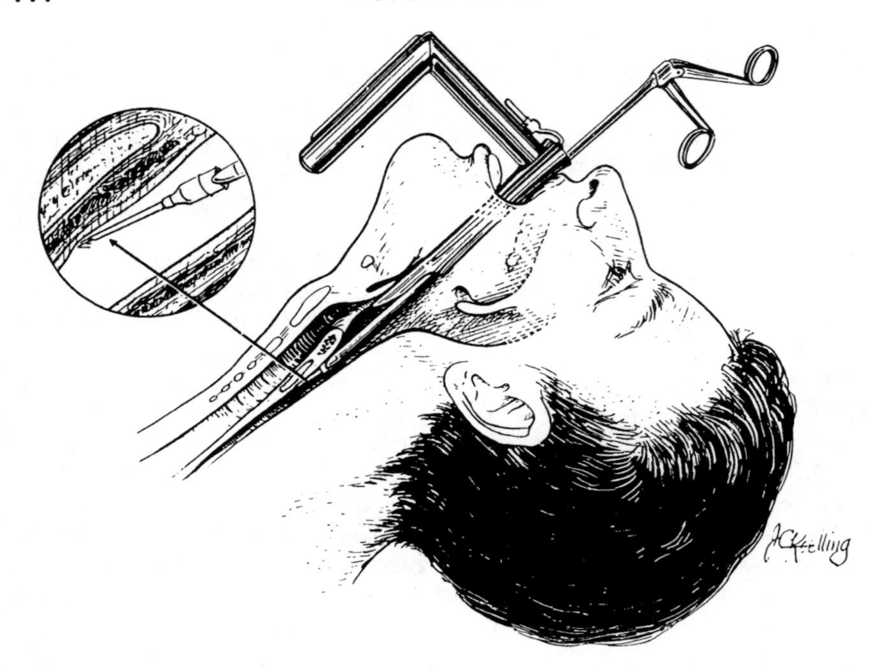

Figure 19.9. A technique for introducing bipolar wire electrodes into the posterior cricoarytenoideus muscle through a laryngoscope. (From T. Shipp et al, ©1970, *Journal of the Acoustical Society of America.*

Hiroto et al (1967) of Kurume, Japan, found little or no activity in posterior cricoarytenoideus during phonation while the cricothyroid, thyroarytenoid, lateral cricoarytenoid, and interarytenoid muscles were very active. The change in electrical activity begins before the onset of speech sounds (with occasional exceptions for voiceless consonant syllables). The time interval from the onset of EMG changes to the onset of sound was least in cricothyroid and for consonant syllables.

Hirose and Gay (1972, 1973) of the Haskins Laboratories in New Haven found that the posterior cricoarytenoideus participates in the production of voiceless consonants but not voiced consonants. A reciprocal pattern occurs in the interarytenoideus. They found three types of vocal attack that are characterized by the coordinated actions of the abductor and adductor muscles which act reciprocally for each type of attack.

The group associated with Thomas Shipp showed that with normal adults there is greater airflow and greater activity in cricothyroid and interarytenoid muscles in modal phonation, compared with those in vocal fry—the nonpathologic phonational register encompassing a range of frequencies below those of the modal register (McGlone and Shipp, 1971). Frequency through the vocal range is changed chiefly by the activity in cricothyroid and thyroarytenoid muscles, and their activity is highly correlated with measures of subglottal pressure. However, no systematic correlation occurs between (1) activity in posterior cricoarytenoid and interarytenoid muscles and air flow measures, and (2) frequency changes (Shipp and McGlone, 1971).

A series of papers from Oslo in 1970 added to our understanding of laryngeal function (Kotby and Haugen, 1970a–d). They described a basic postural ('resting') activity in the posterior cricoarytenoid which fluctuates with breathing, rising most with deep inspiration. In phonation, it becomes very active as it does during sphincteric actions. Kotby and Haugen suggest that the concept of antagonists—abductors *vs.* adductors—is false.

The elegant study of Gay, Strome, Hirose and Sawashima, who represent the Universities of Connecticut, Harvard, and Tokyo, is especially noteworthy (Gay et al, 1972). Both peroral and percutaneous insertion of wire electrodes was done in five subjects with apparently meticulous technique and computer analysis. They, too, showed that increases in fundamental frequency on phonation were accompanied progressively by increases of EMG activity in cricothyroid muscle—but they were able to show this *in vocalis also*. Moreover, to a lesser extent the increase occurs in *all* the other intrinsic muscles too, but the posterior cricoarytenoid increase comes at the highest pitch level.

Gay et al, disagreeing with Faaborg-Andersen (cited above), find that the posterior cricoarytenoid can also act as a tensor of the vocal folds, following the curve of activity of the cricothyroid. They warn, however, that there are individual differences between subjects—which also may account for earlier confusions in laryngeal EMG research. They were unable to demonstrate a strict linear relationship of the activity in cricothyroid and vocalis to fundamental frequency, and they discounted adductor muscle actions as "probably secondary."

Changes in intensity seem to be related more to expiratory muscle force than laryngeal factors; there may be "a trading relationship between laryngeal and subglottal forces depending on whether a specific frequency level is aimed for or not" (Gay et al, 1972).

SOME CLINICAL APPLICATIONS OF LARYNGEAL EMG

Knutsson et al (1969) and Sutton et al (1972) demonstrated the possibility of studying the characteristics of individual motor units in various laryngeal muscles by means of percutaneous and peroral electrodes. Kotby and Haugen (1970) and Dedo (1970, 1972) took up the problem of diagnosing various vocal fold mobility disorders and paralysis. Ueda et al (1971, 1972), Ohyama et al (1972), and Lauerma et al (1972) investigated experimental artificial voice production, while Shipp (1970) studied the function of the pharyngeal inferior constrictor and crico-pharyngeus during alaryngeal voice production in laryngectomized patients.

Stuttering Larynx

Combining fiber optics and EMG techniques, Freeman and Ushijima (1974) studied stuttering in three patients. Recordings were made with percutaneous fine-wire electrodes in four of the five intrinsic laryngeal muscles. The so-called Wingate hypothesis was justified: thus laryngeal activity patterns differ between the stuttering moments and when the patients could be induced to speak fluently. The abnormal patterns at the stuttering moments reveal (1) disruption of the normal reciprocity of the abductors and adductors, (2) disruption of the normal synchrony between adductors, and (3) generally higher activity in the intrinsic muscles.

Laryngeal Reflexes

A number of experimental studies employing electromyography have been reported. Thus, Kurozumi et al (1971) have shown the existence of bilateral motor innervation from the medulla oblongata to the laryngeal muscles. Murakami and Kirchner (1971, 1972) investigated the effects of different frequencies of electrical stimulation on the various muscles and described the EMG activity of intrinsic muscles during different reflexes and laryngeal closure. Abo-El-Enein and Wyke (1966, 1969) and Wyke (1968) described the effects of anesthesia and nerve-sectioning on the myotatic reflexes. The laryngeal initiation of swallowing via sensory endings in the mucosa of the larynx has been described by Storey (1968a and b) of Toronto.

Laryngeal Muscle Training

Hardyck et al (1966) first reported that EMG feedback could be used (by the motor-unit training technique) in laryngeal muscles for treating persons whose silent reading was hampered by subvocalization. These matters are discussed in Chapter 6.

CHAPTER 20

Muscles of Mastication, Face, and Neck

MUSCLES OF MASTICATION

The first concerted effort to apply electromyography to problems of orthodontics and normal temporomandibular physiology was made by a dentist, Robert Moyers, while working as a graduate student at Iowa and as Professor of Orthodontics at the University of Toronto in 1949 (where another early electromyographer—J.V.B.—was at work). Following Moyer's lead, other dental scientists published a number of good studies, although some of them have disagreed with certain details of his earliest work (1949 et seq). The following account is a review of the published reports in the field, and a number of special sources including the research of graduate students.

Also valuable are the newer reports of work with inserted electrodes by Ahlgren (1966, 1967) and Møller (1975, 1976) in Sweden, Munro (1974) and his colleagues in Sydney, and Vitti (1969a and b, 1970, 1971 et seq with others) in Brazil. The good fortune of having Munro and later Vitti as visiting colleagues with the Basmajian group through 1972 to 1974 greatly enhanced general understanding of mandibular and facial mechanisms. Another area that has emerged with vigor is that of mandibular reflexes. A discussion of the various reflexes now makes up a later section of this chapter.

Mandibular Rest Position

While the biomechanics and orthodontics of the mandibular rest position must remain the province of dental clinical research specialists, a few words on the contributions of electromyography are appropriate here. Møller (1975, 1976) correctly ascribed the confusion in the literature to the wide differences in equipment and technique, and he pleads for better quantification, as did Frame et al (1973).

Is the rest position controlled actively or determined passively? There now appears to be no doubt that in deliberately relaxed but fully upright postures, even with the lips together (but the teeth not together), there is virtually no EMG activity (Vitti and Basmajian, 1975), except with occlusal interference (Funakoshi et al, 1976). However, where the force of gravity acts on the jaw of an unrelaxed person, low-level activity is the rule in the anterior part of temporalis; greater alertness and functional disorders both raise the level substantially (Møller, 1976). During sleep, intermittent increases and decreases of activity occurred throughout the night in the study of Fuchs (1975), as might be expected. However,

Fuchs was able to elicit the patterns of change, and his paper should be consulted for details.

Temporalis

RESTING TONUS

As noted above, wire electrodes and the best surface-electrode techniques have clearly shown that at normal and deliberate rest (with the lips but not the teeth occluded) there is no activity in temporalis. However, the controversy began when Moyers (1949, 1950) reported for normal subjects a "remarkably even state of *tonus*" in all three parts of the muscle when it is at rest, stating also that the normal maintenance of mandibular posture is shared by all the parts. Carlsöö (1952), although he agreed that temporalis is the main postural muscle in the habitual rest position, insisted that the posterior part of temporalis was the more important part in this position. MacDougall and Andrew (1953) also agreed that resting postural tonus was obtainable, but they were less precise.

Latif (1957), while working in Basmajian's Toronto laboratory, made a definitive EMG study of both temporalis muscles in 25 normal teenage children. In the physiological resting position of the mandible in the upright subject, both the anterior and posterior fibers of temporalis were continuously active in almost all the subjects. However, this activity was much greater in the posterior fibers (Fig. 20.1), as hinted at by Carlsöö's

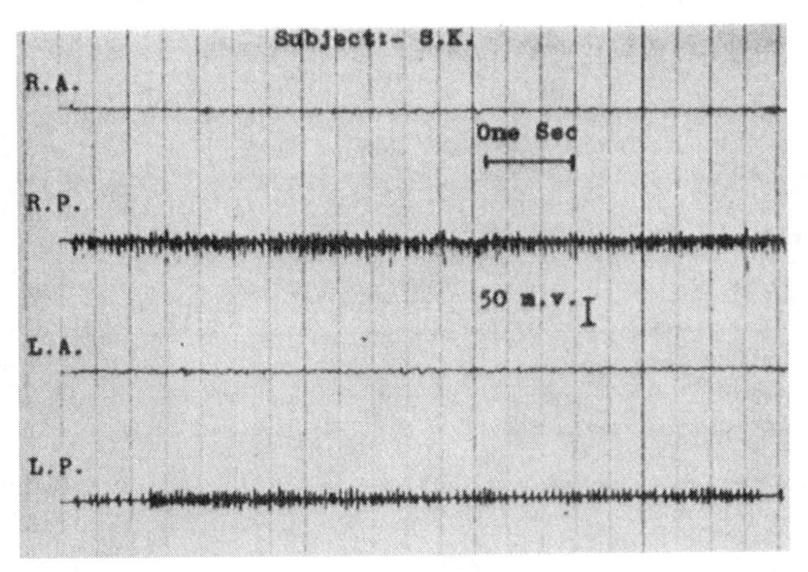

Figure 20.1. EMG records of anterior and posterior parts of right and left temporalis muscles at rest show much greater postural activity in posterior parts (*R.P.* and *L.P.*) compared with anterior (*R.A.* and *L.A.*).

earlier findings and in contradiction to those of Moyers. In the same year, Latif's finding was duplicated independently by Kawamura and Fujimoto (1957) of Osaka. However, with coaxial needle electrodes in the anterior, middle, and posterior part of the temporalis of 57 patients, Vitti (1969a) found no resting postural tonus in the majority, and later Vitti and Basmajian (1975, 1977) thoroughly documented this finding in children and in adults. Vitti found some relationship between the state of dentition and the presence of activity in the posterior fibers. His similar study (Vitti, 1969b) of retraction of the angle of the mouth showed activity in the posterior fibers in most subjects and the reverse for the middle and anterior fibers. Christensen et al (1969) of Copenhagen and Osaka indicate that actually there may be four distinct functional areas of the temporalis in cats.

END-TO-END OCCLUSION (INCISOR BITE)

Latif found all parts of temporalis were active, the greater activity being somewhat more frequently in the anterior fibers (40% of muscles), but in many (22%) the posterior fibers predominated, while in a third of the muscles the activity was equal throughout (Fig. 20.2). In several muscles, in contrast, the temporalis was inactive with incisor bite, even though this condition was considered to be the norm (quite erroneously) by Keith (1920). These results of Latif confirmed the findings of MacDougall and Andrew (1953). Vitti (1970) reported that the anterior

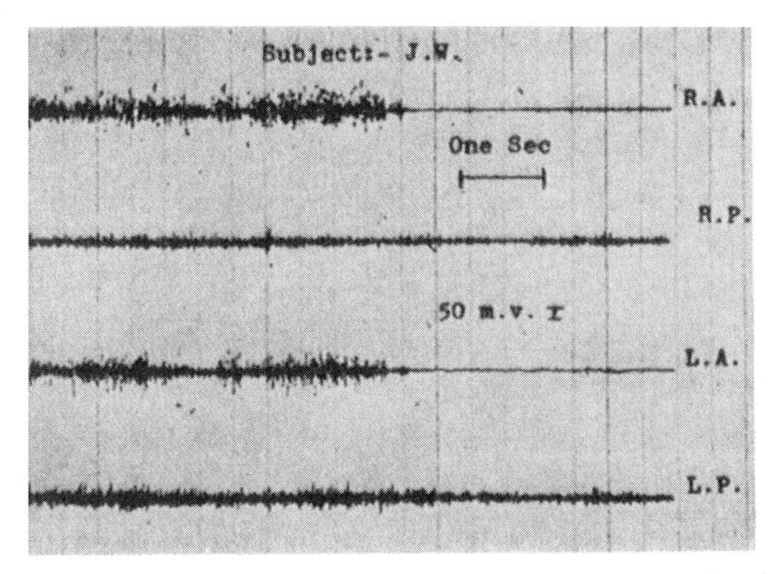

Figure 20.2. EMG records of temporalis muscles during end-to-end occlusion from right anterior (*R.A.*), right posterior (*R.P.*), left anterior (*L.A.*), and left posterior (*L.P.*) parts (From A. Latif, ©1957, *American Journal of Orthodontics.*)

and middle fibers were active with normal dentition and incomplete dentition with molar support; in edentulous patients, all three parts were active during incisive bite.

MOLAR OCCLUSION

All the fibers of the temporalis showed marked activity in all subjects (Latif, 1957; Ahlgren, 1967; Vitti, 1970; Vitti and Basmajian, 1975, 1977), as would be expected (Fig. 20.3). This is the chief function of the temporalis.

RETRACTION OF THE JAW

A universal finding was a marked activity in the posterior fibers of temporalis with lesser activity in the anterior fibers during the drawing back of the jaw from the protruded (protracted) position. This is in keeping with the accepted teaching (Latif, 1957; Vitti, 1971; Vitti and Basmajian, 1975, 1977).

PROTRACTION

Latif's findings were not in agreement with the opinion of McCollum (1943) and the findings of Moyers that the anterior fibers are active during protraction. He found *no* activity, as did Carlsöö et al, Woelfel et al (1960), and Vitti (1971). Indeed the activity even dropped from the resting tonus (Fig. 20.4). Apparently the temporalis shifts the burden of

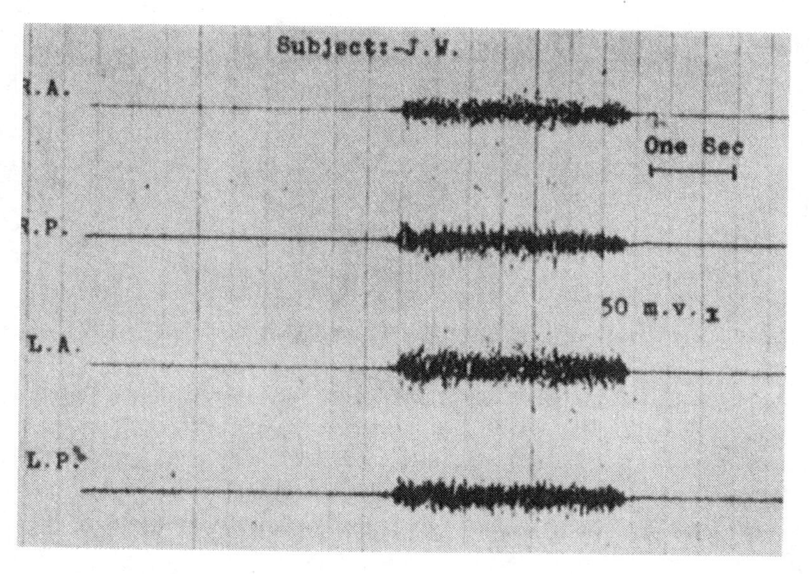

Figure 20.3. EMG records of temporalis muscles during molar occlusion from right anterior (*R.A.*), right posterior (*R.P.*), left anterior (*L.A.*), and left posterior (*L.P.*) parts. (From A. Latif, ©1957, *American Journal of Orthodontics.*)

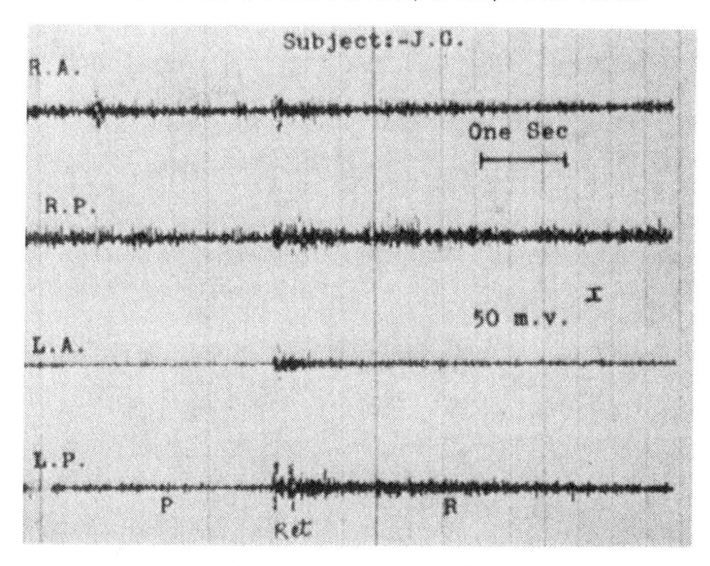

Figure 20.4. EMG records of temporalis muscles during protraction of mandible (*P.*) and at rest (*R.*). *Ret*, retraction. R.A. and R.P., from right anterior and posterior; L.A. and L.P., from left. (From A. Latif, ©1957, *American Journal of Orthodontics.*)

supporting the jaw to the muscles that protrude it, chiefly the lateral pterygoids.

LATERAL MOVEMENTS

Moyers' findings that the temporalis abducts the mandible were clearly confirmed by the observation that this action is almost universal (Fig. 20.5). In repetitive side-to-side movements, first one temporalis and then the other acts, each pulling the mandible only to its own side (Latif, 1957; Vitti, 1971; Vitti and Basmajian, 1975, 1977).

DEPRESSION

During ordinary opening of the mouth, temporalis is inactive (MacDougall and Andrew, 1953, Latif, 1957; Vitti, 1971). When the mandible is forcibly depressed (maximal opening of the mouth), some irregular muscle potentials do appear, suggesting that temporalis acts then as a protector against dislocation of the jaw.

The following paragraph summarizes the functions of temporalis. It maintains mandibular posture in the physiological resting position, the posterior fibers taking a more active part than the anterior. Its chief function is molar occlusion, and it is an ipsilateral abductor (and therefore a contralateral adductor) of the mandible. During maximal opening of the mouth, the temporalis acts as a preventative to dislocation, but it plays no role either in ordinary opening of the mouth or in protraction

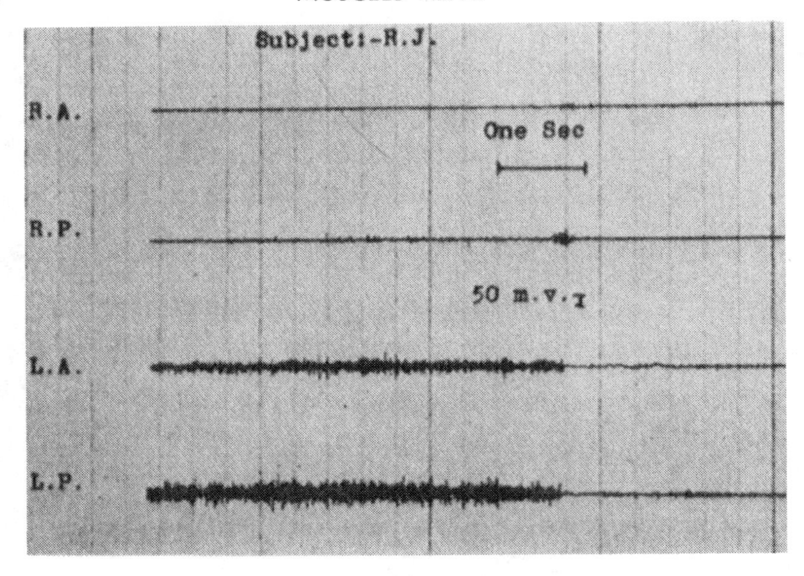

Figure 20.5. EMG records of temporalis muscles during "left lateral position" of mandible. Note greatest activity in left posterior (*L.P.*) fibers, and least in right anterior (*R.A.*). (From A. Latif, ©1957, *American Journal of Orthodontics.*)

of the jaw. In end-to-end occlusion (incisor bite) the anterior fibers are more active. The temporalis retracts the protruded jaw, the posterior fibers being especially active (Vitti and Basmajian, 1977).

Finally, Biggs et al (1968) showed that the EMG activity of monozygotic twins was similar in the duration of activity in their temporalis (and masseter) muscles.

MASSETER

As would be expected, during forceful centric occlusion the masseter muscle is very active (Pruzansky, 1952; Moyers, 1950; Ahlgren, 1967; Vitti and Basmajian, 1975, 1977). During chewing movements the maximal activity occurs in the masseter at about the time the jaw reaches the temporary position of centric occlusion which accounts for about a fifth of the chewing cycle (Gibbs, 1975). Masseter is not an important muscle in the habitual resting position (Carlsöö, 1952; Vitti and Basmajian, 1975, 1977), although it does show some activity in its superficial part during protrusion (Carlsöö, 1952; Vitti and Basmajian, 1977) and with increasing weights on the chin (Carlsöö, 1952) (Fig. 20.6). It also acts in certain types of ipsilateral and contralateral movements of the mandible (Vitti and Basmajian, 1977). MacDougall and Andrew found that its deep fibers are occasionally active during retraction.

Hannam (1976) used on-line computer sampling and cross-correlation analysis of the bilateral myoelectric activities of the masseter muscles. Cross-correlation showed no significant patterns of difference from bite to bite when the 10 subjects were taken as a whole. However, individuals showed some slight differences. In almost half his sample, the muscle contralateral to the side chewed upon started to contract first, while the rest acted in miscellaneous ways.

In awake rabbits, Schärer and Pfyffer (1970) were able to stimulate cortical centers while recording from the masseter through wire electrodes. The results were identical to normal chewing in masseters.

Medial Pterygoid

During simple protraction, Moyers always found strong activity with needle electrodes in medial pterygoid. Others agree (Vitti and Basmajian, 1977; Fortinguerra and Vitti, 1979). This decreased slightly if the mandible was depressed while being protracted. Unilateral contractions of the muscle accompanied (caused?) contralateral abduction of the chin and were particularly impressive when there was an added element of protraction (Moyers, 1950; Carlsöö, 1952). In the rabbit and rat, O'Dell et al (1969) found with wire electrodes that medial pterygoid opposes forced lateral mandibular movement.

Lateral Pterygoid and Digastricus

During mandibular depression, Moyers (1950) found that the first large potentials to appear are in the lateral pterygoid, and he believed that the activity reaches its peak even before the other muscles become active (Fig. 20.7). Furthermore, it continues throughout the movement. Woelfel et al found the lateral pterygoid very active in contralateral excursions, uncontrolled openings, and protrusions of the mandible. However, it was inactive during hinge openings of up to 1 cm. Apparently its function is to draw forward the articular disc in the temporomandib-

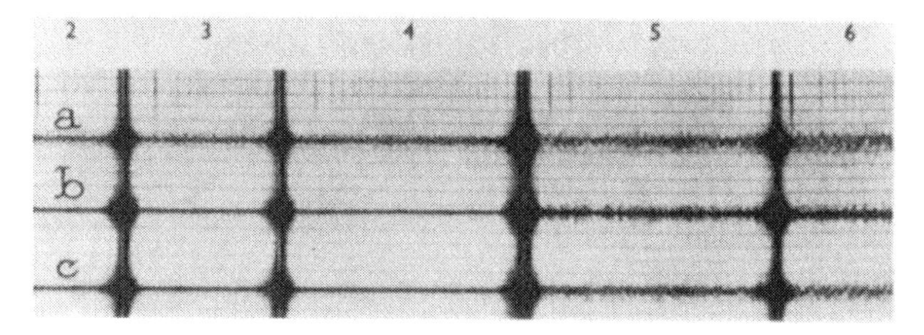

Figure 20.6. EMG records from posterior part of temporalis (a), superficial part of masseter (*b*) and medial pterygoid (c), with progressively increasing loads starting at "2." (From S. Carlsöö, ©1952, *Acta Odontologica Scandinavica*.)

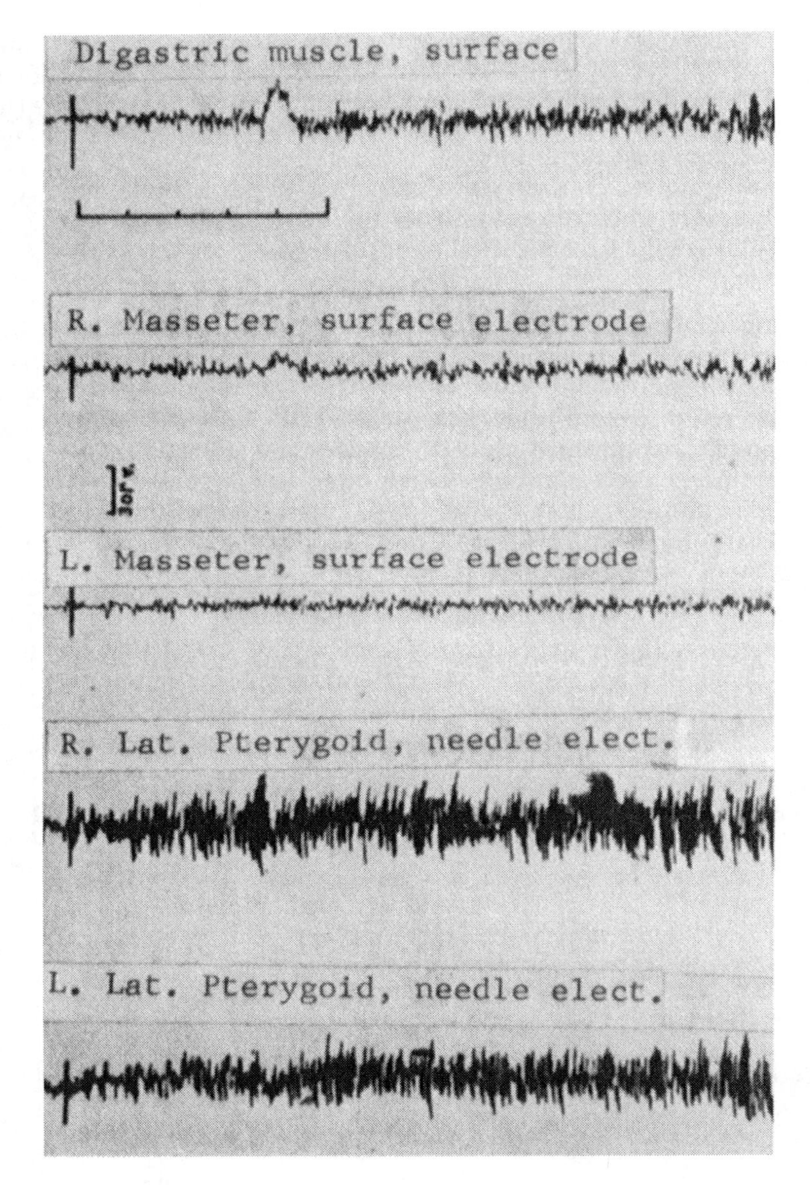

Figure 20.7. EMG records of mandibular depression. (From R.E. Moyers, ©1950, *American Journal of Orthodontics.*)

ular joint, along with the head of the mandible. In rhesus monkeys (which have the same morphological arrangement as man in this muscle) McNamara (1973) found that the inferior head of the lateral pterygoid is active only during opening and protrusive movements. The superior head was active only during such closing movements as chewing, swallowing, and clenching of the teeth. Supporting McNamara, Grant (1973) insists that the muscle is two muscles, both functionally and structurally.

Gross and Lipke (1979) and Juniper (1981, 1983) confirmed this is the case for human beings in a study with inserted fine-wire electrodes. Also using wire electrodes (but not in separate parts of the muscle) Lehr and Owens (1980) disagreed because there was consistent activity for both elevation and depression of the mandible. Owens et al (1975) had earlier reported that the silence of the lateral pterygoid coincides with the elevating and retruding of the mandible into the centric position. Lehr and Owens (1980) summarized the general activities of the whole muscle as:

1. The lateral pterygoid muscles are the prime movers in the protrusive movements, whether in an isolated protrusion or as an initiator for the incisor clench.
2. Retrusion involves only minimal activity, and it is not clear as to whether this activity is one of a mechanical stimulation, or a reflex pattern due to an occlusal abnormality.
3. Isolated right or left transversions rely heavily on the contralateral lateral pterygoid muscle for the total movement. The ipsilateral activity previously reported is not consistently present. This finding casts doubt on the function of this muscle as a stabilizer for the condyle during ipsilateral transversion.
4. The molar clench does not involve the lateral pterygoid muscles to a significant degree.
5. The lateral pterygoid muscles are partially responsible for the maximal depression of the mandible.
6. In the chewing cycle the lateral pterygoids are bilaterally active, and this activity is both in alternation with and minimally overlapping the elevator muscles. The sequence of lateral pterygoid activity in the chewing cycle is lead by the ipsilateral lateral pterygoid muscle.

During mandibular depression, Moyers wrote that the digastric muscle comes into action after the lateral pterygoid and is not as important. However, its action is essential for the maximum depression of forced or complete opening of the mouth. However, with modern needle-electrode techniques, König et al (1978) found consistent activity from the onset of opening, increasingly marked with resistance. In 20 normal subjects they found practically no activity during jaw closing. As soon as the mandible was being raised, there was a decrease in the digastric muscles activity.

During the incisor bite, right and left molar bite, the digastric muscle showed slight to moderate activity in the majority of the individuals, and the activity always occurred in antagonism with the agonist muscles, showing that the digastric muscles are usually active, only in the opening phase, which confirms the findings of some authors.

In the protraction of the jaw with and without occlusal contact, their findings showed moderate activity of the digastric muscles in almost all the cases, which supported earlier findings.

In the retraction of the jaw from protraction, a negligible (almost *nil*) activity was observed in some individuals. It is possible that this activity occurred due to slight depression of the jaw. When the retraction of the jaw was performed from the normal position, practically all the individuals showed marked activity. In this movement, the effort to retract the jaw was greater than when the retraction was performed starting from protraction and the jaw depressed; on the other hand, during the return of the jaw from retraction to normal position, the muscles were inactive, except in one case in which the activity was classified as negligible.

During lateral movements to either side, König et al found marked activity in most of the individuals in both muscles. This activity observed in the digastric muscles can be interpreted as resulting from the slight depression and protrusion of the jaw.

The right and left digastric muscles do not function as individuals. They contract bilaterally with greatest activity during uncontrolled openings and retrusions of the mandible (Woelfel et al, 1960). Munro (1974) agreed but added a new dimension: when tooth contact occurs during elevation of the mandible in chewing, the posterior belly contracts without activity in the anterior belly. Apparently this is related to positioning of the hyoid bone, not the jaw.

Other work on the digastric muscles has dealt with their general functions—e.g., coughing, breathing, and swallowing (Lesoine and Paulsen, 1968; Hrycyshyn and Basmajian, 1972) and mandibular reflexes (Munro and Basmajian, 1971). Coughing and swallowing strongly recruit the activity of the digastric muscles.

König et al (1978) also found substantial activity in swallowing, pointing out that the hyoid bone rises to its highest position during the activity and for most of the period that dental contact is achieved.

Coordination during Chewing

Using subcutaneous wire electrodes, Ahlgren (1966) studied the muscular coordination and EMG activity during chewing of peanuts and of chewing gum. Considerable variation of patterns from subject to subject was a salient finding. The commonest pattern during gum chewing (in

44%) was one in which the ipsilateral temporalis contracts first, then the contralateral temporalis and both masseters contract simultaneously. Vitti and Basmajian (1977) also found marked activity in the medial pterygoid muscles bilaterally. The peak of activity and the ending of activity occurs simultaneously most of the time. With peanut chewing, the commonest pattern was both masseters and temporalis contracting simultaneously.

Comparing the activity of masseter and temporalis muscles while chewing hard and soft foods, Steiner et al (1974) found marked variations between normal subjects. Masseter always led the action. With hard foods (carrots) the delay before recruitment of temporalis was shorter, and the total activity was considerably greater.

In a series of experiments with dogs, Vitti (1965) found a constant activity of temporalis, masseter, and medial pterygoid during mandibular elevation in chewing towards either side; there was a decrease with "incisive chewing," in which the greatest activity was in the anterior and middle fibers of temporalis, middle and deep part of masseter, and medial pterygoid. Unlike Vitti's finding in dogs, bats (*myotis lucifugus*) have clear alternations of activity between the muscles of the two sides (Kalen and Gans, 1970).

In freely feeding domestic cats, Gorniak and Gans (1980) found a "working side" and a "nonworking side." The adductors of the working side were generally most active while chewing foods of tougher consistency. The digastrics and lateral pterygoids reflected vertical and lateral displacements.

Summary of Mandibular Movements

The various movements of the jaw are produced by cooperative activity of several muscles bilaterally or unilaterally. Mandibular elevation is performed by the temporalis, masseter, and medial pterygoids and depression by the lateral pterygoids and digastrics. The digastrics show their greatest activity in forceful opening of the mouth at the limit of depression of the mandible. Lateral movements are performed by the ipsilateral temporalis and masseter and the contralateral medial pterygoid (and, to a lesser extent, the lateral pterygoid). Protraction is performed by the medial and lateral pterygoids while retraction is by the temporalis, chiefly its posterior fibers, and perhaps the deep part of masseter.

Mandibular Reflexes

Electromyographic investigations by several observers have shown that following tooth contact in man, there is an inhibition of the mandibular elevator muscles (Ahlgren, 1969; Beaudreau et al, 1969; Griffin and Munro, 1969; Kloprogge, 1975; Nagasawa et al, 1976; Lund, 1976). It is believed that this inhibition is induced reflexly by stimulation of periodontal end organs. Griffin and Munro (1969) found that, at the time of inhibition of the mandibular elevators, the lateral pterygoid muscle was inactive, whereas localized activity of the digastric muscle occurred, mainly towards the end of the inhibitory phase of the mandibular elevators. They also found localized activity of the digastric muscle with inhibition of the masseter muscle after the tapping of a single tooth; these effects were abolished by local anesthesia of the tooth. Beaudreau et al (1969) also found a motor pause in the temporalis and masseter muscles following the hitting of teeth, but the digastric muscle in their series also exhibited a motor pause at the time of inhibition of the mandibular elevators.

Munro and Basmajian (1971) investigated the six mandibular muscles with fine wire electrodes in normal persons performing the open-close-open cycle. A microphone applied to the skin over the zygoma was used to determine the time of initial tooth contact.

A period of inhibition of 12 to 20 ms mean duration was found in each mandibular elevator, commencing after a mean interval of 10 to 14 ms from initial tooth contact. In each cycle the inhibitory periods of the six mandibular elevators were found to be almost synchronous. The mean duration of the overlap of inhibition of the mandibular elevators was 11.9 ms, commencing after a mean interval of 14.6 ms from initial tooth contact.

In a further five subjects the activity of the anterior temporalis muscles was recorded with that of the anterior bellies of the digastric muscles. The activity of the digastric muscles at the time of inhibition of the mandibular elevators was analyzed. No activity was seen in 13% of digastric muscles at this time; a further 25% showed activity continuing through part or all of the period of inhibition of the mandibular elevators. In the remaining 62% of cycles there was a localized burst of digastric muscle activity at this time, usually either throughout or towards the later end of the period of inhibition of the mandibular elevators. It has previously been shown that the lateral pterygoid muscle is inactive at this time.

This pattern of muscular activity would be expected to result in jaw opening (or at least in a reduction in occlusal pressure) in the interval from 15 to 27 ms after initial tooth contact.

Quinn et al (1982) recently described an "action tonic stretch reflex" of masseter muscle and evidence that it can be voluntarily inhibited.

A lateral jaw movement reflex has been described by Lund et al (1971) and central coordination of rhythmical mastication by Dellow and Lund (1971). Anderson and Mahan (1971) at Emory University described the interaction of receptors in tooth pulp and periodontal ligaments in the jaw-depression reflex. Klineberg et al (1970) emphasize the role of mechanoreceptors in the temporomandibular joint capsule in the reflex control of chewing, while Bratzlavky (1972) explains the reflex pauses in the elevators entirely on spindle afferent discharges. Further work has been done by many investigators on the reflex activities of the mandibular elevators to elucidate postural and muscle-spindle neuromotor functions (see for example: Clark and Wyke, 1965; Lamarre and Lund, 1975; Övall and Elmqvist, 1975; Godaux and Desmedt, 1975a and b; Amano and Funakoshi, 1976; Ongerboer de Visser and Goor, 1976; Hagbarth et al, 1976; Widmalm and Hedegård, 1976; van Steenberghe, 1981; Felli and McCall, 1979; van der Glas and van Steenberghe, 1982). A most elegant monograph by Widmalm (1976), his thesis at the University of Göteborg, Sweden, offers an excellent review and illustrations (Fig. 20.8).

MUSCLES OF FACIAL EXPRESSION

Systematic EMG of the muscles of expression was badly neglected until a few years ago, although clinical electromyographers are constantly concerned with facial palsies. Worthy of at least a brief mention here is the ongoing study in various medical centers of the "postauricular response," the EMG reflex following acoustic stimulation recorded from the postauricular group of facial muscles, e.g., the report of De Grandis and Santoni (1980). (Orbicularis oculi has been investigated with the eye muscles; see p 478.)

Covert Emotions and the Facial Muscles

It is increasingly well known that the human facial musculature lacks the postural reflex mechanisms of the limbs where muscle spindles and the γ-loop respond to stretches induced by gravity (Hultborn, 1976; Bratzlavsky, 1981). However, facial muscles respond both visibly and invisibly to emotions in quite predictable ways which are becoming an important part of behavioral research in both the U.S. and Japan (Ekman, 1973; Schwartz et al, 1976; Sumitzuji et al, 1977; Fridlund and Izard, 1982). The responses are under exquisite controls, a great part of which responds to normal and distorted emotional states. The use of EMG techniques to diagnose and to study covert states of affect is becoming quite important. For an excellent discussion of the state of its art and science, see Fridlund and Izard's recent reviews (1982, 1983).

The techniques developed to date indicate that low-level facial EMG activity will become a useful tool for studying a wide range of emotion states and, in the hands of experts, for diagnosing and tracking depressive disorders. The multivariate pattern classification which is basic to the method appears to permit the disentanglement of arousal states from those due to distinct emotions.

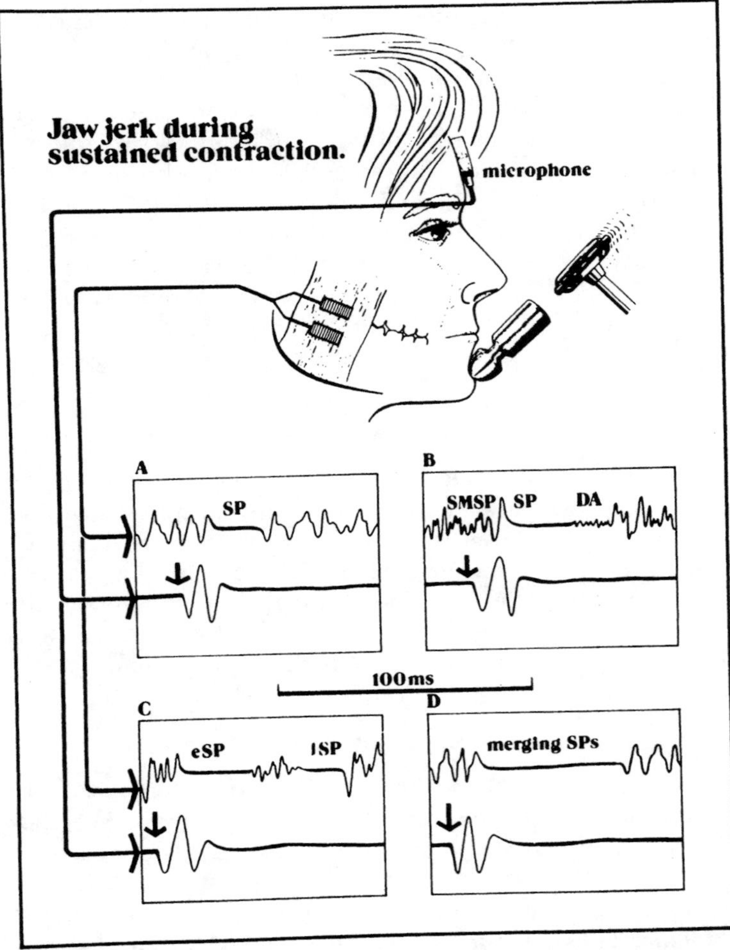

Figure 20.8. Widmalm's method of eliciting jaw jerks during sustained contraction of masseter.

Mentalis

Mentalis muscle, which wrinkles the chin, was the subject of an investigation primarily concerned with its reactions during sleep. Hishikawa et al (1965) of Osaka showed a fairly continuous "tonic" activity of mentalis in man, except during deep sleep, when it relaxes completely for short periods. More concerned with dental problems, Schlossberg and Harris (1956) described considerable activity in mentalis muscle during swallowing, as did Jacob et al (1971). Isley and Basmajian (1973) found only slight activity in mentalis during gentle cheek puffing and with various gentle grimacings with the lips. When these were more forceful, the activity rose; during forceful pulling of the corners of the mouth downward and pursing the lips, activity rose to "marked."

Buccinator

Although, in some respects this muscle lining the cheek is more related to chewing, it has many other functions too. The first major paper on buccinator was by de Sousa and Vitti (1965) at São Paulo. They reported that, with needle electrodes, the muscle consistently showed activity in retracting the mouth laterally and in various complex functions (which has been described with the *Mouth*, (p 429). However, distension of the cheeks left buccinator silent. Work with wire electrodes (Isley and Basmajian, 1973; Basmajian and Newton, 1973) convincingly shows its complexity of function. Buccinator undoubtedly is recruited in individualized patterns and even exhibits variations within the same subject, depending on the amount of force and on slight shifts of emphasis in lip movement. Gentle puffing fails to recruit buccinator activity; more forceful effort increases it to varying degrees with various subjects but not to a marked level. Pulling the corners of the mouth back—or any movements involving such a component—recruits variable (but rarely marked) activity, even when done forcefully. A wide forced smile is most effective, however, resulting in marked activity in almost half of the subjects (Isley and Basmajian, 1973). Blanton et al (1970) found rather similar results, although their subjects showed more activity in blowing. They also demonstrated activity during mandibular movements, which they ascribed to sympathetic behavior and not prime action. Sucking movements were also quite effective, they said, in recruiting buccinator; however, results with thumb-sucking showed very little activity in buccinator (Vitti et al, 1975).

A related study revealed the marvelous control man has over various parts of buccinator (Basmajian and Newton, 1973). With minimal feedback training, a series of subjects rapidly learned to consciously control contractions in different parts of the muscle while maintaining silence in the remainder—and they could do this while performing on the clarinet!

Lip Muscles

Two muscles are included entirely within the confines of the lips—orbicularis oris superior and inferior (OOS and OOI); they really are two muscles and not just a continuous sphincter, as textbooks imply. Others radiate outward to bone and fascia to move the corners of the mouth; upward, levator anguli oris (LAO); upward and laterally, two zygomaticus muscles, major and minor; laterally, risorius and buccinator (see above); and downward, depressor anguli oris (DAO). The upper lip has an "elevator," the levator labii superioris (LLS) and the lower lip a "depressor," the depressor labii inferioris (DLI). There are also various wisps here and there which vary with the individual. All facial muscles (including others higher on the face) are supplied by branches of the same nerve—the facial or 7th cranial nerve.

As noted before, the complexity of the arrangement and the superficial nature of their disposal has retarded EMG progress of these muscles until lately. Speech pathologists and musicians could not wait for anatomists and physiologists and have taken the initiative for obvious practical reasons. Here, let us consider first the general kinesiologic studies; second, the applied speech studies; third, the musician-related work; and finally, clinical (surgical) applications.

LIP KINESIOLOGY

A number of investigators have now emphasized that OOS and OOI are separate muscles (Jacob et al, 1971; Basmajian and White, 1973; Isley and Basmajian, 1973; White and Basmajian, 1973). They act in complex fashions during a multitude of lip positionings. Thus, forceful puffing, pulling the corners of the mouth in different directions, pursing the lips, and curling the lips over the teeth all recruit varying levels of activity, and there again they vary from subject to subject (Isley and Basmajian, 1973). DAO has been investigated separately by Vitti et al (1972) and by Sales and Vitti (1979), who find similar rational results.

Zygomaticus major acts intermittently, and only a wide forced smile recruits maximal activity. The corner-moving muscles are aptly named—they do what their names proclaim. The same is true of the LLS. All of these muscles are most active during broad smiling. Puffing the cheeks out brings out the least response in this group.

While only slight myoelectric activity can be picked up in orbicularis oris during the rest position, during aberrant oral activity, such as thumb-sucking, there is marked activity in the lips (Vitti et al, 1975). Kelman and Gatehouse (1975) found that diffuse surface-electrode pickup of the lip muscles remained surprisingly consistent in individuals repeating the same utterances. Garrity (1975) reported a high correlation between surface-electrode pickups from the lips and from the chin and she then made the wrong conclusion (that one was as good as the other to reveal lip functions). All she was doing was recording cross-talk between pickup sites, no serious problem in the general type of speech research she represents. See also the section below on *Musical Performance*, which must be much more precise. Vitti and his colleagues in Brazil (Farret et al, 1982a and b) have further elucidated functions of the orbicularis oris, mentalis, and depressor labii inferioris in speech.

LEVATOR LABII SUPERIORIS ALAEQUE NASI

Vitti et al (1974) confirmed the assumption that this muscle raises the upper lip and dilates the nostril, but he found it remains silent when the face is resting and composed. In various facial grimaces it springs into activity, but in other facial expressions it remains relaxed, e.g., raising the eyebrows, closing the eyes, opening the mouth, and blowing without distension of the cheeks.

Procerus and Frontalis Muscles

These two muscles have been studied and compared with surface electrodes and wire electodes simultaneously (Vitti and Basmajian, 1976). This was done because they are being often used by psychologists as a reference area for muscular tension studies. In the resting state, as with other facial muscles, we found no activity, contrary to the report of Sumitsuji et al (1967) that there are high-amplitude EMG discharges in the frontalis. Akamatsu (1960) studied the EMG activity of frontalis in schizophrenic patients; he reported a relationship between symptom pattern and patient's posture on the one hand and EMG activity on the other. Earlier, Tokizane (1954) reported that except for frontalis and levator labii superioris, facial muscles showed electrical silence in the resting state. Although they found no significant change in frontalis due to postural variation, Sumitsuji et al found that frontalis activity decreased when their subjects lie down. Not only frontalis but also mentalis and other facial muscles decreased their EMG amplitude in the supine posture. The resting silence from sensitive surface and wire electrodes in studies of normal subjects agrees with wide clinical EMG experience that striated muscle at rest shows complete relaxation (Vitti and Basmajian, 1976). The reports to the contrary, mentioned above, apparently arise from varying levels of tension in the face over a period of some time. To reiterate: at rest, frontalis and procerus are silent and become active on demand or in response to specific (perhaps uncontrolled) emotional states or expressions.

Elevation of the eyebrows, frowning with the forehead, closing the eyes normally, and whistling all did not recruit procerus muscle. The same silence resulted from the frontalis muscle in the movements of pulling the eyebrows downward; closing the eyes either normally or strongly; in the expressions of sadness, gladness, and revulsion; transverse wrinkling over the bridge of the nose; whistling; and winking either the right or the left eye.

The Vitti/Basmajian data showed that the frontalis has a very marked activity only in elevating the eyebrows and frowning with the forehead. It showed its antagonism to the procerus, which is inactive in these movements, and confirmed the muscle-stimulation experiments of Duchenne, who showed the procerus is a direct antagonist of the frontalis.

Confirming the classical descriptions given by some anatomy textbooks and by Martone (1962), the procerus muscle shows a very marked activity during the production of transverse wrinkles over the bridge of the nose and moderate activity in pulling the eyebrows downward. Closing the eyes strongly recruits a marked activity in the muscle. However, in this movement the Vitti/Basmajian subjects always provoked transverse wrinkles over the bridge of the nose, which, as already noted, always makes the muscle active. In winking the right and left eyes, the procerus was active with a little less intensity. This was also true during facial expressions of revulsion and aggression. In all these movements, subjects always showed transverse wrinkles over the bridge of the nose.

Finally, in the facial expression of sadness and gladness, some subjects showed negligible activity in procerus muscle. These expressions having been performed by the subjects voluntarily and not "naturally" could lead to some confusion. Furthermore, as subjects performed called-for expressions, these sometimes did not correspond to the usual concept of spontaneous or involuntary expressions. However, the videotape record of those facial expressions and movements always showed that whenever the subjects provoked transverse wrinkles over the bridge of the nose, the procerus was active while the frontalis was not.

Speech Research

The lip muscles have been the subject of research by a growing number of investigators. Only the group at Haskins Laboratory and the University of Connecticut has used fine-wire electrodes in a limited fashion for speech research on the lips (Gay and Hirose, 1973; Gay et al, 1974; Gay, 1975), a natural extension of their laryngeal research technique. Most speech research has avoided intramuscular electrodes. At the University of North Carolina, Lubker, with Parris (1970), extended his speech studies of the palate (p 434) to lip muscles. Some progress has emerged in improving electrode techniques when fine-wire inserted electrodes are unacceptable to the investigator; thus, Allen et al (1973) have combined fine-wire techniques with paint-on techniques (using the NASA spray-on electrode method for EKGs) with promising early results. Concentrating on orbicularis oris (with surface electrodes) are a group at Essex University in Colchester, England (Tatham and Morton, 1968a, b; Tatham, 1971), and Fibiger (1971) in Stockholm. These investigators are concerned with articulated speech and with disturbances such as stuttering. Fibiger has spread his interest to other muscles of the lip (1972), as has still another group in Stockholm (Persson et al, 1969; Leanderson et al, 1970, 1971; Leanderson, 1972).

The last named group has shown among other applied findings that linguistic context considerably influences the recruitment of muscles used to produce any given phoneme, i.e., the EMG activity pattern associated with a phoneme is not constant. When, for instance, *i* was preceded by a rounded vowel *y* there was a considerable activity in depressor labii inferioris, while this activity was much reduced when *i* was preceded by ϵ. A difference in the amount of motor unit activity was observed in one and the same muscle when *p* and *b* were spoken in the same context. The activity was more vigorous at the closure of the *p* than at that of the *b*. These differences between voiceless and voiced consonants can probably be explained on the basis of a difference in the time course of the intraoral pressure during the closure, which has been established by previous measurements. Orbicularis oris has been studied effectively during articulated speech with tiny paint-on electrodes connected to fine wires to conduct the potentials (McAllister et al, 1974). At the Haskins Laboratories wire hook electrodes have been used quite effectively also (Gay et al, 1974).

At the Speech Research Laboratory of the University of Wisconsin, Ronald Netsell and his colleagues are also exploring the use of electromyography in lip function during normal and distorted speech (Netsell and Cleeland, 1973). Daniel and Guitar (1973) of that group have studied the effects of EMG feedback on facial and speech gestures during recovery following neural anastomosis. In Seattle, Folkins and Abbs (1975) found that the muscles of the lips and jaw are capable of on-line compensatory motor reorganization in response to resistive loading of the jaw during speech. In Kyoto and Osaka, the pioneer EMG work of Koyama (1982) under the tutelage of Professor Kumamoto has analyzed the difficulties of Japanese people learning English pronunciation.

Particularly useful in future studies of articulatory function of the facial and jaw muscles has been the recent work of O'Dwyer et al (1981) in Sydney. EMG recordings were taken from 18 orofacial and mandibular muscles while gestures believed to be specific to each muscle were performed. The anatomic criteria for the placement of the electrodes, the specificity of interference patterns obtained, and the degree of differentiation of the temporal sequence of activity from that in neighboring muscles were used to decide on the degree of differentiation of the temporal sequence of activity from that in neighboring muscles, and they were used to decide on the degree of certainty that a particular muscle was being recorded. The appropriateness of each gesture as a stimulus to any muscle was determined on the basis of the level of activation occurring with the gesture relative to other muscles and its degree of variability between subjects. In their six subjects they found sufficient variability to sound a warning to others with less refined techniques limited to surface electrodes.

Musical Performance

A group of music professors from Appalachian State University joined Basmajian from time to time to pioneer the use of wire electrodes in the lips and cheeks for elucidating the part played by these muscles in the *embouchure* of wind instrument performance. The details are reported elsewhere in a series of papers (Basmajian and White, 1973; White and Basmajian, 1973; Isley and Basmajian, 1973; Basmajian and Newton, 1973). Here the main results are described very briefly.

Musical performance, like facial expressions, recruits muscles in patterns that differ from person to person. No absolute pattern is demonstrably more efficient than another to produce desired expressions or notes. However, significant correlation was found between certain patterns of lip muscle contraction and proved proficiency in a long series of performers. Also found was the fact that performers can deliberately change their patterns during a performance if they are given artificial EMG feedback *with no apparent alteration* of the sounds produced.

Surgical Application

DePalma et al (1958) made a brief EMG study of the lip musculature in full thickness flaps that had been rotated by plastic surgery from the upper to the lower lip to fill a wedge-shaped defect. In four patients, there was regeneration and reinnervation of the motor nerve to the flap with return of apparently normal function. A similar study has been reported in Sweden by Isaksson et al (1962).

NECK MUSCLES

Platysma

The first and probably the only careful study of this broad sheet of subcutaneous muscle was done by de Sousa (1964), who investigated 20 men using needle electrodes. The obvious actions of this rather obvious muscle produced the greatest activity (pulling the thoracic skin up and the angle of the mouth down).

During abrupt inspiration (but not expiration) there was activity that de Sousa interprets as helping to reduce the constricting effect of the skin on the subcutaneous veins of the neck. Widening the opening of the mouth ("buccal rima") elicited marked activity, but the natural opening of mouth and jaws did not. Active and resisted movements (flexion, extension, and rotation) of the neck did not recruit platysma, nor did swallowing.

Sternomastoid, Scalenes, and Longus Colli

The longus colli muscle is unique; it is the only spinal muscle that both lies in front of the spinal column and has attachments confined to the vertebrae. Therefore, its effects must be directly upon and restricted to the cervical spine—unlike other anterior muscles of the head and neck which exert their effects more or less indirectly. Other than the single case study by Fountain et al (1966), no EMG reports were available on the role of this interesting but deeply located muscle until the 1970s.

In quick succession, two studies were completed on the sterno(cleido)-mastoid. In São Paulo, de Sousa et al (1973) and then Vitti et al (1973) in Basmajian's Atlanta laboratory reported results that overlap and agree; these studies probably will remain definitive. The latter study also included the longus colli muscle. Using bipolar wire electrodes, Vitti et al examined the right and left longus colli (LC) and sternocleidomastoideus (StM) muscles electromyographically in 10 healthy young adults. Action potentials were recorded on FM magnetic tape, and each experiment was also videotaped. The head-neck motions were recorded using a special neck goniometer. The muscles were studied in sitting, supine, prone, and lateral positions both during free movements and against resistance.

Vitti et al (1973) found complete inactivity in both sternomastoid and longus colli muscles in relaxed sitting, normal breathing, deep expiration, and wet and dry swallowing. There was very marked synchronous EMG activity of the LC and StM muscles during resisted forward flexion and marked activity during neck flexion against head weight in the supine position and during resisted right and left side-bending. Variable activity was found in both muscles during deep breathing, coughing, forceful blowing, loading on top of the head, resisted backward extension, neck holding against head weight in the prone position, and in twisting movements downwards and upwards. During free flexion-extension movements, LC and StM act synchronously. During free lateral bending they work homolaterally but during free rotation to the right, the right LC works with the left StM and *vice versa.*

As noted on p 467, Fruhling and Basmajian studied the *scalenus* anterior muscles of eight young men with inserted wire electrodes (unpublished). Figure 20.9 summarizes the findings for deliberate movements of the head and neck. On p 409, the *primary respiratory role* of the scalenes was pointed out, and it is re-emphasized here.

Longissimus Cervicis

Only Fountain et al (1966) have examined this deep muscle in one human subject. It was a strong extensor of the cervical spine but had no lateral-bending effect. In relaxed sitting, it was silent but acted in synchrony with longus colli during rotation.

When the human subject is sitting or standing in a relaxed position, both longus and longissimus cervicis are almost completely inactive. This is in keeping with most postural responses in the human erect trunk. The same is not true in the dog unless the head is supported externally.

Longus colli is a strong flexor of the cervical spine, acting reciprocally with the longissimus, a strong extensor. Both muscles are synchronously active during rotation. Influences on lateral bending of the neck appear to be minimal, contrary to widespread belief. Most surprising and interesting is a pronounced increase in activity in longus colli during talking, coughing, and swallowing. Apparently this represents a reflex stabilizing of the neck during pharyngeal contractions.

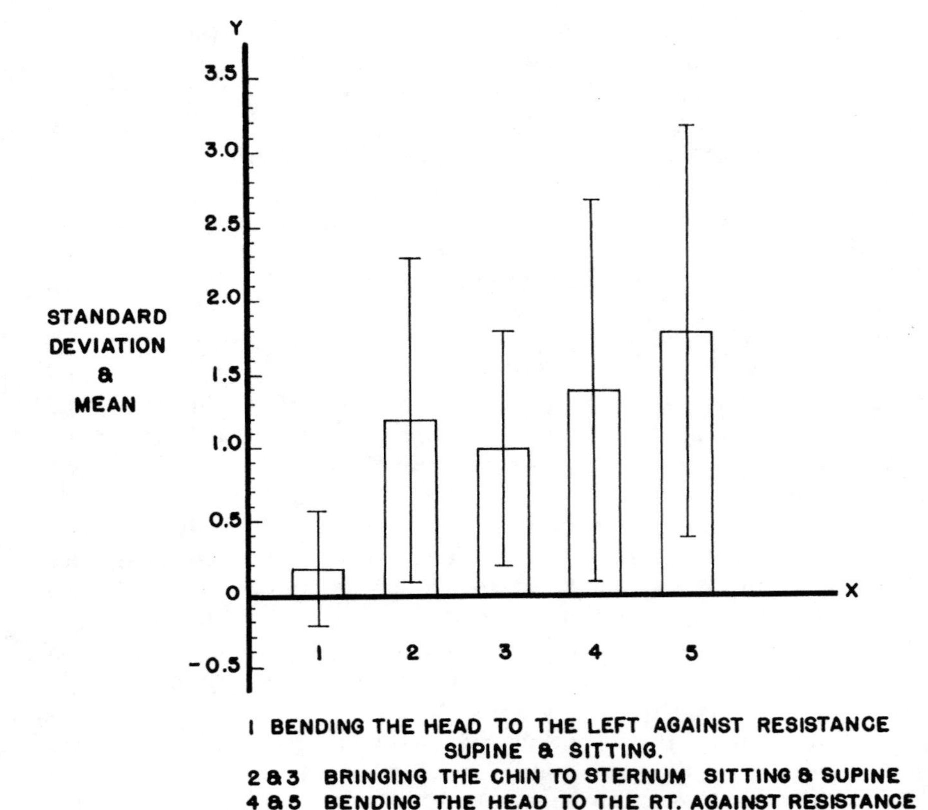

Figure 20.9. EMG results of scalenus anterior—movements of neck.

Semispinalis Capitis and Splenius Capitis

Although Tourney and Paillard (1952) gave a general account of the splenius, probably the only EMG study with modern techniques and intramuscular electrodes of the specific functions of these two huge, but obscure neck muscles was reported by Takebe et al, (1974). Fifteen normal adult subjects had bipolar fine-wire electrodes placed in the middle of the bellies of these muscles (determined by repeated dissecting room experiments). Myopotentials were integrated and summated by computer analysis.

Semispinalis is almost limited to one main function: extension of the head on the neck. No doubt it also has antigravity functions when one leans forward; but in well-balanced, erect postures it falls economically silent, contrary to widespread opinion. It is not a rotator.

Splenius capitis is equal to semispinalis in its exuberance during extensor activity; but it is also just as active during rotation of the head to its own side. It is, then, an important extensor-rotator. Probably it is as important in neck rotation as the better known sternomastoid.

NECK MOVEMENTS

The slightest attempt of a supine subject to raise his head from the couch is accompanied by marked activity in the sternomastoid and scalenes. This finding of Campbell (1955) has often been confirmed in scattered observations by Basmajian.

No other systematic reports are available on the activity of these muscles as demonstrated electromyographically. However, Hellebrandt et al (1956) published an account of their research on the tonic neck reflexes in "exercises of stress" of human subjects. This group had noted previously that spontaneous ipsilateral head-turning occurred regularly during unilateral wrist-extension in hypertonic neurological conditions. From their study they conculded that exercise against resistance calls upon synergists which may be far removed from the part exercised. The pattern of concurrent action in the neck that developed as a result of stress in exercising the upper limb is sufficient to modify the position of the head. Following this discovery, the use of strong voluntary head-movement in the direction of the spontaneous positioning was tested; it appeared to augment the output of work. Therefore, this reflex positioning of the head does, in fact, appear to help the normal work of the upper limb.

Other Neck Muscles

Trapezius is discussed with the upper limb in Chapter 12 (p 265), the oral diaphragm (floor of the mouth) in Chapter 19 (p 431), the laryngeal muscles in Chapter 19 (p 441) and digastricus earlier in the present chapter with the muscles of mastication (p 455). No systematic studies have been reported on the infrahyoid muscles except for cricothyroid, which was studied (along with the anterior belly of the digastric) by Lesoine and Paulsen (1968). The greatest intensity of contraction in cricothyroid occurs in swallowing. The next level of intensity occurs during phonation; then, in reducing order: during strong expiration and coughing (equal); strong inspiration; and, finally, quiet breathing (very slight).

CHAPTER 21

Extraocular Muscles and Muscles of Middle Ear

EXTRAOCULAR MUSCLES

Although Scandinavian, Japanese, and Italian workers have also published in this field, most of the ocular electromyography that has received wide attention was done by several small groups of investigators in the U.S.A. The two most prolific groups were located in San Francisco and New York City. The former includes Marg, Tamler, Jampolsky, and other occasional associates; the latter was centered on one investigator, Goodwin M. Breinin. Of necessity, this chapter will be entirely a review and synthesis of the work of these and other investigators.

Before proceeding further, it must be clear that *electro-oculography*, which is sometimes confused with electromyography, is distinctly a different technique and therefore is not the subject of this chapter. Electro-oculography is recorded through superficial skin electrodes in the region of the orbit and consists of potentials arising in the retina, extraocular muscles, and other sources in the orbital cavity. With a venerable but rather dull history dating back about a century, it has contined to show more promise than results. For an excellent early summary of electro-oculography, see Marg (1951). More recent papers with more than passing interest are: a brief up-to-date review by Peters (1971) and a description of the basic aspects by Bicas (1972). These should serve as a jumping-off place for readers interested in exploring electro-oculography. The widest use of the technique is in sleep studies—where the emphasis is on the presence or absence of rapid eye movements (REM) (see p 474).

Electronystagmography is related to electro-oculography. It depends on the existence of a corneoretinal electrical potential that varies with the eye movements during nystagmus. It, too, is beyond the scope of this chapter (for a good review, see Milojevic, 1965).

Electromyography of the muscles that move the eyeball uses direct recording of motor unit action potentials from these muscles. Jampolsky et al (1959) warn, with proper alarm, against attempts to measure relative strength of extraocular muscles by using only the amplitude of potentials as the criterion, because there are many sources of EMG artifact. Variations in the structure of the muscles being tested and in the positioning of the electrodes can also lead to errors. Jampolsky (1970) continued to advise caution, flatly stating that electromyography ". . . is of no practical usefulness in management of the usual strabismus patient." He conceded that it is ". . . a very valuable laboratory and investigative tool" to study ocular rotations.

Normal Position and Movement of the Eyeball

Several groups have shown that the recti exhibit fairly strong, persistent activity to maintain the position of the eye during waking hours. Changing visual inputs constantly affect the level of motor unit activity, apparently without imbalance resulting (Serra, 1970). We may conclude, then, that positioning results from a balance of activity among these muscles, Björk and Kugelberg (1953) demonstrated a persistence of this activity even in darkness.

These last-named workers also showed that each change of ocular fixation is accompanied by a gradual increase of activity in the prime mover or agonist with a reciprocal decrease in the antagonist whether the movements were slow or fast. With extreme positions of gaze, e.g., far to one or other side, the antagonist is usually completely inhibited. Kadefors et al (1974) found that the average rectified EMG signal of the lateral rectus tends to be an exponential rather than a linear function of the degree of ocular deviation. Muscle length is not a major function.

During fast movements, there is a complete inhibition of the antagonist accompanying the sharp short burst of activity in the agonist (Fig. 21.1). During a saccadic eye movement, Tamler et al (1959) found a heightened burst of activity in the agonist, inhibition of the antagonist, and coactivity of the auxiliary muscles. A saccadic eye movement is a type that occurs in changing the gaze from one point in space to another and includes small terminal oscillations. Tamler et al confidently assert that rapid movements of this sort are not ballistic in nature (see Chapter 9, p 225). Miller's (1958) independent findings would seem to agree (Fig. 21.2), although he stated that for large movements one or more bursts of

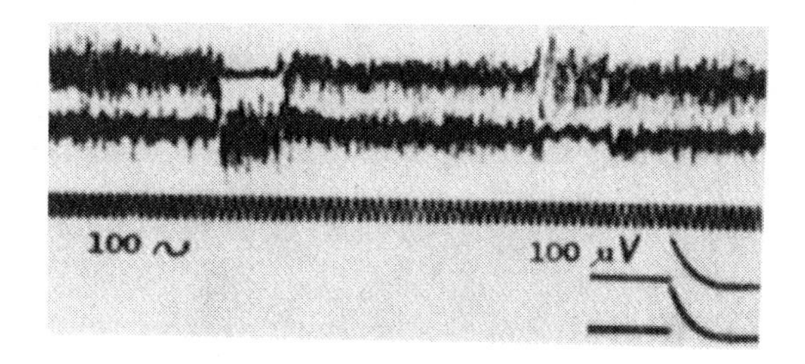

Figure 21.1. EMG records from lateral rectus (*upper tracing*) and medial rectus of one eye during quick changes of fixation from 10° outward to center of scale and back to 10° outward. There is abrupt inhibition in the antagonist simultaneous with increase in agonist activity. (From Å. Björk and Å. Kugelberg, ©1953, *Electroencephalography and Clinical Neurophysiology*.)

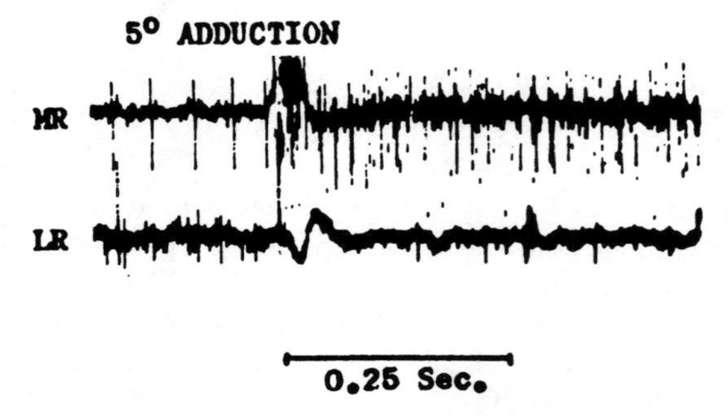

Figure 21.2. EMG records from medial and lateral rectus during a 5° saccadic movement. (From J.E. Miller, ©1958, *American Journal of Ophthalmology.*)

activity, in addition to the initial burst, may be found. Nonetheless, the antagonist remains completely inhibited. Yamazaki (1968) found essentially the same (but limited to 13 ms in duration) in the medial and lateral recti as part of tiny horizontal flick movements when monocular fixation is performed.

Robinson (1965) has shown that both smooth pursuit and saccadic movements may occur with completely temporal independence. Smooth movements may occur just before, after, or with saccadic ones.

There seems to be no doubt, then, that the muscles of the eye are in more-or-less continuous activity, except during sleep (see below). Indeed, their role may be best compared to dynamic guy-wires.

Position of Ocular Rest

Among oculists, the position of the eyes during sleep and anesthesia has been the subject of much interest, speculation, and investigation. Reported observations have been conflicting and sometimes confusing. During anesthesia, for example, different degrees of vergence have been reported, as well as versions during the induction stage. In surgical anesthesia, the movements cease, and during the asphyxial stage of anesthesia, a convergent depressed position may be assumed. A similar behavior may be noted during fainting (Wójtowicz, 1966) and sleep (Breinin, 1957 *et seq*).

According to Breinin (1957b), the disappearance of esotropia during anesthesia is a frequent observation, with straight eyes or divergence taking the place of esotropia. While investigating the electromyographic changes during anesthesia and sleep, Breinin found that the "innervation" to all horizontal recti rapidly decreased during the induction of general anesthesia with intravenous Pentothal in a patient with intermittent exotropia. When the surgical plane was reached, all nerve impulses were repressed, and the eyes occupied a moderately divergent position.

As the level of anesthesia lightened, individual motor units began to discharge. When the plane of anesthesia was again deepened, the motor unit activity disappeared promptly. Therefore, the position of the eyeball during surgical anesthesia suggests that the motor nerve impulses to the extraocular muscles cease completely.

The position of the eyeball during anesthesia is apparently dictated by anatomical-mechanical factors—bony, fascial, and muscular—and represents the anatomical position of rest. It is, then, the *basic* position which is modified by the normal factors of motor innervation during consciousness. If this basic position is one of divergence, then a basic stress toward divergence must underlie whatever position the eyes assume in consciousness, according to Breinin. "It opposes esodeviation and facilitates exodeviation." The frequency of divergence under surgical anesthesia supports the concept of a basic anatomical divergence, to which, in exotropia, may be added innervational divergence.

Eye Movements during Sleep

During the light dozing phase preceding ordinary sleep, very irregular discharges and bursts of potentials appeared in Breinin's investigations. He reports that a run of double and single MUAP occurred at intervals, as well as rhythmical trains of potentials (Fig. 21.3). During what appeared to be lightening of sleep induced by particularly cacophonous snores, a single burst of MUAPs occurred, accompanied by an obvious movement of the eyeball.

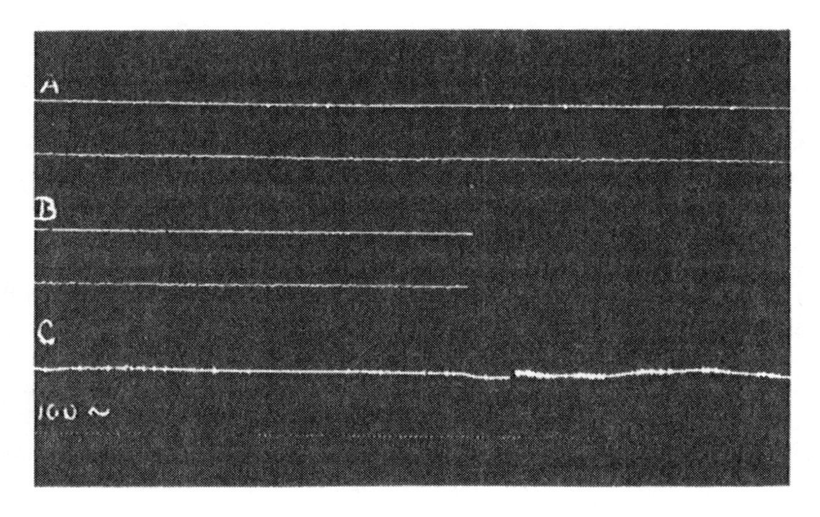

Figure 21.3. (*A* and *B*) EMG records (*upper tracing in each case*) with simultaneous electro-oculograms of medial rectus during sleep. *Tracing C* shows resumption of normal waking pattern (toward its right end). (From G.M. Breinin, ©1957b, *AMA Archives of Ophthalmology.*)

The commoner method of sleep researchers is the much less precise DC electro-oculography. Initially, eyelid closure induces upward movement of the globes; there they remain for 55 to 85% of the time spent in stages 2, 3, and 4. With the onset of the stage now known as the rapid eye movement (REM) sleep, they move down to the central position and begin to flick rapidly—usually in oblique directions but sometimes in the horizontal or vertical directions (Jacobs et al, 1971). Eye movements do not occur in infants during quiet sleep but appear as both slow rolling movement and REMs (in varying amounts) during irregular sleep (Prechtal and Lenard, 1967).

Asymmetric Convergence

Asymmetric convergence may be defined as convergence of the eyes occurring in any direction other than along the median plane. For example, this would occur if an object approached one eye directly from the front. This would require a much greater convergent movement of the other eye to fixate on the approaching object. Breinin (1955) observed that whenever an eye turned in either a version or a vergence movement, there was an increase in the electrical activity of the agonist accompanied by a reciprocal inhibition of the antagonist. When he noted no change in activity in either horizontal rectus of the stationary eye during asymmetric convergence, he concluded that the antagonistic innervations for a version and vergence movement cancelled each other in the brain and never reached the muscles. If the motor impulse to that eye were to change, it would have to move. However, fixation precludes such movement. Miller (1959) seems to agree. He found that sudden asymmetric convergence is accompanied by a burst in the yoke muscles (i.e., pairs of muscles which are agonists in performing versions) followed by a convergent pattern in both medial recti. No increase in coactivity to horizontal antagonists was found during slow asymmetric convergence on an approaching target until the near point of convergence was approached.

An early EMG study by Blodi and Van Allen (1957) also agreed with that of Breinin. They found that there was no change in the myoelectric signal from the horizontal muscles of the apparently stationary eye during asymmetric convergence. They also suggested that the sum total of innervation of the two horizontal muscles has to remain the same, regardless of the position of the eye.

On the other hand, Tamler et al (1958) presented electromyographic data which support Hering's concept of peripheral receipt and adjustment of opposing stimuli to the apparently stationary eye during asymmetric convergence. According to them, the reason the eye remains stationary is due to cocontraction of opposing horizontal recti, that is, there is a simultaneous increase in innervation of the lateral and medial rectus muscles. One possible reason why other investigators have not found this by EMG was, they suggest, a failure to induce a sufficient amount of angular convergence in the moving eye in order to register observable changes on the electromyogram of the stationary eye.

Hering's law appears to be supported by both of two test methods of asymmetric convergence. The first is that of smooth, binocular convergence along the axis of one eye, and the second is that of uncovering one eye to force a fusional convergent movement while the other eye continues to fixate.

In "breaking fusion" Tamler et al often found simultaneous decrease in innervation of the horizontal recti of the stationary eye. They explain that if it did not occur, then continued refusion movements with repeated covering and uncovering of an eye would cause greater and greater buildup of electrical activity in the horizontal muscles of the stationary eye.

Blodi and Van Allen (1960) finally laid to rest the theory that vergence movements by the medial recti were somehow controlled by the sympathetic nervous system. After anesthetizing the cervical sympathetic trunk in human subjects, they found no EMG change and no directly observable change from the normal response of convergence on command.

Cocontraction

Cocontraction as applied to extraocular muscles (as with synergists in other skeletal muscles) is defined as the simultaneous increased contraction of extraocular muscles which are normally antagonistic in their primary field of action. One suggestion has been that cocontraction occurs when the eye moves from its primary position to any secondary or tertiary position. According to this hypothesis, as the lateral rectus abducts an eye, the vertical recti as well as the superior and inferior obliques cocontract to steady the eye in its horizontal path. At the same time, the cocontraction is believed to prevent undue torsion of the globe and perhaps helps to maintain the abduction because the lateral rectus loses its mechanical advantage as the movement progresses. The same reasoning would apply to adduction, supraduction, and infraduction (Tamler et al, 1959).

Tamler et al (1959) studied the electrical activity of four extraocular muscles simultaneously in many subjects. They concluded that there is little or no increased coactivity of auxiliary extraocular muscles in adduction, abduction, supraduction, and infraduction during slow, vertical, and horizontal following movements in planes through the primary position (Fig. 21.4). This does not necessarily mean that they do not contribute to the movement. Apparently, the effect of muscle activity on eye movement depends upon the position of the eye at that particular moment. Hence, they suggest that the "primary position innervational

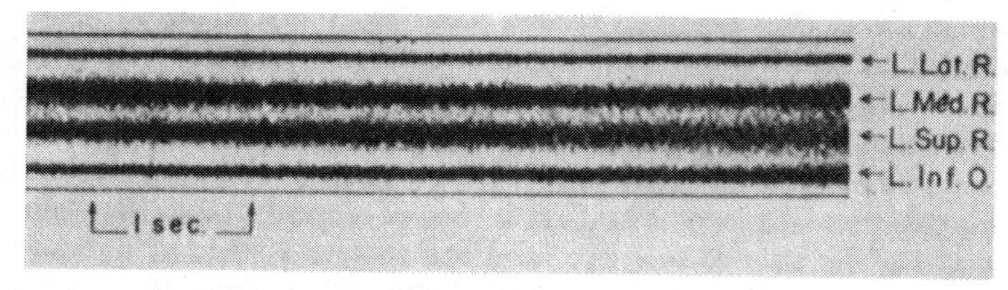

Figure 21.4. EMG records from various muscles of "up gaze" in sagittal plane of left eye. No obvious change in lateral and medial recti. (From E. Tamler et al, ©1959, *AMA Archives of Ophthalmology.*)

tonus" of the auxiliary muscles, which does not appear to change during these movements, may indeed contribute to the movement after the eye has gone a certain distance. Also, the primary position tonus may be all that is necessary to steady the eye and prevent torsion during the movement. Tamler and his group conclude that although one can deduce muscle contraction with certainty from the electrical activity of a muscle, one cannot firmly state the function of the muscle at that moment.

Furthermore, they insist that normal muscle activity, as found in the primary position, is present in the auxiliary muscles during movement but that systematic changes in coactivity do not generally occur. Occasionally, they did find barely perceptible changes in the recordings from those muscles which are auxiliary during movements. These changes, when elicited, had no systematic pattern and therefore are not consistent with a cocontraction hypothesis that requires systematic changes with a given direction of movement. Therefore, these authors, quoted here so extensively, are in basic agreement with Breinin's conclusion regarding the same problem.

Innocenti (1971) studied the EMG activity of extraocular muscles in unanesthetized rabbits during oculocompensatory positions. During inclination around a longitudinal axis, units belonging to the superior and medial rectus and to the superior oblique were seen to be activated by ipsilateral and inhibited by contralateral inclination; the units belonging to the inferior rectus, inferior oblique, and lateral rectus showed an opposite pattern.

During upward inclination around a bitemporal axis, the units of medial and inferior rectus and superior oblique of both eyes were activated while the units belonging to the other extraocular muscles were inhibited and often unaffected. An opposite pattern was exhibited by the same units during downwards incliniation. A linear ratio was found between the degree of inclination and the increase in discharge rate of extraocular units.

Stretch and Proprioceptive Effects

Sears et al (1959) were concerned about the proprioceptive function of the tension in the extraocular muscles. Previously, muscle spindles had been observed in the muscles of the eye by Cooper and Daniel (1949). Using EMG techniques, Sears et al investigated the possible existence of stretch mechanisms in human extraocular muscles using patients prior to surgical enucleation of the eye. Under topical anesthesia they inserted a concentric needle electrode into one of the horizontal recti muscles. Recordings were made with the eyes both in primary and in horizontal gaze, the latter about 30° beyond the midline. A suture was placed through the tendon of one horizontal rectus muscle which was then cut away from its insertion distal to this ligature. The effects of disinsertion and of gentle manual pulling on the suture were recorded from the agonist muscle, while the muscle was acting as either agonist, antagonist, or yoke muscle. With the eyes in primary gaze, they found that the action potentials were unchanged by disinsertion or manual pull. During agonist contraction, however, stretch of the agonist itself, of the antagonist, or of the yoke muscle produced a decrease in the frequency and amplitude of the discharges. This inhibitory effect was not observed in a patient with oculomotor palsy when stretch was applied to the muscles of an eye which had a previous retrobulbar alcohol injection.

We can only agree with Sears and his colleagues that probably the extraocular muscles of man do have receptors which can be stimulated by passive stretch. Breinin (1957a) reached the same conclusion, stating: ". . . it appears that a proprioceptive mechanism must be postulated for the extraocular muscles. It is not implied that such a mechanism provides muscle position sense or awareness. Failure to make this distinction has resulted in confusion in the past."

Bearing on the problem of control mechanisms is the possibility of innervational plasticity of the oculomotor system. This was investigated experimentally by Metz and Scott (1970). Studies in 14 rhesus monkeys utilizing superior and lateral rectus muscle cross-transplantation resulted in return of conjugate eye movements within 10 days. Electromyographic recordings from transplanted and normal extraocular muscles indicated no change in innervational patterns. Central sixth nerve stimulation and postmortem orbital examination demonstrated that observed ocular movement can be explained mechanically and need not result from innervational plasticity.

Eye-Head Coordination

Eye-head coordination was investigated by recording from the neck and eye muscles in monkeys by Bizzi et al (1971). The results show that (1) during eye-head turning, neural activity reaches the neck muscles before the eye muscles; and (2) all agonist neck muscles are activated simultaneously, regardless of the initial head position. Since overt movement of the eyes precedes that of the head, it was concluded that the central neural command initiates the eye-head sequence but does not specify its serial order. Furthermore, it was determined that the compensatory eye movement is not initiated centrally but instead is dependent upon reflex activation arising from movement of the head.

Eyelids

Björk and Kugelberg (1953) performed the first and definitive studies on the levator palpebrae superioris and orbicularis oculi. They found the levator to be continually active during waking hours, except when looking sharply downwards or when the eyes are closed. Lowering of the upper lid is accompanied by progressive decrease in levator activity but with little or no activity in the orbicularis. In other words, the lid is lowered passively by the relaxing of the levator. The lowering of the lid in the act of blinking, however, is caused by activity in orbicularis with immediate cessation of activity in the levator (Fig. 20.5). This is followed immediately by the reverse (inhibition of orbicularis and activity in levator).

Blinking was also studied by Van Allen and Blodi (1962). They too found a quick, well-controlled, reciprocal relationship between the orbicularis oculi and levator palpebrae superioris. This occurred as expected in normal people, but it also was preserved in the presence of gross neurological conditions affecting the region.

Though not an appropriate topic here, the blink reflex may interest some readers. Since the early work of Kugelberg (1952), scattered papers

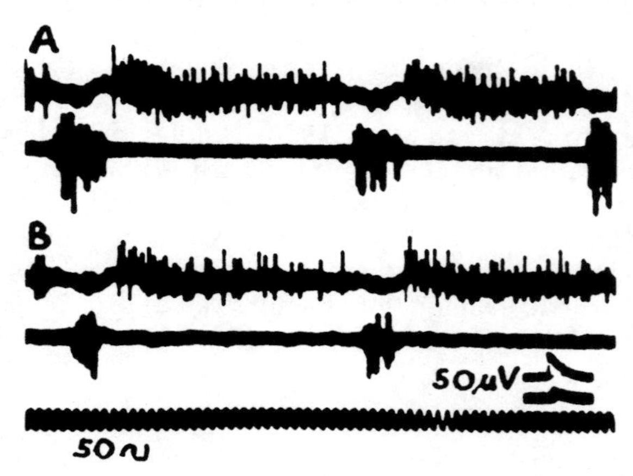

Figure 21.5. (*A*) EMG signals of blinking from levator palpebrae superioris (*upper trace*) and orbicularis oculi. (*B*) Spontaneous blinking. (From Å. Björk and E. Kugelberg, ©1953, *Electroencephalography and Clinical Neurophysiology*.)

have appeared (e.g., Wójtowicz and Gwóźdź, 1968; Young and Shahani, 1969; Ferrari and Messina, 1972).

In view of the indecision surrounding the innervation of the levator palpebrae and arising from the drooping or ptosis of the lid that is part of Horner's syndrome, Blodi and Van Allen (1960) investigated this muscle further. Duplicating a Horner's syndrome by producing a block in the cervical sympathetic trunk with local anesthetic, they found no significant effect on the levator. Even with marked ptosis of the upper lid, the activity in levator was not reduced. Blodi and Van Allen also found normal EMG signals in the levators of patients with long-standing Horner's syndrome. Therefore, taking all this together, we must conclude that the sympathetic nervous system does not innervate the striated levator palpebrae. The drooping of the lids in Horner's syndrome is not due to any direct paralytic effect on the striated fibers but appears to be due to paralysis of smooth muscle.

MUSCLES OF THE MIDDLE EAR

As the reader would suspect, the tiny stapedius and tensor tympani muscles have not been studies extensively. However, Berlendis and De Caro (1955) in Italy, Wersäll (1958) in Stockholm, Kirikae (1960) of Tokyo, Carmel and Starr (1963) of Bethesda, MD, Dewson et al (1965) at Stanford University, Henson (1965) at Yale, and Teig (1973) in Oslo have performed EMGs within the middle ear of experimental animals. In Copenhagen, Salomon and Starr (1963) and in Zurich, Fisch and Schulthess (1963) made direct recordings from human patients. But most studies of tensor tympani and stapedius in man have been made by an ingenious *indirect* method now gaining fairly wide acceptance. This depends on measurements of changes in acoustic impedance in the external ear canal. Interested readers should consult Møller (1961) for a technical description.

Wersäll's beautiful experiments, which were performed on rabbits and cats and were described in his monograph, established several important points. He showed that there is continuous activity at rest, i.e., resting tonus, in these middle ear muscles. When sound reaches the eardrum there is a small delay (less than 0.1 s) before a reflex contraction of the muscles becomes evident. This delay is constant regardless of the intensity of the sound. There is, however, a threshold above which the sound must rise to elicit the reflex. It is generally higher in the tensor tympani than the stapedius. Nonetheless, contrary to widely held opinion, protective action against sudden loud noises is no more developed than against a moderate noise.

Wersäll showed a characteristic difference between the two muscles in the dependence of reflex tension on sound intensity. Stapedius, but not tensor tympani, reaches a plateau before the sound is so intense as to become injurious. Under the stimulation of steady sound, the reflex contractions of both muscles often showed undulatory fluctuations. During strong sustained stimulation, the tension drops to a steady state in both muscles. Wersäll ascribed this fall in tension to fatigue.

Kirikae (1960) measured the latency of the acoustic reflex more precisely than Wersäll. For stapedius it was 3.4 ms and is shorter than it is for tensor. Activating time, which requires muscle contraction, is 2 to 3 ms. Maximal contraction takes it a further several ms. Fatigue or adaptation phenomenon were also prominent in his studies.

In unrestrained cats, Carmel and Starr (1963) added a number of new observations or confirmed Wersäll's findings. As in his investigation, the middle ear muscles were found to relax gradually during prolonged sound stimulation. Both muscles contract during certain bodily movements and vocalization. Reflexes are modified by previous exposure to loud noises. Application of electric shock concurrently with test signals also modifies the responses of the muscles. Thus they concluded that a wide variety of dynamic processes influence the activity of middle ear muscles, not just the physical characteristics of sound impinging on the ear.

The limited studies on man reported by Salomon and Starr (1963) revealed the middle ear muscles to be active with general motor events, such as closing the eyes, movements of the face and head, vocalization, yawning, swallowing, coughing, and laughing. Thus they proposed the theory that central controls for these functions are closely integrated. Dewson et al (1965) found (in cats) a close similarity or relationship in the EMG reaction of the middle ear muscles and the stage of rapid eye movements during sleep. During surgery et al (1963) recorded EMG activity of the stapedius muscles in 10 patients (six with facial palsy, two with otosclerosis, and one each with Menière's disease and chronic otitis media). Only two who had normal ossicles were reported on. Both had spontaneous resting activity at the noise level of the operating room, but general anesthesia abolished it completely. Reflex latency studies confirmed results obtained by indirect methods.

The activity and function of middle-ear muscles in echo-locating bats forms the subject of an intriguing paper by Henson (1965). He found that stapedius begins to contract just before vocalization and reaches its maximum as each pulse begins, apparently to dampen the ossicles during the sound-emission stage. Relaxation starts immediately and continues during the echo phase, apparently to allow its maximum perception. Tensor tympani seems to play no part in this system. Jen and Suga (1976) have provided elegant expansion to Henson's earlier work.

Finally, mention should be made of a related technique for the study of reflex contractions of stapedius and tensor tympani by the indirect method of acoustic impedance at the tympanic membrane. The Zwislocki acoustic impedance bridge in the external ear canal is used in both experimental animals and man (e.g., Borg, 1972; Deutsch, 1972).

CHAPTER 22

Motor Nerve Conduction Velocity and Residual Latency

Inseparably wrapped up in function with the skeletal muscles are their motor nerves. The use of human nerve conduction velocity in studying abnormal states was introduced by Harvey and Masland (1941). Their technique, which has been followed with minor variations by others, consisted of recording action potentials from the hypothenar muscles following electrical stimulation of the ulnar nerve (Figs. 22.1, *top* and *bottom*; 22.2). A simple and clear description of technique is given by Lundervold et al (1965), and the entire field is reviewed in book form by Smorto and Basmajian (1979). That book deals at great length with all aspects of nerve conduction tests, both motor and sensory, with an emphasis on clinical applications. Here in this chapter only a general consideration is given to the most significant features of normal electroneurography.

Hodes et al (1948) adapted this technique to the study of nerve conduction velocity as well as the action potentials of normal and abnormal nerves. Stimulating the ulnar nerve first at the elbow and then at the wrist while recording action potentials of the abductor digiti minimi, they determined the difference between the two latencies of response. The distance between the two points of stimulation divided by the difference in latencies yielded an accurate measure of the conduction velocity of the most rapidly conducting fibers in the nerve. Assuming the speed was constant, Hodes et al calculated the time that should be taken by the impulse to reach the muscle and found that it was always less than the measured latency of the muscle action potential. Therefore, there is a small "residual latency" caused by either a slower velocity in the finer terminal portions of the nerve, or a delay at the neuromuscular junction, or a combination of these two factors. Indeed, Trojaborg (1964) has shown that the conduction velocity falls in the distal part of the median and ulnar nerve trunks. Spiegel and Johnson (1962) hold the opposite view.

In a series of papers, Magladery et al (1950a–c) extensively analyzed patterns of electrical activity evoked by stimulation of mixed peripheral nerves in human limbs. Their results indicate a slowing of the impulse in distal portions of both motor and afferent nerve fibers, which accounts for a portion of the residual latency as described by Hodes et al.

Norris et al (1953) determined ulnar nerve conduction velocities and

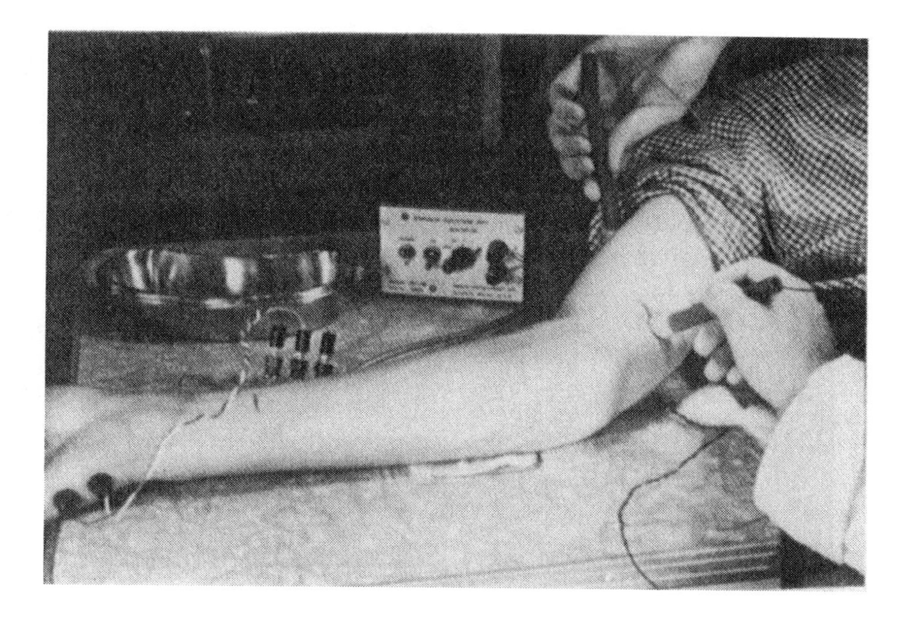

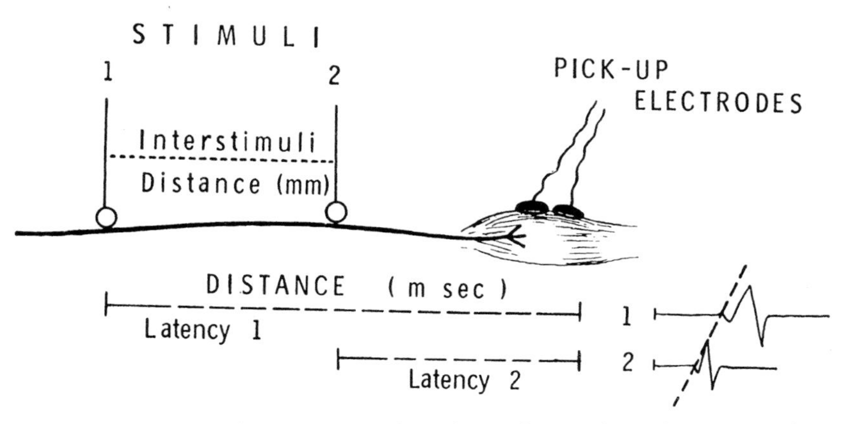

Figure 22.1. (*Top*) Electromyographic electrodes on hypothenar muscles and stimulating cathode on ulnar nerve above the elbow. *Black mark* above wrist is site for other stimulation point. (From M.D. Low et al ©1962, *American Journal of the Medical Sciences*.) (*Bottom*) Diagram showing method of computing conduction speed in a segment of nerve. (From M.P. Smorto and J.V. Basmajian, ©1979, Williams & Wilkins, Baltimore.)

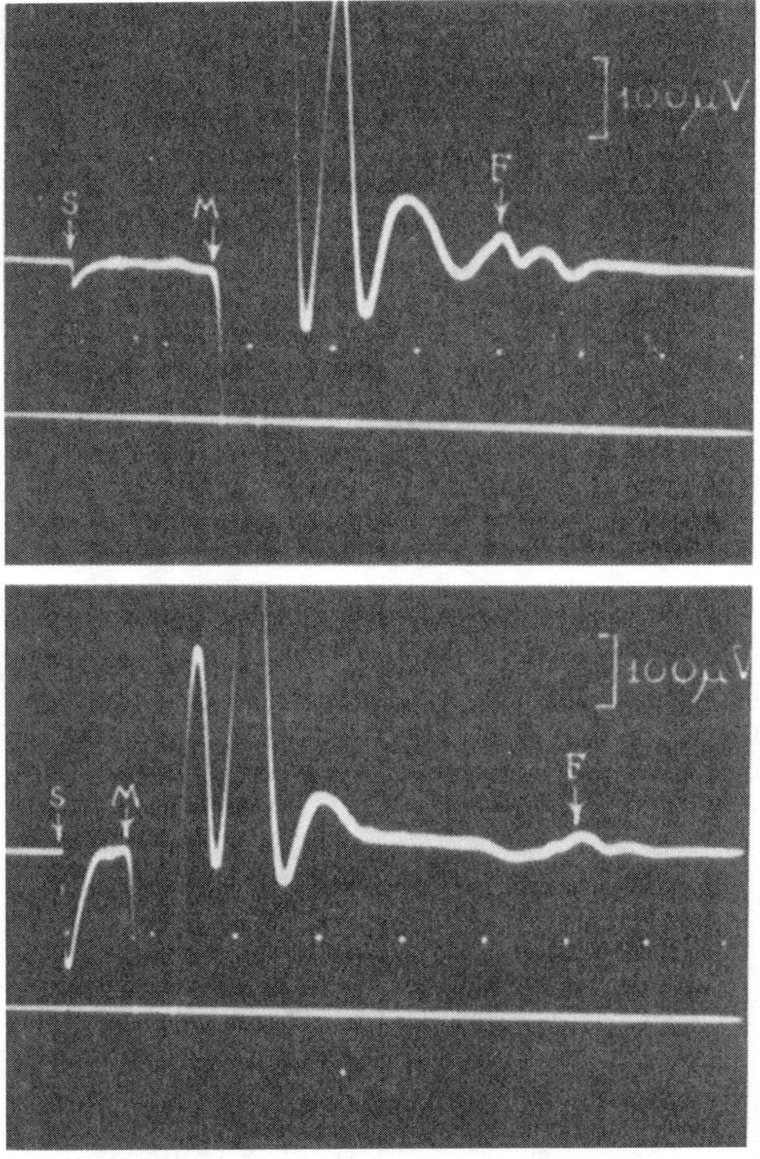

Figure 22.2. (*Upper record*) Stimulus above elbow. (*Lower record*) Stimulus above wrist. Conduction time from stimulus (*S*) to the start of a muscle twitch (*M*) can be measured by comparing with the time signal (series of *white dots* at 5-ms intervals). (From M.D. Low et al, ©1962, *American Journal of the Medical Sciences.*)

"residual latencies" in 25 subjects ranging in age from 20 to 90 years. They found a decrease in velocity but no change in residual latency with increasing age.

Normal minimum values in average adults (rounded out, in meters per s) for various human nerves are: median and ulnar, 47 m/s; deep peroneal, 38 m/s; posterior tibial, 39 m/s (Schubert, 1963). The mean velocity for the radial nerve is 56 m/s according to Downie and Scott (1964), but it is 66 to 74 m/s according to Gassel and Diamantopoulos (1964). The mean velocity in the sciatic nerve is 51 m/s when abductor hallucis is recorded, and 56 m/s when a shorter length of nerve is studied, i.e., when gastrocnemius is used for recording (Gassel and Trojaborg, 1964) (see Table 22.1).

Mayer (1963) gives the following motor conduction velocities for the age group 10 to 35 years, which are rounded out to the closest whole number:

Median: wrist-elbow—59 ± 4, and elbow-axilla—66 ± 5 m/s
Ulnar: wrist-elbow—59 ± 4, and elbow-axilla—64 ± 3 m/s
Common peroneal: 50 ± 6 m/s
Posterior tibial: 46 ± 4 m/s

In a much older age group (51 to 80 years) the velocities were lower by about 4 or 5 m/s in almost all the above categories.

One always must apply "average normal" and "minimum" velocities with caution to individual cases. A wide fluctuation is possible during one session or from day to day, as shown by our experiments described

Table 22.1
Normal Readings for Motor Nerve Conduction Velocity (m/s)

	Mean	Range	S.D.	Source
Ulnar	55.1		6.4	Johnson & Olsen
	58.7	50.8–66.7	4.0	Abramson et al
	59.1	49.1–65.5		Henriksen
	60.4	47.0–73.0	5.8	Thomas & Lambert
Median	56.1	46.8–68.4	4.5	Abramson et al
	56.4	47.9–68.3	5.4	Jebsen
	58.5	53.0–64.3		Henriksen
Radial	58.4	45.4–82.5	6.7	Jebsen
	72.0		6.1	Gassel et al
Tibial	46.2	37.4–58.9	3.3	Jebsen
	50.2		9.3	Johnson & Olsen
Peroneal	47.3	40.2–57.0	4.3	Jebsen
	50.1		7.2	Johnson & Olsen
	51.0		3.3	Thomas & Lambert
	51.5	45.6–56.3		Henriksen

[a] From M.P. Smorto and J.V. Basmajian, © 1979, Williams & Wilkins, Baltimore.

below and those of Christie and Coomes (1960). With variable physical factors playing a very real role, considerable error can creep into the results (Schubert, 1964).

LaFratta and Smith (1964) have shown that, on the average, conduction velocities are greater in women than in men, although the overlap shown in their graph seems quite large. They also found a slight decline in velocity with the older age groups.

The use of these EMG methods to determine slowing in nerves affected by injury and various neuropathies has been discussed by Johnson and Olsen (1960) and Bastron and Lambert (1960), and by others. In thorough articles, Thomas (1961) and Dunn et al (1964) reviewed the progress made in the clinical applications of these techniques. The purpose of this chapter (and, indeed, this book) is to avoid purely clinical discussions while providing a solid foundation of information about normal neuromuscular function as revealed by this application of electromyography. One does this reluctantly, for the recent clinical literature on conduction velocity is profuse. (See *Clinical Electroneurography*, Smorto and Basmajian, 1979).

Detailed study of normal conduction velocity and residual latency, carried out on volunteers over a period of hours, showed that conduction velocity in the motor fibers of the ulnar nerve between the elbow and the wrist fluctuates considerably, sometimes rather widely from time to time in any individual (Low et al, 1962). Therefore, single clinical estimations must be viewed with caution. Residual latency, a function of the delay in the terminal neuromuscular apparatus, also shows a fluctuation with time. Nevertheless, the study of the residual latency offered a new tool in the investigation of neuromuscular function. It deserves wider investigation, but to date it has been neglected.

The most striking observation in individual experiments was the fluctuation in conduction velocity. In individual subjects, velocities measured 15 minutes apart fluctuated by as much as 8 m/s. Generally, the first two or three determinations (during the first 30 minutes of an experiment) tended to be higher than those that followed. Aside from this (inconstant) tendency for the velocity to decrease as the experiment progressed, values did not vary in a predictable way.

Because the temperature of the limbs was not controlled, temperature variations may account for some of the differences observed between individuals and for some of the differences from time to time in the same person. Such variations were considered and deliberately accepted from the start so that the results might be directly comparable to determinations made on patients in routine diagnostic clinics where elaborate temperature control is impractical.

The mean control value for the conduction velocity in the ulnar nerve, which was 56.3 ± 4.2 m/s, agrees closely with values quoted in the literature, those given by Thomas et al (1959) for the ulnar nerve, being 56 ± 4.6 m/s.

No one had commented previously on the fluctuation in conduction velocity with the passage of time in normal subjects. Reasons for the irregular fluctuations are not evident in the literature. Undoubtedly, progressive cooling of the forearm occurs as it lies exposed during an experiment but, as explained before, temperature changes can be only a part of the explanation.

Changes in blood flow, variations in metabolic processes of the nerve or nearby muscle, or changing influences from the central nervous system all must contribute to the fluctuations. An important implication of their occurrence is that we should be cautious in making only one or two determinations in a brief period to compare with "normal" or "abnormal" values. Johnson and Olsen (1960) state that their whole procedure takes ". . . five or ten minutes in a co-operative patient." None of the other writers reports exactly what time is involved in the determination of their velocities. Sources or error in routine "diagnostic" tests are many and complex. Gassel (1964) incriminates the following: recordings of potentials arising from muscles at a distance from electrodes; anomalous innervation; and spread of stimuli other than that over which the recording electrodes are placed. Simpson (1964) also warns against facile acceptance of results.

Nerve conduction velocities may be influenced by various conditions other than neurological disease and by pharmacological influences. For detailed reports (which would be out of place in this book) one should refer to the following: the effects of chronic alcoholic polyneuropathy (Mawdsley and Mayer, 1965); the acute effects of ethanol (Low et al, 1962); effects of acute drug intoxication (Pihkanen et al, 1965); and of cold (de Jesus et al, 1973).

Aging affects nerve conduction velocity, especially during the early years. From 1 month of age to 4 years, mean velocity rises from 33 m/s to over 50 m/s (Wagner and Buchthal, 1972). In adults, Trojaborg (1976) reported a drop of 2 m/s in the musculocutaneous nerve for each decade of increase in age. However, Nielson (1973) has shown that there is no significant difference due to the sex of the subject.

RESIDUAL LATENCY

The mean of control residual latencies of 1.52 ± 0.21 ms (Low et al, 1962) is somewhat lower than the other three mean values of residual latency in adults reported in the literature. Hodes et al (1948), Norris et al (1953), and Bolzani (1955) gave values of 2.2, 1.7, and 2.68 ms, respectively.

Like the values for conduction velocity, the residual latencies determined in each person fluctuate throughout an experiment lasting several hours. They tend to drop slightly, probably for the same sort of reasons that cause variations in conduction velocity. Only one paper comments on any positive change in residual latencies during experiments. Hodes et al (1948) reported that in regenerating nerves (ulnar and median) following suture, the residual latencies were greater than the values for the same nerves under normal conditions. At the same time, they found that nerve conduction velocity was slowed. In a study by Norris et al (1953) performed on a large number of subjects from 50 to 90 years of age, the velocity of conduction in the ulnar nerve decreased with increasing age, but the residual latencies remained unchanged.

Before concluding this chapter, a more comprehensive definition of "residual latency" than that offered by other writers must be made by discussing its elements. An important component of the residual latency must be the time taken for the muscle to respond to depolarization of the endplate. The residual latency must also include a short time taken by the muscle action potential to pass beneath the proximal recording electrode because muscles have a slow conduction velocity ranging from 1.3 to 4.7 m/s, as shown by Ramsay (1960) and Eccles and O'Connor (1939). With surface electrodes, this factor is extremely small because they are placed over the area of innervation and they gather potentials from a relatively wide area.

Another component of the residual latency, the slowing of the impulse in the final terminal fibers of the nerve, has been calculated by Eccles and O'Connor (1939) to be about 0.2 ms. There is a fourth and perhaps the most significant component: the time consumed by neuromuscular transmission or the "synaptic delay." Direct measurements of this delay give values of about 0.5 ms (Eccles and O'Connor, 1939). According to Nachmansohn (1959) and Fatt (1959), this delay must be due to the time taken by electrical events in the nerve ending causing the release of acetylcholine and the corresponding buildup of the postsynaptic potential. The transmission of acetylcholine across the 500 Å gap cannot be a factor in the delay since Eccles and O'Connor estimated that this process should take no more than 10 μs.

Thus the latent period from the arrival of the impulse at the fine terminal nerve branches to the beginning of the muscle action potential is composed of the four factors discussed above. This time has been measured directly; in mammalian striated muscle it is approximately 0.85 ms (Eccles and O'Connor, 1939). The values for residual latency determined in our experiments are of the order of 1.5 ms; therefore, they must include factors in addition to the four previously mentioned.

According to some authors (Magladery and McDougal, 1950; Gilliatt and Thomas, 1959) there is a decrement in conduction velocity along a nerve in an extremity. Thus the distal portions conduct more slowly than proximal segments of the same nerve (Copack et al, 1975). If this is true, then the assumption that the velocity of the impulse from elbow to wrist is the same as that from wrist to muscle must introduce an error in calculation of the residual latency by the accepted method (i.e., subtracting the calculated latency of response in the hypothenar muscles applied at the elbow from the observed latency). This is mentioned by Magladery and McDougal, and probably it contributes most of the remainder of the residual latency. In our series this portion would be about 0.67 ms.

SENSORY NERVE CONDUCTION

Although this subject is fully considered in our other book (Smorto and Basmajian, 1979), here it might be briefly reviewed as an introduction for some readers. Direct recording of nerve potentials dates from 1949 when Dawson and Scott developed the first clinical approach to the study of the sensory nerve latencies in man. They showed that by summation of single responses, potentials may be recorded from the skin. They first applied the method to clinical diagnosis of peripheral nerve lesions.

The early experiments were concerned with sensory latencies obtained by applying electrical stimuli to the ulnar and median nerves at the wrist. The stimuli evoked afferent volleys of potentials in the sensory fibers of the nerve trunk. A pair of surface electrodes were placed over the course of the nerve at different levels of the arm and elbow. Dawson and Scott analysed these waves, which were small, di-, or triphasic and were less than 50 μV in amplitude. The durations were 2 to 3 ms. This method elicited not only orthodromic ascending sensory waves, but also antidromic stimulation of the motor fibers of the same nerve.

Seven years later Dawson (1956) modified his technique in order to eliminate the antidromic stimulation; instead of applying the stimulus at the wrist (where both sensory and motor fibers are present) he stimulated the nerve at the fingers where no motor fibers are present. The sensory nerve potentials evoked in this way were recorded at the wrist. However, the evoked potentials were much smaller in amplitude (5 to 15 μV). Nevertheless, Dawson successfully recorded the small potentials in 14 normal subjects. (The sensory latency between the fingers and the wrist averaged 4.73 ms.) The motor latencies from the wrist to hypothenar muscles or wrist to thenar muscle were about equal—5.12 ms. Details of methodologies and results are given in Smorto and Basmajian (1979).

Shagass (1961) described a method for recording cerebral evoked potentials after stimulation of major peripheral nerves of the body by using an averaging technique. In 1963 Liberson and Kim, by using the mnemotron, were able to record evoked potentials from the median, ulnar, radial, tibial, and peroneal nerves. They also concluded that while the method of cerebral evoked potentials permits assertion of the continuity of a sensory nerve, a precise determination of the conduction velocity may be somewhat difficult.

Determination of conduction velocities in sensory fibers elicited by H and F reflex responses has held considerable interest, but systematic study which early depended on the work of Liberson et al (1966) has now expanded to involve many clinical investigators. The chief usefulness of the technique is in the presence of pathology; therefore, a review of this field may be obtained from Smorto and Basmajian (1979).

References

Aarons L: Diurnal variations of muscle action potentials and word associations related to psychological orientation. *Psychophysiology* 5:77–91, 1968.

Abo-El-Enein MA, Wyke B: Laryngeal myotatic reflexes. *Nature* 209:682–686, 1966.

Adler N, Perry J, Kent B, Robertson K: Electromyography of the vastus medialis oblique and vasti in normal subjects during gait. *Electromyogr Clin Neurophysiol* 23:643–649, 1983.

Adrian ED, Bronk DW: The discharge of impulses in motor nerve fibres. Part I. Impulses in single fibres of the phrenic nerve. *J Physiol* 66:81–101, 1928.

Adrian ED, Bronk DW: The discharge of impulses in motor nerve fibres. Part I. Impulses in single fibres of the phrenic nerve. *J Physiol* 66:81–101, 1928.

Adrian ED, Bronk DW: The discharge of impulses in motor nerve fibres. Part II. The frequency of discharge in reflex and voluntary contractions. *J Physiol* 67:19–151, 1929.

Agarwal GC, Berman BM, Lohnberg P, Stark L: Studies in postural control systems. II. Tendon jerk input. *IEEE Trans Syst Sci Cybernet* 6:122–126, 1970.

Agarwal GC, Gottlieb GL: An analysis of the electromyogram by Fourier, simulation and experimental techniques. *IEEE Trans Biomed Eng* 22:225–229, 1975.

Agostoni E: Diaphragm contraction as a limiting factor to maximum expiration. *J Appl Physiol* 17:427–428, 1962.

Agostoni E: Diaphragm activity during breath holding: factors related to onset. *J Appl Physiol* 18:30–36, 1963.

Agostoni E, Sant'Ambrogio G, Carrasco H del P: Elettromiografia del diaframma e pressione transdiaframmatica durante la tosse, lo sternuto ed il riso. *Atti Accad Nazion Lincei* 28:493–496, 1960.

Agostoni E, Sant'Ambrogio G, Garrasco H del P: Electromyography of the diaphragm in man and transdiaphragmatic pressure. *J Appl Physiol* 15:1093–1097, 1960.

Ahlborg B, Bergstrom J, Ekelund LG, Guarnieri G, Harris RC, Hultman E, Nordesjo LO: Muscle metabolism during isometric exercise performed at constant force. *J Appl Physiol* 33:224–228, 1972.

Ahlgren J: Mechanism of mastication: a quantitative cinematographic and electromyographic study of masticatory movements in children, with special reference to occlusion of the teeth. *Acta Odont Scandinav* 24:suppl. 44, 109 pp., 1966.

Ahlgren J: Kinesiology of the mandible: an EMG study. *Acta Odont Scandinav* 25:593–611, 1967.

Ahlgren J: The silent period in the EMG of the jaw muscles during mastication and its relationship to tooth contact. *Acta Odontol Scandinav* 3:(27)219–227, 1969.

Ahlgren J, Lipke DP: Electromyographic activity in digastric muscles and opening force of mandible during static and dynamic conditions. *Scand J Dent Res* 85:152–154, 1977.

Akamatsu K: Electromyographic studies on the expressionless faces of schizophrenics. *Hiroshima Med J* 8:2249, 1960.

Åkerblom B: *Standing and Sitting Posture* (tr. by Ann Synge). Stockholm, A.-B. Nordiska Bokhandeln., 1948.

Åkerblom B: Anatomische und Physiologische Grundlagen zur Gestaltung von Sitzen. *Ergonomics* 12:120–131, 1969.

Alexander AB: An experimental test of assumptions relating to the use of electromyographic biofeedback as a general relaxation training technique. *Psychophysiology* 12:656–662, 1975.

Alexander MA, Season EH: Ideopathic scoliosis: an electromyographic study. *Arch Phys Med Rehab* 59:314–315, 1978.

Allen CEL: Muscle action potentials used in the study of dynamic anatomy. *Br J Phys Med* 11:66–73, 1948.

Allen G, Lubker J, Harrison E: New paint-on electrode for surface electromyography. *J Acoust Soc Am Suppl* 1:52, S124, 1972.

Allen GD, Lubker JF, Turner DT: Adhesion to mucous membrane for electromyography. *J Dent Res* 52:394, 1973.

Allert ML: Anal reflex in the electromyogram of bladder and intestinal sphincter muscles. *Wien Z Nervenh* 27:281–287, 1969.

Amano N, Funakoshi M: Tonic periodontal-masseteric reflex in man. *J Gifu Dent Soc* 3:182–183, 1976.

Amato A, Hermsmeyer CA, Kleinman KM: Use of electromyographic feedback to increase inhibitory control of spastic muscles. *Phys Ther* 53:1063–1066, 1973.

Anch AM, Remmers JE, Sauerland EK, De Groot WJ: Oropharyngeal patency during waking and sleep in the Pickwickian syndrome: Electromyographic activity of the tensor veli palatini. *Electromyogr Clin Neurophysiol* 21:317–330, 1981.

Andersen P, Hendriksson J: Capillary supply of quadriceps femoris muscle of non-adaptive response to exercise. *J Physiol* 270:677–690, 1977.

Anderson GBJ, Örtengren R, Nachemson A: Intradiskal pressure, intra-abdominal pressure and myoelectric back muscle activity related to posture and loading. *Clin Orthop Related Res* 129:156–164, 1977.

Anderson KV, Mahan PE: Interaction of tooth pulp and periodontal ligament receptors in a jaw-depression reflex. *Exp Neurol* 23:295–302, 1971.

Andersson JG, Jonsson B, Örtengren R: Myoelectric activity in individual lumbar erector spinae muscles in sitting: a study with surface and wire electrodes. *Scand J Rehabil Med* 3(suppl):91–108, 1974.

Andersson JG, Örtengren R, Nachemson AL, Elfstrom G, Broman H: The sitting posture: an electromyographic and discometric study. *Orthop Clin North Am* 6:105–120, 1975.

Andersson S, Stener B: Experimental evaluation of the hypothesis of ligamento-muscular protective reflexes. II. A study in cat using the medial collateral ligament of the knee joint. *Acta Physiol Scand* 48(suppl): 27–49, 1959.

Andreassen S: Interval Pattern of Single Motor Units. Ph.D. dissertation, Technical University of Denmark.

Andreassen S, Rosenfalck A: Recording from a single motor unit during strong effort. *IEEE Trans BME* 25:501–508, 1978.

Andreassen S, Rosenfalk A: Regulation of the firing patterns of single motor units. *J Neurol Neurosurg Psychiatry* 43:897–906, 1980.

Angel RW: Electromyographic patterns during ballistic movement of normal and spastic limbs. *Brain Res* 99:387–392, 1975.

Angel RW: Antagonist muscle activity during rapid arm movements: central versus proprioceptive influences. *J Neurol Neurosurg Psychiatry* 40:683–686, 1977.

Anggard L, Ottoson D: Observations on the functional development of the neuromuscular apparatus in fetal sheep. *Exp Neurol* 7:294–304, 1963.

Antonelli D, Beckey GA, Hary D, Zeman B: Development and evaluation of a reference EMG signal acquisition system. Report from Pathokinesiology Lab., Rancho Los Amigos Hospital, CA, 1982.

Aranda L, Jara O: Estudio electromiografico del reflejo bulbocavernoso en condicion normal y patologica. *Neurocirurgia (Santiago)* 27:248–250, 1969.

Archibald KC, Goldsmith EI: Sphincteric electromyography. *Arch Phys Med* 48:387–392, 1967.

Arienti A: Estudos electromiográficos da locomoção humana (Portuguese text). *Resenha ClinCient Instit Lorenz* 17:175–178, 1948a.

Arienti A: Analyse oscillographique de la marche de l'homme (French text). *Acta Physiother Rheumat Belg* 3:190–192, 1948b.

Arlazoroff A, Rapoport Y, Shanon E, Streifler M: Observations on the electromyogram of the oesophagus at rest and during Valsalva's manoeuvre. *Electroencephalogr Clin Neurophysiol* 33:110–113, 1972.

Ashworth B, Grimby L, Kugelberg E: Comparison of voluntary and reflex activation of motor units: functional organization of motor neurones. *J Neurol Neurosurg Psychiatry* 30:91–98, 1967.

Asmussen E: The weight-carrying function of the human spine. *Acta Orthop Scandinav* 29:276–290, 1960.

Asmussen E, Klausen K: Form and function of the erect human spine. *Clin Orthop* 25:55–63, 1962.

Avni J, Chaco J: Objective measurements of postural changes in depression. Paper presented to Annual Meeting of American Psychological Association, Boston, MA, 1972.

Avon G, Schmitt L: Electromyographie du trapèze dans diverses positions de travail à la machine à écrire. *Ergonomics* 18:619–626, 1975.

Bäckdahl M, Carlsöö S: Distribution of activity in muscles acting on the wrist (an electromyographic study). *Acta Morph Neerl-Scand* 4:136–144, 1961.

Backhouse KM: The mechanics of normal digital control in the hand and analysis of the ulnar drift of rheumatoid arthritis. *Ann Royal Coll Surg Engl* 43:154–173, 1968.

Backhouse KM, Catton WT: An experimental study of the functions of the lumbrical muscles in the human hand. *J Anat* 88:133–141, 1954.

Bahoric A, Chernick V: Electrical activity of phrenic nerve and diaphragm in utero. *J Appl Physiol* 39:513–518, 1975.

Bailey JA, Powers JJ, Waylonis GW: A clinical evaluation of electromyography of the anal sphincter. *Arch Phys Med* 51:403–409, 1970.

Bajd T, Stanič U, Tomšič M, Krajnik J: Symposium and Seminars, "Informatica 74," Bled, Jugoslavia, October 7–12, 1974.

Baker M, Regenos E, Wolf SL, Basmajian JV: Developing strategies for biofeedback: applications in neurologically handicapped patients. *Phys Ther* 57:402–408, 1977.

Baker PS: A method for obtaining muscle action potentials from a locust during free turning in flight. Proceedings of Physiol Soc, November 1972, 2 P–3 P.

Baldissera F, Gustafsson B, Parmiggiani F: Saturating summation of the after-hyperpolarization conductance in spinal motoneurones: a mechanism for secondary range repetitive firing. *Brain Res* 146:69–82, 1978.

Ballesteros MLF: See Fernandez-Ballesteros.

Ballintijn CM: Muscle coordination of the respiratory pump of the carp (cyprinus carpio L.) *J Exp Biol* 50:569–591, 1969.

Ballintijn CM, Hughes GM: The muscular basis of the respiratory pumps in the trout. *J Exp Biol* 43:349–362, 1965a.

Ballintijn CM, Hughes GM: The muscular basis of the respiratory pumps in the dogfish (*scyliorhinus canicula*). *J Exp Biol* 43:363–383, 1965b.

Balshan ID (Goldstein): Muscle tension and personality in women. *Arch Gen Psychiatry* 7:436–448, 1962.

Bankov S, Jørgensen K: Maximum strength of elbow flexors with pronated and supinated forearm. *Comm Dan Nat Ass Infant Paral* No. 29, 1969.

Banks RW, Barker D, Stacey MJ: *Muscle Receptors and Movement*, Taylor A, and Prochazka (eds). London, Macmillan, 1981, p 5.

Baranova-Krylov IN: Testing of the excitability of the spinal centres of a pair of antagonistic muscles (Russian text). *Zh Vyssh Nerv Diat Pavlov* 16:889–891, 1969.

Bárány M: ATPase activity of myosin correlated with speed of muscle shortening. *J Gen Physiol* 50:197, 1967.

Barnett CH, Harding D: The activity of antagonist muscles during voluntary movement. *Ann Phys Med* 2:290–293,

Barnett CH, Richardson A: The postural function of the popliteus muscle. *Ann Phys Med* 1:177–179, 1953.

Barrett DM, Wein AJ: Flow evaluation and simultaneous external sphincter electromyography in clinical urodynamics. *J Urol* 125:538–541, 1981.

Bartoshuk AK, Kaswich JA: Electromyographic gradients as a function of tracking cues. *Psychon Sci* 6:43–44, 1966.

Basmajian JV: Electromyography. *Univ Toronto Med J* 30:10–18, 1952.

Basmajian JV: *Primary Anatomy, ed 3.* Baltimore, Williams & Wilkins, 1955a.

Basmajian JV: Letters to the Editor. *Lancet* 2:1140, 1955b.

Basmajian JV: Electromyography of two-joint muscles. *Anat Rec* 129:371–380, 1957a.

Basmajian JV: New views on muscular tone and relaxation. *Can Med Assoc J* 77:203–205, 1957b.

Basmajian JV: A new six-channel electromyograph for studies on muscle. *I.R.E. Trans Med Electronics* PGME-11:45–47, 1958a.

Basmajian JV: Electromyography of iliopsoas. *Anat Rec* 132:127–132, 1958b.

Basmajian JV: "Spurt" and "shunt" muscles: an electromyographic confirmation. *J Anat* 93:551–553, 1959.

Basmajian JV: Weight-bearing by ligaments and muscles. *Can J Surg* 4:166–170, 1961.

Basmajian JV: Control and training of individual motor units. *Science* 141:440–441, 1963a.

Basmajian JV: Conscious control of single

nerve cells. *New Scientist* 20:662–664, 1963b.
Basmajian JV: *Primary Anatomy,* ed 6. Baltimore, Williams & Wilkins, 1970.
Basmajian JV: Electromyography comes of age. *Science* 176:603–609, 1972.
Basmajian JV: A simple method to improve performance of fine-wire emg electrodes. *Am J Phys Med* 53:269–270, 1975.
Basmajian JV: Facts *vs.* myths in emg biofeedback. *Biofeedback Self-Reg* 1:396–371, 1976.
Basmajian JV: The human bicycle: an ultimate biological convenience. *Orthop Clin North Am* 7:1027–1029, 1976b.
Basmajian JV: Motor learning and control: a working hypothesis. *Arch Phys Med Rehab* 58:38–41, 1977.
Basmajian JV, Baeza M, Fabrigar C: Conscious control and training of individual spinal motor neurons in normal human subjects. *J New Drugs* 5:78–85, 1965.
Basmajian JV, Bazant FJ: Factors preventing downward dislocation of the adducted shoulder joint: an electromyographic and morphological study. *J Bone Joint Surg* 41A:1182–1186, 1959.
Basmajian JV, Bentzon JW: An electromyographic study of certain muscles of the leg and foot in the standing position. *Surg Gynecol Obstet* 98:662–666, 1954.
Basmajian JV, Boyd WH: (Motion picture film) Electromyography of the Diaphragm: Direct Recording Technique in the Rabbit. Exhibited at Am. Assoc. of Anatomists Annual Meeting, Chicago, March 21–23, 1961; abstract in *Anatomical Record* 139:337, 1961.
Basmajian JV, Clifford HC, McLeod WD, Nunnally HN: *Computers in Electromyography.* London, Butterworths, 1975.
Basmajian JV, Cross GL: Duration of motor unit potentials from fine-wire electrodes. *Am J Phys Med* 50:144–148, 1971.
Basmajian JV, Dutta CR: Electromyography of the pharyngeal constrictors and soft palate in rabbits. *Anat Rec* 139:443–449, 1961a.
Basmajian JV, Dutta CR: Electromyography of pharyngeal constrictors and levator palati in man. *Anat Rec* 139:561–563, 1961b.
Basmajian JV, Forrest WJ, Shine G: A simple connector for fine-wire electrodes. *J Appl Physiol* 21:1680, 1966.
Basmajian JV, Greenlaw RK: Electromyography of iliacus and psoas with inserted fine-wire electrodes (abstr). *Anat Rec* 160:310, 1968.
Basmajian JV, Griffin WR: Function of anconeus muscle: an electromyographic study. *J Bone Joint Surg* 54A:1712–1714, 1972.
Basmajian JV, Harden TP, Regenos EM: Integrated actions of the four heads of quadriceps femoris: an electromyographic study. *Anat Rec* 172:15–20, 1972.
Basmajian JV, Hudson JE: Miniature source-attached differential amplifier for electromyography. *Am J Phys Med* 53: 234–236
Basmajian JV, Kukulka CG, Narayan MG, Takebe K: Biofeedback treatment of foot-drop after stroke compared with standard rehabilitation technique: effects on voluntary control and strength. *Arch Phys Med Rehab* 56:231–236, 1975.
Basmajian JV, Latif A: Integrated actions and functions of the chief flexors of the elbow: a detailed electromyographic analysis. *J Bone Joint Surg* 39A:1106–1118, 1957.
Basmajian JV, Lovejoy JF, Jr: Functions of the popliteus muscle in man: a multifactorial electromyographic study. *J Bone Joint Surg* 53A:557–562, 1971.
Basmajian JV, Newton WJ: Feedback training of parts of buccinator muscle in man. *Psychophysiology* 92, 1973.
Basmajian JV, Samson J: Standardization of methods in single motor unit training. *Am J Phys Med* 52:250–256, 1973.
Basmajian JV, Simard TG: Effects of distracting movements on the control of trained motor units. *Am J Phys Med* 46:1427–1449, 1967.
Basmajian JV, Spring WB: Electromyography of the male (voluntary) sphincter urethrae. *Anat Rec* 121:388, 1955.
Basmajian JV, Stecko GA: A new bipolar indwelling electrode for electromyography. *J Appl Physiol* 17:849,
Basmajian JV, Stecko G: Role of muscles in arch support of the foot. *J Bone Joint Surg* 45A:1184–1190, 1963.
Basmajian JV, Super G: Dantrolene sodium in the treatment of spasticity. *Arch Phys Med* 54:60–64, 1973.
Basmajian JV Szatmari A: Effect of largactil (chlorpromazine) on human spasticity and electromyogram. *AMA Arch Neurol* 73:224–231, 1955a.
Basmajian JV, Szatmari A: Chlorpromazine and human spasticity: an electromyographic study. *Neurology* 5:856–860, 1955b.
Basmajian JV, Travill A: Electromyography of the pronator muscles in the forearm.

Anat Rec 139:45–49, 1961.
Basmajian JV, White ER: Neuromuscular control of trumpeters' lips. *Nature* 241:70, 1973.
Bass L, Moore WJ: The role of protons in nerve conduction. *Prog Biophys Mol Biol* 27:143–147, 1973.
Baštecký J, Vinař O, Roth Z: Delayed auditory feedback and EMG of mimic muscles in schizophrenia. *Activ Nerv Suppl* (Praha), 10:212–214, 1968.
Baštecký J, Vinař O, Roth Z: Lee effect and EMG of the mimic muscles in schizophrenia (Czech. text; English summary) *Bratislavské Lekaraké Listy* 49:618–622, 1969.
Bastian J: Neuro-muscular mechanisms controlling a flight maneuver in the honeybee. *J Comp Physiol* 77:126–140, 1972.
Bastron JA, Lambert EH: The clinical value of electromyography and electrical stimulation of nerves. *Med Clin North Am* 44:1025–1036, 1960.
Battye CK, Joseph J: An investigation by telemetering of the activity of some muscles in walking. *Med Biol Eng* 4:125–135, 1966.
Baumann F: Zur Frage der muskulären Verspannung bei der Koxarthrose auf Grund elektromyographischer Untersuchungen. *Z Orthop Grenz* 106:500–508, 1969.
Baumann F: Die Elektromyographie in der Orthopädie. *Z Orthop Grenz* 109:326–338, 1971.
Baumann F: Der muskuläre Entspannungseffekt bei Hüftoperationen. In Otte P, Schlegel K-F: *Bücherei des Orthopäden*, Stuttgart, Ferdinand Enke Verlag, 1972, vol 9, pp 43, 103.
Baumann F, Behr O: Elektromyographische Untersuchungen der Hüftmuskulatur nach Arthrodese. *Arch Orthop Unfall-Chir* 66:1–17, 1969.
Bauwens P: Electromyography. *Br J Phys Med* 11:130–136, 1948.
Bauwens P: The analysis of action potentials in electromyography. *Proc Inst Electr Eng* 97(107):217–222, 1950.
Bauwens P, Richardson AT: Electrodiagnosis. In *British Encyclopaedia of Medical Practice 2.* London, Butterworth, 1951.
Bawa P, Calancie BW: Repetitive doublet firing in the human flexor carpi radialis. *J Physiol* 332:33P, 1982.
Baykushev S: Two variants of feedback artificially created in central motor disturbances on the base of EMG. *Electromyogr Clin Neurophysiol* 13:477–483, 1973.
Bearn JG: The significance of the activity of the abdominal muscles in weight lifting. *Acta Anat* 45:83–89, 1961a.
Bearn JG: An electromyographic study of the trapezius, deltoid, pectoralis major, biceps and triceps muscles, during static loading of the upper limb. *Anat Rec* 140:103–108, 1961b.
Beatty CH, Peterson RD, Bocek RM: Metabolism of red and white muscle fiber groups. Am J Physiol 204(5):939–942, 1963.
Beaudreau DE, Daugherty WF Jr, Masland WS: Two types of motor pause in masticatory muscles. *Am J Physiol* 216:16–21, 1969.
Beck A: Elecktromyographische Untersuchungen am Sphincter ani (German text). *Pflüger's Arch Ges Physiol* 224:278–292, 1930.
Becker RO: The electrical response of human skeletal muscle to passive stretch. *J Bone Joint Surg* 42A:1091–1103, 1960.
Becker RO, Chamberlin JT: A modified coaxial electrode for electromyography. *Arch Phys Med* 41:149–151, 1960.
Beevor CE: Croonian lectures on muscular movements and their representation in the central nervous system. *Lancet* 1:1715–1724, 1903.
Beevor CE: *Croonian Lectures on Muscular Movements and Their Representation in the Central Nervous System.* London, Adlard and Son, 1904.
Bekoff A: Ontogeny of leg motor output on the chick embryo. *Brain Res* 106:271–291, 1976.
Belanger AY, McComas AJ: Extent of motor unit activation during effort. *J Appl Physiol* 51:1131–1135, 1981.
Belcastro AN, Bonen A: Lactic acid removal rates during controlled and uncontrolled recovery exercise. *J Appl Physiol* 39:932–936, 1975.
Belenkii VY, Gurfinkel VS, Paltsev YI: Elements of control of voluntary movements. *Biofizika* 12:135–141, 1967.
Bellemare F, Grassino A: The fatigue and recovery of the human diaphragm. In *Proc 4th Congr of the Int Soc of Electrophysiol Kinesiol* 74–79, 1979.
Bellemare F, Grassino A: Evaluation of human diaphragm fatigue. *Appl Physiol* 53:1196–1206, 1982.
Bell-Berti F: Control of pharyngeal cavity size for English voiced and voiceless stops. *J Acoust Soc Am* 57:456–461, 1974.
Bell-Berti F: An electromyographic study of velopharyngeal function in speech. *J Speech Hearing Res* 19:225–240, 1976.
Bell-Berti F, Hirose H: Patterns of palato-

glossal activity and their implications for speech organization. Status Report on Speech Research. Report SR-34 Haskins Laboratories, New Haven, CN, 1973.

Bentley DR, Hoy RR: Postembryonic development of adult motor patterns in crickets: a neural analysis. *Science* 170:1409–1411, 1970.

Beránek R, Novotný I: Spontaneous electrical activity and excitability of the denervated insect muscle. *Physiol Bohemoslovenica* 8:87–92, 1959.

Berger SM, Hadley SW: Some effects of a model's performance on an observer's electromyographic activity. *Am J Psychol* 88:263–276, 1973.

Berger W, Quintern J, Dietz V: Pathophysiology of gait in children with cerebral palsy. *Electroencephalogr Clin Neurophysiol* 53:538–548, 1982.

Bergey GE, Squires RD, Sipple WC: Electrocardiogram recording with pasteless electrodes. *IEEE Trans BME* 18:206, 1971.

Bergström RM: The relation between the number of impulses and the integrated electric activity in electromyogram. *Acta Physiol Scand* 45:97–101, 1959.

Bergström RM: The relation between the integrated kinetic energy and the number of action potentials in the electromyogram during voluntary muscle contraction. Ann Acad Scient Fennicae (Series A):V. Medica 93 Suomalainen Tiedeakatemia, Helsinki, Finland, 1962.

Bergström RM, Kerttula Y: On the neural control of breathing as studied by electromyography of the intercostal muscles of the rat. Ann Acad Scient Fennicae, Series A, V. Medica, 79.

Berlendis PA, DeCaro LG: L'unitá motoria del muscolo stapedio (Italian text). *Boll Soc Med Chir Pavia* 69:33–36, 1955.

Bernshtein VM: Statistical parameters of the electric signal of a muscle model. *Biophysics* 12:693–703, 1967.

Bernstein NA: *Investigations on the Biodynamics of Walking, Running, and Jumping. Part II.* Central Scientific Institute of Physical Culture, Moscow.

Bernstein NA, Salzberger O, Pavelnko P, Popova T, Sadchikov N, Osipov L: Untersuchungen über die Biodynamik der Locomotion. I Teil. Verlag des Instituts für Experimentelle Medizin der Soviet-Union, Moscow, 1935.

Bertoz A, Metral S: Behavior of a muscular group subjected to a sinusoidal and trapezoidal variation of force. *J Appl Physiol* 29:378–384, 1970.

Betts RP, Brown BH: Method for recording electrocardiograms with dry electrodes applied to unprepared skin. *Med Biol Eng* 14:313–315, 1976.

Bicas HEA: Electro-oculography in the investigation of ocular imbalance. I. Basic aspects. *Vision Res* 12:993–1010, 1972.

Biederman W: *Electrophysiology.* London, 1898.

Bierman W, Ralston HJ: Electromyographic study during passive and active flexion and extension of the knee of the normal human subject. *Arch Phys Med* 46:71–75, 1965.

Bierman W, Yamshon LJ: Electromyography in kinesiologic evaluations. *Arch Phys Med* 29:206–211, 1948.

Biggs NL, Pool HE, Blanton PL: An electromyographic study of the temporalis and masseter muscles of monozygotic twins. Trans of Int Soc EMG Kinesiol, Montreal *Electromyography,* 8(suppl):163–171, 1968.

Bigland B, Lippold OCJ: The relation between force, velocity and integrated electrical activity in human muscles. *J Physiol* 123:214–224, 1954a.

Bigland B, Lippold OCJ: Motor unit activity in the voluntary contraction of human muscle. *J Physiol* 125:322–335, 1954b.

Bigland B, Lippold OCJ, Wrench A: The electrical activity in isotonic contractions of human calf muscle. Proc of the Physiol Society, March 20–21. *J Physiol* 120:40P–41P, 1953.

Bigland-Ritchie B: EMG and fatigue of human voluntary and stimulated contractions. Human muscle fatigue: physiological mechanisms. London, Pitman Medical, 1982.

Bigland-Ritchie B, Woods JJ: Integrated EMG and oxygen uptake during dynamic contractions of human muscles. *J Appl Physiol* 36:475–479, 1974.

Bigland-Ritchie B, Donovan EF, Roussos CS: Conduction velocity and EMG power spectrum changes in fatigue of sustained maximal efforts. *J Appl Physiol* 51:1300–1305, 1981.

Bills AG: *The Psychology of Efficiency.* New York, Harper, 1943.

Binder MD, Kroin JS, Moore GP, Stauffer EK, Stuart DG: Correlation analysis of muscle spindle responses to single motor unit contractions. *J Physiol* 257:325–336, 1976.

Binder MD, Kroin JS, Moore GP, Stuart DG: The response of Golgi tendon organs to single motor unit contractions. *J Physiol* 271:337–349, 1977.

Binder MD, Stuart DG: Responses of Ia and spindle group II afferents to single motor-units contractions. *J Neurophysiol* 43:621–629, 1980.

Biro G, Partridge LD: Analysis of multiunit spike records. *J Appl Physiol* 30:521–526, 1971.

Bishop B: Reflex control of abdominal muscles during positive-pressure breathing. *J Appl Physiol* 19:224–232, 1964.

Bizzi E, Chapple W, Hogan N: Mechanical properties of muscles. Implications for motor control. *TINS* 5(11):395–398, 1982.

Bizzi E, Kalil RE, Tagliasco V: Eye-head coordination in monkeys: evidence for centrally patterned organization. *Science* 173:452–454, 1971.

Bjelle A, Hagberg M, Michaelson G: Occupational and individual factors in acute shouder-neck disorders among industrial workers. *Br J Indust Med* 38:356–363, 1981.

Björk Å, Kugelberg E: The electrical activity of the muscles of the eye and eyelids in various positions and during movement. *Electroencephalog Clin Neurophysiol* 5:595–602, 1953.

Björk A, Wåhlin Å: The effect of succinylcholine on the cat diaphragm: An electromyographic study. *Acta Anaesth Scand* 4:13–20, 1960.

Blanton PL, Biggs NL, Perkins RC: Electromyographic analysis of the buccinator muscle. *J Dent Res* 49:389–394, 1970.

Blinowska A, Verroust J, Cannet G: The determination of motor electromyographic power spectrum. *Electromyogr Clin Neurophysiol* 19:281–290, 1979.

Blinowska A, Verroust J, Cannet G: An analysis of synchronization and double discharge effects on low frequency electromyographic power spectrum. *Electromyogr Clin Neurophysiol* 20:465–480, 1980.

Blodi FC, Van Allen MW: Electromyography of extraocular muscles in fusional movements: I. Electric phenomena at the breakpoint of fusion. *Am J Ophthalmol* 44(4), Special Part II (not I) of the October issue):136–142, 1957.

Blodi FC, Van Allen MW: The effect of paralysis of the cervical sympathetic system on the electromyogram of extraocular muscles in the human. *Am J Ophthalmol* 49:679–683, 1960.

Boeëthius J, Knutsson E: Resting membrane potential in chick muscle cells during ontogeny. *J Exp Zool* 174:281–286, 1970.

Bole CT: Electromyographic kinesiology of the genioglossus muscles in man. M.S. Thesis, Ohio State University, Columbus, OH, 1965.

Bolzani L: La curva intensitádurata (i/t) ela velocitá di conduzione del nervo nell'uomo. *Arch Suisses Neurol Psychiatr* 74:148–154, 1955.

Bonde-Petersen F, Mork AL, Nielsen E: Local muscle blood flow and sustained contractions of human arm and back muscle. *Eur J Appl Physiol* 34:43–50, 1975.

Bonde-Petersen F, Robertson CH Jr: Blood flow in "red" and "white" calf muscles in cats during isometric and isotonic exercise. *Acta Physiol Scand* 112:243–251, 1981.

Booker HE, Rubow RT, Coleman PJ: Simplified feedback in neuromuscular retraining: an automated approach using electromyographic signals. *Arch Phys Med* 50:621–625, 1969.

Borden GJ, Harris KS: Oral feedback. II. An electromyographic study of speech under nerveblock anesthesia. *J Phonet* 1:297–308, 1973.

Borg E: Excitability of the acoustic m. stapedius and m. tensor tympani reflexes in the nonanesthetized rabbit. *Acta Physiol Scand* 85:374–389, 1972.

Borg JL, Grimby L, Hannerz J: Axonal conduction velocity and voluntary discharge properties of individual short toe extensor motor units in man. *J Physiol* 277:143–152, 1978.

Bors E: Ueber das Zahlenverhältnis zwischen Nerven- und Muskelfasern. *Anat Anz* 60:14–416, 1926.

Bors EJ, Blinn KA: Spinal reflex activity from the vesical mucosa in paraplegic patients. *AMA Arch Neurol* 78:339–354, 1957.

Bors E, Blinn DA: Bulbocavernosus reflex. *J Urol* 82:128–130, 1959.

Bors E, Blinn KA: Abdominal electromyography during micturition. *Calif Med* 102:17–22, 1965.

Botelho SY, Steinberg SA: Electromyography in canine fetus (Abstr). *BullAm Assoc EMG Electrodiag* 12:17, 1965.

Boter J, Den Hertog A, Kuiper J: Disturbance-free skin electrodes for persons during exercise. *Med Biol Eng* 4:91, 1966.

Bouisset S, Denimal J, Soula C: Relation entre l'accélération d'un raccourcissement musculaire et l'activité electromyographique intégrée. *J Physiol Paris* 55:203, 1963.

Bouisset S, Goubel F: Relation entre l'activité électromyographique intégrée et la vitesse d'execution de mouvements monoarticulaires simples. *J Physiol* 59:359, 1967.

Bouisset S, Goubel F: Interdependence of relations between integrated EMG and diverse biomechanical quantities in normal voluntary movements. *Activitas Nervosa Superior* 13:23–31, 1971.

Bouisset S, Maton B: Quantitative relationship between surface EMG and intramuscular electromyographic activity in voluntary movement. *Am J Phys Med* 51:285–295, 1972.

Bouisset S, Maton B: Comparison between surface and intramuscular EMG during voluntary movement. In Desmedt JE: *New Developments in EMG and Clinical Neurophysiology*, Basel, Karger, 1973, vol 1, pp 533–539.

Bouisset S, Zattara M: A sequence of postural movements precedes voluntary movements. *Neurosci Lett* 22:263–270, 1981.

Boyd IA, Roberts TDM: Proprioceptive discharges from stretchreceptors in the knee-joint of the cat. *J Physiol* 122:38–58, 1953.

Boyd WH: The electrical activity of intercostal muscles during quiet breathing in rabbits. *Can J Physiol Pharm* 46:749–755, 1968.

Boyd WH: Coordinated electrical activity of diaphragm and intercostal muscles in rabbits during quiet breathing. *Arch Phys Med* 50:127–132, 1969.

Boyd WH, Basmajian JV: Electromyography of the diaphragm in rabbits. *Am J Physiol* 204:943–948, 1963.

Boyd WH, Blincoe H, Hayner JC: Sequence of action of the diaphragm and quadratus lumborum during quiet breathing. *Anat Rec* 151:579–582, 1965.

Braithwaite F, Channell GD, Moore FT, Whillis J: The applied anatomy of the lumbrical and interosseous muscles of the hand. *Guy's Hosp Rep* 97:185–195, 1948.

Brandell BR: An electromyographic-cinematographic study of the muscles of the index finger. *Arch Phys Med* 51:278–285, 1970.

Brandell BR: Functional roles of the calf and vastus muscles in locomotion. *Am J Phys Med* 56:59–74, 1977.

Brandell BR, Huff GJ, Spark GJ: An electromyographic-cinematographic study of the thigh muscles using M.E.R.D. (muscle electronic recording device). I. *Trans Int Soc EMG Kines,* Montreal, 1968, pp 67–76.

Brandstater ME, Lambert EH: Motor unit anatomy. In Desmedt JE: *New Developments in Electromyography and Clinical Neurophysiology.* Basel, Karger, 1973, vol 1, pp 14–22.

Brandt K, Fenz WD: Specificity in verbal and physiological indicants of anxiety. *Percept Motor Skills* 29:663–675, 1969.

Brantner JN, Basmajian JV: Effects of training on endurance in hanging by the hands. *J Motor Behav* 7:131–134, 1975.

Bratanova Ts Kh: On bioelectric activity of muscles-antagonists in the course of elaboration of a motor habit (Russian text). *Zh Vyssh Nerv Diat Pavlov* 16:411–416, 1966.

Bratzlavsky M: Pauses in activity of human jaw closing muscle. *Exp Neurol* 36:160–165, 1972.

Bratzlavsky M: Reciprocal innervation of motoneurons lacking Ia input in men. *Exp Neurol* 73:517–524, 1981.

Braun: *Ann Phys Chem* 60:552, 1897.

Braune CW, Fischer O: Der Gang des Menschen. I Teil. Versuche unbelasten und belasten Menschen. *Abhandl. d. Math.-Phys. Cl. d. k. Sächs, Gesellsch. Wissensch.*, 21:153–322, 1895.

Breinin GM: The nature of vergence revealed by electromyography. *AMA Arch Ophthalmol* 54:407–409, 1955.

Breinin GM: Electromyographic evidence for ocular muscle proprioception in man. *AMA Arch Ophthalmol* 57:76–180, 1957a.

Breinin GM: The position of rest during anesthesia and sleep: electromyographic observations. *AMA Arch Ophthalmol* 57:323–326, 1957b.

Breinin GM: Analytic studies of the electromyogram of human extraocular muscle. *Am J Ophthalmol* 46:123–142, 1958.

Brennan JB: Clinical method of assessing tonus and voluntary movement in hemiplegia. *Br Med J* 1:767–768, 1959.

Briggs DI, Landau B, Akert K, Youmans WB: Kinesiology of the abdominal compression reaction (Abstr). *Physiologist* 3:30, 1960.

Briscoe G: The muscular mechanism of the diaphragm. *J Physiol* (Lond), 54:46–53,

1920.

Briscoe JC: The mechanism of post-operative massive collapse of the lungs. *Q J Med* 13:293–334, 1920.

Broadbent TR, Swinyard CA: The dynamic pharyngeal flap, its selective use and electromyographic evaluation. *Plast Reconstruct Surg* 23:301–312, 1959.

Brody G, Balasubramanian R, Scott RN: A model for myo-electric signal generation. *Med Biol Eng* 12:29–41, 1974.

Broman H: An investigation on the influences of a sustained contraction on the succession of action potentials from a single motor unit. Ph.D. dissertation, Chalmers University of Technology, Göteborg, Sweden.

Broman H: An investigation on the influence of a sustained contraction on the succession of action potentials from a single motor unit. *Electromyogr Clin Neurophysiol* 17:341–358, 1977.

Broman H, Kadefors R: A spectral moment analyzer for quantification of electromyograms. Proc 4th Cong Int Soc Electrophysiol Kinesiol, 1979, pp 90–91.

Broman H, Magnusson R, Petersén I, Ortengren R: Vocational electromyography. *Dev EMG Clin Neurophysiol* 1:656–664, 1973.

Broman H, Mambrito B, De Luca CJ: Peripheral regulation of muscle output: Interplay between motor unit recruitment and firing rates. In press, 1985.

Brooke MH, Kaiser KK: Three myosin adenosine triphosphatase systems: The nature of their pability and sulfhydryl dependence. *J Histochem Cytochem* 18:670–672, 1970.

Brooks JE: Intracellular electromyography: resting and action potentials in normal human muscle. *Arch Neurol* 18:291–300, 1968.

Brooks JE: Disuse atrophy of muscle: intracellular electromyography. *Arch Neurol* 22:27–30, 1970.

Broome HL, Basmajian JV: The function of the teres major muscle: an electromyographic study. *Anat Rec* 170:309–311, 1971a.

Broome HL, Basmajian JV: Survival of iliopsoas muscle after Sharrard procedure. *Am J Phys Med* 50:301–302, 1971b.

Brown DM, Basmajian FV: A bioconverter for upper extremity rehabilitation. *Am J Phys Med,* in press, 1985.

Brown ME, Long C, Weiss G: Electromyographic kinesiology of the hand. Part 1. Method. *Phys Ther Rev* 40:453–458, 1960.

Brown SH, Cooke JD: Responses to force perturbations preceding voluntary arm movements. *Brain Res* 220:350–355, 1981a.

Brown SH, Cooke JD: Amplitude and instruction-dependent modulation of movement-related electromyogram activity in humans. *J Physiol* 316:97–107, 1981b.

Brown WEK, Harland R, Basmajian JV: Electromyography of quadriceps femoris. Unpublished work at the University of Toronto, 1957.

Brudny J, Korein J, Grynbaum BB, Friedman FW, Weinstein S, Sachs-Frankel G, Belandres PV: EMG feedback therapy: review of treatment of 114 patients. *Arch Phys Med Rehab* 57:55–61, 1976.

Bruno LJJ, Davidowitz J, Heferline RF: EMG waveform duration: a validation method for the surface electromyogram. *Behav Res Meth Instr* 2:211–219, 1970.

Buchthal F: The functional organization of the motor unit: a summary of results. *Am J Phys Med* 38:125–128, 1959.

Buchthal F, Clemmesen S: On differentiation of muscle atrophy by electromyography. *Acta Psychiatr Neurol* 16:143–181, 1941.

Buchthal F, Engbaek L: Refractory period and conduction velocity of the striated muscle fibre. *Acta Physiol Scand* 59:199–219, 1963.

Buchthal F, Erminio F, Rosenfalck P: Motor unit territory in different human muscles. *Acta Physiol Scand* 45:72–87, 1959.

Buchthal F, Faaborg-Andersen K: Electromyography of laryngeal and respiratory muscles; correlation with phonation and respiration. *Ann Otol Rhinol Laryngol* 73:118–124, 1964.

Buchthal F, Fernandez-Ballesteros ML: Electromyographic study of the muscles of the upper arm and shoulder during walking in patients with Parkinson's disease. *Brain* 88:875–896, 1965.

Buchthal F, Guld C, Rosenfalck P: Action potential parameters in normal human muscle and their dependence on physical variables. *Acta Physiol Scand* 32:200–218, 1954.

Buchthal F, Guld C, Rosenfalck P: Multielectrode study of the territory of a motor unit. *Acta Physiol Scand* 39:83–103, 1957.

Buchthal F, Guld C, Rosenfalck P: Volume conduction of the spike of the motor unit potential investigated with a new type of multielectrode. *Acta Physiol Scand* 38:331–354, 1957b.

Buchthal F, Madsen EE: Synchronous activity in normal and atrophic muscle. *Electroencephalogr Clin Neurophysiol* 2:425–444, 1950.

Buchthal F, Pinelli P, Rosenfalck P: Action potential parameters in normal human muscle and their physiological determinants. *Acta Physiol Scand* 32:219–229, 1954.

Buchthal F, Rosenfalck P: On the structure of motor units. In Desmedt JE: *New Developments in EMG and Clinical Neurophysiology*, Basel, Karger, 1973, vol 1, pp 71–85.

Buchthal F, Schmalbruch H: Contraction time and fibre types of normal and diseased human muscle. In *Muscle Disease* (proceedings of an International Congress, Milan, May, 1969). Excerpta Medica International Congress Series No. 199.

Buchthal F, Schmalbruch H: Contraction times and fiber types in intact human muscle. *Acta Physiol Scand* 79:435–452, 1970.

Budingen HJ, Freund HJ: The relationship between the rate of isometric tension and motor unit recruitment in a human forearm muscle. *Pflugers Arch* 362:61–67, 1976.

Budzynski TH, Stoyva JM: An instrument for producing deep muscle relaxation by means of analog information feedback. *J Appl Behav Anal* 2:231–237, 1969.

Budzynski T, Stoyva J, Adler C: Feedback-induced muscle relaxation: application to tension headache. *J Behav Exp Psychiatry* 1:205–211, 1970.

Buller AJ, Eccles JC, Eccles RM: Interactions between motoneurones and muscles in respect of the characteristic speeds of their responses. *J Physiol* 150:417–439, 1960.

Burbank DP, Webster JG: Reducing skin potential motion artifact by skin abrasion. *Med Biol Eng* 16:38, 1978.

Burdet A, Taillard W, Blanc Y: Stehen und Gehen mit Gehhilfen: eine elektromyokinesiographische Untersuchung. *Z Orthop* 117:247–259, 1979.

Burke RE, Levine DN, Zajac FE: Mammalian motor units: physiological-histochemical correlation in three types in cat gastrocnemius. *Science* 174:709, 1971.

Burke RE, Rudomin P, Zajac III FE: The effect of activation history on the tension production by individual muscle units. *Brain Res* 109:515–529, 1976.

Burke RE, Tsairis P: Anatomy and innervation ratios in motor units of cat gastrocnemius. *J Physiol* (Lond), 234:749–765, 1975.

Cacioppo JT, Petty RE: Electromyograms as measures of extent and affectivity of information-processing. *Am Psychologist* 36:441–456, 1981.

Cain WC: Nature of perceived effort and fatigue: roles in strength and blood flow in muscle contractions. *J Motor Behav* 5:33–47, 1973.

Caldwell CW, Reswick JB: A percutaneous wire electrode for chronic research use. *IEEE Trans Biomed Eng* 22:429–432, 1975.

Cameron WE, Binder MD, Botterman BR, Reinking RM, Stuart DG: Motor unit-muscle spindle interaction in active muscles of decerebrate cats. *Neurosci Lett* 19:55–60, 1980.

Cameron WE, Binder MD, Botterman BR, Reinking RM, Stuart DG: "Sensory partitioning" of cat medial gastrocnemius muscle by its muscle spindle and tendon organs. *J Neurophysiol* 46:32–47, 1981.

Campbell EJM: An electromyographic study of the role of the abdominal muscles in breathing. *J Physiol* 117:222–233, 1952.

Campbell EJM: The muscular control of breathing in man. Ph.D. Thesis, University of London; quoted in his 1958 monograph (see reference below), 1954.

Campbell EJM: An electromyographic examination of the role of the intercostal muscles in breathing in man. *J Physiol* 129:12–26, 1955a.

Campbell EJM: The role of the scalene and sternomastoid muscles in breathing in normal subjects. An electromyographic study. *J Anat* 89:378–386, 1955b.

Campbell EJM: The functions of the abdominal muscles in relation to the intra-abdominal pressure and the respiration. *Arch Middlesex Hosp* 5:87–94, 1955c.

Campbell EJM: *The Respiratory Muscles and the Mechanics of Breathing*, London, Lloyd-Luke (Medical Books), 1958.

Campbell EJM, Green JH: The expiratory function of the abdominal muscles in man, an electromyographic study. *J Physiol* 120:409–418, 1953a.

Campbell EJM, Green JH: The variations in intra-abdominal pressure and the activity of the abdominal muscles during breathing; a study in man. *J Physiol* 122:282–290, 1953b.

Campbell EJM, Green JH: The behaviour of the abdominal muscles and the intra-abdominal pressure during quiet breathing and increased pulmonary ventilation: a study in man. *J Physiol* 127:423–426, 1955.

Campbell KM, Biggs NL, Blanton PL, Leher RP: Electromyographic investigation of the relative activity among four components of the triceps surae. *Am J Phys Med* 52:30–41, 1973.

Canter A, Kondo CY, Knott JR. A comparison of EMG feedback and progressive muscle relaxation training in anxiety neurosis. *Br J Psychiatry* 127:470–477, 1975.

Cappozzo A, Leo T, Pedotti A: A general computing method for the analysis of human locomotion. *J Biomech* 8:307–320, 1975.

Car A: La commande corticale du centre déglutiteur bulbaire. *J Physiol* (Paris), 62:361–386, 1970.

Cardus D, Quesada EM, Scott FB: Studies on the dynamics of the bladder. *J Urol* 90:425–433, 1963.

Carlson KE, Alston W, Feldman DJ: Electromyographic study of aging in skeletal muscle. *Am J Phys Med* 43:141–145, 1964.

Carlsöö S: Nervous coordination and mechanical function of the mandibular elevators: an electromyographic study of the activity, and an anatomic analysis of the mechanics of the muscles. *Acta Odontol Scand* 10(suppl. 11):1–132, 1952.

Carlsöö S: The static muscle load in different work positions: an electromyographic study. *Ergonomics* 4:193–211, 1961.

Carlsöö S: A method of studying walking on different surfaces. *Ergonomics* 5:271–274, 1962.

Carlsöö S: Influence of frontal and dorsal loads on muscle activity and on the weight distribution in the feet. *Acta Orthop Scand* 34:299–309, 1964.

Carlsöö S: The initiation of walking. *Acta Anat* 65:1–9, 1966.

Carlsöö S: *How Man Moves.* Translated from Swedish edition by W.P. Michael. London, Heinemann; and New York, Crane Russak, 1973.

Carlsöö S, Edfeldt AW: Attempts at muscle control with visual and auditory impulses as auxiliary stimuli. *Scand J Psychol* 4:231–235, 1963.

Carlsöö S, Fohlin L: The mechanics of the two-joint muscles rectus femoris, sartorius and tensor fasciae latae in relation to their activity. *Scand J Rehab Med* 1:107–111, 1969.

Carlsöö S, Guharay AR: A study of the muscle activity and blood flow in the muscles of the upper arm in supporting heavy loads. *Acta Physiol Scand* 72:366–369, 1968.

Carlsöö S, Johansson O: Stabilization of and load on the elbow joint in some protective movements: an experimental study. *Acta Anat* 48:224–231, 1962.

Carlsöö S, Mayr J: A study of the load on joints and muscles in work with a pneumatic hammer and a bolt gun. *Work Environ Health* 11:32–38, 1974.

Carlsöö S, Molbech S: The function of certain two-joint muscles in a closed muscular chain. *Acta Morph NeerScand* 4:377–386, 1966.

Carlsöö S, Norstrand A: The coordination of the knee muscles in some voluntary movements and in the gait in cases with and without knee joint injuries. *Acta Chir Scand* 134:423–426, 1968.

Carlsöö S, Skoglund G: En rörelseananatomisk studie av fall med gångrubbningar (Swedish text). *Lakartidningen* 66:3966–3972, 1969.

Carlsson SG, Gale EN, Ohman A: Treatment of temporomandibular joint syndrome with biofeedback training. *J Am Dent Assoc* 91:602–605, 1975.

Carman DJ, Blanton PL, Biggs NL: Electromyography of the anterolateral abdominal musculature (abstr). *Anat Rec* 169:474, 1971.

Carmel PW, Starr A: Acoustic and nonacoustic factors modifying middle-ear muscle activity in waking cats. *J Neurophysiol* 26:598–616, 1963.

Carvalho CAF, Garcia OS, Vitti M, Berzin F: Electromyographic study of the m. tensor fasciae latae and m. sartorius. *Electromyogr Clin Neurophysiol* 12:387–400, 1972.

Carvalho CAF, König B Jr, Vitti M: Electromyographic study of the muscles "extensor digitorum brevis" and "extensor hallucis brevis." *Rev Hosp Fac Med São Paulo* 22:65–72, 1967.

Carvalho CAF, Vitti M: Estudo Eletromiográfico do músculo extensor longo do halux. *Folio Clin Biol* 34:77–91, 1965.

Casarin G, Puricelli R, Villa A: Una tecnica telemetrica via radio nello studio elettromiografico della deambulazione. *Riabilitazione* 7:177–189,

Catton WT, Gray JE: Electromyographic study of the action of the serratus anterior muscle in respiration (abstr). *J Anat* 85:412, 1951.

Chaco J, Wolf E: Subluxation of the glenohumeral joint in hemiplegia. *Am J Phys Med* 50:139–143, 1971.

Chaco J, Yules RB: Velopharyngeal incompetence post tonsilloadenoidectomy: an electromyographic study. *Acta Otolaryngol* 68:276–278, 1969.

Chaffin DB: Localized muscle fatigue—definition and measurement. *J Occup Med* 15:346–354, 1973.

Chantraine A: Exploration électrique des sphincters striés urétral et anal. *J Urol Nephr* 80:207–213, 1974.

Chapman AJ: An electromyographic study of social facilitation: a test of the 'mere presence' hypothesis. *Br J Psychol* 65:123–128, 1974.

Cheney PD, Kasser R, Holsapple J: Reciprocal effect of single corticomotoneuronal cells on wrist extensor and flexor muscle activity in primate. *Brain Res* 247:164–168, 1982.

Cheng IS, Koozekanani SH, Fatchi MT: A simple computer-television interface system for gait analysis. *IEEE Trans Biomed Eng* 22:259–260, 1975.

Christensen E: Topography of terminal motor innervation in striated muscles from stillborn infants. *Am J Phys Med* 38:17–30, 1959.

Christensen EH: Muscular work and fatigue. Delivered at "Muscle as a Tissue" International Conference, Lankenau Hospital, Philadelphia in 1960. pp. 176–189 in *Muscle as a Tissue* Rodahl K, Horvath SM (eds). New York, McGraw-Hill; 1962.

Christensen J, Hauser RL: Circumferential coupling of electrical slow waves in circular muscle of cat colon. *Am J Physiol* 221:1033–1037, 1971.

Christensen J, Takata M, Kawamura Y: Electrophysiologic analysis of innervation of temporalis muscle in the cat. *J Dent Res* 48:327, 1969.

Christie BGB, Coomes EN: Normal variation of nerve conduction in three peripheral nerves. *Ann Phys Med* 5:303–309, 1960.

Ciriello VM: A longitudinal study of the effects of two training regimens on muscle strength and hypertrophy of fast twitch and slow twitch fibers. Ph.D. dissertation, Boston University, Boston, MA, 1982.

Clamann HP: A quantitative analysis of the firing pattern of single motor units of a skeletal muscle of man and their utilization in isometric contractions. Ph.D. Thesis, Johns Hopkins University, 1968.

Clamann HP: Statistical analysis of motor unit firing patterns in a human skeletal muscle. *Biophys J* 9:1233–1251, 1969.

Clamann HP: Activity of single motor units during isometric tension. *Neurology* 20:254–260, 1970.

Clamann HP, Broecker MS: Relation between force and fatiguability of red and pale skeletal muscles in man. *Am J Phys Med* 58:70–85, 1979.

Clamann PH, Gillies DJ, Henneman E: Effects of inhibitory inputs on critical firing level and rank order of motoneurons. *J Neurophysiol* 37:1350–1360, 1974.

Clark RFK, Wyke BD: Temporomandibular articular reflex control of the mandibular musculature. *Intern Dent J* 25:289–296, 1975.

Clark RW, Luschel ES, Hoffman DS: Recruitment order, contractile characteristics, and firing patterns of motor units in temporalis muscle of monkeys. *Exp Neurol* 61:31–52, 1978.

Clarke AM: Relationship between the electromyogram and isometric reflex response in pre-strain conditions in normal humans. *Nature* 214:1114–1115, 1967.

Cleeland CS: Behavioral technics in the modification of spasmodic torticollis. *Neurology* 23:1241–1247, 1973.

Clemmesen S: Some studies of muscle tone. *Proc Roy Soc Med* 44:637–646, 1951.

Clendenin MA, Szumski AJ: Influence of cutaneous ice application on single motor units in humans. *J Am Phys Ther Assoc* 51:166–175, 1971.

Cleveland S, Kuschmierz A, Ross H-G: Static input-output relations in the spinal recurrent inhibitory pathway. *Biol Cybern* 40:223–231, 1981.

Close JR: *Motor Function in the Lower Extremity: Analysis by Electronic Instrumentation.* Charles C Thomas, Springfield, IL, 1964.

Close JR, Nickle ED, Todd FN: Motor-unit action-potential counts: their significance in isometric and isotonic contractions. *J Bone Joint Surg* 42A:1207–1222, 1960.

Cnockaert JC: Comparaison electromyographique du travail en allongement et en raccourcissement au course de mouvement de va-et-vient. *Electromyogr Clin Neurophysiol* 15:477–489.

Cobb S, Forbes A: Electromyographic studies of muscle fatigue in man. *Am J Physiol*

65:234–251, 1923.

Coërs C, Woolf, AL: *The Innervation of Muscle, a Biopsy Study.* Oxford, Blackwell Scientific Publications; Springfield, IL, 1959.

Coggshall JC, Bekey GA: A stochastic model of skeletal muscle based on motor unit properties. *Math Biosci* 7:405–419, 1970.

Cohen AH, Gans C: Muscle activity in rat locomotion: movement analysis and electromyography of the flexors and extensors of the elbow. *J Morphol* 146:177–196, 1975.

Cohen L: Interaction between limbs during bimanual voluntary activity. *Brain* 93:259–272, 1970.

Cohen MJ: The relation between heart rate and electromyographic activity in a discriminated escape-avoidance paradigm. *Psychophysiology* 10:8–20, 1973.

Cohen MJ, Johnson HJ: Relationship between heart rate and muscular activity within a classical conditioning paradigm. *J Exp Psychol* 90:222–226, 1971.

Comtet JJ, Auffray Y: Physiologie des muscles élévateurs de l'épaule. *Rev Chirurg Orthop (Paris)* 56:105–117, 1970.

Cook WA: Antagonistic muscles in the production of clonus in man. *Neurology* 17:779–782, 1967.

Cooper RR: Alterations during immobilization and regeneration of skeletal muscle in cats. *J Bone Joint Surg* 54A:919–953, 1972.

Cooper S, Daniel PM: Muscle spindles in human extrinsic eye muscles. *Brain* 72:1–24, 1949.

Copack PB, Felman E, Lieberman JS, Gilman S: Differences in proximal and distal conduction velocity of efferent nerve fibers to the medial gastrocnemius muscle. *Brain Res* 91:147–150, 1975.

Cotton FJ: Subluxation of the shoulder—downward. *Boston Med Surg J* 185:405–407, 1921.

Coulmance M, Gahery Y, Massion J, Swett JE: The placing reaction in standing cat: a model for the study of the coordination of posture and movement. *Exp Brain Res* 37:265–281, 1979.

Coursey RS: Electromyograph feedback as a relaxation technique. *J Consult Clin Psychol* 43:825–834, 1975.

Cox DJ, Freundlich A, Meyer RG: Differential effectiveness of electromyographic feedback, verbal relaxation instructions, and medication placebo with tension headaches. *J Consult Clin Psychol* 43:892–899, 1975.

Crochetiere WJ, Vodovnik L, Reswick JB: Electrical stimulation of skeletal muscle. A study of muscle as an actuator. *Med Biol Eng* 5:111–125, 1967.

Cunningham DP, Basmajian JV: Electromyography of genioglossus and geniohyoid muscles during deglutition. *Anat Rec* 165:401–410, 1969.

Currier DP: Measurement of muscle fatigue. *J Am Phys Ther Assoc* 49:724–730, 1969.

Currier DP: Maximal isometric tension of the elbow extensors at varied positions. Part 2. Assessment of extensor components by quantitative electromyography. *Phys Ther* 52:1265–1276, 1972.

Czéh G, Székely G: Muscle activities recorded simultaneously from normal and supernumerary forelimbs in ambystoma. *Acta Physiol Acad Sci Hungar* 40:287–301, 1971.

Czipott Z, Herpai S: Elektromyographische Untersuchungen bie Meniskus-, Knie-, und Bandverletzungen. *Z Orthop* 109:768–778, 1971.

Da Hora B: O "musculus anconeus." Contribuição ao estudo da sua arquitetura e das suas funções (Portuguese text). Thesis, University of Recife, Recife, Brazil, 1959.

Daniel B, Guitar B: EMG feedback and recovery of facial and speech gestures following neural anastomosis. Unpublished Document (No. 7), Speech Research Laboratory, University of Wisconsin, 1973.

Daniel EE: The electrical activity of the alimentary tract. *Am J Digest Dis* 13:297–319, 1968.

Danielsson C-O, Franksson C, Petersén I: Stress incontinence following the Manchester operation for prolapse. *Acta Obstet Gynecol Scand* 35:335–344, 1956.

Davis MH, Saunders DR, Creer TL, Chai H: Relaxation training facilitated by biofeedback apparatus as a supplemental treatment in bronchial asthma. *J Psychosom Res* 17:121–128, 1973.

Dawson GD: The relative excitability and conduction velocity of sensory and motor nerve fibers in man. *J Physiol* 131:436–451, 1956.

Dawson GD, Scott JW: The recording of nerve action potentials through the skin in man. *J Neurol Neurosurg Psychiatry* 12:259–267, 1949.

Day BL, Marsden CD, Obeso JA, Rothwell JC, Traub MM: Manual motor function in a deafferented man. *J Physiol* 320:23–24P, 1981.

deAndrade JR, Grant C, Dixon ASTJ: Joint distension and reflex muscle inhibition in the knee. *J Bone Joint Surg* 47A:313–322, 1965.

Dedo HH: The paralyzed larynx: an electromyographic study in dogs and humans. *Laryngoscope* 80:1455–1517, 1970.

Dedo HH: Electromyographic and visual evaluation of recurrent laryngeal nerve anastomosis in dogs. *Ann Otol Rhinol Laryngol* 80:664–669, 1971.

Dedo H, Hall WN: Electrodes in laryngeal electromyography: reliability comparison. *Otol Rhinol Laryngol* 78:172–181, 1969.

Dedo HH, Ogura JH: Vocal cord electromyography in the dog. *Laryngoscope* 75:201–311, 1965.

de Freitas V, Vitti M: Electromyographic study of the trapezius (pars media) and rhomboideus major muscles in movements of the arm (Parts 1 and 2). *Electromyogr Clin Neurophysiol* 21:469–478 and 479–485, 1981a and b.

de Freitas V, Vitti M, Furlani J: Electromyographic analysis of the levator scapulae and rhomboideus major muscles in movements of the shoulder. *Electromyogr Clin Neurophysiol* 19:335–342, 1979.

de Freitas V, Vitti M, Furlani J: Análise eletromiográfica dos músculos elevador da escápula e rombóide maior em movimentos da cabeça e tronco. *Ciência e Cultura* 32:218–220, 1980.

de Girardi-Quirion C: Lá rétroaction biologique en physiothérapie: une expérience vécue en pédiatrie. *Physiother Can* 28:14–19, 1976.

De Grandis D, Santoni P: The post-auricular response. A single motor unit study. *Electroencephalogr Clin Neurophys* 50: 437–440, 1980.

de Jesus PV, Hausmanowa-Petrusewicz I, Barchi RL: The effect of cold on nerve conduction of human slow and fast nerve fibers. *Neurology* 23:1182–1189, 1973.

Dejonckere P: La technique externe d'électromyographie laryngée en clinique laryngologique et phoniatrique. *Acta Otolaryngol Belg* 29(4):677–691, 1975.

Delhez L: Evolution de l'activité électrique intégrée du diaphragme durant l'hyperventilation. *Compte Rend Séances Soc Biol* 158:2496–2500, 1964.

Delhez L, Bottin R, Damoiseau J, Petit JM: Examen comparatif des électromyogrammes des piliers du diaphragme dérivés au moyen de trois modéles de-sondes-électrodes. *Electromyography* 4: 5–14, 1964a.

Delhez L, Bottin-Thonon A, Petit JM: Influence de l'entraînement sur la force maximum des muscles respiratoires. *Trav Soc Méd Belge d'Educ Phys Sports* 20:52–63, 1968.

Delhez L, Damoiseau J, Deroanne R: Comportement electrique du diaphragme et des muscles abdominaux durant la respiration sous pression positive intermittente chez des sujets normaux et emphysémateux. *Rev Electrodiag Thérap* 1:197–209, 1964b.

Delhez L, Petit JM: Données actuelles de l'électromyographie respiratoire chez l'homme normal. *Electromyography* 6: 101–146, 1966.

Delhez L, Petit JM, Deroanne R, Pirnay F, Sneppe R: Evolution de l'activité électrique integrée de quatre muscles locomoteurs durant la marche et la course sur tapis roulant. *Electromyography* 9:417–431, 1969.

Delhez L, Troquet J, Damoiseau J, Petit JM: Activité antagoniste du diaphragme à la fin de l'expiration forcée. *J Physiol Paris* 55:241–242, 1963.

Delhez L, Troquet J, Damoiseau J, Petit JM: Influence des modalités d'activité électrique des muscles abdominaux et du diaphragme sur les diagrammes volume/pression de relâchement du thorax et des poumons. *Arch Intern Physiol Biochim* 71:175–194, 1963.

Delhez L, Troquet J, Damoiseau J, Petit JM: Nécessité de l'électromyographie dans les mesures d'elastance thoraco-pulmonaire. *Rev Electrodiag-Thérap* 1:39–45, 1964c.

Delhez L, Troquet J, Damoiseau J, Pirnay F, Deroanne R, Petit JM: Influence des modalités d'exécution des manoeuvres d'expiration forcée et d'hyperpression thoracoabdominale sur l'activité électrique du diaphragme. *Arch Intern Physiol Biochim* 72:76–94, 1964d.

Dellow PG, Lund JP: Evidence for central timing of rhythmical mastication. *J Physiol* 215:1–13, 1971.

De Luca CJ: Myoelectric analysis of isometric contractions of the human biceps brachii. M. Sc. Thesis, University of New Brunswick, Fredericton, New Brunswick, Canada, 1968.

De Luca CJ: A model for a motor unit train recorded during constant force isometric contractions. *Biol Cybernet* 19:159–167, 1975.

De Luca CJ: Physiology and mathematics of myoelectric signals. *IEEE Trans Biomed Eng* 26:313–325, 1979.

De Luca CJ, Berenberg W: A Polar technique for displaying EMG signals. *Proc 28th Ann Conf Engineering in Medicine and Biology*, 1975, p 21.

De Luca CJ, Forrest WJ: An electrode for recording single motor unit activity during strong muscle contractions. *IBEEE Trans Biomed Eng* BME-19:367–372, 1972.

De Luca CJ, Forrest WJ: Some properties of motor unit action potential trains. *Kybernetics* 12:160–168, 1973a.

De Luca CJ, Forrest WJ: Force analysis of individual muscles acting simultaneously on the shoulder joint during isometric abduction. *J Biomech* 6:385–393, 1973b.

De Luca CJ, Forrest WJ: Probability distribution function of the inter-pulse intervals of single motor unit action potentials during isometric contraction. In Desmedt JE: *New Developments in Electromyography and Clinical Neurophysiology.* Basel, Karger, 1973c, vol 1, 638–647.

De Luca CJ, LeFever RS, McCue MP, Xenakis AP: Behaviour of human motor units in different muscles during linearly varying contractions. *J Physiol* 329:113–128, 1982a.

De Luca CJ, LeFever RS, McCue MP, Xenakis AP: Controls scheme governing concurrently active human motor units during voluntary contractions. *J Physiol* 329:129–142, 1982b.

De Luca CJ, LeFever RS, Stulen FB: Pasteless electrodes for clinical use. *Med Biol Eng Comput* 17:387–390, 1979.

De Luca CJ, Sabbahi MA, Stulen FB, Bilotto G: Some properties of the median frequency of the myoelectric signal during localized muscle fatigue. In Knuttgen HK et al: *Biochemistry of Exercise*, 1983, vol 13, pp 175–186.

De Luca CJ, van Dyk EJ: Derivation of some parameters of myoelectric signals recorded during sustained constant force isometric contractions. *Biophys J* 15:1167–1180, 1975.

Dempster WT, Finerty JC: Relative activity of wrist moving muscles in static support of the wrist joint: an electromyographic study. *Ann J Physiol* 150:596–606, 1947.

Denny-Brown D: On the nature of postural reflexes. *Proc Roy Soc* 104B:252–301, 1929.

Denny-Brown D, Foley JM: Myokymia and the benign fasciculation of muscular cramps. *Trans Assoc Am Physicians* 61:88–96, 1948.

Denny-Brown D, Pennybacker JB: Fibrillation and fasciculation in voluntary muscle. *Brain* 61:311–334, 1938.

Denslow JS, Gutensohn OR: Neuromuscular reflexes in response to gravity. *J Appl Physiol* 23:243–247, 1967.

DePalma AT, Leavitt LA, Hardy SB: Electromyography in full thickness flaps rotated between upper and lower lips. *Plast Reconstruct Surg* 21:448–452, 1958.

Desmedt JE, Godaux E: Fast motor units are not preferentially activated in rapid voluntary contractions in man. *Nature* 267:717–719, 1977a.

Desmedt JE, Godaux E: Ballistic contractions in man: characteristics recruitment pattern of single motor units of the tibialis anterior muscle. *J Physiol* 264:673–693, 1977b.

Desmedt JE, Godaux E: Ballistic contractions in fast or slow human muscles: discharge patterns of single motor units. *J Physiol* 285:185–196, 1978.

Desmedt JE, Godaux E: Ballistic skilled movements: load compensation and patterning of motor commands. *Progress Clin Neurophysiol* 4:21–55, 1978a.

Desmedt JE, Godaux E: Spinal motoneuron recruitment in man: rank deordering with direction but not with speed of voluntary movement. *Science* 214:933–936, 1981.

de Sousa OM: Aspectos da arquitetura e da ação dos músculos estriados, baseada na electromiografia (Portuguese text). *Folia Clin Biol* 28:12–42, 1958.

de Sousa O Machado: Estudo electromiográfico do m. platysma (Portuguese text with English summary). *Folia Clin Biol (Brazil)* 33:42–52, 1964.

de Sousa OM, Berzin F, Berardi AC: Electromyographic study of the pectoralis major and latissimus dorsi muscles during medial rotation of the arm. *Electromyography* 9:407–416, 1969.

de Sousa OM, de Moraes JL, de Morais Vieira FL: Electromyographic study of the brachioradialis muscle. *Anat Rec* 139:125–131, 1961.

de Sousa OM, de Morais WR, Ferraz EF: Observações anatômicas e electromiográficas sôbre o "m. pronator quadratus" (Portuguese text). *Folia Clin Biol* 27:214–219, 1957.

de Sousa OM, de Morais WR, Ferraz EC de F: Estudo electromiográfico de alguns músculos do antebraço durante a pronação (Portuguese text). *Rev Hosp Clin* 13:346–354, 1958.

de Sousa OM, Furlani J: Electromyographic study of the m. rectus abdominis. *Acta Anat* 88:281–298, 1974.

de Sousa OM, Furlani J, Garcia OS: Atividade de músculos antagonistas: estudo electromiográfico. *Rev Hosp Clin Fac Med S Paulo* 30:471–473, 1975.

de Sousa OM, Furlani J, Vitti M: Étude électromyographique du m. sternocleidomastoideus. *Electromyogr Clin Neurophysiol* 13:93–106, 1973.

de Sousa OM, Vitti M: Estudo electromiográfico do m. buccinator. *O Hospital (São Paulo)* 68:105–117, 1965.

de Sousa OM, Vitti M: Estudio electromiográfico de los músculos adductores largo y mayor (abstr). *Arch Mex Anat* 7:52–53, 1966.

Deutsch LJ: The threshold of the stapedius reflex for pure tone and noise stimuli. *Acta Otolaryngol* 74:248–251, 1972.

deVries HA: Muscle tonus in postural muscles. *Am J Phys Med* 44:275–291, 1965.

deVries HA: Efficiency of electrical activity as a physiological measure of the functional state of muscle tissue. *Am J Phys Med* 47:10–22, 1968.

deVries HA: Method for evaluation of muscle fatigue and endurance from electromyographic fatigue curves. *Am J Phys Med* 47:125–135, 1968.

deVries HA, Burke RK, Hopper RT, Sloan JH: Relationship of resting EMG level to total body metabolism with reference to the origin of "tissue noise." *Am J Phys Med* 55:139–147, 1976.

deVries H, Burke RK, Hopper R, Sloan JH: Effect of EMG biofeedback in relaxation training. *Am J Phys Med* 56:75–81, 1977.

Dewson JH III, Dement WC, Simmons FB: Middle ear muscle activity in cats during sleep. *Exp Neurol* 12:1–8, 1965.

Di Benedetto A, Siebens AA, Cincotti JJ, Grant AR, Glass P: A study of diaphragmatic and crural innervation by the direct recording of action potentials in the dog. *J Thorac Cardiovas Surg* 38:104–107, 1959.

Dietz V, Bischofberger E, Wita C, Freund HJ: Correlation between the discharges of two simultaneously recorded motor units and physiological tremor. *Electroencephalogr Clin Neurophysiol* 40:97–105, 1976.

Dill JD, Lockemann PC, Naka KI: An attempt to analyze multiunit recordings. *Electroencephalogr Clin Neurophysiol* 28: 79–82, 1972.

Dixon HH, Dickel HA: Tension headache. *Northw Med* 66:817–820, 1967.

Donisch EW, Basmajian JV: Electromyography of deep muscles in man. *Am J Anat* 133:25–36, 1972.

Dorai Raj BS: Diversity of crab muscle fibers innervated by a single motor axon. *J Cell Comp Physiol* 64:41–54, 1964.

Doty RW, Bosma JF: An electromyographic analysis of reflex deglutition. *J Neurophysiol* 19:44–60, 1956.

Downie AW, Scott TR: Radial nerve conduction studies. *Neurology* 14:839–843, 1964.

Doyle AM, Mayer RF: Studies of the motor unit in the cat. *Bull Sch Med Univ Md* 54:11–17, 1969.

Draper MH, Ladefoged P, Whitteridge D: Expiratory muscles involved in speech. *J Physiol* 138:17–25, 1957.

Duarte Cintra AI, Furlani J: Electromyographic study of quadriceps femoris in man. *Electromyogr Clin Neurophysiol* 21:539–554, 1981.

Dubo HIC, Peat M, Winter DA, Quanbury AO, Hobson DA, Steinke T, Reimer G: Electromyographic temporal analysis of gait: normal human locomotion. *Arch Phys Med Rehab* 57:415–418, 1976.

DuBois-Reymond E: Untersuchungen ueber thierische elektricitaet, Vol. II, second part. Berlin, Reimer, Verslag von G., 1849.

Duchenne GBA: *Physiologie des Mouvements*, transl. by E. B. Kaplan (reissued in 1959). Philadelphia and London, W. B. Saunders, 1949 (originally published 1867).

Dunn HG, Buckler WSJ, Morrison GCE, Emery AW: Conduction velocity of motor nerves in infants and children. *Pediatrics* 34:708–727, 1964.

Duthie HL, Watts JM: Contribution of the external anal sphincter to the pressure zone in the anal canal. *Gut* 6:64–68, 1965.

Dutta CR, Basmajian JV: Gross and histological structure of the pharyngeal constrictors in the rabbit. *Anat Rec* 137:127–134, 1960.

Duval A: Analyse functionelle des muscles de la paroi antero-laterale de l'abdomen par électromyocartographié. *Bull Assoc Anat* 59:743–755, 1975.

Eason RG: Electromyographic study of local and generalized muscular impairment. *J Appl Physiol* 15:479–482, 1960.

Eastman MC, Kamon E: Posture and subjective evaluation of flat and slanted desks. *Human Factors* 18:15–26, 1976.

Eberhart HD, Inman VT, Bresler B: The principal elements in human locomotion. In Klopsteg PE, Wilson PD: *Human Limbs and Their Substitutes*, New York, McGraw-Hill, 1954.

Eberhart HD, Inman VT, Saunders JBdeCM, Levens AS, Bresler B, Cowan TD: Fundamental studies of human locomotion and other information related to the design of artificial limbs. A Report to the N.R.C. Committee on Artificial Limbs. University of California, Berkeley, 1947.

Eble JN: Reflex relationships of paravertebral muscles. *Am J Physiol* 200:939–943, 1961.

Eccles JC, O'Connor WJ: Responses which nerve impulses evoke in mammalian striated muscles. *J Physiol* 97:44–102, 1939.

Edelwejn Z: Effect of high temperature on muscle bioelectric activity. *Acta Physiol Polon* 15:433–439, 1964.

Edström L, Kugelberg E: Histological composition, distribution of fibres and fatiguability of single motor units. *J Neurol Neurosurg Psychiatry* 31:424–433, 1968.

Edwards RG, Lippold OC: The relation between force and integrated electrical activity in fatigued muscle. *J Physiol* 132:677–681, 1956.

Edwards RHT, Hyde S: Methods of measuring muscle strength and fatigue. *Physiotherapy* 63:51–55, 1977.

Ekman P: *Darwin and Facial Expression.* New York, Appleton-Century-Crofts, 1973.

Ekstedt J: Human single muscle fiber action potentials. *Acta Physiol Scandinav* 61(suppl 226):1964.

Ekstedt J, Lindholm B, Ljunggren S, Stålberg E: The jittermeter: a variability calculator for use in single fiber electromyography. *Electroencephalogr Clin Neurophysiology* 30:154–158, 1971.

Ekstedt J, Nilsson G, Stålberg E: Calculation of the electromyographic jitter. *J Neurol Neurosurg Psychiatry* 37:526–539, 1974.

Elble RJ, Randall JE: Motor-unit activity responsible for 8- to 12-Hz component of human physiological finger tremor. *J Neurophysiol* 39:370–383, 1976.

Ekstedt J, Stålberg E: Single fiber EMG for the study of the microphysiology of the human muscle. In Desmedt JE: *New Development in EMG and Clinical Neurophysiology.* 1:84–112, 1973.

Elftman H: The function of the arms in walking. *Human Biol* 11:529–535, 1939.

Elftman H: Biomechanics of muscle: with particular application to studies of gait. *J Bone Joint Surg* 48A:363–377, 1966.

Elkus R, Basmajian JV: Endurance: why do people hanging by their hands let go? *Am J Phys Med* 52:124–127, 1973.

Ellaway PH: Recurrent inhibition of fusimotor neurons exhibiting background discharges in the decerebrate and spinal cat. *J Physiol* 216:419–439, 1971.

Elliott BC, Blanksby BA: Reliability of averaged integrated electromyograms during running. *J Hum Mov Studies* 2:28–35, 1976.

Elorante V, Komi PV: Function of the quadriceps femoris muscle under the full range of forces and different contraction velocities of concentric work. *Electromyogr Clin Neurophysiol* 21:419–431, 1981.

Emanuel M: The pathophysiology of the urinary sphincter. *Surg Clin North Am* 45:1467–1480, 1965.

Eng GD: Spontaneous potentials in premature and full-term infants. *Arch Phys Med Rehab* 57:120–121, 1976.

Engberg I: Reflexes to foot muscles in the cat. *Acta Physiol Scand* 62(suppl 235):1964.

Engel WK: The essentiality of histo- and cytochemical studies of skeletal muscle in the investigation of neuromuscular disease. *Neurology* 12:778–794, 1962.

Engel WK: Fiber-type nomenclature of human skeletal muscle for histochemical purposes. *Neurology* 24:344–348, 1974.

Engelhardt JK, Ishikawa K, Lisbin SJ, Mori J: Neurotrophic effects on passive electrical properties of cultured chick skeletal muscles. *Brain Res* 110:170–174, 1976.

Epstein BR, Foster KR: Anisotropy in the dielectic properties of skeletal muscle. *Med Biol Eng Comput* 21:51–55, 1983.

Ertekin C, Reel F: Bulbocavernosus reflex in normal men and in patients with neurogenic bladder and/or impotence. *J Neurol Sci* 28:1–15, 1976.

Faaborg-Andersen KL: Electromyographic investigation of intrinsic laryngeal muscles in humans: an investigation of subjects with normally movable vocal cords and patients with vocal cord paresis. *Acta Physiol Scand* 41(suppl 140):1–148, 1957.

Fairbank TJ: Fracture-subluxations of the shoulder. *J Bone Joint Surg* 30-B:454–460,

1948.

Farrar WB: Using electromyographic biofeedback in treating orofacial dyskinesia. *J Prosthet Dent* 35:384–387, 1976.

Farret SM, Vitti M, Farret MMB: Electromyographic analysis of the upper and lower orbicularis muscles in the production of speech. *Electromyogr Clin Neurophysiol* 22:125–136, 1982a.

Farret SM, Vitti M, Farret MMB: Electromyographic analysis of the mentalis and depressor labii inferior muscles in the production of speech. *Electromyogr Clin Neurophysiol* 22:137–148, 1982b.

Fatt P: Skeletal neuromuscular transmission. In *Handbook of Physiology, Section 1: Neurophysiology, Vol. 1.* Washington, American Physiological Society, 1959, pp 199–212.

Feinstein B, Lindegård B, Nyman E, Wohlfart G: Morphological studies of motor units in normal human muscles. *Acta Anat* 23:127–142, 1955.

Feldkamp L, Abbink F, Güth V: Elektromyographische studie des bewegungsverhaltens bei sänglingen mit zerebraler bewegungsstörung. *Z Orthop* 114:32–37, 1976.

Felli AJ, McCall WD Jr: Jaw muscle silent periods before and after rapid palatal expansion. *Am J Orthod* 76:676–681, 1979.

Fényes I, Gergely C, Tóth S: Clinical and electromyographic studies of "spinal reflexes" in premature and full-term infants. *J Neurol Neurosurg Psychiatry* 23:63–68, 1960.

Fernandez-Ballesteros ML, Buchthal F, Rosenfalck P: The pattern of muscular activity during the arm swing of natural walking. *Acta Physiol Scand* 63:296–310, 1964.

Ferrari E, Messina C: Blink reflexes during sleep and wakefulness in man. *Electroencephalogr Clin Neurophysiol* 32:55–62, 1972.

Ferraz ECF, de Moraes JL, Parolari JB: Atividade dos músculos fibulares longo e curto (Portuguese text). *Folia Clin Biol* 28:140–142, 1958.

Fetz EE, Finocchio DV: Operant conditioning of neural and muscular activity. *Science* 174:431–435, 1971.

Fibiger S: I. Communication disorders. A. Stuttering explained as a physiological tremor. Report from the Dept. of Speech Communication, Royal Instit. Technol. (KTH), Stockholm, privately printed, 1971.

Fibiger S: Further discussion on "stuttering explained as a physiological tremor." Report from the Dept. of Speech Communication, Royal Instit. Technol. (KTH) Stockholm, privately printed, 1972.

Fick R: *Handbuch der Anatomie und Mechanik der Gelenke*, vol 3. Jena, Germany, Gustav Fischer, 1911.

Fink BR: A method of monitoring muscular relaxation by the integrated abdominal electromyogram. *Anesthesiology* 21:178–185, 1960.

Fink BR, Hanks EC, Holaday DA, Ngai SH: Monitoring of ventilation by integrated diaphragmatic electromyogram. *JAMA* 172:1367–1371, 1960.

Finley FR, Cody KA, Finizie RV: Locomotion pattern in elderly women. *Arch Phys Med* 50:140–146, 1969.

Fisch U, Schulthess GV: Electromyographic studies of the human stapedial muscle. *Acta Oto-laryngol* 56:287–297, 1963.

Fleck H: Action potentials from single motor units in human muscle. *Arch Phys Med* 43:99–107, 1962.

Fleshman JW, Munson JB, Sypert GW, Friedman WA: Rheobase, input resistance, and motor-unit type in medial gastrocnemius motoneurons in the cat. *J Neurophysiol* 46:1326–1338, 1981.

Flint MM: Abdominal muscle involvement during performance of various forms of sit-up exercise. *Am J Phys Med* 44:224–234, 1965a.

Flint MM: An electromyographic comparison of the function of the iliacus and the rectus abdominis muscles. A preliminary report. *J Am Phys Ther Assoc* 45:248–253, 1965b.

Flint MM, Drinkwater BL, McKittrick JE: Shoulder dynamics subsequent to a radical mastectomy. *Electromyography* 10: 171–182, 1970.

Flint MM, Gudgell J: Electromyographic study of abdominal muscular activity during exercise. *Res Quart* 36:29–37, 1965.

Floyd WF, Negus VE, Neil E: Observations on the mechanism of phonation. *Acta Otolaryngol* 48:16–25, 1957.

Floyd WF, Silver PHS: Electromyographic study of patterns of activity of the anterior abdominal wall muscles in man. *J Anat* 84:132–145, 1950.

Floyd WF, Silver PHS: Function of erector spinae in flexion of the trunk. *Lancet Jan* 20:133–138, 1951.

Floyd WF, Silver PHS: The function of the erectores spinae muscles in certain movements and postures in man. *J Physiol*

129:184–203, 1955.

Floyd WF, Walls EW: Electromyography of the sphincter ani externus in man. *J Physiol* 122:599–609, 1953.

Folkins JW, Abbs JH: Lip and jaw motor control during speech: responses to resistive loading of the jaw. *J Speech Hearing Res* 18:207–220, 1975.

Forbes A: The interpretation of spinal reflexes in terms of present knowledge of nerve conduction. *Physiol Rev* 2:361–414, 1922.

Forbes A, Thatcher C: *Am J Physiol* 52:409, 1920.

Forrest WJ, Basmajian JV: Function of human thenar and hypothenar muscles: an electromyographic study of twenty-five hands. *J Bone Joint Surg* 47A: 1585–1594, 1965.

Forrest WJ, Khan MA: Electromyography of the flexor pollicis brevis and adductor pollicis in twenty hands: a preliminary report. (Transactions of the Internat. Soc. EMG. Kines., Montreal 1968.) *Electromyography* 8(suppl 1) 49–53, 1968.

Fortinguerra CRH, Vitti M: Estudo eletromiográfico da ação do m. pterigoideu medial em movimentos mandibulares. *Rev Assoc Paul Cirurg Dent* 33:501–508, 1979.

Fountain FP, Minear WL, Allison RD: Function of longus colli and longissimus cervicis muscles in man. *Arch Phys Med* 47:665–669, 1966.

Frame JW, Rothewell PS, Duxbury AJ: The standardization of electromyography of the masseter muscle in man. *Arch Oral Biol* 18:1419–1423, 1973.

Franksson C, Petersén I: Electromyographic investigations of disturbances in the striated muscles of the urethral sphincter. *Br J Urol* 27:154–161, 1955.

Freeman F, Ushijima T: The stuttering larynx: an EMG, fiberoptic study of laryngeal activity accompanying the moment of stuttering. Status Report SR-41, Haskins Laboratories, New Haven, CN, 1974.

Freeman MAR, Wyke B: Articular contributions to limb muscle reflexes: the effects of partial neurectomy of the knee-joint on postural reflexes. *Br J Surg* 53:61–69, 1966.

Freeman MAR, Wyke B: Articular reflexes at the ankle joint: an electromyographic study of normal and abnormal influences of ankle-joint mechanoreceptors upon reflex activity in the leg muscles. *Br J Surg* 54:990–1001, 1967.

Frenckner B, Euler CV: Influence of pudendal block on the function of the anal sphincters. *Gut* 16:482–489, 1975.

Freund HJ, Budingen HJ, Dietz V: Activity of single motor units from human forearm muscles during voluntary isometric contractions. *J Neurophysiol* 38:933–946, 1975.

Freund H-J, Dietz V, Wita CW, Kapp H: Discharge characteristics of single motor units in normal subjects and patients with supraspinal motor disturbances. *New Dev Electromyogr Clin Neurophysiol* 3:242–250, 1973.

Fridlund AJ, Cottam GL, Fowler SC: In search of the general tension factor: tensional patterning during auditory stimulation. *Psychophysiology* 19:136–145, 1982.

Fridlund AJ, Fowler SC, Pritchard DA: Striate muscle tensional patterning in frontalis EMG biofeedback. *Psychophysiology* 17:47–55, 1980.

Fridlund AJ, Izard CE: Electromyographic studies of facial expressions of emotions and patterns of emotions. In *Social Psychophysiology*. New York, Guilford Press, 1982.

Fridlund AJ, Izard CE: Electromyographic studies of facial expressions of emotions and patterns of emotions. In: Cacioppo JT, Petty RE: *Social Psychology: A Sourcebook*. New York, Plenum, 1983.

Friedebold G: Die aktivität normaler Rückenstreckmuskulatur im Elektromyogramm unter verschiedenen Haltungsbedingungen; eine Studie zur Skelettmuskelmechanik (German text). *Z Orthop* 90:1–18, 1958.

Friedman DH: Detection of signals by template matching. Baltimore, Johns Hopkins Press, 1968.

Friedman WA, Sypert GW, Munson JB, Fleshman JW: Recurrent inhibition in type-identified motoneurons. *J Neurophysiol* 46:1349–1359, 1981.

Fritzell B: An electromyographic study of the movements of the soft palate in speech. *Folia Phoniatr (Basel)* 15:307–311, 1963.

Fritzell B, Kotby MN: Observations on thyroarytenoid and palatal levator activation for speech. *Folia Phoniatr (Basel)* 28:1–7, 1976.

Fruhling M, Basmajian JV, Simard TG: A note on the conscious controls of motor units by children under six. *J Motor Behav*

1:65–68, 1969.

Fuchs P: The muscular activity of the chewing apparatus during night sleep. *J Oral Rehab* 2:35–48, 1975.

Fudel-Osipova SI, Grishko FE: Features specific to electromyograms taken during voluntary muscle contraction in old age. *Biull Eksp Biol Med* 3:9–14, 1962.

Fudema JJ, Fizzell JA, Nelson EM: Electromyography of experimentally immobilized skeletal muscles in cats. *Am J Physiol* 200:963–967, 1961.

Fujiwara M, Basmajian JV: Electromyographic study of two-joint muscles. *Am J Phys Med* 54:234–242, 1975.

Funakoshi M, Fujita N, Takehana S: Relations between occlusal interference and jaw muscle activities in response to changes in head position. *J Dent Res* 55:684–690, 1976.

Furlani J: Electromyographic study of the m. biceps brachii in movements of the glenohumeral joint. *Acta Anat* 96:270–284, 1976.

Furlani J, Bérzin F, Vitti M: Electromyographic study of the gluteus maximus muscle. *Electromyogr Clin Neurophysiol* 14:377–388, 1974.

Furlani J, Vitti M, Bérzin F: Estudo eletromiográfico do m. biceps femoral. *Folia Clin Biol* 1:188–192, 1973.

Furlani J, Vitti M, Bérzin F: Musculus biceps femoris, long and short head: an electromyographic study. *Electromyogr Clin Neurophysiol* 17:13–19, 1977.

Gaarder K: Control of states of consciousness, Parts I and II. *Arch Gen Psychiatry* 25:429–435; 430–441, 1971.

Gallai V, Firenze C, Mazzotta G, Agostini L: Electromyographic study of bulbo-cavernosus muscle in women according to parity and labor. *Electromyogr Clin Neurophysiol* 23:195–201, 1983.

Galvani L: *DeViribus Electricitatis*, transl by Green R, MA, Cambridge, 1953.

Gans C, Gorniak GC: Electromyograms are repeatable: precautions and limitations. *Science* 210:795–797, 1980.

Gans C, Hughes GM: The mechanism of lung ventilation in the tortoise *testudo graeca linné*. *J Exp Biol* 47:1–20, 1967.

Gardner E, Gray DJ, O'Rahilly R: *Anatomy: a Regional Study of Human Structure*. Philadelphia, W. B. Saunders, 1960.

Garrity LI: Measurement of subvocal speech: correlations between two muscle leads and between the recording methods. *Percept Motor Skills* 40:327–330, 1975.

Gassel MM: Sources of error in motor nerve conduction studies. *Neurology* 14:825–835, 1964.

Gassel MM, Diamantopoulos E: Patterns of conduction times in the distribution of the radial nerve: a clinical and electrophysiological study. *Neurology* 14:222–231, 1964.

Gassell MM, Trojaborg W: Clinical and electrophysiological study of the pattern of conduction times in the distribution of the sciatic nerve. *J Neurol Neurosurg Psychiatry* 27:351–357, 1964.

Gasser HS, Erlanger J: The nature of conduction of an impulse in the relatively refractory period. *Am J Physiol* 73:613, 1925.

Gasser HS, Newcomer HS: Physiological action currents in the phrenic nerve. An application of the thermionic vacuum tube to nerve physiology. *Am J Physiol* 57:1–26, 1921.

Gatev V: Studies of the electrical activity of the antagonistic muscles of the arm in normal children aged between 1 and 5 months. *Compte Rend l'Acad Bulg Sci* 20:743–747, 1967.

Gath I: Analysis of point process signals applied to motor unit firing patterns. I. Superposition of independent spike trains. *Math Biosci* 22:211–222, 1974.

Gath I, Stalberg E: The calculated radial decline of the extracellular action potential compound with *in situ* measurements in the human brachial biceps. *EEG Clin Neurophysiol* 44:547–552, 1978.

Gay T: Some electromyographic measures of coarticulation in VCV utterances. Proc Fifth Essex Phonetics Symposium, 1975, pp 15–28.

Gay T, Hirose H: Effect of speaking rate on labial consonant production: a combined electromyographic/high-speed motion picture study. *Phonetica* 22:44–56, 1973.

Gay T, Strome M, Hirose H, Sawashima M: Electromyography of the intrinsic laryngeal muscles during phonation. *Ann Otol Rhinol Laryngol* 81:401–410, 1972.

Gay T, Ushijima T, Hirose H, Cooper FS: Effect of speaking rate on labial consonant-vowel articulation. *J Phonet* 2:47–63, 1974.

Geddes LA: *Electrodes and the Measurement of Bioelectric Events*. New York, John Wiley & Sons, 1972.

Gelfand IM, Gurfinkel' VS, Tsetlin ML, Shik ML: Problems in analysis of movements. In Gelfand IM, Gurfinkel' VS, Fomin SV, Tsetlin ML: Models of the structural functional organization of certain biological systems. Am. translation, 1971. Cambridge, MA, M.I.T., 1966, pp 330–345.

Gellhorn E: Patterns of muscular activity in man. *Arch Phys Med* 28:568–574, 1947.

Gellhorn E: In Johnson WR: *Science and Medicine of Exercise and Sports.* New York, Harper & Bros, 1960, ch 7, pp 108–122.

Germana J: Patterns of autonomic and somatic activity during classical conditioning of a motor response. *J Comp Physiol Psychol* 69:173–178, 1969.

Gerstein GL, Clark WA: Simultaneous studies of firing patterns in several neurons. *Science* 143:1325–1327, 1964.

Gettrup E: Sensory regulation of wing twisting in locusts. *Exp Biol* 44:1–16, 1966.

Ghez C, Martin JH: The control of rapid limb movement in the cat. *Exp Brain Res* 45:115–125, 1982.

Gibbs CH: Electromyographic activity during the motionless period of chewing. *J Prosthet Dent* 34:35–40, 1975.

Gilliatt RW, Thomas PK: Changes in nerve conduction with ulnar lesions at the elbow. *J Neurol Neurosurg Psychiatry* 23:312–321, 1960.

Gillies JD: Motor unit discharge patterns during isometric contraction in man. *J Physiol (Lond)* 223:36–37P, 1972.

Gilson AS, Mills WB: Single responses of motor units in consequence of volitional effort. *Proc Soc Exp Biol Med* 45:650–652, 1940.

Gilson AS, Mills WB: Activities of single motor units in man during slight voluntary efforts. *Am J Physiol* 133:658–669, 1941.

Giovine GP: Premesse al trattamento neurochirurgico della disfunzioni vescicali neurogene. I. Studi sulla funzione dello sfintere striato dell'-uretra: l'elettrosfinterografia (Italian text). *Chirurgia* 14:39–62, 1959.

Girton DG, Kamiya J: A very stable electrode system for recording human scalp potentials with direct-coupled amplifiers. *Electroencephalogr Clin Neurophysiol* 37:85, 1974.

Givens MW, Teeple JB: Myoelectric frequency changes in children during static force production. *Electroencephalogr Clin Neurophysiol* 45:173–177, 1978.

Glaser EM, Marks WB: The on-line separation of interleaved neuronal pulse sequences. Rochester Conf. on Data Acquisition in Biology and Medicine, 1966, pp 137–156.

Gleason TF, Goldstein WM, Ray RD: Letter to the Editor. *J Bone Joint Surg* 65A:1031, 1983.

Godaux E, Desmedt JE: Evidence for a monosynaptic mechanism in the tonic vibration reflex of the human masseter muscle. *J Neurol Neurosurg Psychiat* 38:161–168, 1975a.

Godaux E, Desmedt JE: Exteroceptive suppression and motor control of the masseter and temporalis muscles in normal man. *Brain Res* 85:447–458, 1975b.

Godfrey KE, Kindig LE, Windell EJ: Electromyographic study of duration of muscle activity in sit-up variations. *Arch Phys Med Rehab* 58:132–135, 1977.

Goldberg LJ, Derfler B: Relationship among recruitment order, spike amplitude, and twitch tension of single motor units in human masseter muscle. *J Neurophysiol* 40:879–890, 1977.

Gollnick PD, Armstrong RB, Saltin B, Saubert IV CW, Sembrowich WL, Shepherd RE: Effect of training on enzyme activity and fiber composition of human skeletal muscle. *J Appl Physiol* 1:107–111, 1973.

Golstein I, Balshan ID: The relationship of muscle tension and autonomic activity to psychiatric disorders. *Psychosom Med* 27:39–52, 1965.

Goldstein ID (see also Balshan ID).

Gomez Oliveros L: Estudios anatómicos y electromigráficos de la fonación. *Acta Otorrinolar Esp* 19(6):37–88, 1969.

Gorniak GC, Gans C: Quantitative assay of electromyograms during mastication in domestic cats (*Felis catus*). *J Morphol* 163:253–281, 1980.

Goss CM (ed): *Gray's Anatomy of the Human Body*, ed 27. Philadelphia, Lea & Febiger, 1959.

Goto Y, Kumamoto M, Okamoto T: Electromyographic study of the function of the muscles participating in thigh elevation in the various planes. (Japanese text; English abstract). *Res J Phys Ed* 18:269–276, 1974.

Goto Y, Matsushita K, Tsujino A, Okamoto T: Hadoru-Soh no Kineshiorogiteki Kohsatsu (A kinesiological study of the hurdle running). In *The Science of Human Movement.* Tokyo, Kyorin Shorin, 1976, pp 145–158.

Gottlieb GL, Agarwall GC: Dynamic relationship between isometric muscle tension and the electromyogram in man. *J Appl Physiol* 30:345–351, 1971.

Gottlieb GL, Agarwal GC, Stark L: Interactions between voluntary and postural mechanism of the human motor system. *J Neurophysiol* 33:365–381, 1970.

Gottlieb GL, Myklebust BM, Penn RD, Agarwal GC: Reciprocal excitation of muscle antagonists by the primary afferent pathway. *Exp Brain Res* 46:454–456, 1982.

Goubel F, Bouisset S: Relation entre l'activité électromyographique intégrée et la travail mécanique effectué au cours d'un mouvement monoarticulaire simple. *J Physiol* 59:241, 1967.

Goubel F, Lestienne F, Bouisset S: Détermination dynamique de la compliance musculaire *in situ. J Physiol* 60:255, 1968.

Granit R: Neuromuscular interaction in postural tone of the cat's isometric soleus muscle. *J Physiol* 143:387–402, 1958.

Granit R: During discussion of his paper, Muscle tone and postural regulations, in "Muscle as a Tissue" International Conference, Lankenau Hospital, Philadelphia, 1960.

Granit R: The gamma (γ) loop in the mediation of muscle tone. *Clin Pharmacol Ther* 5:837–847, 1964.

Granit R, Henatsch HD, Steg G: Tonic and phasic ventral horn cells differentiated by post-tetanic potentiation in cat extensors. *Acta Physiol Scand* 37:114–126, 1956.

Granit R, Phillips CG, Skoglund S, Steg G: Differentiation of tonic from phasic alpha ventral horn cells by stretch, pinna and crossed extensor reflexes. *J Neurophysiol* 20:470–481, 1957.

Grant JCB, Basmajian JV: *Grant's Method of Anatomy: By Regions Descriptive and Deductive,* ed 7. Baltimore, Williams & Wilkins, 1965.

Grant PG: Lateral pterygoid: two muscles? *Am J Anat* 138:1–10, 1973.

Grassino AE, Whitelaw WA, Milic-Emili J: Influence of lung volume and electrode position on electromyography of the diaphragm. *J Appl Physiol* 40:971–975, 1976.

Gray ER: The role of leg muscles in variations of the arches in normal and flat feet. *J Am Phys Ther Assoc* 49:1084–1088, 1969.

Gray ER: Conscious control of motor units in a tonic muscle. *Am J Phys Med* 50:34–40, 1971a.

Gray ER: Conscious control of motor units in a neuromuscular disorder. *Electromyography* 11:515–517, 1971b.

Gray ER, Basmajian JV: Electromyography and cinematography of leg and foot ("normal" and flat) during walking. *Anat Rec* 161:1–16, 1968.

Green EE, Green AM, Walters ED: Voluntary control of internal states: psychological and physiological. *J Transpersonal Psychol* 2:1–26, 1970.

Green EE, Walters ED, Green AM, Murphy G: Feedback technique for deep relaxaton. *Psychophysiology* 6:371–377, 1969.

Green JG, Neil E: The respiratory function of the laryngeal muscles. *J Physiol* 129:134–141, 1953.

Greenlaw RK: *Function of Muscles about the Hip during Normal Level Walking.* Ph.D. Thesis, Queen's University, Canada, 1973.

Greenwood R, Hopkins A: Muscle responses during sudden falls in man. *J Physiol (Lond)* 254:507–518, 1976a.

Greenwood R, Hopkins A: Landing from an unexpected fall and a voluntary step. *Brain* 99:375–386, 1976b.

Greenwood R, Hopkins A: Monosynaptic reflexes in falling man. *J Neurol Neurosurg Psychiatry* 40:448–454, 1977.

Gregg RA, Mastellone AF, Gersten JW: Cross exercise—a review of the literature and study utilizing electromyographic techniques. *Am J Phys Med* 36:269–280, 1957.

Gresczyk EG: *Electromyographic Study of the Effect of Leg Muscles on the Arches of the Normal and Flat Foot.* M.S. Thesis, University of Vermont, 1965.

Grieve DW, Pheasant ST: Myoelectric activity, posture and isometric torque in man. Electromyogr Clin Neurophysiol 16:3–21, 1976.

Griffin CJ, Munro RR: Electromyography of the jaw-closing muscles in the open-close-clench cycle in man. *Arch Oral Biol* 14(2):141–150, 1969.

Grim P: Anxiety change produced by self-induced muscle tension and by relaxation with respiration feedback. *Behav Ther* 2:11–17, 1971.

Grimby L: Normal plantar response: integration of flexor and extensor reflex components. *J Neurol Neurosurg Psychiatry* 26:39–50, 1963a.

Grimby L: Pathological plantar response: disturbances of the normal integration of flexor and extensor reflex components. *J Neurol Neurosurg Psychiatry* 26:314–321, 1963b.

Grimby L, Hannerz J: Recruitment order of motor units on voluntary contraction: changes induced by proprioceptive afferent activity. *J Neurol Neurosurg Psychiatry* 31:565–573, 1968.

Grimby L, Hannerz J: Differences in recruitment order of motor units in phasic and tonic flexion reflex in 'spinal man.' *J Neurol Neurosurg Psychiatry* 33:562–570, 1970.

Grimby L, Hannerz J: Differences in recruitment order and discharge pattern of motor units in the early and late flexion reflex components in man. *Acta Physiol Scand* 90:555–564, 1974.

Grimby L, Hannerz J: Disturbances in the voluntary recruitment order of anterior tibial motor units in bradykinesia of parkinsonism. *J Neurol Neurosurg Psychiatry* 37:47–54, 1974b.

Grimby L, Hannerz J: Firing rate and recruitment order of toe extensor motor units in different modes of voluntary contraction. *J Physiol* 264:865–879, 1977.

Grimby L, Hannerz J, Hedman B: The fatigue and voluntary discharge properties of single motor units in man. *J Physiol* 316:545–554, 1981.

Grimby L, Hannerz J, Rånlund T: Disturbances in the voluntary recruitment order of anterior tibial motor units in spastic paraparesis upon fatigue. *J Neurol Neurosurg Psychiatry* 37:40–46, 1974.

Grønbaek P, Skouby AP: The activity pattern of the diaphragm and some muscles of the neck and trunk in chronic asthmatics and normal controls: a comparative electromyographic study. *Acta Med Scand* 168:413–425, 1960.

Gross BD, Lipke DP: A technique for percutaneous lateral pterygoid electromyography. *Electromyogr Clin Neurophysiol* 19:47–55, 1979.

Gross D, Grassino A, Ross WRD, Macklem PT: Electromyogram pattern of diaphragmatic fatigue. *J Appl Physiol* 46:1–7, 1979.

Grossman WI, Weiner H: Some factors affecting the reliability of surface electromyography. *Psychosom Med* 28:78–83, 1966.

Grundy M, Tosh PA, McLeish RD, Smidt L: An investigation of the centres of pressure under the foot while walking. *J Bone Joint Surg* 57B:98–103, 1975.

Grynbaum BB, Brudny J, Korein J, Belandres PV: Sensory feedback therapy for stroke patients. *Geriatrics* 31:43–47, 1976.

Guitar B: Reduction of stuttering frequency using analog electromyographic feedback. *J Speech Hearing Res* 18:672–685, 1975.

Guld C, Rosenfalck A, Willison RG: Report of the committee on emg instrumentation. *Electroencephalogr Clin Neurophysiol* 28:399–413, 1970.

Gurfinkel' VS, Ivanova AN, Kots YM, Pyatetskii-Shapiro IM, Shik ML: Quantitative characteristics of the work of motor units in the steady state. *Biofizika* 9(5):636–638, 1964.

Gurfinkel' VS, Levik YS: Forming an unfused tetanus. *Hum Physiol (Russian)* 2:914–924, 1976.

Gurfinkel' VS, Pal'tsev EI: Coactivation of antagonist muscles by the tendon reflex. *Agressologie* 13:37–43, 1972.

Gurfinkel' VS, Pal'tsev EI: Reflex response of antagonist muscles during elicitation of the tendon reflex. *Neurophysiology* 5:57–62, 1973.

Gurfinkel' VS, Surguladze TD, Mirskii ML, Tarko AM: Work of human motor units during rhythmic movements. *Biofizika* 15:1090–1095, 1970.

Gurkow HJ, Bast TH: Innervation of striated skeletal muscle. *Am J Phys Med* 37:269–277, 1958.

Guttmann L, Silver JR: Electromyographic studies on reflex activity of the intercostal and abdominal muscles in cervical cord lesions. *Paraplegia* 3:1–22, 1965.

Gydikov A, Kosarov D: Studies on the activity of alpha motoneurons in man by means of a new electromyographic method, neurophysiology studied in man. Amsterdam, Excerpta Medica, 1972, pp 321–329.

Gydikov A, Kosarov D: Influence of various factors on the length of the summated depolarized area of the muscle fibres in voluntary activating of motor units and in electrical stimulation. *Electromyogr Clin Neurophysiol* 14:79–93, 1974a.

Gydikov A, Kosarov D: Some features of different motor units in human biceps brachii. *Pflugers Arch* 347:75–88, 1974b.

Hagbarth KE, Hellsing G, Löfstedt L: TVR and vibration-induced timing of motor impulses in the human jaw elevator muscles. *J Neurol Neurosurg Psychiatry* 39:719–728, 1976.

Hagberg M: The amplitude distribution of surface EMG in static and intermittent static muscular performance. *Eur J Appl Physiol* 40:265–272, 1979.

Hagberg M: The elevated arm: myoelectric amplitude and spectral changes in some shoulder muscles. Proc 4th Congr Int Soc Electrophysiol Kinesiol, 1979, pp 70–71.

Hagberg M: Electromyographic signs of shoulder muscular fatigue in two elevated arm positions. *Am J Phys Med* 60:111–121, 1981a.

Hagberg M: Muscle endurance and surface electromyogram in isometric and dynamic exercise. *J Appl Physiol* 51:1–7, 1981b.

Hagberg M, Ericson BE: Myoelectric power spectrum dependence on muscular contraction level of elbow flexors. *Eur J Appl Physiol* 48:147–156, 1982.

Hagg G: Electromyographic fatigue analysis based on the number of zero crossings. *Pflugers Arch* 391:78–80, 1981.

Haines RW: The laws of muscle and tendon growth. *J Anat* 66:578–585, 1932.

Haines RW: On muscles of full and short action. *J Anat* 69:20–24, 1934.

Hairston LE, Sauerland EK: Electromyography of the human pharynx: discharge patterns of the superior pharyngeal constrictor during respiration. *Electromyogr Clin Neurophysiol* 21:299–306, 1981.

Håkansson CH: Conduction velocity and amplitude of the action potential as related to circumference in the isolated fibre of frog muscle. *Acta Physiol Scand* 37:14–34, 1956.

Håkansson CH: Action potentials recorded intra- and extracellularly from the isolated frog muscle fibre in Ringer's solution and in air. *Acta Physiol Scand* 39:291–318, 1957a.

Håkansson CH: Action potential and mechanical response of isolated cross striated frog muscle fibres at different degrees of stretch. *Acta Physiol Scand* 41:199–216, 1957b.

Håkansson CH, Toremalm NG: Studies of the physiology of the trachea, part IV. *Ann Otolaryngol Rhinol Laryngol* 76:873–885, 1967.

Hakkinen K, Komi PV: Electromyographic and mechanical characteristics of human skeletal muscle during fatigue under voluntary and reflex conditions. *EEG Clin Neurophysiol* 53:436–444, 1983.

Hallén LG, Lindahl O: Muscle function in knee extension: an emg study. *Acta Orthop Scand* 38:434–444, 1967.

Hallett M, Marsden CD: Ballistic flexion movements of the human thumb. *J Physiol* 294:33–50, 1979.

Hallett M, Shahani BT, Young ER: EMG analysis of stereotyped voluntary movements in man. *J Neurol Neurosurg Psychiatry* 38:1154–1162, 1975a.

Hallet M, Shahani BT, Young ER: EMG analysis of patients with cerebellar deficits. *J Neurol Neurosurg Psychiatry* 38:1163–1169, 1975b.

Hamilton WJ, Appleton AB: In *Textbook of Human Anatomy*, Boyd JD, Le Gros Clark WE, Hamilton WJ, Yoffey JM, Zuckerman S, Appleton AB (eds). London, Macmillan, 1956, p 206.

Hannam AG: Effect of voluntary contraction of the masseter and other muscles upon the masseteric reflex in man. *J Neurol Neurosurg Psychiatry* 35:66–71, 1972.

Hannam AG: Computer analysis of the correlation between the activity of the masseter muscles during unilateral chewing in man. *Electromyogr Clin Electrophysiol* 16:165–175, 1976.

Hannerz J: Discharge properties of motor units in man. *Experientia* 29:45–46, 1973.

Hannerz J: An electrode for recording single motor unit activity during strong muscle contractions. *Electroencephalogr Clin Neurophysiol* 37:179–181, 1974.

Hanson RJ Jr, Sussman HM, MacNeilage PF: Single motor unit potentials in speech musculature. In *Proc VII Intern Congr Phon Sci*. The Hague, Mouton, 1971. pp 316–319.

Hara T: Evaluation of recovery from local muscle fatigue by voluntary test contractions. J Hum Ergol 9:35–46, 1980.

Hardy RH: A method of studying muscular activity during walking. Med Biol Illustration 9:158–163, 1959.

Hardyck CD, Petrinovich LF, Ellsworth DW: Feedback of speech muscle activity during silent reading: rapid extinction. *Science* 154:1467–1468, 1966.

Harris RC, Hultman E, Sahlin E: Glycolytic intermediates in human muscle offer isometric contraction. *Pflugers Arch* 389:277–282, 1981.

Harris KS: Action of the extrinsic musculature in the control of tongue position: preliminary report. Report SR-25/26. New Haven, CN, Haskins Laboratories, pp 87–96.

Harris KS, Rosov R, Cooper FS, Lysaught GF: A multiple suction electrode system. *Electroencephalogr Clin Neurophysiol* 17:698–700, 1964.

Harris RI, Beath T: Hypermobile flat-foot with short tendo achillis. *J Bone Joint Surg* 30A:116–140, 1948.

Harrison VF, Koch WB: Voluntary control

of single motor unit activity in the extensor digitorum muscle. *Phys Ther* 52:267–272, 1972.

Harrison VF, Mortensen OA: Identification and voluntary control of single motor unit activity in the tibialis anterior muscle. *Anat Rec* 144:109–116, 1962.

Hart BL, Kitchell RL: Penile erection and contraction of penile muscles in the spinal and intact dog. *Am J Physiol* 210:257–262, 1966.

Harvey AM, Masland RL: Method for study of neuromuscular transmission in human subjects. *Bull Johns Hopkins Hosp* 68:81–93, 1941.

Hary D, Belman MJ, Propst J, Lewis S: A statistical analysis of the spectral moments used in EMG tests of endurance. *J Appl Physiol* 53:779–783, 1982.

Hasan Z, Stuart D: Mammalian muscle receptor. In *Handbook of the Spinal Cord.* New York, Marcel Dekker, 1984, vol 3.

Haskell B, Rovner H: Electromyography in the management of the incompetent anal sphincter. *Dis Colon Rectum* 10:81–84, 1970.

Hayes KJ: Wave analyses of tissue noise and muscle action potentials. *J Appl Physiol* 15:749–752, 1960.

Hefferline RF, Perera TB: Proprioceptive discrimination of a covert operant without its observation by the subject. *Science* 139:834–835, 1963.

Hellebrandt FA, Houtz SJ, Partridge MJ, Walters CE: Tonic neck reflexes in exercises of stress in man. *Am J Phys Med* 35:144–159, 1956.

Hellebrandt FA, Waterland JC: Indirect learning: the influence of unimanual exercise on related muscle groups of the same and opposite side. *Am J Phys Med* 41:45–55, 1962a.

Hellebrandt FA, Waterland JC: Expansion of motor patterning under exercise stress. *Am J Phys Med* 41:56–66, 1962b.

Henderson RL: Remote action potentials at the moment of response in a simple reaction-time situation. *J Exp Psychol* 44:238–241, 1952.

Henneman E, Olson CB: Relation between structure and function in the design of skeletal muscles. *J Neurophysiol* 28:581–598, 1965.

Henneman E, Somjen G, Carpenter DO: Excitability and inhibitability of motoneurons of different sizes. *J Neurophysiol* 28:599–620, 1965.

Henson OW Jr: The activity and function of the middle-ear muscles in echo-locating bats. *J Physiol* 180:871–887, 1965.

Herberts P, Kadefors R: A study of painful shoulder in welders. *Acta Orthop Scand* 47:381–387, 1976.

Herberts P, Kadefors R, Broman H: Localized muscle fatigue in shoulder muscles: A preliminary study employing the spectral moment analyzer. *Proc 4th Cong of the Int Soc of Electrophysiol Kinesiol*, 1979, pp 72–73.

Herberts P, Kaiser E, Magnusson R, Petersén I: Power spectra of myoelectric signals in muscles of arm amputees and healthy normal controls. *Acta Orthop Scand* 39:1–32, 1969.

Herman R, Bragin SJ: Function of gastrocnemius and soleus muscles. *J Am Phys Ther Assoc* 47:105–113, 1967.

Hermann GW: An electromyographic study of selected muscles involved in the shot put. *Res Q* 33:1–9, 1962.

Hermann L: Vebereine wirkung galvanischer strome auf muskeln und nervern. *Pflugers Arch Ges Physiol* 5:223–275, 1871.

Hermansen L, Osnes JB: Blood and muscle pH after maximal exercise in man. *J Appl Physiol* 32:304–308, 1972.

Hershler C, Milner M: Kinematic analysis of gait by stroboscopic flash photography and computer. In *Human Engineering Program Progress Report No. 1.* Hamilton, Canada, Chedoke Hospitals, 1976.

Hicks JH: The function of the plantar aponeurosis. *J Anat* 85:414–415, 1951.

Hicks JH: The mechanics of the foot. II. The plantar aponeurosis and the arch. *J Anat* 88:25–31, 1954.

Hinson MM: An electromyographic study of the push-up for women. *Res Q* 40(2): 1969.

Hirose H, Gay T: The activity of the intrinsic muscles in voicing control: an electromyographic study. *Phonetica*, 25:140–160, 1972.

Hirose H, Gay T: Laryngeal control in vocal attack: an electromyographic study. *Folio Phoniatr* 25:203–213, 1973.

Hirose K, Uono M, Sobue I: Quantitative electromyography comparison between manual values and computer ones on normal subjects. *Electromyogr Clin Neurophysiol* 14:315–320, 1974.

Hiroto I, Hirano M, Toyozumi Y, Shin T: Electromyographic investigation of the intrinsic laryngeal muscles related to speech sounds. *Otol Rhinol Laryngol*

76:861–873, 1967.

Hirschberg GG: Electromyographic evidence of the role of intercostal muscles in breathing. *News Lett Am Assoc EMG Electrodiagn*, 4:2–3, 1957.

Hirschberg GG, Adamson JP, Lewis L, Robertson KJ: Patterns of breathing of patients (with) poliomyelitis and respiratory paralysis. *Arch Phys Med* 43:529–533, 1962.

Hirschberg GG, Dacso MM: The use of electromyography in the study of clinical kinesiology of the upper extremity. *Am J Phys Med* 32:13–21, 1953.

Hirschberg GG, Nathanson M: Electromyographic recording of muscular activity in normal and spastic gaits. *Arch Phys Med* 33:217–225, 1952.

Hishikawa Y, Sumitsuji N, Matsumoto K, Kaneko Z: H-reflex and EMG of the mental and hyoid muscles during sleep, with special reference to narcolepsy. *Electroencephalogr Clin Neurophysiol* 18:487–492, 1965.

Hixon TJ, Siebens AA, Minifie FD: An EMG electrode for the diaphragm. *J Acoust Soc Am* 46:1588–1589, 1969.

Hobart DJ, Kelley DL, Bradley LS: Modification occurring during acquistion of a novel throwing task. *Am J Phys Med* 54:1–24, 1975.

Hodes R, Gribetz L, Moskowitz JA, Wagman IH: Low threshold associated with slow conduction velocity. *Arch Neurol* 12:510–526, 1965.

Hodes R, Larrabee MG, German W: The human electromyogram in response to nerve stimulation and the conduction velocity of motor axons. *Arch Neurol Psychiatry* 60:340–365, 1948.

Hoeffer PFA: Physiological mechanisms in spasticity. *Br J Phys Med* (n.s.) 15:88–90, 1952.

Hof, AL, van den Berg JW: Linearity between the weighted sum of the EMGs of the human triceps surae and the total torque. *J Biomech* 10:529–539, 1977.

Hogan NJ: Myoelectric prosthesis control: optimal estimation applied to EMG and the cybernetic considerations for its use in a man-machine interface. Ph.D. Dissertation, M.I.T., Cambridge, MA, 1976.

Hogan NJ, Mann RW: Myoelectric signal processing: Optimal estimation applied to electromyography. Part I. Derivation of the optimal myoprocessor. *IEEE Trans BME* 27:382–395, 1980.

Hogue RE: Upper-extremity muscular activity at different cadences and inclines in normal gait. *J Am Phys Ther Assoc* 49:963–972, 1969.

Holbrook SH: *The Golden Age of Quackery.* New York, MacMillan, 1959.

Holliday TA, Van Meter JR, Julian LM, Asmundson VS: Electromyography of chickens with inherited muscular dystrophy. *Am J Physiol* 209:871–876, 1965.

Hollinshead WH: *Anatomy for Surgeons.* New York, Hoeber-Harper, 1958, vol 3, p 388.

Holonen JP: Motor unit firing with different feedback methods. *EEG Clin Neurophysiol* 21:83–93, 1981.

Holonen JP, Falk B, Kalimo H: The firing rate of motor units in neuromuscular disorders. *J Neurol* 225:269–276, 1981.

Holt KS: Facts and fallacies about neuromuscular function in cerebral palsy as revealed by electromyography. *Dev Med Child Neurol* 8:255–268, 1966.

Holt KS, Pollack M, Sheridan MD: Discussion on developmental paediatrics. *Practitioner* 202:433–436, 1969.

Hoogmartens MJ, Basmajian JV: Postural tone in the deep spinal muscles of idiopathic scoliosis patients and their siblings: an etiologic study based on vibration-induced electromyography. *Electromyogr Clin Neurophysiol* 16:93–114, 1976.

Hoover F: The functions and integration of the intercostal muscles. *Arch Int Med* 30:1–33, 1922.

Hopf HC, Schlegel HJ, Lowitzsch K: Irradiation of voluntary activity to the contralateral side in movements of normal subjects and patients with central motor disturbances. *Eur Neurol* 12:142–147, 1974.

Horn CV: Electromyographic investigation of muscle imbalance in patients with paralytic scoliosis. *Electromyography* 9:447–455, 1969.

Hoshikawa T, Matsui H, Miyashita M: In Jokl: *Medicine and Sport, vol 8: Biomechanics III.* Basel, Karger, 1973, pp 342–348.

Hoshiko M: Sequence of action of breathing muscles during speech. *J Speech Hearing Res* 3:291–297, 1960.

Hoshiko M: Electromyographic investigation of the intercostal muscles during speech. *Arch Phys Med* 43:115–119, 1962.

Houk JC, Henneman E: Responses of Golgi tendon organs to active contractions of the soleus muscle of the cat. *J Neurophysiol* 30:466–481, 1967.

Houk JC, Rymer WZ: Neural control of muscle length and tension. In Brookhart

JM, Mountcastle VB, Brooks VB, Geiger SR: *Handbook of Physiology.* Section 1. *The Nervous System*, Bethesda, MD, American Physiological Society, vol 2, part 1, ch 8, pp 257–323, 1981.

Houk JC, Rymer WZ, Crago PE: Dependence of dynamic response of spindle receptors on muscle length and velocity. *J Neurophysiol* 46:143–166, 1981.

Houtz SJ, Fischer FJ: An analysis of muscle action and joint excursion during exercise on a stationary bicycle. *J Bone Joint Surg* 41A:123–131, 1959.

Houtz SJ, Fischer FJ: Function of leg muscles acting on foot as modified by body movements. *J Appl Physiol* 16:597–605, 1961.

Houtz SJ, Walsh FP: Electromyographic analysis of the function of the muscles acting on the ankle during weight-bearing with special reference to the triceps surae. *J Bone Joint Surg* 41A:1469–1481, 1959.

Hoyle G: Exploration of neuronal mechanisms underlying behaviour in insects. In Reiss R: *Neural Theory and Modeling.* Stanford CA, Stanford University Press, 1964, pp 346–376.

Hoyle G, Willows AOD: Neuronal basis of behavior in *tritonia. J Neurobiol* 4:239–254, 1973.

Hrycyshyn AW, Basmajian JV: Electromyography of the oral stage of swallowing in man. *Am J Anat* 133:333–340, 1972.

Hubbard AW: In Johnson WR: *Science and Medicine of Exercise and Sports.* New York, Harper Brothers, 1960, ch2, pp 7–39.

Hufschmidt HI, Hufschmidt T: Antagonist inhibition as the earliest sign of a sensory-motor reaction. *Nature (Lond.)* 174:607, 1954.

Hughes GM, Ballintijn CM: Electromyography of the respiratory muscles and gill water flow in the dragonet. *J Exp Biol* 49:583–602, 1968.

Hultborn H, Pierrot-Deseilligny E: Input-output relations in the pathway of recurrent inhibition to motoneurones in the cat. *J Physiol* 297: 267–287, 1979.

Hultborn H, Lindstrom S, Wigstrom H: On the function of recurrent inhibition in the spinal cord. *Exp Brain Res* 37:399–403, 1979.

Huntington DA, Harris KS, Sholes GN: An electromyographic study of consonant articulation in hearing-impaired and normal speakers. *J Speech Hearing Res* 11:147–158, 1968.

Husson R: Etude des phénomenes physiologiques et acoustiques fondamentaux de la voix chantée. Thesis, Faculty of Sciences, Paris, 1950.

Hustert R: Neuromuscular coordination and proprioceptive control of rhythmical abdominal ventilation in intact *locusta migratoria migratorioides. J Comp Physiol* 97:159–179, 1975.

Hutch JA, Elliott HW: Electromyographic study of electrical activity in the paraurethral muscles prior to and during voiding. *J Urol* 99:759–765, 1968.

Inbar GF, Allin J, Golos E, Koehler W, Kranz H: EMG spectral shift with muscle length, tension and fatigue. *Proc IEEE Melecon. Conf.*, vol 8, pp 2 and 3, 1981.

Ihre T: Studies on anal function in continent and incontinent patients. *Scand J Gastroenterol, Suppl* 25, 1974.

Iida M, Basmajian JV: Electromyography of hallux valgus. *Orthop Related Res* 101:220–224, 1974.

Iida M, Basmajian JV: Electromyography of plantaris muscle. *Electromyogr Clin Neurophysiol* 15:311–316, 1975.

Iida M, Viel E, Iwasaki T, Ito H, Yazaki K: Activité E.M.G. des muscles superficiels et profonds du dos pendant les exercices de rééducation couramment utilisés. *Electrodiagnostic-therapie* 13:55–67, 1976.

Ikai M: Crossed reflexes of limbs observed in healthy man. *Jap J Physiol* 6:29–39, 1956.

Ingjer F: Effect of endurance training on muscle fiber ATPase activity, capillary supply and mitochondrial content in man. *J Physiol* 294:419–432, 1979.

Inglis J, Campbell D, Donald MW: Electromyographic biofeedback and neuromuscular rehabilitation. *Can J Behav Sci* 8:299–323, 1976a.

Inglis J, Sproule M, Leicht M, Donald MW, Campbell D: Electromyographic biofeedback treatment of residual neuromuscular disabilities after cerebrovascular accident. 28:260–264, 1976b.

Inman VT: Functional aspects of the abductor muscles of the hip. *J Bone Joint Surg* 29:607–619, 1947.

Inman VT, Ralston HJ, Saunders JBCM, Feinstein B, Wright EW Jr: Relation of human electromyogram to muscular tension, Advisory Committee on Artifical Limbs, N.R.C., Series 11, Issue 18.

Inman VT, Ralston HJ, Saunders JBCM, Feinstein B, Wright EW Jr: Relation of human electromyogram to muscular tension. *Electroencephalogr Clin Neurophysiol*

4:187–194, 1952.

Inman VT, Saunders JBCM, Abbott LC: Observations on the function of the shoulder joint. *J Bone Joint Surg* 26:1–30, 1944.

Innocent GM: Electrical activity of single extraocular muscles during the oculocompensatory positions. *Electromyography* 11:25–38, 1971.

Inouye T, Shimizu A: The electromyographic study of verbal hallucination. *J Nerv Ment Dis* 151:415–422, 1970.

Ioffe ME, Andreyev AE: Inter-extremities coordination in local motor conditioned reactions of dogs (in Russian). *Zh Vyssh Nerv Deyat* 19:557–565, 1969.

Isaksson I, Johanson B, Petersén I, Sellden U: Electromyographic study of the Abbe-and fan flaps. *Acta Chir Scand* 123:343–350, 1962.

Ishida H, Kimura T, Okada M: Symp. 5th Cong. Int. Primat. Soc. Tokyo, Japan Science Press, 1974.

Isley CL, Basmajian JV: Electromyography of the human cheeks and lips. *Anat Rec* 176:143–148, 1973.

Ismail AH, Barany JW, Manning KR: *Assessment and Evaluation of Hemiplegic Gait.* Technical Report for the National Institutes of Health, Purdue University, Lafayette, IN, 1965.

Ito H, Iwasaki T, Yamada M, Yasaki K, Tanaka S, Iida M: Electromyography of latissimus dorsi muscle. *J Jpn Ph Assoc* 3:23–36, 1976.

Jacob PP, Haridas R, Ammal PJ: An electromyographic study of the behaviour of orbicularis oris and mentalis muscle. *Ind J Med Res* 59:311–320, 1971.

Jacobs A, Fenton GS: Visual feedback of myoelectric output to facilitate muscle relaxation in normal persons and patients with neck injuries. *Arch Phys Med* 50:34–39, 1969.

Jacobs L, Feldman M, Bender MB: Eye movement during sleep. I. The pattern in the normal human. *Arch Neurol* 25:151–159, 1971.

Jacobs MB: *Antagonist EMG Temporal Patterns during Rapid Voluntary Movement.* Ph.D. Dissertation, University of Toledo, Ohio, 1976.

Jacobs MB, Andrews LT, Iannone A, Greniger L: Antagonist EMG temporal patterns during rapid voluntary movement. *Neurology* 30:36–41, 1980.

Jacobson A, Kales A, Lehmann D, Hoedemaker FS: Muscle tonus in human subjects during sleep and dreaming. *Exp Neurol* 10:418–424, 1964.

Jacobson E: *Progressive Relaxation.* Chicago, University of Chicago Press, 1929.

Jacobson E: Electrical measurements concerning muscular contraction (tonus) and the cultivation of relaxation in man: studies on arm flexors. *Am J Physiol* 107:230–248, 1933.

Jampolsky A: What can electromyography do for the ophthalmologist? *Invest Ophthalmol* 9:570–599, 1970.

Jampolsky A, Tamler E, Marg E: Artifacts and normal variations in human ocular electromyography. *AMA Arch Ophthalmol* 61:402–413, 1980.

Janda V, Kozák P: Zur Funktion der motorischen Einheit unter Ischämie. *Dtsch Z Nervenheilkd* 185:598–605, 1964.

Janda V, Stará V: The role of thigh adductors in movement patterns of the hip and knee joint. *Courrier* (*Centre Internat de l'Enfance*), 15:1–3, 1965.

Janda V, Véle F: A polyelectromyographic study of muscle testing with special reference to fatigue. Proc. of IX World Rehab. Congress, Copenhagen, pp 80–84, 1963.

Jankowska E, McCrea D, Mackel R: Pattern of "non-reciprocal" inhibition of motoneurones by impulses in group Ia muscle spindle afferents in the cat. *J Physiol* 316:393–409, 1981a.

Jankowska E, McCrea D, Mackel R: Oligosynaptic excitation of motoneurones by impulses in group Ia muscle spindle afferents in the cat. *J Physiol* 316:441–425, 1981b.

Jarcho LW, Eyzaguirre C, Berman B, Lilienthal JL Jr: Spread of excitation in skeletal muscle: some factors contributing to the form of the electromyogram. *Am J Physiol* 168:446–457, 1952.

Jarcho LW, Vera CL, McCarthy CG, Williams PM: The form of motor-unit and fibrillation potentials. *Electroencephalog Clin Neurophysiol* 10:527–540, 1958.

Jasper HH, Ballem G: Unipolar electromyograms of normal and denervated human muscle. *J Neurophysiol* 12:231–244, 1949.

Jasper HH, Forde WO: The R. C. A. M. C. electromyograph mark III. *Can J Res* 25:100–110, 1947.

Jefferson NC, Ogawa T, Syleos C, Zambetoglou A, Necheles H: Restoration of respiration by nerve anastomosis. *Am J Physiol* 198:931–933, 1960.

Jefferson NC, Phillips CW, Necheles H: Observations on diaphragm and stomach of the dog following phrenicotomy. *Proc Soc Exp-Biol Med* 72:482–485, 1949.

Jen PHS, Suga N: Coordinated activities of middle-ear and laryngeal muscles in echolocating bats. *Science* 191:950–952, 1976.

Jenerick H: An analysis of the striated muscle fibre action current. *Biophysical J* 4:77–91, 1964.

Jennische E: Relation between membrane potential and lactate in gastrocnemius and soleus muscle in the cat during tourniquet ischemia and postischemia reflow. *Pflügers Arch* 394:329–332, 1982.

Jesel M, Isch Treussard C, Isch F: EMG of the anal and urethral sphincters in the diagnosis of lesions of the cauda equina and lumbar section of the spinal cord. *Rev Neurol* 122:431–434, 1970.

Johansson S, Larsson LE, Ortengren R: An automatic method for the frequency analysis of myoelectric signals evaluated by an investigation of the spectral changes following strong sustained contractions. *Med Biol Eng* 8:257–264, 1970.

Johnson CE, Basmajian JV, Dasher W: Electromyography of sartorius muscle. *Anat Rec* 173:127–130, 1972.

Johnson CP: Analysis of five tests commonly used in determining the ability to control single motor units. *Am J Phys Med* 55:113–121, 1976.

Johnson DR: *An Electromyographic Study of Extrinsic and Intrinsic Muscles of the Thumb.* M.Sc. Thesis, Queen's University, Canada, 1970.

Johnson DR, Forrest WJ: An electromyographic study of the abductors and flexors of the thumb in man (abstr). *Anat Rec* 166:325, 1970.

Johnson EW, Olsen KJ: Clinical value of motor nerve conduction velocity determination. *JAMA* 172:2030–2035, 1960.

Johnson HE, Garton WH: Muscle re-education in hemiplegia by use of electromyographic device. *Arch Phys Med* 54:320–322, 1973.

Johnston R, Lee KH: Myofeedback: a new method of teaching breathing exercises in emphysematous patients. *Phys Ther* 56:826–829, 1976.

Johnston TB, Davies DV, Davies F: *Gray's Anatomy: Descriptive and Applied*, ed 32. London, Longmans, Green, 1958.

Johnston TB, Whillis J: (eds): *Gray's Anatomy: Descriptive and Applied*, ed 31. London, Longmans, Green and Co., 1954.

Jones DS, Beargie RJ, Pauly JE: An electromyographic study of some muscles of costal respiration in man. *Anat Rec* 117:17–24, 1953.

Jones DS, Pauly JE: Further electromyographic studies on muscles of costal respiration in man. *Anat Rec* 128:733–746, 1957.

Jones FW: *The Principles of Anatomy as Seen in the Hand*, ed 2. London, Bailliére, Tindall & Cox, 1942, pp 258–259.

Jones FW: *Structure and Function as Seen in the Foot*, ed 2. London, Bailliére, Tindall & Cox, 1949, pp 246–256.

Jones GM, Watt DGD: Observations on the control of stepping and hopping movements in man. *J Physiol* 219:709–727, 1971.

Jones NB, Lago PJA: Spectral analysis and the interference EMG. *IEEE Proc* 129: 673–678, 1982.

Jones RL: The human foot. An experimental study of its mechanics, and the role of its muscle and ligaments in the support of the arch. *Am J Anat* 68:1–39, 1941.

Jones RL: The functional significance of the declination of the axis of the subtalar joint. *Anat Rec* 93:151–159, 1945.

Jonsson B: The functions of individual muscles in the lumbar part of the erector spinae muscle. *Electromyography* 10:5–21, 1970.

Jonsson B: Electromyography of the erector spinae muscle. In Jokl E: *Medicine and Sport, vol 8: Biomechanics III.* Basel, Karger, 1973, pp 294–300.

Jonsson B, Bagge UE: Displacement, deformation and fracture of wire electrodes for electromyography. *Electromyography* 8:328–347, 1968.

Jonsson B, Hagberg M: The effect of different working heights on the deltoid muscle: a preliminary methodological study. *Scand J Rehab Med* 3(suppl):26–32, 1974.

Jonsson S, Jonsson B: Function of the muscles of the upper limb in car driving. Part I. The deltoid muscle. Part II. The trapezius muscle. Part III. The brachialis, brachioradialis, biceps brachii and triceps brachii muscles. IV. The pectoralis major, serratus anterior and latissimus dorsi muscles. *Ergonomics* 18:375–388 & 643–649, 1975.

Jonsson S, Jonsson B: Function of the muscles of the upper limb in car driving. Part V. The supraspinatus, infraspinatus, teres minor and teres major muscles. *Ergonomics* 19:711–717, 1976.

Jonsson B, Olofsson BM, Steffner LC: Function of the teres major, latissimus dorsi and pectoralis major muscles: a preliminary study. *Acta Morphol. Neerl-Scand* 9:275–280, 1972.

Jonsson B, Reichmann S: Reproducibility in kinesiologic EMG-investigations with intramuscular electrodes. *Acta Morphol Neerl-Scand* 7:73–90, 1968.

Jonsson B, Reichmann S: Radiographic control in the insertion of emg electrodes in the lumbar part of the erector spinae muscle. *Z Anat Entwickl-Gesch* 130:192–206, 1970.

Jonsson B, Rundgren A: The peroneus longus and brevis muscles: a roentgenologic and electromyographic study. *Electromyography* 11:93–103, 1971.

Jonnson B, Steen B: Function of the gracilis muscle. An electromyographic study. *Acta Morphol Neerl-Scand* 4:325–341, 1966.

Jonsson B, Synnerstad B: Electromyographic studies of muscle function in standing: a methodological study. *Acta Morphol Neerl-Scand* 6:361–370, 1967.

Joseph J: *Man's Posture: Electromyographic Studies,* Springfield IL, Charles C Thomas, 1960.

Joseph J: Electromyography of posture and gait in man (abstr). *Bull Am Assoc EMG Electrodiag* 12:24, 1965.

Joseph J: The pattern of activity of some muscles in women walking in high heels. *Ann Phys Med* 9:295–299, 1968.

Joseph J, Nightingale A: Electromyography of muscles of posture: leg muscles in males. *J Physiol* 117:484–491, 1952.

Joseph J, Nightingale A: Electromyography of muscles of posture: thigh muscles in males. *J Physiol* 126:81–85, 1954.

Joseph J, Nightingale A: Electromyography of muscles of posture: leg and thigh muscles in women, including the effects of high heels. *J Physiol* 132:465–468, 1956.

Joseph J, Nightingale A, Williams PL: A detailed study of the electric potentials recorded over some postural muscles while relaxed and standing. *J Physiol* 127:617–625, 1955.

Joseph J, Watson R: Telemetering electromyography of muscles used in walking up and down stairs. *J Bone Joint Surg* 49B:774–780, 1967.

Joseph J, Williams PL: Electromyography of certain hip muscles. *J Anat* 91:286–294, 1957.

Jouffroy FK, Jungers WL, Stern JT: Téléélectromyographie des divers faisceaux du muscle quadriceps femoris au cours de la locomotion chez un Lémurien de Madagascar (*Lemur fulvus*). *CR Acad Sci Paris* 288:1627–1630, 1979.

Jüde HD, Drechsler F, Neuhauser B: Elementare elektromyographische Analyse and Bewegungstmuster des M. mylo-glossus. *Dtsch Zahärztl Z* 30:457–461, 1975.

Jungers WJ, Jouffroy FK, Stern JT Jr: Gross structure and function of the quadriceps femoris in *Lemur fulvus*: an analysis based on telemetered electromyography. *J Morphol* 164:287–299, 1980.

Juniper RP: The superior pterygoid muscle. *Br J Oral Surg* 19:121, 1981.

Juniper RP: Electromyography of the two heads of external pterygoid muscle via the intra-oral route. *Electromyogr Clin Neurophysiol* 23:21–33, 1983.

Kadefors R, Kaiser E, Petersén I: Dynamic spectrum analysis of myo-potentials with special reference to muscle fatigue. *Electromyography* 8:39–74, 1968.

Kadefors R, Petersén I: Spectral analysis of myo-electric signals from muscles of the pelvic floor during voluntary contraction and during reflex contractions connected with ejaculation. *Electromyography* 10:45–68, 1970.

Kadefors R, Petersén I, Herberts P: Muscular reaction to welding work. *Ergonomics* 19:543–558, 1976.

Kadefors R, Petersén I, Tengroth B: Quantitative analysis of EMG from m rectus lateralis oculi. *Scand J Rehab Med* 6(suppl)3:115–120, 1974.

Kahn SD: Comparative advantages of bipolar abraded skin surface electrodes over bipolar intramuscular electrodes for single motor unit recording in psychophysiological research. *Psychophysiology* 8:635–647, 1971.

Kaiser E, Petersén I: Frequency analysis of muscle action potentials during tetanic contraction. *Electromyography* 3:5–17, 1963.

Kaiser E, Petersén I: Muscle action potentials studied by frequency analysis and duration measurement. *Acta Neurol Scand* 41:19–41, 1965.

Kaiser E, Petersén I: Muscle action potentials studied by frequency analysis and duration measurement. *Acta Neurol Scand Suppl* 13:213–236, 1965.

Kalen FC, Gans C: How does the bat chew? Evidence from electromyography (abstr). *Anat Rec* 166:327, 1970.

Kamei S, Matsui H, Miyashita M: An electromyographic analysis of Japanese archery (Japanese text: English abstr). *Res J Phys Ed* 15:39–46, 1971.

Kamen G: The acquisition of maximal isometric plantar flexor strength: a force-time curve analysis. *J Motor Behav* 15:63–73, 1983.

Kammer AE, Heinrich B: Neural control of bumblebee fibrillar muscles during shivering. *J Comp Physiol* 337–345, 1972.

Kamon E: Electromyography of static and dynamic postures of the body supported on the arms. *J Appl Physiol* 21:1611–1618, 1966.

Kamon E, Gormley J: Muscular activity pattern for skilled performance and during learning of a horizontal bar exercise. *Ergonomics* 11:345–357, 1968.

Kamp A, Kok ML, de Quartel FW: A multiwire cable for recording from moving subjects. *Electroencephalogr Clin Neurophysiol* 18:422–423, 1965.

Kanosue K, Yoshida M, Akazawa K, Fujii K: The number of active motor units and their firing rates in voluntary contractions of the human brachialis muscle. *Jpn J Physiol* 29:427–443, 1979.

Kaplan M, Kaplan T: Flat foot. A consideration of the anatomy and physiology of the normal foot, the pathology and mechanism of flat foot, with the resulting roentgen manifestations. *Radiology* 25: 485–491, 1935.

Karlins M, Andrews LM: *Biofeedback: Turning on the Power of Your Mind*. New York, Warner Paperback Library, 1973. Originally published by J. B. Lippincott, 1972.

Karlsson E, Jonsson B: Function of the gluteus maximus muscle: an electromyographic study. *Acta Morphol Neerl-Scand* 6:161–169, 1965.

Kaseda Y, Nomura S: Electromyographic studies on the swimming movement of carp. I. Body movement. *Jpn J Vet Sci* 35:335–342, 1973.

Kasvand T, Milner M, Quanbury A, Winter DA: Computers and the kinesiology of gait. *Comput Biol Med* 6:111–120, 1976.

Kasvand T, Milner M, Rapley LF: A computer-based system for the analysis of some aspects of human locomotion. In *Transactions of Conference on Human Locomotor Engineering*, University of Sussex, Institute of Mechanical Engineers, Publishers, London, pp 297–306, 1971.

Kato M, Murakami S, Takahashi K, Hirayama H: Motor unit activities during maintained voluntary muscle contraction at constant levels in man. *Neurosci Lett* 25:149–154, 1981.

Kato M, Tanji J: Volitionally controlled single motor units in human finger muscles. *Brain Res* 40:345–357, 1972a.

Kato M, Tanji J: Cortical motor potentials accompanying volitionally controlled single motor unit discharges in human finger muscles. *Brain Res* 47:103–111, 1972b.

Kawamura Y, Fujimoto J: Some physiologic considerations on measuring rest position of the mandible. *Med J Osaka Univ* 8:247–255, 1957.

Kawasaki M, Ogura JH, Takenouchi S: Neurophysiologic observations of normal deglutition. *Laryngoscope* 74:1747–1780, 1964.

Kazai N, Kumamoto M, Okamoto T, Yamashita N, Goto Y, Maruyama H: Yakyu no Toh-Dohsa (Oba-hando Suroh) ni okeru Johshi Johshitaikingun no Sayo-Kijyo (Electromyographic study of the overhand pitching in terms of the functional mechanism of the upper extremity and the shoulder girdle muscles). *Res J Phys Ed* 21:137–144, 1976.

Kazai N, Okamoto T, Kumamoto M: Electromyographic study of supported walking of infants in the initial period of learning to walk. In Komi PV et al: *Biomechanics V*. Baltimore, University Park Press, 1976.

Keagy RD, Brumlik J, Bergan JJ: Direct electromyography of the psoas major muscle in man. *J Bone Joint Surg* 48A:1377–1382, 1966.

Kear M, Smith RN: A method of recording tendon strain in sheep during locomotion. *Acta Orthop Scand* 46:896–905, 1975.

Keehn DG: An iterative spike separation technique. *IEEE BME* 13:19–28, 1966.

Keith A: In a discussion of a paper by L. H. Buxton: The teeth and jaws of savage man. *Trans Br Soc Orthodon 1916–1920*:79–88, 1920.

Keith A: The history of the human foot and its bearing on orthopaedic practice. *J Bone Joint Surg* 11:10–32, 1929.

Kelly KA, Code CF, Elveback LR: Patterns of canine gastric electrical activity. *Am J Physiol* 217:461–470, 1969.

Kelman AW, Gatehouse S: A study of the electromyographic activity of the muscle orbicularis oris. *Folia Phoniatr* 27:177–189, 1975.

Kelsen SG, Altose MD, Stanley NN, Levinson RS, Cherniack NS, Fishman AP: Electromyographic response of respiratory muscles during elastic loading. *Am J Physiol* 230:675–683, 1976.

Kelton LW, Wright RD: The mechanism of easy standing by man. *Austr J Exp Biol*

Med Sci 27:505–515, 1949.

Kendall HO, Kendall FP, Wadsworth GE: Muscles: testing and function, ed 2. Baltimore, Williams & Wilkins, 1971.

Kenney WE, Heaberlin PC Jr: An electromyographic study of the locomotor pattern of spastic children. *Clin Orthop* 24:139–151, 1962.

Kernell D: High-frequency repetitive firing of cat lumbosacral motoneurons stimulated by long-lasting injected currents. *Acta Physiol Scand* 65:74–86, 1965.

Kernell D, Monster AW: Threshold current for repetitive impulse firing in motoneurones innervating muscle fibres of different fatigue sensitivity in the cat. *Brain Res* 229:193–196, 1981.

Kernell D, Monster AW: Motoneurone properties and motor fatigue. *Exp Brain Res* 46:197–204, 1982.

Khan MA: *Morphology and Electromyography of the Flexor Pollicis Brevis and Adductor Pollicis Muscles* M.Sc. Thesis. Queen's University, Canada, 1969.

Kiesswetter H: EMG-patterns of pelvic floor muscles with surface electrodes. *Urol Int* 31:60–69, 1970.

Kimm J, Sutton D: Foreperiod effects on human single motor unit reaction times. *Physiol Behav* 10:539–542, 1973.

Kinsman RA, O'Banion K, Robinson S, Staudenmayer H: Continuous biofeedback and discrete posttrial verbal feedback in frontalis muscle relaxation training. *Psychophysiology* 12:30–35, 1975.

Kirikae I: *The Structure and Function of the Middle Ear*. Tokyo, The University of Tokyo Press, 1960.

Kirikae I, Hirose H, Kawamura S, Sawashima M, Kobayashi T: An experimental study of central motor innervation of the laryngeal muscles in the cat. *Ann Otol Rhinol Laryngol* 71:222–242, 1962.

Kiviat MD, Zimmermann TA, Donovan WH: Sphincter stretch: a new technique resulting in continence and complete voiding in paraplegics. *J Urol* 114:895–897, 1975.

Klausen K: The form and function of the loaded human spine. *Acta Physiol Scand* 65:176–190, 1965.

Kleppe D, Groendijk HE, Huijing PA, Van Wieringen PC: Single motor unit control in the human mm. abductor pollicis brevis and mylohyoideus in relation to the number of muscle spindles. *Electromyogr Clin Neurophysiol* 22:21–22, 1982.

Klineberg IJ, Greenfield BE, Wyke BD: Contributions to the reflex control of mastication from mechanoreceptors in the temporomandibular joint capsule. *Dent Pract* 21:73–83, 1970.

Kloprogge, MJGM: Reflex control of the jaw muscles by stimuli from receptors in the periodontal membrane. *J Oral Rehab* 2:259–272, 1975.

Knowlton GC, Bennett RL, McClure R: Electromyography of fatigue. *Arch Phys Med* 32:648–652, 1951.

Knutsson B, Lindh K, Telhag H: Sitting—an electromyographic and mechanical study. *Acta Orthop Scand* 37:415–428, 1966.

Knuttson E, Mårtensson A, Martensson B: The normal electromyogram in human vocal muscles. *Acta Otolaryngol* 68:526–536, 1969.

Koczocik-Przedpelska J, Tobola S, Gruszczyński W: Aktywność miolektryczna podczas czynności zautomatyzowanej oraz ruchn dowolnego (Polish text; English abstr). *Acta Physiol Polon* 17:593–599, 1966.

Koepke GH, Smith EM, Murphy AJ, Dickinson DG: Sequence of action of the diaphragm and intercostal muscles during respiration. I. Inspiration. *Arch Phys Med* 39:426–430, 1958.

Kogi K, Hakamada T: Frequency analysis of the surface electromyogram in muscle fatigue. *J Sci Labour* (Tokyo) 38:519–528, 1962a.

Kogi K, Hakamada T: Slowing of surface electromyogram and muscle strength in muscle fatigue. *Rep Inst Sci Lab* 60:27–41, 1962b.

Kollberg S, Petersén I, Stener I: Preliminary results of an electromyographic study of ejaculation. *Acta Chir Scand* 123:478–483, 1962.

Komi PV: Relationship between muscle tension, EMG and velocity of contraction under concentric and eccentric work. In Desmedt JE: *New Developments in Electromyography and Clinical Neurophysiology* Basel, Karger, 1973.

Komi PV, Buskirk ER: Reproducibility of electromyographic measurements with inserted wire electrodes and surface electrodes. *Electromyography* 10:357–367, 1970.

Komi PV, Karlsson J: Skeletal muscle fibre types, enzyme types, enzyme activities and physical performance in young males and females. *Acta Physiol Scand* 103:210–218, 1978.

Komi PV, Rusko H: Quantitative evaluation of mechanical changes during fatigue loading of eccentric and concentric work. *Scand J Med* 3(suppl):121–126, 1974.

Komi PV, Tesch P: EMG frequency spectrum, muscle structure and fatigue during dynamic contractions in man. *Eur J Appl Physiol* 42:41–50, 1979.

Komi P, Viitasalo JHT: Signal characteristics of EMG at different levels of muscle tension. *Acta Physiol Scand* 96:267–276, 1976.

König B Jr, Vitti M, Bérzin F, De Camargo AM, Fortinguerra CRH: Electromyographic analysis of the digastric muscle. *Ciência e Cultura* 30:463–465, 1978.

Konopacki RA, Cole KJ: Evaluation of electrodes for speech muscle electromyography. *J Acoust Soc Am Suppl* 1:71, S33, 1982.

Kopec J, Hausman-Petrusewicz I: Application of harmonic analysis to the electromyogram evaluation. *Acta Physiol Polon* 17:598–608, 1966.

Korein J, Brudny J, Grynbaum B, Sachs-Frankel G, Weisinger M, Levidow L: Sensory feedback therapy of spasmodic torticollis and dystonia: results in treatment of 55 patients. *Adv Neurol* 14:375–402, 1976.

Kotby MN: Percutaneous laryngeal electromyography: standardization of the technique. *Folio Phoniatr* 27:116–127, 1975.

Kotby MN, Haugen LK: The mechanics of laryngeal function. *Acta Otolaryngol* 70:203–211, 1970a.

Kotby MN, Haugen LK: Critical evaluation of the action of the posterior crico-arytenoid muscle, utilizing direct emg-study. *Acta Otolaryngol* 70:260–268, 1967b.

Kotby MN, Haugen LK: Attempts at evaluation of the function of various laryngeal muscles in the light of muscle and nerve stimulation experiments in man. *Acta Otolaryngol* 70:419–427, 1970c.

Kotby MN, Haugen LK: Clinical application of electromyography in vocal fold mobility disorders. *Acta Otolaryngol* 70:428–437, 1970d.

Koyama S, Okamoto T, Yoshizawa M, Kumamoto M: Electromyographic study of English pronuciation by Japanese. Privately published, Dept. of Physiology, Kyoto University, 1981.

Kozmyan EI: Time relations of excitation and inhibition of antagonist muscles (Russian text). *Zh Vssh Nerv Deiat Pavlov* 17:125–133, 1965.

Kramer H, Frauendorf H, Küchler G: Die Beeinflussung des mittels Oberflächenelektroden abgeleiteten Elektromyogramms durch ableittechnische Variablen. II. Zum Einfluss von Abstand, Flächengrösse und Andruck der Electroden auf die ableitbare electrische Muskelaktivität. *Acta Biol Med Germ* 28:489–496, 1972.

Kramer JF, Reid DC: Backward walking: a cinematographic and electromyographic pilot study. *Physiother Can* 33:77–86, 1981.

Kranz H: Control of motoneuron firing during maintained voluntary contraction in normals and in patients with cerebral lesions. In JE Desmedt (ed): *Progress in Clinical Neurophysiology* 10:358–367, 1981.

Kranz H, Baumgartner G: Human alpha motoneurone discharge, a statistical analysis. *Brain Res* 67:324–329, 1974.

Kranz H, Chan H, Caddy DJ, Williams AM: Factors contributing to the decrease in the frequency content of the electromyogram during muscle contration. In *New Approaches to Nerve and Muscle Disorders, Basic and Applied Contributions.* Proc. of the 2nd Symposium on the Foundation for Life Science, pp 104–113, 1981.

Kranz H, Williams AM, Cassell J, Caddy DJ, Silberstein RB: Factors determining the frequency content of the electromyogram. *J Appl Physiol* 55:392–399, 1983.

Krnjević K, Miledi R: Motor units in the rat diaphragm. *J Physiol* 40:427–439, 1958a.

Krnjević K, Miledi R: Failure of neuromuscular propagation in rats. *J Physiol* (Lond), 140:440–461, 1958b.

Kreifeldt JG, Yao S: A signal-to-noise investigation of nonlinear electromyographic processors. *IEEE Trans BME* 21:298–308, 1974.

Kudina LP: Reflex effects of muscle afferents on antagonists studied on single firing motor units in man. *Electroencephalogr Clin Neurophysiol* 50:214–221, 1980.

Kuffler SW, Vaughan Williams EM: The distribution of small motor nerves to frog skeletal muscle, and the membrane characteristics of the fibers they innervate. *J Physiol* 121:289–317, 1953.

Kugelberg E: Clinical electromyography. *Prog Neurol Psychiatry* 8:264–282, 1953.

Kugelberg E: Electromyograms in muscular disorders. *J Neurol Neurosurg Psychiatry* 10:122–133, 1947.

Kukulka CG, Clamann PH: Comparison of the recruitment and discharge properties of motor units in human brachial biceps and adductor pollicis during isometric contractions. *Brain Res* 219:45–55, 1981.

Kukulka CG, Brown DM, Basmajian JV: Biofeedback training for early finger joint mobilization. *Am J Occup Ther* 29:469–470, 1975.

Kumar S: Physiologic responses to weight lifting in different planes. *Ergonomics* 23:987–993, 1980.

Kuroda E, Klissouras V, Mulsum JH: Electrical and metabolic activities and fatigue in human isometric contraction. *J Appl Physiol* 29:358–367, 1970.

Kurozumi Tashiro T, Harady Y: Laryngeal responses to electrical stimulation of the medullary respiratory centers in the dog. *Laryngoscope* 81:1960–1967, 1971.

Kwatney E, Thomas DH, Kwatny HG: An application of signal processing techniques to the study of myoelectric signals. *IEEE Trans Biomed Engng* 17:303–313, 1970.

LaBan MM, Raptou AD, Johnson EW: Electromyographic study of function of iliopsoas muscle. *Arch Phys Med* 46:676–679, 1965.

Ladd HW, Broman H, O'Riain MD: Regenerative processes in peripheral nerve injury: a new method for their evaluation. *Arch Phys Med Rehab* 63:124–129, 1982.

Ladd H, Jonsson B, Lindegren U: The learning process for fine neuromuscular controls in skeletal muscles of man. *Electromyogr Clin Neurophysiol* 12:213–223, 1972.

Ladd HW, Simard TG: Bilaterally controlled neuromuscular activity in congenitally malformed children—an electromyographic study. *Interclin Inform Bull* 11:9–16, 1972.

LaFratta CW, Smith OH: A study of the relationship of motor nerve conduction velocity in the adult to age, sex and handedness. *Arch Phys Med* 45:407–412, 1964.

Lagasse PP: Prediction of maximum speed of human movement by two selected muscular coordination mechanisms and by maximum static strength. *Percept Motor Skills* 49:151–161, 1979.

Lago P, Jones NB: Effect of motor-unit firing time statistics on e.m.g. spectra. *Med Biol Eng Comput* 15:648–655, 1977.

La Joie WJ, Cosgrove MD, Jones WG: Electromyographic evaluation of human detrusor muscle activity in relation to abdominal muscle activity. *Arch Phys Med Rehab* 57:382–386, 1976.

Lake N: The arches of the foot. *Lancet* 2:872–873, 1937.

Lake LF: An electromyographic study of the action of the second and third dorsal interosseous muscles and their interaction with extensor digitorum communis and flexor digitorum sublimis of the normal hand. M.A. Thesis, Department of Anatomy, Washington University, St. Louis, MO, 1954.

Lake LF: An electromyographic study of finger movement. *Anat Rec* 127:322–323, 1957.

Lam HS, Morgan DL, Lampard DG: Derivation of reliable electromyograms and their relation to tension in mammalian skeletal muscles during synchronous stimulation. *Electroencephalogr Clin Neurophysiol* 46:72–80, 1979.

Lamarre Y, Lund JP: Load compensation in human masseter muscles. *J Physiol (Lond)* 253:21–35, 1975.

Landa J: Shoulder muscle activity during selected skills on the uneven parallel bars. *Res Q* 45:120–127, 1974.

Landau WM: Comparison of different needle leads in EMG recording from a single site. *Electroencephalogr Clin Neurophysiol* 3:163–168, 1951.

Landau WM, Clare MH: The plantar reflex in man: with special reference to some conditions where the extensor response is unexpectedly absent. *Brain* 82:321–355, 1959.

Lane RH: Clinical application of anorectal physiology. *Proc Roy Soc Med* 68:28–30, 1975.

Lapides J, Ajemian EP, Stewart BH, Breakey BA, Lichtwardt JR: Further observations on the kinetics of the urethrovesical sphincter. *J Urol* 84:86–94, 1960.

Lapides J, Sweet RB, Lewis LW: Role of striated muscle in urination. *J Urol* 77:247–250, 1957.

Larson JD, Foulkes D: Electromyographic suppression during sleep, dream recall, and orientation time. *Psychophysiology* 5:548–555, 1969.

Larsson LE, Linderholm H, Ringqvist T: The effect of sustained and rhythmic contractions on the electromyogram (EMG). *Acta Physiol Scand* 65:310–318, 1965.

Larsson LE: On the relation between the EMG frequency spectrum and the duration of symptoms in lesions of the peripheral motor neuron. *Electroencephalogr Clin Neurophysiol* 38:69–78, 1975.

Lashley KS: The problem of serial order in

behavior. In Jeffress LA: *Cerebral Mechanism in Behavior*, The Hixon Symposium. New York, John Wiley & Sons, 1951.

Last RJ: *Anatomy Regional and Applied*. London, J. & A. Churchill, 1954.

Latif A: An electromyographic study of the temporalis muscle in normal persons during selected positions and movements of the mandible. *Am J Orthodont* 43:577–591, 1957.

Lauerma KSL, Harvey JE, Ogura JH: Cricopharyngeal myotomy in subtotal supraglottic laryngectomy: an experimental study. *Laryngoscope* 82:447–453, 1972.

Lawrence JH, De Luca CJ: Myoelectric signal vs. force relationship in different human muscles. *J Appl Physiol* 54:1653–1659, 1983.

Leanderson R: *On the Functional Organization of Facial Muscles in Speech*. Published in book form by Acta Laryngol., Stockholm, 1972.

Leanderson R, Meyerson BA, Persson A: Effect of L-dopa on speech in parkinsonism: an EMG study of labial articulatory function. *J Neurol Neurosurg Psychiatry* 34:679–681, 1971.

Leanderson R, Persson A, Öhman, S: Electromyographic studies of the function of the facial muscles in dysarthria. *Acta Otolaryngol* 263:89–94, 1970.

Leavitt LA, Beasley WC: Clinical application of quantitative methods in the study of spasticity. *Clin Pharmacol Therap* 5:918–941, 1964.

Le Bozec S, Maton B: The activity of anconeus during voluntary elbow extension: the effect of lidocaine blocking of the muscle. *EMG Clin Neurophysiol* 22:265–275, 1982.

Le Bozec S, Maton B, Cnockaert JC: The synergy of elbow extensor muscles during static work in man. *Eur J Appl Physiol* 43:57–68, 1980.

Le Bozec S, Maton B, Cnockaert JC: The synergy of elbow extensor muscles during dynamic work in man. Part I. Elbow extension. *Eur J Appl Physiol* 44:255–269, 1980.

LeFever RS: Statistical analysis of concurrently active human motor units. Ph.D. dissertation, M.I.T., Cambridge, MA, 1980.

LeFever RS, De Luca CJ: The contribution of individual motor units to the EMG power spectrum. *Proc 29th ACEMB*:56, 1976.

LeFever RS, De Luca CJ: Decomposition of superimposed action potential trains. *Proc of 8th Annual Meeting of the Soc for Neuroscience*, 299, 1978.

LeFever R, De Luca CJ: A procedure for decomposing the myoelectric signal into its constituent action potentials. Part I. Technique, theory and implementation. *IEEE Trans Biomed Eng* 29:149–157, 1982.

LeFever RS, Xenakis AP, De Luca CJ: A procedure for decomposing the myoelectric signal into its constituent action potentials. Part II. Execution and test for accuracy. *IEEE BME* 29:158–164, 1982.

Lehr RP, Blanton PL, Biggs NL: An electromyographic study of the mylohyoid muscle. *Anat Rec* 169:651–660, 1971.

Lehr RP Jr, Owens SE Jr: An electromyographic study of the human lateral pterygoid muscles. *Anat Rec* 196:441–448, 1980.

Leibrecht BE, Lloyd AJ, Pounder S: Auditory feedback and conditioning of the single motor unit. *Psychophysiology* 10:1–7, 1973.

Leifer LJ: Characterization of single muscle fiber discharge during voluntary isometric contraction of the biceps brachii muscle in man. Ph.D. Thesis, Stanford University, Stanford, CA, 1969.

Lenman JAR, Potter JL: Electromyographic measurement of fatigue in rheumatoid arthritis and neuromuscular disease. *Ann Rheumat Dis* 25:76–84, 1966.

Lesage Y, Le Bar R: Étude électromyographique simultanée des différent chefs du quadriceps. *Ann Méd Phys* 13:292–297, 1970.

Lesoine W, Paulsen H-J: Elektromyographische Unterschungen am Musculus cricothyreodeus und Musculus digastricus venter anterior. *Z Laryngol Rhinol Otol* 47:543–55, 1968.

Lestienne F, Bouisset S: Pattern temporel de la mise en jeu d'un agoniste et d'un antagoniste en fonction de la tension de l'agoniste. *Rev Neurol Paris* 118:550–554, 1968.

Lestienne F, Goubel F: Contribution relatif de deux agonistes à un travail avec et sans raccourcissement. *J Physiol* 61:342–343, 1969.

Levens AS, Inman VT, Blosser JA: Transverse rotation of the segments of the lower extremity in locomotion. *J Bone Joint Surg* 30A:859–872, 1948.

Levine IM, Jossmann PB, Tursky B, Meister M, De Angelis V: Telephone telemetry of bioelectric information. *JAMA* 188:794–798, 1964.

Levitt MN, Dedo HH, Ogura JH: The cricopharyngeus muscle, an electromyographic study in the dog. *Laryngoscope* 75:122–136, 1965.

Levy R: The relative importance of the gastrocnemius and soleus muscles in the ankle jerk of man. *J Neurol Neurosurg Psychiatry* 26:148–150, 1963.

Lewes D: Multipoint electrocardiography without skin preparation. *Lancet,* July 3, p 17, 1965.

Lewis RS, Basmajian JV: Electromyographic studies in embryos and fetuses. *Proc Can Fed Biol Soc* 2:41–42, 1959.

Li CL, Lundervold A: Electromyographic study of cleft palate. *Plast Reconstr Surg* 21:427–432, 1958.

Liberson WT: Une nouvelle application de Quartz pièzoélectrique: Pièzoélectrographie de la marche et des mouvements volontaires. *Le Travail Humain (Paris)* 4:1–7, 1936.

Liberson WT: New development in the study of conduction time in peripheral neuropathies. *Electroencephalogr Clin Neurophysiol* 12:264, 1960.

Liberson WT: Normal and pathological gaits (abstr). *Bull Am Assoc EMG Electrodiagn* 12:25, 1965a.

Liberson WT: Biomechanics of gait: a method of study. *Arch Phys Med* 46:37–48, 1965b.

Liberson WT, Dondey M, Asa MM: Brief repeated isometric maximal exercises. *Am J Phys Med* 41:3–14, 1962.

Liberson WT, Gratzer M, Zalis A, Grabinski B: Comparison of conduction velocities of motor and sensory fibers determined by different methods. *Arch Phys Med* 47:17–23, 1966.

Liberson WT, Kim KC: Mapping evoked potentials elicited by stimulation of the median and peroneal nerves. *Electroencephalogr Clin Neurophysiol* 15:721, 1963.

Liberson WT, Holmquest HJ, Halls A: Accelerographic study of gait. *Arch Phys Med* 43:547–551, 1962.

Libkind MS: II. Modelling of interference bioelectrical activity. Biofizika 13:685–693, 1968.

Libkind MS: III. Modelling of interference bio-electrical activity. *Biophysics* 14:395–398, 1969.

Lieb FJ, Perry J: Quadriceps function: an anatomical and mechanical study using amputated limbs. *J Bone Joint Surg* 50A:1535–1538, 1968.

Lieb FJ, Perry J: Quadriceps function: an electromyographic study under isometric conditions. *J Bone Joint Surg* 53A:749–758, 1971.

Lindqvist C: The motor unit potential in severely paretic muscles after acute anterior poliomyelitis: an electromyographic study using fatigue experiments. *Acta Psychiatr Neurol Scand* 34(suppl 131):1–72, 1959.

Lindsley DB: Electrical activity of human motor units during voluntary contraction. *Am J Physiol* 114:90–99, 1935.

Lindström LR: *On the Frequency Spectrum of EMG Signals.* Technical Report-Research Laboratory of Medical Electronics, Chalmers University of Technology, Göteborg, Sweden, 1970.

Lindström LR, Magnusson R, Petersén I: Muscular fatigue and action potential conduction velocity changes studied with frequency analysis of EMG signals. *Electromyography* 4:341–353, 1970.

Lindström L, Broman H, Magnusson R, Petersén I: On the inter-relation of two methods of EMG analysis, EEG Clin. *Neurophysiology* 7:801, 1973.

Lindström L, Kadefors R, Petersén I: An electromyographic index for localized muscle fatigue. *J Appl Physiol* 43:750–754, 1977.

Lindström L, Magnusson R: Interpretation of myoelectric power spectra: a model and its applications. *Proc IEEE* 65:653–662, 1977.

Lindström L, Magnusson R, Petersén I: Muscle load influence on myoelectric signal characteristics. *Scand J Rehab Med Suppl* 3:27–148, 1974.

Lindström L, Petersén I: Power spectra of myoelectric signals: motor-unit activity and muscle fatigue. In Stalberg E, Young RR (eds): *Clinical Neurophysiology.* London, Butterworth, 1981, pp 66–87.

Lippold OCJ: The relation between integrated action potentials in a human muscle and its isometric tension. *J Physiol* 117:492–499, 1952.

Lippold OCJ: Physiological tremor. *Sci Am* 224:65–73, 1971.

Lippold OCJ, Redfearn JWT, Vučo J: The rhythmical activity of groups of motor units in the voluntary contraction of muscle. *J Physiol* 137:473–487, 1957.

Lippold OCJ, Redfearn JWT, Vučo J: The electromyography of fatigue. *Ergonomics* 3:121–131, 1970.

Littler JW: The physiology and dynamic function of the hand. *Surg Clin North Am*

40:259–266, 1960.

Livingston RB, Paillard J, Tournay A, Fessard A: Plasticité d'une synergie musculaire dans l'exécution d'un mouvement volontaire chez l'homme. *J Physiol Paris* 43:605–619, 1951.

Lloyd AJ: Muscle activity and kinesthetic position responses. *J Appl Physiol* 25:659–663, 1968.

Lloyd AJ: Surface electromyography during sustained isometric contractions. *J Appl Physiol* 30:713–719, 1971.

Lloyd AJ, Caldwell LS: Accuracy of active and passive positioning of the leg on the basis of kinesthetic cues. *J Comp Physiol Psychol* 60:102–106, 1965.

Lloyd AJ, Leibrecht BC: Conditioning of a single motor unit. *J Exp Psychol* 88:391–395, 1971.

Lloyd AJ, Shurley JT: The effect of sensory perceptual isolation on single motor unit conditioning. *Psychophysiology* 13:340–344, 1976.

Lloyd AJ, Voor JH: The effect of training on performance efficiency during a competitive isometric exercise. *J Motor Behav* 5:17–24, 1973.

Lockhart RD: *Cunningham's Textbook of Anatomy*, ed 9, Brash JC (ed). London, Oxford University Press, New York, Toronto, 1951.

Lockhart RD, Hamilton GF, Fyfe F: *Anatomy of the Human Body.* London, Faber & Faber, 1959, pp 92, 216.

Long C, Brown ME: Electromyographic kinesiology of the hand. Part III. Lumbricalis and flexor digitorum profundus to the long finger. *Arch Phys Med* 43:450–460, 1962.

Long C, Brown ME: Electromyographic kinesiology of the hand: muscles moving the long finger. *J Bone Joint Surg* 46A:1683–1706, 1964.

Long C, Brown ME, Weiss G: Electromyographic kinesiology of the hand. Part II. Third dorsal interosseus and extensor digitorum of the long finger. *Arch Phys Med* 42:559–565, 1960.

Long C, Conrad PW, Hall EW,, Furler SL: Intrinsic-extrinsic muscle control of the hand in power grip and precision handling. *J Bone Joint Surg* 52A:853–867, 1970.

Lopata M, Evanich MJ, Lourenço RV: The electromyogram of the diaphragm in the investigation of human regulation of ventilation. Chest 70 (suppl. 162S–165S), 1976.

Lopez A, Richardson PC: Capacitive electrocardiographic and bioelectric electrodes. *IEEE Trans BME* 16:99, 1969.

Louis-Sylvestre J, MacLeod P: Le muscle vocal humain est-il asynchrone? *J Physiol Paris* 60:373–389, 1968.

Loofbourrow GN: Electrographic evaluation of mechanical response in mammalian skeletal muscle in different conditions. *J Neurophysiol* 11:153–168, 1948.

Lourenço RV, Cherniak NS, Malm JR, Fishman AP: Nervous output from the respiratory center during obstructed breathing. *J Appl Physiol* 21:527–533, 1966.

Low MD, Basmajian JV, Lyons GM: Conduction velocity and residual latency in the human ulnar nerve and the effects on them of ethyl alcohol. *Am J Med Sci* 244:720–730, 1962.

Lowe AA, Johnston WD: Tongue and jaw muscle activity in response to mandibular rotations in a sample of normal and anterior open-bite subjects. *Am J Orthodont* 76:565–576, 1979.

Lubker JF: An electromyographic-cinefluorographic investigation of velar function during normal speech. *Cleft Palate J* 5:1–18, 1968.

Lubker JF: Normal velopharyngeal function in speech. *Clin Plast Surg* 2:249–259, 1975.

Lubker J, Fritzell B, Lindqvist J: Speech production. Velopharyngeal function: and electromyographic study. Privately printed, 1970.

Lubker J, May K: Palatoglossus function in normal speech production. In *Papers from the Institute of Linguistics, 1973.* University of Stockholm, Sweden, 1973, pp 17–26.

Lubker JF, Parris PJ: Simultaneous measurements of intraoral pressure, force of labial contact, and labial electromyographic activity during production of the stop consonant cognates p and b . *J Acoust Soc Am* 47:625–633, 1970.

Lugaresi E, Coccagna G, Farneti P, Montovani M, Cirignotta F: Snoring. *Electroencephalogr Clin Neurophysiol* 39:59–64, 1975.

Lukas JS, Peeler DJ, Kryter KD: Effects of sonic booms and subsonic jet flyover noise on skeletel muscle tension and a paced tracing task. Report to N.A.S.A. (CR-1522), 1970.

Lund JP: Oral-facial sensation in the control of mastication and voluntary movements of the jaw. In Sessle BJ, Hannam AG: *Mastication and Swallowing*, Toronto, University of Toronto Press, 1976.

Lund JP, McLachlan RS, Dellow PG: A

lateral jaw movement reflex. *Exp Neurol* 31:189–199, 1971.

Lundervold AJS: *Electromyographic Investigations of Position and Manner of Working in Typewriting.* Oslo, W. Brøggers Boktrykkeri A/S, 1951.

Lundervold A, Bruland H, Stensrud P: Conduction velocity in peripheral nerves: a general introduction. *Acta Neurol Scand* 41(suppl. 13):259–262, 1965.

Lundervold A, Li CL: Motor units and fibrillation potentials as recorded with different kinds of needle electrodes. *Acta Psychiatr Neurol* 28:201–211, 1953.

Luschei E, Saslow C, Glickstein M: Muscle potentials in reaction time. *Exp Neurol* 18:429–442, 1967.

Luscher R, Ruenzel P, Henneman E: How the size of motoneurons determines their susceptibility to discharge. *Nature* 282:859–861, 1979.

Lynn PA, Bettles ND, Hughes AD, Johnson SW: Influence of electrode geometry on bipolar recordings of the surface electromyogram. *Med Biol Eng Comput* 16:651–660, 1978.

Lynne-Davies P, Coker M, Simkins D, Widrow B: Adaptive filtering in respirator electromyogram processing. *Proc 4th Congr Int Soc Electrophysiol Kinesiol* 170–171, 1979.

Lyons K, Perry J, Gronley JK, Barnes L, Antonelli D: Timing and relative intensity of hip extensor and abductor muscle action during level and stair ambulation. *Phys Ther* 63:1597–1605, 1983.

MacConaill MA: Some anatomical factors affecting the stabilizing functions of muscles. Irish JM, ed. *Sci* 6:160–164, 1946.

MacConaill MA: The movements of bones and joints. 2. Function of the musculature. *J Bone Joint Surg* 31B:100–104, 1949.

MacConaill MA, Basmajian JV: *Muscles and Movements: A Basis for Human Kinesiology.* Baltimore, Williams & Wilkins, 1969.

MacDougall JDB, Andrew BL: An electromyographic study of the temporalis and masseter muscles. *J Anat* 87:37–45, 1953.

Machado, see de Sousa, O. Machado.

MacNeilage PF: Preliminaries to the study of single motor unit activity in speech musculature. *J Phonet* 1:55–71, 1973.

MacNeilage PF, Rootes TO, Chase RA: Speech production and perception in a patient with severe impairment of somesthetic perception and motor control. *J Speech Hearing Res* 10:449–467, 1967.

MacNeilage PF, Szabo RK: Frequency control of single motor units in upper articulatory musculature. Paper presented at 83rd Meeting, Acoustical Soc. of Am., Buffalo, New York.

MacNeilage PF, Sussman HM, Sussman RJ: Parametric study of single motor unit waveforms in upper articulatory musculature. Paper presented at 83rd Meeting, Acoustical Soc. of Am., Buffalo, New York, 1972.

Magladery JW, McDougal DB Jr: Electrophysiological studies on nerve and reflex activity in normal man (Part I). *Bull Johns Hopkins Hosp* 86:265–290, 1950a.

Magladery JW, McDougal DB Jr, Stoll J: Electrophysiological studies of nerve and reflex activity in normal man. Parts II and III. *Bull Johns Hopkins Hosp* 86:291–312 and 313–339, 1950b and c, 1950.

Magora A, Rozin R, Robin GC, Gonen B, Simkin A: Investigation of gait. I. A technique of combined recording. *Electromyography* 10:385–396, 1970.

Magoun HW, Rhines R: *Spasticity: the Stretch-Reflex and Extrapyramidal Systems.* Springfield, IL, Charles C Thomas, 1947.

Malmo RB, Shagass C, Davis JF: Electromyographic studies of muscular tension in psychiatric patients under stress. *J Clin Exp Psychopathol* 12:45–66, 1951.

Malmo RB, Smith AA: Forehead tension and motor irregularities in psychoneurotic patients under stress. *J Personality* 23:391–406, 1955.

Mambrito B: Motor control of two muscle joints. Ph.D. dissertation, M.I.T., Cambridge, MA, 1983.

Mambrito B, De Luca CJ: Acquisition and decomposition of the EMG signal. *New Dev EMG Clin Neurophysiol* 10:52–72, 1983.

Mann R, Inman VT: Phasic activity of intrinsic muscles of the foot. *J Bone Joint Surg* 66A:469–481, 1964.

Marciniak W: Czynnośc̀ bioelektryczna mieśni goleni w chodzie poziomym u dziece zdrowyck (Bioelectric activity of leg muscles of healthy children during level walking). *Chir Narz Ruchu Ortop Pol* 15:643–650, 1975.

Marey EJ: *Movement,* transl. by E Pritchard. New York, D. Appleton and Co., 1895.

Marg E: Development of electrooculography. *AMA Arch Ophthalmol* 45:169–185, 1951.

Marg E, Tamler E, Jampolsky A: Activity of a human oculorotary muscle unit. *Elec-*

troencephalogr Clin Neurophysiol 14:754–757, 1962.

Marinacci AA: Dynamics of neuromuscular diseases. *Arch Neurol* 1:243–257, 1959.

Marinacci AA: *Applied Electromyography.* Philadelphia, Lea & Febiger, 1968.

Markee JE, Logue JT, Williams M, Stanton WB, Wrenn RN, Walker LB: Two-joint muscles of the thigh. *J Bone Joint Surg* 37A:125–142, 1955.

Marsden C, Meadows J, Hodgson H: Observations on the reflex response to muscle vibration in man and its voluntary control. *Brain* 92:829–846, 1969.

Marsden CD, Meadows JC, Lange GW, Watson RS: (a) The role of the ballistocardiac impulse in the genesis of physiological tremor. *EEG Clin Neurophysiol* 27:169–178. (b) Variations in human physiological finger tremor, with particular reference to changes with age. *Brain* 92:647–662, 1969.

Marsden CD, Meadows JC, Merton PA: Fatigue in human muscle in relation to the number and frequency of motor impulses. *J Physiol (Lond)* 258:94P, 1976.

Marsden CD, Merton PA, Morton HB: Servo action in the human thumb. *J Physiol (Lond)* 1–44, 1976.

Mårtensson A, Skoglund CR: Contraction properties of internal laryngeal muscles. *Acta Physiol Scand* 60:318–336, 1964.

Martin HN, Hartwell EM: On the respiratory function of the internal intercostal muscles. *J Physiol* 2:24–27, 1879.

Martin JD, De Troyer A: The behaviour of the abdominal muscles during inspiratory mechanical loading. *Resp Physiol* 50:63–73, 1982.

Martin JG, Habib M, Engel LA: Inspiratory muscle activity during induced hyperinflation. *Resp Physiol* 39:303–313, 1980.

Martone AL: Anatomy of facial expression and its prosthodontic significance. *J Prosthet Dent* 12:1020–1042, 1962.

Masland WS, Sheldon D, Hershey CD: The stochastic properties of individual motor unit interspike intervals. *Am J Physiol* 217(5):1384–1388, 1969.

Mason RR, Munro RR: Relationship between EMG potentials and tension in abduction of the little finger. *Electromyography* 9:185–199, 1969.

Masuda T, Miyano H, Sadoyama T: The measurement of muscle fiber conduction velocity using a gradient threshold zero-crossing method. *IEEE Trans Biomed Eng* 29:673–678, 1982.

Mathews AM, Gelder MG: Psycho-physiological investigations of brief relaxation training. *J Psychosom Res* 13:1–12, 1960.

Maton B: Human motor unit activity during the onset of muscle fatigue in submaximal isometric isotonic contractions. *Eur J Appl Physiol* 46:271–281, 1981.

Maton B, Bouisset S: Variation de l'intervalle moyen. *Rev Electroencephalogr Clin Neurophysiol* 2:340–341, 1972.

Maton B, Bouisset S: Motor unit activity and preprogramming of movement in man. *Electroencephalogr Clin Neurophysiol* 38:658–660, 1975.

Maton B, Le Bozec S, Cnockaert JC: The synergy of elbow extensor muscles during dynamic work in man. Part II. Braking of elbow flexion. *Eur J Appl Physiol* 44:271–278, 1980.

Matsushita K, Goto Y, Okamoto T, Tsujino A, Kumamoto M: Soh no Kindenzuteki Kenkyu (An electromyographic study of spring running). *Res J Phys Ed* 19:147–156, 1974.

Matteucci C: *Traites des Phenomenen Electrophysiologiques.* Paris, 1844.

Matthews PBC: *Mammalian Muscle Receptors and Their Central Actions.* London, Arnold, 1972.

Mawdsley C, Mayer RF: Nerve conduction in alcohol polyneuropathy. *Brain* 88(Part II):335–356, 1965.

Mayer RF: Nerve conduction studies in man. *Neurology* 13:1021–1030, 1963.

McAllister R, Lubker J, Carlson J: An emg study of some characteristics of the Swedish rounded vowels. *J Phonet* 2:267–278, 1974.

McCann GD, Ray CB: An information processing and control system for biological research. *Ann NY Acad Sci* 128:830–848, 1966.

McCollum BB: Oral diagnosis. *J Am Dent A* 30:1218–1233, 1943.

McFarland GB, Krusen UL, Weathersby HT: Kinesiology of selected muscles acting on the wrist: electromyographic study. *Arch Phys Med* 43:165–171, 1962.

McGlone RE, Shipp T: Some physiologic correlates of vocal-fry phonation. *J Speech Hearing Res* 14:769–775, 1971.

McGregor AL: *Synopsis of Surgical Anatomy,* ed 7. Baltimore, Williams & Wilkins, 1950.

McGuigan FJ: Covert oral behavior and auditory hallucinations. *Psychophysiology* 3:73–80, 1966.

McGuigan FJ: Covert oral behavior during

the silent performance of language tasks. *Psychol Bull* 74:309–326, 1970.

McGuigan FJ, Rodier WI, III: Effects of auditory stimulation on covert oral behavior during silent reading. *J Exp Psychol* 76:649–655, 1968.

McKeon B, Burke D: Muscle spindle discharge in response to contraction of single motor units. *J Neurophysiol* 49:291–302, 1983.

McLeod WD, Nunnally HN, Cantrell PE: Dependence of emg power spectra on electrode type. *IEEE Trans Biomed Eng* (March) 172–175, 1976.

McLeod WD, Thysell R: Cortically evoked potentials co-related with single motor units, unpublished, 1973.

McNamara JA: The independent functions of the two heads of the lateral pterygoid muscle. *Am J Anat* 138:197–206, 1973.

McPhedran AM, Wuerker RB, Henneman E: Properties of motor units in a homogeneous red muscle (soleus) of the cat. *J Neurophysiol* 28:71–84, 1965.

Meijers LMM, Teulings JLHM, Eijkman EGJ: Model of the electromyographic activity during brief isometric contractions. *Biol Cybernetics* 25:7–16, 1976.

Mendell LM, Henneman E: Input to motoneuron pools and its effects. In Mountcastle BM: *Medical Physiology*, ed 14. St. Louis, C.V. Mosby, 1980, vol 1, pp 742–761.

Merletti R, Sabbahi MA, De Luca CJ: Median frequency of the myoelectric signal: effects of ischemia and cooling. *Eur J Appl Physiol*, 1984.

Merton PA: Voluntary strength and fatigue. *J Physiol* 123:553–564, 1954.

Metz HS, Scott AB: Innervational plasticity of the oculomotor system. *Arch Ophthalmol* 84:86–91, 1970.

Middaugh SJ: Single motor unit control with emg feedback, non-feedback performance, and discrimination of motor unit level events. (abstr); EMG feedback as a muscle reeducation technique: a controlled study (abstract). Two papers presented to the Society of Psychophysiological Research, 1976. *Psychophysiology* 14:102–103, 1976.

Miles M, Mortensen OA, Sullivan WE: Electromyography during normal voluntary movements. *Anat Rec* 98:209–218, 1947.

Miller AJ: Characteristics of the swallowing reflex induced by peripheral nerve and brain stem stimulation. *Exp Neurol* 34:210–222, 1972a.

Miller AJ: Significance of sensory inflow to the swallowing reflex. *Brain Res* 43:147–159, 1972b.

Miller JE: Electromyographic pattern of saccadic eye movements. *Am J Ophthalmol* 46:(No. 5, special Part II (not I) of the November issue):183–186, 1958.

Miller JE: The electromyography of vergence movement. *AMA Arch Ophthalmol* 62:790–794, 1959.

Mills KS: Power spectral analysis of electromyogram and compound muscle action potential during muscle fatigue and recovery. *J Physiol* 362:401–409, 1982.

Milner M, Basmajian JV, Quanbury AO: Multifactorial analyses of walking by electromyography and computer. *Am J Phys Med* 50:235–258, 1971.

Milner-Brown HS, Brown WF: New methods of estimating the number of motor units in a muscle. *J Neurol Neurosurg Psychiatry* 39:258–265, 1976.

Milner-Brown HS, Stein RB: The relation between the surface electromyogram and muscular force. *J Physiol (Lond.)* 246:549–569, 1975.

Milner-Brown HS, Stein RB, Lee RG: Synchronization of human motor units: possible roles of exercise and supraspinal reflexes. *Electroencephalogr Clin Neurophysiol* 38:245–254, 1975.

Milner-Brown HS, Stein RB, Yemm R: Mechanisms for increasing force during voluntary contractions. *J Physiol (Lond)* 226:18–19P, 1972.

Milner-Brown HS, Stein RB, Yemm R: Changes in firing rate of human motor units during linearly changing voluntary contractions. *J Physiol* 230:371–390, 1973a.

Milner-Brown HS, Stein RB, Yemm R: The contractile properties of human motor units during voluntary isometric contractions. *J Physiol* 228:285–306, 1973b.

Milojevic B: Electronystagmography. *Laryngoscope* 75:243–258, 1965.

Milojevic B, Hast M: Cortical motor centers of the laryngeal muscles in the cat and dog. *Ann Otol Rhinol Laryngol* 73:979–989, 1964.

Miranda JM, Lourenço RV: Influence of diaphragm activity on the movement of total chest compliance. *J Appl Physiol* 24:741–746, 1968.

Mishelevich DJ: On-line real-time digital computer separation of extracellular neuroelectric signals. *IEEE BME* 17:147–150, 1970.

Missiuro W: Studies on developmental stages of children's reflex reactivity. *Child Dev* 34:33–41, 1963.

Missiuro W, Kozlowski S: Investigations on adaptative changes in reciprocal innervation of muscle. *Arch Phys Med* 44:37–41, 1961.

Missiuro W, Kirschner H, Kozlowski S: Electromyographic manifestations of fatigue during work of different intensity. *Acta Physiol Polon* 13:11–23, 1962a.

Missiuro W, Kirschner H, Kozlowski S: Contribution à l'étude de la fatigue au cours de travaux musculaires d'intensités différentes. *J Physiol Paris* 54:717–727, 1962b.

Mitolo M: Studio elettromiografico nel corso dell'allenamento all'esercizio fisico (Italian text). *Boll Soc Ital Biol Sper* 32:1413–1415, 1956.

Mitolo M: Studio elettromiografico nel corso dell'allenamento all'esercizio fisico (Italian text). *Lav Umano* 9:2–23, 1957.

Mitolo M: L'allenamento del muscolo all'esercizio fisico in vecchiaia (ricerche dinamometriche, ergografiche ed elettromiografiche). *Lavoro Umano* 16:371–390, 1964.

Miwa N, Matoba M: Relation between the muscle strength and electromyogram (Japanese text). *Orthop Traumatol* 8:121–124, 1959.

Miwa N, Tanaka T, Matoba M: Electromyography in kinesiologic evaluations: subjects on the two joint muscle and the relation between the muscular tension and electromyogram. *J Jpn Orthop Assoc* 36:1025–1035, 1963.

Miyashita M, Miura M, Matsui H, Minamitate K: Measurement of the reaction time of muscular relaxation. *Ergonomics* 15:555–562, 1972.

Miyoshi K: The mechanism of arch support in the foot (Japanese text; English summary). *J Jpn Orthop Assoc* 40:977–984, 1966.

Mizote M: The effect of digital nerve stimulation on recruitment order of motor units in the first deep lumbrical muscle of the cat. *Brain Res* 248:245–255, 1982.

Mohr TM, Allison JD, Patterson R: Electromyographic analysis of the lower extremity during pedaling. *J Orthop Sports Phys Ther* 2:163–170, 1981.

Møller AR: Bilateral contraction of the tympanic muscles in man: examined by measuring acoustic impedance-change. *Ann Otol Rhinol Laryngol* 70:735–752, 1961.

Møller E: Quantitative features of masticatory muscle activity. In Rowe NH: *Occlusion: Research in Form and Function*, Ann Arbor, University of Michigan Press, 1975.

Møller E: Evidence that the rest position is subject to servo-control. In Anderson DJ, Mathews B: *Mastication.* Bristol, Wright & Sons, 1976.

Monster AW, Chan H: Isometric force production by motor units of extensor digitorum communis muscle in man. *J Neurophysiol* 40:1432–1443, 1977.

Monster AW: Firing rate behavior of human motor units during isometric voluntary contraction: relation to unit size. *Brain Res* 171:349–354, 1979.

Moore AD: Synthesized EMG waves and their implications. *Am J Phys Med* 46:1302–1316, 1967.

Moore AD: Synthetic EMG waves. In Basmajian JV: *Muscles Alive*, ed 3. Baltimore, Williams & Wilkins, 1966, pp 53–70.

Moore JC: Fabrication of suction-cup electrodes for electromyography. *Electroencephalogr Clin Neurophysiol* 20:405–406, 1966.

Moore JC: Excitation overflow: an electromyographic investigation. *Arch Phys Med Rehab* 56:115–120, 1975.

Mori S: Discharge patterns of soleus motor units with associated changes in force exerted by foot during quiet stance in man. *J Neurophysiol* 36:458–471, 1973.

Mori S, Ishida A: Synchronization of motor units and its simulation in parallel feedback system. *Biol Cybernetics* 21:107–11, 1976.

Morin C, Pierrot-Deseilligny E, Bussel B: Role of muscular afferents in the inhibition of the antagonist motor nucleus during a voluntary contraction in man. *Brain Res* 103:373–376, 1976.

Moritani T, deVries HA: Reexamination of the relationship between the surface integrated electromyogram and force of isometric contraction. *Am J Phys Med* 57:263–277, 1978.

Morosova IA, Shik LL: Action potentials of respiratory muscles in patients with respiratory deficiencies (Russian text). *Byull Eksper Biol Med* 63:61–65, 1957.

Morris JM, Benner G, Lucas DB: An electromyographic study of the intrinsic muscles of the back in man. *J Anat* 96:509–520, 1962.

Morris JM, Lucas DB, Bresler B: The role of the trunk in stability of the spine. Biomechanics Laboratory, University of California. Publication no. 42, 1961.

Mortimer JT, Magnusson R, Petersén I: Conduction velocity in ischemic muscle: effect on EMG frequency spectrum. *Am J Physiol* 219:1324–1329, 1970a.

Mortimer JT, Magnusson R, Petersén I: Isometric contraction, muscle blood flow, and the frequency spectrum of the electromyogram. In Spring E: *Proceedings of the First Nordic Meeting on Medical and Biological Engineering.* Helsinki, Finland, Soc. Med. & Biol. Eng., Publishers, 1970b.

Mortimer JT, Kerstein MD, Magnusson R, Petersén I: Muscle blood flow in the human biceps as a function of developed muscle force. *Arch Surg* 103:376–377, 1971.

Morton DJ: *The Human Foot,* New York, Columbia University Press, 1935, p 119.

Morton DJ: *Human Locomotion and Body Form. A Study of Gravity and Man.* Baltimore, Williams & Wilkins, 1952.

Mosher CB, Gerlach RL, Stuart DG: Soleus and anterior tibial motor units of the cat. *Brain Res* 44:1–11, 1972.

Moyers RE: Temporomandibular muscle contraction patterns in angle class II, division I malocclusions: an electromyographic analysis. *Am J Orthodont* 35:837–857, 1949.

Moyers RE: An electromyographic analysis of certain muscles involved in temporomandibular movement. *Am J Orthodont* 36:481–515, 1950.

Muellner SR: The voluntary control of micturition in man. *J Urol* 80:473–478, 1958.

Muller N, Gulston G, Cade C, Whittan J, Bryan MH, Bryan AC: Respiratory muscle fatigue in infants. *Clin Res* 25:17a, 1978.

Munro RR: Activity of the digastric muscle in swallowing and chewing. *J Dent Res* 53:530–537, 1974.

Munro RR, Adams C: Electromyography of the intercostal muscles in connected speech. *Electromyography* 11:365–378, 1971.

Munro RR, Adams C: The patterns of internal intercostal muscle activity used by some non-native speakers of English. *Electromyogr Clin Neurophysiol* 13:271–288, 1973.

Munro RR, Basmajian JV: The jaw opening reflex in man. *Electromyography* 11:191–206, 1971.

Murakami Y, Fukuda H, Kirschner JA: The cricopharyngeus muscle: an electrophysiological and neuropharmacological study. *Acta Otolaryngol* (Suppl 311):1972.

Murakami Y, Kirschner JA: Electrophysiological properties of laryngeal reflex closure. *Acta Otolaryngol* 71:416–425, 1971.

Murakami Y, Kirschner JA: Mechanical and physiological properties of reflex laryngeal closure. *Ann Otol Rhinol Laryngol* 81:59–72, 1972.

Murphey DL, Blanton PL, Biggs NL: Electromyographic investigation of flexion and hyperextension of the knee in normal adults. *Am J Phys Med* 50:80–90, 1971.

Murphy AJ, Koepke GH, Smith EM, Dickinson DG: Sequence of action of the diaphragm and intercostal muscles during respiration. II. Expiration. *Arch Phys Med* 40:337–342, 1959.

Murray MP, Drought AB, Kory RC: Walking patterns of normal men. *J Bone Joint Surg* 46A:335–360, 1964.

Murray MP, Seirge AA, Sepic SB: Normal postural stability and steadiness: quantitative assessment. *J Bone Joint Surg* 57A:510–516, 1975.

Muscio B: Is a fatigue test possible? *Br J Psychol* 12:31–46, 1921.

Mustard WT: Iliopsoas transfer for weakness of the hip abductors. *J Bone Joint Surg* 34A:647–650, 1952.

Mustard WT: Personal communications, 1958.

Muybridge E: *The Human Figure in Motion.* Reprinted in 1955 by Dover Publications Inc., New York, 1887.

Myers SJ, Sullivan WP: Effect of circulatory occlusion on time of muscular fatigue. *J Appl Physiol* 24:54–59, 1968.

Myklebust BM, Gottlieb GL, Agarwall GC, Kaufman B, Poyezdala J, Baldwin L, Vaccaro D, Penn R: Reciprocal excitation: observations in normal adult, infants, and cerebral palsy patients, *11th Ann. Meet Soc Neurosci,* Abstr 559, 1981.

Nachemson A: Electromyographic studies of the vertebral portion of the psoas muscle. *Acta Orthop Scand* 37:177–190, 1966.

Nachmansohn D: *Chemical and Molecular Basis of Nerve Activity.* New York, Academic Press, 1959.

Naeije M, Zorn H: Relation between EMG power spectrum shifts and muscle fiber action potential conduction velocity changes during local muscular fatigue in man. *Eur J Appl Physiol* 50:23–33, 1982.

Nafpliotis H: Electromyographic feedback to improve ankle dorsiflexion, wrist extension, and hand grasp. *Phys Ther* 56:821–825, 1976.

Nagasawa T, Sasaki H, Tsuru H: Masseteric silent period after tooth contact in full denture wearers. *J Dent Res* 55:314, 1976.

Nakaarai K: Electromyographic study on the bulbocavernosus muscle of male subjects with normal and abnormal urination. *Electromyography* 8:367–381, 1968.

Napier JR: The foot and the shoe. *Physiotherapy* 43:65–74, 1957.

Naponiello LV: An electromyographic study of certain muscles in the easy standing position (abstr). *Anat Rec* 127:339, 1957.

Nashner LM: Balance adjustments of humans perturbed while walking. *J Neurophysiol* 44:650–664, 1980.

Negus VE: *The Comparative Anatomy and Physiology of the Larynx.* London, William Heinemann, 1949.

Nesbit RM, Lapides J: The physiology of micturition. *J Mich Med Soc* 58:384–388, 1959.

Netsell R, Cleeland CS: Modification of lip hypertonia in dysarthria using EMG feedback. *J Speech Hearing Dis* 38:131–140, 1973.

Nielson VK: Sensory and motor nerve conduction in the median nerve in normal subjects. *Acta Med Scand* 194:435–443, 1973.

Nieporent HJ: An electromyographic study of the function of the respiratory muscles in normal subjects. Dissertation for M.D. degree, University of Zurich, 30 pp, 1956.

Nordh E, Jonsson B, Ladd H: The learning process for fine neuromuscular controls in skeletal muscles of man. VII. interelectrode distance. *Electromyogr Clin Neurophysiol* 14:475–483, 1974.

Norris EH Jr, Gasteiger EL: Action potentials of single motor units in normal muscle. *Electroencephalogr Clin Neurophysiol* 7:115–126, 1955.

Norris FH Jr, Gasteiger EL, Chatfield PO: An electromyographic study of induced and spontaneous muscle cramps. *Electroencephalogr Clin Neurophysiol* 9:139–147, 1960.

Norris FH Jr, Irwin RL: Motor unit area in a rat muscle. *Am J Physiol* 200:944–946, 1960.

Norris AH, Shock NW, Wagman IH: Age changes in the maximum conduction velocity of motor fibers of human ulnar nerves. *J Appl Physiol* 5:589–593, 1953.

Novozamsky V, Buchberger J: Die Fusswölbung nach Belastung durch einin 100 km-Marsch. *Z Anat Entwickl-Gesch* 131:243–248, 1970.

Obrist PA: Heart rate and somatic coupling during classical aversive conditioning in humans. *J Exp Psychol* 77:180–193, 1968.

O'Connell AL: Electromyographic study of certain leg muscles during movements of the free foot and during standing. *Am J Phys Med* 37:289–301, 1958.

O'Connell AL: Effect of sensory deprivation on postural reflexes. *Electromyography* 11:519–527, 1971.

O'Connell AL, Gardner EB: The use of electromyography in kinesiological research. *Res Quart* 34:166–184, 1963.

O'Connell AL, Mortensen OA: An electromyographic study of the leg musculature during movements of the free foot and during standing. *Anat Rec* 127:342, 1957.

O'Dell NL, Topp GL, III, Bernard GR: Musculoskeletal arrangements for lateral mandibular movements in the rabbit and rat: electromyographic and other analyses. *J Dent Res* 49:1111–1117, 1970.

O'Donnell TF Jr: Measurement of percutaneous muscle surface pH. *Lancet* 2:533, 1975.

O'Dwyer NJ, Quinn PT, Guitar BE, Andrews G, Neilson PD: Procedures for vertification of electrode placement in EMG studies of orofacial and mandibular muscles. *J Speech Hearing Res* 24:273–288, 1981.

Oganisyan AA, Ivanova SN: A new method of implantation of electrodes into the muscles of a dog's extremities for EMG recording during free movement (Russian text). *Byull Eksper Biol Med* 57:136–138, 1964.

Ogata T: The differences in some labile constituents and some enzymatic activities between the red and white muscle. *J Biochem* 47:726–732, 1960.

Ogawa T, Jefferson NC, Toman JE, Chiles T, Zambetoglou A, Necheles H: Action potentials of accessory respiratory muscles in dogs. *Am J Physiol* 199:569–572, 1960.

Ohtsuki T: Decrease in grip strength induced by simultaneous bilateral exertion with reference to finger strength. *Ergonomics* 24:37–48, 1981.

Ohyama M, Ueda N, Harvey JE, Mogi G, Ogura JH: Electrophysiologic study of reinnervated laryngeal motor units. *Laryngoscope* 82:237–251, 1972.

Oka H, Okamoto T, Kumamoto M: Electromyographic and cinematographic study of the volleyball spike. In *International Series on Biomechanics, Vol. 1A, Biomechanics, Vol. V-A*. Baltimore, University Park Press, 1976, pp 326–331.

Okada M: An electromyographic estimation of the relative muscular load in different human postures. *J Human Ergol* 1:75–93, 1972.

Okada M: Quantitative studies on the bearing of the anti-gravity muscles in human postures with special references to electromyographic estimation of the postural muscle load. *J Fac Sci Univ Tokyo* 4(Sec V):471–530, 1975.

Okamoto T: Electromyographic study of the function of m. rectus femoris. *Res J Phys Ed* (Japan) (English text) 12:175–182, 1968a.

Okamoto T: A study of the variation of discharge pattern during flexion of the upper extremity. *J Lib Arts Dept Kansai Med School* 2:111–122, 1968.

Okamoto T: Electromyographic study of the learning process of walking in 1- and 2-year-old infants. *Medicine and Sports: Biomechanics III*, 8:328–333, 1973.

Okamoto T: Yoshoji no Suiei-Shidokatei no Bunseki (A kinesiological study of swimming). *J Health Phys Ed Rec* 26:409–414, 1976.

Okamoto T, Kumamoto M: Electromyographic study of the process of acquisition of proficiency in gymnastic kip. *Res J Phys Ed* 17:385–394, 1971.

Okamoto T, Kumamoto M: Electromyographic study of the learning process of walking in infants. *Electromyography* 12:149–158, 1972.

Okamoto T, Kumamoto M, Takagi K: Electromyographic study of the function of m. adductor longus and m. adductor magnus. *Jpn J Phys Fitness* 15:43–48, 1966.

Okamoto T, Takagi K, Kumamoto M: Electromyographic study of elevation of the arm. *Res J Phys Ed (Jap Soc Phys Ed)* 11:127–136, 1967.

Okamoto T, Tokuyama H, Yoshizawa M, Dodaira A, Tsujino A, Kumamoto M: Yohshoji no Suiei no Kindenzuteki Kenkyu (An electromyographic study of swimming in children). In *The Science of Human Movement*. Tokyo, Kyorin Shoin, 1976, pp 115–126.

Okhnyanskaya LG, Yusevich Yu S, Nikiforova NA: Some questions relating to electromyographic investigations of synergies in man. *Electromyogr Clin Neurophysiol* 14:391–414, 1974.

Olsen CB, Carpenter DO, Henneman E: Orderly recruitment of muscle action potentials: motor unit threshold and EMG amplitude. *Arch Neurol* 19:591–597, 1968.

Önal E, Lopata M, Evanich J: Effects of electrode position on esophageal diaphragmatic EMG in humans. 47:1236–1238, 1979.

Ongerboer de Visser BW, Goor C: Cutaneous silent period in masseter muscles: a clinical and diagnostic evaluation. *J Neurol Neurosurg Psychiatry* 39:674–679, 1976.

Ono K: Electromyographic studies of the abdominal wall muscles in visceroptosis. I. Analysis of patterns of activity of the abdominal wall muscles in normal adults. *Tohoku J Exp Med* 68:347–354, 1958.

Oota Y: Electromyography in the study of clinical kinesiology. *Kyushu J Med Sci* 7:75–91, 1956a.

Oota Y: Electromyography in the practice of orthopedic surgery. *Kyushu J Med Sci* 7:49–62, 1956b.

Orchardson R: The generation of nerve impulses in mammalian axons by changing the concentrations of the normal constituents of extracellular fluid. *J Physiol* 275:177–189, 1978.

Ortengren R, Lindstrom L, Petersén I: Electromyographic evaluation and modelling of localized fatigue during periodic muscular work. *Proc 4th Congr Int Soc Electrophysiol Kinesiol* 96–97, 1979.

Öwall B, Elmqvist D: Motor pauses in emg activity during chewing and biting. *Odontol Rev* 26:17–38, 1975.

Owens SE, Lehr RP Jr, Biggs NL: The functional significance of centric relation as demonstrated by electromyography of the lateral pterygoid muscles. *J Prosthet Dent* 33:6–9, 1975.

Panin N, Lindenauer HJ, Weiss AA, Ebel A: Electromyographic evaluation of the "cross exercise" effect. *Arch Phys Med* 42:47–52, 1961.

Palla S, Ash MM: Effects of bite force on the power spectrum of the surface electromyogram of human jaw muscles. *Arch Oral Biol* 26:287–295, 1981.

Parker PA, Scott RN: Statistics of the myoelectric signal from monopolar and bipolar electrodes. *Med Biol Eng* 11:591–596, 1973.

Partridge MJ, Walters CE: Participation of

the abdominal muscles in various movements of the trunk in man: an electromyographic study. *Phys Ther Rev* 39:791–800, 1959.

Paton WDM, Wand DR: The margin of safety of neuromuscular transmission. *J Physiol* (Lond) 191:59–90, 1967.

Patton NJ, Mortensen OA: A study of some mechanical factors affecting reciprocal activity in one-joint muscles. *Anat Rec* 166:360, 1970.

Patton NJ, Mortensen AO: An electromyographic study of reciprocal activity of muscles. *Anat Rec* 170:255, 1971.

Paul GL: Physiological effects of relaxation training and hypnotic suggestion. *J Abnorm Psychol* 74:425–437, 1969.

Pauly JE: Electromyographic studies of human respiration. *Chicago Med School Q* 18:80–86, 1957.

Pauly JE: An electromyographic analysis of certain movements and exercises. Part I. Some deep muscles of the back. *Anat Rec* 155:223–234, 1966.

Pauly JE, Steele RW: Electromyographic analysis of back exercises for paraplegic patients. *Arch Phys Med* 47:730–736, 1966.

Payne JK, Higenbottam T, Guindi GM: A surface electrode for laryngeal electromyography. *J Neurol Neurosurg Psychiatry* 43:853–854, 1980.

Payton OD: Electrical correlates of motor skill development in an isolated movement: the triceps and anconeus as agonists. *Proc World Conf Phys Ther,* 1974, pp 132–139.

Payton OD, Kelley DL: Electromyographic evidence of the acquisition of a motor skill. *Phys Ther* 52:261–266, 1972.

Payton OD, Su S, Meydrich EF: Abductor digiti shuffleboard: a study in motor learning. *Arch Phys Med Rehab* 57:169–174, 1976.

Pearson K: The control of walking. *Sci Am* 235:72–87, 1976.

Peat M, Grahame RE: Shoulder function in hemiplegia: a method of electromyographic and electrogoniometric analysis. *Physiother Can* 29:1–7, 1977.

Pedotti A: A study of motor coordination and neuromuscular activities in human locomotion. *Biol Cybernetics* 26:53–62, 1977.

Perritt RQ, Milner M: A simple, inexpensive, 8-channel multiplexer for electromyography in human locomotion. *Med Biol Eng* (Jan):104–106, 1976.

Person RS: Electromyographical study of coordination of the activity of human antagonist muscles in the process of developing motor habits (Russian text). *Jurn Vys'cei Nervn Dejat* 8:17–27, 1958.

Person RS: *Antagonistic Muscles in Human Movements.* (Russian text; English Abstract). Moscow, Acad Science, U.S.S.R. (publishers), 1965.

Person RS: *Electromyography in Human Study* (Russian text). Moscow, Acad Science, U.S.S.R. (publishers), 1969.

Person RS: Rhythmic activity of a group of human motoneurones during voluntary contractions of a muscle. *Electroencephalogr Clin Neurophysiol* 36:585–595, 1974.

Person RS, Kudina LP: Pattern of human motoneuron activity during voluntary muscular contraction. *Neurophysiology* 3(6):455–462, 1971.

Person RS, Kudina LP: Discharge frequency and discharge pattern in human motor units during voluntary contractions of muscle. *Electroencephalogr Clin Neurophysiol* 32:471–483, 1972.

Person RS, Libkind MS: Modelling of interference bio-electrical activity. *Biophysics* 12:145–153, 1967.

Person RS, Mishin LN: Auto- and cross-correlation analysis of the electrical activity of muscles. *Med Electron Biol Eng* 2:155–159, 1964.

Person RS, Roshtchina NA: Electromyographic investigation of coordinated activity of antagonistic muscles in movements of fingers of the human hand (Russian text). *J Physiol USSR* 94:455–462, 1958.

Persson A, Leanderson R, Öhman S: Electromyographic studies of facial muscle activity in speech. Proc 6th IFSECN Congress. *Electroencephalogr Clin Neurophysiol* 27:641–735, 1969.

Pertuzon E, Lestienne F: Caractère électromyographique d'un mouvement monoarticulaire executé à vitesse maximale. *J Physiol* 60:513, 1968.

Petajan JH, Philip BA: Frequency control of motor unit action potentials. *Electroencephalogr Clin Neurophysiol* 27:66–72, 1969.

Petajan JH, Williams DD: Behavior of single motor units during pre-shivering tone and shivering tremor. *Am J Phys Med* 51:16–22, 1972.

Peter JB, Bernard RJ, Edgerton VR: Metabolic profiles of three fiber types of skeletal muscle in guinea pigs and rabbits.

Biochemistry 11:2627, 1972.

Peters JF: Eye movement recording: a brief review. *Psychophysiology* 8:414–416, 1971.

Petersén I, Franksson C: Electromyographic study of the striated muscles of the male urethra. *Br J Urol* 27:148–153, 1955.

Petersén I, Franksson C, Danielsson CO: Electromyographic study of the muscles of the pelvic floor and urethra in normal females. *Acta Obstet Gynecol Scand* 34:273–285, 1955.

Petersén I, Kadefors R, Person J: Neurophysiologic studies of welders in shipbuilding work. *Environ Res* 11:226–236, 1976.

Petersén I, Kugelberg E: Duration and form of action potential in the normal human muscle. *J Neurol Neurosurg Psychiatry* 12:124–128, 1949.

Petersén I, Stener B: Experimental evaluation of the hypothesis of ligamento-muscular protective reflexes. III. A study in man using the medial collateral ligament of the knee joint. *Acta Physiol Scand* 48(Suppl 166):51–61, 1959.

Petersén I, Stener I: An electromyographical study of the striated urethral sphincter, the striated anal sphincter, and the levator ani muscle during ejaculation. *Electromyography* 10:23–44, 1970.

Petersén I, Stener I, Selldén U, Kollberg S: Investigation of urethral sphincter in women with simultaneous electromyography and micturition urethro-cystography. *Acta Neurol Scand* 38(Suppl 3):145–51, 1962.

Petit JM, Delhez L, Deroanne R, Pirnay F, Boccar M: Régulation de la ventilation pendant l'exercise musculaire. *Trav Soc Méd Belge d'Educ Phys Sports* 20:82–87, 1968.

Petit JM, Delhez L, Troquet J: Aspects actuels de la mécanique ventilatoire. *J Physiol Paris* 57:7–113, 1965.

Petit JM, Milic-Emili G, Delhez L: Role of the diaphragm in breathing in conscious normal man: an electromyographic study. *J Appl Physiol* 15:1101–1106, 1960.

Petrofsky SJ: Filter bank analyzer for automatic analysis of the EMG. *Med Biol Eng Comp* 18:585–590, 1980.

Petrofsky SJ, Lind AR: The influence of temperature on the amplitude and frequency components of the EMG during brief and sustained isometric contractions. *Eur J Appl Physiol* 44:189–200, 1980a.

Petrofsky SJ, Lind AR: Frequency analysis of the surface electromyogram during sustained isometric contractions. *Eur J Appl Physiol* 43:173–182, 1980b.

Petrofsky JS, Phillips CA, Sawka MN, Hanpeter D, Strafford D: Blood flow and metabolism during isometric contractions in cat skeletal muscle. *J Appl Physiol* 50:493–502, 1981.

Phillips CG: In Laying the ghost of 'muscles versus ligaments', a lecture to the Xth Can. Congr. of Neurol Science, 1975.

Phuon-Monich: La capacité de travail statique intermittent. (M.D. thesis), Paris, Editions A.G.E.M.P. (Publishers), 1963.

Pierce DS, Wagman IH: A method of recording from single muscle fibers or motor units in human skeletal muscle. *J Appl Physiol* 19:366–368, 1964.

Pierce JM Jr, Roberge JT, Newmann MM: Electromyographic demonstration of bulbocavernosus reflex. *J Urol* 83:319, 1960.

Pierrot-Deseilligny E, Bussel B: Evidence of recurrent inhibition by motoneurons in human subjects. *Brain Res* 88:105–108, 1975.

Piette P: L'exploration électromyographique du muscle transversaire épinaux. *Electromyogr Clin Neurophysiol* 14:69–77, 1974.

Pihkanen T, Harenko A, Huhmar E: Observations on the conduction velocity in peripheral nerves in states of drug intoxication: studies of the ulnar nerve in acute drug intoxication. *Acad Neurol Scand* 41(Suppl 13):267–271, 1965.

Piper H: Uber den willkurlichen Muskeltetanus. *Pflugers Arch Ges Physiol Mensch Tiere* 119:301–338, 1907.

Piper H: Neue Versuche uber den willkurlichen Tetanus der quergestreiften Muskeln. *Z Biol* 50:393–420, 1908.

Piper H: *Electrophysiologie Menschlicher Muskeln* (German text). Berlin, Springer-Verlag, 1912.

Pishkin V, Shurley JT: Electrodermal and electromyographic parameters in concept identification. *Psychophysiology* 5:112–118, 1968.

Pishkin V, Shurley JT, Wolfgang A: Assessment of drugs in decision-making under stress. *Dis Nerv Syst* 29:841–843, 1968.

Pivik T, Dement WC: Phasic changes in muscular and reflex activity during non-REM sleep. *Exp Neurol* 27:115–124, 1970.

Pocock GS: Electromyographic study of the quadriceps during resistive exercises. *J*

Am Phys Ther Assoc 43:427–434, 1963.

Podivinsky F: Factors affecting the course and the intensity of crossed motor irradiation during voluntary movement in healthy human subjects. *Physiol Bohemoslov* 13:172–178, 1961.

Polgar J, Johnson MA, Weightman D, Appleton D: Data on fiber size in thirty-six human muscles: an autopsy study. *J Neurol Sci* 19:307–318, 1973.

Polit A, Bizzi E: Characteristics of motor programs underlying arm movements in monkeys. *J Neurophysiol* 41:542–556, 1979.

Pollak V: The waveshape of action potentials recorded with different types of electromyographic needles. *Med Biol Eng* 9:657–664, 1971.

Pollock LJ, Davis L: Reflex activities of the decerebrate animal. *J Comp Neurol* 50:377–411, 1930.

Pommeranke WT: A study of the sensory areas eliciting the swallowing reflex. *Am J Physiol* 84:36–41, 1928.

Pompeiano O, Wand P, Sontag K-H: The relative sensitivity of Renshaw cells to orthodromic group Ia volleys caused by static stretch and vibration of extensor muscles. *Arch Ital Biol* 113:238–279, 1975.

Porter NH: (unpublished) Reported in Todd IP: Some aspects of the physiology of continence and defecation. *Arch Dis Childh* 37:181–183, 1960.

Portmann G: Myographies des cordes vocales chez l'homme (French text). *Rev Laryngol* 77:1–10, 1956.

Portmann G: The physiology of phonation (the Semon lecture for 1956) *J Laryngol Otol* 71:1–15, 1957.

Portmann G, Robin JL, Laget P, Husson R: La myographie des cordes vocales (French text). *Acta Otolaryngol* 46:250–263, 1956.

Portnoy H, Morin F: Electromyographic study of postural muscles in various positions and movements. *Am J Physiol* 186:122–126, 1956.

Potter A, Menke L: Capacitive type of biomedical electrode. *IEEE Trans BME* 17:350, 1970.

Poudrier C, Knowlton GC: Command-force relations during voluntary muscle contraction. *Am J Phys Med* 43:109–116, 1964.

Powers WR: *Conscious Control of Single Motor Units in the Preferred and Non-Preferred Hand.* Ph.D. Thesis, Queen's University Kingston, Ontario, Canada, 1969.

Pózniak-Patewicz E: Electromyographic investigations of nape and temporal muscles in headaches of various origin (Polish text; English abstr). *Neurol Neurochir Pol* 9:461–468, 1975.

Prabhu VG, Oexter YT: Electromyographic studies of skeletal muscle of rat given cortisone. *Arch Neurol* 24:253–258, 1971.

Prechtl HFR, Lenard HG: A study of eye movements in sleeping newborn infants. *Brain Res* 5:477–493, 1967.

Proebster R: Uber Muskelaktionsstrome am gesunden und Kranken Menschen. *Orthop Clin* 50:1, 1928.

Pruzansky S: The application of electromyography to dental research. *J Am Dent Assoc* 44:49–68, 1952.

Quinn PT, Neilson PD, McCaughey J, O'-Dwyer NJ: Action tonic stretch reflex in human masseter—evidence for voluntary suppression. *EMG Clin Neurophysiol* 22:3–8, 1982.

Rack PMH: Limitations of somatosensory feedback in control of posture and movement. In Brookhart JM, Mountcastle VB, Brooks VB, Geiger SR: *Handbook of Physiology,* The Nervous System, Bethesda, MD, American Physiological Society, 1981, vol 2, part 1, ch 7, pp 229–256.

Radcliffe CW: The biomechanics of below-knee prostheses in normal, level, bipedal walking. *Artif Limbs* 6:16–24, 1962.

Rainaut JJ: Étude de la marche par electromyographie télémétrique. *Rev Chir Orthop Repat App Moteur* (Paris) 57:427–437, 1971.

Ralston HJ: Uses and limitations of electromyography in the quantitative study of skeletal muscle function. *Am J Orthodontol* 47:521–530, 1961.

Ralston HJ: Effects of immobilization of various body segments on the energy cost of human locomotion. *Ergonomics* 7:53–60, 1964.

Ralston HJ, Libet B: The question of tonus in skeletal muscle. *Am J Phys Med* 32:85–92, 1953.

Ralston HJ, Todd FN, Inman VT: Comparison of electrical activity and duration of tension in the human rectus femoris muscle. *Electromyogr Clin Neurophysiol* 16:277–286, 1976.

Ramsey RW: Some aspects of the biophysics of muscle. In Bourne GH: *The Structure and Function of Muscle.* New York, Academic Press, 1960, vol 2, pp 323–327.

Ranney DA, Basmajian JV: Electromyography of the rabbit fetus. *J Exp Zool* 144:179–186, 1960.

Rao VR: Interesting electromyographic changes induced by smoking. *J Postgrad Med* 9:138–139, 1963.

Rao VR: Reciprocal inhibition: inapplicability to tendon jerks. *J Postgrad Med* 11:123–125, 1965.

Rao VR, Rindani TH: The influence of smoking on electromyograms. *J Postgrad Med* 8:170–172, 1962.

Raper AJ, Thompson WT Jr, Shapiro W, Patterson JL Jr: Scalene and sternomastoid muscle function. *J Appl Physiol* 21:497–502, 1966.

Raphael LJ: The physiological control of durational differences between vowels preceding voiced and voiceless consonants in English. *J Phonet* 3:25–33, 1975.

Raphael LJ, Bell-Berti F: Tongue musculature and the feature of tension in English vowels. *Phonetica* 32:61–73, 1975.

Rasch PJ, Burke RK: *Kinesiology and Applied Anatomy: the Science of Human Movement,* ed 6. Philadelphia, Lea & Febiger, 1978.

Rattner WH, Gerlaugh BL, Murphy JJ, Erdman WJ II: The bulbocavernosus reflex. I. Electromyographic study of normal patients. *J Urol* 80:140–141, 1958.

Ravaglia M: Sulla particolare attività dei capi muscolari del quadricipite femorale nell'uomo (indagine elettromiografica) (Italian text). *Circ Org Movimento* 44:498–504, 1957.

Ravaglia M: Indagine elettromiografica della funzione dei muscoli pettorali nella meccanica respiratoria (Italian text). *Ginnastica Med* 5:1–3, 1958.

Ray RD, Johnson RJ, Jameson RM: Rotation of the forearm: an experimental study of pronation and supination. *J Bone Joint Surg* 33A:993–996, 1951.

Redford JB, Butterworth TR, Clements EL: Use of electromyography as a prognostic aid in the management of idiopathic scoliosis. *Arch Phys Med* 50:433–438, 1969.

Reesor LA, Susman RL, Stern JT Jr: Electromyographic studies of the human foot: experimental approaches to hominid evolution. *Foot Ankle* 3:391–407, 1983.

Reinking RM, Stephens JA, Stuart DG: The motor units of cat medial gastrocnemius: problem of their categorisation on the basis of mechanical properties. *Exp Brain Res* 23:301–313, 1975.

Reis DJ, Wooten GF, Hollenberg M: Differences in nutrient blood flow of red and white skeletal muscle in the cat. *Am J Physiol* 213:592–596, 1967.

Reis FP, de Camargo AM, Vitti M, de Carvalho CAF: Electromyographic study of the subclavius muscle. *Acta Anat* 105:284–290, 1979.

Reis FP, Carvalho CAF: Electromyographic study of the popliteus muscle. *Electromyogr Clin Neurophysiol* 13:445–455, 1973.

Reiter R: Quoted by Podlusky, MU, and Mann RW. In Letters to the Editor "Forum" IEEE Spectrum, February 1969.

Renshaw B: Collateral effects of centripetal impulses in axons of spinal ventral roots. *J Neurophysiol* 9:191, 1946.

Richards C, Knutsson E: Evaluation of abnormal gait patterns by intermittent-light photography and electromyography. *Scand J Rehab Med* 3(Suppl):61–68, 1974.

Richter J: Vorlaufige Mitteilung über elektromyographische Untersuchungen am unbelasteten und belasteten Fuss. *Arch Orthop Unfall-Chir* 59:168–176, 1966.

Riddle HFV, Roaf R: Muscle imbalance in the causation of scoliosis. *Lancet* 1(June 18):1245–1247, 1955.

Rideau Y, Duval A: Function of the anterior thigh muscles. *Anat Clin* 1:29–42, 1978.

Rideau Y, von Torklus V, Duval A. et al: Electromyocartographie de la paroi postérieure et inférieure de l'abdomen lors de l'extension du tronc. *Bull Assoc Anat Nancy* 59:757–767, 1975.

Robinson DA: The mechanics of human smooth pursuit eye movement. *J Physiol* 180:569–591, 1965.

Roman C, Car A: Déglutitions et contractions oesophagiennes réflexes obtenues par la stimulation des nerfs vague et laryngé supérieur. *Exp Brain Res* 11:48–74, 1970.

Romeny BM, Denier van der Gon JJ, Gielen CCAM: Changes in recruitment order of motor units in the human biceps muscle. *Exp Neurology* 78:360–368, 1982.

Rosemeyer B: Elektromyographische Untersuchungen der Rückenund Schultermuskulatur im Stehen und Sitzen unter Berücksichtigung der Haltung des Autofahrers. *Arch Orthop Unfall-Chir* 69:59–70, 1971.

Rosenfalck A: Evaluation of the electromyogram by mean voltage recording. In *Medical Electronics: Proceedings of the Second International Conference on Medical Electronics, Paris June 24–27, 1959,* London, Iliffe & Sons, 1960, pp 9–12.

Rosenfalck P: *Intra- and Extracellular Potential Fields of Active Nerve and Muscle Fibers.* Kobenhavn, Academisk Forlag, 1969.

Rosenfalck A, Andreassen S: Impaired regulation of force and firing pattern of single motor units in patients with spasticity. *J Neurol Neurosurg Psychiatry* 43:907–916, 1980.

Rosenthal RG, Sabbahi MA, Merletti R, De Luca CJ: Possible fiber typing by analysis of surface EMG signals. Proc of 11th Annual Meeting of the Soc for Neuroscience, 1981, p 683.

Rositano SA: Ultraflexible bioelectrodes and wires. Proc. 23rd ACEMB, 1970, p 149.

Ross HG, Cleveland S, Haase J: Contribution of single motoneurons to Renshaw cell activity. *Neurosci Lett* 1:105–108, 1975.

Ross HG, Cleveland S, Haase J: Quantitative relation between discharges frequencies of a Renshaw cell and an intracellularly depolarized motoneuron. *Neurosci Letters* 3:129–132, 1976.

Rubin HJ: Further observations on the neurochronaxic theory of voice production. *AMA Arch Otolaryngol* 72:207–211, 1960.

Ruch TC, Patton HD, Woodbury JW, Towe AL: *Neurophysiology.* Philadelphia, W.B. Saunders, 1963.

Ruëdi L: Some observations on the histology and function of the larynx (Semon lecture for 1958). *J Laryngol Otol* 73:1–20, 1959.

Rummel RM: An electromyographic analysis of patterns used to reproduce muscular tension. *Res Q Am Assoc Health Phys Ed* 45:64–71, 1974.

Ruskin AP: Anal sphincter electromyography. *Electromyography* 10:425–428, 1970.

Ryall RW: Renshaw cell mediated inhibition of Renshaw cells: patterns of excitation and inhibition from impulses in motor axon collaterals. *J Neurophysiol* 33:257–270, 1970.

Ryall RW, Piercey M: Excitation and inhibition of Renshaw cells by impulses in peripheral afferent nerve fibers. *J Neurophysiol* 34:242–251, 1971.

Ryall RW, Piercey M, Polosa C, Goldfarb J: Excitation of Renshaw cells in relation to orthodromic and antidromic excitation of motoneurons. *J Neurophysiol* 35:137–148, 1972.

Sacco G, Buchthal F, Rosenfalck P: Motor unit potentials at different ages. *Arch Neurol* 6:366–373, 1962.

Sabbahi MA, De Luca CJ, Powers WR: The effect of ischemia, cooling and local anesthesia on the median frequency of the myoelectric signal. *Proc 4th Cong Int Soc Electrophysiol Kinesiol* 1979, pp 94–95.

Sabbahi MA, De Luca CJ: Topical anesthesia: H-reflex recovery changes by desensitization of the skin. *EEG Clin Neurophysiol* 52:328–335, 1981.

Sabbahi MA, De Luca CJ: Topical anesthesia: modulation of the monosynaptic reflexes by densensitization of the skin. *EEG Clin Neurophysiol* 54:677–688, 1982.

Sadoyama T, Masuda T, Miyano H: Relationship between muscle fiber conduction velocity and frequency parameters of surface EMG during sustained contraction. *Eur J Appl Physiol* 51:247–256, 1983.

Sadoyama T, Miyano H: Frequency analysis of surface EMG to evaluation of muscle fatigue. *Eur J Appl Physiol* 47:239–246, 1981.

Sahlin K, Alvestrand A, Brandt R, Hultman E: Intracellular pH and bicarbonate concentration in human muscles during recovery from exercise. *J Appl Physiol* 45:474–480, 1978.

Sahlin K, Harris RC, Hultman E: Creatine kinase equilibrium and lactate content compated with muscle pH in tissue samples obtained after isometric exercise. *Biochem J* 152:173–180, 1975.

Saito M, Matsui H, Miyamura M: Effect of physical training on the calf and thigh blood flows. *Jpn J Physiol* 30:955–959, 1980.

Sala E: Studio elettromiografico dell'innervazione dei muscoli flessore breve ed opponente del pollice (Italian text). *Riv Pat Nerv* 80:131–147, 1959.

Sales RD, Vitti M: Analisé eletromiográfica dos mm. orbicularis oris em indivíduos portadores de maloclusão classe I. *Rev Assoc Paul Cirurg Dent* 33:399–411, 1979.

Salmons S, Henriksson J: The adaptive response of skeletal muscles to increased use. *Muscle Nerve* 4:94–105, 1981.

Saltin B, Gollnick PD: Skeletal muscle adaptability: significance for metabolism and performance. *Handbook of Physiology.* Bethesda, MD, American Physiological Society, 1981.

Salomon G, Starr A: Electromyography of middle ear muscles in man during motor activities. *Acta Neurol Scand* 39:161–168, 1963.

Samilson RL, Morris JM: Surgical improvement of the cerebral-palsied upper limb:

electromyographic studies and results in 128 operations. *J Bone Joint Surg* 46A:1203–1216, 1964.

Samson J: *The Acquisition, Retention and Transfer of Single Motor Unit Control.* Ph.D. Dissertation, Univ. of Illinois at Urbana-Champaign, 1971.

Sano E, Ando K, Katori I, Yamada H, Samperi H, Sugahara R: Electromyographic studies on the forearm muscle activities during finger movements. *J Jpn Orthop Assoc* 51:331–337, 1977.

Sant'Ambrogio G, Frazier MF, Wilson MF, Agostoni E: Motor innervation and pattern of activity of cat diaphragm. *J Appl Physiol* 18:43–46, 1963.

Sant'Ambrogio G, Widdicombe JG: Respiratory reflexes acting on the diaphragm and inspiratory intercostal muscles of the rabbit. *J Physiol* 180:766–779, 1965.

Sato M: Electromyographical study of skilled movement. *J Fac Sci Univ Tokyo Sec V, Vol II, Part 4:*323–369, 1963.

Sato M: Some problems in the quantitative evaluation of muscle fatigue by frequency analysis of the electromyogram. *J Anthro Soc Japan* 73:20–27, 1965.

Sato M: Muscle fatigue in the half rising posture (English text). *Zinruigaku Zassi J Anthropol Soc (Nippon)* 74:13–19, 1966.

Sato M, Hayami A, Sato H: Differential fatiguability between the one- and two-joint muscles. *Zinruigaku Zassi (J Anthropol Soc Nippon)* 73:82–90, 1965.

Sauerland EK, Harper RM: The human tongue during sleep; electromyographic activity of the genioglossus muscle. *Exp Neurol* 51:160–170, 1976.

Sauerland EK, Mitchell SP: Electromyographic activity of intrinsic and extrinsic muscles of the human tongue. *Texas Rep Biol Med* 33:445–455, 1975.

Sauerland EK, Sauerland BAT, Orr WC, Hairston LE: Non-invasive electromyography of human genioglossal (tongue) activity. *Electromyogr Clin Neurophysiol* 21:279–286, 1981a.

Sauerland EK, Orr WC, Hairston LE: EMG patterns of oropharyngeal muscles during respiration in wakefulness and sleep. *Electromyogr Clin Neurophysiol* 21:307–316, 1981b.

Saunders JBCM, Inman VT, Eberhart HD: The major determinations in normal and pathological gait. *J Bone Joint Surg* 35A:543–558, 1953.

Schärer P, Pfyffer GL: Comparison of habitual and cerebrally stimulated jaw movements in the rabbit. *Helv Odontol Acta* 14:6–10, 1970.

Scherr MS, Crawford PL, Sergent B, Scherr CA: Effect of bio-feedback techniques on chronic asthma in a summer camp environment. *Ann Allergy* 35:289–295, 1975.

Scherrer J, Bourguignon A: Changes in the electromyogram produced by fatigue in man. *Am J Phys Med* 38:170–180, 1959.

Scherrer J, Bourguignon A, Marty R: Evaluation electromyographique du travail statique (French text). *J Physiol Paris* 49:376–378, 1957.

Scherrer J, Bourguignon A, Samson M, Marty R: Sur un caractère particulier de la contraction isométrique maximum au cours de la fatigue chez l'homme (French text). *J Physiol Paris* 48:704–707, 1956.

Scherrer J, Lefebvre J, Bourguignon A: Activité électrique du muscle strié squelettique et fatigue (French text). In *Fourth International Congress of Electroencephalography and Clinical Neurophysiology,* Brussels, 1957, pp 99–123.

Scherrer J, Monod H: Le travail musculaire local et la fatigue chez l'homme (French text). *J Physiol Paris* 52:419–501, 1960.

Scheving LE, Pauly JE: An electromyographic study of some muscles acting on the upper extremity of man. *Anat Rec* 135:239–246, 1959.

Schlapp M: Observations on a voluntary tremor—violinist's vibrato. *Q J Exp Physiol* 58:357–368, 1973.

Schloon H, O'Brien MJ, Scholten CA, Prechtl HFR: Muslce activity and postural behaviour in newborn infants. *Neuropädiatrie* 7:384–415, 1976.

Schlossberg L, Harris SC: An electromyographic investigation of the functioning perioral and suprahyoid musculature in normal occlusion and Class I division I dysplasia cases. *Am J Orthod* 42:153, 1956.

Schmidt EM: An instrument for separation of multiple unit neuroelectric signals. *IEEE BME* 18:155–157, 1971.

Schmidt EM, Stromberg MW: Computer dissection of peripheral nerve bundle activity. *Comput Biomed Res* 185:446–455, 1969.

Schmidt RS: Action of intrinsic laryngeal muscles during release calling in leopard frog. *J Exp Zool* 181:233–244, 1972.

Schoolman A, Fink BR: Permanently implanted electrode for electromyography of the diaphragm in the waking cat. *Electroencephalogr Clin Neurophysiol* 15:127–128, 1963.

Schubert HA: A study of motor nerve conduction: determination of velocity. *South Med J* 56:666–668, 1963.

Schubert HA: Conduction velocities along course of ulnar nerve. *J Appl Physiol* 19:423–426, 1964.

Schulte FJ, Schwenzel W: Motor control and muscle tone in the newborn period. Electromyographic studies. *Biol Neonat* 8:198–215, 1965.

Schuster MM: The riddle of the sphincters. *Gastroenterology* 69:249–262, 1975.

Schwartz GE, Fair PL, Salt P, Mandel MR, Klerman GL: Facial expression and imagery in depression: an electromyographic study. *Psychosom Med* 38:337–347, 1976a.

Schwartz GE, Fair PH, Salt P, Mandel MR, Klerman GL: Facial muscle patterning to affective imagery in depressed and nondepressed subjects. *Science* 192:489–491, 1976b.

Schwartz MS, Stålberg E, Schiller HH, Thiele B: The reinnervated motor unit in man: a single fiber EMG multielectrode investigation. *J Neurol Sci* 27:303–312, 1976.

Schwartz RP, Trautman O, Heath AL: Gait and muscle function recorded by electrobasograph. *J Bone Joint Surg* 18:445–454, 1936.

Schweitzer TW, Fitzgerald JW, Bowden JA, Lynne-Davies P: Spectral analysis of human inspiratory diaphragmatic electromyograms. *J Appl Physiol* 46:152–165, 1979.

Schwestka R, Windhorst U, Schaumberg R: Patterns of parallel signal transmission between multiple alpha efferents and multiple Ia afferents in the cat semitendinosus muscle. *Exp Brain Res* 43:34–46, 1981.

Scott RN: A method of inserting wire electrodes for electromyography. *IEEE Trans Bio-Med Eng* BME-12:46–47, 1965.

Scott RN, Thompson GB: An improved biopolar wire electrode for electromyography. *Med Biol Eng* 7:677–678, 1969.

Scripture EW, Smith TL, Brown EM: On the education of muscular control and power. *Studies Yale Psychol Lab* 2:114–119, 1894.

Scully HE, Basmajian JV: Motor-unit training and influence of manual skill. *Psychophysiology* 5:625–632, 1969.

Sears ML, Teasdall RD, Stone HH: Stretch effects in human extraocular muscle: an electromyographic study. *Bull Johns Hopkins Hosp* 104:174–178, 1959.

Serra C: Neuromuscular studies on various oto-laryngological problems. *Electromyography* 4:254–294, 1964.

Serra C: Motor driving aptitude evaluation: an electromyographic contribution. *Eur Medicophys* 4:1–5, 1968.

Serra C: Electromyographic contributions to human visual perception mechanisms. *Int J Neurol* 8:62–69, 1970.

Serra C, Covello L: Elettromiografia clinica (Italian text). *Acta Neurol (Naples)* 20:1–507, 1959.

Serra C, Lambiase M: Fumo e sistema nervoso. II. Modificazioni dei potenziali d'azione muscolari da fumo (Italian text). *Acta Neurol (Naples)* 8:494–506, 1957.

Serra C, Pasanisi EJ, De Natale G: L'attività elettrica muscolare del coniglio in condizione di ipotermia controllata. *Il Cardarelli (Rev Ospedali Riun di Napoli)* 5:1–8, 1963.

Settineri LIC, Rodriguez RB: Estudo electromiográfico da mobilizacão ativa e passiva do cotovelo. *Med Esporte Porto Alegre (Brazil)* 1:161–166, 1974.

Seyfert S, Kunkel H: Analysis of muscular activity during voluntary contraction of different strengths. *Electromyogr Clin Neurophysiol* 14:323–330, 1974.

Seyffarth H: *The Behaviour of Motor-Units in Voluntary Contraction*, I. Kommisjon Hos Jacob Dybwad, A.W. Brøggers Boktrykkeri A/S, Oslo, 1940.

Shagass C: Evoked cortical potentials and sensations in man. *J Neuropsychiatry* 2:262–270, 1961.

Shafik A: A new concept of the anatomy of the anal sphincter mechanism and the physiology of defecation. *Invest Urol* 12:412–419, 1975.

Shahani BT, Young RR: Physiological and pharmacological aid in the differential diagnosis of tremor. *J Neurol Neurosurg Psychiatry* 39:772–783, 1976.

Shambes GM: Static postural control in children. *Am J Phys Med* 55:221–252, 1976.

Shankweiler D, Harris KS, Taylor ML: Electromyographic studies of articulation in aphasia. *Arch Phys Med* 49:1–8, 1968.

Sharp JT, Druz W, Danon J, Kim MJ: Respiratory muscle function and the use of respiratory muscle electromyography in the evaluation of respiratory regulation. *Chest* 70(suppl):150S–154S, 1976.

Sharrard WJW: Posterior iliopsoas transplantation in treatment of paralytic dislocation of the hip. *J Bone Joint Surg* 46B:426–44, 1964.

Sheffield FJ: Electromyographic study of the abdominal muscles in walking and other movements. *Am J Phys Med* 41:142–147, 1962.

Sheffield FJ, Gersten JW, Mastellone AF: Electromyographic study of the muscles of the foot in normal walking. *Am J Phys Med* 35:223–236, 1956.

Sherrington CS: The integrative action of the nervous system, ed 2. New Haven, Yale University Press, 1906.

Sherrington CS: Reciprocal innervation of antagonist muscles. Fourteenth note. On double reciprocal innervation. *Proc Roy Soc Lond Ser B* 91:249–268, 1909.

Sherrington CS: Ferrier Lecture. Some functional problems attaching to convergence. *Proc Roy Soc* 105B:332–362, 1929.

Shevlin MG, Lehmann JF, Lucci JA: Electromyographic study of the function of some muscles crossing the glenohumeral joint. *Arch Phys Med* 50:264–270, 1969.

Shiavi R: Control of and interaction between motor units in a human skeletal muscle during isometric contraction. Ph.D. thesis, Drexel Univ., Philadelphia, PA, 1972.

Shiavi R: A wire multielectrode for intramuscular recording. *Med Biol Eng*:721–723, 1974.

Shiavi R, Negin M: The effect of measurements errors on correlation estimates in spike-interval sequences. *IEEE BME* 9:374–378, 1973.

Shiavi R, Negin M: Stochastic properties of motoneuron activity and the effect of muscular length. *Biol Cybernetics* 19:231–237, 1975.

Shik ML, Orlovsky GN: Neurophysiology of locomotor automatism. *Physiol Rev* 56:465–501, 1976.

Shimazu H, Hongo T, Kubota K, Narabayashi H: Rigidity and spasticity in man: electromyographic analysis with reference to the role of the globus pallidus. *Arch Neurol* 6:10–17, 1962.

Shipp T: A technique for examination of laryngeal muscles during phonation. Proceedings of I.S.E.K., Montreal, 1968. *Electromyography* 8(suppl):21–26, 1968.

Shipp T: EMG of pharyngoesophageal musculature during alaryngeal voice production. *J Speech Hearing Res* 13:184–192, 1970.

Shipp T, Deatsch WW, Robertson K: A technique for electromyographic assessment of deep neck muscle activity. *Laryngoscope* 78:418–432, 1968.

Shipp T, Fishman BV, Morrissey P: Method and control of laryngeal emg electrode placement in man. *J Acoust Soc Am* 48:429–430, 1970.

Shipp T, McGlone RE: Laryngeal dynamics associated with voice frequency change. *J Speech Hearing Res* 14:761–768, 1971.

Shwedyk E et al: A nonstationary model for the electromyogram. *IEEE Trans Biomed Eng* BME-24:1977.

Sills FD, Olsen AL: Action potentials in unexercised arm when opposite arm exercised. *Res Q AAHPE* 29:213–221, 1958.

Simard T: Fine sensorimotor control in healthy children: an electromyographic study. *Pediatrics* 43:1035–1041, 1969.

Simard T: Unusual electromyographic motor unit action potentials in human healthy subjects. *Electromyography* 11:83–91, 1971.

Simard TG, Basmajian JV: Factors influencing motor unit training in man. *Proc Can Fed Biol Soc* 8:63, 1965.

Simard TG, Basmajian JV: Methods in training the conscious control of motor units. *Arch Phys Med* 48:12–19, 1967.

Simard TG, Basmajian JV, Janda V: Effect of ischemia on trained motor units. *Am J Phys Med* 47:64–71, 1968.

Simard TG, Ladd HW: Conscious control of motor units with thalidomide children: an electromyographic study. *Dev Med Child Neurol* 11:743–748, 1969.

Simon W: The real-time sorting of neuroelectric action potentials in multiple unit studies. *Electroencephalogr Clin Neurophysiol* 18:192–195, 1965.

Simonson E, Weiser P (eds): Psychological aspects and physiological correlates of work and fatigue. Springfield, IL, Charles C Thomas, 1976.

Simoyama M, Tanaka R: Reciprocal Ia inhibition at the onset of voluntary movements in man. *Brain Res* 82:334–337, 1974.

Simpson HM, Climan MH: Pupillary and electromyographic changes during an imagery task. *Psychophysiology* 8:483–490, 1971.

Simpson JA: Fact and fallacy in measurement of conduction velocity in motor nerves. *J Neurol Neurosurg Psychiatry* 27:381–385, 1964.

Sjodin B: Lactate dehydrogenase in human skeletal muscles. *Acta Physiol Scand Suppl*:436, 1976.

Sloan AW: Electromyography during fatigue of healthy rectus femoris. *South Afr*

Med J 39:395–398, 1965.

Smidt GL: Methods of studying gait. *Phys Ther* 54:14–17, 1974.

Smidt GL, Arora JS, Johnston, RC: Accelerographic analysis of several types of walking. *Am J Phys Med* 50:285–300, 1971.

Smith HM Jr, Basmajian JV, Vanderstoep SF: Inhibition of neighboring motoneurons in conscious control of single spinal motoneurons. *Science* 183:975–976, 1974.

Smith JW: Muscular control of the arches of the foot in standing: an electromyographic assessment. *J Anat* 88:152–163, 1954.

Smith OC: Action potentials from single motor units in voluntary contraction. *Am J Physiol* 108: 629–638, 1934.

Smith RP Jr: Frontalis muscle tension and personality. *Psychophysiology* 10:311–312, 1973.

Smorto MP, Basmajian JV: *Clinical Electroneurography: an Introduction to Nerve Conduction Tests.* Baltimore, Williams & Wilkins, 1972.

Soechting JF, Roberts WJ: Transfer characteristics between emg activity and muscle tension under isometric conditions in man. *J Physiol Paris* 70:779–793, 1975.

Solomon S, Ladd HW, Bradley BC, Macklem PT: A Long-term electromyographical evaluation of inspiratory function in quadriplegic patients. *Proc of 4th Cong of the Int Soc of Electrophysiol Kinesiol*, 1979, pp 92–93.

Solomonow M, Scopp R: Frequency response of isometric muscle force during recruitment. Proc Fifth Annual Conference of IEEE Eng in Med and Biol Society, 1983, pp 179–183.

Soto RA, Sanz OP, Sica REP, Chorny D: Facilitation of muscle activity by contralateral homonymous muscle action in man. *Medicina* 34:481–484, 1974.

Spiegel MH, Johnson EW: Conduction velocity in the proximal and distal segments of the motor fibers of the ulnar nerve of human beings. *Arch Phys Med* 43:57–61, 1962.

Spoor A, Van Dishoeck HAE: Electromyography of the human vocal cords and the theory of Husson. *Pract Oto-rhinolaryngol* 20:353–360, 1960.

Spruit R: *Een Analyse van Vorm en Ligging van de MM. Glutaei en de Adductoren* (An Analysis of Form and Location of Gluteal and Adductor Muscles) (Dutch text with English summary). Leiden Drukkerij Albani-Den. Haag, 1965.

Stålberg E: Propagation velocity in human muscle fibers in situ. *Acta Physiol Scand* 70(suppl):287, 1967.

Stålberg E: Macro EMG, a new recording technique. *J Neurol Neurosurg Psychiatry* 43:475–482, 1980.

Stålberg E, Ekstedt J: Single fiber EMG and microphysiology of the motor unit in normal and diseased muscle. In Desmedt JE: *New Developments in Electromyography and Clinical Neurophysiology.* Basel, Karger, 1973, pp 113–129.

Stålberg E, Schwartz MS, Thiele B, Schiller HH: The normal motor unit in man. *J Neurol Sci* 27:291–301, 1976.

Stålberg E, Schwartz MS, Trontelj JV: Single fiber electromyography in various processes affecting the anterior horn cell. *J Neurol Sci* 24:403–415, 1975.

Steendijk R: On the rotating function of the ilio-psoas muscle. *Acta Neerl Morphol* 6:175–183, 1948.

Stein RB, Bawa P: Reflex responses of human soleus muscle to small perturbations. *J Neurophysiol* 39:1105–1116, 1976.

Steiner JE, Michman J, Litman A: Time sequences of the activity of the temporal and masseter muscles in healthy young human adults during habitual chewing of different test foods. *Arch Oral Biol* 19:29–34, 1974.

Steiner TJ, Thexton AJ, Weber WV: An EMG electrode system for selected-site recording from a muscle. *J Appl Physiol* 32:531–532, 1972.

Steindler A: *Kinesiology of the Human Body under Normal and Pathological Conditions.* Springfield, IL, Charles C Thomas, 1955.

Stener B: Experimental evaluation of the hypothesis of ligamentomuscular protective reflexes. I. A method of adequate stimulation of tension receptors in the medial collateral ligament of the knee joint of the cat, and studies of the innervation of the ligament. *Acta Physiol Scand* 48(suppl 166):5–26, 1959.

Stephens JA, Garnett R, Bulli NP: Reversal of recruitment order of single motor units produced by cutaneous stimulation during voluntary muscle contraction in man. *Nature* 272:362–364, 1978.

Stephens JA, Taylor A: Fatigue of maintained voluntary muscle contraction in man. *J Physiol* 220:1–18, 1972.

Stephens JA, Usherwood TP: The fatigability of human motor units. *J Physiol*

250:37–38, 1975.

Stern JT: Investigations concerning theory of 'spurt' and 'shunt' muscles. *J Biomechanics* 4:437–453, 1971.

Stern JT Jr, Wells JP, Vangor AK, Fleagle JG: Electromyography of some muscles of the upper limb in *Ateles* and *Lagothrix*. *Yearbk Phys Anthropol* 20:498–507, 1977.

Stern JT Jr, Wells JP, Jungers WJ, Vangor AK: An electromyographic study of the pectoralis major in atelines and *Hylobates*, with special reference to the evolution of pars clavicularis. *Am J Phys Anthropol* 52:13–25, 1980a.

Stern JT Jr, Wells JP, Jungers WJ, Vangor AK: An electromyographic study of serratus anterior in atelines and *Alouatta*: implications for hominoid evolution. *Am J Phys Anthropol* 52:323–334, 1980b.

Stern MM: A model that relates low frequency EMG power to fluctuations in the number of active motor units: application to local muscle fatigue. Ph.D. Dissertation, Univerisity of Michigan, 1971.

Stetson RH: Speech movements in action. *Tr Am Laryngol Assoc* 55:29–41, 1933.

Stevens A, Slijins H, Reybrouck T, Bonte G, Michels A, Rosselle N, Roelandts P, Krauss E, Verheyen G: A polyelectromyographical study of the arm muscles at gradual isometric loading. Electromyogr Clin Neurophysiol 13:465–476, 1973.

Stiles RN: Frequency and displacement amplitude relations for normal hand tremor. *J Appl Physiol* 40:44–54, 1976.

Stolov WC: The concept of normal muscle tone, hypotonia and hypertonia. *Arch Phys Med* 47:156–168, 1966.

Storey AT: Laryngeal initiation of swallowing. *Exp Neurol* 20:359–365, 1968a.

Storey AT: A functional analysis of sensory units innervating epiglottis and larynx. *Exp Neurol* 20:366–383, 1968b.

Stotz S, Heimstädt P: Uber die Rollwirking des M. Iliopsoas in Abhängigkeit von Schenkelhalsund Antetorsionswinkel bie der infantilen Zerebralparese. *Z Orthop Grenz* 108:229–243, 1970.

Stratford P: Electromyography of the quadriceps femoris muscles in subjects with normal knees and acutely effused knees. *Phys Ther* 62:279–283, 1982.

Straus WL, Weddell G: Nature of the first visible contractions of the forelimb musculature in rat fetuses. *J Neurophysiol* 3:358–369, 1940.

Stribley RF, Alberts JW, Tourtellotte WW, Cockrell JL: A quantitative study of stance in normal subjects. *Arch Phys Med Rehab* 55:74–80, 1974.

Stulen FB: A technique to monitor localized muscular fatigue using frequency domain analysis for the myoelectric signal. Ph.D. dissertation, Massachusetts Inst. Technol., Cambridge, MA, 1980.

Stulen FB, De Luca CJ: The relation between the myoelectric signal and physiological properties of constant force isometric contractions. *Electroencephalogr Clin Neurophysiol* 45:681–698, 1978a.

Stulen FB, De Luca CJ: A non-invasive device for monitoring metabolic correlates of myoelectric signals. *Proc 31st Ann Conf on Engng in Med and Biol*:264, 1978b.

Stulen FB, De Luca CJ: A non-invasive device for monitoring localized muscular fatigue, *Proc 14th Ann Meeting Assoc Advancement in Medical Instrumentation*:268, 1979.

Stulen FB, De Luca CJ: Frequency parameters of the myoelectric signal as a measure of muscle conduction velocity. *IEEE Trans Biomed Engng* 28:515–523, 1981.

Stulen FB, De Luca CJ: Muscle fatigue monitor: a non-invasive device for observing localized muscular fatigue. *IEE Trans on Biomed Engng* 29:760–768, 1982.

Struppler A: Silent period (s.p.). *Electromyogr Clin Neurophysiol* 5:163–168, 1965.

Stuart DG, Eldred E, Hemingway A, Kawamura Y: Neural regulation of the rhythm of shivering. In *Temperature—Its Measurement and Control in Science and Industry*, New York, Rheinhold, 1963, pp 545–557.

Styck J, Hoogmartens M: A suitable type of wire for wire-electrodes. *Electromyogr Clin Neurophysiol* 15:291, 1975.

Sullivan WE, Mortensen OA, Miles M, Greene LS: Electromyographic studies of m. biceps brachii during normal voluntary movement at the elbow. *Anat Rec* 107:243–252, 1950.

Sumitsuji N, Matsumoto K, Tanaka M, Kashiwagi T, Kaneko Z: Electromyographic investigation of the facial muscles. *Electromyography* 7:77–96, 1967.

Sumitsuji N, Matsumoto K, Tanaka M, Kashiwagi T, Kaneko Z: In *Proceedings of the 4th International College of Psychosomatic Medcine*, Kyoto, Japan, 1977.

Sunderland S: The actions of the extensor digitorum communis, interosseous and lumbrical muscles. *Am J Anat* 77:189–209, 1945.

Šurina I, Jágr J: Action potentials of levator

and tensor muscles in patients with cleft palate. *Acta Chir Plast* 11:21–30, 1969.

Susman RL, Stern JT Jr: Telemetered electromyography of flexor digitorum profundus and flexor digitorum superficialis in *Pan troglodytes* and implications for interpretation of the O.H. 7 hand. *Am J Phys Anthropol* 50:565–574, 1979.

Susman RL, Stern JT Jr: EMG of the interosseous and lumbrical muscles in the chimpanzee (*Pan traglodytes*) hand during locomotion. *Am J Anat* 157:389–397, 1980.

Susset JG, Rabinovitch H, Mackinnon KJ: Parameters of micturition: clinical study. *J Urol* 94:113–121, 1965.

Sussman HM, Hanson RJ, MacNeilage PF: Studies of single motor units in the speech musculature: methodology and preliminary findings. *J Acoust Soc Am* 51:1372–1374, 1972.

Sutherland DH: An electromyographic study of the plantar flexors of the ankle in normal walking on the level. *J Bone Joint Surg* 48A:66–71, 1966.

Sutherland DH, Schottstaedt ER, Larsen LJ, Ashley RK, Callander JN, James PM: Clinical and electromyographic study of seven spastic children with internal rotation gait. *J Bone Joint Surg* 51A:1070–1082, 1969.

Sutton DL: Surface and needle electrodes in electromyography. Dent Prog 2:127–131, 1962.

Sutton D, Kimm J: Reaction time of motor units in biceps and triceps. *Exp Neurol* 23:503–515, 1969.

Sutton D, Kimm J: Alcohol effects on human motor unit reaction time. *Physiol Behav* 5:889–892, 1970.

Sutton D, Larson CR, Farrell DM: Cricothyroid motor units. *Acta Otolaryngol* 74:145–151, 1972.

Suzuki R: Function of the leg and foot muscles from the viewpoint of the electromyogram (English text). *J Jpn Orthop Surg Soc* 30:775–786, 1956.

Swezey RL, Fiegenberg DS: Inappropriate intrinsic muscle action in the rheumatoid hand. *Ann Rheum Dis* 30:619–625, 1971.

Takebe K, Basmajian JV: Gait analysis in stroke patients to assess treatment of drop-foot. *Arch Phys Med Rehab* 57:305–310, 1976.

Takebe K, Vitti M, Basmajian JV: The functions of semispinalis capitis and splenius capitis muscles: an electromyographic study. *Anat Rec* 179:477–480, 1974.

Tam H, Webster JG: Minimizing electrode motion artifact by skin abrasion. *IEEE Trans* BME-24:134, 1977.

Tamler E, Jampolsky A, Marg E: An electromyographic study of asymmetric convergence. *Am J Ophthalmol* 46 (No. 5, special Part II (not I) of the November issue):174–181, 1958.

Tamler E, Marg E, Jampolsky A: An electromyographic study of coactivity of human extraocular muscles in following movements. *AMA Arch Ophthalmol* 61:270–273, 1959.

Tamler E, Marg E, Jampolsky A, Nawratzki I: Electromyography of human saccadic eye movements. *AMA Arch Ophthalmol* 62:657–661, 1959.

Tanji J, Kato M: Volitionally controlled single motor unit discharges and cortical motor potentials in human subjects. *Brain Res* 29:243–246, 1971.

Tanji J, Kato M: Discharges of single motor units at voluntary contraction of abductor digiti minimi muscle in man. *Brain Res* 45:590–593, 1972.

Tanji J, Kato M: Recruitment of motor units in voluntary contraction of a finger muscle in man. *Exp Neurol* 40:759–770, 1973a.

Tanji J, Kato M: Firing rate of individual motor units in voluntary contraction of abductor digiti minimi muscle in man. *Exp Neurol* 40:771–783, 1973b.

Tasaki I, Singer I, Takenaka T: Effects of internal and external ionic environment on exitability of squid giant axon. *J Gen Physiol* 48:1095–1123, 1967.

Tatham MAA, Morton K: Some electromyography data towards a model of speech production. "*Occasional Papers,*" Language Centre, Univ. of Essex, Colchester, Engl 1:24, 1968a.

Tatham MAA, Morton K: Further electromyography data towards a model of speech production. "*Occasional Papers*", Language Centre, Univ. of Essex, Colchester, Engl 1:25–59, 1968b.

Tatham MAA: Speech synthesis and models of speech production. *Interim Report to Science Research Council.* Essex Univ., Colchester, Engl, privately printed, 1971.

Tauber ES, Coleman RM, Weitzman ED: Absence of tonic electromyographic activity during sleep in normal and spastic nonmimetic skeletal muscles in man. *Ann Neurol* 2:66–68, 1977.

Taylor A: The contribution of the intercostal muscles to the effort of respiration in man. *J Physiol* 151:390–402, 1960.

Teig E: Differential effect of graded contraction of middle ear muscles on the sound transmission of the ear. *Acta Physiol Scand* 88:382–391, 1973.

Teng EL, McNeal DR, Kralj A, Waters RL: Electrical stimulation and feedback training: effects on the voluntary control of paretic muscles. *Arch Phys Med Rehab* 57:228–233, 1976.

Tergast P: Ueber das Verhaltnis von Nerve und Muskel (German text) *Arch Mikr Anat* 9:36–46, 1873.

Terzuolo CA, Soechting JF, Viviani P: Studies on the control of some simple motor tasks. I. Relations between parameters of movements and EMG activities. *Brain Res* 58:212–216, 1973a.

Terzuolo CA, Soechting JF, Viviani P: Studies on the control of some simple motor tasks. II. On the cerebellar control of movements in relation to the formulation of intentional commands. *Brain Res* 58:217–222, 1973b.

Terzuolo CA, Viviani P: Parameters of motion and EMG activities during some simple motor tasks in normal subjects and cerebellar patients. In Cooper IS, Riklan M, Snider RS: *The Cerebellum, Epilepsy and Behavior*. New York, Plenum Press, 1974, pp 173–215.

Tesch P, Karlsson J: Lactate in fast and slow twitch skeletal muscle fibers of man during isometric contraction. *Acta Physiol Scand* 99:230–236, 1977.

Tesch P, Sjodin B, Thorstensson A, Karlsson J: Muscle fatigue and its relation to lactate accumulation and LDH activity in man. *Acta Physiol Scand* 103:413–420, 1978.

Thom H: Elektromyographische Untersuchungen zur Funktion des M. trapezius. *Elektromedizin* 10:65–72, 1965.

Thomas G: The function of the extensor muscles of the back. *German Med Monthly* 14:564–566, 1969.

Thomas JS, Schmidt EM, Hambrecht FT: Limitations of volitional control of single motor unit recruitment sequence (abstr). Society of Neuroscience, 1976, Annual Meeting, Toronto, Canada, 1976, p 536.

Thomas JS, Schmidt EM, Hambrecht FT: Facility of motor unit control during tasks defined directly in terms of unit behaviors. *Exp Neurol* 59:384–395, 1978.

Thomas LJ, Tiber N, Schireson S: The effects of anxiety and frustration of muscular tension related to the temporomandibular joint syndrome. *Oral Surg Oral Med Oral Pathol* 36:763–768, 1973.

Thomas PK: Recent advances in the clinical electrophysiology of muscle and nerve. *Postgrad M J* 37:377–384, 1961.

Thomas PK, Sears TA, Gilliatt RW: The range of conduction velocity in normal motor nerve fibers to the small muscles of the hand and foot. *J Neurol Neurosurg Psychiatry* 22:175–181, 1959.

Thomson SA: Hallux varus and metatarsus varus. In *Clinical Orthopaedics* Philadelphia, Lippincott, 1960, no. 16, pp 109–118.

Thorstensson A, Karlsson J, Viitasalo JHT, Luhtanen P, Komi PV: Effect of strength training on EMG of human skeletal muscle. *Acta Physiol Scand* 98:232–236, 1976.

Thysell RV: Reaction time of single motor units. *Psychophysiology* 6:174–185, 1969.

Tichauer ER: A pilot study of the biomechanics of lifting in simulated industrial work situations. *J Safety Res* 3:98–115, 1971.

Tilney F, Pike FH: Muscular coordination experimentally studied in its relation to the cerebellum. *Arch Neurol Psychiatry* 1:289–334, 1925.

Tokita T, Tashiro K, Kato K: Electromyography of the esophagus and its clinical applications. *Acta Otolaryngol* 70:269–278, 1970.

Tokizane H: Electromyographic study of facial muscles. *Ochyanomizu Med J* 2:1, 1954.

Tokizane T, Kawamata K, Tokizane H: Electromyographic studies on the human respiratory muslces. *Jpn J Physiol* 2:232–247, 1952.

Tokizane T, Shimazu H: *Functional Differentiation of Human Skeletal Muscle*. Springfield, IL, Charles C Thomas, 1964a.

Tokizane T, Shimazu H: *Functional Differentiation of Human Skeletal Muscle* Tokyo, University of Tokyo Press, 1964b.

Tokuyama H, Okamoto T, Kumamoto M: Electromyographic study of swimming in infants and children. In *International Series on Biomechanics, Vol. 1B, Biomechanics V-B* Baltimore, University Park Press, 1976, pp 215–220.

Tokuyama H, Okamoto T, Yoshizawa M, Kumamoto M: Suieiundoh no Kisoteki Kenkyu (A basic study of swimming—An electromyographic study of the flutter kick in childhood). In *Reports of the Japanese Research Committee of Sports Sciences*, no.4, 1977, pp 25–27.

Tomaszewska J: Badania patokinetyki miésni po amputacji uda (Studies on the pathokinesiology of muscles after above-knee amputation) (Polish text with French and English summaries). *Roczniki Nauk WSWF W Poznan* 8:3–55, 1964.

Tönnis D: Elektromyographische Untersuchungen über die Rollwirkung des M. iliopsoas. *Z Orthop Grenzg* 102:61–68, 1966a.

Tönnis D: Ekektromyographische Untersuchungen über die Beterlegung des M. iliopsoas an der Adund Abduktion des Oberschenkels. *Z Orthop Grenzg* 102:68–74, 1966b.

Török Z, Hammond PH: On the performance of single motor units as sources of control signals. Proceedings of the Conf. on Human Locomotor Engineering, U. of Sussex, 1971.

Tournay A, Fessard A: Etude électromyographique de la synergie entre l'abducteur du pouce et le muscle cubital postérieur (French text). *Rev Neurol* 80:631, 1948.

Tournay A, Paillard J: Electromyographie des muscles radiaux à l'état normal (French text). *Rev Neurol* 89:277–279, 1953.

Townsend MA, Shiavi R, Lainhart SP, Caylor J: Variability in synergy patterns of leg muscles during climbing, descending and level walking of highly-trained athletes and normal males. *Electromyogr Clin Neurophysiol* 18:69–80, 1978.

Townsend RE, House JF, Addario D: A comparison of biofeedback-mediated relaxation and group therapy in the treatment of chronic anxiety. *Am J Psychiatry* 132:589–601, 1975.

Toyoshima S, Matsui H, Miyashita M: An electromyographic study of the upper arm muscles involved in throwing (Japanese text, English abstract). *Res J Phys Ed* 15:103–110, 1971.

Travill AA: Electromyographic study of the extensor apparatus of the forearm. *Anat Rec* 144:373–376, 1962.

Travill A, Basmajian JV: Electromyography of the supinators of the forearm. *Anat Rec* 139:557–560, 1961.

Trelease RB, Sieck GC, Harper RM: A new technique for acute and chronic recordings of crural diaphragm EMG in cats. *EEG Clin Neurophysiol* 53:459–462, 1982.

Trojaborg W: Motor nerve conduction velocities in normal subjects with particular reference to the conduction in proximal and distal segments of median and ulnar nerve. *Electroencephalogr Clin Neurophysiol* 17:314–321, 1964.

Trojaborg W: Motor and sensory conduction in the musculocutaneous nerve. *J Neurol Neurosurg Psychiatry* 39:890–899, 1976.

Trontelj JV, Pečak F, Dimitrijevîć MR: Segmental neurophysiological mechanisms in scoliosis. *J Bone Joint Surg* 61-B:309–313, 1979.

Trusgnich DJ, Agarwal GC, Gottlieb GL: Modeling of the electromyogram: effect of motor unit synchronization. *Proc of the 4th Cong of ISEK* 1979, pp 196–197.

Tsurumi N: An electromyographic study of the gait of children. (Japanese text; English abstract). *J Jpn Orthop Assoc* 43:611–628, 1969.

Tursky B: Integrators as measuring devices of bioelectric output. *Clin Pharmacol Ther* 5:887–892, 1964.

Tusiewicz K, Moldofsky H, Bryan AC, Bryan MH: Mechanics of the rib cage and diaphragm during sleep. *J Appl Physiol* 43:600–602, 1977.

Tuttle R, Basmajian JV: Electromyography of knuckle-walking: results of four experiments on the forearm of *Pan gorilla. Am J Phys Anthropol* 37:255–266.

Tuttle R, Basmajian JV: Electromyographic studies of brachial muscles in *Pan gorilla* and hominoid evolution. *Am J Phys Anthropol* 41:71–90, 1974.

Tuttle RH, Basmajian JV: Electromyography of *Pan gorilla*: an experimental approach to the problem of hominization. In *Proc. Symp. 5th Congr. Int'l Primat. Soc. (1974)*. Tokyo, Japan Science Press, 1975.

Tuttle R, Basmajian JV: Electromyography of the pongid shoulder muscles and hominid evolution. *Am J Phys Anthropol* 44:212, 1976.

Tuttle RH, Basmajian JV: Electromyography of pongid shoulder muscles and hominoid evolution. I. Retractors of the humerus and rotators of the scapula. *Yearbk Phys Anthropol* 20:491–497, 1977.

Tuttle RH, Basmajian JV: Electromyography of pongid shoulder muscles. II. Deltoid, rhomboid and "rotator cuff." III. Quadrupedal positional behavior. *Am J Phys Anthropol* 49:47–56; 57–70, 1978.

Tuttle R, Basmajian JV, Ishida H: Electromyography of the gluteus maximus muscle in gorilla and the evolution of bipedalism. In Tuttle RH: *Antecedents of Man*, The Hague, Mouton, 1975, vol 1.

Tuttle RH, Velte MJ, Basmajian JV: Electromyography of brachial muscles in *Pan troglodytes* and *Pan pygmaeus. Am J Phys Anthropol* 61:75–83, 1983.

Ueda N, Ohyama M, Harvey JE, Mogi G, Ogura JH: Subglottal pressure and induced live voices of dogs with normal, reinnervated and paralyzed larynges. I. On voice function of the dog with a normal larynx. *Laryngoscope* 81:1948–1959, 1971.

Ueda N, Ohyama M, Harvey JE, Ogura JH: Influence of certain extrinsic laryngeal muscles on artificial voice production. *Laryngoscope* 82:468–482, 1972.

Ushijima T, Hirose H: Electromyographic study of the velum during speech. *J Phonet* 2:315–326, 1974.

Valbo AB, Hagbarth KE, Torebjork HE, Wallin BG: Somatosensory, proprioceptive, and sympathetic activity in human peripheral nerves. *Physiol Rev* 59:919–957, 1979.

Van Allen MW, Blodi FC: Electromyographic study of reciprocal innervation in blinking. *Neurology* 12:371–377, 1962.

van Boxtel A, Goudswaard GM, von der Molen, von der Bosch EJ: Changes in electromyogram power spectra of facial and jaw-elevator muscles during fatigue. *J Appl Physiol* 54:51–58, 1983.

van der Glas HW, van Steenberghe D: A fully automatic analysis method for the various silent period parameters of masticatory muscles in man. *Arch Int Physiol Biochem* 90:28–30, 1982.

van der Straaten JHM, Lohman AHM, van Linge B: A combined electromyographic & photographic study of the muscular control of the knee during walking. *J Hum Movement Studies* 1:25–32, 1975.

Vanderstoep SF: *A Comparison of the Ability to Control Single Motor Units in Selected Human Skeletal Muscles.* Ph.D. Dissertation, Univ. of Southern CA, 1971.

Vandervoort AA, Quinlan J, McComas AJ: Twitch potentiation after voluntary contraction. *Exp Neurol* 81:141–152, 1983.

van Gijn J: Babinski response: stimulus and effector. *J Neurol Neurosurg Psychiatry* 38:180–186, 1975.

van Gijn J: Equivocal plantar responses: a clinical and electromyographic study. *J Neurol Neurosurg Psychiatry* 39:275–282, 1976.

van Gool JD, de Ridder ALA, Kuijten RH, Donckerwolcke RAMG, Tiddens HA: Measurement of intravesical and rectal pressures simultaneously with electromyography of anal sphincter in children with myelomeningocele. *Dev Med Child Neurol* 18:287–301, 1976.

van Harreveld A: The structure of the motor units in the rabbit's m. sartorius. *Arch Néerl Physiol* 28:408–412, 1946.

van Harreveld A: On the force and size of motor units in the rabbit's sartorius muscle. *Am J Physiol* 151:96–106, 1947.

van Hoecke J, Perot C, Goubel F: Contribution des muscles biceps brachii et pronator teres a l'effort de prono-supination. I. Travail statique. *Eur J Appl Physiol* 38:83–91, 1978.

van Linge B, Mulder JD: Function of the supraspinatus muscle and its relation to the supraspinatus syndrome. *J Bone Joint Surg* 45B:750–754, 1963.

van Steenberghe D, van der Glas HW, Grisar PR, Vande Putte KM: The effects of acoustic masking of the silent period in the masseter electromyogram in man during sustained isometric contraction. *Electromyogr Clin Neurophysiol* 21:611–625, 1981.

Van Trees HL: *Detection, Estimation, and Modulation Theory, Part I.* New York, John S. Wiley, 1968.

Varnauskas E, Bjorntorp P, Fahlen M, Prerovsky I, Stenberg J: Effects of physical training on exercise blood flow and enzymatic activity in skeletal muscle. *Cardiovas Res* 4:418–422, 1970.

Vasilescu C, Dieckmann G: Electromyographic investigations in torticollis. *Appl Neurophysiol* 38:153–160, 1965.

Vereecken RL, Derluyn J, Verduyn H: Electromyography of the perineal striated muscles during cystometry. *Urol Int* 30:92–98, 1975.

Verroust J, Blinowska A, Cannet G: Functioning of the ensemble of motor units of the muscle determined from global EMG signal. *Electromyogr Clin Neurophysiol* 21:11–24, 1981.

Vigreaux B, Cnockaert JC, Pertuzon E: Factors influencing quantified surface EMGs. *Eur J Appl Physiol* 41:119–129, 1979.

Viitasalo JHT, Komi PV: Signal characteristics of EMG during fatigue. *Eur J Appl Physiol* 37:111–121, 1977.

Viitasalo JT, Komi PV: Effects of fatigue on isometric force- and relaxation-time characteristics in human muscle. *Acta Physiol Scand* 111:87–95, 1981.

Viljanen AA: The relation between the electrical and mechanical activity of human intercostal muscles during voluntary inspiration. *Acta Physiol Scand* Suppl:296, 1967.

Viljanen AA, Poppius H, Bergström RM, Hakumäki M: Electrical and mechanical activity in human respiratory muscles. *Acta Neurol Scand* 41(suppl 13):237–239, 1964.

Visser SL, de Rijke W: Influence of sex and age on emg contraction pattern. *Eur Neurol* 12:229–235, 1974.

Vitti M: Estudo electromiográfico dos músculos mastigadores no cão. *Folia Clin Biol* 34:101–114, 1965.

Vitti M: Analise electromiográfica do m. temporal no posição de repousa da mandíbula. *O Hospital (São Paulo)* 76:339–349, 1969a.

Vitti M: Comportamento electromiográfica do m. temporal no repuxamento da comissura bucal. *O Hospital (São Paulo)* 75:349–354, 1969b.

Vitti M: Comportamento eletromiográfica do músculo temporal nas mordidas incisiva e molares homo e heterolaterais. *O Hospital (São Paulo)* 78:207–214, 1970.

Vitti M: Electromyographic analysis of the musculus temporalis in basic movements of the jaw. *Electromyography* 11:389–403, 1971.

Vitti M, Bankoff ADP: Simultaneous EMG of latissimus dorsi and sternocostal part of pectoralis major muscles during classic natatory stroke. *Electromyogr Clin Neurophysiol* 19:505–510, 1979.

Vitti M, Basmajian JV: Muscles of mastication in small children: an electromyographic analysis. *Am J Orthod* 68:412–419, 1975.

Vitti M, Basmajian JV: Electromyographic investigation of procerus and frontalis muscles. *Electromyogr Clin Neurophysiol* 16:227–236, 1976.

Vitti M, Basmajian JV: Integrated actions of masticatory muscles: simultaneous EMG from eight intramuscular electrodes. *Anat Rec* 187:173–189, 1977.

Vitti M, Basmajian JV, Ouelette PL, Mitchell DL, Eastman WP, Seaborn RD: Electromyographic investigations of the tongue and circumoral muscular sling with fine-wire electrodes. *J Dent Res* 54:844–849, 1975.

Vitti M, Corrêa ACF, Fortinguerra CRH, Bérzin F, König B Jr: Electromyographic study of the "muscle depressor anguli oris." *Electromyography Clin Neurophysiol* 12:119–125, 1972.

Vitti M, Fortinguerra CRH, Corrêa ACF, König B Jr, Bérzin F: Electromyographic behavior of the levator labii superioris alaeque nasi. *Electromyogr Clin Neurophysiol* 14:37–43, 1974.

Vitti M, Fujiwara M, Iida M, Basmajian JV: The integrated roles of longus colli and sternocleidomastoid muscles: an electromyographic study. *Anat Rec* 177:471–484, 1973.

Vogt AT: Electromyograph responses and performance success estimates as a function of internal-external control. *Percept Motor Skills* 41:977–978, 1975.

Volta A: An account of some discoveries made by M. Galvani. *Philos Trans* 83:10, 1973.

Volta A: Collezione dell'opere del Cavagliere Conte Allessandro Volta, Florence, 1816.

von Meyer H: *Die Wechslnde Lage des Schwerpunktes im Menschlichen Köper.* Leipzig, Englemann, 1868.

Vrbová G: The effect of motoneurone activity on the speed of contraction of striated muscle. *J Physiol* 169:513–526, 1963.

Vredenbregt J, Rau G: Surface electromyography in relation to force, muscle length and endurance. In Desmedt JE: *New Developments in EMG and Clinical Neurophysiology* Basel, Karger, 1973, pp 607–622.

Vreeland RW, Sutherland DH, Dorsa JJ, Williams LA, Collins CC, Schottsteadt ER: A three-channel electromyography with synchronized slow-motion photography. *IRE Trans Bio-med Electronics* BME-8:4–6, 1961.

Wachholder K, McKinley C: Über die innervation und tätigheit der atemmuskeln. *Arch Ges Physiol* 222:575–588, 1929.

Wagman IH, Pierce DS, Burger RE: Proprioceptive influence in volitional control of individual motor units. *Nature* 207:957–958, 1965.

Wagner AL, Buchthal F: Motor and sensory conduction in infancy and childhood: reappraisal. *Dev Med Child Neurol* 14:189–216, 1972.

Wait JV, Atwater AE, Wetzel MC: Computer-aided approach to gait analysis. *Am J Phys Med* 53:229–233, 1974.

Walmsley RP: Electromyographic study of the phasic activity of peroneus longus and brevis. *Arch Phys Med Rehab* 58:65–69, 1977.

Walshe FMR: On certain tonic or postural

reflexes in hemiplegia with special reference to the so-called "associated movements." *Brain* 46:1–37, 1923.

Walters CE, Partridge MJ: Electromyographic study of the differential action of the abdominal muscles during exercise. *Am J Phys Med* 36:259–268, 1957.

Warmolts JR, Engels WK: Open biopsy electromyography: correlation of motor unit behavior with histochemical muscle fiber type in human limb muscle. *Arch Neurol* 27:512–517, 1972.

Waterland JC, Hellebrandt FA: Involuntary patterning associated with willed movement performed against progressively increasing resistance. *Am J Phys Med* 43:13–29, 1964.

Waterland JC, Munson N: Reflex association of head and shoulder girdle in nonstressful movements of man. *Am J Phys Med* 43:98–108, 1964a.

Waterland JC, Munson N: Involuntary patterning evoked by exercise stress: radioulnar pronation and supination. *J Am Phys Ther Assoc* 44:91–97, 1964b.

Waterland JC, Shambes GM: Head and shoulder linkage: stepping in place. *Am J Phys Med* 49:279–289, 1970.

Waters P, Strick PL: Influence of "strategy" on muscle activity during ballistic movements. *Brain Res* 207:189–194, 1981.

Waters RL, Frazier J, Garland DE, Jordan C, Perry J: Electromyographic gait analysis before and after operative treatment for hemiplegic equinus and equinovarus deformity. *J Bone Joint Surg* 64-A:284–288, 1982.

Waters RL, Morris JM: Effect of spinal support on the electrical activity of muscles of the trunk. *J Bone Joint Surg* 52-A:51–60, 1970.

Waters RL, Morris JM: Electrical activity of muscles of the trunk during walking. *J Anat* 111:191–199, 1972.

Watt D, Jones GM: On the functional role of the myotatic reflex in man. *Proc Can Fed Biol Soc* 9:13, 1966.

Watt DGD: Analysis of muscle receptor connections with spike-triggered averaging. 1. Spindle primary and tendon organ afferents. *J Neurophysiol* 39:1375–1392, 1976.

Waylonis GW, Aseff JN: Anal sphincter electromyography in the first two years of life. *Arch Phys Med Rehab* 54:525–527, 1973.

Weathersby HT: Electromyography of the thenar muscles. *Anat Rec* 127:386, 1957.

Weathersby HT: The forearm stabilisers of the thumb: an electromyographic study. *Anat Rec* 154:439, 1966.

Weathersby HT, Sutton LR, Krusen UL: The kinesiology of muscles of the thumb: an electromyographic study. *Arch Phys Med* 44:321–326, 1963.

Weddell G, Feinstein B, Pattle RE: The electrical activity of voluntary muscle in man under normal and pathological conditions. *Brain* 67:178–257, 1944.

Weiss MA: Electromyography in the selection of muscle reeducation methods. Polish Medical History and Science, July 1959 Bulletin, 1959.

Weiss MA: The neurophysiological aspects of the immediate post myoplastic amputation fitting from the neurophysiological point of view. (Unpublished manuscript of talk given in Washington), 1966.

Wells JG, Morehouse LE: Electromyographic study of the effect of various headward accelerative forces upon the pilot's ability to perform standardized pulls on the aircraft control stick. *J Aviation Med* 21:48–54, 1950.

Wersäll R: The tympanic muscles and their reflexes: physiology and pharmacology with special regard to noise generation by the muscles. *Acta Otolaryingol* 139(suppl):1–112, 1958.

Wertheimer LG, Ferraz ECDF: Observações electromiográficas sôbre as funções dos músculos supra-espinhal e deltóide nos movimentos do ombro, (Portuguese text). *Folia Clin Biol* 28:276–289, 1958.

Whatmore GB, Kohli DR: Dysponesis: a neurophysiologic factor in functional disorders. *Behav Sci* 13:102–124, 1968.

Wheatley MD, Jahnke WD: Electromyographic study of the superficial thigh and hip muscles in normal individuals. *Arch Phys Med* 32:508–515, 1951.

White ER, Basmajian JV: Electromyography of lip muscles and their role in trumpet playing. *J Appl Physiol* 35:892–897, 1973.

Whitehead WF, Orr WC, Engel BT, Schuster MM: External anal sphincter response to rectal distention: learned response or reflex. *Psychophysiology* 19:57–62, 1981.

Widmalm SE: *Reflex Activity of the Masseter Muscle in Man and EMG Study* (Suppl. 72 vol. 34 of *Acta Odontol Scand*) Doctoral Thesis, Faculty of Odontology, Göteborg University, 1976.

Widmalm SE, Hedegård B: Reflex activity in the masseter muscle of young individuals. Part I. Experimental procedures—results: & Part II. (no subtitle). *J Oral Rehab* 3:41–55; 167–180, 1976.

Wiedenbauer MM, Mortensen OA: An electromyographic study of the trapezius muscle. *Am J Phys Med* 31:363–372, 1952.

Wiesendanger M, Schneider P, Villoz JP: Elektromyographische Analyse der raschen Wilkürbewegung. Schweiz. *Arch Neurol Neurochir Psychiatr* 100:88–99, 1967.

Wiesendanger M, Schneider P, Villoz JP: Electromyographic analysis of a rapid volitional movement. *Am J Phys Med* 48:17–24, 1969.

Wilkins BR: A theory of position memory. *J Theor Biol* 7:374–387, 1964.

Williams JGL: A resonance theory of "microvibrations." *Psychol Rev* 70:547–558, 1963.

Willison RG: A method for measuring motor unit activity in human muscle. *J Physiol* 168:35–36, 1963.

Wilson A, Wilson AS: Psychophysiological and learning correlates of anxiety and induced muscle relaxation. *Psychophysiology* 6:740–748, 1970.

Wilson DM: Proprioceptive leg reflexes in cockroaches. *J Exp Biol* 43:397–409, 1965.

Windhorst U, Schwestka R: Interaction between motor units in modulating discharge patterns of primary muscle spindle endings. *Exp. Brain Res* 45:417–427, 1982.

Windhorst U, Schwestka R, Koehler W: The effect of double activation of single motor units on the discharge patterns of primary muscle spindle endings in the cat. *Neurosci Lett* 28:303–307, 1982.

Windle WF: *Physiology of the Fetus*. Philadelphia, W.B. Saunders, 1940.

Windle WF, Minear WL, Austin MF, Orr DM: The origin and early development of somatic behaviour in the albino rat. *Physiol Zool* 8:156–185, 1935.

Winter DA: Camera speed for normal and pathological gait analysis. *Med Biol Eng Comput* 20:408–412, 1982.

Winter DA, Quanbury AO: Multichannel biotelemetry system for use in emg studies, particularly in locomotion. *Am J Phys Med* 54:142–147, 1975.

Winter DA, Quanbury AO, Hobson DA, Sidwall HG, Reimer G, Trenholm BG, Steinke T, Shlosser H: Kinematics of normal locomotion—statistical study based on T.V. data. *J Biomech* 7:479–486, 1974.

Winter DA, Rau G, Kadefors R, Broman H, De Luca CJ: Units, Terms and Standards in the Reporting of EMG Research. Published by the International Society of Electrophysiological Kinesiology, 1980.

Woelfel JB, Hickey JC, Stacy RW, Rinear L: Electromyographic analysis of jaw movements. *J Prosthet Dent* 10:688–697, 1960.

Wójtowicz S: Reciprocal innervation of directly and indirectly synergic and antagonistic external eye muscles. *Polish Med J* 5:656–661, 1966.

Wójtowicz S, Gwoździ E: Protective function of the orbicularis oculi muscle in electromyographic investigations. *Polish Med J* 7:465–470, 1968.

Wolf SL, Basmajian JV: A rapid cooling device for controlled cutaneous stimulation. *Am Phys Therap Assoc J* 53:25–27, 1973.

Wolf SL, Basmajian JV: Assessment of paraspinal electromyographic activity in normal subjects and in chronic back pain patients using a muscle biofeedback device. In Asmussen E, Jørgensen K, *Biomechanics VI-B*, Baltimore, University Park Press, 1980.

Wolf SL, Basmajian JV, Russe CTC, Kutner M: Normative data on low back mobility and activity levels: implications for neuromuscular re-education. *Am J Phys Med* 58:217–229, 1979.

Wolf SL, Letbetter WD: Effect of skin cooling on spontaneous emg activity in triceps surae of the decerebrate cat. *Brain Res* 91:151–155, 1975.

Wolf SL, Letbetter WD, Basmajian JV: Effects of a specific cutaneous cold stimulus on single motor unit activity of medial gastrocnemius muscle in man. *Am J Phys Med* 55:177–183, 1976.

Wolpert R, Wooldridge CP: The use of electromyography as biofeedback therapy in the management of cerebral palsy: a review and case study. *Physiother Can* 27:5–10, 1975.

Woodburne RT: *Essentials of Human Anatomy*. New York, Oxford University Press, 1957.

Woods JJ, Bigland-Ritchie B: Linear and non-linear surface EMG/force relationships in human muscles. *Am J Phys Med* 62:287–299, 1983.

Wright G, Wasserman E, Pottala E, Dukes-Dobos R: The effect of general fatigue on isometric strength-endurance measurements and the electromyogram of the biceps brachii. *Am Indust Hyg Assoc J* 37:274–279, 1976.

Wyke B: Effects of anesthesia upon intrinsic laryngeal reflexes. *J Laryngol Otol* 82:603–612, 1968.

Wyke B: Deus ex machina vocis: an analysis of the laryngeal reflex mechanisms of speech. *Br J Disorders Communic* 4:3–25, 1969.

Wyman R: Patterns of frequency variation in dipteran flight motor units. *Comp Biochem Physiol* 35:1–16, 1970.

Wyrick W, Duncan A: Electromyographical study of reflex, premotor, and simple reaction time of relaxed muscle to joint displacement. *J Motor Behav* 6:1–10, 1974.

Yabe K, Tamaki Y: Inhibitory effect of unilateral contraction on the contralateral arm. *Percept Motor Skills* 43:979–982, 1976.

Yamaji K, Misu A: Kinesiologic study with electromyography of low back pain. *Electromyography* 8:189, 1968.

Yamashita N: The mechanism of generation and transmission of forces in leg extension. *J Hum Ergol* 4:43–52, 1975.

Yamashita N, Kumamoto M, Okamoto T: Electromyographic study of the giant swing on the horizontal bar. In *Sixth International Congress of Physical Medicine*, 1972, vol 2, pp 361–363.

Yamashita N, Takagi K, Okamoto T: Tetsubo-Undoh ni okeru Junte-Sharin no Kindenzuteki Kenkyu (Electromyographic study on the giant swing [forward] on the horizontal bar). *Res Phys Ed* 15:93–102, 1971.

Yamazaki A: Electrophysiological study on "flick" eye movements during fixation (Japanese text; English abstract). *Acta Soc Ophthalmol Jpn* 71: 2446–2459, 1968.

Yamshon LJ, Bierman W: Kinesiologic electromyography. II. The trapezius. *Arch Phys Med* 29:647–651, 1948.

Yamshon LJ, Bierman W: Kinesiologic electromyography. III. The deltoid. *Arch Phys Med* 30:286–289, 1949.

Yemm R: Variations in the electrical activity of the human masseter muscle occurring in association with emotional stress. *Arch Oral Biol* 14:873–878, 1969a.

Yemm R: Masseter muscle activity in stress: adaptation of response to a repeated stimulus in man. *Arch Oral Biol* 14:1437–1439, 1969b.

Yemm R: Temporomandibular dysfunction and masseter muscle response to experimental stress. *Br Dent J* 127:508–510, 1969c.

Yoshi H, Takagi K, Kumamoto M, Ito M, Ito K, Yamashita N, Okamoto T, Nakagawa H: Suiso ni yoru Kayak Soho no Kindenzugakuteki Kenkyu (Electromyographic study of kayak paddling in the paddling tank). *Res J Phys Ed* 18:191–198, 1974.

Yoshizawa M, Tokuyama H, Okamoto T, Kumamoto M: Electromyographic study of the breaststroke. *International Series on Biomechanics, Vol B, Biomechanics V-B*, Baltimore, University Park Press, 1976, pp 222–229.

Youmans WB, Murphy QR, Turner JK, Davis LD, Briggs DI, Hoye AS: *The Abdominal Compression Reaction: Activity of Abdominal Muscles Elicited from the Circulatory System.* Baltimore, Williams & Wilkins, 1963.

Youmans WB, Tjioe DT, Tong EY: Control of involuntary activity of abdominal muscles. *Am J Phys Med* 53:57–74, 1974.

Young RR, Shahani B: An EMG study of cutaneous blink reflexes (abstr.). *Electroencephalogr Clin Neurophysiol* 26:630–636, 1969.

Zappalá A: Influence of training and sex on the isolation and control of single motor units. *Am J Phys Med* 49:348–361, 1970.

Zernicke RD, Waterland JC: Single motor unit control in m. biceps brachii. *Electromyogr Clin Neurophysiol* 12:225–241, 1972.

Zhukov EK, Zakharyants JZ: Electrophysiological data concerning certain mechanisms of overcoming fatigue (Russian text) *J Physiol U.S.S.R.* 46:819–827, 1960.

Zinnir NR, Sterling AM: *Female Incontinence.* New York, A.R. Liss, 1981.

Zipp P: Effect of electrode parameters on the bandwidth of the surface EMG power-density spectrum. *Med Biol Comput* 116:537–541, 1978.

Żuk T: Badania elektromiograficzne w skoliozach (Polish text; Russian and English summaries). *Chir Narz Ruchu i Ortop Polska* 25:589–595, 1960.

Żuk T: Etiopatogeneze skoliózy na podkladě elektromyografických záznamů (Czech test; Russian and English summaries). *Acta Chir Orthop Traumat Chechoslov* 24:69–74, 1962a.

Żuk T: The role of spinal and abdominal muscles in the pathogenesis of scoliosis. *J Bone Joint Surg* 44-B:102–105, 1962b.

Zwaagstra B, Kernell D: The duration of after-hyperpolarization in hindlimb alpha motoneurones of different sizes in the cat. *Neurosci Lett* 19:303–307, 1980.

Index

Page numbers in italics denote figures; those followed by "t" or "f" denote tables or footnotes, respectively.

F

G

H

N

O

P